Lecture Notes in Computer Science 2778

Edited by G. Goos, J. Hartmanis, and J. van Leeuwen

Springer-Verlag Berlin Heidelberg GmbH

Peter Y. K. Cheung   George A. Constantinides
Jose T. de Sousa (Eds.)

Field-Programmable Logic and Applications

13th International Conference, FPL 2003
Lisbon, Portugal, September 1-3, 2003
Proceedings

Springer

Series Editors

Gerhard Goos, Karlsruhe University, Germany
Juris Hartmanis, Cornell University, NY, USA
Jan van Leeuwen, Utrecht University, The Netherlands

Volume Editors

Peter Y. K. Cheung
George A. Constantinides
Imperial College of Science, Technology, and Medicine
Dept. of Electrical and Electronic Engineering
Exhibition Road, London SW7 2 BT, UK
E-mail: p.cheung@ic.ac.uk; george.constantinides@ieee.org

Jose T. de Sousa
Technical University of Lisbon
INESC-ID/IST
R. Alves Redol, 9, Apartido 13069, 1000-029, Lisboa, Portugal
E-mail: jts@inesc-id.pt

Cataloging-in-Publication Data applied for

A catalog record for this book is available from the Library of Congress

Bibliographic information published by Die Deutsche Bibliothek
Die Deutsche Bibliothek lists this publication in the Deutsche Nationalbibliografie;
detailed bibliographic data is available in the Internet at <http://dnb.ddb.de>.

CR Subject Classification (1998): B.6-7, C.2, J.6

ISSN 0302-9743
ISBN 978-3-540-40822-2   ISBN 978-3-540-45234-8 (eBook)
DOI 10.1007/978-3-540-45234-8

http://www.springer.de

© Springer-Verlag Berlin Heidelberg 2003
Originally published by Springer-Verlag Berlin Heidelberg New York in 2003.

Typesetting: Camera-ready by author, data conversion by Steingräber Satztechnik GmbH
Printed on acid-free paper   SPIN 10931431   06/3142   5 4 3 2 1 0

Preface

This book contains the papers presented at the 13th International Workshop on Field Programmable Logic and Applications (FPL) held on September 1–3, 2003. The conference was hosted by the Institute for Systems and Computer Engineering-Research and Development of Lisbon (INESC-ID) and the Department of Electrical and Computer Engineering of the IST-Technical University of Lisbon, Portugal.

The FPL series of conferences was founded in 1991 at Oxford University (UK), and has been held annually since: in Oxford (3 times), Vienna, Prague, Darmstadt, London, Tallinn, Glasgow, Villach, Belfast and Montpellier. It brings together academic researchers, industrial experts, users and newcomers in an informal, welcoming atmosphere that encourages productive exchange of ideas and knowledge between delegates.

Exciting advances in field programmable logic show no sign of slowing down. New grounds have been broken in architectures, design techniques, run-time reconfiguration, and applications of field programmable devices in several different areas. Many of these innovations are reported in this volume.

The size of FPL conferences has grown significantly over the years. FPL in 2002 saw 214 papers submitted, representing an increase of 83% when compared to the year before. The interest and support for FPL in the programmable logic community continued this year with 216 papers submitted. The technical program was assembled from 90 selected regular papers and 56 posters, resulting in this volume of proceedings. The program also included three invited plenary keynote presentations from LSI Logic, Xilinx and Cadence, and three industrial tutorials from Altera, Mentor Graphics and Dafca.

Due to the inclusive tradition of the conference, FPL continues to attract submissions from all over the world. The accepted contributions were submitted by researchers from 32 different countries:

USA	42	Belgium	6	Brazil	2	Estonia	1
Spain	33	Netherlands	6	Canada	2	Norway	1
UK	29	Mexico	5	Hungary	2	India	1
Germany	14	Greece	4	Iran	2	Slovakia	1
Japan	13	Poland	4	Korea	2	Slovenia	1
Portugal	12	Switzerland	4	Romania	2		
Italy	9	Australia	3	Singapore	2		
Czech Rep.	8	Ireland	3	Austria	1		
France	7	China	2	Egypt	1		

We would like to thank all the authors for submitting their first versions of the papers and the final versions of the accepted papers. We also gratefully acknowledge the reviewing work done by the Program Committee members and many additional reviewers who contributed their time and expertise towards the compilation of this volume. The members of our Program Committee and all other reviewers are listed on the following pages. We are particularly pleased that of the 1029 reviews sought, 95% were completed.

We would like to thank QuickSilver Technology for their sponsorship of the Michal Servit Award, Celoxica for sponsoring the official FPL website www.fpl.org, Xilinx and Synplicity for their early support of the conference, and Coreworks for help in registration processing. We are indebted to Richard van de Stadt, the author of CyberChair. This excellent free software made our task of managing the submission and reviewing process much easier. We are grateful for the help and advice received from Wayne Luk and Horácio Neto. In addition, we acknowledge the help of the following research students from Imperial College London in checking the integrity of the manuscripts: Christos Bouganis, Wim Melis, Gareth Morris, Andy Royal, Pete Sedcole, Nalin Sidahao, and Theerayod Wiangtong.

We are grateful to Springer-Verlag, particularly Alfred Hofmann and Anna Kramer, for their work in publishing this book.

June 2003

Peter Y.K. Cheung
George A. Constantinides
Jose T. de Sousa

Organization

Organizing Committee

Program Chair	Peter Y.K. Cheung, Imperial College London, UK
Program Co-chair	George A. Constantinides, Imperial College London, UK
General Chair	Jose T. de Sousa, INESC-ID/IST, Technical University of Lisbon, Portugal
Publicity Chair	Reiner Hartenstein, University of Kaiserslautern, Germany
Local Chair	Horácio Neto, INESC-ID/IST, Technical University of Lisbon, Portugal
Finance Chair	Fernando Gonçalves, INESC-ID/IST, Technical University of Lisbon, Portugal
Exhibition Chair	João Cardoso, INESC-ID/UA, University of Algarve, Portugal

Program Committee

Nazeeh Aranki	Jet Propulsion Laboratory, USA
Jeff Arnold	Stretch, Inc., USA
Peter Athanas	Virginia Tech, USA
Neil Bergmann	Queensland University of Technology, Australia
Dinesh Bhatia	University of Texas, USA
Eduardo Boemo	University of Madrid, Spain
Gordon Brebner	Xilinx, Inc., USA
Andrew Brown	University of Southampton, UK
Klaus Buchenrieder	Infineon Technologies AG, Germany
Charles Chiang	Synopsys, Inc., USA
Peter Cheung	Imperial College London, UK
George Constantinides	Imperial College London, UK
Andre DeHon	California Institute of Technology, USA
Jose T. de Sousa	Technical University of Lisbon, Portugal
Carl Ebeling	University of Washington, USA
Hossam ElGindy	University of New South Wales, Australia
Manfred Glesner	Darmstadt University of Technology, Germany
Fernando Goncalves	Technical University of Lisbon, Portugal
Steven Guccione	Quicksilver Technology, USA
Reiner Hartenstein	University of Kaiserslautern, Germany
Scott Hauck	University of Washington, USA
Brad Hutchings	Brigham Young University, USA
Tom Kean	Algotronix Consulting, UK
Andreas Koch	University of Braunschweig, Germany
Dominique Lavenier	University of Montpellier II, France
Philip Leong	Chinese University of Hong Kong, China
Wayne Luk	Imperial College London, UK
Patrick Lysaght	Xilinx, Inc., USA
Bill Mangione-Smith	University of California at Los Angeles, USA
Reinhard Männer	University of Mannheim, Germany
Oskar Mencer	Bell Labs, USA
George Milne	University of Western Australia
Toshiyaki Miyazaki	NTT Network Innovation Labs, Japan
Fernando Moraes	PUCRS, Brazil
Horacio Neto	Technical University of Lisbon, Portugal
Sebastien Pillement	ENSSAT, France
Dhiraj Pradhan	University of Bristol, UK
Viktor Prasanna	University of Southern California, USA
Michel Renovell	University of Montpellier II, France
Jonathan Rose	University of Toronto, Canada
Zoran Salcic	University of Auckland, New Zealand
Hartmut Schmeck	University of Karlsruhe, Germany
Rainer Spallek	Dresden University of Technology, Germany

Adrian Stoica	Jet Propulsion Laboratory, USA
Jürgen Teich	University of Paderborn, Germany
Lothar Thiele	ETH Zürich, Switzerland
Liones Torres	University of Montpellier II, France
Stephen Trimberger	Xilinx, Inc., USA
Milan Vasilko	Bournemouth University, UK
Ranga Vemuri	University of Cincinnati, USA
Roger Woods	Queen's University of Belfast, UK

Steering Committee

Jose T. de Sousa	Technical University of Lisbon, Portugal
Manfred Glesner	Darmstadt University of Technology, Germany
John Gray	Independent Consultant, UK
Herbert Grünbacher	Carinthia Technical Institute, Austria
Reiner Hartenstein	University of Kaiserslautern, Germany
Andres Keevallik	Tallinn Technical University, Estonia
Wayne Luk	Imperial College London, UK
Patrick Lysaght	Xilinx, Inc., USA
Michel Renovell	University of Montpellier II, France
Roger Woods	Queen's University of Belfast, UK

Additional Reviewers

Anuradha Agarwal
Ali Ahmadinia
Seong-Yong Ahn
Rui Aguiar
Miguel A. Aguirre
Bashir Al-Hashimi
Ferhat Alim
Jose Alves
Hideharu Amano
Jose Nelson Amaral
David Antos
António José Araújo
Miguel Arias-Estrada
Rubén Arteaga
Armando Astarloa
José Augusto
Shailendra Aulakh
Vicente Baena
Zachary Baker
Jonathan Ballagh
Sergio Bampi

Francisco Barat
Jorge Barreiros
Marcus Bednara
Peter Bellows
Mohammed Benaissa
AbdSamad BenKrid
Khaled BenKrid
Pascal Benoit
Manuel Berenguel
Daniel Berg
Paul Berube
Jean-Luc Beuchat
Rajarshee Bharadwaj
Unai Bidarte
Bob Blake
Brandon Blodget
Jose A. Boluda
Vanderlei Bonato
Andrea Boni
Marcos R. Boschetti
Ignacio Bravo

Ney Calazans
Danna Cao
Francisco Cardells-Tormo
Joao Cardoso
Dylan Carline
Luigi Carro
Nicholas Carter
Gregorio Cappuccino
Joaquín Cerdà
Abhijeet Chakraborty
François Charot
Seonil Choi
Bobda Christophe
Alessandro Cilardo
Christopher Clark
John Cochran
James Cohoon
Stuart Colsell
Katherine Compton
Pasquale Corsonello
Tom Van Court
Octavian Cret
Damian Dalton
Alan Daly
Martin Danek
Klaus Danne
Eric Debes
Martin Delvai
Daniel Denning
Arturo Diaz-Perez
Pedro Diniz
Peiliang Dong
Cillian O'Driscoll
Mark E. Dunham
Alireza Ejlali
Tarek El-Ghazawi
Peeter Ellervee
Wilfried Elmenreich
Rolf Enzler
Ken Erickson
Roberto Esper-Chaín Falcón
Béla Fehér
Michael Ferguson
Marcio Merino Fernandes
Viktor Fischer

Toshihito Fujiwara
Rafael Gadea-Girones
Altaf Abdul Gaffar
Federico Garcia
Alberto Garcia-Ortiz
Ester M. Garzon
Manjunath Gangadhar
Antonio Gentile
Raul Mateos Gil
Rafael Gadea-Girones
Federico Garcia
Jörn Gause
Fahmi Ghozzi
Guy Gogniat
Richard Aderbal Gonçalves
Gokul Govindu
Gail Gray
Jong-Ru Guo
Manish Handa
Frank Hannig
Jim Harkin
Martin Herbordt
Antonin Hermanek
Fabiano Hessel
Teruo Higashino
Roland Höller
Renqiu Huang
Ashraf Hussein
Shuichi Ichikawa
José Luis Imaña
Minoru Inamori
Preston Jackson
Kamakoti
Parivallal Kannan
Irwin Kennedy
Tim Kerins
Jawad Khan
Sami Khawam
Daniel Kirschner
Tomoyoshi Kobori
Fatih Kocan
Dirk Koch
Zbigniew Kokosinski
Andrzej Krasniewski
Rohini Krishnan

Georgi Kuzmanov
Soonhak Kwon
David Lacasa
John Lach
Jesus Lazaro
Barry Lee
Dong-U Lee
Gareth Lee
Jirong Liao
Valentino Liberali
Bossuet Lilian
Fernanda Lima
John Lockwood
Andrea Lodi
Robert Lorencz
Michael G. Lorenz
David Rodriguez Lozano
Shih-Lien Lu
Martin Ma
Usama Malik
Cesar Augusto Marcon
Theodore Marescaux
Eduardo Marques
L.J. McDaid
Paul McHardy
Maire McLoone
Bingfeng Mei
Mahmoud Meribout
Uwe Meyer-Baese
Yosuke Miyajima
Sumit Mohanty
Gareth Morris
Elena Moscu
Francisco Moya-Fernandez
Madhubanti Mukherjee
Tudor Murgan
Ciaron Murphy
Takahiro Murooka
Kouichi Nagami
Ulrich Nageldinger
Jeff Namkung
Ángel Grediaga Olivo
Pilar Martinez Ortigosa
Selene Maya-Rueda
Wim Melis

Allen Michalski
Maria Jose Moure
John Nestor
Jiri Novotny
John Oliver
Eva M. Ortigosa
Fernando Ortiz
Damjan Oseli
Jingzhao Ou
Marcio Oyamada
Chris Papachristou
Fernando Pardo
Stavros Paschalakis
Kolin Paul
Cong Vinh Phan
Juan Manuel Sanchez Perez
Stefania Perri
Mihail Petrov
Thilo Pionteck
Marco Platzner
Jüri Põldre
Dionisios N. Pnevmatikatos
Kara Poon
Juan Antonio Gomez Pulido
Federico Quaglio
Senthil Rajamani
Javier Ramirez
Juergen Reichardt
Javier Resano
Fernando Rincón
Francisco Rodríguez-Henríquez
Nuno Roma
Eduardo Ros
Gaël Rouvroy
Andrew Royal
Giacinto Paolo Saggese
Marcelino Santos
Gilles Sassatelli
Toshinori Sato
Sergei Sawitzki
Pascal Scalart
Bernd Scheuermann
Jan Schier
Clemens Schlachta
Klaus Schleisiek

Herman Schmit
David Schuehler
Ronald Scrofano
Pete Sedcole
Peter-Michael Seidel
Shay Seng
Sakir Sezer
Naoki Shibata
Tsunemichi Shiozawa
Nalin Sidahao
Reetinder Sidhu
Valery Sklyarov
Iouliia Skliarova
Gerard Smit
Raphael Some
Ioannis Sourdis
Lionel Sousa
Ludovico de Souza
François-Xavier Standaert
Henry Styles
Qing Su
Vijay Sundaresan
Noriyuki Takahashi
Shigeyuki Takano
Kalle Tammemäe
Konstantinos Tatas
Raoul Tawel
John Teifel

Yann Thoma
Tim Todman
Jon Tombs
Cesar Torres-Huitzil
Kuen Tsoi
Marek Tudruj
Richard Turner
Fabrizio Vacca
Sudhir Vaka
Eduardo do Valle Simoes
József Vásárhelyi
Joerg Velten
Felip Vicedo
Tanya Vladimirova
Markus Weinhardt
Theerayod Wiangtong
Juan Manuel Xicotencatl
Andy Yan
Keiichi Yasumoto
Pavel Zemcik
Xiaoyang Zeng
Yumin Zhang
Jihan Zhu
Ling Zhuo
Peter Zipf
Claudiu Zissulescu-Ianculescu
Mark Zwolinski

Table of Contents

Technologies and Trends

Communications Applications

High Level Design Tools 1

Reconfigurable Architectures

Cryptographic Applications 1

Place and Route Tools

Multi-context FPGAs

Cryptographic Applications 2

Low-Power Issues 1

Run-Time Configurations

Cryptographic Applications 3

Compilation Tools

Asynchronous Techniques

Biology-Related Applications

Codesign

Reconfigurable Fabrics

Dynamic Reconfiguration

SoC Architectures

Emulation

Cache Design

Arithmetic 1

Biologically Inspired Designs

Low-Power Issues 2

SoC Designs

Cellular Applications

Arithmetic 2

Fault Analysis

Network Applications

High Level Design Tools 2

Technologies and Trends (Posters)

Applications (Posters)

Tools (Posters)

FPGA Implementations (Posters)

Video and Image Applications (Posters)

Reconfigurable and Low-Power Systems (Posters)

Design Techniques (Posters)

Neural and Biological Applications (Posters)

Codesign and Embedded Systems (Posters)

Reconfigurable Systems and Architectures (Posters)

DSP Applications (Posters)

Dynamic Reconfiguration (Posters)

Arithmetic (Posters)

Design and Implementations 1 (Posters)

Design and Implementations 2 (Posters)

Reconfigurable Circuits
Using Hybrid Hall Effect Devices

Steve Ferrera and Nicholas P. Carter

Department of Electrical and Computer Engineering, University of Illinois,
Urbana-Champaign, Illinois 61801, USA

Abstract. Hybrid Hall effect (HHE) devices are a new class of reconfigurable logic devices that incorporate ferromagnetic elements to deliver non-volatile operation. A single HHE device may be configured on a cycle-by-cycle basis to perform any of four different logical computations (OR, AND, NOR, NAND), and will retain its state indefinitely, even if the power supply is removed from the device. In this paper, we introduce the HHE device and describe a number of reconfigurable circuits based on HHE devices, including reconfigurable logic gates and non-volatile table lookup cells.

1 Introduction

Over the last two decades, CMOS circuitry has become the dominant implementation technology for reconfigurable logic devices and semiconductor systems in general, because advances in fabrication processes have delivered geometric rates of improvement in both device density and speed. However, CMOS circuits suffer from the disadvantage that they require power to maintain their state. When power is removed from the system, all information about the state of a computation and the configuration of a reconfigurable circuit is lost, requiring that the reconfigurable device be configured and any information about the ongoing computation be reloaded from non-volatile storage each time the system containing it is powered on.

Magnetoelectronic circuits [1] overcome this limitation of CMOS systems by incorporating ferromagnetic materials, similar to those used in conventional hard disks. The magnetization state of these materials remains stable when power is removed from the device, allowing them to retain their state without a power supply and to provide "instant-on" operation when power is restored.

Much of the previous work on magnetoelectronic circuits has focused on the use of magnetoelectronic devices to implement non-volatile memory. In this paper, we describe a new class of magnetoelectronic device, the hybrid Hall effect device [2,4], that can be reconfigured on a cycle-by-cycle basis to implement a variety of logic functions, and present two initial applications for these devices: reconfigurable gates and non-volatile lookup table elements.

In the next section, we describe the HHE device and its basic operation. Section 3 presents two reconfigurable gate designs based on HHE devices, while Section 4

P.Y.K. Cheung et al. (Eds.): FPL 2003, LNCS 2778, pp. 1–10, 2003.

illustrates how the HHE device could be used to provide non-volatile storage for conventional lookup-table based reconfigurable systems. In Section 5, we present simulation results for our circuits. Related work is mentioned in Section 6, and Section 7 presents conclusions and our plans for future work.

2   HHE Device Description and Operation

The hybrid Hall effect device [2] is a semiconductor structure that contains a ferromagnetic element for non-volatile storage. Fig. 1 shows the physical structure of an HHE device along with a functional block diagram. The input to the device is a current along the input wire, which is at the top of Fig. 1(a). As shown in the figure, the current along the input wire creates a magnetic field in the ferromagnetic element beneath it. If the magnitude of the current is high enough, the induced magnetic field in the ferromagnetic element will magnetize it in the direction of the magnetic field, and the magnetization will remain stable once the input current is removed. An input current of sufficient magnitude in the opposite direction will magnetize the ferromagnetic element in the opposite direction, creating two stable states that can be used to encode binary values. If the magnitude of the input current is below the value required to change the magnetization state of the ferromagnetic element, which is a function of the dimensions of the device and the material used to implement the ferromagnetic element, the ferromagnetic element will retain its old magnetization state indefinitely.

The output voltage of the device is generated by passing a bias current through the insulated conductor at the bottom of Fig. 1(a). According to the Hall effect [3], the interaction of this bias current with the magnetic field generated by the magnetized ferromagnetic element produces a voltage perpendicular to the bias current. The sign of this voltage is determined by the magnetization of the ferromagnetic element and its magnitude is proportional to the magnitude of the bias current. Depending on the intended use of the device, a fabrication offset voltage may be added [4], making the

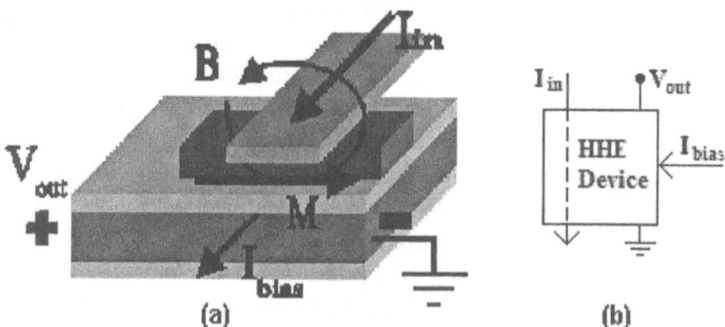

Fig. 1. HHE diagrams. (a) Physical structure. From top to bottom, the blocks represent an input wire, ferromagnetic element, insulator, conducting output channel, and bottom insulator. (b) Functional block.

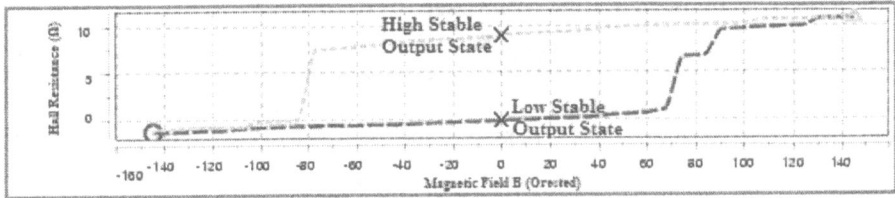

Fig. 2. Hysteresis loop for an HHE device

output voltage approximately 0V for one magnetization state and VDD for the other. Adding this fabrication offset makes it significantly easier to integrate the HHE device with CMOS circuits. Previous experiments [4] have fabricated HHE devices a small number of microns on a side, and the technology is expected to scale to significantly smaller devices in the near future, which will also reduce the amount of current required to set the magnetization state of the device.

The behavior of the HHE device is summarized by the hysteresis graph in Fig. 2. The Hall resistance, which relates the magnitude of the output voltage to that of the bias current, is plotted as a function of the magnetic field generated by the current along the input wire. As shown in the figure, there are two stable states when the input current, and thus the magnetic field, is 0, which correspond to the two magnetization states of the ferromagnetic element. A magnetic field of approximately ±90 Oersted is required to change the magnetization state of the ferromagnetic element and shift the Hall resistance from one half of the hysteresis curve to the other.

3 HHE as a Reconfigurable Gate

Fig. 3 shows a high-level block diagram of the circuitry required to implement a reconfigurable gate using an HHE device. Signals A and B are the inputs to the gate, while signals G0 and G1 select the logic function to be performed by the gate. In the following subsections, we present two designs for the interface circuitry that converts the CMOS-compatible gate inputs into appropriate input currents for the HHE device.

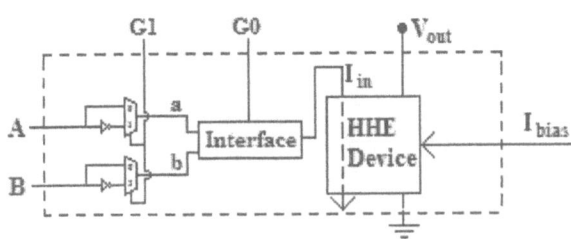

Fig. 3. HHE reconfigurable gate

3.1 HHE Reconfigurable Gate with Reset

Our first reconfigurable gate design uses a reset-evaluate methodology similar to that used in dynamic CMOS circuits. In the reset phase of each clock cycle, an input current of fixed magnitude and direction is applied to the device to set its magnetization state. In the evaluate phase, a current in the opposite direction whose magnitude is determined by the inputs to the reconfigurable gate is applied, possibly switching the magnetization state to the opposite value. An HHE-based gate using this clocking methodology has been demonstrated in [4].

Fig. 4 illustrates the interface logic for a 2-input HHE reconfigurable gate using this methodology. To simplify the interface logic, we assume the use of an HHE gate with two input wires that are vertically stacked in different metal layers. Our conversations with researchers working at the device level indicate that such an extension to the base HHE device is possible, and we are initiating efforts to fabricate a test device with this design.

As shown in the figure, one of the input wires is only used in the reset phase. During this phase, the RESET signal is high, causing a current to flow upward through transistor MR and setting the magnetization state of the HHE device in the direction that corresponds to a logical 0. The PULSE input is held low during this period to ensure that no current flows through the other input wire.

During gate evaluation, the PULSE input is pulled high while the RESET input remains low. Depending on the inputs to the gate, any or all of transistors Ma, Mb, and MG0 may be turned on, allowing a current I_{in1} to flow downward through the input wire. These three transistors are sized such that at least two of them must be on in order for I_{in1} to be large enough to reverse the magnetization state of the HHE device, creating a majority function. Depending on the value of the G0 configuration input, this causes the device to compute either the AND or OR of its other inputs. Similarly, the value of the G1 input shown in Fig. 3 determines whether or not the inputs to the gate are inverted before they connect to the HHE device, allowing the gate to compute the NAND and NOR of its inputs as well.

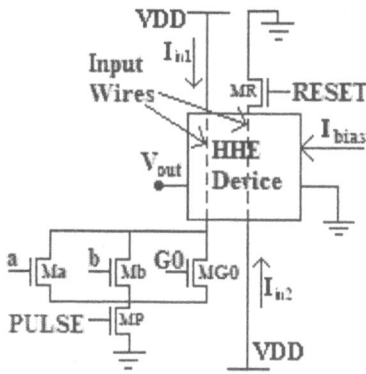

Fig. 4. Interface logic for HHE reconfigurable gate with reset. Current in the left input wire may only flow downwards, while current in the right input wire may only flow upwards.

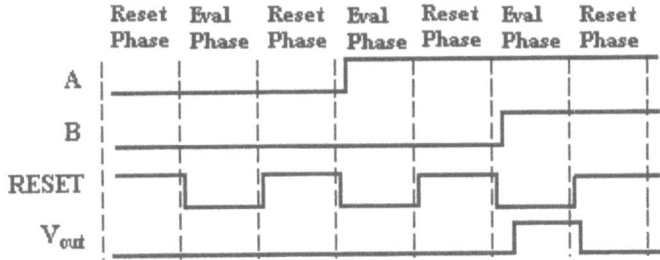

Fig. 5. Waveform of HHE reconfigurable gate operation with reset

Fig. 5 illustrates the operation of the reconfigurable gate when configured to compute the AND of its inputs (G0 = 0, G1 = 0). During each reset phase, the magnetization state of the HHE device is configured in the direction that represents a logical 0 by the reset path of the gate. During each evaluation phase, the magnetization state of the gate is conditionally set in the direction that represents a logical 1 based on the value of inputs A and B.

The circuit shown in Fig. 4 can be extended to compute functions of additional inputs by adding additional transistors to the pull-down chain shown in the figure and appropriately sizing the transistors in the pull-down chain. In Section 5, we present simulation results for a four-input reconfigurable gate of the type described in the next subsection. In addition, structures that connect additional configuration inputs in parallel with the transistor MG0 are also possible, allowing the gate to compute threshold or symmetric functions [5,6].

3.2 HHE Reconfigurable Gate with Output Feedback

One drawback to the circuit shown in Fig. 4 is that it consumes power each time the RESET signal is asserted, regardless of the value of its inputs and configuration. If the output of the gate remains constant from one cycle to the next, this can result in significant wasted power. To address this limitation, we have designed the static reconfigurable gate shown in Fig. 6, which uses output feedback to eliminate the need for a reset phase. Rather than resetting the magnetization state of the HHE device to a logical 0 on each cycle, this design provides two conditional pull-down chains, one of which allows current to flow in the direction that sets the device to a logical 0, and one of which allows current to flow in the direction that corresponds to a logical 1. The PULSE input to each pull-down chain prevents static power consumption by only allowing current flow during the time required to evaluate the output of the device. (approximately 2ns for current HHE devices) Feedback from the output of the device to the pull-down chains disables the chain that corresponds to the current output value, preventing power from being consumed on input changes that do not change the output of the device.

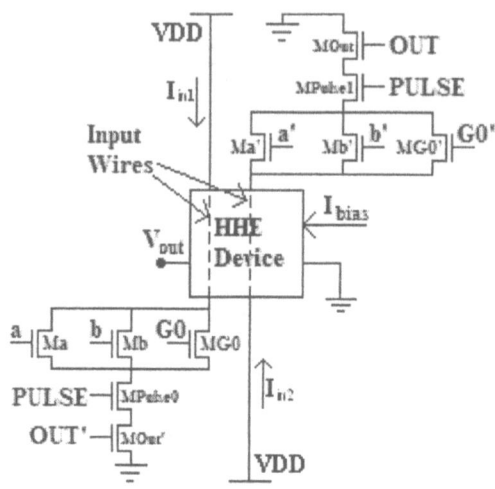

Fig. 6. Interface logic for HHE reconfigurable gate with output feedback

To demonstrate the operation of the gate with output feedback, consider the case where the gate is configured to compute the AND of its inputs (G0 = 0, G1 = 0). In this case, the a and b inputs to the circuit receive the uninverted values of the A and B inputs to the gate, while the a' and b' inputs receive the complement of A and B. Assume that the output of the gate starts at a logical 0. In this case, the left-hand pull-down chain in the figure is enabled, while the right-hand chain is disabled. Since G0 is set to logical 0, both the A and B inputs to the gate must be high for enough current to flow through the left-hand pull-down chain to flip the magnetization state of the HHE device and change the output of the gate to "1." If the output of the gate starts at a logical 1, however, the right-hand pull-down chain is enabled while the left-hand one is disabled. Because G0 = 0, G0' = 1, and only one of the A or B inputs to the circuit must be 0 for enough current to flow to set the output of the gate to logical 0. Thus, the circuit computes the logical AND of its inputs.

This gate design requires somewhat more configuration logic than the reset-evaluate design, because it is necessary to provide both the true and inverted values of each input signal and the G0 configuration bit. The set of logic functions that can be computed by this style of gate can be expanded by including configuration circuitry that allows each input to be inverted or disabled individually, reducing the number of gates required to implement a circuit at the cost of increased gate complexity.

HHE reconfigurable gates with input inversion and input enabling may be incorporated into non-volatile reconfigurable logic devices such as PLAs and CPLDs. Currently, EEPROM transistors are the underlying technology of these systems. EEPROMs are useful for realizing product terms with wide-AND operations such as those commonly used in state machines and control logic. HHE-based logic for these devices will be more efficient than EEPROM-based logic because of its greater flexibility. For example, an HHE-based device would allow either two-level AND-OR or two-level OR-AND implementation of a given logical function, depending on which

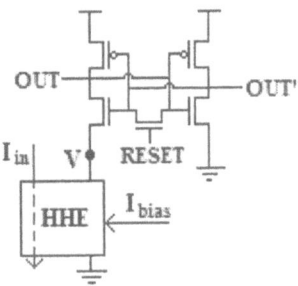

Fig. 7. HHE device incorporated into a logic block LUT cell

resulted in the fewest number of HHE gates used (i.e. fewest number of product/sum terms.) For complete non-volatile operation, the configuration bits for these HHE gates may be stored in non-volatile HHE LUT cells as described in the next section.

4 Non-volatile LUTs Using HHE Devices

HHE devices may also be used to add non-volatile operation to FPGAs based on more-conventional SRAM lookup tables, as shown in Fig. 7. In this circuit, an HHE device fabricated without an offset voltage is used to store the state of each SRAM cell in a lookup table by applying an appropriate current I_{in} to the HHE device during configuration. The HHE device will retain its state without requiring a power supply, acting as non-volatile storage for the configuration of the device.

To copy the state stored in the HHE device into the lookup table, the RESET signal is asserted to equalize the values of OUT and OUT'. When RESET goes low, a bias current is applied to the HHE device, causing it to generate either a positive or a negative voltage on terminal V depending on the magnetization state of its ferromagnetic element (since the HHE device has been fabricated without an offset voltage.) The cross-coupled inverters in the SRAM cell then act as a differential amplifier, bringing the output voltages of the SRAM cell to full CMOS levels. By applying RESET and the bias current to each SRAM cell in an FPGA simultaneously, the entire device can be reconfigured extremely quickly at power-on.

Although only a single HHE device is depicted in Fig. 7, one more may be added to the right leg of the LUT cell. In this manner, one of two configurations may be dynamically loaded into the LUT cell by applying the appropriate read bias current I_{bias} through the desired HHE device.

5 Simulation Results

Using the HSPICE™ circuit simulator, we created a circuit model of the HHE device based on the techniques presented in [7], and have simulated HHE designs for recon-

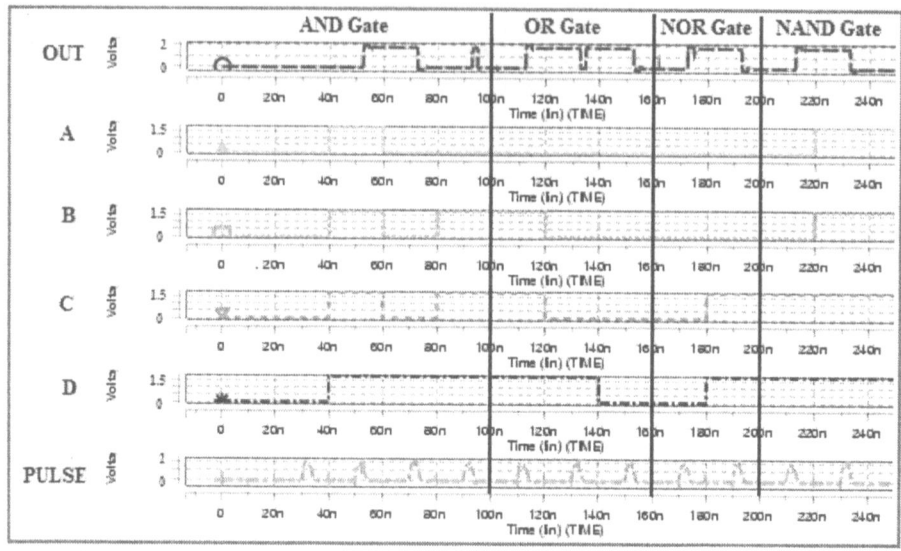

Fig.8. Simulations for HHE reconfigurable gate

figurable gate structures and non-volatile LUTs. Simulations have also been performed comparing power consumption between a reconfigurable gate with output feedback against one that uses reset pulses. The designs incorporate .18u CMOS transistors in a 1.8V technology.

In Fig. 8, we illustrate the operation of a 4-input HHE reconfigurable gate with output feedback. The gate is configured to compute different functions of its inputs over the course of the simulation, and inputs are allowed to change on multiples of 20ns. 10ns after each input change, the PULSE input to the gate is asserted to cause the gate to compute its output. The simulated device requires 2ns to compute its outputs, matching current experiments with prototype HHE devices.

One may notice that the HHE gate output attempts to switch at 92ns and 132ns. However, the output does not fully switch because not all of the inputs are logic 1 or logic 0 respectively. This indicates that input currents did not exceed the switching threshold of the ferromagnetic element, so the output reverts back to its previous state when PULSE goes low.

Simulations were also performed to compare the power dissipation of a reconfigurable gate using output feedback against one that uses reset pulses. In Fig. 9, we show the input current pulses associated with the input vectors from Fig. 8 for both types of gate, illustrating that the reset-based design requires more and larger current pulses than the static gate with output feedback. For these input vectors, simulations show an average power consumption of 4.09mW for the reset pulse design. Average power consumption for the design with output feedback is 1.69mW, an improvement of 2.42x, although power consumption for both gates will scale with clock frequency.

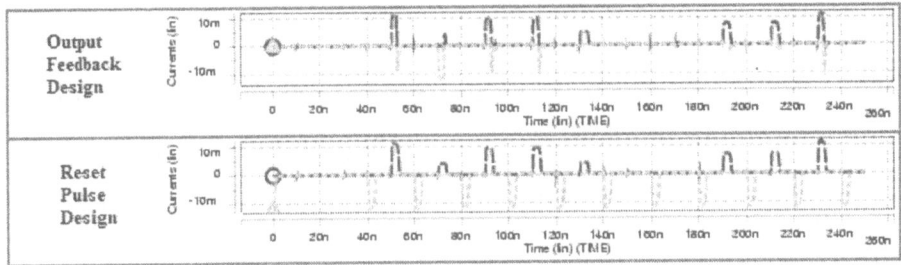

Fig. 9. Input current pulses for output feedback and reset pulse designs for HHE reconfigurable gate. Top curves represent current through left HHE input wire. Bottom curves represent current through right HHE input wire

In Fig. 10 we illustrate the operation of a non-volatile LUT using HHE devices. During the first 40ns, the output state of the HHE device is initialized to logic 1 and the RESET signal is high. At 40ns, the RESET signal is removed, and the LUT output becomes the same as that of the HHE device. At 60ns, the power is turned off (VDD=0), and the LUT output decreases exponentially due to discharge effects. At 140ns, power is restored, and the RESET signal is asserted. At 150ns, the RESET signal is disabled, and the LUT output is restored its pre-shutdown value.

6 Related Work

A number of other technologies exist that provide non-volatile storage in reconfigurable devices. Anti-fuses have the benefit of small area, but they are one-time programmable (OTP) and are mainly used for programming interconnections and not logic. EPROMs/EEPROMs are reprogrammable, but consume constant static power since they realize functions using wired-AND logic. Giant-magnetoresistive (GMR) devices are another type of magnetoelectronic device that can easily be integrated into LUT cells [8]. GMR-based designs have the disadvantage that two devices are required to hold the state of each LUT, as opposed to one HHE device, potentially making them less attractive, although this will depend on how well each type of device scales with fabrication technology.

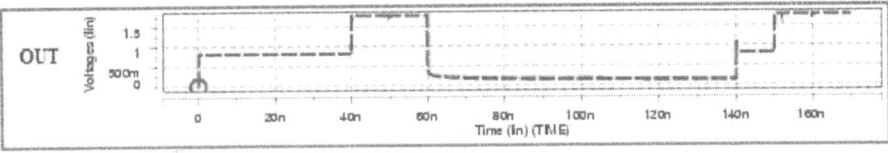

Fig. 10. Simulations for non-volatile LUT using HHE devices

7 Conclusions and Future Work

Hybrid Hall effect devices are a new class of magnetoelectronic circuit element that can be used to implement non-volatile reconfigurable logic and storage. In this paper, we have presented circuit-level designs for reconfigurable gates and non-volatile lookup table cells based on these devices, demonstrating the potential of these devices. We are currently working with researchers at the Naval Research Lab to fabricate prototypes of these circuits.

Future studies of HHE-based reconfigurable logic will focus on the system issues involved in building large-scale reconfigurable logic systems based on magnetoelectronic devices. In particular, the small size and fine-grained reconfigurablility of these devices makes them very attractive and is leading us towards the design of systems based on simple logic blocks and regular interconnect patterns, trading reduced logic block utilization for reductions in interconnect area and complexity.

Acknowledgements

The authors would like to thank Mark Johnson of the NRL for information regarding the properties of HHE devices and for the experimental data used in our models. This material is based on work supported by the ONR under award No. N00014-02-1-1038. Any opinions, findings, and conclusions or recommendations expressed in this publication are those of the authors and do not necessarily reflect the views of the ONR.

References

1. Prinz, G.A.: Magnetoelectronics. Science 282 (1998) 1660-1663
2. Johnson, M., Bennett, B.R., Yang, M.J., Miller, M.M., Shanabrook B.V.: Hybrid Hall Effect Device. App. Phys. Lett. 71 (1997) 974-976
3. Streetman, B.G., Banerjee S.: Solid State Electronic Devices. 5th edn. Prentice Hall, Upper Saddle River, New Jersey (2000)
4. Johnson, M., Bennett, B.R, Hammar, P.R., Miller, M.M.: Magnetoelectronic Latching Boolean Gate. Solid-State Electronics 44 (2000) 1099-1104
5. Aoyama, K., Sawada, H., Nagoya, A., Nakajima, K.: A Threshold Logic-Based Reconfigurable Logic Element with a New Programming Technology. FPL (2000) 665-674
6. Muroga, S.: Threshold Logic and Its Applications. John Wiley & Sons, Inc. (1971)
7. Das, B., Black, Jr., W.C.: A Generalized HSPICE^{TM} Macro-Model for Pinned Spin-Dependent-Tunneling Devices. IEEE Transactions on Magnetics 35 (1999) 2889-2891
8. Das, B., Black, Jr., W.C.: Programmable Logic Using Giant-Magnetoresistance and Spin-Dependent Tunneling Devices. J. Appl. Phys. 87 (2000) 6674-6679

Gigahertz FPGA by SiGe BiCMOS Technology for Low Power, High Speed Computing with 3-D Memory

Chao You*, Jong-Ru Guo*, Russell P. Kraft, Michael Chu, Robert Heikaus,
Okan Erdogan, Peter Curran, Bryan Goda, Kuan Zhou, and John F. McDonald

youc@rpi.edu, guoj@rpi.edu

Abstract. This paper presents an improved Xilinx XC6200 FPGA using IBM SiGe BiCMOS technology. The basic cell performance is greatly enhanced by eliminating redundant signal multiplexing procedures. The simulated combinational logic result has a 30% shorter gate delay than the previous design. By adjusting and properly shutting down the CML current, this design can be used in lower-power consumption circuits. The total saved power is 50% of the first SiGe FPGA developed in the same group. Lastly, the FPGA with a 3-D stacked memory concept is described to further reduce the influence of parasitics generated by the memory banks. The circuit area is also reduced to make dense integrated circuits possible.

1 Introduction

Field Programmable Gate Arrays (FPGAs) have exclusive applications in many areas such as high-speed networking and digital signal processing. The maximum speed of most current FPGAs is around 300 MHz. The first gigahertz FPGA is a Current Mode Logic (CML) version of the Xilinx XC6200 implemented using IBM's SiGe BiCMOS technology [1], [2]. The XC6200 architecture has been selected because of its open source bit stream an available programming tools. Driven by the developing technology, the primary goal of this work is to design high-speed FPGAs while simultaneously focusing on reduced power applications.

The power consumption of this first gigahertz FPGA has a total cell power calculated as following:

$$P_{total} = N_{cell} \times N_{CML-tree} \times V_{supply} \times I_{tree}$$

Where P_{total} is the total cell power, N_{cell} is the total number of cells, $N_{CML-tree}$ is the total number of CML trees inside one basic cell and I_{tree} is the amount of current in each CML tree.

To obtain more power saving, all efforts are focusing on those four factors by reducing the voltage supply, amount of current in CML tree, number of trees in each cell and total number of cells used in an application.

---

* Both authors have contributed equally to this work.

P.Y.K. Cheung et al. (Eds.): FPL 2003, LNCS 2778, pp. 11–20, 2003.
© Springer-Verlag Berlin Heidelberg 2003

The layout of the paper is this. Section 2 introduces the SiGe BiCMOS technology. Section 3 elaborates on an improved multiplexer structure to reduce the power supply from 3.4 V to 2 V. Section 4 presents an improved basic cell structure that eliminates redundant multiplexing procedures and thus increases performance. Section 5 illustrates the dynamic routing circuits that shut down unused circuits. Section 6 presents the concept of the 3-D FPGA aiming at reducing the area and the influence of parasitic effects. All the results of different FPGAs utilizing different currents are summarized and compared in Section 7. Finally, a brief future plan is described in Section 8.

2   SiGe BiCMOS Technology from IBM

The FPGA design is implemented by the IBM SiGe BiCMOS 7HP technology. The technology has all features of Si-base transistors, such as polysilicon base contact, polysilicon emitter contact, self-alignment and deep-trench isolation. With the use of a linearly graded Ge profile in the base region, three device factors, current gain, early voltage and base transit time, are improved [3]. Figure 1 shows an I_c and f_T curve in the IBM 7HP technology. The peak f_T point is at 1 mA. Later part of this paper shows a trade off between the power and performance based on the data from this curve.

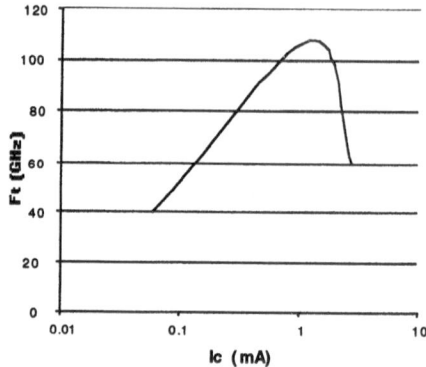

Fig. 1. I_c versus f_T in the IBM 7HP SiGe BiCMOS

3   Improved Multiplexer Structure

The XC6200 is a multiplexer-based FPGA instead of a LUT-based FPGA. As the building block of a basic cell, the single multiplexer design determines the supply voltage, gate delay and total power consumption of the basic cell.

The previous design uses a multiplexer design as shown in Figure 2 (a 4:1 multiplexer is shown as an example). The selection bits come in complementary pairs. Each time, only one branch of the tree is turned on. The corresponding transistor pair will be selected to pass the input signal through. For the NFET gate to work properly, the voltage supply must be large enough. The tallest CML tree in the design, which is an 8:1 multiplexer in the basic cell, determines the chip voltage.

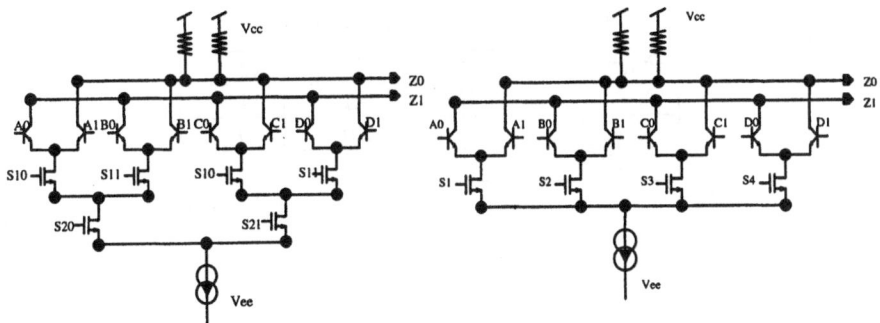

Fig. 2. Previous multiplexer   **Fig. 3.** New multiplexer

An improved multiplexer design is shown in Figure 3. Instead of using complementary selection bits for decoding, a separate decoder is used in the new multiplexer. During operation, only one NFET obtains a SET signal from the decoder. The corresponding transistor pair will be selected to pass the input signal through. There are two advantages for using this new multiplexer structure.

1. The new multiplexer has a single-level selection tree throughout the basic cell. The power supply is reduced to a minimum. For example, the previous uses a 3.4 V power supply, while the tested new chip uses a power supply of 2.0 V.
2. The previous multiplexer can't be turned off since at least one branch is on no matter what the selection bits are. The new multiplexer can be completely turned off by turning off the decoder. When one multiplexer is not used in an application, it may be turned off to save power. The later part of this paper has a detail description about how dynamic routing works.

By implement the basic cell with the new multiplexer structure, the power consumption can be reduced by 40% without any modification to the other parts of design. The extra decoders added to the circuits can be implemented with CMOS. The programming bits only add negligible leakage power consumption.

4   Improved Basic Cell

A basic cell is the prime element of the XC6200. The original XC6200 basic cell and the function unit are shown in Figure 4. Each cell obtains two signals in each direction. One signal is from the cell's neighbor in corresponding N, S, E and W directions. The other is the FastLANE® signal, which is a shared signal by four basic cells in the same row or column. The FastLANE provides more routing capability to an XC6200 cell.   Each basic cell performs two-variable logic functions. The two-variable inputs are selected from the above eight signals. An example of the logic function table is shown in Table 1.

Logic function result "C" (for combination logic) is sent out to a CS multiplexer (S for sequential logic). The selected signal "F" from the Function Unit is selected again at the output multiplexers before it can reach the next neighbor cell. To provide more routing capability, each basic cell also routes one signal in one direction to another. The function is called a redirection function.

Table 1. Example of Logic Function Table

Function	X1	X2	X3
INV	XX	A1	A1
A1 AND B1	A1	B0	A0
A1 XOR B1	A1	B1	B0

XX: Don't care. "1" is for signal. "0" is for compliment signal

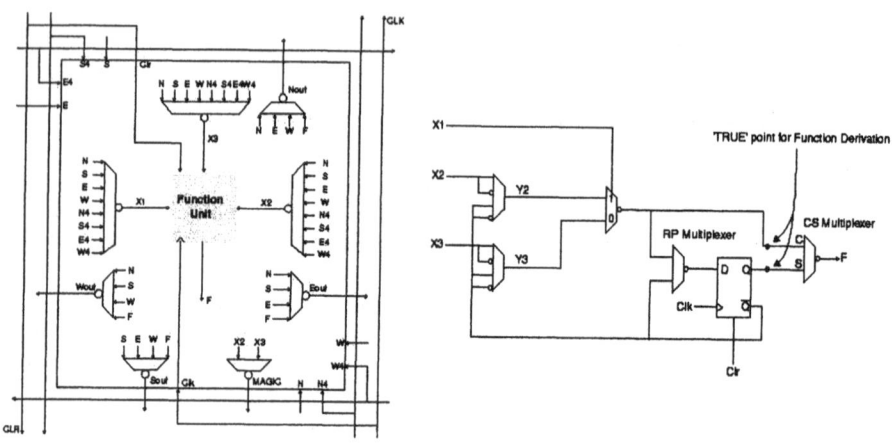

Fig. 4. XC6200 and Function Unit

When its neighbor cell finally receives the logic result, the neighbor cell will select this result again among other signals. The neighbor cell will determine which signal to use: combinational logic result, sequential logic result, redirected signal or FastLANE signal. Because of this, all multiplexing procedures at the output stage of a basic cell can be considered as redundant, which must be removed to increase performance.

Figure 5 shows an improved basic cell structure (BCII) and a function unit without the superfluous multiplexing circuits. The combinational logic result and sequential logic result are delivered to neighbor cells directly. The diamond arrow stands for a combinational logic result and the round arrow stands for a sequential logic result in Figure 5. To preserve the routing capability the XC6200 provided, the redirection function is sustained. Instead of sending one output in each direction, BCII sends out three outputs in each direction. They are the combinational logic result, sequential logic result and redirection function result.

With more outputs in each direction, each BCII now receives three inputs from its neighbor cell in one direction. The input multiplexers and redirection multiplexers must be modified to adapt to this change. As shown in Figure 5, the input multiplexers now receive three inputs from each neighbor cell and one signal from FastLANE. The X2 and X3 multiplexers in Figure 4 also receive a feed back signal from the MS-Latch. The implementation of the X2 and X3 multiplexer in BCII are shown in Figure 6. The 16:1 multiplexer and 17:1 multiplexers are implemented by a two-level multiplexing process as shown in the figure. The redirection multiplexers now receive nine signals in three directions. 9:1 multiplexers are used for the

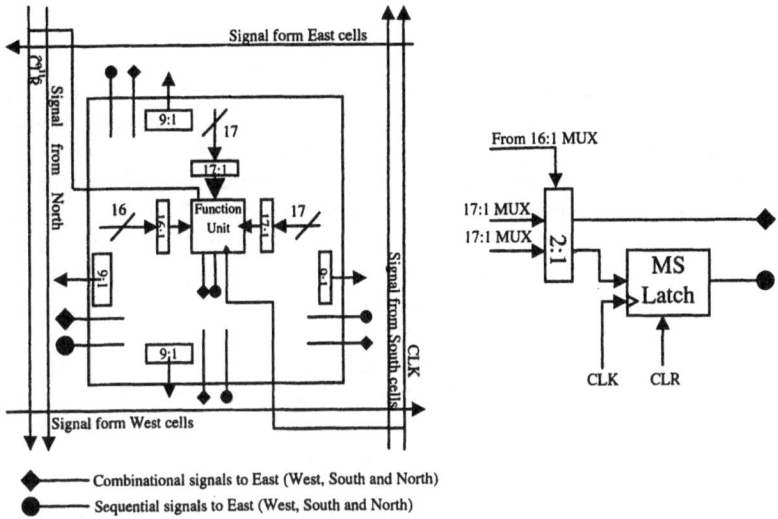

Fig. 5. BCII (Left) and Function Unit (Right)

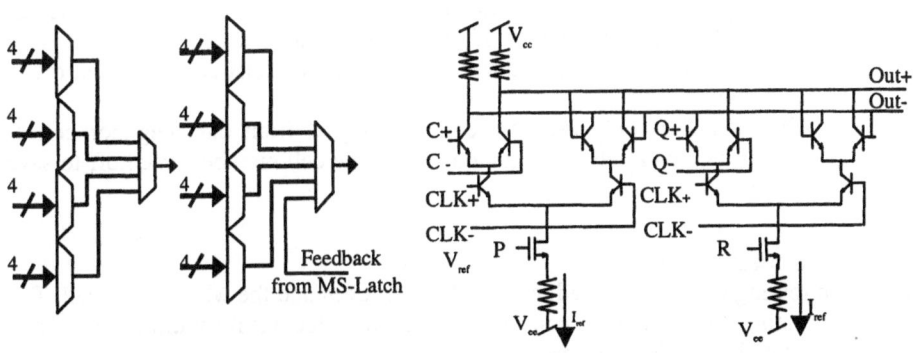

Fig. 6. 16:1 input MUX (Left) and 17:1 input MUX (Right)

Fig. 7. Improved First Stage MS-Latch

redirection function as shown in Figure 5. 9:1 multiplexers are implemented by the new multiplexer structure described in Section 2.

Besides the modification made on the input multiplexers and redirection multiplexers, the Master-Slave Latch (MS-Latch) can be modified to save power as well. The first stage MS-latch can be combined with the RP multiplexer in Figure 4. The revised circuit is shown in Figure 7. The selection bits for the RP multiplexer are used as enable bits for the two CML trees. In operation, only one CML tree is turned on to pass the signal to the second stage. The benefit of using this revised first stage MS-latch is that it can be turned off by setting both R and P to zero.

The major advantage of BCII is the shorter gate delay on the critical signal path. With the IBM SiGe technology, the current in CML trees can be reduced while still maintaining a shorter or comparable gate delay with the original design.

5   Dynamic Routing

Other factors that can be altered in the power equation are the number of cells and the number of CML trees in each basic cell. One primary notion is to turn off unused cells or unused CML trees when an application is loaded into the FPGA.

In a chip scale, for example, when a loaded application uses only 12 cells in a 4 x 4 gate array, turning off the unused 4 cells will save 25% power. The cells that need to be turned off must be determined by the application. Thus, the turning-off scheme is dynamically controlled by the application instead of hard-wired design. One way to turn off an entire cell is by introducing another enable bit for the cell. The enable bit will enable/disable all CML trees in a single cell.

In a cell scale, there are more complicated schemes to turn off unused CML trees. When one cell is configured to perform a certain function, the incoming signals pass through specific path with several multiplexers involved. All other circuits, which are not involved in the path, have to be turned off to save power. The circuits that have to be turned off are listed as follows.

1.   When a cell only performs the combination logic, the sequential logic circuit and all redirection multiplexers may be turned off to save power.
2.   Each cell only accepts two signals to generate a logic function. Then at least two output-drivers and the redirection multiplexers in its neighbor cells can be turned off to save power. If the input signal of a cell is selected from FastLANE, all four output-drivers in all neighbor cells can be turned off to save power.
3.   When a cell is only used to redirect a signal from one neighbor cell to another neighbor cell, the function unit of the cell may be turned off to save power.

Dynamic Routing Circuits
Dynamic routing circuits are used to turn off CML trees and the whole basic cell. The dynamic routing circuits are the control circuit of the select bit for multiplexers. They are implemented by CMOS to save layout area.

Figure 8 shows the 17:1 multiplexer with dynamic routing circuits. As shown in the figure, the decoder has an enable bit. A reset on the decoder enable bit will turn off all multiplexers in the figure. The Selection-Indicator and the Complement-Selector works as the decoder for the second level multiplexer. Either a signal or its complement signal will be selected as the output. The Sequential Logic Selection Control circuit indicates the feedback from sequential logic will be selected at the second level multiplexer. The Safety Circuit is used to prevent the case that both the input and feedback signal are enabled.

Figure 9 shows a redirection multiplexer. The redirection multiplexer is only needed for redirecting signals. One 4-bit decoder's outputs are used as selection bits. The first design has ten outputs from the decoder. The first nine outputs work as selection bits for the multiplexer. The last output is asserted when the output multiplexer needs to be turned off. The remaining six outputs are unused by any other circuits. Other RAM bits are needed as the Master Key (the master enable for the whole cell), FU Enable and MS latch enable. Further research shows that these RAM bits can be connected to the otherwise unused outputs of the decoder.

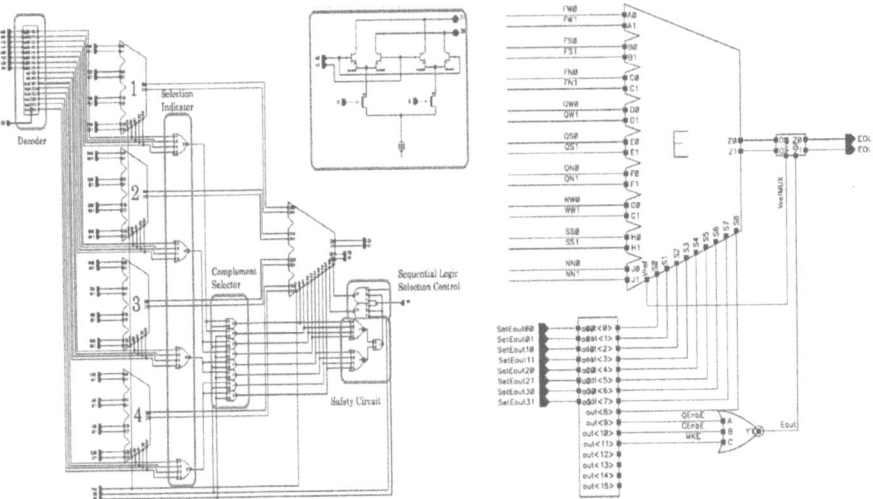

Fig. 8. Dynamic Routing Circuit   **Fig. 9.** Output MUX with control circuit

A revised redirecting multiplexer with its control circuits is shown in Figure 9. Output nine is used for the sequential enable. Output ten is used for the combinational enable. Output eleven is used for the master key.

In the same figure, The QEnbE is the enable bit for the driver of sequential logic facing the east side. The CEnbE is the enable bit for the driver of combinational logic facing the east side. The MKE is the master key from the east side. If any of the first nine signals are selected, the output MUX driver is on. Otherwise, the redirection multiplexer and the driver are turned off. The following table shows the actual power reduction in the new design for three cases.

Table 2. Dynamic Routing Power Usage

Design		Tree #	Usage
BCII Maximum Usage		21	100%
Case I (Comb./Sequential. Logic)		10/12	47.6%/57.1%
Case II	Sequential, One Redir.	15	71.4%
	Sequential, Two Redir.	18	85.7%
	Sequential, Three Redir	21	100%
Case III		3 tree/dir	14.2%/dir

Case I: Only combinational logic or sequential logic is used.
Case II: Sequential logic and redirection function are used.
Case III: Only redirection function is used

6   3-D Stack Memory Concept for FPGA

In the FPGA, memory is used to program logic function. However, the considerable amount of memory and wire has occupied a significant area of the chip and reduced the operating frequency due to the longer interconnect. 3-D integration is one of the solutions to reduce the area and alleviate the parasitic effects.

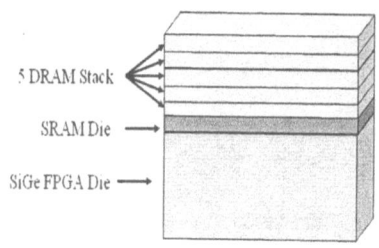

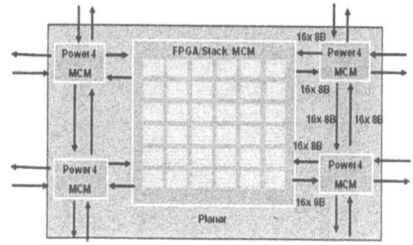

Fig. 10. The 3-D integration of the FPGA/ stack memory.

Fig. 11. The block diagram of the 3-D FPGA chip

In the 3-D design concept, similar circuits are grouped together, for example memory, and manufactured on the same wafer. By using "chemical adhesives", these wafers can be glued and stacked. To connect the different wafer layer circuits, 3-D vias are used. Figure 10 shows the cross section of a 3-D structure.

On the top is the DRAM stacks which provide the FPGA different bit streams to the configuration memory below it (the SRAM die). After selection, the desired bit stream is stored in the SRAM die then passed to the SiGe FPGA die to perform the logic function. It is easy to observe the size of the FPGA is greatly reduced and the parasitic effects can be eased due to the shorter length of the interconnect wires.

Figure 11 shows the block diagram of the 3-D FPGA chip. The 3-D FPGA/Stack Memory block is located in the center and the cells around it are the processors. The FPGA/ stack memory is used to route signals between processors that can broaden the multi-processor applications by programming different personalities in the DRAM. The DSP chip, A/D and D/A converters or communication circuits can replace the processor blocks. Thus, the FPGA chip can serve as the reconfigurable interface between blocks for different applications.

7   Simulation Results

The first gigahertz FPGA chip opens a gate to fast reconfigurable computing. Its continuing work involves performance improvement, lower power consumption design and curbing the basic cell layout area to a scaleable size. High performance is still the primary goal while keeping the power consumption as low as possible. With the IBM SiGe 7HP technology, the performance and power consumption can be balanced by varying the amount of current in CML trees. By moving the current from the peak f_T point, the power consumption is reduced with the lost of performance. Table 3 exhibits the power and delay relationship for a simulated AND gate. Figure12 shows the power and delay trade-off of BCII in the IBM SiGe 7HP process. The best trade-off point is at 0.4 mA CML tree. A current larger than 0.8 mA will have a short gate delay at the sacrifice of power savings.

Table 3. Power and Delay Chart for Designs

Design	Power (mW)	Delay (ps)
BC 0.6 mA	53	80
BCII 0.8 mA	16	46
BCII 0.6 mA	12	55
BCII 0.4 mA	8	70
BCII 0.2 mA	4	120

An AND gate is simulated for design comparison
BC has 28 CML trees. BCII has 21 CML trees (10 trees for Combinational Logic)

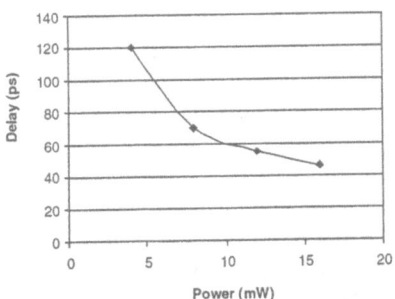

Fig. 12. Power Delay Trade-off in the IBM 7HP Process

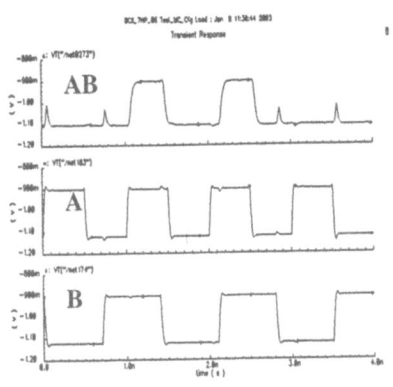

Fig. 13. Simulation result of an AND gate   **Fig. 14.** BCII IBM 7HP layout

Figure 13 shows a simulated AND gate result. The current in the CML tree is 0.6 mA. The gate delay is 55 ps at the temperture of 25°C with a voltage swing of 250 mV.

8   Conclusion and Future Work

With the improved basic cell structure, the performance of a basic cell is improved by 30%. Adjusting the current in CML trees can result in different power savings. The dynamic routing method reduces the power consumption further. One design with the BCII cell has been shipped out for fabrication. The layout of the chip is shown in Figure 14. Future work involves chip test and measurement. The IBM SiGe 8HP technology will be released around August 2003. Further research result will be forth coming after implementing the BCII structure with the faster IBM SiGe 8HP technology.

References

[1] John F. McDonald and Bryan S. Goda, "Reconfigurable FPGA's in the 1-20GHz Bandwidth with HBT BiCMOS", Proceedings of the first NASA/ DoD Workshop on Evolvable Hardware, pp. 188-192.

[2] Bryan S. Goda, John F. McDonald, Stephen R. Carlough, Thomas W. Krawczyk Jr. and Russell P. Kraft, "SiGe HBT BiCMOS FPGAs for fast reconfigurable computing," IEE Proc.-Compu. Digi. Tech, vol.147, no. 3 pp. 189-194.

[3] "IBM SiGe Designer's manual", (IBM Inc. Burlington Vermont. 2001).

[4] D. Harme, E. Crabbe, J. Cressler, J. Comfort, J. Sun, S. Stiffler, E. Kobeda, M. Burghartz, J. Gilbert, A. Malinowski, S. Dally, M. Rathanphanyarat, W. Saccamango, J. Cotte, C. Chu, and J. Stork, "A High Performance Epitaxial SiGe-Base ECL BiCMOS Technology," IEEE IEDMTech Digest, 1992, pp. 2.1.1-2.1.4.

Implementing an OFDM Receiver
on the RaPiD Reconfigurable Architecture*

Carl Ebeling[1], Chris Fisher[1], Guanbin Xing[2], Manyuan Shen[2], and Hui Liu[2]

[1] Department of Computer Science and Engineering
University of Washington
[2] Department of Electrical Engineering
University of Washington

Abstract. Reconfigurable architectures have been touted as an alternative to ASICs and DSPs for applications that require a combination of high performance and flexibility. However, the use of fine-grained FPGA architectures in embedded platforms is hampered by their very large overhead. This overhead can be reduced substantially by taking advantage of an application domain to specialize the reconfigurable architecture using coarse-grained components and interconnects. This paper describes the design and implementation of an OFDM Receiver using the RaPiD architecture and RaPiD-C programming language. We show a factor of about 6x increase in cost-performance over a DSP implementation and 15x over an FPGA implementation.

1 Introduction

Current SOC platforms provide a mix of programmable processors and ASIC components to achieve a balance between the flexibility and ease-of-use of a processor and the cost/performance advantage of ASICs. Although ASIC designs will continue to provide the best implementation for point problems, typically several orders of magnitude better than processors, platforms require flexibility that ASICs cannot provide. Relying on ASIC components reduces a platform's coverage to only those computations the ASICs can handle. It also reduces the platform's longevity by restricting its ability to adapt to changes caused by new standards or improvements to algorithms.

There has been a recent move to include configurable components on platforms in the form of FPGA blocks. So far this has met with only modest success. The main obstacles have been the large overhead of traditional fine-grained FPGAs and a lack of high-productivity programming tools for would-be users of configurable architectures. It is difficult to find a use for configurable logic when the cost-performance penalty approaches two orders of magnitude over ASIC components.

Our research has focused on showing that coarse-grained configurable architectures specialized to an application domain can substantially reduce the overhead of reconfigurable architectures to the point where they can be a viable alternative to DSP and ASIC components. We have defined a coarse-grained configurable architecture called RaPiD, designed a programming language called RaPiD-C, and implemented a

---

* This research was funded by the National Science Foundation.

P.Y.K. Cheung et al. (Eds.): FPL 2003, LNCS 2778, pp. 21–30, 2003.
© Springer-Verlag Berlin Heidelberg 2003

compiler that maps RaPiD-C programs to the RaPiD architecture. This paper begins with a brief description of the RaPiD architecture and programming tools. We then describe the OFDM application and describe in detail the implementation of the compute-intensive parts of this application in RaPiD. Using an emulator of the RaPiD architecture, we have demonstrated the real-time execution of this application. We then compare the performance and cost of this implementation to that of DSP, FPGA and ASIC implementations. Finally we discuss the system issues of constructing a complete implementation of OFDM in a platform that contains RaPiD components.

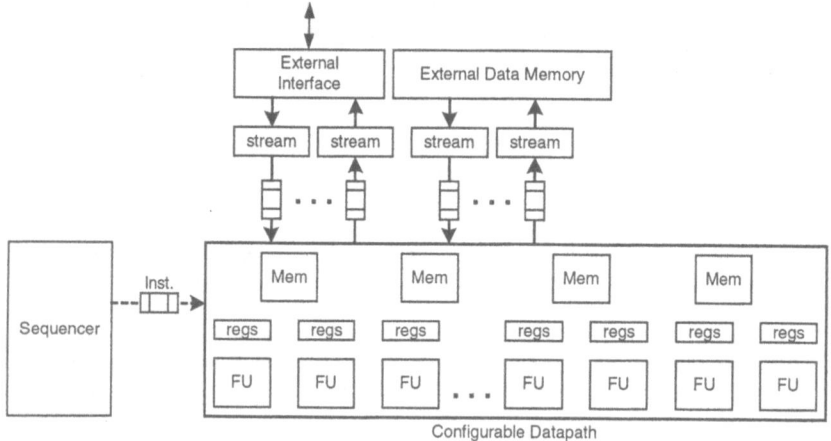

Fig. 1. Architecture of a RaPiD component.

1.1 The RaPiD Architecture

Fig. 1 gives an overview of the RaPiD architecture. RaPiD has been described in detail in previous papers [1,2,3,4] and will only be summarized here. The configurable datapath, which is the main part of a RaPiD component, contains a set of specialized functional units such as ALUs and multipliers that are appropriate to the application domain. The datapath also contains a large number of registers distributed throughout the datapath, which provide the data bandwidth required to keep all the functional units busy on every cycle. There are no register files: Instead, a number of small, embedded memories provide storage for data that is used repeatedly.

The components of the datapath are connected together via a configurable interconnection network comprising segmented busses and multiplexers. The sequencer runs a program that executes a computation on the configured datapath. Part of the interconnection network is statically configured before execution, while the rest is controlled dynamically by the sequencer. The compiler determines which part of the interconnect is statically configured and which is controlled by the program.

The datapath accesses data in external memory and other nodes using a streaming data model. Data streams are created by independent programs that operate on behalf of the datapath to read or write data in memory or communication buffers. The stream programs are decoupled from the datapath using FIFOs and executed ahead/behind the datapath to prefetch/poststore data. Communication with other components, which can be processors, smart memories or ASIC components, is done using data streams in the style of Unix pipes.

Programs for RaPiD are written in the RaPiD-C programming language, which is an assembly-level language for the parallel RaPiD datapath. The programmer writes programs using datapath instructions, each of which specifies a parallel computation using a data-parallel style of programming. The compiler performs the pipelining and retiming necessary to execute datapath instructions at the rate of one per clock cycle. For more details on the RaPiD-C language and compiler, refer to [5].

In [4] we define a "benchmark cell" which comprises 3 ALUs, 3 embedded memories, one multiplier, and 6 datapath registers, connected using a set of 14 tracks of interconnect buses. The benchmark RaPiD array, comprised of 16 benchmark cells, was shown to perform well across a range of applications.

2   Multiple-Antenna OFDM Application

Orthogonal frequency-division multiplexing (OFDM) is a form of modulation that offers a significant performance improvement over other modulation schemes on broadband frequency selective channels. These inherent advantages make OFDM the default choice of a variety of broadband applications, ranging from digital video broadcasting (DVB-T), to wireless LAN, to fixed broadband access IEEE 802.16a. Currently, OFDM is also regarded as the top candidate for the 4th generation cellular network. OFDM can be combined with an antenna array to perform spatially selective transmission/reception so that information can be delivered to the desired user in an efficient manner. This combination provides an elegant solution to high performance, high data rate multiple-access networks.

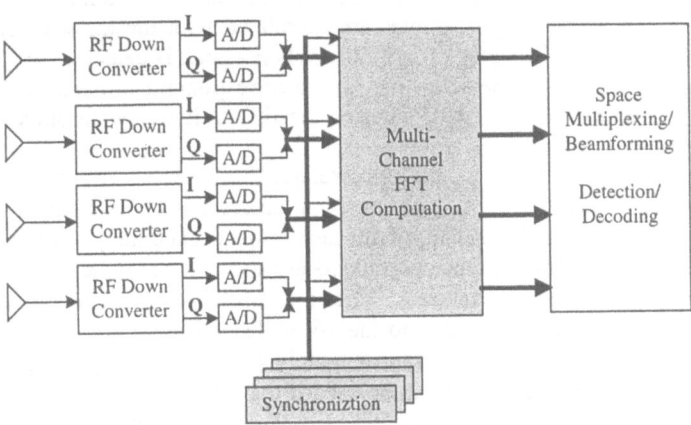

Fig. 2. Antenna Array OFDM Receiver

Fig. 2 shows the block diagram of an antenna array OFDM receiver. The signals from the antennas are first down-converted to base-band and digitized. The OFDM frame timing and carrier frequency are then detected using the synchronization module. Following that, the FFT is computed for each antenna input to convert the incoming signal back to frequency domain, allowing information on each individual

ing signal back to frequency domain, allowing information on each individual sub-carrier to be demodulated.

The shaded portions are the computationally intensive units, namely, synchronization and multi-channel FFTs, which were implemented using the RaPiD architecture. Demodulation and beam-forming typically comprise a small part of the total computation and can be realized using ASIC or programmable DSP components.

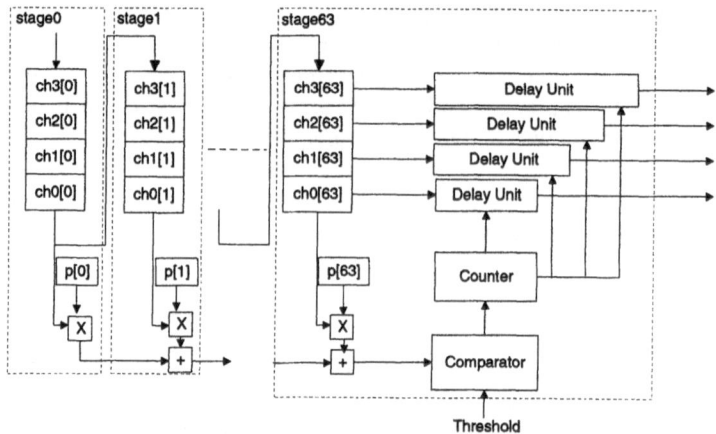

Fig. 3. OFDM Synchronization Module

Synchronization. Synchronization is achieved by correlating the received signal with a known pilot embedded in the transmitted signal. Peaks at the outputs reveal not only the timing of the OFDM frames but also the carrier information. As shown in Fig. 3, the signals from the four antennas are interleaved into a single stream entering the array. Successive windows of 64 samples are correlated with the known pilot signal and the result is compared against a threshold. When a peak is detected, the time offset is used to set the delay of the delay line for the corresponding antenna. This initializes the delay lines so that the signals entering the FFT unit are synchronized to each other and to the start of the frame.

We implemented a maximally parallel correlator that searches for the peak signal over a 64-sample window. This synchronizer can detect a frame-to-frame synchronization offset of +/- 32 samples and apply this new offset immediately to the data symbols in the same frame. Since all the coefficients are +1/-1, the complex multiplies are implemented by a complex add/subtract. Thus a total of 128 16-bit ALUs and about 520 registers are required, in addition to the 16 embedded memories used to implement the delays. This corresponds to approximately 43 RaPiD benchmark cells.

A less expensive implementation can be used if the offset changes by at most +/-4 samples from one frame to the next. In this case, the requirements are reduced to 22 ALUs and 16 embedded memories, corresponding to 8 cells of the benchmark RaPiD array. We will call these two different implementations "parallel" and "tracking".

Multi-Channel FFT. We chose the Radix-4-decimated-in-frquency architecture proposed by Gold & Bially [9], which is particularly efficient when the FFT is computed simultaneously on interleaved data streams. Fig. 4 shows one radix-4 stage, which is composed of three main blocks: a) a delay commutator that delays and skews the input; b) a 4-point butterfly computation unit that requires only adders and I/Q swaps; and c) a twiddle factor multiplier. The initial delay unit is implemented as part of the delay lines in the synchronizer. The commutator is also implemented using datapath memories, while the butterfly computation and twiddle factor multiplication are implemented by using the ALU and the multiplier function units, respectively. All calculations are performed using 16-bit fixed-point numbers.

Each radix-4 stage has slightly different memory requirements for the commutator, but the entire 64-way FFT uses a total of 31 datapath memories, 18 ALUs, 12 multipliers and about 120 registers. This fits comfortably in the 16 cells of the benchmark RaPiD array.

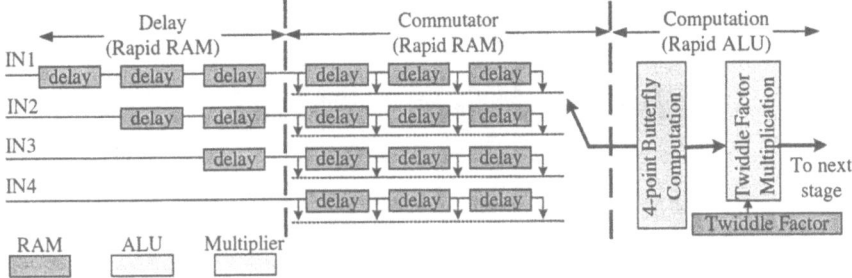

Fig. 4. One stage of the multi-channel FFT implementation

2.1 Implementation Details

This multi-channel OFDM front-end comprising the synchronizer and FFT was programmed using RaPiD-C, compiled using the RaPiD-C compiler [5], and run on the RaPiD architecture emulator [3]. The emulator comprises a set of FPGAs that implement the sequencer, datapath and data streams, and a 64 MByte streaming memory system. This memory system is used to simulate the high-rate data streams that occur in communications systems and can provide 4 I/O streams at an average data rate of close to 1 data value per cycle each. The emulator currently runs at 25 MHz.

To run the OFDM emulation, the external memory is loaded with 200 frames of 4-channel OFDM signals, interleaved, as they would appear in the data stream from the antenna. Each frame consists of one pilot symbol followed by 6 data symbols and one guard interval, as shown below. Both the pilot and data symbol have 64 samples.

Pilot	Data Symbol 1	Data Symbol 2	• • •	Data Symbol 6	Guard Interval

The RaPiD array then executes the four-channel OFDM demodulation program described above at a clock rate of 25 MHz. The input data stream is read at the rate of one complex value per cycle using two parallel 16-bit streams, and the output values are captured to memory at the same rate. The output values can then be analyzed and displayed using standard analysis modules. Fig. 5 shows a screen shot of data generated during the experiment.

The modulation scheme used in the experiment is QPSK. The current 25MHz RaPiD emulator implementation yields an effective data rate of 4ch x (6 / 8) x

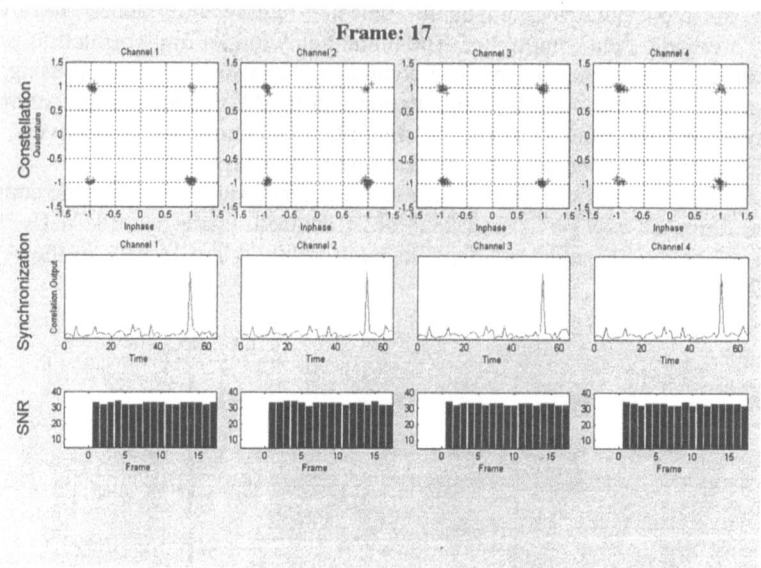

Fig. 5. Screen shot of the experimental results

2bits/symbol x 25Mhz = 150Mbps when 4 channels transmit different bit streams; or a data rate of 1ch x (6/8) x 2bits/symbol x 25Mhz = 37.5Mbps when 4 channels are combined in a low SNR scenario. In either case, the achieved data rate far exceeds that of any existing standard.

2.2 System Issues and Implications

Our OFDM implementation executes the synchronizer and interleaved FFT in a single RaPiD component. This results in a very complex program and a very large array. The correlator and FFT functions have very different requirements: the correlator has no need of multipliers and memories (unless it is time-multiplexed), while the FFT relies heavily on multipliers and memories. Moreover, the correlator and FFT alternate their computation: The correlator operates only on the pilot frames, while the FFT operates only on the data frames. Finally, there are different performance options for the correlator. Thus, the OFDM receiver can be implemented in a number of different ways in a system comprising multiple reconfigurable components. These alternatives are shown in Fig. 6 and described below. The numbers in the lower right-hand corner of each array indicates the size of the array in terms of the number of RaPiD benchmark cells. We have assumed that arrays are sized in multiples of 8 cells. Thus the 43-cell implementation of the correlator fits into a 48-cell array.

One, high-performance array (a, b): The single, high-performance array implemented as described above (a) is very expensive. If it is implemented as a homogenous array with evenly distributed resources, then it has a low utilization because of the varying requirements of the correlator and interleaved FFT. Using the more constrained tracking synchronizer reduces the cost of the array dramatically (b).

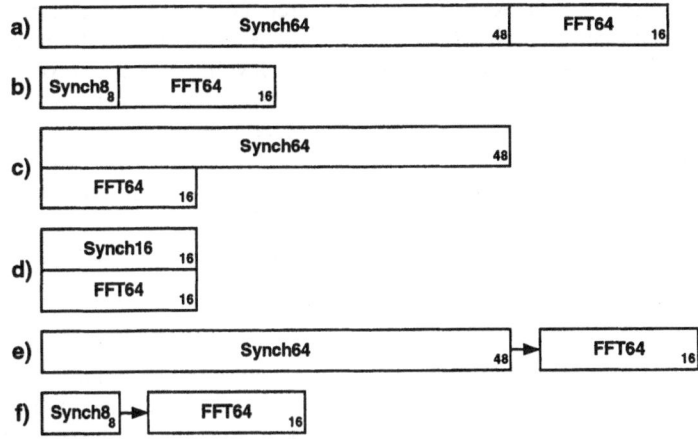

Fig. 6. Alternative OFDM receiver implementations

One array that alternates between the synchronizer and FFT (c, d): This implementation time-multiplexes the array between the two functions. To make this feasible, the datapath must be able to switch between two tasks very quickly, in a few microseconds rather than the many milliseconds required by FPGAs. One advantage of a coarse-grained architecture like RaPiD is that it has many fewer configuration bits. Even so, downloading a new configuration still takes too long for this application. The alternative is to use multiple configuration contexts, allowing reconfiguration in only a few clock cycles. However, the extra context increases the cost by about 10%.

Rapid switching between the correlator and the FFT makes more efficient use of the array since the two computations are not active at the same time. This is especially true if a cheaper tracking synchronizer is used that matches the size of the FFT (d).

Two arrays, one executing the synchronizer and one the FFT (e, f): Instead of providing a single reconfigurable array, platforms typically will provide multiple components with different sizes and capabilities. This permits the functionality of a component to be matched to the requirements of the function. In this case, the synchronizer can be mapped to a component with just ALUs and registers, while the FFT is mapped to a component with multipliers and memories. If a tracking synchronizer is used, then a much smaller array (f) can be used for the correlator.

3 Performance and Area Comparison

In this section we compare the performance and cost of the RaPiD implementation of the OFDM receiver to implementations in a DSP, ASIC and FPGA. We first describe how we obtained the numbers for each of these technologies before presenting the results. Area estimates are given in terms of λ^2 where λ is half the minimum feature size of the given technology. Performance is normalized to a .18μ process technology. Although we have been very careful in reaching the area and performance estimates, we must stress that they are only estimates. However, even if our estimates are off even by a factor of 2, the overall conclusions do not change.

RaPiD - Our study of the RaPiD benchmark architecture [4] included detailed layouts of a benchmark cell in a .5μ 3-layer metal CMOS technology and our area estimates for the RaPiD implementation are based on this layout. We also developed a timing model using this layout, and showed that except for some recursive filters, applications can be pipelined to achieve a 100 MHz clock rate. Scaling to a .18μ technology would push the clock rate above 300 MHz.

Table 1 gives the performance and area results for several alternative RaPiD implementations of the OFDM receiver. The letters in this table refer to the corresponding implementations in Fig. 6. Note that the first implementation (a) is used as the comparison for the ASIC and FPGA implementations, (b) is used to compare against the DSP. Implementations (b) and (f) are the implementations that would most likely be used in practice. Implementation (d) is the most cost-effective option, but it has the disadvantage of requiring rapid run-time reconfiguration, which may reduce the performance slightly, cause increased power dissipation and increase system complexity. Implementations (b) and (f) are almost as good and have better system characteristics in that they use arrays that are more generally useful.

Table 1. Performance and Area Results

Implementation (Synchronization window size)	Performance Msamples/sec	Area (Mλ^2)
(a) Single, homogenous array (+/- 32)	75x4 = 300	3485
(b) Single, homogenous array (+/- 4)	75x4 = 300	1413
(d) Shared, homogenous array (+/- 8)	75x4 = 300	1055
(e1) Two, homogenous arrays (+/- 32)	75x4 = 300	3554
(e2) Two, heterogeneous arrays (+/- 32)	75x4 = 300	2184
(f1) Two, homogeneous arrays (+/- 4)	75x4 = 300	1482
(f2) Two, heterogeneous arrays (+/- 4)	75x4 = 300	1264
(f3) Two, heterogeneous arrays (+/- 8)	75x4 = 300	1513
TI C6203 DSP, 300MHz 3 antennas, tracking Correlator (+/- 4)	8x3 = 24	750
Standard-cell ASIC, 600 MHz (+- 32)	150x4 = 600	1020
Custom ASIC, 600MHz (+/- 32)	150x4 = 600	490
FPGA, 100MHz (+/- 32)	25x4 = 100	19,920 (2938 CLBs)

DSP - We chose a TI C6203 DSP [10] running at 300MHz for comparison. This DSP is based on a VLIW architecture with 6 ALUs and 2 multipliers. This DSP can execute the FFT for 3 antennas with tracking synchronization at a sample rate of 8MHz. For area, we used the published die size of 14.89 mm2 in a .18μ 5-level metal CMOS technology. In the absence of a die photo, we estimated the core of this DSP to be about 40% of the total die area, which corresponds to approximately 750 Mλ^2. This 40% excludes the pads and some of the memory that would not be used with a DSP component implementation.

FPGA - We implemented the OFDM receiver using the same algorithm we used for the RaPiD array and mapped it to the Virtex2 architecture. The FPGA implementation uses the same number of multipliers and ALUs since there are no constant multiplications besides the +1/-1 multiplications in the correlator. We used CLB-based memories instead of block RAMs because they are a more efficient implementation of the small memories used by this algorithm. We did not use the multipliers in the Virtex2 architecture since we wanted to compare to traditional fine-grained FPGA architectures. Using these multipliers would have reduced the area by less than 10%. We estimated the area of the FPGA implementation by mapping the circuit to Virtex2 CLBs. We have estimated the size of a Virtex2 CLB at $6.78M\lambda^2$ using a published die photo for the Virtex2 1000 [11], which is implemented in a .15μ 8-level metal CMOS technology, and the architecture description of the Virtex2. We estimate the clock rate of the Virtex2 FPGA implementation for this application, normalized to .18μ technology, at 100 MHz.

ASIC - We estimated the area of an ASIC implementation using two methods, one for a custom implementation, and one for a standard cell implementation. First, we started with the algorithm implemented used with RaPiD and added up the area of the custom function units and memories used by this implementation. We then multiplied this by a factor of 1.5 to account for interconnect, registers and control to get a conservative area estimate for a custom layout. The second method used the equivalent gate count given by the Xilinx tools for this implementation, and converted this to ASIC area using Toshiba's published density for .18μ technology of 125,000 gate/mm², and a typical 70% packing factor. We estimated the ASIC clock rate at 600MHz.

4 Conclusions

The results of the experimental OFDM implementation clearly show that there is a role for coarse-grained configurable architectures. The RaPiD implementation has about 6 times the cost-performance of a DSP implementation, while an ASIC has about 7 times the cost-performance of RaPiD. Thus RaPiD falls right between the programmable and ASIC alternatives. Finally, the RaPiD implementation has about 15 times the cost-performance of the FPGA implementation, demonstrating the advantage of a specializing a configurable architecture to a problem domain. It is important to remember, of course, that DSPs and FPGAs have much more flexibility than RaPiD, which trades some flexibility for higher performance and lower cost.

Future system-on-chip platforms will have to provide components that have both high-performance and programmability. Processors and ASICs can provide only one, and FPGAs are ruled out for reasons of cost and power except for a very narrow range of bit-oriented applications. The question is whether coarse-grained configurable architectures will really be able to fill the gap as indicated by our research.

We believe that coarse-grained configurable architectures will become increasingly important as new very-high NRE technologies move the world away from ASICs and towards platforms. High-end FPGAs, which are beginning to take on a distinct platform look, now incorporate specialized coarse-grained components like multipliers, ALUs, and memories in addition to processor cores. It is still not clear what type of coarse-grained architecture will be the most effective. RaPiD achieves very high performance via highly pipelined datapaths and is appropriate for many of the intensive computations that must be offloaded to a hardware implementation. However, there is still much research to be done to explore this architecture space. The programming and compiler tools required to do this exploration are either non-existent or relatively primitive. Our research is now focused on building a set of architecture-independent programming and compiler tools for coarse-grained configurable architectures.

References

[1] C. Ebeling, D.C. Cronquist, P. Franklin, J. Secosky and S.G. Berg. "Mapping applications to the RaPiD configurable architecture," *Proceedings of the IEEE Symposium on FPGAs for Custom Computing Machines* pp. 106-115, 1997.

[2] C. Ebeling, D. C. Cronquist, P. Franklin, J. Secosky, and S. G. Berg. "Mapping Applications to the RaPiD Configurable Architecture," *Field-Programmable Custom Computing Machines 1997*

[3] C. Fisher, K. Rennie, G. Xing, S. Berg, K. Bolding, J. Naegle, D. Parshall, D. Portnov, A. Sulejmanpasic, and C. Ebeling. "An Emulator for Exploring RaPiD Configurable Computing Architectures," In *Proceedings of the 11th International Conference on Field-Programmable Logic and Applications* (FPL 2001), Belfast, pp. 17-26, August, 2001

[4] D. Cronquist, C. Fisher, M. Figueroa, P. Franklin and C. Ebeling. "Architecture design of reconfigurable pipelined datapaths," *Proceedings of the Conference on Advanced Research in VLSI* pp. 23-40, 1999.

[5] D. Cronquist, P. Franklin, S.G. Berg and C. Ebeling. "Specifying and Compiling Applications for RaPiD," *IEEE Symposium on FPGAs for Custom Computing Machines* pp. 116-125, 1998.

[6] Richard Van Nee and Ramjee Prasad, *OFDM for Wireless Multimedia Communications,* Artech, 1999

[7] Guanbin Xing, Hui Liu and Shi. R, "A multichannel MC-CDMA demodulator and its implementation", *in Proc. Asilomar '99*

[8] D. Cronquist. "Reconfigurable Pipelined Datapaths", Ph.D. Thesis, Department of Computer Science and Engineering, University of Washington, Seattle, WA, 1999.

[9] B. Gold and T. Bially, "Parallelism in Fast Fourier Transform Hardware", IEEE Trans. on Audio and Electroacoustics, 21:5-16, Feb. 1985.

[10] Texas Instruments Web Page, focus.ti.com/docs/military/catalog/general/general.jhtml?templateId=5603&path=templatedata/cm/milgeneral/data/dsp_die_rev

[11] Chipworks, www.chipworks.com/Reports/Flyers/Xilinx_XC2V1000.htm

Symbol Timing Synchronization in FPGA-Based Software Radios: Application to DVB-S

Francisco Cardells-Tormo[1], Javier Valls-Coquillat[2], and Vicenc Almenar-Terre[3]

[1] Inkjet Commercial Division (ICD) R&D Lab, Hewlett-Packard,
08190 Sant Cugat del Valles, Barcelona, Spain,
francisco.cardells@hp.com
[2] Department of Electronic Engineering, Polytechnic University of Valencia (UPV),
Camino de Vera s/n, 46022 Valencia, Spain,
jvalls@eln.upv.es
[3] Department of Telecommunications, Polytechnic University of Valencia (UPV),
Camino de Vera s/n, 46022 Valencia, Spain,
valmenar@dcom.upv.es

Abstract. The design of all-digital symbol timing synchronizers for FP-GAs is a complex task. There are several architectures available for VLSI wireless transceivers but porting them to a software defined radio (SDR) platform is not straightforward. In this paper we report a receiver architecture prepared to support demanding protocols such as satellite digital video broadcast (DVB-S). In addition, we report hardware implementation and area utilization estimation. Finally we present implementation results of a DVB-S digital receiver on a Virtex-II Pro FPGA.

1 Introduction

Software-defined radio (SDR) enables the consumer to roam from a wireless standard to another in a seamless and transparent way. A hardware capable of supporting SDR must have: flexibility and extensibility (to accommodate various physical layer formats and network protocols), high speed of computation (to support broadband protocols) and low power consumption (to be mobile). With the advent of 90 nm and beyond processes, application specific integrated circuits (ASICs) are becoming too expensive to miss a single market. Due to this fact, field-programmable gate arrays (FPGAs) are the hardware platform of choice for SDR: FPGAs can quickly be adapted to new emerging standards or to cope with the last minute changes of the specifications. In particular, this work will focus on a certain subset of FPGA architectures: look-up table (LUT)-based FPGAs. We have targeted commercial devices belonging to this set: Xilinx Virtex-II Pro. Area estimations are to be done in logic cells (LCs), consisting of a 4-input LUT (4-LUT) and a storage element. According to this definition, Xilinx slices consist of two LCs.

As we have previously stated, SDR must support all kinds of wireless physical layers in a flexible way. In transmission, the physical layer must perform data-to-symbol encoding and carrier modulation, whereas in reception the operations are reversed to recover the data stream. Yet, the recovery process is not straightforward due to the fact that we must cope with signal distortions due to the transmission channel. Therefore the demodulator must be able to estimate the carrier phase shift and the symbol timing delay incurred; and ultimately it must correct those effects. Otherwise, symbols would not correctly be recovered, data detection would fail and in the end, we would obtain an unacceptable high bit-error rate (BER). In this paper we will focus on the latter distortion. Due to the fact that in reception symbols are not aligned with the receiver clock edges, we must know when symbols begin and end to achieve symbol lock (i.e., symbol synchronization). In addition to channel effects, the analog signal is sampled in the digital-to-analog converter (DAC) with a clock independent of the symbol clock, and this must be compensated too. Symbol lock can be achieved in many ways, they all can be found in the literature [1].

The goal of this paper is to report the most suitable architecture of a receiver on FPGAs and to prove the feasibility (in terms of area and clock rates) of performing timing recovery on FPGAs. Although we will base our discussion on a particular standard, the results and ideas found here can be extended to other designs and protocols. Our standard of choice is the satellite digital video broadcasting (DVB-S), an European protocol for satellite communications [2]. The features of this standard are the following: modulation is based on quaternary phase shift keying (QPSK) with absolute mapping (no differential encoding); satellite ground stations could use an intermediate frequency (IF) of 70 MHz; it requires a symbol rate between 20.3 and 42.2 Mbauds (for a satellite transponder bandwidth-symbol rate ratio (BW/Rs) of 1.28).

The structure of the paper is as follows. First we will present the receiver architecture suitable for FPGAs. Secondly we will give implementation details of each block and area estimations. Thirdly, we will report implementation results on Virtex-II Pro FPGAs. Finally, we will present the conclusions of this paper.

2   FPGA-Based Architecture of SDR Receivers

The architecture of an FPGA-based receiver is depicted in figure 1. We have accommodated each functional block of the receiver in its appropriate clock domain according to their throughput requirements. We have chosen to decouple the functionality as much as possible to be able to do this partitioning: the resulting architecture corresponds to a non-data aided synchronizer. Otherwise, we could have used the output of the symbol detector for producing a timing estimate (i.e., a decision-directed architecture).

First of all, $r(t)$ (i.e., the bandlimited signal coming from the RF downconverter) is sampled in the ADC every T'_s seconds. Secondly, the samples $r(mT'_s)$ are down-converted from IF to baseband and then low-pass filtered. The output is downsampled to provide a data rate $1/T_s$ slightly higher than two samples per symbol (e.g., 100 MHz). There are two ways to perform down-conversion. The first technique, covered in detail in [3], multiplies the signal by a carrier

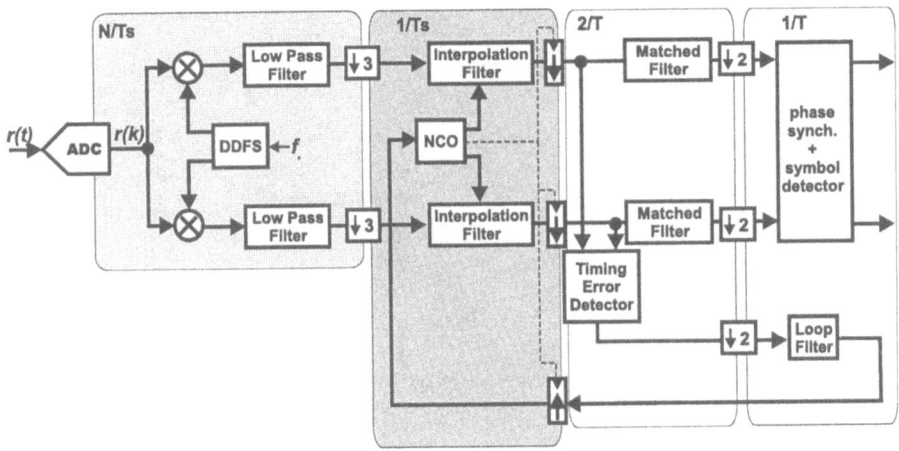

Fig. 1. Receiver architecture: downconverter, timing synchronization loop, matched filter, symbol detector.

centered in IF. Logic must be run at 1.25 times the Nyquist frequency (i.e., at least 232 MHz). In this first stage there are no feed-back loops and therefore pipelining can be used. In addition the output signal must be downsampled to $1/T_s$ using an integer ratio (N), in particular for N=3 then $1/T'_s$ would become 300 MHz. The second technique is based on subsampling, it can be applied to low transponder bandwidths but it will not be considered in this discussion.

Thirdly, we have the symbol timing synchronizer. It works at a data rate slightly higher than two samples per symbol (i.e., $1/T_s = 100\,MHz$). Its working principle can be seen in figure 2. The sampling instants, mT_s, of the incoming data, $x(mT_s)$, are fixed by the ADC sampling clock. New data, $y(kT_i)$, is only generated for the instants kT_i, with T_i the interpolator clock period which is an integer fraction of the symbol period T (e.g. $T_i = T/2$). The ADC sampling rate is asynchronous with the symbol clock (i.e., T/T_s is irrational). Figure 2 shows that the sampling clock and the symbol clock are accommodated using a delayed version of the nearest sample.

$$y(k \cdot T_i) = x((m_k + \mu_k) \cdot T_s) \tag{1}$$

The time delay is defined as $\tau_k = \mu_k \cdot T_s$, where μ_k is the fractional delay. The degree of quantization of μ determines the resolution in the time axis. Besides, we would like to compensate the fact that the receiver and the transmitter clock are not in phase (i.e., they not transition in the same instants). Consequently, the time delay should be slightly shifted from the position shown in figure 2. This correction is achieved using a combinational loop. The error is estimated solely from $y(kT_i)$ and by the timing error detector (TED).

Finally, we have one synchronized sample per symbol. But we must still correct any carrier frequency or phase errors and detect the symbol. This block must have a combinational loop for error detection. Guidelines for the design of

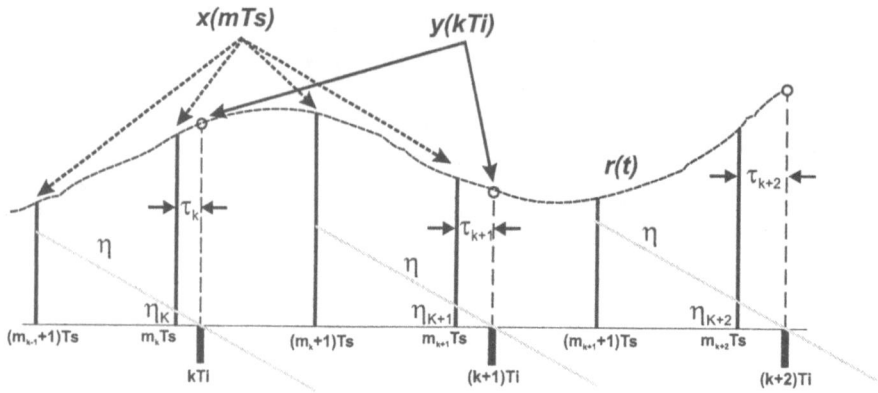

Fig. 2. Sample time relations

this block can be found in [4] and [5]. Current $0.13\mu m$ FPGA technology (e.g., Virtex II Pro) supports bit-rates up to 86 Mbps for DVB-S.

3   Mapping the Symbol Synchronizer in Logic Cells

In this section we will cover the implementation details of each module the symbol synchronizer consists of. Except for the TED, all modules are built in the same way in any all-digital symbol synchronization system, thus making the following discussion fully extensible. In addition to the hardware description, we will estimate area requirements in terms of LCs for our target FPGA. In order to do this in an effective way, we have reduced the design space to a smaller subset by only allowing to set a few parameters. In particular, we can only adjust the wordlength of the signals in table 1. The rest of the signals involved in the design will be affected by those parameters, and they can be found in figures 3, 4, 5, 6 and 7. For our particular case study, we will also use the default values depicted in table 1.

Table 1. Design Parameters.

Parameter	Value	Description
C1	14	Input data wordlength
C2	14	TED output wordlength
C3	28	Loop IIR Filter inner wordlength
C4	4	Fractional delay resolution

3.1   Interpolator

The task of the interpolator is to compute the value of the y signal at the time the interpolator clock transitions (i.e., intermediate values between signal samples $x(m \cdot T_s)$). This can be done using a fractional delay filter [6]. In this paper we will focus on FIR filter implementations [7–9]. The FPGA-implementation of a timing synchronizer using IIR filters can be found in [10]. We must place an interpolation filter in each arm. The number of taps is reduced to four for our normalized bandwidth [9].

$$x[(m_k + \mu_k)T_s] = \sum_{n=-I_1}^{I_2} x[(m_k - n)T_s] \cdot h_n(\mu_k) \tag{2}$$

The fractional delay is variable, and so are the coefficients of the interpolator filter. There are two ways of generating the coefficients, either we pre-compute their values and we store them in a ROM, or we compute them on-the-fly. The two approaches will be discussed next.

ROM-Based Architecture. The direct implementation of equation 2 provides the architecture depicted in figure 3. The filter could not be implemented in its transposed form because we are not using constant coefficients. The filter coefficients can be pre-computed in many ways such as by windowing the impulse response of the ideal interpolator, using optimality criterions and so forth.

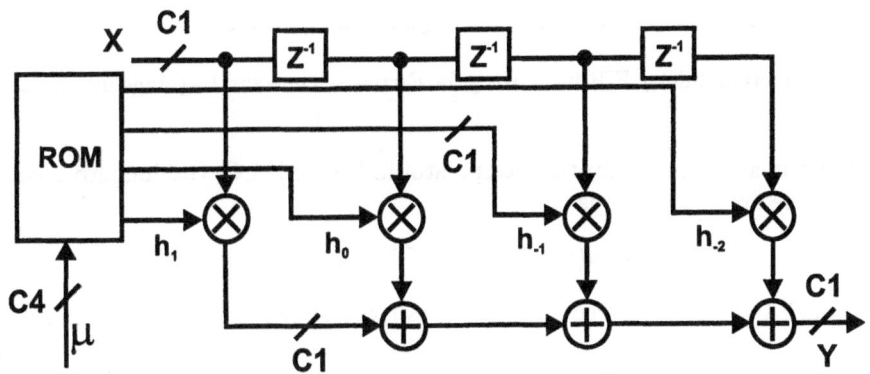

Fig. 3. Fractional-delay FIR filter using for ROM-based coefficients

The area estimation for the interpolator in a Virtex-II Pro (considering the I and Q datapaths) in LCs is

$$Area(C1, C4) = \left(4 \cdot C1 \cdot \frac{2^{C4}}{15}\right) + 8 \cdot C1^2 + 12 \cdot C1 \tag{3}$$

The ROM-based interpolator requires 114 slices and 8 embedded multipliers.

Farrow Architecture. If we use a polynomial interpolation, then coefficients could be stored in a memory as previously discussed or they could be generated on the fly. The 4 intermediate points can be interpolated by a Lagrange polynomial. By reordering of equation 2 as explained in [9] in chapter 9, we obtain the architecture depicted in figure 4. Each filter bank is a constant coefficient FIR filter and therefore we can use its transposed form and shorten the combinational path. The fact that coefficients are constant also helps with the implementation of multipliers for they can be optimized using canonic signed digit code (CSDC).

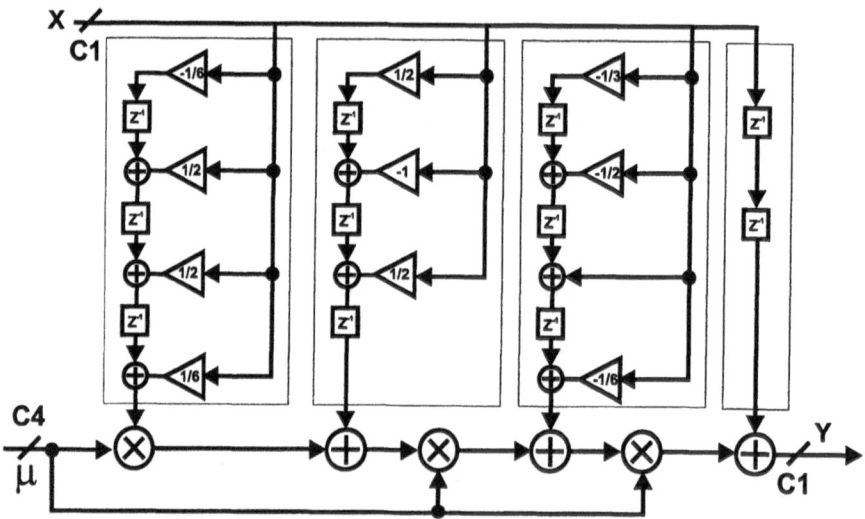

Fig. 4. Fractional-delay FIR filter using the Farrow scheme and Lagrange interpolation

The area estimation for the interpolator in Virtex-II LCs (taking into account that there are two arms) is

$$Area(C1, C4) = 8 \cdot C1 \cdot C4 + 26 \cdot C1 \qquad (4)$$

Although the Farrow interpolator requires fewer logic resources than the ROM-based interpolator, the biggest disadvantage of this technique is the long critical path. The combinational delay is approximately two times longer. Indeed, the ROM-based interpolator has 5 logic levels (an embedded multiplier and four adders), while the Farrow interpolator has 5 logic levels (four adders and three embedded multipliers) plus 3 logic levels in the multiplication with a fixed coefficient.

Polyphase Filters. Although we have only covered interpolation filters, there are other options for performing interpolation such as polyphase filters [11]. A polyphase filter bank consists of M matched filters operating in parallel, the symbol timing synchronizer selects a sample from one filter each cycle. We have not considered this option in our architecture due to the fact that this architecture is not suitable in FPGAs. The filter bank works at a clock rate as high as M times twice the symbol rate, taking into account that M defines the time precision and should be made as high as possible, it is undeniably true that for high symbol rates the required clock frequency for the banks is not achievable in FPGAs.

3.2 Timing Error Detector

In this paper we have focused on non-data-aided TEDs, thus the estimation will be performed using interpolated samples. The method we will use is the discrete form of the maximum likelihood timing synchronization: the so-called early-late gate detector. This technique consists in finding the point where the slope of the interpolator filter output is zero. If the current timing estimation is too early, then the slope of the interpolator filter output is positive indicating that the timing phase should be advanced. If the current timing estimate is too late, then the slope of the interpolator filter output is negative indicating that the timing phase should be retarded.

The slope is calculated using a derivative. If it is approximated with a finite difference, the TED for QPSK is performed using the following expression [12] (chapter 8):

$$e(k) = Re \left\{ y^* \left(k \cdot T + \hat{\varepsilon}_k \right) \cdot \left[y \left((k-1) \cdot T + T/2 + \hat{\varepsilon}_{k-1} \right) - y \left(k \cdot T + T/2 + \hat{\varepsilon}_k \right) \right] \right\} \tag{5}$$

Yet this approximation requires sampling at twice the symbol rate. That is the reason why the symbol timing synchronizer uses this data rate. In figure 5 we depict the hardware implementation of the TED. The output of both TEDs is added before being sent to the loop filter. This TED requires 77 slices and 2 multipliers.

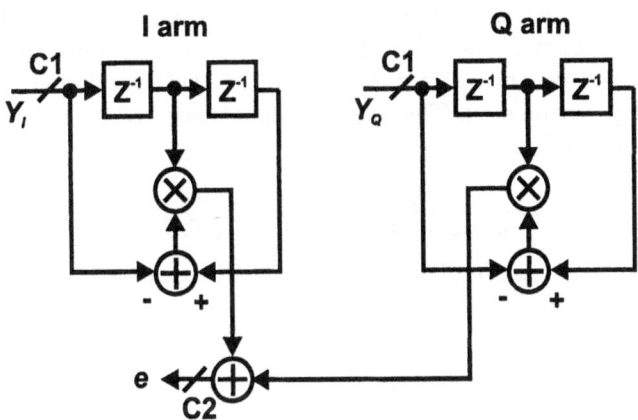

Fig. 5. Timing Error Detector

3.3   Loop Filter

The loop filter implements a standard proportional-integral control as it can be seen in figure 6. Using Virtex II embedded multipliers this module requires 28 slices and 2 multipliers.

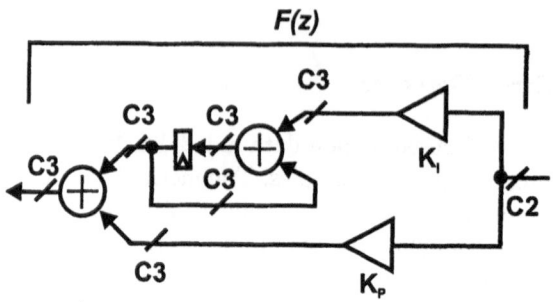

Fig. 6. Loop Filter

3.4   Controller

The controller consists of two timers driven by a clock of frequency of $1/T_s$. In steady-state one timer underflows at the symbol rate while the other underflows at twice the symbol rate. On rollover, the former latches the fractional delay (i.e., the loop filter output), while the latter latches the interpolator output (i.e., the timing error detector input). Therefore a fractional delay value is provided to the interpolator with T-periodicity, and the interpolator outputs two samples per symbol. Timers consists of module-1 counters that are decremented by their nominal period values divided by the clock period as in figure 7. In addition both timers compensate the nominal cycle with a slight shift depending on the timing error-detector output.

In addition to the necessary control signals, the timer with symbol periodicity, generates the fractional interval in the cycle previous to rollover. In figure 2, the relationship between the counter value μ in the cycle previous to rollover and the time delay can be seen. The following equation provides the mathematical formulation as in [1]:

$$\mu_k = \frac{\eta(m_k)}{\eta(m_{k+1}) - 1 + \eta(m_k)} \approx \left(\frac{T_i}{T_s}\right) \cdot \eta(m_k) \tag{6}$$

The two counters are implemented using 56 slices and one extra multiplier to obtain the fractional delay.

4   FPGA-Implementation

The proposed architecture for the symbol timing synchronizer has been fully described in VHDL. We have implemented the design of the synchronizer in a

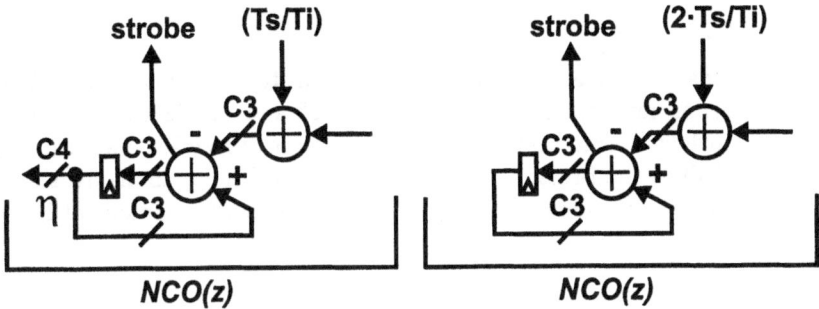

Fig. 7. Receiver architecture showing the timing synchronization loop

Xilinx Virtex-II Pro FPGA. We have used Synplify-Pro 7.1 as synthesis tool and FPGA-vendor design tools, Xilinx ISE 5.1., for place and route. We report the implementation results in table 2.

Table 2. Implementation results of a DVB-S receiver on Virtex-II Pro, speed grade 7.

FPGA		Implementation	
Receiver module	Clock rate	Slices	Multipliers
Down-converter	329 MHz	325	2
Symbol Timing Synchronizer	124 MHz	370	13
Symbol detector	43 MHz	134	6

According to this paper and to [3, 4], the overall area requirements for a DVB-S receiver are: 829 slices and 21 multiplier blocks. We also report the maximum clock rate of each module in the receiver. It can be verified that we are able to run the system fast enough to process DVB-S satellite data. Indeed, the down-converter can be run faster than the Nyquist frequency. The symbol timing synchronizer can be run at twice the maximum symbol rate, i.e. 84 MHz. The symbol detector can be run at the maximum symbol required by DVB-S.

5   Conclusions

In this paper we have presented an SDR receiver architecture adapted for FP-GAs. It is based on decoupling carrier phase and symbol timing recovery loops and on performing the functionality in the most suitable clock domain. We have discussed the mapping non-data aided timing synchronizers on FPGAs, and we have found out that FIR ROM-based interpolation filters are the right choice. We have proved that we can meet the timing requirements and that whole DVB-S receiver can be mapped in Virtex-II Pro FPGAs.

Acknowledgements

This work was supported by the Spanish "Ministerio de Ciencia y Tecnologia" under grant number "TIC2001-2688-C03-02" and "TIC2001-2688-C03-02". Francisco Cardells-Tormo acknowledges the support of Hewlett-Packard, Barcelona R& D Lab in the presentation of these results. The opinions expressed by the authors are theirs alone and they do not represent the opinions of HP.

References

1. Gardner, F.M.: Interpolation in Digital Modems – Part I: Fundamentals. IEEE Transactions on Communications **41** (1993) 501–507
2. E.T.S.I.: Digital Video Broadcasting (DVB). framing structure, channel coding and modulation for 11/12 Ghz satellite services. European Standard EN 300 421 (1997)
3. Cardells-Tormo, F., Valls-Coquillat, J.: High Performance Quadrature Digital Mixers for FPGAs. In: 12th International Conference on Field Programmable Logic and Applications (FPL2002). (2002) 905–914
4. Cardells-Tormo, F., Valls-Coquillat, J., Almenar-Terre, V., Torres-Carot, V.: Efficient FPGA-basd QPSK Demodulation Loops: Aplication to the DVB Standard. In: 12th International Conference on Field Programmable Logic and Applications (FPL2002). (2002) 102–111
5. Dick, C., Harris, F., Rice, M.: Synchronization in software radios - carrier and timing recovery using FPGAs. In: IEEE Symposium on Field-Programmable Custom Computing Machines. (2000)
6. Laakso, T.I., Valimaki, V., Karjalainen, M., Laine, U.K.: Splitting the Unit Delay. IEEE Signal Processing Magazine **13** (1996) 30–60
7. Erup, L., Gardner, F.M., Harris, R.A.: Interpolation in Digital Modems – Part II: Implementation and Performance. IEEE Transactions on Communications **41** (1993) 998–1008
8. Kootsookos, P.J., Williamson, R.C.: FIR Approxiation of Fractional Sample Delay Systems. IEEE Transactions on Circuits and Systems II: Analog and Digital Signal Processing **43** (1996) 269–271
9. Meyr, H., Moeneclaey, M., Fechtel, S.A.: Digital Communication Receivers. Synchronization, Channel Estimation and Signal Processing. John Wiley & Sons, (New York)
10. Wu, Y.C., Ng, T.S.: FPGA Implementation of Digital Timing recovery in Software Radio Receiver. In: IEEE Asia-Pacific Conference on Circuits and Systems (APCCAS 2000). (2000) 703–707
11. Harris, F.J., Rice, M.: Multirate Digital Filters for Symbol Timing Synchronization in Software Defined Radios. IEEE Journal on Selected Areas in Communications **19** (2001) 2346–2357
12. Mengali, U., D'Andrea, A.N.: Synchronization Techniques for Digital Receivers. Plenum Press, (New York and London)

An Algorithm Designer's Workbench for Platform FPGAs*

Sumit Mohanty and Viktor K. Prasanna

Electrical Engineering Systems,
University of Southern California, CA, USA,
{smohanty,prasanna}@usc.edu

Abstract. Growing gate density, availability of embedded multipliers and memory, and integration of traditional processors are some of the key advantages of Platform FPGAs. Such FPGAs are attractive for implementing compute intensive signal processing kernels used in wired as well as wireless mobile devices. However, algorithm design using Platform FPGAs, with energy dissipation as an additional performance metric for mobile devices, poses significant challenges. In this paper, we propose an algorithm designer's workbench that addresses the above issues. The workbench supports formal modeling of the signal processing kernels, evaluation of latency, energy, and area of a design, and performance tradeoff analysis to facilitate optimization. The workbench includes a high-level estimator for rapid performance estimation and widely used low-level simulators for detailed simulation. Features include a confidence interval based technique for accurate power estimation and facility to store algorithm designs as library of models for reuse. We demonstrate the use of the workbench through design of matrix multiplication algorithm for Xilinx Virtex-II Pro.

1 Introduction

High processing power, programmability, and availability of a processor for control intensive tasks are some of the unique advantages for Platform FPGAs which integrate traditional processors into the FPGA fabric [13]. Therefore, Platform FPGAs are attractive for implementing complex and compute intensive applications used in wired and wireless devices [13]. Adaptive beam forming, multi-rate filters and wavelets, software defined radio, image processing, etc. are some of the applications that target reconfigurable devices [11]. Efficient design of an application requires efficient implementation of constituent kernels. In this paper, kernel refers to a signal processing kernel such as matrix multiplication, FFT, DFT, etc. Algorithm design for a kernel refers to the design of a datapath and the data and control flow that implements the kernel.

Even though FPGA based systems are not designed for low-power implementations, it has been shown that energy-efficient implementation of signal

---

* This work is supported by the DARPA Power Aware Computing and Communication Program under contract F33615-C-00-1633.

P.Y.K. Cheung et al. (Eds.): FPL 2003, LNCS 2778, pp. 41–50, 2003.

processing kernels is feasible using algorithmic techniques [3], [5]. However, the major obstacle to the widespread use of Platform FPGAs is the lack of high-level design methodologies and tools [7]. Moreover with energy dissipation as an additional performance metric for the mobile devices, algorithm design using such FPGAs is difficult. In this paper, we discuss the design of a workbench for the algorithm designers based on the domain specific modeling technique [3] that addresses the above issues. A *workbench* in the context of algorithm design refers to a set of tools that aid an algorithm designer in designing energy-efficient algorithms. Although the workbench supports algorithm design for any kernel, we primarily target kernels that can be implemented using regular loop structures enabling maximum exploitation of hardware parallelism.

The workbench enables parameterized modeling of the datapath and the algorithm. Modeling using our workbench follows a hybrid approach, which starts with a top-down analysis and modeling of the datapath and the corresponding algorithm followed by analytical formulation of cost functions for various performance metrics such as energy, area, and latency. The constants in these functions are estimated through a bottom-up approach that involves profiling individual datapath components through low-level simulations. The model parameters can be varied to understand performance tradeoffs. The algorithm designer uses the tools integrated in the workbench to estimate performance, analyze the effect of parameter variation on performance, and identify optimization opportunities. Our focus in designing the workbench is not to develop new techniques for compilation of high-level specifications onto FPGAs, rather to formalize some of the design steps such as modeling, tradeoff analysis, estimation of various performance metrics using the available solutions for simulation and synthesis.

The workbench facilitates both high-level estimation and low-level simulation. High-level refers to the level of abstraction where performance models of the algorithm and the architecture can be defined in terms of parameters and cost functions. In contrast, low-level refers to the level of modeling suitable for cycle-accurate or RT-level simulation and analysis. The workbench is developed using Generic Modeling Environment, GME (GME 3), a graphical tool-suite, that enables development of a modeling language for a domain, provides graphical interface to model specific problem instances for the domain, and facilitates integration of tools that can be driven through the models [4].

The following section discusses some related work. Section 3 describes our algorithm design methodology. Section 4 discusses the design of the workbench. Section 5 demonstrates the use of the workbench. We conclude in Section 6.

2   Related Work

Several tools are available for efficient application design using FPGAs. Simulink and Xilinx System Generator provide a high-level interface to design applications using pre-compiled libraries of signal processing kernels [12]. Other tools such as [6] and [13] provide integrated design environments that accept high-level specification as VHDL and Verilog scripts, schematics, finite state machines, etc. and provide simulation, testing, and debugging support for the designer to implement a design using FPGAs. However, these tools start with a single conceptual design. Design space exploration is performed as part of implementation or through local optimizations to address performance bottlenecks identified during synthesis. For Platform FPGAs, the ongoing design automation efforts use available support for design synthesis onto the FPGA and processor and focus on communication synthesis through shared memory. However, a high-level programming model suitable for algorithmic analysis still remains to be addressed.

Several C/C++ language based approaches such as SystemC, Handel-C, SpecC are primarily aimed at making the C language usable for hardware and software design and allow efficient and early simulation, synthesis, and/or verification [1], [2]. However, these approaches do not facilitate modeling of kernels at a level of abstraction that enables algorithmic analysis. Additionally, generating source code and going through the complete synthesis process to evaluate each algorithm decision is time consuming.

In contrast, our approach enables high-level parameterized modeling for rapid performance estimation and efficient tradeoff analysis. Using our approach, the algorithm designer does not have to synthesize the design to verify design decisions. Once a model has been defined and parameters have been estimated, design decisions are verified using the high-level performance estimator. Additionally, parameterized modeling enables exploration of a large design space in the initial stages of the design process and hence is more efficient when the final design needs to meet strict latency, energy, or area requirements.

3   Algorithm Design Methodology

Our algorithm design methodology is based on domain specific modeling, a technique for high-level modeling of FPGAs, developed by Choi et al. [3]. This technique has been demonstrated successfully for designing energy efficient signal processing kernels [3], [5]. A domain, in the context of domain specific modeling, refers to a class of architectures such as uniprocessor, linear array, etc. and the corresponding algorithm that implements a specific kernel [3].

In our methodology, the designer initially creates a model using the domain specific modeling technique for the kernel for which an algorithm is being designed. This model consists of RModules, Interconnects, component specific parameters and power functions, component power state matrices, and a system-wide energy function. We have extended the model to include functions for latency and area as well. A *Relocatable Module (RModule)* is a high-level architecture abstraction of a computation or storage module. *Interconnect* represents the resources used for data transfer between the RModules. A component (building block) can be a RModule or an Interconnect. *Component Power State (CPS) matrices* capture the power state for all the components in each cycle.

Parameter estimation refers to the estimation of area and power functions for the components. Power and area functions capture the effect of component specific parameters on the average power dissipation and area of the component respectively. Latency is implicit in the algorithm specification and is also 0specified as a function. Ultimately, component specific area and power functions and the latency function are used to derive system-wide (for the kernel) energy, area, and latency functions.

Following parameter estimation, the designs are analyzed for performance tradeoffs as follows: a) each component specific parameter is varied to observe the effect on different performance metrics, b) if there exists alternatives for a building block then each alternate is evaluated for performance tradeoffs, and c) fraction of total energy dissipated by each type of component is evaluated to identify candidates for energy optimization. Based on the above analysis, the algorithm designer identifies the optimizations to be performed.

The workbench is used to assist designing using this methodology. Various supports include a graphical modeling environment based on domain specific modeling, integration of widely used simulators, curve-fitting for cost function estimation, and tradeoff analysis. In addition, the workbench integrates a high-level estimator, kernel performance estimator, that rapidly estimates latency, energy, and area for a kernel to enable efficient tradeoff analysis. The workbench also supports storage of various components as a library for reuse and a confidence interval based technique for statistically accurate power estimation.

4   Algorithm Designer's Workbench

We have used GME to create the modeling environment for the workbench [4]. GME provides support to define a metamodel (modeling paradigm). A metamodel contains syntactic, semantic, and visualization specifications of the target domain. GME configures itself using the metamodel to provide a graphical interface to model specific instances of the domain. GME enables integration of various tools to the modeling environment. Integration, in this context refers, to being able to drive the tools from the modeling environment. In the following, we discuss various aspects of the workbench in detail.

4.1   Modeling

We provide a hierarchical modeling support to model the datapath. The hierarchy consists of three types of components; micro, macro, and library blocks. A micro block is target FPGA specific. For example, as shown in Figure 1 (b), the micro blocks specific to Xilinx Virtex II Pro are LUT, embedded memory cell, I/O Pad, embedded multiplier, and interconnects [13]. In contrast, for Actel ProASIC 500 series of devices, there will be no embedded multiplier. Macro blocks are basic architecture components such as adders, counters, multiplexers, etc. designed using the micro blocks. A library block is an architecture component that is used by some instance of the target class of architectures associated with the domain. For example, if linear array of processing elements (PE) is our target architecture, a PE is a library block. Both macro and library blocks are also referred to as composite blocks. We have developed a basic metamodel using GME that provides templates for different kind of building blocks and associated parameters. Once the target device and the kernel are identified, prior

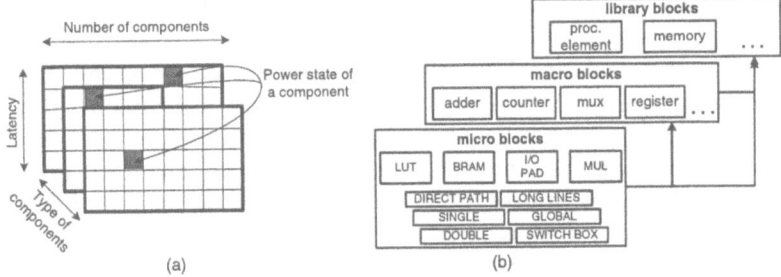

Fig. 1. Component Power State Matrices and Hierarchical Modeling

to using the workbench, the designer needs to modify the basic metamodel to support the specific instances of the building blocks. Being able to modify the modeling paradigm is a considerable advantage over many design environments where the description of a design is limited by the expressive power of the modeling language. Each building block is associated with a set of component specific parameters (Section 3). States is one such parameter which refer to various operating states of each building block. Currently, we model only two states, ON and OFF for each micro block. In ON state the component is active and in OFF state it is clock gated. However, for composite blocks it is possible to have more than 2 states due to different combination of states of the constituent micro blocks. Power is always specified as a function with switching activity as the default parameter.

Given an algorithm, Component Power State (CPS) matrices capture the operating state for all the components in each cycle. For example, consider a design that contains k different types of components $(C_1, \ldots, C_k)$ with n_i components of type i. If the design has the latency of T cycles, then k two dimensional matrices are constructed where the i-th matrix is of size $T \times n_i$ (Figure 1(a)). An entry in a CPS matrix represents the operating state of a component during a specific cycle and is determined by the algorithm.

Modeling based on the technique described above has the following advantages: a) the model separates the low-level simulation and synthesis tools from high-level algorithmic analysis techniques, b) various parameters are exposed at high-level thus enabling algorithm-level tradeoff analysis, c) performance models for energy, area, and latency are generated in the form of parameterized functions which allows rapid performance estimation, and d) using GME's ability to store models in a database, models for various kernels are stored and reused during the design of other kernels that share the same building blocks.

4.2   Performance Estimation

Our workbench supports performance evaluation through rapid high-level estimation as well as low-level simulation using third-party simulators. We have developed a high-level estimation tool, a kernel performance estimator, to estimate different performance metrics associated with the algorithm. The input to the kernel performance estimator is the model of the datapath and the CPS matrices associated with the algorithm. The designer has the option of providing switching activity for each building block. The default assumption is a switching activity of 12.5% (default value used by Xilinx XPower [13]).

Area of the design is evaluated as the sum of the individual areas of the building blocks that constitute the datapath. Latency estimate is implicit in the CPS matrices. Energy dissipation is estimated as a function of the CPS matrices and datapath. Overall energy dissipation is modeled as energy dissipated by each component during each cycle over the complete execution period. As energy can be evaluated as *power × time*, for a given component in a specific cycle, we use the power function for the operating state specified in the CPS matrices and duration of each cycle to evaluate the energy dissipated. Extending the above analysis for each component over the complete execution period, we evaluate the overall energy dissipation for the design. The kernel performance estimator is based on the above technique and is integrated into the workbench. Once a model is defined, it is possible to automatically invoke the estimator to generate the performance estimates and update the model based on these estimates. The estimator operates at a level of abstraction where the domain specific modeling technique is defined. Therefore, the estimator is not affected by the modifications to the basic metamodel as described in Section 4.1.

Low-level simulation is used in our methodology to estimate various component specific parameters. Low-level simulation is supported in the workbench through widely used simulators available for the FPGAs such as ModelSim, Xilinx XPower, Xilinx Power Estimator, etc. [13]. Different simulators have different input requirements. For example, Power Estimator requires a list of different modules used in a design, expected switching activity, area in CLB slices, and frequency of operation to provide a rapid and coarse estimate of average power. In contrast, ModelSim and XPower accept placed and routed design and input waveforms to perform fairly accurate estimation of power and latency. Therefore, we have added capability in our modeling environment to specify VHDL source code and input vectors as files. Derivation of cost function involves simulation of the design at different instances where each instance refers to a combination of parameter values and curve-fitting to generate functions.

Energy dissipation depends on various factors such as voltage, frequency, capacitance, and switching activity. Therefore, simulating once using a random test vector may not produce statistically accurate results. We use confidence intervals to estimate average power and confidence in the estimate to generate statistically significant results. The approach uses results from multiple simulations performed using a set of randomly generated inputs (Gaussian distribution is assumed) and computes the mean of the performance values and the confidence associated with the mean. Given a synthesized design and set of input test vectors, the workbench automatically performs multiple simulations and evaluates the mean and the confidence interval and updates the model using the results.

4.3 Modeling and Simulating the Processors

Kernels that are control intensive are potential candidates for design using both the FPGA and the processor. We assume shared memory communication between the FPGA and the processor. We model computations on the processor as functions. A function is implemented using high-level languages such as C and is compiled and loaded into the instruction memory. It is assumed that the processor is aware of the memory location to read input and write output. Once a function is invoked the FPGA stalls until the results are available.

Functions are specified in the CPS matrices. A function specified in a cell (i, j) refers to invoking the function in the ith cycle. We model each function as time taken and energy dissipated during execution. Therefore, whenever a function is encountered in the CPS matrices, the kernel performance estimator includes the latency and energy values associated with the function to the overall energy or latency estimates. We use the available simulators for the processors to estimate latency and energy values for each function. For example, ModelSim, which uses the SWIFT model for PowerPC, can be used to simulate the processor and estimate latency [13]. We evaluate energy dissipation by scaling the latency value based on the vendor provided estimates. For example, PowerPC on Virtex-II Pro consumes 0.9 mW per MHz. Using this estimate, a program that executes over 9×10^8 cycles will have a latency of 3 seconds and power dissipation of approximately 810 mW when the processor is operating at 300 MHz.

4.4 Design Flow

We briefly describe the design flow for algorithm design using our workbench. We assume that designer has already identified a suitable domain for the kernel to be designed and the target Platform FPGA.

Workbench Configuration (1): In this step, the designer analyzes the kernel to define a domain specific model. However, the designer does not derive any of the functions associated with the model. The designer only identifies the RModules and Interconnects and classifies them as micro, macro, and library blocks and also identifies the associated component specific parameters. Following this, the designer modifies the basic metamodel to configure GME for modeling using the workbench. The building blocks identified for one kernel can also be used for other kernels implemented using the same target FPGA. Therefore, modifications to the metamodel are automatically saved for future usage.

Modeling (2): The model of the datapath is graphically constructed in this step using GME. GME provides the ability to drag and drop modeling constructs and connect them appropriately to specify the structure of the datapath. If appropriate simulators are integrated, the designer can specify high-level scripts for the building blocks to be used in the next step. In addition, CPS matrices for the algorithm are also specified. The workbench facilitates building a library of components to save models of the building blocks and associated performance estimates (see *Step 3*) for reuse during the design of other algorithms.

Parameter Estimation (3): Estimation of the cost functions for power and area involves synthesis of a building block, low-level simulations, and in case of power, the use of confidence intervals to generate statistically significant power estimates. The simulations are performed off-line or, if required simulator is integrated, automatically using specified high-level scripts. Latency functions are estimated using the CPS matrices. System-wide energy and area functions are estimated using the latency function and component specific power and area functions.

Tradeoff Analysis and Optimization (4): In this step, the designer uses the workbench to understand various performance tradeoffs as described in Section 3. While the workbench facilitates generation of the comparison graphs, the designer is responsible for specific design decisions based on the graphs. Once the design is modified based on the decisions, kernel performance estimator is used for rapid verification. Tradeoff analysis and optimization is performed iteratively till the designer meets the performance goals.

5 Illustrative Example: Energy-Efficient Design of Matrix Multiplication Algorithm

A matrix multiplication algorithm for linear array architectures is proposed in [10]. We use this algorithm to demonstrate modeling, high-level performance estimation, and performance tradeoff analysis capabilities of the workbench. The focus is to generate an energy efficient design for matrix multiply using Xilinx Virtex-II Pro. The matrix multiplication algorithm and the architecture are shown in Figure 2. In this experiment, only FPGA is considered while designing the algorithm.

In *step 1*, the architecture and the algorithm were analyzed to define the domain specific model. Various building blocks that were identified are register, multiplexer, multiplier, adder, processing element (PE), and interconnects between the PEs. Among these building blocks only the PE is a library block and the rest of the components are micro blocks. Component specific parame-

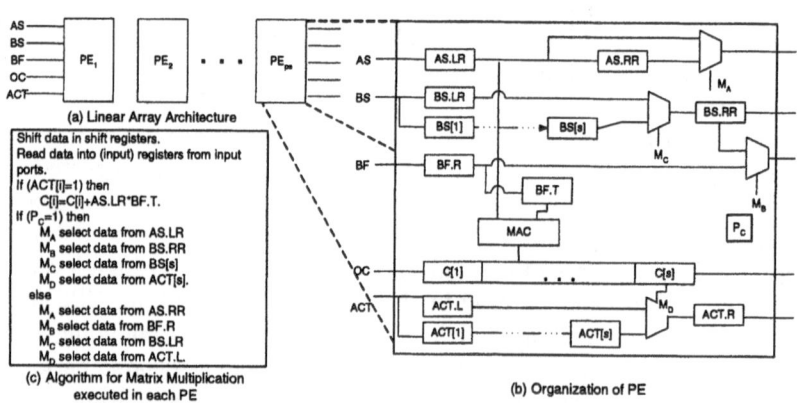

Fig. 2. Architecture and Algorithm for Matrix Multiplication [10]

ters for the PE include number of register (s) and power states ON and OFF. ON refers to the state when the multiplier (within the PE) is in ON state and OFF refers to the state when the multiplier is in OFF state. Additionally, for the complete kernel design number of PEs (pe) is also a parameter. For $N \times N$ matrix multiplication, the range of values for s is $1 \leq s \leq N$ and for pe it is $1 \leq pe \leq N(\lceil N/s \rceil)$. For matrix multiplication with larger size matrices (large values of N) it is not possible to synthesize the required number of PEs due to area constraint. In such cases, block matrix multiplication is used. Therefore, block-size (bs) is also a parameter. Based on the above model, the basic meta-model was enhanced to support modeling of linear array architecture for matrix multiply.

Step 2 involved modeling using the configured GME. Once the datapath was modeled we generated the cost function for power and area for the different components. Switching activity was the only parameter for power functions. To define the CPS matrices, we analyzed the algorithm to identify the operating state of each component in different cycles. As per the algorithm shown in Figure 2 (c), in each PE, the multiplier is in ON state for $T/(\lceil n/s \rceil)$ cycles and is in OFF state for $T \times (1 - 1/\lceil n/s \rceil)$ cycles [10]. All other components are active for the complete duration.

In *Step 3*, we performed simulations to estimate the power dissipated and area occupied by the building blocks. Currently, we have integrated ModelSim, Xilinx XPower, and Xilinx Power Estimator to the workbench [13]. The latency (T) of this design using $N \lceil N/s \rceil$ PEs and s storage per PE [10] is $T = (N^2 + 2N \lceil N/s \rceil - \lceil N/s \rceil + 1)$. Using the latency function, component specific power functions, and CPS matrices, we derived the system-wide energy function.

Finally, we performed a set of tradeoff analyses to identify suitable optimizations. Figure 3 (a) shows the variation of energy, latency, and area for different block sizes for 16×16 matrix multiplication. It can be observed that energy is minimum at a block size of 4 and area and latency are minimum at block size 2 and 16 respectively. This information is used to identify a suitable design (block size) based on latency, energy, or area requirements. Figure 3 (b) shows energy distribution among multipliers, registers, and I/O pads for three different designs. Design 1 corresponds to the original design described in [10] and Design 2 and 3 are low energy variants discussed in [5]. Using Figure 3, we identify

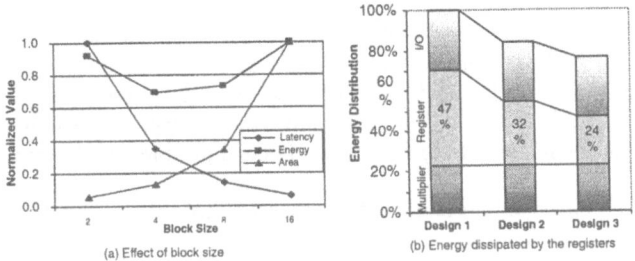

(a) Effect of block size   (b) Energy dissipated by the registers

Fig. 3. Analysis of Matrix Multiplication Algorithm

that the registers dissipate the maximum energy and select them as candidates for optimization. Optimizations considered include reduction of number of registers through analysis of data movements (Design 2) and use of CLB based SRAMs instead of registers to reduce energy dissipation (Design 3). Details of the optimized algorithm are available in [5], [9].

Using the workbench, an optimized matrix multiplication algorithm was designed and compared against the Xilinx IP core for matrix multiplication. The best design obtained using our approach achieved 68% reduction energy dissipation, 35% reduction in latency while occupying 2.3 × area when compared with the design provided by Xilinx [9].

6   Conclusion

We discussed an algorithm designer's work bench suitable for a general class of FPGA and Platform FPGA devices. The work discussed in this paper is part of the MILAN project. MILAN addresses the issues related to the design of end-to-end applications [8]. In contrast, the workbench addresses the issues pertaining to the design of application kernels only. The workbench is developed as a stand-alone tool with plans for integration into the MILAN environment.

References

1. Cai, L., Olivarez, M., Kritzinger, P., and Gajski, D: C/C++ Based System Design Flow Using SpecC, VCC, and SystemC. Tech. Report 02-30, UC, Irvine, June 2002.
2. The Handel-C language. http://www.celoxica.com/
3. Choi, S., Jang, J., Mohanty, S., and Prasanna, V.K.: Domain-Specific Modeling for Rapid System-Wide Energy Estimation of Reconfigurable Architectures. Engineering of Reconfigurable Systems and Algorithms, 2002.
4. Generic Modeling Environment. http://www.isis.vanderbilt.edu/ Projects/gme/
5. Jang, J., Choi, S., and Prasanna, V.K.: Energy-Efficient Matrix Multiplication on FPGAs. Field Programmable Logic and Applications, 2002.
6. Mentor Graphics FPGA Advantage. http://www.mentor.com/fpga-advantage/
7. McGregor, G., Robinson, D., and Lysaght, P.: A Hardware/Software Co-design Environment for Reconfigurable Logic Systems. Field-Programmable Logic and Applications, 1998.
8. Model-based Integrated Simulation. http://milan.usc.edu/
9. Mohanty, S., Choi, S., Jang, J., and Prasanna, V.K.: A Model-based Methodology for Application Specific Energy Efficient Data Path Design using FPGAs. Conference on Application-Specific Systems, Architectures and Processors, 2002.
10. Prasanna, V.K. and Tsai, Y.: On Synthesizing Optimal Family of Linear Systolic Arrays for Matrix Multiplication. IEEE Tran. on Computers, Vol. 40, No. 6, 1991.
11. Srivastava, N., Trahan, J., Vaidyanathan, R., and Rai, S.: Adaptive Image Filtering using Run-Time Reconfiguration. Reconfigurable Architectures Workshop, 2003.
12. System Generator for Simulink. http://www.xilinx.com/products/software/ sysgen/product_details.htm.
13. Xilinx Virtex-II Pro and Xilinx Embedded Development Kit (EDK). http://www.xilinx.com/

Prototyping for the Concurrent Development of an IEEE 802.11 Wireless LAN Chipset

Ludovico de Souza, Philip Ryan, Jason Crawford, Kevin Wong,
Greg Zyner, and Tom McDermott

Cisco Systems, Wireless Networking Business Unit, Sydney, Australia,
ludi@cisco.com

Abstract. This paper describes how an FPGA based prototype environment aided the development of two multi-million gate ASICs: an IEEE 802.11 medium access controller and an IEEE 802.11a/b/g physical layer processor. Prototyping the ASICs on a reconfigurable platform enabled concurrent development by the hardware and software teams, and provided a high degree of confidence in the designs. The capabilities of modern FPGAs and their development tools allowed us to easily and quickly retarget the complex ASICs into FPGAs, enabling us to integrate the prototyping effort into our design flow from the start of the project. The effect was to accelerate the development cycle and generate an ASIC which had been through one pass of beta testing before tape-out.

1 Introduction

The IEEE 802.11 standards [1] describe high-speed wireless local area networks. We have implemented an 802.11 solution as two SoCs, corresponding to the low-level physical communication layer (the PHY), and the high-level medium access control layer (the MAC), as shown in Figure 1.

The complexity and evolving nature of the 802.11 standards make a degree of programmability essential for the MAC. Most MACs achieve this by embedding a processor coupled to hardware acceleration blocks, such as a packet engine or a cryptography module. However, it is difficult a-priori to make good decisions as to what areas require hardware acceleration. The ability to run firmware on real hardware in a real network allows design decisions to be based on real data.

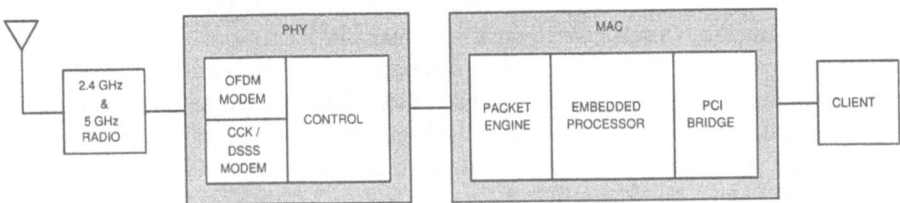

Fig. 1. An 802.11 Wireless LAN, showing the PHY and MAC SoCs.

P.Y.K. Cheung et al. (Eds.): FPL 2003, LNCS 2778, pp. 51–60, 2003.

Conversely, the PHY standard is stable and well understood, and the computational requirements of the PHY require dedicated hardware [2]; but this makes it vulnerable to its own problems. There is little, if any, scope to fix design faults after tape out. Mistakes are expensive, fabrication NRE costs are in the order of a million dollars. Simulation is too slow to exhaustively test the complex design. It is very difficult to envisage and plan for all the ways in which the system might be excited in practice. The vagaries of the medium mean that it is very hard to simulate - the only way to verify is to do it "virtual" real time.

Our FPGA prototype addressed both of these problems. Interfacing the two developing systems with each other and with external entities provided each system with the necessary real-world stimulus to develop and verify its implementation. The ability to execute firmware in a real system environment, months in advance of first silicon, accelerated the development of the product by allowing both hardware and software to be developed simultaneously rather than sequentially, improving confidence and reducing risk in the complete design.

An FPGA based prototype was the best solution to meet our needs and fit in with our existing development process. The capacity and speed of modern FPGAs offered us a near full-speed implementation; a realistic prototype running at one quarter of the final system speed. The maximum benefit of prototyping was achieved by implementing the core functionality of the prototype directly from the ASIC Verilog source, without modification. The pushbutton process from synthesis to bitstream of our multi-million transistor designs was achievable on a time-scale of hours, without user intervention, allowing many design/test/debug iterations per day.

In this paper we share our experience of how modern FPGA tools and hardware allowed us to implement a successful solution, and the positive impact it has had on our ASIC development flow.

2   Requirements

The prototype served three different purposes; as a platform for firmware development, as a testbed, and to interface with real hardware to expose the design to real world conditions. The following sections address these issues.

2.1   MAC Hardware/Firmware Co-development

The attraction of a prototype implementation of the MAC was the ability to execute firmware, running code on the actual platform, months before real silicon was available. When developing software for an embedded processor on a yet to be built SoC, one could make use of simulators, such as the Armulator in the ARM developer suite [3], or one could attempt to integrate a physical processor into the prototype environment, such as using the ARM Integrator Core Module, or using the PowerPC embedded in the Xilinx Virtex-II Pro FPGAs. However our experience is that this is somewhat cumbersome; there is a lack of observability in the implementation, and we are subject to constraints specific to that implementation, such as differences in the cache configuration. The concern is that we are not using the actual hardware that will be implemented in the final ASIC.

A much more accurate and insightful analysis of the firmware execution is achieved by synthesising the processor soft-core into the prototype, with the rest of the system. Our FPGA implementation did this, implementing the complete MAC SoC within an FPGA, and so conserved the timing relationship between the processor and the rest of the system. An in-house developed RISC processor, a V8 SPARC written in synthesizable behavioural Verilog without any FPGA specific timing optimisation, was used on this project. The FPGA implementation was quite acceptable, synthesizing to over 50MHz while using about 10 percent of the XC2V6000 FPGA.

2.2 PHY Hardware Testability

The motivation for a prototype implementation of the PHY was increased testability; the ability to throw huge amounts of data at a complex design, to run the design.

The PHY ASIC supports the OFDM, CCK, and DSSS modulations used in the 802.11a, 802.11b, and 802.11g physical layers. This entails considerable computational effort, covering operations such as synchronisation and frequency correction, transforms such as the FFT and IFFT, equalisation, Viterbi decoding, interleaving, deinterleaving, scrambling, filtering, and so on. The magnitude of this processing, coupled with a limited power budget, demands a highly optimised data-flow implementation, providing little scope for post-silicon design fixes. Simulation of the complex design is time consuming, having a large number of variables that influence the performance of the system. This means that only a limited test coverage is possible in simulation, which is usually sufficient to highlight critical design faults, but may leave many lesser problems undetected. The real-time nature of the prototype allows for instantaneous feedback and analysis. Experimentation and verification are performed with the system running in a much more realistic environment than could be feasibly achieved in any reasonable time in simulation. A packet, lasting 200 microseconds in real time, corresponds to 2 minutes of simulation time, an incredible speed difference - on the prototype, 5000 packets can be demodulated every second, a task that takes almost 3 hours in simulation.

2.3 Why Emulation Wasn't an Option

A requirement common to the PHY and the MAC was to interface with real hardware. We wanted the prototype to be part of a real system, to expose the design to real world conditions. This meant interfacing the PHY with a radio, in our case an existing 5GHz radio, and interfacing the MAC with a host device, in our case an existing Cisco Access Point.

This doesn't imply that a full speed real time prototype is necessary. Because a full speed prototype was unachievable given the speed of current FPGAs, the prototype had to be a time scaled implementation. A quarter speed implementation was achieved, meaning that the prototype PHY operates on signals of one quarter the bandwidth used in real systems. This is not a problem for communicating with other prototype PHY implementations, or with other appropriately modified equipment. It does however prevent communication with regular 802.11 systems. We allowed for this by having a mechanism to bypass the FPGA PHY implementation, to interface the MAC with a silicon PHY implementation, to allow the prototype to communicate with any 802.11 system, permitting more extensive MAC level testing.

These requirements demanded a real implementation, not just emulation.

3   Implementation

3.1   Integrating with the ASIC Design Flow

Our group has a standard process for the development of its ASICs. Designs are coded in synthesizable behavioural Verilog, simulated with tools such as Cadence's NC-Verilog [4], and synthesized and implemented as an ASIC using Cadence's physical design tools, for 0.13μm, 0.18μm and 0.25μm standard CMOS processes.

We decided to start the prototype implementation flow from the behavioural Verilog code base, the same code used for ASIC synthesis. This allows the FPGA synthesis tool to perform FPGA-centric optimisations and infer instantiations of embedded blocks such as memories and multipliers in the FPGAs. All FPGA specific details, such as the instantiation of the clock buffers and the partitioning of the design between FPGAs, were contained within high level wrapper modules. These wrapper modules together instantiated the complete, unmodified, ASIC core. By confining the FPGA implementation details entirely within these wrapper files we avoided inconsistencies associated with maintaining parallel source trees; the FPGA and ASIC implementations used an identical source base, giving us a very high confidence in the match between ASIC and FPGA implementations. As a side effect, synthesis warnings from the FPGA tools often provided meaningful feedback to the designers for coding changes in advance of the ASIC synthesis.

3.2   Choosing a Platform

An enormous amount of logic was required for the implementation, the ASICs contain more than 10 million transistors. A similar approach to wireless baseband modelling has been employed by Berkeley Wireless Research Center with their custom BEE FPGA processor [5]. BWRC chose to implement their own system whereas while we originally did this about four years ago, we found it more cost effective to buy off-the-shelf units; the capacity and speed of modern FPGAs met our needs. We chose the Dini Group's DN3000K10 [6] populated with 3 Xilinx Virtex-II 6 million gate parts in the FF1152 package [7].

The Virtex-II parts provide clock gating and embedded hardware multipliers, both of which our PHY uses. Gated clocking allows us to reduce power consumption by removing the clock to inactive circuits. Because the Virtex-II parts have support for gated clocking our ASIC Verilog was directly implementable. The computationally intensive nature of the PHY meant that we made significant use of the embedded hardware multipliers in the Virtex-IIs.

There were several clocks used in the project, principally the 33MHz PCI clock and a 20MHz PHY clock. The clocks were distributed to all FPGAs simultaneously, to avoid difficulties with skew across the partition.

3.3   Partitioning

We partitioned the two ASICs into three FPGAs, as shown in Figure 2. The MAC was sufficiently small to fit within a single FPGA. The PHY, however, had to be split across two FPGAs. One problem faced was the large number of configuration and status registers; the PHY contains roughly a thousand configuration and status registers, each up to 16 bits in length, which are accessed through a narrow configuration interface. It was infeasible to carry this large number of registers across the partition, so the interface and configuration blocks were duplicated, appearing in both PHY FPGAs. Having done so, there was then a natural split within the ASIC that supported a partition given the available IOs. Because we could partition satisfactorily by hand, no effort was made to use an automated partitioning tool.

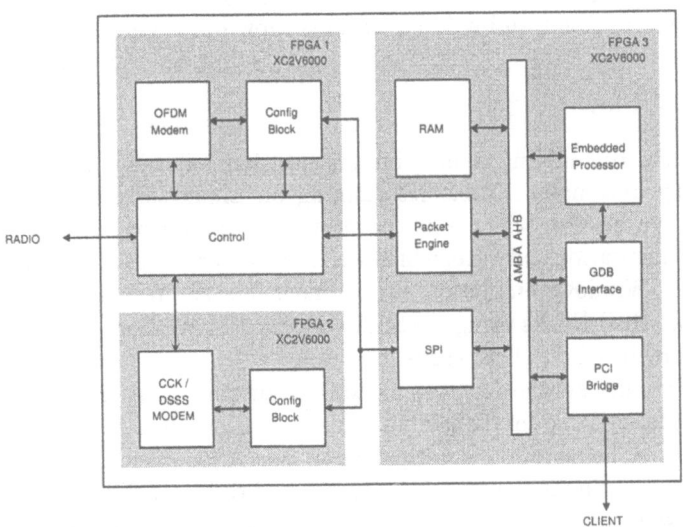

Fig. 2. The partitioned implementation, with the PHY in FPGAs 1 and 2 on the left, and the MAC in FPGA 3 on the right.

3.4   Interfaces

The full benefit of the prototype could only be achieved if the system was able to interface with existing hardware, forming a complete system. For one interface the PCI connector was used, for the other interface a custom interposer board was constructed. The abundance of FPGA IOs were necessary to meet this need.

The Cisco Access Point that was used for product development interfaces with our MAC across a MiniPCI interface. Using a passive PCI to MiniPCI adapter we were able to connect the prototyping board directly to the Access Point. Having the PCI bus directly connected to an FPGA was very important, a number of FPGA development boards have inbuilt PCI implementations, but our MAC SoC already contains a PCI implementation that we wanted to exercise.

A PCI solution is also capable of being inserted into any standard PC, opening up interesting debugging opportunities. We used the PC as a platform for running development support software, such as GUIs to control and observe the configuration of the PHY, tools to monitor the execution of software on the embedded processor, and even to act as an 802.11 client, sending and receiving network traffic through our development system.

Two interposer cards were developed to provide an interface to an existing radio. For development involving the prototype PHY the interposer card included the missing analog features of the PHY, the DAC and ADC. For development requiring a full speed PHY the interposer card included a previous generation silicon PHY. There were also test headers for devices such as logic analysers and oscilloscopes.

A photo of the completed system, showing the prototype interfacing between a Cisco Access Point and a 5GHz radio is shown in Figure 3.

3.5   Tool Flow

The process of building a new implementation following design changes was completely automated. Makefiles running on Linux servers executed first the Synplify synthesiser, which was the native Linux version, and then the Xilinx ISE implementation software, which was the PC version executed in Linux under the WINE Windows emulation environment [8]. The resulting bitstreams were loaded into the FPGAs using the Smart Media flash memory based configuration system present on the Dini cards.

The Synplify synthesizer was capable of automatically inferring instances of Xilinx features from the behavioural Verilog. Block RAMs were automatically inferred from behavioural descriptions of our ASIC memories, although some initial non-functional coding-style changes had to be made to the ASIC Verilog. The hardware multiplier units were also automatically inferred. There were no modifications required to the ASIC design for FPGA synthesis, which was critical to automating the process.

The Xilinx ISE flow was also completely automated. No hand placement was necessary; even as the designs stretched the capacity of the chips, the automatic placement was always satisfactory.

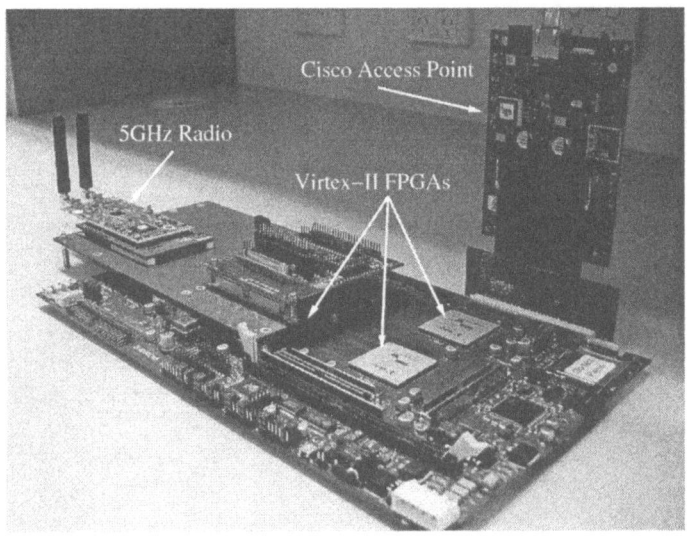

Fig. 3. The prototyping system, showing the 5GHz radio and Access Point.

The utilisation and build times for the three Xilinx Virtex-II XC2V6000 parts are shown below. We found that executing the tools on Linux Pentium 4 servers was many times quicker than executing the same sequence using the Solaris versions of the tools on our high-end Sun servers; this meant saving hours per re-implementation.

Table 1. FPGA Usage.

FPGA	LUTs	Registers	Block RAMs	Multipliers	Build Time
1 (PHY)	57,000 (83%)	32,000 (46%)	48 (33%)	56 (38%)	6 Hours
2 (PHY)	40,000 (59%)	11,000 (15%)	9 (6%)	48 (33%)	3 Hours
3 (MAC)	10,000 (14%)	7,000 (9%)	93 (64%)	0 (0%)	1 Hour

3.6   Custom Development Tools

Being able to plug the development board into a PCI slot and have a simple, high bandwidth interface from the PC into the system offered great development opportunities. Having software to take advantage of this interface as the opportunities became apparent was well worth the effort. The PHY chip contains around a thousand configuration and status registers. Providing a simple interface to manage this configuration was critical to the success of the system. We ported the existing powerful graphical user interface [10] used to configure our group's existing ASICs. This program was written in Python [11], an interpreted programming language. This allowed the hardware developers to write their own scripts to communicate with the prototype and debug their hardware. An example view is shown in Figure 4. Because this is the same interface that we will use to configure and debug the silicon ASICs, it will be a transparent switch for the developers to move from working with the prototype to the ASIC, they are already experienced with the test environment.

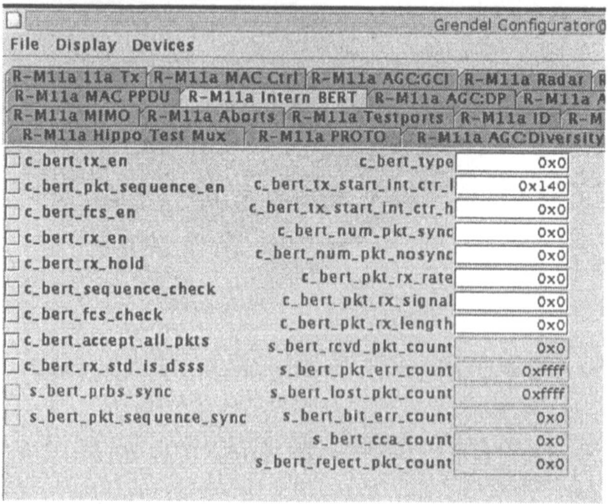

Fig. 4. A screen capture showing the software used to configure and debug the implementation.

Firmware development for the embedded processor also made use of the PCI interface. Firmware is loaded into the processor memory across the PCI bus. Hardware support was added to the embedded processor to provide an interface with the GNU GDB debugger [12] running on the host PC. This allowed for the observation of code execution in the embedded processor, and provided features such as stepping, breakpoints, and watchpoints.

4   Experiences and Lessons

4.1   Revision Control

A difficulty that became apparent from having a rapid re-implementation was keeping the software and the hardware in synchronisation. Modifications to the register map in the hardware would cause existing software to break. A method was implemented to automatically generate a hardware and software implementation of the memory map from a common abstract view, meaning that any changes to the mapping were simultaneously updated in both the hardware Verilog and software map files.

Given the 6 hour re-implementation time, and the fact that the source database was actively being updated, there was sometimes confusion about the versions of files that had been taken to construct the implementation. The build scripts were modified to automatically embed checks such as time stamps and revision numbers into the implementation. These were accessible from our development software, which could detect if the hardware and software were not using the same version.

4.2   Simulating vs Prototyping

When making enhancements to the hardware it was often quicker to test changes through the prototype implementation than to extensively simulate them. Still, the 6 hour implementation time was long enough to warrant a careful analysis before re-implementation, guessing the cause of unexpected behaviour could waste this time.

Simulation and prototyping complemented each other; bugs found in simulation were examined on the prototype, and bugs found in the prototype were examined in simulation. Using both methods increased our understanding and confidence in the design.

4.3   Tool Problems

We found that the FPGA tools were lacking in their ability to handle partial resynthesis. There are large blocks in our design, such as the Viterbi decoder and the processor core, that are not actively developed, remaining constant between each resynthesis. Having blocks such as these reused from previous implementations could save on re-implementation time. But it was non-trivial to make the tools do this, and having done so the time saved was not great.

The FSM optimisations performed by Synplify were, after much head scratching, found to be broken. Several FSMs were observed to lock up in unexpected states, a behaviour inconsistent with the source code. The problem was that the FSMs were being recoded as one-hot state machines, but the synthesizer failed to add logic to check when the FSM was in an illegal state. Disabling the FSM optimisations resolved this problem. This highlighted the need for equivalence checking of the synthesis output.

4.4   Concurrent Development

Having a firmware development proceed concurrently with the development of the MAC and PHY hardware allowed for an evolution and fine tuning of the hardware / software interface. The ability to tightly integrate observability, such as trace-logs, instruction counters, and timers woven deep into the processor, provided us with useful profiling data.

Having the prototype PHY interface directly with a real radio, allowing exposure to real and simulated radio channels with issues such as multipath, allowed us to verify and fine-tune our algorithms. The system continues to be used to implement experimental ideas.

Having the prototype MAC interact with other 802.11 MACs allowed for a better understanding of how these implementations behave and affect our performance.

When using the system we often discovered ways to improve observability. There were test ports that we had not anticipated, that only became apparent once we used the system. Some functionality, such as a trace buffer, was found to be incredibly useful in practice, and so was added to the ASIC code base.

5 Conclusion

The FPGA implementation of our MAC and PHY has become an extensively used, powerful tool. It is vital to the success of our chips. Its ability to transparently fit into the existing ASIC development process, and to automatically produce a re-implementation, without user intervention, have been key to its success.

It has provided great confidence to the hardware and software developers to extensively exercise their implementation, beta-testing in a very realistic environment before tape-out, enhancing the success of the first spin silicon. It has allowed development and experience with our test environment, saving time when the ASICs return from fab.

The system continues to be used for firmware and hardware development.

References

1. The IEEE 802.11 Standards, http://grouper.ieee.org/groups/802/11/
2. Philip Ryan et al., "A Single Chip COFDM Modem for IEEE 802.11a with Integrated ADCs and DACs", ISSCC 2001
3. ARM Developer Suite, ARM Ltd, http://www.arm.com/devtools/ads/
4. Cadence Design Systems, http://www.cadence.com/
5. Berkeley Wireless Research Center, Berkeley Emulation Engine, http://bwrc.eecs.berkeley.edu/Research/BEE/index.htm
6. The Dini Group, http://www.dinigroup.com/
7. Xilinx, http://www.xilinx.com/
8. WINE, http://www.winehq.org/
9. Synplicity, Inc, http://www.synplicity.com/
10. Tom McDermott et al., "Grendel: A Python Interface to an 802.11a Wireless LAN Chipset", The Tenth International Python Conference, 2002
11. The Python Programming Language, http://www.python.org/
12. GDB, The GNU Project Debugger, http://www.gnu.org/software/gdb/gdb.html

ADRES: An Architecture with Tightly Coupled VLIW Processor and Coarse-Grained Reconfigurable Matrix

Bingfeng Mei[1,2], Serge Vernalde[1], Diederik Verkest[1,2,3], Hugo De Man[1,2], and Rudy Lauwereins[1,2]

[1] IMEC vzw, Kapeldreef 75, Leuven, B-3001, Belgium
[2] Department of Electrical Engineering, Katholieke Universiteit Leuven, Belgium
[3] Department of Electrical Engineering, Vrije Universiteit Brussel, Belgium

Abstract. The coarse-grained reconfigurable architectures have advantages over the traditional FPGAs in terms of delay, area and configuration time. To execute entire applications, most of them combine an *instruction set processor* (ISP) and a reconfigurable matrix. However, not much attention is paid to the integration of these two parts, which results in high communication overhead and programming difficulty. To address this problem, we propose a novel architecture with tightly coupled *very long instruction word* (VLIW) processor and coarse-grained reconfigurable matrix. The advantages include simplified programming model, shared resource costs, and reduced communication overhead. To exploit this architecture, our previously developed compiler framework is adapted to the new architecture. The results show that the new architecture has good performance and is very compiler-friendly.

1 Introduction

Coarse-grained reconfigurable architectures have become increasingly important in recent years. Various architectures were proposed [1][2][3][4]. These architectures often comprise a matrix of functional units (FUs), which are capable of executing word- or subword-level operations instead of bit-level ones found in common FPGAs. This *coarse* granularity greatly reduces the delay, area, power and configuration time compared with FPGAs, however, at the expense of flexibility. Other features include predictable timing, a small configuration storage space, flexible topology, etc. However, the reconfigurable matrix alone is not capable of executing entire applications. Most coarse-grained architectures are coupled with processors, typically RISCs. The execution model of such hybrid architectures is based on the well-known 90/10 locality rule[5], i.e., *a program spends 90% of its execution time in only 10% of the code.* Some computational-intensive kernels are mapped to the matrix, whereas the rest code is executed by the processor. So far not much attention is paid to the integration of the two parts of the system. The coupling between the processor and the reconfigurable matrix is often loose, which is essentially two separated parts connected by a

P.Y.K. Cheung et al. (Eds.): FPL 2003, LNCS 2778, pp. 61–70, 2003.

communication channel. This results in programming difficulty and communication overhead. In addition, the coarse-grained reconfigurable architecture consists of components which are similar to those used in processors. This represents a major resource-sharing and cost-saving opportunity, which is not extensively exploited in traditional coarse-grained architectures.

To address the above problems, in this paper we presents a novel architecture called ADRES (*Architecture for Dynamically Reconfigurable Embedded System*), which tightly couples a VLIW processor and a coarse-grained reconfigurable matrix. The VLIW processor and the coarse-grained reconfigurable matrix are integrated into one single architecture but with two virtual functional views. This level of integration has many advantages compared with other coarse-grained architectures, including improved performance, a simplified programming model, reduced communication costs and substantial resource sharing. Nowadays, new programmable architecture can not succeed without good support for mapping applications. In our previous work, we built a compiler framework for a family of coarse-grained architectures [6]. A novel modulo scheduling algorithm was developed to exploit the loop-level parallelism efficiently[7]. In this paper, we present how this compiler framework can be adapted to the ADRES architecture. In addition, some new techniques are proposed to solve the integration problem of the VLIW processor and the reconfigurable matrix.

The paper is organized as follow. Section 2 describes the proposed ADRES architecture and analyzes its main advantages. Section 3 discusses how the compiler framework is ported to the ADRES architecture and some considerations of the compilation techniques. Section 4 reports experimental results. Section 5 covers related work. Section 6 concludes the paper and presents future work.

2   ADRES Architecture

2.1   Architecture Description

Fig. 1 describes the system view of the ADRES architecture. It is similar to a processor with an execution core connected to a memory hierarchy. The ADRES core(fig 3) consists of many basic components, including mainly FUs and register files(RF), which are connected in a certain topology. The FUs are capable of executing word-level operations selected by a control signal. The RFs can store intermediate data. The whole ADRES matrix has two functional views, the VLIW processor and the reconfigurable matrix. These two functional views share some physical resources because their executions will never overlap with each other thanks to the processor/co-processor model. For the VLIW processor, several FUs are allocated and connected together through one multi-port register file, which is typical for VLIW architecture. Compared with the counterparts of the reconfigurable matrix, these FUs are more powerful in terms of functionality and speed. They can execute more operations such as branch operations. Some of these FUs are connected to the memory hierarchy, depending on available ports. Thus the data access to the memory is done through the load/store operation available on those FUs.

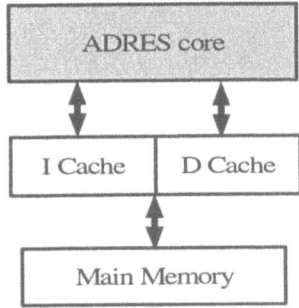

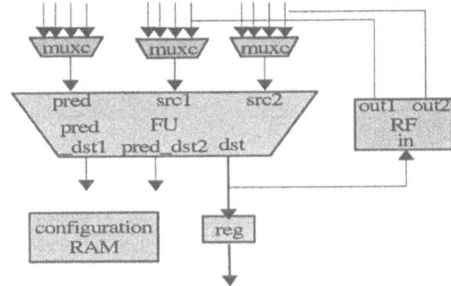

Fig. 1. ADRES system   **Fig. 2.** Example of a Reconfigurable Cell

For the reconfigurable matrix part, apart from the FUs and RF shared with the VLIW processor, there are a number of *reconfigurable cells*(RC) which basically comprise FUs and RFs too(fig. 2). The FUs can be heterogeneous supporting different operation sets. To remove the control flow inside loops, the FUs support predicated operations. The distributed RFs are small with less ports. The multiplexors are used to direct data from different sources. The configuration RAM stores a few configurations locally, which can be loaded on cycle-by-cycle basis. The configurations can also be loaded from the memory hierarchy at the cost of extra delay if the local configuration RAM is not big enough. Like instructions in ISPs, the configurations control the behaviour of the basic components by selecting operations and multiplexors. The reconfigurable matrix is used to accelerate the dataflow-like kernels in a highly parallel way. The matrix also includes the FUs and RF of the VLIW processor. The access to the memory of the matrix is also performed through the VLIW processor FUs.

In fact, the ADRES is a template of architectures instead of a fixed architecture. An XML-based architecture description language is used to define the communication topology, supported operation set, resource allocation and timing of the target architecture [6]. Even the actual organization of the RC is not fixed, FUs and RFs can be put together in several ways, for example, two FUs can share one RF. The architecture shown in fig. 3 and fig. 2 is just one possible instance of the template. The specified architecture will be translated to an internal architecture representation to facilitate compilation techniques.

2.2 Improved Performance with the VLIW Processor

Many coarse-grained architectures consist of a reconfigurable matrix and a relatively slow RISC processor, e.g., TinyRisc in MorphoSys [1] and ARC in Chameleon [3]. These RISC processors execute the unaccelerated part of the application, which only represents a small portion of execution time. However, such a system architecture has some problems due to the huge performance gap between the RISC and the reconfigurable matrix. According to Amdahl's law [5], the performance gain that can be obtained by improving some portion of an application can be calculated as equation 1. Suppose the kernels, representing 90%

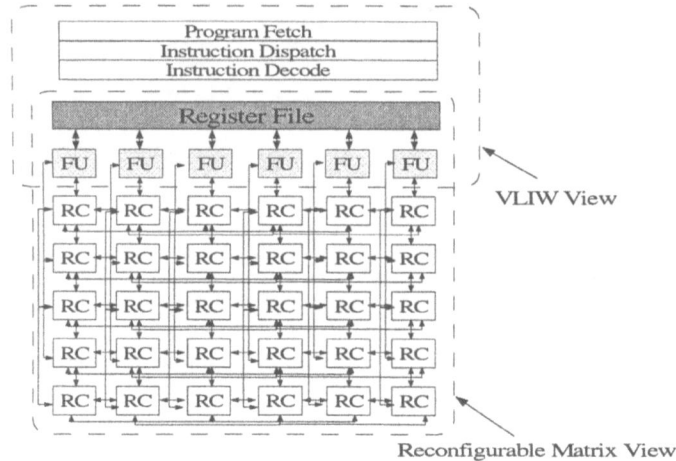

Fig. 3. ADRES core

of execution time, are mapped to the reconfigurable matrix to obtain 30 times of acceleration over the RISC processor, the overall speedup is merely 7.69. Obviously a high kernel speedup is not translated to a high overall speedup. The unaccelerated part, which is often irregular and control-intensive, becomes a bottleneck. Speeding up this part is essential for the overall performance. Although it is hard to exploit higher parallelism for the unaccelerated part on the reconfigurable matrix, it is still possible to discover *instruction-level parallelism* (ILP) using a VLIW processor, where 2-4 times speedup over the RISC is reasonable. If we recalculate the speedup with the assumption of 3 times acceleration for the unaccelerated code, the overall acceleration is now 15.8, much better than the previous scenario. This simple calculation proves the importance of a balanced system. The VLIW can help to improve the overall speedup dramatically in certain circumstances.

$$Speedup_{overall} = \frac{1}{(1 - Fraction_{enhanced}) + \frac{Fraction_{enhanced}}{Speedup_{enhanced}}} \tag{1}$$

2.3 Simplified Programming Model and Reduced Communication Cost

A simplified programming model and reduced communication cost are two important advantages of the ADRES architecture. These are achieved by making the VLIW processor and the reconfigurable matrix share access to the memory.

In traditional reconfigurable architectures, the processor and the reconfigurable matrix are essentially separated. The communication is often through explicit data copying. The normal execution steps are: (1) copy the data from the VLIW memory to that of the reconfigurable matrix; (2) the reconfigurable matrix part computes the kernel; (3) the results are copied back from the memory of the reconfigurable matrix to that of the VLIW processor. Though some techniques are adopted to reduce the data copying, e.g., wider data bus and DMA controller, the overhead is still considerable in terms of performance and energy. From the programming point of view, the separated processor and reconfigurable matrix require significant code rewriting. Starting from a software implementation to map kernels to the matrix, we have to identify the data structures used for communication and replace them with communication primitives. Data analysis should be done to make sure as few as possible data are actually copied. In addition, the kernels and the rest of the code have to be cleanly separated in such a way that no shared access to any data structure remains. These transformations are often complex and error-prone.

In the ADRES architecture, the data communication is performed through the shared RF and memory. This feature is very helpful to map high-level language code such as C to the ADRES architecture without major changes. When a high-level language is compiled to a processor, the local variables are normally allocated in the RF, whereas the static variables and arrays are allocated in the memory space. When the control of the program is transfered between the VLIW processor and the reconfigurable matrix, those variables used for communication can stay in the RF or the memory as they were. The copying is unnecessary because both the VLIW processor and the reconfigurable matrix share access to the RF and memory hierarchy. From programming point of view, this *shared-memory* architecture is more compiler-friendly than the *message-passing* one. Moreover, the RF and memory are alternately shared instead of being simultaneously shared. This eliminates data synchronizing and integrity problems. Code doesn't require any rewriting and can be handled by compiler automatically.

2.4 Substantial Resource Sharing

Since the basic components such as the FUs and RFs of the reconfigurable matrix and those of the VLIW processor are basically the same, one natural thinking is that resources might be shared to have substantial cost-saving. In other coarse-grained reconfigurable architectures, the resources cannot be effectively shared because the processor and the reconfigurable matrix are two separated parts. For example, the FU in the TinyRisc of MorphoSys cannot work cooperatively with the reconfigurable cells in the matrix. In the ADRES architecture, since the VLIW processor and the reconfigurable matrix are indeed two virtual functional views of the same physical entity, many resources are shared among these two parts. Due to its processor/co-processor model, only one of the VLIW processor and the reconfigurable matrix is active at any time. This fact makes the resource sharing possible. Especially, most components of the VLIW processor are reused in the reconfigurable matrix as shown in fig. 3. Although the amount of VLIW resources is only a fraction of those of the reconfigurable matrix, they are generally more powerful. For example, the FUs of the VLIW processor can execute more operations. The register file has much more ports than the counterparts in the reconfigurable matrix. In other words, the resources of the VLIW processor are substantial in terms of functionality. Reusing these resources can help to improve the performance and increase the schedulablity of kernels.

3   Adaptations of Compilation Techniques

Given the ever-increasing pressure of time-to-market and complexity of applications, the success of any new programmable architecture is more and more dependent on good design tools. For example, VLIW processors have gained huge popularity among DSP/multimedia applications although they are neither the most power- or performance-efficient ones. One important reason is that they have mature compiler support. An application written in a high-level programming language can be automatically mapped to a VLIW with reasonable quality. Compared with other coarse-grained reconfigurable architectures, the ADRES architecture is more compiler-friendly due to the simplified programming model discussed in section 2.3. However, some new compilation techniques need to be adopted to fully exploit the potential of the architecture.

3.1   Compilation Flow Overview

Previously, we have developed a compiler framework for a family of coarse-grained reconfigurable architectures [6]. A novel modulo scheduling algorithm and an abstract architecture representation were also proposed to exploit loop-level parallelism [7]. They have been adapted to the ADRES architecture. The overall compilation flow is shown in fig. 4. We use the IMPACT compiler framework [8] as a frontend to parse C source code, do some optimization and analysis, and emit the intermediate representation (IR), which is called *lcode*. Taking *lcode* as input, the compiler first tries to identify the pipelineable loops, which can be accelerated by the reconfigurable matrix. Then, the compilation process is divided into two paths that are for the VLIW processor and the reconfigurable matrix respectively. The identified loops are scheduled on the reconfigurable matrix using the modulo scheduling algorithm we developed [7]. The scheduler takes advantage of the shared resources, e.g., the multi-port VLIW register file,

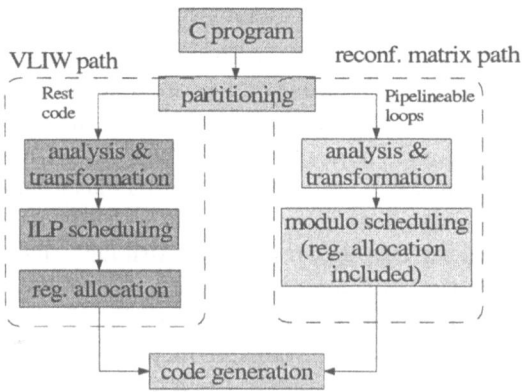

Fig. 4. Compilation Flow for the ADRES architecture

to maximize performance. The remaining code is mapped to the VLIW processor using regular VLIW compilation techniques, including ILP scheduling and register allocation. Afterwards, the two parts of scheduled code are put together, ready for being executed by the ADRES architecture.

3.2   Interface Generation

The compilation techniques for the VLIW architecture are already mature and the main compilation techniques for the coarse-grained architecture were developed in our previous work. Adapted to the ADRES architecture, the most important problem is how to make the VLIW processor and the reconfigurable matrix work cooperatively and communicate with each other. Thanks to ADRES's compiler-friendly features, interface generation is indeed quite simple(fig. 5).

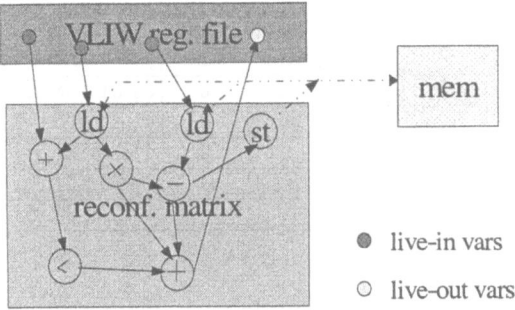

Fig. 5. Interfacing between the VLIW processor and the Reconfigurable matrix

Each loop mapped to the matrix has to communicate with the rest of application, e.g., taking input data and parameters, and writing back results. As mentioned in section 2.3, the communication of the ADRES architecture is performed through shared register file and shared memory. Using dataflow analysis, the *live-in* and *live-out* variables are identified, which represents the input and output data communicated through the shared register file. These variables will be allocated in the VLIW register file. Since these variables will occupy some register space throughout the lifetime of the loops, they are subtracted from the capacity of the VLIW register file. Therefore the scheduler won't overuse the VLIW register file for other tasks. As to the variables mapped to the memory, we don't need to do anything. The mapped loop can access the correct address through the load/store operations available on some FUs of the VLIW processor.

4   Experimental Results

For the purpose of experiment, an architecture resembling the topology of MorphoSys [1] is instantiated from the ADRES template. In this configuration, a total of 64 FUs are divided into four tiles, each of which consists of 4x4 FUs. Each FU is not only connected to the 4 nearest neighbor FUs, but also to all FUs within the same row or column in this tile. In addition, there are row buses and column buses across the matrix. The first row of FUs is also used by the VLIW processor, and are connected to a multi-port register file. Only the FUs in the first row are capable of executing memory operations, i.e., load/store operations.

The testbench consists of 4 programs, which are derived from C reference code of TI's DSP benchmarks and MediaBench [9]. The *idct* is a 8x8 inverse discrete cosine transformation, which consists two loops. The *adpcm-d* refers to an ADPCM decoder. The *mat_mul* computes matrix multiplication. The *fir_cpl* is a complex FIR filter. They are typical multimedia and digital signal processing applications with abundant inherent parallelism.

Table 1. Schedule results

loop	no. of ops	live-in vars	live-out vars	II	IPC	sched. density
idct1	93	4	0	3	31	48.4%
idct2	168	4	0	4	42	65.6%
adpcm-d	55	9	2	4	13.8	21.5%
mat_mul	20	12	0	1	20	31.3%
fir_cpl	23	9	0	1	23	35.9%

The schedule results are shown in table 1. The second column refers to the total number of operations within the loop body. The II is *initiation interval*, meaning the loop starts a new iteration every II cycles [10]. The live-in and live-out variables are allocated in the VLIW register file. The instructions-per-cycle (IPC) reflects how many operations are executed in one cycle on average. Scheduling density is equal to $IPC/No.ofFUs$. It reflects the actual utilization of all FUs for computation. The results show the IPC is pretty high, ranging from 13.8 to 42. The FU utilization is ranged from 21.5% to 65.6%. For kernels such as *adpcm-d*, the results are constrained by achievable minimal II(MII).

The table 2 shows comparisons with the VLIW processor. The tested VLIW processor has the same configuration as the first row of the tested ADRES architecture. The compilation and simulation results for VLIW architecture are obtained from IMPACT, where aggressive optimizations are enabled. The re-

Table 2. Comparisons with VLIW architecture

app.	total ops (ADRES)	total cycles (ADRES)	total ops (VLIW)	total cycles (VLIW)	speed-up
idct	211676	6097	181853	38794	6.4
adpcm_d	8150329	594676	5760116	1895055	3.2
mat_mul	20010518	1001308	13876972	2811011	2.8
fir_cpl	69126	3010	91774	18111	6.0

sults for the ADRES architecture are obtained from a developed co-simulator, which is capable of simulating the mixed VLIW and reconfigurable matrix code. Although these testbenches are small applications, the results already reflect the integration impact of the VLIW processor and the reconfigurable matrix. The speed-up over the VLIW is from 2.8 to 6.4, showing pretty good performance.

5   Related Work

Many coarse-grained reconfigurable architectures have been proposed in recent years. MorphoSys [1] and REMARC [4] are typical ones consisting of a RISC processor and a fabric of reconfigurable units. For MorphoSys the communication is performed through a DMA controller and a so-called frame buffer. In REMARC, the coupling is tighter. The matrix is used as a co-processor next to the MIPS processor. Neither of these architectures has compiler support for the matrix part. Chameleon [3] is a commercial architecture that comprises an ARC processor and a reconfigurable processing fabric as well. The communication is through a 128-bit bus and a DMA controller. The data has to be copied between the two memory spaces. Compiler support is limited to the processor side.

Another category of reconfigurable architectures presents much tighter integration. Examples are ConCise [11], PRISC [12] and Chimaera [13]. In these architectures, the reconfigurable units are deeply embedded into the pipeline of the processor. Customized instructions are built with these reconfigurable units. The programming model is simplified compared with the previous category because resources such as memory ports and register file are exposed to both the processor and the reconfigurable units. This leads to good compiler support. However, these architectures do not have much potential for performance, constrained by limited exploitable parallelism.

6   Conclusions and Future Work

Coarse-grained reconfigurable architectures have been gaining importance recently. Many new architectures are proposed, which normally comprise a processor and a reconfigurable matrix. In this paper, we address the integration problem between the processor and the reconfigurable matrix, which has not received enough attention in the past. A new architecture called ADRES is proposed, where a VLIW processor and a reconfigurable matrix are tightly coupled in a single architecture and many resources are shared. This level of integration brings a lot of benefits, including increased performance, simplified programming model, reduced communication cost and substantial resource sharing.

get fpl03.bblOur compiler framework was adapted to the new without much difficulty. It proves that the ADRES architecture is very compiler-friendly. The VLIW compilation techniques and the compilation techniques for the reconfigurable matrix can be applied to the two parts of the ADRES architecture respectively. The partitioning and interfacing of the accelerated loops and the rest of code can be handled by the compiler without requiring code rewriting.

However, we have not implemented the ADRES architecture at the circuit level yet. Therefore, many detailed design problems have not been taken into account and concrete figures such as area and power are not available. Hence, to implement the ADRES design is in the scope of our future work. On the other hand, we believe the compiler is even more important than the architecture. We will keep developing the compiler to refine the ADRES architecture from the compiler point of view.

References

1. Singh, H., Lee, M.H., Lu, G., Kurdahi, F.J., Bagherzadeh, N., Filho, E.M.C.: Morphosys: an integrated reconfigurable system for data-parallel and computation-intensive applications. IEEE Trans. on Computers **49** (2000) 465–481
2. C. Ebeling, D. Cronquist, P.F.: RaPiD - reconfigurable pipelined datapath. In: Proc. of Field Programmable Logic and Applications. (1996)
3. : Chameleon Systems Inc. (2002) http://www.chameleonsystems.com.
4. Miyamori, T., Olukotun, K.: REMARC: Reconfigurable multimedia array coprocessor. In: FPGA. (1998) 261
5. Patterson, D.A., Hennessy, J.L.: Computer Architecture: A Quantitative Approach. Morgan Kaufmann Publishers, Inc. (1996)
6. Mei, B., Vernalde, S., Verkest, D., Man, H.D., Lauwereins, R.: DRESC: A retargetable compiler for coarse-grained reconfigurable architectures. In: International Conference on Field Programmable Technology. (2002)
7. Mei, B., Vernalde, S., Verkest, D., Man, H.D., Lauwereins, R.: Exploiting loop-level parallelism for coarse-grained reconfigurable architecture using modulo scheduling. In: Proc. Design, Automation and Test in Europe (DATE). (2003)
8. Chang, P.P., Mahlke, S.A., Chen, W.Y., Warter, N.J., Hwu, W.W.: IMPACT: An architectural framework for multiple-instruction-issue processors. In: Proceedings of the 18th International Symposium on Computer Architecture (ISCA). (1991)
9. Lee, C., Potkonjak, M., Mangione-Smith, W.H.: Mediabench: A tool for evaluating and synthesizing multimedia and communicatons systems. In: International Symposium on Microarchitecture. (1997) 330–335
10. Rau, B.R.: Iterative modulo scheduling. Technical report, Hewlett-Packard Lab: HPL-94-115 (1995)
11. Kastrup, B.: Automatic Synthesis of Reconfigurable Instruction Set Accelerations. PhD thesis, Eindhoven University of Technology (2001)
12. Razdan, R., Brace, K., Smith, M.D.: Prisc software acceleration techniques. In: Proc. 1994 IEEE Intl. Conf. on Computer Design. (1994)
13. Hauck, S., Fry, T.W., Hosler, M.M., Kao, J.P.: The Chimaera reconfigurable functional unit. In: Proc. the IEEE Symposium on FPGAs for Custom Computing Machines. (1997)

Inter-processor Connection Reconfiguration Based on Dynamic Look-Ahead Control of Multiple Crossbar Switches

Eryk Laskowski[1] and Marek Tudruj[1,2]

[1] Institute of Computer Science Polish Academy of Sciences
ul. Ordona 21, 01-237 Warsaw, Poland
[2] Polish-Japanese Institute of Information Technology
ul. Koszykowa 86, 02-008 Warsaw, Poland
{laskowsk, tudruj}@ipipan.waw.pl

Abstract. A parallel system architecture for program execution based on the look-ahead dynamic reconfiguration of inter-processor connections is discussed in the paper. The architecture is based on inter-processor connection reconfiguration in multiple crossbar switches that are used for parallel program execution. Programs are structured into sections that use fixed inter-processor connections for communication. The look-ahead dynamic reconfiguration assumes that while some inter-processor connections in crossbar switches are used for current section execution, other connections are in advance configured for execution of further sections. Programs have to be decomposed into sections for given time parameters of reconfiguration control, so, as to avoid program execution delays due to connection reconfiguration. Automatic program structuring is proposed based on the analysis of parallel program graphs. The structuring algorithm finds the partition into sections that minimizes the execution time of a program executed with the look-ahead created connections. The program execution time is evaluated by simulated program graph execution with reconfiguration control modeled as an extension of the basic program graph.

1 Introduction

A new kind of parallel program execution environment based on dynamically reconfigurable connections between processors [1-4] is the main interest of this paper. The proposed environment assumes new paradigm of point-to-point inter-processor connections. Link connection reconfiguration involves time overheads in program execution. These overheads cannot be completely eliminated by an increase of the speed of communication/reconfiguration hardware. However, they can be neutralized by a special method applied at the level of system architecture and at the level of program execution control. This special method is called the look-ahead dynamic link connection reconfigurability [5,6]. Special architectural solutions for the look-ahead reconfigurability consist in increasing the number of hardware resources used for link connection setting (multiple crossbar switches) and using them interchangeably in parallel for program execution and run-time look-ahead reconfiguration. Application programs are partitioned into sections executed by clusters of processors whose mu-

P.Y.K. Cheung et al. (Eds.): FPL 2003, LNCS 2778, pp. 71–80, 2003.
© Springer-Verlag Berlin Heidelberg 2003

tual connections are prepared in advance. The connections are set in crossbar switches and remain fixed during section execution. At sections boundaries, processor's communication links are switched to look-ahead configured crossbar switches.

If a program can be partitioned into sections whose connection reconfiguration does not delay communication, we obtain the quasi-time transparency of reconfiguration control since connection reconfiguration overlaps with program execution. Then, the multi-processor system behaves as a fully connected processor structure.

An algorithm for program partitioning into sections for execution with the look-ahead prepared connections was designed. Sections are defined by program graph analysis performed at compile time. The algorithm finds the graph partition and the number of crossbar switches that provide time transparent connection control. It is based on list scheduling and iterative task clustering heuristics. The optimization criterion is the total execution time of a program. This time is determined by program graph symbolic execution, which takes into account time parameters of the system.

The paper consists of three parts. In the first part, the look-ahead inter-processor connection reconfiguration principles are discussed. In the second part, main features of the applied graph partitioning algorithm are discussed. In the third part, experimental results of efficiency measures of program execution based on the look-ahead dynamic reconfiguration are shown and discussed.

2   The Principle of the Look-Ahead Connection Reconfiguration

The look-ahead dynamic connection reconfiguration assumes anticipated inter-processor connection setting in communication resources provided in the system. An application program is partitioned into sections, which assume fixed direct inter-processor connections, Fig. 1. Connections for next sections are prepared while current sections are executed. Before execution of a section, the prepared connections are enabled for use in parallel and in a very short time. Thus, this method can provide inter-processor connections with almost no delay in the linear program execution time. In other words it can provide a time transparent control for dynamic link connection

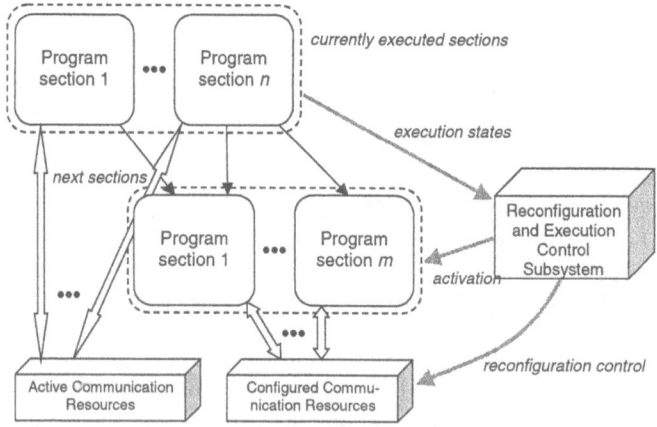

Fig. 1. Program execution based on the look-ahead connection reconfiguration

reconfiguration. The redundant resources used for anticipated connection setting can be link connection switches (crossbar switches, multistage connection networks), processor sets and processor link sets [5].

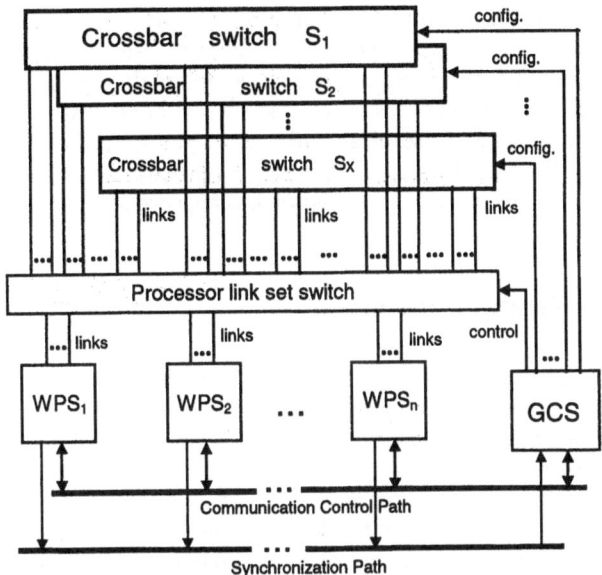

Fig. 2. Look-ahead reconfigurable system with multiple connection switches

The general scheme of a look-ahead dynamically reconfigurable system based on redundancy of link connection switches is shown in Fig. 2. Worker processor subsystems (WPSi) can consist of a single processor or can include a data processor and a communication processor sharing a common memory. WPSis have sets of communication links connected to crossbar switches S_1, ..., S_x, which are interchangeably used as active and configured communication resources. The subsystem links can be switched between the switches by the Processor Link Set Switch. This switch is controlled by the Global Control Subsystem (GCS). CGS collects messages on the section execution states in worker subsystems (link use termination) sent via the Control Communication Path. The simplest implementation of such path is a bus. Depending on the availability of links in the switches S_1, ..., S_x, GCS prepares connections for execution of next program sections. In parallel with reconfiguration, synchronization of states of all processors in clusters for next sections is performed using a hardware Synchronization Path [6]. When all connections for a section are ready and the synchronization has been reached, GCS switches all links of processors, which will execute the section, to the look-ahead configured connections in a proper switch. Then, it enables execution of the section in involved worker processors.

Program sections for execution with the look-ahead connection reconfiguration are defined at compile time by an analysis of the program graph. Three basic program execution strategies can be identified which differ in granularity of control:

1) synchronous, with inter-section connection switching controlled at the level of all worker processors in the system,

2) asynchronous processor-restrained, where inter-section connection switching is controlled at the level of dynamically defined worker processor clusters,

3) asynchronous link-restrained, with granularity of control at the level of single processor links.

The strategies require synchronization of process states in processors in different ways. In the synchronous strategy, processes in all processors in the system have to be synchronized when they reach the end of section points. In the asynchronous processor-restraint strategy, process states in selected subsets of processors are synchronized when they reach the end of use of all links in a section. In the asynchronous link-restraint strategy, the end of use of links in pairs of processors has to be synchronized. In this paper we will be interested in the asynchronous processor-restrained strategy.

3   Program Graph Partitioning

The initial program representation is a weighted Directed Acyclic Graph (DAG) with computation task nodes and communication i.e. node data dependency edges. Programs are executed according to the macro-dataflow model governed by arrivals of all data from all node predecessors. A task executes without pre-emption and, after completion, it sends data to all successor tasks. The graph is static and deterministic.

A two-phase approach is used to tackle the problem of scheduling and graph partitioning in the look-ahead reconfigurable environment [9]. In the first phase, a scheduling algorithm reduces the number of communications and minimizes program execution time, based on program DAG. The program schedule is defined as task-to-processor and communication-to-link assignment with specification of starting time of each task and communication. Assignment preserves the precedence constraints coming from the DAG. The scheduling algorithm is an improved version of ETF /Earliest Task First/ scheduling [8]. In the second phase, scheduled program graph is partitioning into sections for the look-ahead execution in the assumed environment.

A program with specified schedule is expressed in terms of the Assigned Program Graph (APG), Fig. 3. This graph is composed of the non-communicating code nodes (rectangles) and communication instruction nodes (circles). There are activation edges and communication edges in the graph. Weights of nodes represent execution times. Communication is done according to the synchronous model. All nodes are assigned to processors and all communication edges to processor link connections. The graph is assumed static and acyclic. All control in the graph is deterministic.

Program sections correspond to such sub-graphs in the APG for which the following conditions are fulfilled:

i.   section sub graphs are mutually disjoint in respect to communication edges connecting processes allocated to different processors,

ii.   sub-graphs are connected in respect to activation and communication edges,

iii.   inter-processor connections do not change inside a section sub-graph,

iv.   section sub-graphs are complete in respect to activation paths and include all communication edges incident to all communication nodes on all activation paths between any two communication nodes which belong to a section,

v.   all connections for a section are prepared before this section activation, which simultaneously concerns all processors involved in section execution.

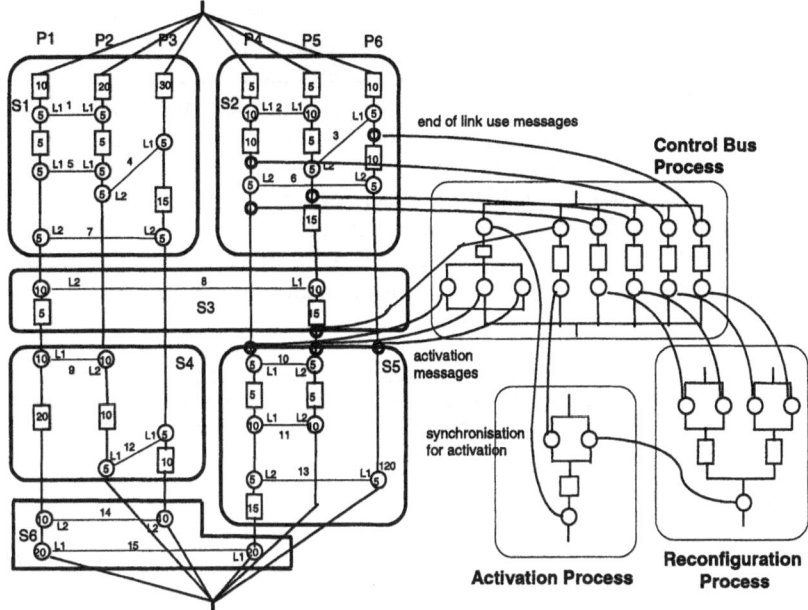

Fig. 3. Modeling the program and the reconfiguration control as a common graph

Program execution time is the optimization criterion. It is estimated by simulated execution of the program graph in a modeled look-ahead reconfigurable system. A partitioned APG graph is automatically extended at all section boundaries by sub-graphs, which model the look-ahead reconfiguration control, Fig. 3. The Communication Control Path, Synchronization Path and Global Control Subsystem are modeled as PAR structures executed on virtual additional processors. Connection setting is modeled by pairs of parallel nodes with input communications ("end of links use" messages) from worker subsystems. After their synchronization, a reconfiguration process is executed by a "Reconfiguration Processor". Weights of nodes correspond to latencies of control actions, such as crossbar switch reconfiguration, bus latency, etc. Synchronization of processors for link switching between crossbars is modeled in the "Activation Processor". Activation of next section, is done by multicast transmissions over control bus to all involved worker subsystems.

The algorithm finds program graph partition into sections assigned to crossbar switches. It also finds the minimal number of switches that allow reconfiguration time transparency. Its outline is given in Fig. 4. To simplify graph analysis, a Communication Activation Graph (CAG) is introduced, which contains nodes and edges that correspond to communication edges and activation paths between communication edges of the initial APG graph, respectively. The algorithm starts with all communications assigned to the same crossbar switch. In each step, a vertex of CAG is selected and then, the algorithm tries to include this vertex to a union of existing sections determined by edges of the current vertex. The heuristics tries to find such a union of sections, which doesn't break rules of graph partitioning. The union, which gives the shortest program execution time, is selected. The choice of the vertex for visiting depends on the following APG and CAG graph parameters used in heuristic rules: the critical path of APG, the delay of vertex of CAG, reconfiguration criticality function for the investigated vertex, and the dependency on link use between communications, see [10] for details. When section clustering doesn't give any execution time improvement, the section of the current vertex is left untouched and crossbar switch is assigned to it. As with section clustering, the choice of the switch depends on program execution time. When algorithm cannot find any crossbar switch for section that allows creating connections with no reconfiguration time overhead, then current number of used switches (*curr_x* in Fig. 4) is increased by 1 and algorithms is restarted. Vertices can be visited many times. The algorithm stops when with all vertices visited, no further time improvement is obtained. A list of visited vertices (*tabu list*) is used to prevent algorithm from frequent visiting small subset of vertices.

```
Begin
B := initial set of section, each section is composed of
        single communication and assigned to crossbar 1
curr_x := 1     {current number of switches used}
finished := false
While not finished
    Repeat until each vertex of CAG is visited and there is no
                execution time improvement during last β steps {1}
        v := vertex of CAG which maximizes the selection function
                and which is not placed in tabu list
        S := set of sections that contain communications of all
                predecessors of v
        M := Find_sections_for_clustering( v, S )
        If M ≠ ∅ Then
            B := B - M
            Include to B a new section built of v and
            communications that are contained in sections in M
        Else
            s := section that consists of communication v
            Assign crossbar switch (from 1..curr_x) to section s
            If reconfiguration introduces time overheads Then
                curr_x := curr_x + 1
                Break Repeat
            EndIf
        EndIf
    EndRepeat
    finished := true
EndWhile
End
```

Fig. 4. The general scheme of the graph partitioning algorithm

4   Experimental Results

The described algorithm finds optimal program partitions, which correspond to given reconfiguration control parameters. It can be also used for evaluation and comparative studies of execution efficiency of programs for different program execution control strategies and system parameters. The following parameters were used to characterize an application program in the experiments that were performed:

t_R - reconfiguration time of a single connection in a crossbar switch,

t_v – section activation time overhead,
a   – average time interval between connection reconfigurations in a program,
$R = a / (t_R + t_v)$ – reconfiguration control efficiency for a program.

Reconfiguration control efficiency of a given system for the program is a ratio of the average time interval between reconfigurations of processor links in the program to the connection reconfiguration service time in the system. It represents suitability of the reconfiguration control for execution of the program.

Experiments with execution of exemplary program graphs have shown that the look-ahead connection reconfiguration behaves better (positive and large program execution speedup S_q with the look-ahead reconfiguration versus execution with the on-request reconfiguration) for systems in which the reconfiguration efficiency is poor (low values of the parameter **R**), Fig. 5. For systems with sufficiently fast connection reconfiguration control, the on-request reconfiguration can be sufficient (introduces lower reconfiguration control overhead). This confirms the suitability of the look-ahead reconfiguration for fine-grain parallel systems.

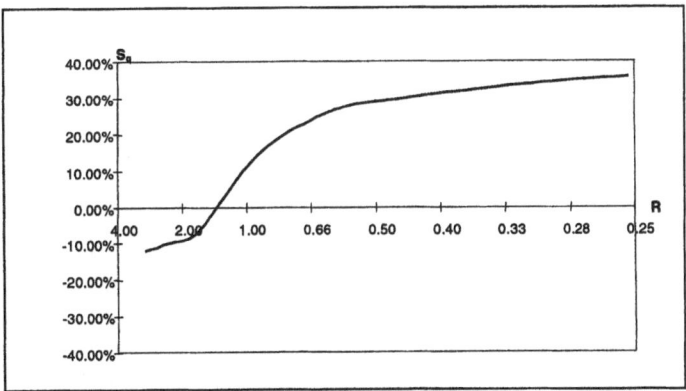

Fig. 5. On-request connection reconfiguration versus look-ahead reconfiguration.

We have examined program execution speedup versus parameters of reconfiguration control (t_R and t_v), the number of crossbar switches, the number of processor links, for the parallel program of Strassen matrix multiplication [11] (two recursion levels). Fig. 6 presents speedup on 12 processors with 2 and 4 links vs. execution on 1 processor with the on-request reconfiguration. The speedup is 7 and 9 in a narrow area of the lowest values of t_v, t_R. With the look-ahead reconfiguration, Fig. 7, the speedup increases to 9.5 in a larger area of lowest t_v, t_R. With an increase in the number of crossbars, the high speedup area increases and is much larger than for the on-request reconfiguration. Fig. 8 shows the reduction of the reconfiguration control time overhead when look-ahead control is used instead of on-request for 2 and 4 crossbars. When reduction is close to 100%, the system behaves as a system with fully-connected inter-processor network. It happens in a narrow area of t_v, t_R that increases when there are more crossbars used. When the number of switches is 6, Fig. 9, the good speedup and reconfiguration overhead area is much larger. So, multiple crossbar switches used with the look-ahead control strongly reduce reconfiguration time overheads. The larger is the number of processor links, the look-ahead method is prevailing vs. on-request for a wider range of reconfiguration and activation time parameters.

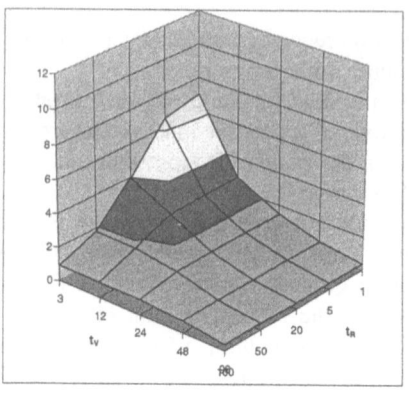 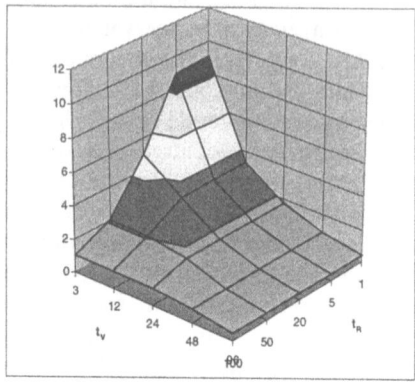

a) 12 processors, 2 links   b) 12 processors, 4 links

Fig. 6. Speed-up for Strassen algorithm in on-request environment, single crossbar switch.

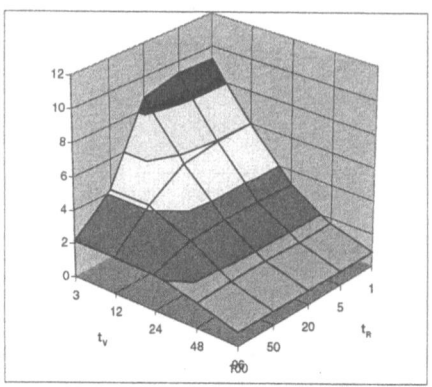 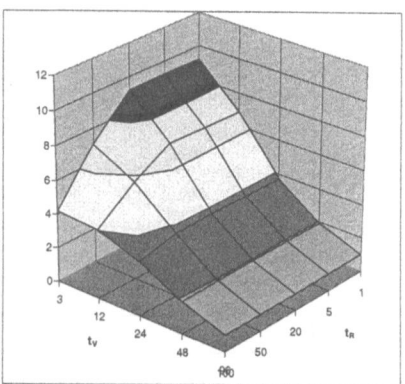

a) 12 processors, 4 links, 2 x-bar switches   b) 12 processors, 4 links, 4 x-bar switches

Fig. 7. Speedup for Strassen algorithm executed in the look-ahead environment.

5   Conclusions

The paper has presented and discussed the concept of the connection look-ahead dynamic reconfigurability in multi-processor systems based on multiple crossbar switches. Program partitioning into sections executed with fixed inter-processor connections is the essential feature of this environment. Programs can be automatically structured for the look-ahead reconfiguration paradigm using the graph partitioning algorithm presented in the paper. For a sufficient number of crossbar switches, the system can behave - from a program point of view- as fully connected multi-processor cluster. This number depends on relation between program time parameters – the granularity of parallelism and time parameters of the dynamic reconfiguration control.

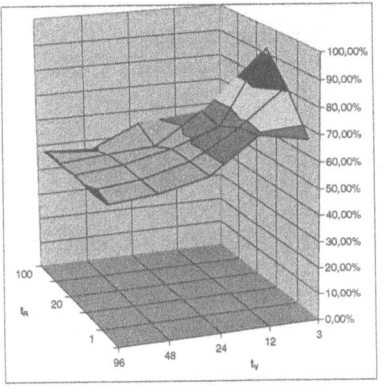
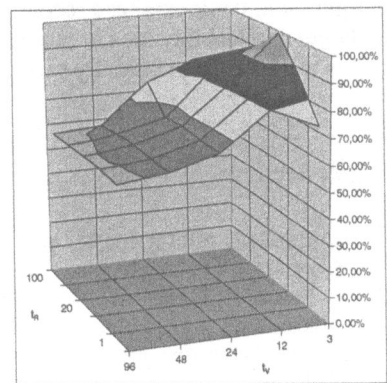

a) 12 processors, 4 links, 2 x-bar switches   b) 12 processors, 4 links, 4 x-bar switches

Fig. 8. Reduction of reconfiguration control overhead for Strassen algorithm with the look-ahead reconfiguration vs. on-request method.

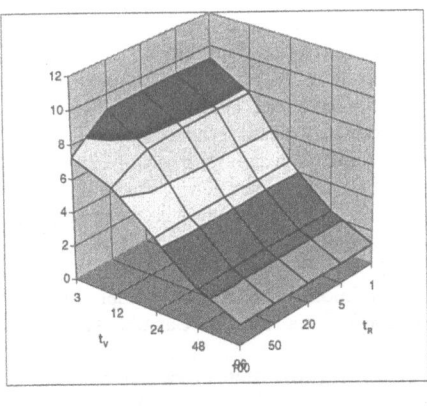
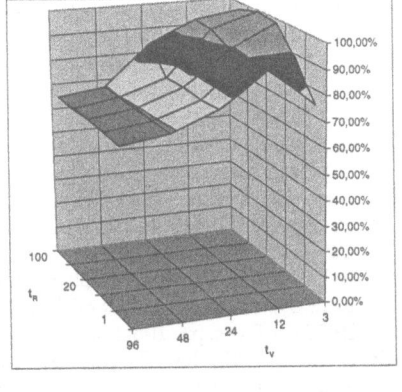

a)   b)

Fig. 9. Speedup (a) and reduction of reconfiguration time overhead (b) for Strassen algorithm executed in the look-ahead environment with 12 processors, 4 links, 6 x-bar switches

This relation can be evaluated using the notion of the reconfiguration control efficiency of a system for a program, introduced in the paper. Experiments show that the lower is the reconfiguration efficiency of the system, the better results are obtained from the application of the look-ahead dynamic connection reconfiguration. It confirms the value of the look-ahead reconfiguration for fine-grained parallel programs, which have time-critical reconfiguration requirements.

Experiments with Strassen algorithm show that quasi time transparency of inter-processor connection setting can be obtained for reconfiguration control time parameters whose area depends on the number of crossbar switches used for dynamic look-ahead connection reconfiguration. This area is narrow for 2 crossbar switches and becomes larger with the bigger number of switches used. Comparing the on-request reconfiguration, the look-ahead reconfiguration gives better program execution speedup and larger applicability area in respect to various system time parameters.

This work has been partially supported by the KBN research grant T11C 007 22.

References

[1]   T. Muntean, SUPERNODE, Architecture Parallele Dynamiquement Reconfigurable de Transputers, *11-emes Journees sur l'Informatique* , Nancy, Janvier 1989.

[2]   A. Bauch, R. Braam, E. Maehle, DAMP - A Dynamic Reconfigurable Multiprocessor System With a Distributed Switching Network, *2-nd European Conf. on Distributed Memory Computing*, Munich, 22-24 April, 1991, pp. 495-504.

[3]   M. Tudruj, "Connection by Communication" Paradigm for Dynamically Reconfigurable Multi-Processor Systems, *Proceedings of the PARELEC 2000*, Trois Rivieres, Canada, August 2000, IEEE CS Press, pp. 74-78.

[4]   M. Tudruj, Embedded Cluster Computing Through Dynamic Reconfigurability of Inter-Processor Connections, *Advanced Environments, Tools and Applications for Cluster Computing*, NATO Advanced Workshop, Mangalia, 1-6 Sept. 2001, Springer Verlag, LNCS 2326.

[5]   M. Tudruj, Multi-transputer architectures with the look-ahead dynamic link connection reconfiguration, *World Transputer Congress '95*, Harrogate, Sept. 1995.

[6]   M. Tudruj, Look-Ahead Dynamic Reconfiguration of Link Connections in Multi-Processor Architectures, *Parallel Computing '95*, Gent, Sept. 1995, pp. 539-546.

[7]   M. Tudruj, O. Pasquier, J.P. Calvez, Fine-grained process synchronization in multi-processors with the look-ahead inter-processor connection setting, *Proc. of 22nd Euromicro Conference: Short Contributions*, Prague, IEEE CS, Sept. 1996.

[8]   Jing-Jang Hwang, Yuan-Chien Chow, Frank D. Angers, Chung-Yee Lee; Scheduling Precedence Graphs in Systems with Inter-processor Communication Times, *Siam J. Comput.*, Vol. 18, No. 2, April 1989, pp. 244-257.

[9]   E. Laskowski, Program Graph Scheduling in the Look-Ahead Reconfigurable Multiprocessor System, *Proceedings of the PARELEC 2000*, Trois Rivieres, Canada, August 2000, IEEE CS Press, pp. 106-110.

[10] E. Laskowski, New Program Structuring Heuristics for Multi-Processor Systems with Redundant Communication Resources, *Proc. of the PARELEC 2002*, Warsaw, Poland, September 2002, IEEE CS Press, pp. 183-188.

[11] B. Wilkinson, M. Allen, *Parallel Programming, Techniques and Applications Using Networked Workstations and Parallel Computers*, Prentice Hall 1999, p. 327.

Arbitrating Instructions in an $\rho\mu$-Coded CCM

Georgi Kuzmanov and Stamatis Vassiliadis

Computer Engineering Lab, Electrical Engineering Dept.,
EEMCS, TU Delft, The Netherlands,
{G.Kuzmanov,S.Vassiliadis}@ET.TUDelft.NL, http://ce.et.tudelft.nl/

Abstract. In this paper, the design aspects of instruction arbitration in an $\rho\mu$-coded CCM are discussed. Software considerations, architectural solutions, implementation issues and functional testing of an $\rho\mu$-code arbiter are presented. A complete design of such an arbiter is proposed and its VHDL code is synthesized for the VirtexII Pro platform FPGA of Xilinx. The functionality of the unit is verified by simulations. A very low utilization of available reconfigurable resources is achieved after the design is synthesized. Simulations of an MPEG-4 case study suggest considerable performance speed-up in the range of 2,4-8,8 versus a pure software PowerPC implementation.

1 Introduction

Numerous design concepts and organizations have been proposed to support the Custom Computing Machine (CCM) paradigm from different prospectives [6, 7]. An example of a detailed classification of CCMs can be found in [8]. In this paper we propose a design of a potentially performance limiting unit of the MOLEN $\rho\mu$-coded CCM: the arbiter [11]. We discuss all design aspects of the arbiter, including software considerations, architectural solutions, implementation issues and functional testing. A synthesizable VHDL code of the arbiter has been developed and simulated. Performance has been evaluated theoretically and by experimentation using Virtex II Pro technology of Xilinx. Synthesis results and performance evaluation for an MPEG-4 case study suggest:

- Less than 1% of the reconfigurable hardware resources available on the selected FPGA (xc2vp20) chip are spent for the implementation of the arbiter.
- Considerable speed-ups in the range of 2,4-8,8 of the MPEG-4 encoder are feasible when the SAD function is implemented in the proposed framework.

The remainder of this paper is organized as follows. The section to follow contains brief background on the MOLEN $\rho\mu$-coded processor and describes the requirements to the design of a general arbiter. In Section 3, software-hardware considerations for a particular arbiter design for PowerPC and Virtex II Pro FPGA are presented. Section 4 discusses functional testing, performance analysis and an MPEG-4 case study with experimental results obtained from a real chip implementation of the arbiter. Finally, we conclude the paper with Section 5.

P.Y.K. Cheung et al. (Eds.): FPL 2003, LNCS 2778, pp. 81–90, 2003.

2 Background

This section presents the MOLEN $\rho\mu$-coded Custom Computing Machine organization, introduced in [11] and described in Fig. 1. The ARBITER performs a partial decoding on the instructions in order to determine where they should be issued. Instructions implemented in fixed hardware are issued to the core processor. Instructions for custom execution are redirected to the *reconfigurable unit*. The reconfigurable unit consists of a custom computing unit (CCU) and the $\rho\mu$-code unit. An operation, executed by the reconfigurable unit, is divided into two distinct phases: **set** and **execute**. The **set** phase is responsible for reconfiguring the CCU hardware enabling the execution of the operation. This phase may be divided into two subphases - partial set (**pset**) and complete set (**cset**). In the **pset** phase the CCU is partially configured to perform common functions of an application (or group of applications). Later, the **cset** sub-phase only reconfigures that blocks in the CCU, which are not covered in the **pset** sub-phase in order to *complete* the functionality of the CCU.

General Requirements to the Arbiter. The arbiter controls the proper co-processing of the GPP and the reconfigurable units. It is closely connected to three major units of the CCM, namely the GPP, the memory and the $\rho\mu$-unit. Each of these parts of the organization has its own requirements, which should be considered when an arbiter is designed. Regarding the core processor, the arbiter should: 1) Preserve the original behavior of the core processor when no reconfigurable instruction is executed. Create the shortest possible critical path penalties. 2) Emulate reconfigurable instruction execution behavior on the core processor using its original instruction set and/or other architectural features.

Regarding the $\rho\mu$-unit the arbiter should: 1) Distribute control signals and the starting microcode address to the $\rho\mu$-unit. 2) Consume minimal hardware resources if implemented in the same FPGA with the $\rho\mu$-unit. Thus more resources will be available for the CCU.

For proper memory management the arbiter should be designed to: 1) Arbitrate the data access between the $\rho\mu$-unit and the core processor. 2) Allow

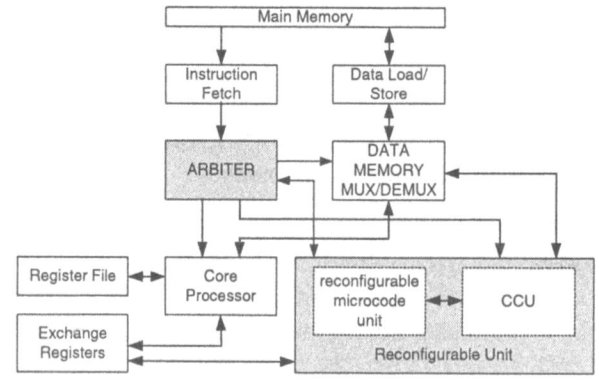

Fig. 1. The MOLEN machine organization

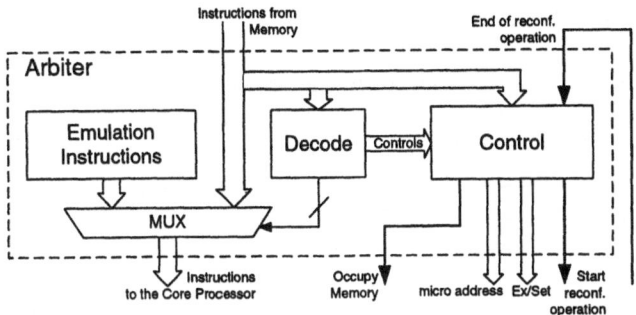

Fig. 2. General Arbiter organization

speeds within the capabilities of the utilized memory technology, i.e., not creating performance bottlenecks in memory transfers.

Reconfigurable instructions should be encoded in consistence with the instruction encoding of the targeted general purpose architecture. The arbiter should also provide proper timing for reconfigurable instruction execution to all units referred above. In Fig.2, a general view of an Arbiter organization is presented. The operation of such an arbiter is entirely based on decoding the input instruction flow. The unit either redirects instructions, or generates an instruction sequence to control the state of the core processor during reconfigurable operations. In such an organization, the critical path penalty to the original instruction flow can be reduced to just one 2-1 multiplexer. Once either of the three reconfigurable instructions has been decoded, the following actions are initiated:

1. Emulation instructions are multiplexed to the processor instruction bus. These instructions drive the processor into wait or halt state.
2. Control signals from the decoder are generated to the control block in Fig.2. Based on these controls, the control block performs the following: a) Redirects the microcode location address of the corresponding reconfigurable instruction to the $\rho\mu$-unit. b) Generates an internal code signal (Ex/Set) for the decoded reconfigurable instruction and delivers it to the $\rho\mu$-unit. c) Initiates reconfigurable operation by generating 'start reconf. operation' signal to the $\rho\mu$-unit. d) Reserves the data memory control for the $\rho\mu$-unit by generating *memory occupy* signal to the (data) memory controller. e) Enters a wait state until signal 'end of reconf. operation' arrives.

An active 'end of reconf. operation' signal initiates the following actions: 1) Data memory control is released back to the core processor. 2) An instruction sequence is generated to ensure proper exiting of the core processor from the wait state. 3) After exiting the wait state, the program flow control is transferred to the instruction immediately after the reconfigurable instruction, executed last.

3   CCM Implementation

The general organization presented in the previous section has been implemented on Virtex II Pro, a platform FPGA chip of Xilinx.

Software Considerations: Because of performance reasons, we decided not to use PowerPC special operating modes instructions (exiting special operating modes is usually performed by interrupt). We employed the *'branch to link register'* (**blr**) to emulate a wait state and *'branch to link register and link'* (**blrl**) to get the processor out of this state. The difference between these instructions is that **blrl** modifies the link register (LR), while **blr** does not. The next instruction address is the effective address of the branch target, stored in LR. When **blrl** is executed, the new value loaded into LR is the address of the instruction following the branch instruction. Thus the emulation instructions, stored into the corresponding block in Fig.2 are reduced to only one instruction for wait and one for 'wake-up' emulation.

Let us assume the following mnemonics for the three reconfigurable instructions: *'pset rm_addr'*, *'cset rm_addr'* and *'exec rm_addr'*. To implement the proposed mechanism, we only need to initialize LR with a proper value, i.e. the address of the reconfigurable instruction. This should be done by the compiler with the *'branch and link'* (**bl**) instruction of PowerPC. In the assembly code of the application program the *'complete set'* instruction should look like this:

$$bl \quad label1 \quad \rightarrow bl \text{ — branch to } label1 \text{ ; LR} = label1$$
$$label1: \textbf{cset } rm\_addr \rightarrow \textbf{blr} \text{ — branch to } label1 \text{ ; LR} = label1$$

Obviously, the processor will execute branch instruction to the same address, because LR remains unchanged and points to an address containing **blr** instruction. Thus we drive the processor into an eternal loop. It is the responsibility of the arbiter to get the processor out of this state. When the reconfigurable instruction is complete, an *'end_op'* signal is generated by the $\rho\mu$-unit to the arbiter, which initiates the execution of **blrl** exactly twice. Thus, the effective address of the next instruction is loaded into the LR, which points to the address of the instruction immediately following the reconfigurable one and the processor exits the eternal loop. Below, the instructions generated by the arbiter to finalize a reconfigurable operation are displayed (instruction alignment is at 4 bytes):

$$label1: \textbf{cset } \quad rm\_addr \rightarrow \textbf{blrl} \text{ — branch to } label1 \quad \text{; LR} = label1+4$$
$$\rightarrow \textbf{blrl} \text{ — branch to } label1+4 \text{ ; LR} = label1+4$$
$$label1+4: \textbf{next instr} \quad \rightarrow .... \text{ — next instruction} \quad \text{; LR} = label1+4$$

This approach allows executions of *blocks of reconfigurable instructions* (BRI): *We define BRI as any sequence of reconfigurable instructions starting with the instruction 'bl' and containing arbitrary number of consecutive reconfigurable instructions. No other instructions can be utilized within a BRI.* Utilizing BRI saves the necessity to initialize LR every time a reconfigurable instruction is invoked, thus saving a couple of **bl** instructions. In this scheme only one **bl** instruction is used to initialize LR in the beginning of the BRI. The time spent for executing a single reconfigurable operation (T_ρ) is estimated to be the time for the *reconfigurable execution* ($T_{\rho E}$), consumed by the $\rho\mu$-unit, plus the time for three *unconditional taken branch instructions* (T_{UTB}) : $T_\rho = 3 \times T_{UTB} + T_{\rho E}$. Assuming the number of reconfigurable instructions in the BRI to be N_{BRI}, the execution time of a reconfigurable instruction within a BRI costs: $T_\rho = 2 \times T_{UTB} + T_{\rho E} + \frac{T_{UTB}}{N_{BRI}}$ In other words, *the time penalty for single reconfigurable instruction execution is* $3 \times T_{UTB}$ *and within a BRI execution - between* $2 \times T_{UTB}$ *and* $3 \times T_{UTB}$. Optionally, the *'instruction synchronization'* instruction (**isync**) can be added before a BRI to avoid out-of-order executions of previous instructions during reconfigurable operation.

Instruction Encoding. To perform the MOLEN processor reconfigurations, the PowerPC Instruction Set Architecture (ISA) is extended with three instructions. To encode these three instructions, we have considered the following: 1.) The encoding scheme should be consistent with the PowerPC instruction format with opcodes (OPC) encoded in the six most-significant bits of the instruction word (see Fig.3). 2.) All three instructions have the same OPC field and same instruction form, which is similar to the I-form. Let us call the new form of the reconfigurable instructions *R-form*. 3.) The OPCodes of the instructions are as close as possible to the OPC of the emulation instructions (shortest Hamming distance), i.e. **blr** and **blrl**. From the free opcodes of the PowerPC architecture, such is opcode '6' ("000110_b"). 4.) Instruction modifiers are implemented in the two least-significant fields of the instruction word, to distinguish the three reconfigurable instructions. 5.) A 24-bit address, embedded into the instruction word, specifies the location of the microcode in memory. A modifier bit R/P (Resident/Pageable), assumed to be a part of the address field, specifies where the microcode is located and how to interpret the address field. If R/P=1 a memory address is specified, otherwise an address of the on-chip storage in the $\rho\mu$-code unit is referred. The address always points to the location of the first microcode instruction. This first address should contain the length or the final address of the microcode. A microprogram is terminated by an *end_op* microinstruction.

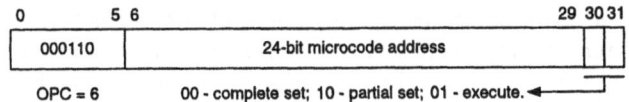

Fig. 3. Reconfigurable instruction encoding: R-form

Hardware Requirements. To implement the instruction bus arbitration, we have considered the following: 1) Information, related to instruction decoding, arbitration and timing is obtained only through the instruction bus. 2) PowerPC instruction bus is 64-bit wide and instructions are fetched in couples. 3) Speculative dummy prefetches should not disturb the correct timing of a reconfigurable instruction execution. 4) Both the arbiter and the $\rho\mu$-unit strobe input signals on rising clock edges and generate output controls on falling clock edges.

The $\rho\mu$-code arbiter for PowerPC has been described in synthesizable VHDL and mapped on the Virtex II Pro FPGA. Fig. 4 depicts the timing of this im-

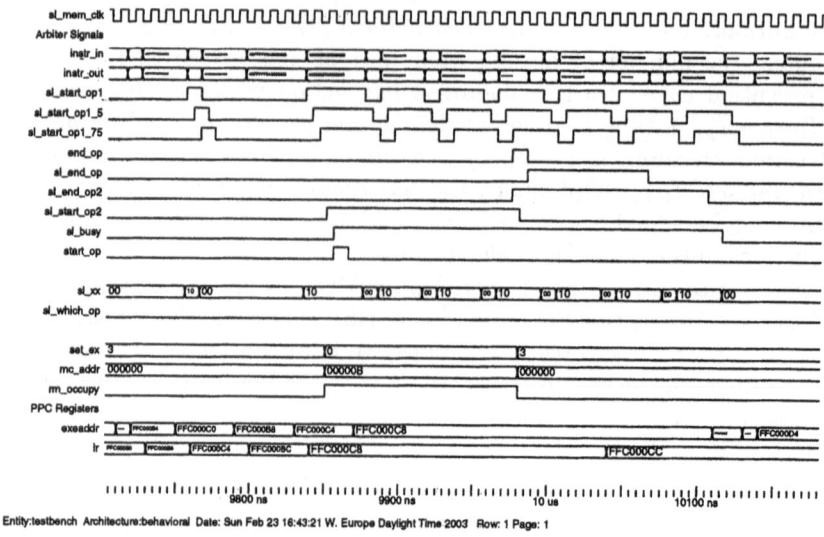

Fig. 4. Reconfigurable instruction execution timing

plementation. The unit uses the same clock ('sl_mem_clk') as the instruction memory, in this case a fast ZBT RAM. The only inputs of the arbiter are the *input instruction* bus ('instr_in') and *end of (reconfigurable) operation* ('end_op').

The *decode unit* of the arbiter (Fig. 2) decodes both OPCodes of the fetched instructions. Non-reconfigurable instructions are redirected (via MUX) to output 'instr_out', directly driving the instruction bus of PowerPC. Alternatively, when either of the decoded two instructions is reconfigurable, the instruction code of **blr** is multiplexed via 'instr_out' from the 'emulation instructions' block. Obviously, the critical path penalty to the original instruction flow is just one 2-1 multiplexer and the decoding logic for a 6-bit value. The decode block generates two internal signals to the control block - sl_start_op1 (explained later) and sl_xx. The latter signal indicates the alignment of the fetched instructions with respect to the reconfigurable ones. A one represents a reconfigurable instruction, a zero - any other instruction. For example, assuming big endian alignment: "sl_xx=10" means a reconfigurable instruction at the least-significant and a non-reconfigurable instruction at the most-significant address.

The control block generates signal *start (reconfigurable) operation* ('start_op') for one clock cycle delayed with two cycles after the moment a reconfigurable operation is prefetched and decoded, thus filtering short (dummy) prefetches. In Fig. 4 the rising edge of the internal signal sl_start_op1 indicates the moment a reconfigurable operation is decoded. One can see that signal ('start_op') is generated only when the reconfigurable instruction is really fetched, i.e. when sl_start_op1 takes longer than one clock cycle. Dummy prefetch filtration has been implemented by two flip-flops, connected in series and clocked by complementary clock edges. The outputs of these flip-flops are denoted by signals sl_start_op1_5 and sl_start_op1_75. The output control to the *ρμ*-unit, sl_start_op is generated between two falling clock edges.

Synchronously with the decoding of a reconfigurable instruction, the two instruction modifier fields (output signal *set_ex*) and *microcode address* (24-bit output *mc_addr*) are registered on rising clock edge (recall Fig.3). The internal flip-flop *sl_which_op* is used only when both of the fetched instructions are reconfigurable (*sl_xx*="11") to ensure the proper timely distribution of *set_ex*, *mc_addr* and controls. In addition, two internal signals (flip-flops) are set when reconfigurable instruction is decoded. These two signals denote that the $\rho\mu$-unit is performing an operation (*sl_start_op2*) and that the arbiter is busy (*sl_busy*) with such an operation, therefore another reconfigurable execution can not be executed. To multiplex the data memory ports to the $\rho\mu$-unit during reconfigurable operations, signal *rm_occupy* is driven to the data memory controller.

When a reconfigurable instruction is over, '*end_op*' is generated by the $\rho\mu$-unit and the *sl_start_op2* flip-flop is reset, thus releasing the data memory (via *rm_occupy*) for access by other units. Now, the control logic should guarantee that the **blrl** instruction is decoded exactly twice. This is done by a counter issuing active *sl_end_op* for precisely two **blrl** cycles, i.e., eight clocks. Instruction codes of **blr** and **blrl** differ only in one bit position. Therefore, redirecting *sl_end_op* via the MUX to this exact position of '*instr_out*' while **blr** is issued, drives **blrl** to the PowerPC. When '*end_op*' is strobed by the arbiter, another counter generates the *sl_end_op2* signal to prevent other reconfigurable operations from starting executions before the current reconfigurable operation has finished properly. The falling edge of signal *sl_end_op2* synchronously resets signal *busy*, thus enabling the execution of reconfigurable operations coming next.

4   Arbiter Testing, Analysis, and Case Study

Testing. To test the operation of the arbiter, we need a program, strictly aligned into memory, which tests all possible sequences of instruction couple alignments. Fig. 5(a) depicts the transition graph of such a test sequence, where a bubble represents an instruction couple alignment (1=reconfigurable instruction, 0 = other instruction). Arrows (transitions) fetch the next aligned instruction couple. Minimum 16 transitions cover all possible situations. The number next to each arrow indicates its position in the program sequence. An extra transition (arrow 0) tests the dummy prefetch filtration. Fig. 5(b) depicts its assembly code.

Performance Analysis. Let us assume T to be the execution time of the original program (say measured in cycles) and T_{SEi} - time to execute kernel i in software, which we would like to speed-up in reconfigurable hardware. With respect to T_ρ from Section 3, $T_{\rho i}$ is the execution time for the reconfigurable implementation of kernel i. Assuming $a_i = \frac{T_{SEi}}{T}$ and $s_i = \frac{T_{SEi}}{T_{\rho i}}$, the speed-up of the program with respect to the reconfigurable implementation of kernel i is:

$$S_i = \frac{T}{T - T_{SEi} + T_{\rho i}} = \frac{1}{1 - (a_i - \frac{a_i}{s_i})} \tag{1}$$

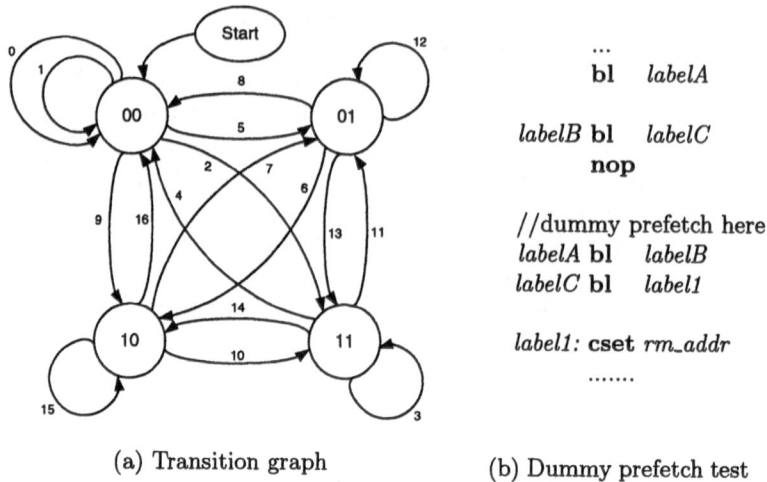

(a) Transition graph   (b) Dummy prefetch test

Fig. 5. Test program

Identically, assuming $a = \sum_i a_i$, all kernels potential candidates for reconfigurable implementation would speed-up the program with:

$$S = \frac{T}{T - \sum_i T_{SEi} + \sum_i T_{\rho i}} = \frac{1}{1 - (a - \sum_i \frac{a_i}{s_i})}, \quad S_{max} = \lim_{\forall s_i \to \infty} S = \frac{1}{1 - a} \quad (2)$$

Where S_{max} is the theoretical maximum speed-up. Parameters a_i may be obtained by profiling the targeted program, or along with s_i, by running an application on the real MOLEN CCM. Further in this Section, we will use (1) and (2) to compare theoretical to actual experimental speed-up.

Experimental Testbench. To prove the benefits of the proposed design we followed an experimental scenario. First, we use *profiling data* for the application to *extract computationally demanding kernels*. Second, we *design hardware engines*, which implement these kernels in performance efficient hardware. Further, we go trough the following steps, to get experimental data for analysis:

1. Describe the MOLEN organization and the hardware kernel designs in VHDL and synthesize them for the selected target FPGA technology.
2. Simulate the pure software implementations of the kernels on a VHDL model of the core processor to obtain performance figures.
3. Simulate the hardware implementations of the same kernels, embedded in the MOLEN organization, and mapped on the target FPGA (i.e., VirtexIIPro).
4. Estimate the speed-ups of each kernel (s_i) and for the entire application (S_i, S), based on data from the previous steps and the initial profiling.
5. Download the FPGA programming stream into a real chip and run the application, to validate the figures from simulations.

Performance Speed-Up: An MPEG-4 Case Study. The application domain of interest in our experimentations is the visual data compression and in particular the MPEG-4 standard. For parameter a_i, we use profiling data

Table 1. Speed-up for pageable $\rho\mu$-code, fixed $\rho\mu$-code, and theoretical maximum.

MPEG-4 SAD	$T_{SEi} = 3404[cyc]$		
	Pag.	Fixed	Theor.
T_{pi},[cyc]	87	51	-
s_i	39	67	∞
$S_i, (a_i = 0,6)$	2,41	2,45	2,50
$S_i, (a_i = 0,66)[3],[4]$	2,80	2,86	2,94
$S_i, (a_i = 0,88)[2],[10]$	7,01	7,51	8,33
$S_i, (a_i = 0,90)[1]$	8,13	8,82	10

Table 2. Synthesis Results for xc2vp20, Speed Grade -5

Slices	84 of 10304	< 1%
Flip Flops	69 of 20608	< 1%
4 input LUTs	150 of 20608	< 1%
Clock period	7.004ns	
Frequency	142.776MHz	

reported in literature [1–4, 10]. Values of some "global" parameters (a_i) regarding overall MPEG-4 performance may be within a standard deviation of 20% [3], with respect to the particular data. On the other hand, "local" parameters regarding implemented kernels (T_{SEi}, T_{pi}, s_i) are less data dependent, thus more predictable (accuracy within 5%). Table 1 contains experimental results for the implementation of the most demanding function in MPEG-4 encoder, the Sum-of-Absolute Differences (SAD), utilizing a design, described in [9] and assuming memory addressing schemes, discussed in [5]. The SAD kernel takes 3404 PowerPC cycles to execute in pure software. For its reconfigurable execution in MOLEN, we run two scenarios: a)worst case, when SAD execution microcode address is pagable and not residing in the $\rho\mu$-unit; and best case, when the microcode is fixed into the $\rho\mu$-unit. Experimental results in Table 1 strongly suggest that considerable speed-up of MPEG-4 encoders in the range of 2,41-8,82 is achievable only by implementing the SAD function as CCU in the MOLEN CCM organization. Both experimental and theoretical results indicate that for great kernel speed-ups ($s_i \gg 1$), the difference in overall performance (S_i) between worst and best case (pageable and fixed μ-code) is diminishing.

FPGA Synthesis Results. The VHDL code of the arbiter has been simulated with Modeltech's ModelSim and synthesized with Project Navigator ISE 5.1 S3 of Xilnx. The target FPGA chip was XC2VP20. Hardware costs obtained by the synthesis tools are reported in Table 2. Post-place-and-route results indicate the total number of slices to be 80 and memory clock frequency of 100 MHz to be feasible. These results strongly suggest that at trivial hardware costs the $\rho\mu$-arbiter design can arbitrate the PowerPC instruction bus without causing severe critical path penalties and frequency decreases. Moreover, virtually all reconfigurable resources of the FPGA remain available for CCUs. Regarding the total number of flip-flops in the arbiter design (69), most of them (52) are used for registering mc_addr and set_ex outputs. Thus only 17 flip-flops are spent for the control block, including the two embedded counters (2×4 flip-flops).

5   Conclusions

In this paper, we proposed an efficient design of a potentially performance limiting unit of an $\rho\mu$-coded CCM: the arbiter. The general $\rho\mu$-coded machine organization MOLEN was implemented on the platform FPGA Virtex II Pro and the arbitration between reconfigurable and fixed PowerPC instructions investigated. All design aspects of the arbiter have been described, including software considerations, architectural solutions, implementation issues and functional testing. Performance has been evaluated analytically and by experimentation. Synthesis results indicate trivial hardware costs for an FPGA implementation. Simulations suggest that considerable speed-ups (in the range of 2,4-8,8) of an MPEG-4 case study are feasible when the SAD function is implemented in the proposed framework. The presented design will be implemented on an FPGA prototyping board.

Acknowledgements

This research is supported by PROGRESS, the embedded systems research program of the Dutch scientific organization NWO, the Dutch Ministry of Economic Affairs and the Technology Foundation STW.

References

1. H.-C. Chang, L.-G. Chen, M.-Y. Hsu, and Y.-C. Chang. Performance analysis and architecture evaluation of MPEG-4 video codec system. In *IEEE International Symposium on Circuits and Systems*, vol. II, pp. 449–452, 28-31 May 2000.
2. H.-C. Chang, Y.-C. Wang, M.-Y. Hsu, and L.-G. Chen. Efficient algorithms and architectures for MPEG-4 object-based video coding. In *IEEE Workshop on Signal Processing Systems*, pp. 13–22, 11-13 Oct 2000.
3. J. Kneip, S. Bauer, J. Vollmer, B. Schmale, P. Kuhn, and M. Reissmann. The MPEG-4 video coding standard - a VLSI point of view. In *IEEE Workshop on Signal Processing Systems,(SIPS98)*, pp. 43–52, 8-10 Oct. 1998.
4. P. Kuhn and W. Stechele. Complexity analysis of the emerging MPEG-4 standard as a basis for VLSI implementation. In *SPIE Visual Comunications and Image Processing (VCIP)*, vol. 3309, pp. 498–509, Jan. 1998.
5. G. Kuzmanov, S. Vassiliadis, and J. van Eijndhoven. A 2D Addressing Mode for Multimedia Applications. In *SAMOS 2001*, vol. 2268 of *Lecture Notes in Computer Science*, pp. 291–306, July 2001. Springer-Verlag.
6. M.Wazlowski, L.Agarwal, T.Lee, A.Smith, E.Lam, H.Silverman, and S.Ghosh. PRISM-II Compiler and Architecture. In *Proc.IEEE Workshop on FPGAs for Custom Computing Machines*, pp. 9–16, 5-7 April, 1993.
7. R.W.Hartenstein, R.Kress, and H.Reining. A new FPGA Architecture for Word-Oriented Datapaths. In *FPL 1994*, pp. 144–155, 1994.
8. M. Sima, S. Vassiliadis, S. Cotofana, J. T. van Eijndhoven, and K. Vissers. Field-Programmable Custom Computing Machines. A Taxonomy. In *FPL 2002.*, vol. 2438 of *Lecture Notes in Computer Science*, pp. 79–88, Sept. 2002. Springer-Verlag.
9. S. Vassiliadis, E. Hakkennes, J. Wong, and G. Pechaneck. The Sum Absolute Difference Motion Estimation Accelerator. In *EUROMICRO 98*, vol. 2, Aug. 1998.
10. S. Vassiliadis, G. Kuzmanov, and S. Wong. MPEG-4 and the New Multimedia Architectural Challenges. In *15th SAER'2001*, 21-23 Sept. 2001.
11. S. Vassiliadis, S. Wong, and S. Cotofana. The MOLEN $\rho\mu$-coded processor. In *FPL 2001*, pp. 275–285, Aug. 2001.

How Secure Are FPGAs
in Cryptographic Applications?*

Thomas Wollinger and Christof Paar

Chair for Communication Security (COSY), Horst Görtz Institute for IT Security,
Ruhr-Universität Bochum, Germany,
{wollinger,cpaar}@crypto.rub.de

Abstract. The use of FPGAs for cryptographic applications is highly attractive for a variety of reasons but at the same time there are many open issues related to the general security of FPGAs. This contribution attempts to provide a state-of-the-art description of this topic. First, the advantages of reconfigurable hardware for cryptographic applications are listed. Second, potential security problems of FPGAs are described in detail, followed by a proposal of a some countermeasure. Third, a list of open research problems is provided. Even though there have been many contributions dealing with the algorithmic aspects of cryptographic schemes implemented on FPGAs, this contribution appears to be the first comprehensive treatment of system and security aspects.

Keywords: cryptography, FPGA, security, attacks, reconfigurable hardware

1   Introduction

The choice of the implementation platform of a digital system is driven by many criteria and heavily dependent on the application area. In addition to the aspects of algorithm and system speed and costs — which are present in most other application domains too — there are crypto-specific ones: physical security (e.g., against key recovery and algorithm manipulation), flexibility (regarding algorithm parameter, keys, and the algorithm itself), power consumption (absolute usage and prevention of power analysis attacks), and other side channel leakages.

Reconfigurable hardware devices, such as Field Programmable Gate Arrays (FPGAs), seem to combine the advantages of SW and HW implementations. At the same time, there are still many open questions regarding FPGAs as a module for security functions. There has been a fair amount of work been done by the research community dealing with the algorithmic and computer architecture aspects of crypto schemes implemented on FPGAs since the mid-1990s (see, e.g., relevant articles in [KP99,KP00,KNP01,KKP02]), often focusing on high-performance implementations. At the same time, however, very little work has

---

* This research was partially sponsored by the German Federal Office for Information Security (BSI).

P.Y.K. Cheung et al. (Eds.): FPL 2003, LNCS 2778, pp. 91–100, 2003.

been done dealing with the system and physical aspects of FPGAs as they pertain to cryptographic applications. It should be noted that the main threat to a cryptographic scheme in the real world is *not* the cryptanalysis of the actual algorithm, but rather the exploration of weaknesses of the implementation. Given this fact, we hope that the contribution at hand is of interest to readers in academia, industry and government sectors.

In this paper we'll start in Section 2 with a list of the advantages of FPGAs in cryptographic applications from a systems perspective. Then, we highlight important questions pertaining to the *security* of FPGAs when used for crypto algorithms in Section 3. A major part of this contribution is a state-of-the-art perspective of security issues with respect to FPGAs, by illuminating this problem from different viewpoints and by trying to transfer problems and solutions from other hardware platforms to FPGAs (Section 4). In Section 5, we provide a list of open problems. Finally, we end this contribution with some conclusions. We would like to stress that this contribution is not based on any practical experiments, but on a careful analysis of available publications in the literature and on our experience with implementing crypto algorithms.

2   System Advantages of FPGAs for Cryptographic Applications

In this section we list the potential advantages of reconfigurable hardware (RCHW) in cryptographic applications. More details and a description to each item can be found in [EYCP01,WP03]. Note that the listed potential advantages of FPGAs for cryptographic applications can only be exploited if the security shortcomings of FPGAs discussed in the following have been addressed.

- Algorithm Agility
- Algorithm Upload
- Architecture Efficiency
- Resource Efficiency
- Algorithm Modification
- Throughput
- Cost Efficiency

3   Security Shortcomings of FPGAs

This section summarizes security problems produced by attacks against given FPGA implementations. First we would like to state what the possible goals of such attacks are.

3.1   Objectives of an Attacker

The most common threat against an implementation of cryptographic algorithm is to learn a confidential cryptographic key. Given that the algorithms applied are publicly known in most commercial applications, knowledge of the key enables the attacker to decrypt future and past communication. Another threat is the one-to-one copy, or "cloning", of a cryptographic algorithm *together* with its key. In some cases it can be enough to run the cloned application in decryption mode to decipher past and future communication. In other cases, execution of a certain cryptographic operation with a presumingly secret key is in most applications the sole criteria which authenticates a communication party. An attacker who can perform the same function can masquerade as the attacked communication party. Yet another threat is given in applications where the cryptographic algorithms are proprietary e.g. pay-TV and government communication. In such scenarios it is already interesting for an attacker to reverse-engineer the encryption algorithm itself.

The discussion above assumes mostly that an attacker has physical access to the encryption device. We believe that in many scenarios such access can be assumed, either through outsiders or through dishonest insiders. In the following we discuss vulnerabilities of modern FPGAs against such attacks.

3.2   Black Box Attack

The classical method to reverse engineer a chip is the so called Black Box attack. The attacker inputs all possible combinations, while saving the corresponding outputs. The intruder is then able to extract the inner logic of the FPGA, with the help of the Karnaugh map or algorithms that simplify the resulting tables. This attack is only feasible if a small FPGA with explicit inputs and outputs is attacked and a lot of processor power is available.

3.3   Readback Attack

Readback is a feature that is provided for most FPGA families. This feature allows to read a configuration out of the FPGA for easy debugging. The idea of the attack is to read the configuration of the FPGA through the JTAG or programming interface in order to obtain secret information (e.g. keys) [Dip]. The readback functionality can be prevented with security bits provided by the manufactures.

However, it is conceivable, that an attacker can overcome these countermeasures in FPGA with fault injection. This kind of attack was first introduced in [BDL97] and it was shown how to break public-key algorithms by exploiting hardware faults. It seems very likely that these attacks can be easily applied to FPGAs, since they are not especially targeted to ASICs. If this is in fact feasible, an attacker is able to deactivate security bits and/or the countermeasures, resulting in the ability to read out the configuration of the FPGA [Kes,Dip].

3.4   Cloning of SRAM FPGAs

In a standard scenario, the configuration data is stored (unprotected) externally in nonvolatile memory (e.g., PROM) and is transmitted to the FPGA at power-up in order to configure the FPGA. An attacker could easily eavesdrop on the transmission and get the configuration file. This attack is therefore feasible for large organizations as well as for those with low budgets and modest sophistication.

3.5   Reverse-Engineering of the Bitstreams

The attacks described so far output the bitstream of the FPGA design. In order to get the design of proprietary algorithms or the secret keys, one has to reverse-engineer the bitstream. The condition to launch the attack is that the attacker has to be in possession of the (unencrypted) bitstream.

FPGA manufactures claim that the security of the bitstream relies on the disclosure of the layout of the configuration data. This information will only be made available if a non-disclosure agreement is signed, which is, from a cryptographic point of view, an extremely insecure situation. This security-by-obscurity approach was broken at least ten years ago when the CAD software company NEOCad reverse-engineered a Xilinx FPGA [Sea]. Even though a big effort has to be made to reverse engineer the bitstream, for large organizations it is quite feasible. In terms of government organizations as attackers, it is also possible that they will get the information of the design methodology directly from the vendors or companies that signed NDAs.

3.6   Physical Attack

The aim of a physical attack is to investigate the chip design in order to get information about proprietary algorithms or to determine the secret keys by probing points inside the chip. Hence, this attack targets parts of the FPGA, which are not available through the normal I/O pins. This can potentially be achieved through visual inspections and by using tools such as optical microscopes and mechanical probes. However, FPGAs are becoming so complex that only with advanced methods, such as Focused Ion Beam (FIB) systems, one can launch such an attack. To our knowledge, there are no countermeasures to protect FPGAs against this form of physical threat. In the following, we will try to analyze the effort needed to physically attack FPGAs manufactured with different underlying technologies.

SRAM FPGAs: Unfortunately, there are no publications available that accomplished a physical attack against SRAM FPGAs. This kind of attack is only treated very superficially in a few articles, e.g. [Ric98]. In the related area of SRAM memory, however there has been a lot of effort by academia and industry to exploit this kind of attack [Gut96,Gut01,AK97,WKM$^+$96,Sch98,SA93,KK99]. Due to the similarities in structure of the SRAM memory cell and the internal structure of the SRAM FPGA, it is most likely that the attacks can be employed in this setting.

Contrary to common wisdom, the SRAM memory cells do not entirely loose the contents when power is cut. The reason for these effects are rooted in the physical properties of semiconductors (see [Gut01] for more details). The physical changes are caused mainly by three effects: electromigration, hot carriers, and ionic contamination. Most publications agree that device can be altered, if 1) threshold voltage has changed by 100mV or 2) there is a 10% change in transconductance, voltage or current.

One can attack SRAM memory cells using the access points provided by the manufactures. An extreme case of data recovery, was described in [AK97], where a cryptographic key war recovered without special equipment. "I_{DDQ} testing" is one of the widely used methods to analyze SRAM cells and it is based on the analysis of the current usage [Gut01,WKM+96,Sch98]. Another possibilities for the attack are also to use the scan path that the IC manufacturers insert for test purposes or techniques like bond pad probing [Gut01].

When it becomes necessary to use access points that are not provided by the manufacturer, the layers of the chip have to be removed. Mechanical probing with tungsten wire with a radius of $0, 1 - 0, 2\mu m$ is the traditional way to discover the needed information. Focused Ion Beam (FIB) workstations can expose buried conductors and deposit new probe points [KK99]. Electron-beam tester (EBT) is another measurement method. EBT measures the energy and amount of secondary electrons that are reflected.

Resulting from the above discussion of attacks against SRAM memory cells, it seems likely that a physical attack against SRAM FPGAs can be launched successfully, assuming that the described techniques can be transfered. However, the physical attacks are quite costly and having the structure and the size of state-of-the-art FPGA in mind, the attack will probably only be possible for large organizations, for example intelligence services.

Antifuse FPGAs: In order to be able to detect the existence or non-existence of the connection one has to remove layer after layer, or/and use cross-sectioning. Unfortunately, no details have been published regarding this type of attack. In [Dip], the author states that a lot of trial-and-error is necessary to find the configuration of one cell and that it is likely that the rest of the chip will be destroyed, while analyzing one cell. The main problem with this analysis is that the isolation layer is much smaller than the whole AF cell. One study estimates that about 800,000 chips with the same configuration are necessary to explore one configuration file of an Actel A54SX16 chip with 24,000 system gates [Dip]. Another aggravation of the attack is that only about 2–5 % of all possible connections in an average design are actually used. In [Ric98] a practical attack against AF FPGAs was performed and it was possible to alter one cell in two months at a cost of $1000.

Flash FPGAs: Flash FPGAs can be analyzed by placing the chip in a vacuum chamber and powering it up. Other possible attacks against flash FPGAs can be found in the related area of flash memory. The number of write/erase cycles are limited to $10,000 - 100,000$, because of the accumulation of electrons in the floating gate causing a gradual rise of the transistors threshold voltage. This fact increases the programming time and eventually disables the erasing of the cell [Gut01]. Another less common failure is the programming disturbance in which unselected erased cells gain charge when adjacent selected cells are written [ASH+93,Gut01]. Furthermore, electron emission causes a net charge loss [PGP+91]. In addition, hot carrier effects build a tunnel between the bands [HCSL89]. Another phenomenon is overerasing, where an erase cycle is applied to an already-erased cell leaving the floating gate positively charged [Gut01].

All the described effects change in a more or less extensive way the cell threshold voltage, gate voltage, or the characteristic of the cell. We remark that the stated phenomenons apply as well for EEPROM memory and that due to the structure of the FPGA cell these attacks can be simply adapted to attack flash/EEPROM FPGAs.

3.7   Side Channel Attacks

Any physical implementation of a cryptographic system might provide a *side channel* that leaks unwanted information. Examples for side channels include in particular: power consumption, timing behavior, and electromagnet radiation. Obviously, FPGA implementations are also vulnerable to these attacks. In [KJJ99] two practical attacks, Simple Power Analysis (SPA) and Differential Power Analysis (DPA) were introduced. Since their introduction, there has been a lot of work improving the original power attacks (see, e.g., relevant articles in [KP99,KP00,KNP01,KKP02]). There seems to be very little work at the time of writing addressing the feasibility of actual side channel attacks against FP-GAs. However, it seems almost certain that the different side channels can be exploited in the case of FPGAs as well.

4   How to Prevent Possible Attacks?

This section shortly summarizes possible countermeasures that can be provided to minimize the effects of the attacks mentioned in the previous section. Most of them have to be realized by design changes through the FPGA manufacturers, but some could be applied during the programming phase of the FPGA.

Preventing the Black Box Attack: The Black Box Attack is not a real threat nowadays, due to the complexity of the designs and the size of state-of-the-art FPGAs. Furthermore, the nature of cryptographic algorithms prevents the attack as well. Todays stream ciphers output a bit stream, with a period length of 128 bits (e.g. w7). Block ciphers, like AES, are designed with a minimum key length of 128 bits. Minimum length in the case of public-key algorithms is 160 bits for elliptic curve cryptosystems and 1024 bits for discrete logarithm and RSA-based systems. It is widely believed that it is infeasible to perform a brute force attack and search a space with 2^{80} possibilities. Hence, implementations of this algorithms can not be attacked with the black box approach.

Preventing the Cloning of SRAM FPGAs: There are many suggestions to prevent the cloning of SRAM FPGAs, mainly motivated by the desire to prevent reverse engineering of general, i.e., non-cryptographic, FPGA designs. One solution would be to check the serial number before executing the design and delete the circuit if it is not correct. Another solution would be to use dongles to protect the design (a survey on dongles can be found in [Kea01]). Both solutions do not provide the necessary security, see [WP03] for more details. A more realistic solution would be to have the nonvolatile memory and the FPGA in one chip or to combine both parts by covering them with epoxy. However, it has to be guaranteed that an attacker is not able to separate the parts.

Encryption of the configuration file is the most effective and practical countermeasure against the cloning of SRAM FPGAs. There are several patents that propose different encryption scenarios [Jef02,Aus95,Eri99,SW99,Alg] and a good number of publications, e.g. [YN00,KB00]. The 60RS family from Actel was the first attempt to have a key stored in the FPGA in order to be able to encrypt the configuration file. The problem was that every FPGA had the same key on board.

An approach in a completely different direction would be to power the whole SRAM FPGA with a battery, which would make transmission of the configuration file after a power loss unnecessary. This solution does not appear practical, however, because of the power consumption of FPGAs. Hence, a combination of encryption and battery power provides a possible solution. Xilinx addresses this with an on-chip 3DES decryption engine in its Virtex II [Xil] (see also [PWF+00]), where the two keys are stored in the battery powered memory.

Preventing the Physical Attack: To prevent physical attacks, one has to make sure that the retention effects of the cells are as small as possible, so that an attacker can not detect the status of the cells. Already after storing a value in a SRAM memory cell for 100–500 seconds, the access time and operation voltage will change [vdPK90]. The solution would be to invert the data stored periodically or to move the data around in memory. Neutralization of the retention effect can be achieved by applying an opposite current [TCH93] or by inserting dummy cycles into the circuit [Gut01]. In terms of FPGA application, it is very costly or even impractical to provide solutions like inverting the bits or changing the location for the whole configuration file. A possibility could be that this is done only for the crucial part of the design, like the secret keys. Counter techniques such as dummy cycles and opposite current approach can be carried forward to FPGA applications.

Antifuse FPGAs can only be protected against physical attack, by building a secure environment around them. If an attack was detected every cell should be programmed in order not to leak any information or the antifuse FPGA has to be destroyed.

In terms of flash/EEPROM memory cell, one has to consider that the first write/erase cycles causes a larger shift in the cell threshold [SKM95] and that this effect will become less noticeable after ten write/erase cycles [HCSL89]. Thus, one should program the FPGA about 100 times with random data, to avoid these effect (suggested for flash/EEPROM memory cells in [Gut01]). The phenomenon of overerasing flash/EEPROM cells can be minimized by first programming all cells before deleting them.

Preventing the Readback Attack: The readback attack can be prevented with the security bits set, as provided by the manufactures, see Section 3.3. If one wants to make sure that an attacker is not able to apply fault injection, the FPGA has to be embedded into a secure environment, where after detection of an interference the whole configuration is deleted or the FPGA is destroyed.

Preventing the Side Channel Attack: In recent years, there has been a lot of work done to prevent side-channel attacks (see, e.g., relevant articles in [KP99,KP00,KNP01,KKP02]). There are "Software" countermeasures that refer primarily to algorithmic changes which are also applicable to implementations in FPGA. Furthermore, there are Hardware countermeasures that often deal either with some form of power trace smoothing or with transistor-level changes of the logic. Neither seem to be easily applicable to FPGAs without support from the manufacturers. However, some proposals such as duplicated architectures might work on today's FPGAs.

5   Open Problems

At this point we would like to provide a list of open questions and problems regarding the security of FPGAs. If answered, such solutions would allow stand-along FPGAs with much higher security assurance than currently available. A more detailed description to all points can be found in [WP03].

- Side channel attacks
- Fault injection
- Key management for configuration encryption
- Secure deletion
- Physical attacks

6   Conclusions

This contribution analyzed possible attack against the use of FPGA in security applications. For black box attacks, we stated that they are not feasible for state-of-the-art FPGAs. However, it seems very likely for an attacker to get the secret information stored in a FPGA, when combining readback attacks with fault injection. Cloning of SRAM FPGA and reverse engineering depend on the specifics of the system under attacked, and they will probably involve a lot of effort, but this does not seem entirely impossible. Physical attacks against FPGAs are very complex due to the physical properties of the semiconductors in the case of flash/SRAM/EEPROM FPGAs and the small size of AF cells. It appears that such attacks are even harder than analogous attacks against ASICs. Even though FPGA have different internal structures than ASICs with the same functionality, we believe that side-channel attacks against FPGAs, in particular power-analysis attacks, will be feasible too.

From the discussion above it may appear that FPGAs are currently out of question for security applications. We don't think that this the right conclusion, however. It should be noted that many commercial ASICs with cryptographic functionality are also vulnerable to attacks similar to the ones discussed here. A commonly taken approach to prevent these attacks is to put the ASIC in a secure environment.

References

[AK97]   R.J. Anderson and M.G. Kuhn. Low Cost Attacks on Tamper Resistant Devices. In *5th International Workshop on Security Protocols*, pages 125–136. Springer-Verlag, 1997. LNCS 1361.

[Alg]   Algotronix Ltd. Method and Apparatus for Secure Configuration of a Field Programmable Gate Array. PCT Patent Application PCT/GB00/04988.

[ASH+93]   Seiichi Aritome, Riichiro Shirota, Gertjan Hemink, Tetsup Endoh, and Fujio Masuoka. Reliability Issues of Flash Memory Cells. *Proceedings of the IEEE*, 81(5):776–788, May 1993.

[Aus95]   K. Austin. Data Security Arrangements for Semicondutor Programmable Devices. United States Patent, No. 5388157, 1995.

[BDL97]   D. Boneh, R. A. DeMillo, and R. J. Lipton. On the Importance of Checking Cryptographic Protocols for Faults. In *EUROCRYPT '97*, pages 37–51. Springer-Verlag, 1997. LNCS 1233.

[Dip]   B. Dipert. Cunning circuits confound crooks. http://www.e-insite.net/ednmag/contents/images/21df2.pdf.

[Eri99]   C. R. Erickson. Configuration Stream Encryption. United States Patent, No. 5970142, 1999.

[EYCP01]   A. Elbirt, W. Yip, B. Chetwynd, and C. Paar. An FPGA-based performance evaluation of the AES block cipher candidate algorithm finalists. *IEEE Transactions on VLSI Design*, 9(4):545–557, August 2001.

[Gut96]   P. Gutmann. Secure Deletion of Data from Magnetic and Solid-State Memory. In *Sixth USENIX Security Symposium*, pages 77–90, July 22–25, 1996.

[Gut01]   P. Gutmann. Data Remanence in Semiconductor Devices. In *10th USENIX Security Symposium*, pages 39–54, August 13–17, 2001.

[HCSL89]   Sameer Haddad, Chi Chang, Balaji Swaminathan, and Jih Lien. Degradations due to hole trapping in flash memory cells. *IEEE Electron Device Letters*, 10(3):117–119, March 1989.

[Jef02]   G. P. Jeffrey. Field programmable gate arrays. United States Patent, No. 6356637, 2002.

[KB00]   S. H. Kelem and J. L. Burnham. System and Method for PLD Bitstram Encryption. United States Patent, No. 6118868, 2000.

[Kea01]   T. Kean. Secure Configuration of Field Programmable Gate Arrays. In *FPL 2001*, pages 142–151. Springer-Verlag, 2001. LNCS 2147.

[Kes]   D. Kessner. Copy Protection for SRAM based FPGA Designs. http://www.free-ip.com/copyprotection.html.

[KJJ99]   P. Kocher, J. Jaffe, and B. Jun. Differential Power Analysis. In *CRYPTO '99*, pages 388–397. Springer-Verlag, 1999. LNCS 1666.

[KK99]   O. Kommerling and M.G. Kuhn. Design Principles for Tamper-Resistant Smartcard Processors. In *Smartcard '99*, pages 9–20, May 1999.

[KKP02]   B. S. Kaliski, Jr., Ç. K. Koç, and C. Paar, editors. *Workshop on Cryptographic Hardware and Embedded Systems — CHES 2002*, Berlin, Germany, August 13-15, 2002. Springer-Verlag. LNCS 2523.

[KNP01]   Ç. K. Koç, D. Naccache, and C. Paar, editors. *Workshop on Cryptographic Hardware and Embedded Systems — CHES 2001*, Berlin, Germany, May 13-16, 2001. Springer-Verlag. LNCS 2162.

[KP99]   Ç. K. Koç and C. Paar, editors. *Workshop on Cryptographic Hardware and Embedded Systems — CHES'99*, Berlin, Germany, August 12-13, 1999. Springer-Verlag. LNCS 1717.

[KP00]   Ç. K. Koç and C. Paar, editors. *Workshop on Cryptographic Hardware and Embedded Systems — CHES 2000*, Berlin, Germany, August 17-18, 2000. Springer-Verlag. LNCS 1965.

[PGP$^+$91]   C. Papadas, G. Ghibaudo, G. Pananakakis, C. Riva, P. Ghezzi, C. Gounelle, and P. Mortini. Retention characteristics of single-poly EEPROM cells. In *European Symposium on Reliability of Electron Devices, Failure Physics and Analysis*, page 517, October 1991.

[PWF$^+$00]   R. C. Pang, J. Wong, S. O. Frake, J. W. Sowards, V. M. Kondapalli, F. E. Goetting, S. M. Trimberger, and K. K. Rao. Nonvolatile/battery-backed key in PLD. United States Patent, No. 6366117, Nov. 28 2000.

[Ric98]   G. Richard. Digital Signature Technology Aids IP Protection. In *EETimes - News*, 1998. http://www.eetimes.com/news/98/1000news/digital.html.

[SA93]   J. Soden and R.E. Anderson. IC failure analysis: techniques and tools for quality and reliability improvement. *Proceedings of the IEEE*, 81(5):703–715, May 1993.

[Sch98]   D.K. Schroder. *Semiconducor Material and Device Characterization*. John Wiley and Sons, 1998.

[Sea]   G. Seamann. FPGA Bitstreams and Open Designs. http://www.opencollector.org/.

[SKM95]   K.T. San, C. Kaya, and T.P. Ma. Effects of erase source bias on Flash EPROM device reliability. *IEEE Transactions on Electron Devices*, 42(1):150–159, January 1995.

[SW99]   C. Sung and B. I. Wang. Method and Apparatus for Securing Programming Data of Programmable Logic Device. United States Patent, Patent Number 5970142, June 22 1999.

[TCH93]   Jiang Tao, Nathan Cheung, and Chenming Ho. Metal Electromigration Damage Healing Under Bidirectional Current Stress. *IEEE Transactions on Elecron Devices*, 14(12):554–556, December 1993.

[vdPK90]   J. van der Pol and J. Koomen. Relation between the hot carrier lifetime of transistors and CMOS SRAM products. In *IRPS 1990*, page 178, 1990.

[WKM$^+$96]   T.W. Williams, R. Kapur, M.R. Mercer, R.H. Dennard, and W. Maly. IDDQ Testing for High Performance CMOS - The Next Ten Years. In *ED&TC'96*, pages 578–583, 1996.

[WP03]   T. Wollinger and C. Paar. How Secure Are FPGAs in Cryptographic Applications? (Long Version). Report 2003/119, IACR, 2003. http://eprint.iacr.org/.

[Xil]   Xilinx Inc. Using Bitstream Encryption. Handbook of the Virtex II Platform. http://www.xilinx.com.

[YN00]   Kun-Wah Yip and Tung-Sang Ng. Partial-Encryption Technique for Intellectual Property Protection of FPGA-based Products. *IEEE Transactions on Consumer Electronics*, 46(1):183–190, 2000.

FPGA Implementations
of the RC6 Block Cipher

Jean-Luc Beuchat

Laboratoire de l'Informatique du Parallélisme, Ecole Normale Supérieure de Lyon,
46, Allée d'Italie, F–69364 Lyon Cedex 07,
Jean-Luc.Beuchat@ens-lyon.fr

Abstract. RC6 is a symmetric-key algorithm which encrypts 128-bit plaintext blocks to 128-bit ciphertext blocks. The encryption process involves four operations: integer addition modulo 2^w, bitwise exclusive or of two w-bit words, rotation to the left, and computation of $f(X) = (X(2X + 1)) \bmod 2^w$, which is the critical arithmetic operation of this block cipher. In this paper, we investigate and compare four implementations of the $f(X)$ operator on Virtex-E and Virtex-II devices. Our experiments show that the choice of an algorithm is strongly related to the target FPGA family. We also describe several architectures of a RC6 processor designed for feedback or non-feedback chaining modes. Our fastest implementation achieves a throughput of 15.2 Gb/s on a Xilinx XC2V3000-6 device.

1 Introduction

In 1997, the National Institute of Standards and Technology (NIST) initiated a process to specify a new symmetric-key encryption algorithm capable of protecting sensitive data. RSA Laboratories submitted RC6 [9] as a candidate for this Advanced Encryption Standard (AES). NIST announced fifteen AES candidates at the First AES Candidate Conference (August 1998) and solicited public comments to select five finalist algorithms (August 1999): MARS, RC6, Rijndael, Serpent, and Twofish. Though the algorithm Rijndael was eventually selected, RC6 remains a good choice for security applications and is also a candidate for the NP 18033 project (via the Swedish ISO/IEC JTC 1/SC 27 member body[1]) and the Cryptrec project initiated by the Information-technology Promotion Agency in Japan[2].

A version of RC6 is more exactly specified as RC6-$w/r/b$, where the parameters w, r, and b respectively express the word size (in bits), the number of rounds, and the size of the encryption key (in bytes). Since all actual implementations are targeted at $w = 32$ and $r = 20$, we use RC6 as shorthand to refer to RC6-32/20/b. A key schedule generates $2r + 4$ words (w bits each) from the b-bytes key provided by the user (see [9] for details). These values (called round

---

[1] http://www.din.de/ni/sc27

[2] http://www.ipa.go.jp/security/enc/CRYPTREC/index-e.html

P.Y.K. Cheung et al. (Eds.): FPL 2003, LNCS 2778, pp. 101–110, 2003.

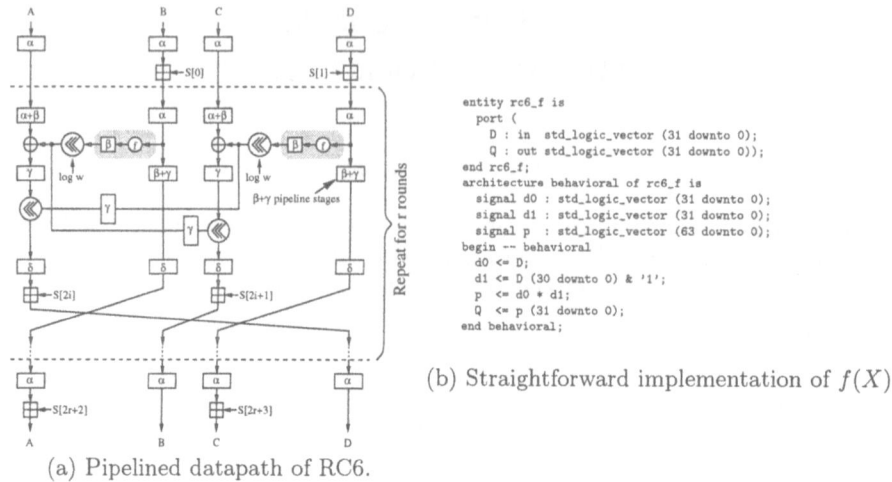

(a) Pipelined datapath of RC6.

(b) Straightforward implementation of $f(X)$.

Fig. 1. Encryption with RC6.

keys) are stored in an array $S[0, \ldots, 2r + 3]$ and are used in both encryption and decryption. The encryption algorithm involves four operations (Figure 1a):

- Integer addition modulo 2^w (denoted by $X \boxplus Y$).
- Bitwise exclusive or of two w-bit words (denoted by $X \oplus Y$).
- Computation of $f(X) = (X(2X + 1)) \bmod 2^w$, where X is a w-bit integer.
- Rotation of the w-bit word X to the left by an amount given by the $\log_2 w$ least significant bits of Y (denoted by $X \lll Y$).

Note that the decryption process requires moreover integer subtraction modulo 2^w and rotation to the right. As the algorithm is similar to encryption, we will not consider it here.

In this paper, we study several hardware architectures of RC6 using Virtex-E and Virtex-II field programmable gate arrays (FPGA). Virtex-E and Virtex-II Configurable Logic Blocks (CLB) provide functional elements for synchronous and combinational logic. Each CLB includes respectively two (Virtex-E) and four (Virtex-II) slices containing basically two 4-input look-up tables (LUT), two storage elements, fast carry logic dedicated to addition and subtraction, and two dedicated AND gates (referred to as MULT_AND) which improve the efficiency of multiplier implementation. Furthermore, Virtex-II devices embed many 18-bit × 18-bit multiplier blocks (also referred to as MULT18x18 blocks) supporting two independent dynamic data input ports: 18-bit signed or 17-bit unsigned. Arithmetic operators dedicated to FPGAs should therefore involve such building blocks.

This paper is organized as follows: Section 2 describes several architectures of a RC6 processor. We then investigate various implementations of $f(X)$ and show that the choice of an algorithm depends on the target FPGA family (Section 3). Finally, Section 4 digests our main results and compare them with recent works on RC6.

2   Architecture of a RC6 Processor

RC6 encrypts plaintext in fixed-size 128-bit blocks. However, messages will often exceed 128 bits and a simple solution, known as Electronic Codebook (ECB) mode, consists in partitioning the plaintext into 128-bit blocks and encrypting each independently. This ECB mode has a major drawback in that identical ciphertext blocks imply identical plaintext blocks and is therefore inadvisable if the secret key is reused for more than one message. More sophisticated chaining modes bring a solution to this problem. For instance, in the Cipher Block Chaining (CBC) mode, a feedback mechanism causes the jth ciphertext block to depend on the first j plaintext blocks and an n-bit initialization vector. Since the entire dependency on preceding blocks is contained in the previous ciphertext block [6], all blocks must be processed sequentially (CBC decryption can however be performed in parallel). This property forbids to pipeline the computation path and implies a slightly different hardware architecture of the block cipher with a lower throughput. The counter (CTR) mode, a non-feedback mode described for example in [3], could remedy the situation if it becomes a standard as recommended in [5]. It is also possible to pipeline the processor in feedback modes if we accept the decomposition of the data stream into d separately encrypted messages, where d is the pipeline depth [11]. Also note that RC6 involves forty 32-bit × 32-bit unsigned multipliers. The implementation of the 20 rounds is therefore only possible on rather large and expensive FPGAs.

Consequently, the hardware architecture of a RC6 processor depends as well on the required chaining mode and the target FPGA. We adopt here a design methodology initially proposed for the hardware implementation of the IDEA block cipher [7, 11]: the simplest RC6 processor contains a single round, the input round, and the output round (Figure 2a). This architecture is tailored to feedback chaining modes: a single plaintext block is encrypted at a time and we can provide a new input block after 21 clock cycles.

Assume now that a non-feedback chaining mode is required or that the data stream is decomposed into several separately encrypted messages. In order to shorten the critical path, each round has a parametric number of internal pipeline stages (parameters α, β, γ, and δ on Figure 1a). Figure 2b depicts an iterative architecture with partial loop unrolling and pipelining. The circuit implements k rounds (k is an integer divisor of the total number of rounds r), the input round, and the output round. Finally, Figure 2c illustrates an architecture with full loop unrolling dedicated to high throughput implementations of the RC6 block cipher.

In addition to the RC6 computation path, each processor contains a subkey memory implemented on CLBs and a control unit. The latter simply consists in a token associated with each plaintext block. This token indicates the validity of the data and selects the correct subkeys in iterative architectures. We have written a C program which generates a structural VHDL description of such a RC6 processor according to several parameters (partial or full loop unrolling, inner-round pipeline stages, and outer-round pipeline stages). Some examples are freely available at `http://www.ens-lyon.fr/~jlbeucha`.

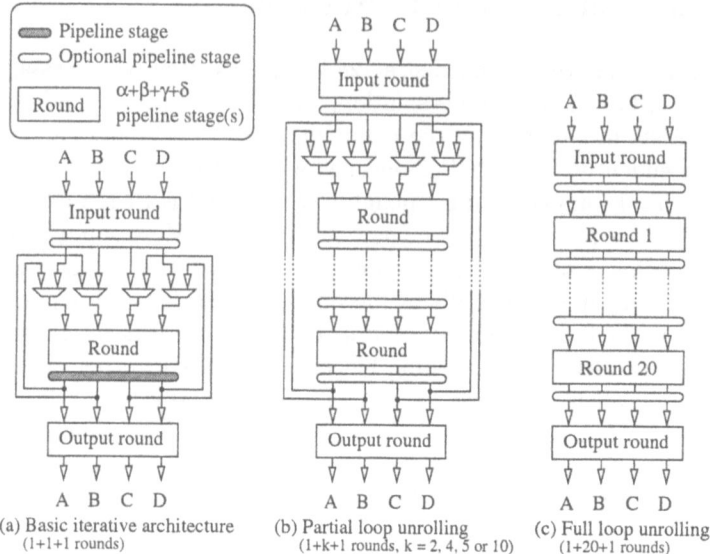

Fig. 2. Some architectures of a RC6 processor.

3   Computation of $f(X)$

The computation of $f(X)$ is the critical arithmetic operation of the block cipher. Therefore, both area and delay of a RC6 processor are closely related to the hardware operator carrying out $f(X) = (X(2X + 1)) \bmod 2^w$. In this section, we investigate and compare a method involving an array of AND gates and carry-propagate adders (CPA), and three algorithms dedicated to FPGA embedding small multiplier blocks. In the following, $X_{q:p}$ denotes $\sum_{i=p}^{q} x_i 2^i$.

3.1   Adder-Based Algorithm

This first algorithm is based on a standard method for squaring described for instance in [8]. Let us consider the problem of computing $f(X)$ when X is a 8-bit unsigned integer. As shown in Figure 3, the partial products can be significantly simplified before performing their addition according to the identities $x_i x_i = x_i$ and $x_i x_j + x_j x_i = 2 x_i x_j$. Finally, based on the well-known relation $x_i x_j + x_i = 2 x_i x_j + x_i \bar{x}_j$, we remove $x_3 x_2$ and x_2 from the leftmost column and replace them by $x_3 \bar{x}_2$. As $f(X)$ is computed modulo 2^8, we ignore the term $2 x_3 x_2$.

Let us formalize the algorithm sketched out in this example. If w is even, the computation of $f(X)$ involves the addition of $\frac{w}{2}$ partial products PP_i defined as follows[3]:

---

[3] A proof of correctness is provided in [1].

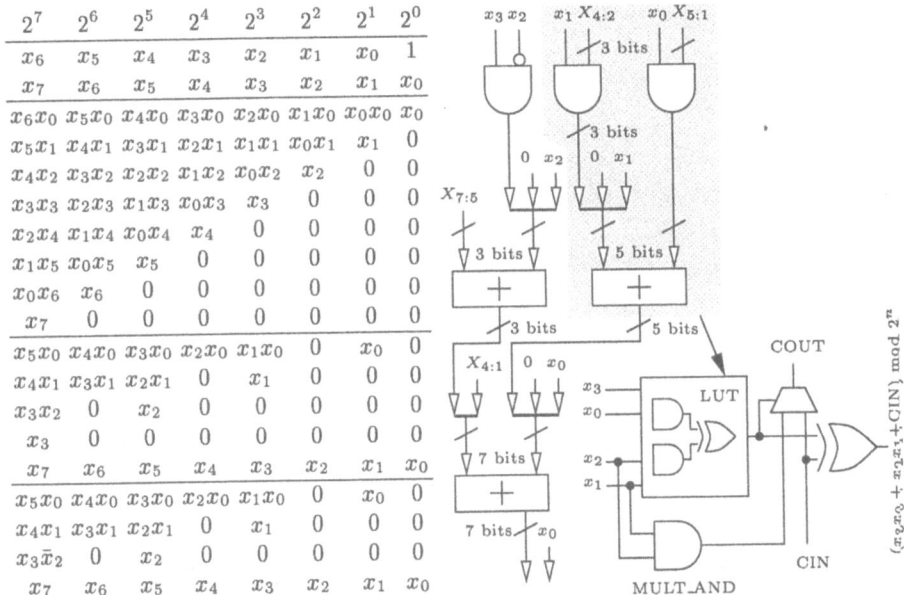

Fig. 3. Algorithm 1: computation of $f(X) = (X(2X+1)) \bmod 2^8$ with AND gates and carry-propagate adders.

$$
PP_i = \begin{cases}
\displaystyle\sum_{j=0}^{w-1} x_j 2^j & \text{if } i = \frac{w}{2} - 1, \\[2ex]
x_{i+1}\bar{x}_i 2^{w-1} + x_i 2^{w-3} & \text{if } i = \frac{w}{2} - 2, \\[2ex]
\displaystyle x_i 2^{2i+1} + \sum_{j=2i+3}^{w-1} x_{j-i-2}x_i 2^j & \text{if } 0 \le i \le \frac{w}{2} - 3.
\end{cases}
$$

The above equation allows to automatically generate the VHDL code of the partial product generator for any even w. Synthesis tools are then able to put to good use the MULT_AND gates in order to generate and add partial products using the same logic resources as a simple multioperand tree adder (Figure 3).

3.2 Multiplier-Based Algorithms

A straightforward algorithm reported in [10] consists in writing the VHDL code depicted by Figure 1b. Since $w = 32$ and Virtex-II devices embed 17-bit $\times$ 17-bit unsigned multipliers, commercial tools like Synplify Pro or XST (Xilinx Synthesis Technology) resort to the well-known divide-and-conquer scheme [8] in order to build the operator (Figure 4).

Each $f(X)$ operator based on this scheme involves three multiplier blocks and two carry-propagate adders. Consequently, a RC6 processor with full loop unrolling requires $2r \cdot 3 = 120$ MULT18x18 blocks and fits in a XC2V4000 device. Let us define the lower and higher words of the operand X as follows:

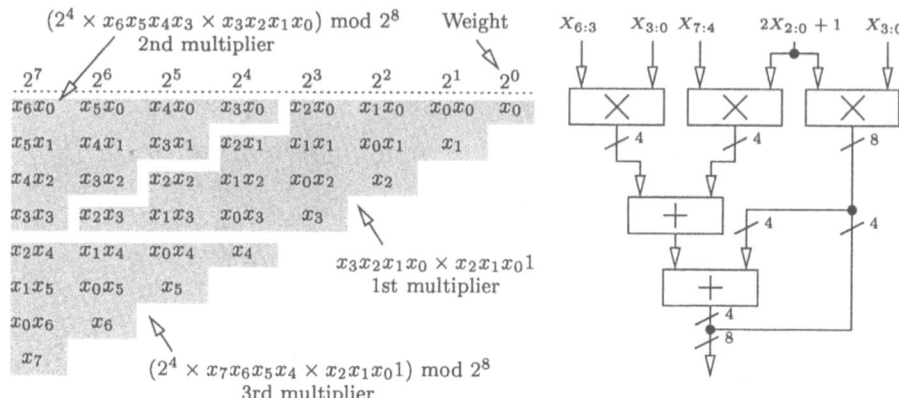

Fig. 4. Algorithm 2: computation of $f(X) = (X \cdot (2X + 1)) \bmod 2^w$ with a divide-and-conquer strategy.

$X_L = \sum_{i=0}^{w/2-1} x_i 2^i$ and $X_H = \sum_{i=0}^{w/2-1} x_{n+i} 2^i$. A solution to reduce the amount of 17-bit × 17-bit unsigned multipliers, and therefore the price of the processor, is to evaluate $f(X)$ according to:

$$f(X) = (2X^2 + X) \bmod 2^w = (2 \cdot (2^{w/2}X_H + H_L)^2 + X) \bmod 2^w$$
$$= (2 \cdot (2^w X_H^2 + 2^{1+w/2}X_H X_L + X_L^2) + X) \bmod 2^w$$
$$= (2^{2+w/2}X_H X_L + 2X_L^2 + X) \bmod 2^w.$$

Figure 5 illustrates how this algorithm works. In this example, we assume that $w = 8$ and that 4-bit × 4-bit multiplier blocks are available. Since $(2 \cdot 2^7 x_0 x_6 + 2 \cdot 2^7 x_2 x_4) \bmod 2^8 = 0$, we can discard these terms. For $w = 32$, this third algorithm involves a 16-bit squarer, a 14-bit × 14-bit multiplier, a 14-bit CPA, and a 31-bit CPA.

Remember that Virtex-II devices embed 17-bit × 17-bit unsigned multipliers. In the following, we describe how to put to good use the most significant bit of their input ports in order to further reduce the area of the $f(X)$ operator. Consider again the computation of $f(X)$ with $w = 8$ (Figure 6). The trick consists in performing the *rectangular* multiplication $(2 \cdot X_{3:0} + 1)X_{3:0}$ and allows to replace the $(w-1)$-bit CPA by a $w/2$-bit CPA.

3.3   Comparison of the Four Algorithms

Table 1 summarizes the main characteristics of the four $f(X)$ operators previously described[4]. From these results, we can conclude that:

---

[4] The VHDL code was synthesized and implemented on Virtex-E and Virtex-II devices with Synplify Pro 7.0.3 and Xilinx Alliance Series 4.1.03i. All input and output signals were routed through the D-type flip-flops available in the Input/Output blocks of Virtex-E or Virtex-II devices. No specific constraints were given to the synthesis tools and it should be possible to improve the results.

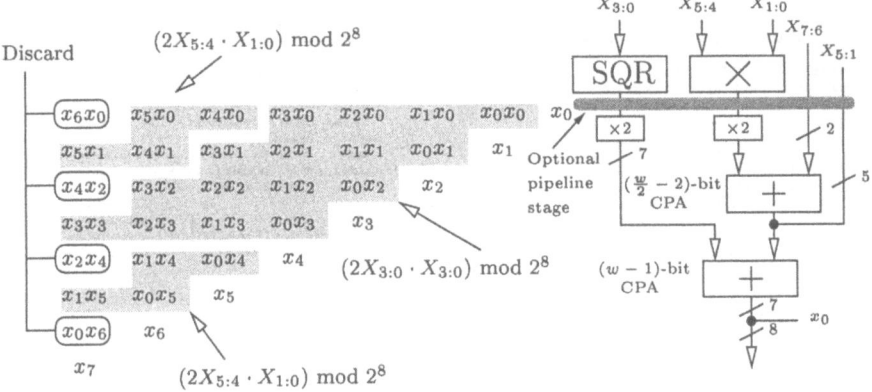

Fig. 5. Algorithm 3: computation of $f(X)$ with a squarer, a multiplier, and two carry-propagate adders.

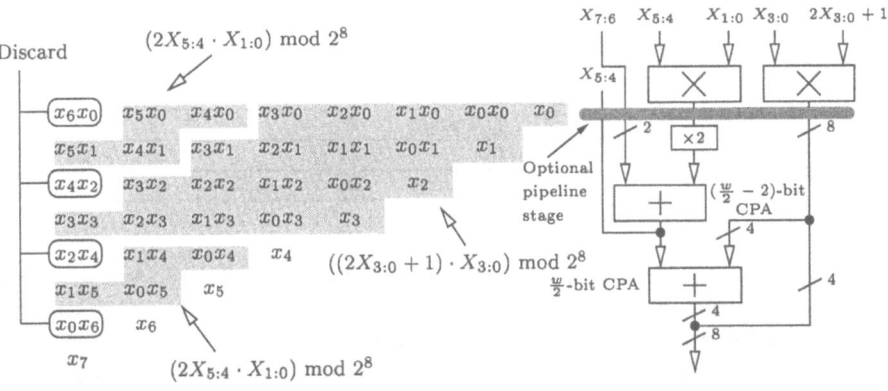

Fig. 6. Algorithm 4: computation of $f(X) = (X(2X+1)) \bmod 2^8$ with two multipliers and two carry-propagate adders.

- On a Virtex-II device, Algorithm 4 leads to the smallest circuit.
- The adder-based operator (Algorithm 1) is the best alternative for the Virtex-E family. This result is not surprising since current synthesis tools are unable to build efficient multipliers from a high-level VHDL description such that shown in Figure 7. We also notice that Algorithm 3 and Algorithm 4 lead to the same circuit area: the fact that both multipliers and adders are now implemented on CLBs explains this result.

In order to improve the frequency of the RC6 processor, our VHDL code generator is able to insert optional pipeline stages in the $f(X)$ operator (parameter β on Figure 1a). For Virtex-II devices, we take advantage of the internal pipeline stage of each MULT18x18 block. Unfortunately, the VHDL coding style depends on both the chosen algorithm and the synthesis tools:

```
entity rc6_f is
  port (
    D  : in  std_logic_vector (31 downto 0);
    Q  : out std_logic_vector (31 downto 0));
end rc6_f;
architecture behavioral of rc6_f is
  signal sqr_q  : std_logic_vector (31 downto 0);
  signal mult_q : std_logic_vector (27 downto 0);
  signal add_q  : std_logic_vector (31 downto 0);
begin  -- behavioral
  sqr_q  <= D (15 downto 0) * D (15 downto 0);
  mult_q <= D (29 downto 16) * D (13 downto 0);
  add_q (17 downto 0)   <= D (17 downto 0);
  add_q (31 downto 18) <= D (31 downto 18) +
                          mult_q (13 downto 0);
  Q (0)          <= add_q (0);
  Q (31 downto 1) <= add_q (31 downto 1) + sqr_q (30 downto 0);
end behavioral;
```

```
entity rc6_f is
  port (
    D  : in  std_logic_vector (31 downto 0);
    Q  : out std_logic_vector (31 downto 0));
end rc6_f;
architecture behavioral of rc6_f is
  signal mult1_q : std_logic_vector (32 downto 0);
  signal mult2_q : std_logic_vector (27 downto 0);
  signal add_q   : std_logic_vector (15 downto 0);
begin  -- behavioral
  mult1_q <= (D (15 downto 0) & '1') * D (15 downto 0);
  mult2_q <= D (29 downto 16) * D (13 downto 0);
  add_q (15 downto 2) <= D (31 downto 18) +
                         mult2_q (13 downto 0);
  add_q (1 downto 0)  <= D (17 downto 16);
  Q (15 downto 0)  <= mult1_q (15 downto 0);
  Q (31 downto 16) <= mult1_q (31 downto 16) + add_q;
end behavioral;
```

(a) VHDL code for Algorithm 3.   (b) VHDL code for Algorithm 4.

Fig. 7. Two VHDL descriptions of the $f(X)$ operator.

Table 1. Comparison of several $f(X)$ operators.

	XCV50E-6		XC2V40-6			
	Slices	Delay [ns]	Pipeline	Slices	Mult. blocks	Delay [ns]
Algorithm 1	134	~ 18	–	148	–	~ 12
Algorithm 2	274	~ 20	–	18	3	~ 12
Algorithm 3	229	~ 20	–	25	2	~ 12
			1 stage	41	2	~ 8
Algorithm 4	230	~ 20	–	17	2	~ 12
			1 stage	25	2	~ 8

- For Algorithms 3 and 4, it sometimes suffices to insert a register after each multiplication. For instance, Synplify Pro is able to infer registered multipliers (option -pipe 1 in the synthesis script). Synthesis tools also provide the designer with libraries containing the basic building blocks of a given FPGA family. It is therefore possible to instantiate a pipelined MULT18x18 block in the VHDL code, instead of expressing the multiplication operator.
- However, if the multiplier does not fit in a single MULT18x18 block, current synthesis tools are unable to simultaneously apply the divide-and-conquer scheme and the retiming algorithm. It is therefore impossible to automatically pipeline the VHDL description of Algorithm 2 depicted on Figure 1. The solution consists here in performing the divide-and-conquer scheme by hand: the VHDL description will then contain three 16-bit × 16-bit multipliers.

The VHDL code illustrated on Figure 7 leads however to poor results on Virtex-E devices. A structural description of the operator (partial product generation, carry-propagate adders, and registers) gives better results.

4   Implementation Results

Table 2 summarizes the main characteristics of several RC6 processors for Virtex-E and Virtex-II devices. For non-feedback chaining modes, processors with full-

Table 2. Some RC6 processors for Virtex-II and Virtex-E devices. CAD tools: Synplify Pro 7.0.3 and Xilinx Alliance Series 4.1.03i (*) or ISE 5.1.03i (†).

	Device	Algo	Rounds	Pipeline α	β	γ	δ	Slices	Mult. blocks	Through-put [Gb/s]
Non-feedback chaining modes	XCV2000E-6†	1	$1+20+1$	1	2	1	1	19198 (99%)	–	~ 10.6
	XCV300E-6*	1	$1+1+1$	1	2	1	1	2068 (67%)	–	~ 0.5
	XC2V3000-6*	4	$1+20+1$	1	1	1	0	8554 (59%)	80 (83%)	~ 15.2
	XC2V3000-6†	4	$1+20+1$	1	1	1	0	10288 (71%)	80 (83%)	~ 14.2
	XC2V3000-6*	1	$1+10+1$	1	1	1	0	7456 (52%)	0 (0%)	~ 4.8
	XC2V1000-6*	4	$1+10+1$	1	1	1	0	4391 (85%)	40 (100%)	~ 7.4
	XC2V500-6*	4	$1+5+1$	1	1	1	0	2365 (76%)	20 (62%)	~ 3.9
	XC2V250-6*	4	$1+4+1$	1	1	1	0	1534 (99%)	16 (66%)	~ 2.8
Feedback chaining modes	XCV300E-6*	1	$1+1+1$	0	0	0	0	1709 (55%)	–	~ 0.16
	XCV300E-6*	4	$1+1+1$	0	0	0	0	1902 (61%)	–	~ 0.15
	XCV400E-6*	1	$1+4+1$	1	0	0	0	3932 (81%)	–	~ 0.16
	XC2V1000†	4	$1+1+1$	0	0	0	0	1560 (30%)	4 (10%)	~ 0.29
	XC2V1000†	4	$1+4+1$	1	0	0	0	2902 (56%)	16 (40%)	~ 0.34

loop unrolling achieve high throughputs. The area and the critical path however depend on the synthesis tools: we have obtained better results with Symplify Pro 7.0.3 and Xilinx Alliance 4.1.03i than with ISE 5.1.03i. Also note that XC2V500 and XC2V250 devices have not enough I/Os to deal with 128-bit words. Our solution consists in defining 64-bit input and output ports and spending two clock cycles for data transmission.

The basic iterative architecture (Figure 2a) seems to be the best one for feedback chaining modes: it requires less slices than systems with partial loop unrolling and achieves the same throughput. As the rounds are combinational, the critical path increases and we obtain very low encryption rates.

A NSA team has implemented RC6 with semi-custom ASICs based on a 0.5 μm CMOS library [10]. Using the architecture depicted by Figure 2c with a pipeline stage between two consecutive rounds and Algorithm 2 to compute $f(X)$, the NSA team reports a throughput of 2.2 Gbits/s. Gaj et al. have proposed an architecture similar to Figure 2 [2, 4]. The main differences lie in the $f(X)$ operator and in the number of pipeline stages per cipher round (3 in our case versus 28 in their system). However, four XCV1000-6 devices are required to implement the algorithm with full loop unrolling and to achieve a throughput of 13.1 Gbits/s. While the throughput is close to ours, this solution is more expensive and requires a more complex PCB.

5   Conclusions

In this paper, several architectures of the RC6 block cipher for Virtex-E and Virtex-II FPGAs have been described. We have also investigated four algorithms computing $f(X)$, which is the critical arithmetic operation of the block cipher. Our experiments indicate that both the choice of an algorithm and the VHDL coding style are strongly related to the target FPGA family. Our VHDL generator allows to quickly explore a wide parameter space and to determine the best architecture for a given set of constraints (feedback or non-feedback chaining mode, FPGA device, ...).

Acknowledgments

The author would like to thank the "Ministère Français de la Recherche", the Swiss National Science Foundation, and the Xilinx University Program for their support.

References

1. J.-L. Beuchat. *Etude et conception d'opérateurs arithmétiques optimisés pour circuits programmables*. PhD thesis, Swiss Federal Institute of Technology Lausanne, 2001.
2. P. Chodowiec, P. Khuon, and K. Gaj. Fast Implementations of Secret-Key Block Ciphers Using Mixed Inner- and Outer-Round Pipelining. In *Proc. ACM/SIGDA International Symposium on Field Programmable Gate Arrays*, pages 94–102, 2001.
3. M. Dworkin. Recommandation for Block Cipher Modes of Operation, 2001. NIST Special Publication 800-38A.
4. K. Gaj and P. Chodowiec. Fast implementation and fair comparison of the final candidates for Advanced Encryption Standard using Field Programmable Gate Arrays. In *Proc. RSA Security Conf. - Cryptographer's Track*, pages 84–99. Springer-Verlag, 2001. Available at http://ece.gmu.edu/crypto/publications.htm.
5. H. Lipmaa, P. Rogaway, and D. Wagner. Comments to NIST concerning AES Modes of Operations: CTR-Mode Encryption.
6. A. J. Menezes, P. C. van Oorschot, and S. A. Vanstone. *Handbook of Applied Cryptography*. CRC Press, 1997.
7. E. Mosanya, C. Teuscher, H. F. Restrepo, P. Galley, and E. Sanchez. Crypto-Booster: A Reconfigurable and Modular Cryprographic Coprocessor. In C. K. Koc and C. Paar, editors, *Proceedings of the First International Workshop on Cryptographic Hardware and Embedded Systems, CHES'99, Worcester, MA*, volume 1717 of *Lecture Notes in Computer Science*, pages 246–256. Springer-Verlag, 1999.
8. B. Parhami. *Computer Arithmetic*. Oxford University Press, 2000.
9. R.L. Rivest, M. J. B. Robshaw, R. Sidney, and Y. L. Yin. The RC6 Block Cipher, 1998.
10. B. Weeks, M. Bean, T. Rozylowicz, and C. Ficke. Hardware Performance Simulations of Round 2 Advanced Encryption Standard Algorithms. Technical report, National Security Agency, 2000.
11. R. Zimmermann, A. Curiger, H. Bonnenberg, H. Kaeslin, N. Felber, and W. Fichtner. A 177 Mbit/s VLSI Implementation of the International Data Encryption Algorithm. *IEEE Journal of Solid-State Circuits*, 29(3):303–307, 1994.

Very High Speed 17 Gbps
SHACAL Encryption Architecture

Máire McLoone and J.V. McCanny

DSiP™ Laboratories, School of Electrical and Electronic Engineering,
The Queen's University of Belfast, Belfast BT9 5AH, Northern Ireland
maire.mcloone@ee.qub.ac.uk, j.mccanny@ee.qub.ac.uk

Abstract. Very high speed and low area hardware architectures of the SHACAL-1 encryption algorithm are presented in this paper. The SHACAL algorithm was a submission to the New European Schemes for Signatures, Integrity and Encryption (NESSIE) project and it is based on the SHA-1 hash algorithm. To date, there have been no performance metrics published on hardware implementations of this algorithm. A fully pipelined SHACAL-1 encryption architecture is described in this paper and when implemented on a Virtex-II X2V4000 FPGA device, it runs at a throughput of 17 Gbps. A fully pipelined decryption architecture achieves a speed of 13 Gbps when implemented on the same device. In addition, iterative architectures of the algorithm are presented. The SHACAL-1 decryption algorithm is derived and also presented in this paper, since it was not provided in the submission to NESSIE.

Keywords: NESSIE, SHACAL

1 Introduction

New European Schemes for Signatures, Integrity and Encryption (NESSIE) was a three-year research project, which formed part of the Information Societies Technology (IST) programme run by the European Commission. The main aim of NESSIE is to present strong cryptographic primitives covering confidentiality, data integrity and authentication, which have been open to a rigorous evaluation and cryptanalytical process. Therefore, submissions not only included private key block ciphers, as with the Advanced Encryption Standard (AES) project [1], but also hash algorithms, digital signatures, stream ciphers and public key algorithms. The first call for submissions was in March 2000 and resulted in 42 submissions. The first phase of the project involved conducting a security and performance evaluation of all the submitted algorithms. The performance evaluation hoped to achieve performance estimates for software, hardware and smartcard implementations of each algorithm. 24 algorithms were selected for further scrutiny in the second phase, which began in July 2001. The National Institute for Standards and Technology's (NIST) Triple-DES and Rijndael private key algorithms and SHA-1 and SHA-2 hash algorithms were also considered for evaluation. The following 7 block cipher algorithms were chosen for further investigation in phase two of the NESSIE project: IDEA, SHACAL,

P.Y.K. Cheung et al. (Eds.): FPL 2003, LNCS 2778, pp. 111–120, 2003.
© Springer-Verlag Berlin Heidelberg 2003

SAFER++, MISTY1, Khazad, RC6 and Camellia. The final selection of the NESSIE cryptographic primitives took place on 27 February 2003. MISTY1, Camellia, SHACAL-2 and Rijndael are the block ciphers chosen to be included in the NESSIE portfolio of cryptographic primitives. Overall, 12 of the original submissions are included along with 5 existing standard algorithms.

Currently, the fastest known FPGA implementations of the NESSIE finalist algorithms are as follows: Leong et al.[2] estimate that their IDEA architecture can achieve a throughput of 2 Gbps on a Virtex XCV1000 FPGA device. Standaert et al.'s [3] Khazad architecture runs at 9.4 Gbps on the XCV1000 FPGA. Their MISTY1 architecture runs at 10 Gbps on the same device. A pipelined Camellia implementation by Ichikawa et al. [4] on a Virtex-E XCV1000E device performs at 6.7 Gbps. The fastest FPGA implementation of the RC6 algorithm is the 15.2 Gbps implementation by Beuchat [5] on the Virtex-II XC2V3000 device. Finally, an iterative architecture of the SAFER++ algorithm by Ichikawa et al. [6] achieves a data-rate of 403 Mbps on a Virtex-E XCV1000E device. For the SHACAL algorithm, only published work on software implementations is currently available. Hence, in this paper, the authors hope to provide a performance evaluation of SHACAL-1 hardware implementations.

Section 2 of the paper provides a description of the SHACAL-1 algorithm. Section 3 outlines the SHACAL-1 architectures for hardware implementation. Performance results are given in section 4 and conclusions are provided in section 5.

2 A Description of the SHACAL-1 Algorithm

The SHACAL algorithm [7], developed by Helena Handschuh and David Naccache, is based on the NIST's SHA hash algorithm. SHACAL is defined as a variable block and key length family of ciphers. There are two versions specified in the submission: SHACAL-1, which is derived from SHA-1 and SHACAL-2, which is derived from SHA-256. However, other versions can easily be derived from the SHA-384 and SHA-512 hash algorithms. Only the SHACAL-1 algorithm is considered for implementation in this paper. However, the SHACAL-1 architecture described can be easily adapted for the other variants.

SHACAL-1 operates on a 160-bit data block utilising a 512-bit key. The key, k, can vary in length within the range, $128 \leq k \leq 512$. If $k < 512$, it is appended with zeros to a length of 512-bits. SHACAL-1 encryption, outlined in Fig. 1, is performed by splitting the 160-bit plaintext into five 32-bit words, A, B, C, D and E. Next, 80 iterations of the compression function are performed. The resulting A, B, C, D and E values are concatenated to form the ciphertext. The compression function is defined as follows:

$$A_{i+1} = ROT_{LEFT\text{-}5}(A_i) + F_i(B_i, C_i, D_i) + E_i + Cnst_i + W_i$$
$$B_{i+1} = A_i$$
$$C_{i+1} = ROT_{LEFT\text{-}30}(B_i)$$
$$D_{i+1} = C_i$$
$$E_{i+1} = D_i \tag{1}$$

where, W_i are the subkeys generated from the key schedule and the $ROT_{LEFT\text{-}n}$ function is defined as a 32-bit word rotated to the left by n positions.

The constants, $Cnst_i$, in hexadecimal, are,

$$
\begin{array}{ll}
Cnst_i = 5a827999 & 0 \le i \le 19 \\
Cnst_i = 6ed9eba1 & 20 \le i \le 39 \\
Cnst_i = 8f1bbcdc & 40 \le i \le 59 \\
Cnst_i = ca62c1d6 & 60 \le i \le 79
\end{array} \tag{2}
$$

and the function $F_i(x, y, z)$ is defined as,

$$
F_i(x, y, z) = \begin{cases}
(x\ AND\ y)\ OR\ (\overline{x}\ AND\ z) & 0 \le i \le 19 \\
x \oplus y \oplus z & 20 \le i \le 39 \\
(x\ AND\ y)\ OR\ (x\ AND\ z)\ OR\ (y\ AND\ z) & 40 \le i \le 59 \\
x \oplus y \oplus z & 60 \le i \le 79
\end{cases} \tag{3}
$$

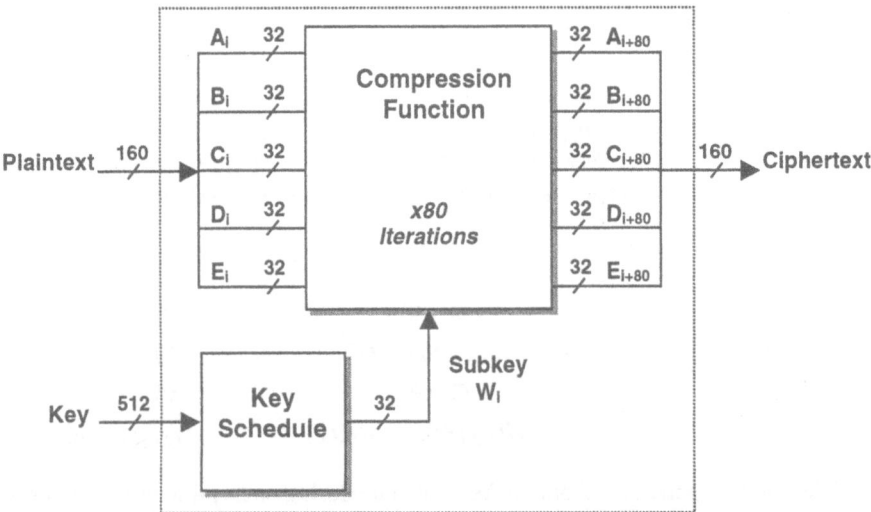

Fig. 1. Outline of SHACAL-1 Encryption Algorithm

In the SHACAL-1 key schedule, the 512-bit input key is expanded to form eighty 32-bit subkeys, W_i, such that,

$$
W_i = \begin{cases}
Key_i & 0 \le i \le 15 \\
ROT_{LEFT-1}(W_{i-3} \oplus W_{i-8} \oplus W_{i-14} \oplus W_{i-16}) & 16 \le i \le 79
\end{cases} \tag{4}
$$

The first 16 subkeys are formed by splitting the input key into sixteen 32-bit values.

2.1 SHACAL-1 Decryption

SHACAL-1 decryption is not defined in the submission. Therefore, it has been
derived and is presented here. It requires an inverse compression function and an
inverse key schedule.

The inverse compression function is as follows:

$$A_{i+1} = B_i$$
$$B_{i+1} = ROT_{RIGHT\text{-}30}(C_i)$$
$$C_{i+1} = D_i$$
$$D_{i+1} = E_i$$
$$E_{i+1} = A_i - [ROT_{LEFT\text{-}5}(B_i) + InvF_i(ROT_{RIGHT\text{-}30}(C_i), D_i, E_i) + InvCnst_i + InvW_i] \qquad (5)$$

where $InvW_i$ are the inverse subkeys generated from the inverse key schedule and the
$ROT_{RIGHT\text{-}n}$ function is defined as a 32-bit word rotated to the right by n positions. The
constants, $InvCnst_i$, in hexadecimal, are,

$$
\begin{aligned}
InvCnst_i &= ca62c1d6 & 0 \le i \le 19 \\
InvCnst_i &= 8f1bbcdc & 20 \le i \le 39 \\
InvCnst_i &= 6ed9eba1 & 40 \le i \le 59 \\
InvCnst_i &= 5a827999 & 60 \le i \le 79
\end{aligned}
\qquad (6)
$$

and the function $InvF_i(x, y, z)$ is defined as,

$$
InvF_i(x, y, z) =
\begin{cases}
x \oplus y \oplus z & 0 \le i \le 19 \\
(x \text{ AND } y) \text{ OR } (x \text{ AND } z) \text{ OR } (y \text{ AND } z) & 20 \le i \le 39 \\
x \oplus y \oplus z & 40 \le i \le 59 \\
(x \text{ AND } y) \text{ OR } (\overline{x} \text{ AND } z) & 60 \le i \le 79
\end{cases}
\qquad (7)
$$

The subkeys generated from the key schedule during encryption are used in reverse
order when decrypting data. Therefore, it is necessary to wait for all of the subkeys to
be generated before beginning decryption. However, an inverse key schedule can be
utilised to generate the subkeys in the order that they are required for decryption. This
inverse key schedule is defined as,

$$
InvW_i =
\begin{cases}
InvKey_i & 0 \le i \le 15 \\
ROT_{RIGHT-1}(InvW_{i-16}) \oplus InvW_{i-13} \oplus InvW_{i-8} \oplus InvW_{i-2} & 16 \le i \le 79
\end{cases}
\qquad (8)
$$

The $InvKey$ is created by concatenating the final 16 subkeys generated during
encryption, such that,

$$
\begin{aligned}
InvKey = \{ &W_{64}, W_{65}, W_{66}, W_{67}, W_{68}, W_{69}, W_{70}, W_{71}, \\
&W_{72}, W_{73}, W_{74}, W_{75}, W_{76}, W_{77}, W_{78}, W_{79} \}
\end{aligned}
\qquad (9)
$$

3   SHACAL-1 Hardware Architectures

The SHACAL-1 algorithm is derived from the SHA hash algorithm. Therefore, the design of a SHACAL-1 hardware architecture can be derived from the design of a SHA hash algorithm architecture. Previous efficient implementations of the SHA-1 and SHA-2 hash algorithms [8], [9], [10] have utilised a shift register design approach. Thus, this methodology has also been used in the iterative and fully pipelined SHACAL-1 architectures described here.

3.1   Iterative SHACAL-1 Architecture

In the iterative encryption and decryption architectures data is passed through the compression or inverse compression function component eighty times. The initial five data blocks are generated from the input block split into five 32-bit blocks. The outputs of the function, A to E form the inputs on consecutive clock cycles. After 80 iterations, the A to E outputs are concatenated to form the 160-bit plaintext/ciphertext.

The main components in both the iterative and fully pipelined architectures are the compression function and the key schedule. The key schedule is implemented using a 16-stage shift register design, as illustrated in Fig. 2.

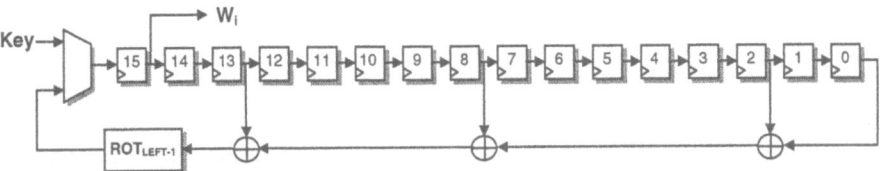

Fig. 2. SHACAL-1 Key Schedule Design

The input 512-bit key is loaded into the registers in 32-bit blocks over 16 clock cycles. On the next clock cycle, the value of register 15 is updated with the result of equation 4. An initial delay of 16 clock cycles is avoided by taking the values, W_i from the output of register 15 and not from the output of register 0. The subkeys, W_p are generated as they are required by the compression function.

The inverse key schedule is designed in a similar manner and is shown in Fig. 3.

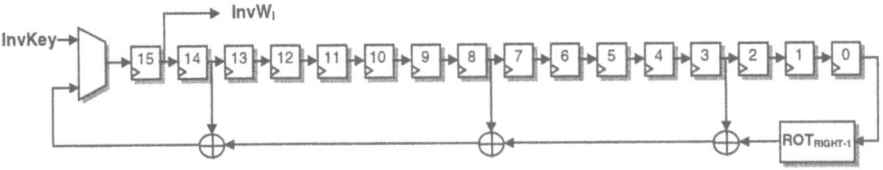

Fig. 3. SHACAL-1 Inverse Key Schedule Design

Typically, for decryption, the sender of the ciphertext will send the receiver the original key used to encrypt the message. Hence, the receiver will have to generate all eighty 32-bit subkeys before commencing decryption. However, this can be avoided if the sender of the ciphertext sends the receiver the final sixteen 32-bit subkeys that were created during encryption of the message as a 512-bit inverse key. Now, the receiver can immediately begin to decrypt the ciphertext, since the subkeys required for decryption can be generated as they are required using this inverse key.

The design of the SHACAL-1 compression function is depicted in Fig. 4. The design requires 5 registers to store the continually updating values of A, B, C, D & E. The values in registers B, C and D are operated on by one of four different functions every 20 iterations, as given in equation 3.

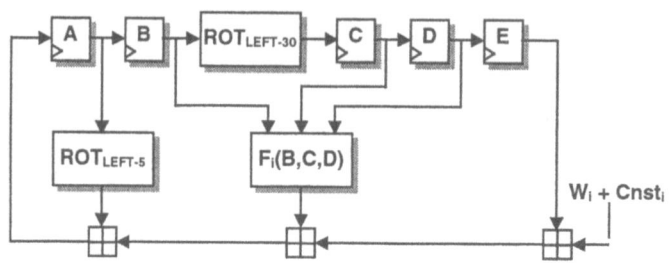

Fig. 4. SHACAL-1 Compression Function Design

The critical path of the overall SHACAL design occurs in the compression function in the calculation of,

$$A_{i+1} = ROT_{LEFT\text{-}5}(A_i) + F_i(B_p C_p D_i) + E_i + Cnst_i + W_i \tag{10}$$

In the architectures described here, this critical path is reduced in 2 ways. Firstly, the addition, $T = Cnst_i + W_i$, is performed on the previous clock cycle and thus, is removed from the critical path. Also, Carry-Save-Adders (CSAs) are utilised. With CSAs, the carry propagation is avoided until the final addition. Therefore, equation 10 is implemented using two CSAs and one Full Adder (FA) rather than three FAs, as shown in Fig. 5. Since the CSAs involve 32-bit additions, this implementation is faster than an implementation using only FAs [11].

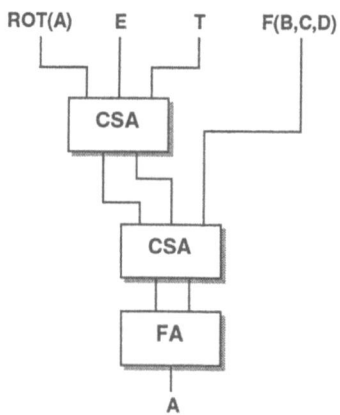

Fig. 5. CSA Implementation in Compression Function

The inverse compression function, outlined in Fig. 6, contains a subtraction. Hence, the throughput of the SHACAL-1 decryption architecture will be slower than that of the encryption architecture. During decryption the critical path of the overall SHACAL design occurs in the inverse compression function, where,

$$E_{i+1} = A_i - [ROT_{LEFT-5}(B_i) + InvF_i(ROT_{RIGHT-30}(C_i), D_i, E_i) + InvCnst_i + InvW_i] \qquad (11)$$

Once again, it can be reduced by performing the addition, $InvCnst_i + InvW_p$ on the previous clock cycle. No significant performance benefit is achieved by using CSAs in the decryption architecture. Since equation 11 includes a subtraction, only one CSA could be used and two full adders would still be required in an implementation.

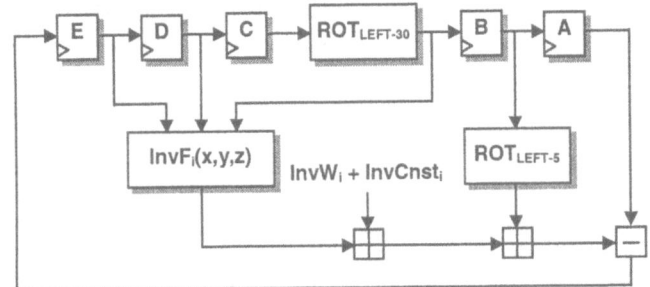

Fig. 6. SHACAL-1 Inverse Compression Function Design

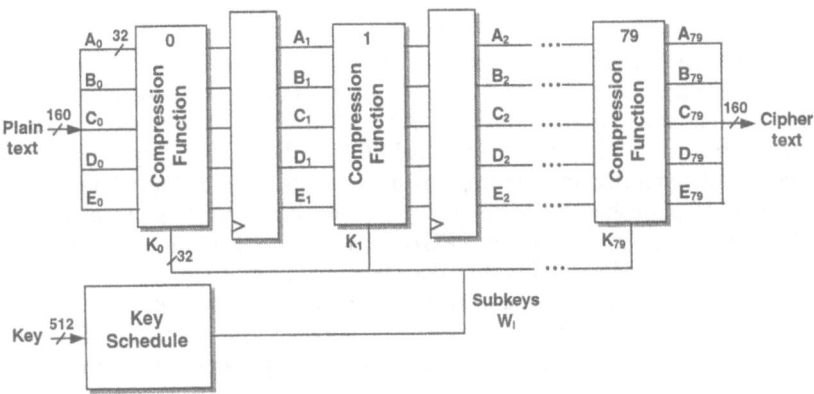

Fig. 7. SHACAL-1 Pipelined Architecture

3.2 Pipelined SHACAL-1 Architecture

In the pipelined SHACAL-1 encryption architecture, the compression function and key schedule are designed as for the iterative architecture. However, the eighty compression function iterations are fully unrolled and registers placed between each component, as depicted in Fig. 7. It is assumed that the same key is used throughput a data transfer session. Every twenty compression function components contains a different function according to equation 3. New plaintext blocks can be accepted on every clock cycle and after an initial delay, the corresponding ciphertext blocks will appear on consecutive clock cycles. This leads to a very high speed design. The pipelined decryption architecture is designed in a similar manner using the inverse compression function and inverse key schedule.

4 Performance Evaluation

To provide a hardware performance evaluation for SHACAL-1, the hardware architectures described in this paper are implemented on Xilinx FPGA devices for demonstration purposes.   The designs were simulated using Modelsim and synthesised using Synplify Pro v7.2 and Xilinx Foundation Series 5.1i software. They were verified using the SHACAL-1 test vectors provided in the submission to NESSIE.

The iterative encryption architecture implemented on the Virtex-II XC2V500 device runs at a clock speed of 110 MHz and hence, achieves a throughput of 215 Mbps. The design utilises just 994 CLB slices. The iterative decryption architecture implemented on the same device has a data-rate of 177 Mbps and requires only 877 slices. The decryption design is slower since its critical path contains a subtraction. However, it is smaller in area since the CSAs used in the design of the compression function in the encryption design incur a slight area penalty. The iterative architectures result in highly- compact yet efficient implementations. In both designs the 160-bit plaintext/ciphertext blocks and 512-bit key are input and output in 32-bit blocks. This implies a lower IOB count and less routing, thus, the designs can be implemented efficiently on smaller FPGA devices.

The pipelined architectures result in very high speed designs. The encryption design when implemented on the Virtex-II XC2V4000 device, operates at 106 MHz with a throughput of 17 Gbps utilising 13,729 CLB slices. The equivalent decryption design metrics are 83 MHz, 13 Gbps and 10,844 slices. Even higher data-rates are attainable if the architectures are implemented on ASIC technology. Table 1 provides a summary of published work on fast, pipelined hardware implementations of the NESSIE block cipher finalists on FPGA devices. The SAFER++ algorithm is not included in the table since an iterative design [6] with a throughput of 403 Mbps is the fastest implementation published to date. For comparison purposes, the table includes performance metrics for a fully pipelined single-chip implementation of the Rijndael algorithm [12]. Overall, the SHACAL-1 algorithm provides the fastest pipelined implementation. However, it is difficult to compare the performance of the algorithms as they are implemented on different devices. The closest in speed to the SHACAL-1 implementation is the 15.2 Gbps RC6 design by Beuchat. Beuchat [5] also carried out an implementation of the pipelined RC6 design on the XCV1600E device, achieving a data-rate of 9.7 Gbps. The SHACAL-1 pipelined design implemented on the same device performs at 10 Gbps. Therefore, the SHACAL-1 algorithm still provides the fastest implementation. Although it is the fastest, the pipelined design is also the largest in area. High-speed, yet lower area implementations are possible by unrolling the algorithm by a lower number of compression function components, while still adopting pipelining. The SHACAL specification provided performance metrics for a software implementation of SHACAL-1 on an 800 MHz Pentium III processor. Encryption was achieved at a data-rate of 52 Mbps, decryption at 55 Mbps and key setup at 180 Mbps [7]. Therefore, even the iterative hardware design outlined in this paper is 4 times faster than their software implementation.

The efficiency of the algorithm implementations is also given in Table 1. Efficiency calculations are not provided for implementations that have been designed specifically to a device, and thus, include features such as multipliers and BRAM components. The MISTY1 implementation is the most efficient pipelined design while the SHACAL-1 implementation on the Virtex-II device is the next most efficient.

Table 1. Summary of NESSIE Algorithm Hardware Implementations

Authors	Algorithm	Device Used	*Area*	Through -put (Mbps)	Efficiency (Mbps/ slices)
Authors	SHACAL-1	XC2V4000	13,729 slices	17021	1.24
Authors	SHACAL-1	XCV1600E	14,768 slices	10123	0.68
Leong et al.[2] (estimated)	IDEA	XCV1000	11,204 slices	2003	0.18
Standaert et al.[3]	Khazad	XCV1000	8,800 slices	9472	1.08
Standaert et al.[3]	MISTY1	XCV1000	6,322 slices	10176	1.6
Icikawa et al. [4]	Camellia	XCV1000E	9,692 slices	6750	0.7
Beuchat[5]	RC6	XC2V3000	8,554 slices 80 multipliers	15200	-
Beuchat[5]	RC6	XCV1600E	14110 slices	9700	0.7
McLoone, et al. [12]	Rijndael	XCV812E	2679 slices 82 BRAMs	6956	-

5   Conclusions

Throughout the NESSIE project, there has been no performance evaluation provided for hardware implementations of the SHACAL algorithm. In this paper, the SHACAL-1 algorithm is studied and both low area iterative and high speed pipelined architectures are presented. These are implemented on Xilinx FPGA devices for demonstration purposes but can readily be implemented on other FPGA or ASIC technologies. The iterative architectures are highly compact, yet efficient, when implemented on Virtex-II devices. A very high speed 17 Gbps design is achieved when the fully pipelined SHACAL-1 architecture is implemented on the Virtex-II XC2V4000 device. Therefore, SHACAL-1 is the fastest algorithm in hardware when compared to pipelined hardware implementations of the other NESSIE block cipher finalists. The other SHACAL algorithm version specified in the submission, SHACAL-2, operates on a 256-bit data block utilising a 512-bit key. The SHACAL-1 architectures described in this paper can be easily adapted for SHACAL-2. SHACAL-2 iterative and pipelined architectures are anticipated to achieve similar speeds to that achieved by SHACAL-1.

The SHACAL-2 algorithm was selected as one of four block ciphers to be included in the NESSIE portfolio of cryptographic primitives and future work will be carried out on SHACAL-2 hardware implementations. The SHACAL algorithm has proven to be very secure with a large security margin. Since it is derived from the SHA hash algorithm, it has already undergone much cryptanalysis. It performs well when implemented in software [13] and as is evident from this paper, very high speed hardware implementations are also possible. Since security and performance were the two main criteria in the NESSIE selection process, the SHACAL algorithm is an ideal candidate for selection.

References

1.   US NIST Advanced Encryption Standard, URL: http://csrc.nist.gov/encryption/aes/
2.   M.P. Leong, O.Y.H. Cheung, K.H. Tsoi, P.H.W. Leong, " A Bit-Serial Implementation of the International Data Encryption Algorithm IDEA", IEEE Symposium on FCCMs 2000, California, April 2000.
3.   F.X. Standaert, G. Rouvroy, "Efficient FPGA Implementation of Block Ciphers Khazad and MISTY1", 3rd NESSIE Workshop, URL:
       http://www.di.ens.fr/~wwwgrecc/NESSIE3/, Germany, November 2002.
4.   T. Ichikawa, T. Sorimachi, T. Kasuya, M. Matsui, "On the criteria of hardware evaluation of block ciphers(1)", Techn report of IEICE, ISEC2001-53, Sept 2001.
5.   J.L. Beuchat, "High Throughput Implementations of the RC6 Block Cipher Using Virtex-E and Virtex-II Devices", INRIA Research Report, URL: http://www.ens-lyon.fr/~jlbeucha/publications.html, July 2002.
6.   T. Ichikawa, T. Sorimachi, T. Kasuya, "On Hardware Implementation of Block Ciphers Selected at the NESSIE Project Phase 1", 3rd NESSIE Workshop, URL:
       http://www.di.ens.fr/~wwwgrecc/NESSIE3/, Germany, November 2002.
7.   H. Handschuh, D. Naccache, "SHACAL", 1st NESSIE Workshop, URL:
       https://www.cosic.esat.kuleuven.ac.be/nessie/workshop/, Belgium, Nov 2000.
8.   K.K. Ting, S.C.L. Yuen, K.H. Lee, P.H.W. Leong, "An FPGA based SHA-256 Processor", 12th International FPL Conference, France, September 2002.
9.   T. Grembowski, R. Lien, K. Gaj, N. Nguyen, P. Bellows, J. Flidr, T. Lehman, B. Schott, "Comparative Analysis of the Hardware Implementations of Hash Functions SHA-1 and SHA-512", Information Security Conference", Oct 2002.
10.   M. McLoone, J.V. McCanny, "Efficient Single-Chip Implementation of SHA-384 & SHA-512", IEEE International FPT Conference, Hong Kong, Dec 2002.
11.   T. Kim, W. Jao, S. Tjiang, "Circuit Optimization Using Carry-Save-Adder Cells", IEEE Transactions on Computer-Aided Design of Integrated Circuits and Systems, Vol. 17, No.10, October 1998.
12.   M. McLoone, J.V. McCanny, "High Performance Single-Chip FPGA Rijndael Algorithm Implementations", 3rd International CHES Workshop, pp 65-77, France, May 2001.
13.   NESSIE, "Performance of Optimized Implementations of the NESSIE Primitives", http://www.cosic.esat.kuleuven.ac.be/nessie/deliverables/D21-v2.pdf February 2003.

Track Placement: Orchestrating Routing Structures to Maximize Routability

Katherine Compton[1] and Scott Hauck[2]

[1] Northwestern University, Evanston, IL USA
kati@ece.northwestern.edu
[2] University of Washington, Seattle, WA USA
hauck@ee.washington.edu

Abstract. The design of a routing channel for an FPGA is a complex process requiring a careful balance of flexibility with silicon efficiency. With a growing move towards embedding FPGAs into SoC designs, and the new opportunity to automatically generate FPGA architectures, this problem is even more critical. The design of a routing channel requires determining the number of routing tracks, the length of the wires in those tracks, and the positioning of the breaks between wires on the tracks. This paper focuses on the last problem, the placement of breaks in tracks to maximize overall flexibility. Our optimal algorithm for track placement finds a best solution provided the problem meets a number of restrictions. Our relaxed algorithm is without restrictions, and finds solutions on average within 1.13% of optimal.

1 Introduction

The design of an FPGA interconnect structure has usually been a hand-tuning process. A human designer, with the aid of benchmark suites and trial-and-error, develops an interconnect structure that attempts to balance flexibility with silicon efficiency. Often, the concentration is on picking the number and length of tracks – long tracks give global communication but with high silicon and delay costs, while short wires can be very efficient only if signals go a relatively short distance.

An area that can sometimes be ignored is the placement of the breaks in these interconnect wires. If we have a symmetric interconnect, with N length-N wires, we simply break one wire at each cell. However, for more irregular interconnects, it can be difficult to determine the best positioning of these breaks.

While a manual solution may be feasible in many cases when only a single architecture is being examined, it is not always practical. For example, track placement becomes extremely critical when we consider automatic generation of custom FPGA architectures for systems-on-a-chip [1]. Here, a track placement may be performed a very large number of times within the inner loop of an architecture generator, and a fast but effective algorithm for automatic track placement becomes a necessity.

In this paper, we address the issue of routing architecture design for reconfigurable architectures with segmented channel routing, such as RaPiD [2] and Garp [3]. We formalize the track placement problem, define a cost metric, and introduce track placement algorithms, including one proven optimal for a subset of these problems.

P.Y.K. Cheung et al. (Eds.): FPL 2003, LNCS 2778, pp. 121–130, 2003.

Achieving the best track placement requires a careful positioning of the breaks on multiple, different-length, routing tracks. For example, in Fig. 1 right, the breaks between wires in the routing tracks are staggered. This helps to provide similar routing options regardless of location in the array. If instead all breaks were lined up under a single unit, as in Fig. 1 left, a signal might become very difficult to route if its source was on one side of the breaks and at least one sink on the other. When large numbers of tracks or tracks of different wire lengths are involved, it can become difficult to find a solution where the breaks are evenly distributed through the array.

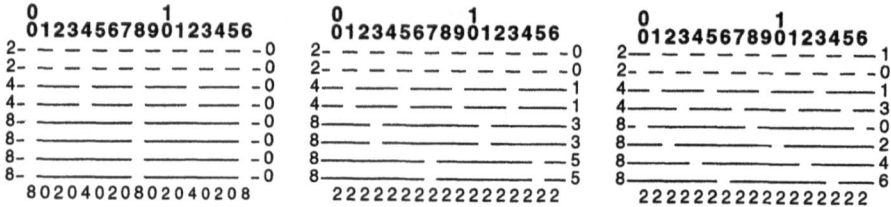

Fig. 1. Three different track placements for the same problem. A very poor one (left), an improved one (middle) and an even better placement (right). In each, the numbers on top indicate position in the architecture, and the numbers on the bottom indicate the number of breaks at the given position. Each track has its associated wire length at left, and offset at right.

The issue of determining the positioning (or offset) of a track within an architecture is referred to as track placement. The goal is to pick the offset that should be used for each track in order to maximize the routability, given a predetermined set of tracks with fixed length wires. For simplicity, each track is restricted to contain wires of only one length, which is referred to as the track's S value, or track length. The actual determination of the quantity and S values of tracks is discussed elsewhere [1], as are issues specific to 2D routing architecture design [4].

2 Problem Description

Finding a good track placement is a complex task. Intuitively, we wish to space tracks with the same S value evenly, such as by placing the length-8 tracks from the problem featured in Fig. 1 at offsets 0, 2, 4, and 6. However, a placement of tracks of one S value can affect the best placement for tracks of another S value. For example, Fig. 1 right shows that the length-4 tracks are placed at offsets 1 and 3 in order to avoid having the breaks of those tracks fall at the same locations as the breaks from the length 8 tracks. This effect is called "correlation" between the affected tracks. Correlations occur between tracks when their S values contain at least one common prime factor. It is these correlations that make track placement difficult.

From the previous example we might conclude that a possible metric for measuring the quality of a track placement would be to compute the "evenness" or "smoothness" of the break distribution throughout the architecture. However, the smoothness metric fails to capture the idea of maintaining similar routing options for every location in the array. The placement in Fig. 1 center has the same smoothness of breaks as the solution at right, but is not an equally good solution. For example, although there are two length-4 tracks, each logic unit position is at the same location along the length-4 wires in both tracks. On the other hand, the architecture at right provides for two different locations along the length-4 wires at every logic unit position. For this reason, we consider the placement at right to be superior in terms of routing options at each position despite the two architectures having the same break smoothness.

Instead, our chosen goal in track placement is to ensure that signals of all lengths have the most possible routes. To quantitatively measure this routability, we examine each possible signal length, and find the region of the interconnect that gives the fewest possible routes for this length signal. Summing across all possible signal lengths yields the "diversity score" of a track placement, as shown in Equation 1. The fewer and smaller the routing bottlenecks, the more routing choices are available throughout the architecture, and the higher the diversity score.

Equation 1. The diversity score for a particular track placement routing problem T and solution set of offsets O for the tracks in T can be calculated using the equation:

$$diversity\_score(T,O) = \sum_{L} \left(\min_{all\ positions} \left(\sum_{Ti \in T} uncut(T_i, O_i, L, position) \right) \right)$$

for all tracks Ti with their given wire lengths and offsets Oi, possible signal lengths L, and all possible positions in the interconnect. The uncut() function returns a binary result that is 0 if there is a break within the range [position, position + L) on track Ti that has been placed at offset O_i, and 1 otherwise.

Fig. 2 shows the diversity score calculation for two different placements of the same track problem. Here we consider a four position window that encapsulates the full repeating pattern of breaks of the placement. The length of the window that needs to be considered can be found by taking the LCM of all S values in the problem. Examining a larger window would simply yield the same results. Within this window we count the number of tracks at each position that can be used to route a signal of the given length towards the right. The different possible signal lengths (L) are listed, and at each position, the number of tracks useable to route that signal length is given.

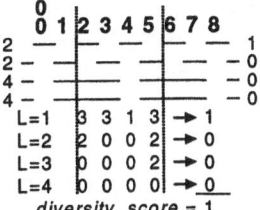
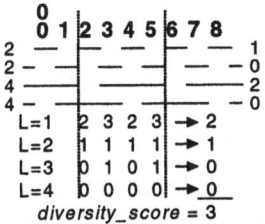

Fig. 2. Diversity score for two different track placements. Positions are given by the top row of numbers. For each track, segment length S is at left, and offset O is at right. The number of routing possibilities is given for each potential signal length L. Diversity score is at bottom.

In Fig. 2 left, two different tracks can be used to route length-2 signals to the right at position 2, but at position 3, no tracks are useable. At right, exactly one track can be used to route length-2 signals from any position in our window. We then take the minimum value at each length (representing a routing bottleneck for that signal length), and sum across L values to arrive at the diversity score.

We have also determined a bound on this value, as described in Theorem 1 (proven elsewhere [5]). Note that comparing diversity scores is only valid across different solutions to the same track placement problem. We discuss the issue of determining the actual quantity and types of tracks that should be used elsewhere [1].

Theorem 1. For all possible offset assignments O to track set T,

$$diversity\_score(T,O) \le \sum_L floor\left(|T| - \sum_{Ti \in T} \min(1, L/S_i)\right)$$

We focus on the worst-case (the regions with the fewest possible routes) instead of the average case. The average number of possible routes for a signal is independent of track placement, and only depends on the number of tracks and their wire lengths. Thus, an average case metric cannot tell the difference between the placements in Fig. 2, while the worst-case will shows that the rightmost version is superior.

3   Proposed Algorithms

We have developed a number of algorithms to solve the track placement problem based on the diversity score cost function.   These track placement algorithms, both optimal and heuristic, are discussed in depth in the next few sections.

3.1   Brute Force Algorithm

Using a brute force approach, we can guarantee finding a solution with the highest possible diversity score. However, examining the entire solution space is a very slow process. The number of cases that the brute force approach must examine for a given problem is the product, over all distinct track lengths in the problem, of a multichoose of the track length and quantity of that length. For example, a modest architecture with 8 length-12 tracks, 4 length-6 tracks and 2 length-4 tracks will require the examination of *over 95 million* distinct cases. Since one of our targeted applications is within an inner loop of an automatic architecture generator [1], this approach is far too slow. Instead, it provides a bound on the diversity score, which is used to verify the results of our optimal algorithm and measure the quality of our other algorithms.

3.2   Simple Spread Algorithm

The Simple Spread algorithm is a simple heuristic for track placement. Tracks are grouped by segment length. For each group, the tracks are spaced evenly within the offsets 0 through S_i-1 (where S_i is the segment length of that group), regardless of the placement of tracks from other groups. This algorithm is simple in design and fast in execution, but disregards correlations between S values. It is included to demonstrate in the results section the necessity of considering these correlations.

3.3   Optimal Factor Algorithm

One of our goals was to develop a fast algorithm that, with some restrictions, would provably find the optimal solution. The Optimal Factor algorithm is the culmination of many theorems and proofs. While many of the theorems will be presented here, due to reasons of space, their proofs are presented elsewhere [5].

Any optimal algorithm must consider the correlations between breaks both within an S group and across S groups. Correlations between two different S values occur when those S values share a common factor. For example, a track with S=6 and one with S=2 (a common factor of 2) will either have breaks at the same location once every 6 positions, or not at all, depending on the offsets chosen for the two tracks. On the other hand, a track with S=2 and one with S=5 will have breaks at the same location once every 10 positions regardless of the offsets chosen, as they are completely uncorrelated (no common factors). The following theorem is therefore used to divide a problem into smaller independent (uncorrelated) problems when possible.

Theorem 2. If T can be split into two sets $G1 \subset T$ and $G2 = T-G1$, where the S values of all tracks in G1 are relatively prime with the S values of all tracks in G2, then *diversity_score*(T,O) = *diversity_score*(G1, O1) + *diversity_score*(G2, O2), where O is the combination of sets O1 and O2. Thus, G1 and G2 may be solved independently.

We can also use the fact that correlation is based on common factors to further simplify the track placement problem. Theorem 3 states that whenever one track's S value has a prime factor that is not shared with any other track in the problem (or a track has a higher quantity of a prime factor than any other track in the problem), the prime factor can be removed. For example, with 2 length-6 tracks and one length-18 track, we can remove a 3 from 18, and essentially have 3 length-6 tracks, which can then be placed evenly using a later theorem. We do not actually change the S value of the length-18 track, just the effective S value used during track placement. After the offsets of tracks are determined, the original S values are restored if necessary.

Theorem 3. If the S_i of unplaced track T_i contains more of any prime factor than the S_j of each and every other track T_j ($I \neq j$), then for all solutions O, *diversity_score*(T, O) = *diversity_score*(T', O), where T' is identical to T, except T'_i has $S'_i = S_i$ with the unshared prime factor removed. This rule can be applied recursively to eliminate all unshared prime factors.

Next the tracks are grouped by S value (which may have been factored using Theorem 3). These groups are then sorted such that the largest S value is considered first. Any full sets (a set of N tracks all with Si = N) are removed from the problem by Theorem 4, placing one track from the set at each possible offset 0 to N-1. Note we are only considering as possible offsets the range 0...|Si|-1 for a track with S=Si. While it is perfectly valid to place a track at an offset greater than Si-1, the fact that all wires within the track are of length Si causes a break to be located every Si locations. Therefore, there will always be a break on track Ti within the range 0 to Si-1.

The basic idea of the remainder of the algorithm is to space the tracks of the largest S value (S_{max}) evenly using Theorems 4 and 5, then remove these tracks from further consideration. However, in order to respect the effect that their breaks have on the placement of tracks from successive S groups, we add pre-placed placeholder tracks (with S = the next largest S value S_{next}) such that they have the same number of breaks at the same positions as the original tracks, but in terms of S_{next} instead of S_{max}. This enables us to determine the best offsets for the real tracks with $S=S_{next}$. This process is repeated within a loop, where S_{next} becomes S_{max}, and we find the new S_{next}.

Theorem 4. Given a set $G \subset T$ of tracks, each with $S=|G|$. If there exists a solution O' for tracks T'=T-G that meets the bound, then there is also solution O for tracks T that meets the bound (and thus is optimal), where each track in G is placed at a different offset $0...|G|-1$.

We do set a number of restrictions, however, in order to ensure that at each loop iteration, (1) we can perfectly evenly space the tracks with $S=S_{max}$, and (2) the breaks from the set of tracks with $S=S_{max}$ can be exactly represented by some integer number of tracks with $S=S_{next}$. These restrictions are outlined in the next two theorems.

Theorem 5. Let S_{max} be the maximum S value currently in the track placement problem, M be the set of all tracks with $S = S_{max}$, N = |M|, and S_{next} be the largest $S \neq S_{max}$. If N>1, S_{max} is a proper multiple of N, and $S_{next} \leq S_{max}*(N-1)/N$, then any solution to T that meets the bound must evenly space out the N tracks with $S = S_{max}$. That is, for each track M_i, with a position O_i, there must be another track M_j with a break at $O_i + S_{max}/N$.

Theorem 6. Given a set of tracks G, all of segment length X, where X is evenly divisible by |G|, and a solution O with these tracks evenly distributed. There is another set of tracks G', all of length Y = |G'| * X / |G|, with a solution O' where the number of breaks at each position is identical for solution O of G and solution O' of G'. If solution in which G has been replaced with G' meets its bound, the solution with the original G also meets its bound.

```
Optimal_Factor(T) {
   Run independently on relatively prime track sets (Th. 2)
   Factor out unshared prime factors from S values (Th. 3)
   While tracks exist without an assigned Oᵢ {
      Place and remove any full sets (Th. 4)
      If all tracks have their Oᵢ assigned, end.
      Let Smax = the largest Sᵢ amongst unplaced tracks
      Let Snext = 2nd largest Sᵢ amongst unplaced tracks
      Let M be the set of tracks with Sᵢ = Smax
      Require: Smax % |M| = 0, Snext ≤ Smax*(|M|-1)/|M| (Th. 5)
      Assign all unassigned tracks in M to Oᵢ, s.t. all tracks in M are
         at a k*Smax/|M|, for all k 0≤k<|M|.
      If all tracks have their Oᵢ assigned, end.
      Require: Snext = c*Smax/|M| for some integer c≥1 (Th. 6), and Snext
         % c = 0 (to make Th. 5 work)
      Use c placeholder tracks w/S=Snext to model existing breaks
   }
}
```

Fig. 3. The pseudocode for the Optimal Factor algorithm. This algorithm has been proven optimal [5] provided all restrictions listed in the pseudocode are met.

Using the theorems we have presented, we can construct an algorithm (Fig. 3) which is optimal [5] provided our restrictions are met. There is one additional restriction that is implied. Because the tracks at a given S value must be evenly spread, the offsets assigned to the placeholder tracks added using Theorem 6 in the previous iteration must fall at offsets calculated using Theorem 5 on the next iteration. This is accomplished by requiring that S_{next} also be evenly divisible by the number of pseudotracks of that length added during the track conversion phase.

3.4   Relaxed Factor Algorithm

While the Optimal Factor Algorithm generates placements that are optimal in diversity score, there are significant restrictions on segment length and track quantity. However, not all architectures may meet these requirements. Therefore we developed a relaxed version, which may not always be optimal, but does demonstrate good results.

The general framework remains basically the same as the optimal version, although the placement and track conversion phases differ in operation. Before we used restrictions on track length and quantity to ensure that our methods are optimal. In the Relaxed Algorithm, when the optimal restrictions are not met, we instead use a number of heuristics where the goal is the even spreading of the breaks in tracks.

The routing architecture can be considered in terms of a topography, where elevation at a given location is equivalent to the number of breaks at that position. Because "mountains" represent areas with many breaks, and therefore fewer potential routing paths, we attempt to avoid their creation. Instead, we focus on placing our tracks to evenly fill the "plains" in our topography, where the breaks from the additional tracks will have a low effect on the overall routability of the resulting architecture. This topography is represented by the breaks[] array, which holds a count of the number of breaks occurring at every position $0...K-1$. In this case, K is the number of positions required to observe the full behavior of the break pattern. As mentioned earlier, this value can be determined by finding the LCM of all S values in the problem.

When each S group is placed, we need to look at the breaks[] array in terms of its effect on that S group. The tracks[] array summarizes the information from the breaks[] array within the potential offsets for a track. Fig. 4 indicates how this is accomplished. Using the tracks[] array, we can choose offsets which place new breaks in slots with the fewest amount of existing breaks. Each time a track is placed, both the breaks[] array and the tracks[] array are updated to reflect the additional breaks. As long as we have at least as many tracks in the current S group as minimum slots in tracks[], this is a simple procedure. However, when there are more minimum locations than tracks, we need to decide which of those minimum offsets should be used.

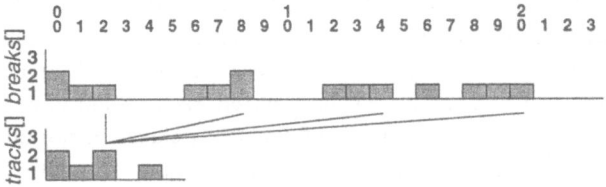

Fig. 4. An example of a breaks[] array for K = 24, and the corresponding tracks[] array for S_{next} = 6. The lines between the two arrays indicate the locations in the breaks[] array used to compute the value of tracks[2].

In that case, we use density-based placement. We first compute the "ideal" density of breaks (ie, if they were placed perfectly smooth) if we were able to achieve a perfect placement of the tracks thus far plus our current S group. This represents our placement goal for the S group. In order to achieve this goal, we consider an increasing region of the array, and attempt to bring its density close to the goal density. This region first begins as a single plain and mountain in the tracks[] array, and the number of tracks needed to bring the density of that region close to the goal density is added. These tracks are spaced evenly in the plain. The region is now increased to include the next plain and mountain in the array (where we treat tracks[] as a circular array). We then add tracks to the new plain in a similar manner to bring the density of the new region close to the goal density. This continues until the entire tracks[] array is included in the region, and we have placed all our tracks in the current S group.

Once we have placed all tracks with $S = S_{max}$, we need to prepare for the next iteration when we place all tracks with $S=S_{next}$. This means that we need to update the tracks[] array to be in terms of S_{next} instead of S_{max}. The tracks[] array is resized to be S_{next} slots in length, and is recomputed using the technique from Fig. 4.

4   Results

A number of terms are used to describe the problems that we have covered in our testing of our algorithms. The value numT refers to the total number of tracks in a particular track placement problem, numS refers to the number of discrete S values in the problem, maxS is the largest S value in the problem, and maxTS is the maximum number of tracks at any one S value in the problem. We have also reduced our very large search space by only considering problems where the number of tracks at each S value is less than the S value itself, since Theorem 4 strongly implies that cases with S or more tracks of a particular S value will yield similar results by placing tracks from any full sets one per potential offset. Cases with only one track, or all track lengths less than 3, are trivial and thus ignored. The three terms, numT, numS, and maxS, along with the restrictions above, define the track placement problems we consider.

Our first test was to verify that the Optimal Factor Algorithm yields a best possible diversity score in practice as well as theory. The results of this algorithm were compared to those of the Brute Force Algorithm for all cases with $2 \leq numTracks \leq 8$, $\leq numS \leq 4$, and $3 \leq maxS \leq 9$, which represents 5236 different routing problems. Note that even with these significant restrictions the runtime of brute-force is over a week of solid computation. In all cases where a solution could be found using the Optimal Factor Algorithm, the resulting diversity score was identical to the Brute Force method. Furthermore, we compared the Relaxed Factor Algorithm to the Optimal Factor Algorithm for this same range and found that Relaxed Factor produces optimal results for all cases that meet the Optimal Factor restrictions within that range.

Next, we compared the performance of the Relaxed and Simple Spread to the results of the Brute Force method for the same search space as above to determine the quality of our algorithms. The results for the heuristics, normalized to Brute Force, are shown categorized by numT, numS, and maxTS in Fig. 5 In these graphs Min is the worst performance of the algorithm at that data point, Max is the best, and Avg is the average performance. In this figure, the Brute Force result is a constant value of 1. Fig. 5 left indicates that both heuristics achieve optimal results in some cases, with the average of the Relaxed algorithm nearly optimal across the entire range. Simple Spread improves with the increasing number of tracks, and both algorithms degrade with an increase in the number of different S values, though only slightly for Relaxed.

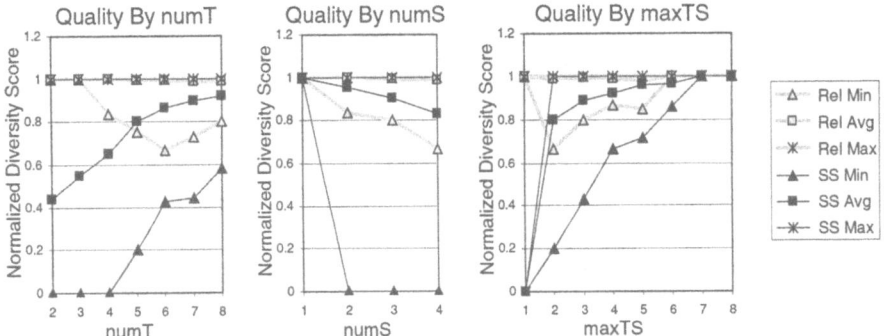

Fig. 5. A comparison of Relaxed and Simple Spread to the Brute Force method, with respect to numT (left), numS (center), and maxTS (right). The diversity scores for each case were first normalized to the Brute Force result, which is represented by a horizontal line at the value 1.

The upswing of both algorithms' minimums towards higher values of numT may be an artifact of our benchmark suite – since we only allow at most 4 unique S values, when there are more than 4 tracks we have at least two tracks with the same S value. Fig. 5 right shows that as the number of tracks per S value increases, the quality of all algorithms improves. The only exception is for the relaxed algorithm when maxTS=1; when there is only one track per S value the Relaxed Algorithm is always optimal. All throughout these tests the Relaxed algorithm is superior to the Simple Spread algorithm. This indicates the critical importance of correlation between S values in track placement. Fig. 5 center also demonstrates how as numS increases, the more difficult it is to solve the problem well, which we expected to be the case. Note that the results for both heuristics are optimal for the case when there is only one S value, as in this case there are no correlations to contend with, and only an even spreading is required.

Next, we used our place and route tool [1] to test the correspondence between diversity score and routability of architectures created using the Relaxed and Simple Spread algorithms. These architectures are based on a tileable coarse-grained architecture similar in structure to RaPiD. This architecture has two length-2 local tracks, four length-4 local tracks, eight length-4 distance tracks, and eight length-8 distance tracks. Note that local tracks do not allow connections between wire segments to form longer wires. We used seven different multi-netlist applications, and created a test case for each application/track placement algorithm. By keeping the proportion of track types constant but varying the total quantity of tracks, the minimum number of tracks required to successfully place and route every netlist in the application onto this target architecture was determined. Fig. 6 lists the results of this experiment. Performing track placement using the Relaxed Algorithm allowed netlists to be routed with on average 27% (and up to 45%) fewer tracks than the Simple Spread Algorithm.

	OFDM	Camera	Radar	Image Processing	Sort	Matrix Multiply	FIR Filters
Simple Spread	27	23	20	23	23	11	20
Relaxed	24	19	13	15	13	10	11

Fig. 6. The number of tracks in our target architecture required to successfully place and route all netlists in an application using the given track placement algorithm.

5 Conclusions

As we have shown, the track placement problem involved fairly subtle choices, including balancing requirements between tracks of the same length, and between tracks of different, but not relatively prime, lengths. We introduced a quality metric, the diversity score, which captures the impact of track placement. Multiple algorithms were presented to solve the track placement problem. One of these algorithms is provably optimal for some situations, though it is complex and works for only a relatively restricted set of cases. We also developed a relaxed version of the optimal algorithm, which appears to be optimal in all cases meeting the restrictions of the optimal algorithm, and whose average appears near optimal overall.

We envision two situations where this presented research can be applied. First, we believe that there is a growing need for automatic generation of FPGA architectures for systems-on-a-chip. Domain-specific FPGAs can achieve much better area, performance, and power than standard FPGAs. Furthermore, because the FPGA subsystem will be within a custom fabricated SoC, the advantage of pre-made silicon of commodity FPGAs is irrelevant. Second, these techniques can be used as a guideline for manual FPGA designers to potentially improve routing architecture quality.

References

1. K. Compton, S. Hauck, "Flexible Routing Architecture Generation for Domain-Specific Reconfigurable Subsystems", *International Conference on Field-Programmable Logic and Applications*, pp. 59-68, 2002.
2 D. C. Cronquist, P. Franklin, C. Fisher, M. Figueroa, C. Ebeling, "Architecture Design of Reconfigurable Pipelined Datapaths", *Twentieth Anniversary Conference on Advanced Research in VLSI*, 1999.
3. J. R. Hauser, "The Garp Architecture", *University of California at Berkeley Technical Report*, 1997.
4. V. Betz, J. Rose, ``Automatic Generation of FPGA Routing Architectures from High-Level Descriptions," *ACM/SIGDA International Symposium on Field Programmable Gate Arrays*, pp. 175 – 184, 2000.
5. K. Compton, S. Hauck, "Track Placement: Orchestrating Routing Structures to Maximize Routability", *University of Washington Technical Report UWEETR-2002-0013*, 2002.

Quark Routing

Sean T. McCulloch[1] and James P. Cohoon[2]

[1] Ohio Wesleyan University, Department of Computer Science, Delaware,
OH, 43015 USA
[2] University of Virginia, Department of Computer Science
Charlottesville, VA 22903, USA

Abstract. With inherent problem complexity, ever increasing instance size and ever decreasing layout area, there is need in physical design for improved heuristics and algorithms. In this investigation, we present a novel routing methodology based on the mechanics of auctions. We demonstrate its efficacy by exhibiting the superior results of our auctionbased FPGA router QUARK on the standard benchmark suite.

1 Introduction

Most typical routers are based on a *sequential* strategy. With this methodology, one of the nets to be routed is chosen first, and is given a path anywhere among the initially unclaimed set of routing resources. Then a second net is routed in the space unused by the first net, and so on, until the last nets must find a path that has not been used by the preceding nets. We feel that the sequential strategy has inherent limitations that impact both solution quality and runtime.

The main contributions of this investigation are a routing methodology based on the mechanics of an auction that supports a simultaneous routing philosophy; and an FPGA router QUARK that implements a version of our methodology. The idea of using an auction for decision-making is not new, and appears with some frequency in the operating systems literature. One particular area is in the allocation of resources in distributed environments. Tasks are given virtual money with which to bid on the various resources in the system. When a resource becomes available, all tasks can bid on that resource. The highest bidder pays for the resource and utilizes it [GAG95].

There are also two similar negotiation-based routers that have some slight resemblance to our methodology—the PathFinder and NC routers [MCMU95, CHAN00]. The routers represent an FPGA as a graph. Initially, each net is assigned an optimal path, even if it conflicts with the paths of other nets. The algorithm iterates until there are no conflicting assignments. The current cost of a shared resource is set to the number of nets that want it. Each net then finds a new shortest path. Because the shared resources become more expensive, nets heuristically end up negotiating an alternate path. Although this process is not dependent on an initial net ordering, it is not a truly simultaneous strategy. In practice, nets are placed down in order of congestion and are typically ripped up in that order if another net overlaps any part of that net's path.

In our methodology, the entire auction of all the routing resources takes place concurrently. This property has led to a different computational model that represents

P.Y.K. Cheung et al. (Eds.): FPL 2003, LNCS 2778, pp. 131–140, 2003.

a novel approach to performing simultaneous routing. In addition, our negotiation processes are both different and more personalized.

2   Basic FPGA Auction Methodology

The concept of our auction-based routing methodology is straightforward—the pins of an FPGA are resources that can be bid upon by the nets. Each net seeks to control a set of pins that realize a complete detailed route for that net. For discussion ease, we define two terms.

- A *pin-auction* is the local auction of a single specific pin. The auction generally consists of several nets bidding for the right to route on that pin.
- A *chip-auction* is the complete auction process over the entire circuit. Thus, this auction comprises all of the various pin-auctions.

Thus, a run of an auction-based router corresponds to a single chip-auction, where a chip auction consists of collection of pin auctions, one for each pin.

Only one net can win a given pin-auction, and thus have the right to be routed upon that pin. The goal of each net, while working independently of the other nets, is to win sufficient pin auctions to realize its detailed routing. The chip auction completes successfully after all nets have achieved a detailed routing. It is important to stress that all of the pin-auctions are taking place simultaneously. The individual pin-auctions start when the chip-auction starts and finish when the chip-auction completes. This requirement gives quick proof to our claim that QUARK is a simultaneous router.

2.1   Income

The algorithm begins with each net being given an initial allocation of funds. These funds are normally the only source of assets available to a net for bidding on various pin-auctions. An alternative method would have been to have periodic "pay periods" for nets. For example, nets that are losing several pin-auctions could be given extra money. However, it is our experience that routinely adding income to the system merely cause the prices in the pin-auctions to increase and hampers the chip-auction's ability to finish. Thus, we recommend limiting the application of additional funds to extreme cases.

After a net has been given its money, the net then places its initial bids. We view this initial bidding as being a distinct process from subsequent bidding actions. The initial bidding process must select the specific path to realize the net. Subsequent bidding processes generally only require checking of pin-auctions and updating of bids if necessary.

Once each net has placed its initial set of bids, the chip-auction enters its iterative main phase. In each iteration of the main phase, all nets are given an opportunity to place or modify bids in the various pin-auctions as they see fit. The nets make bids in such a way as to eventually claim ownership of pins that realize a complete detailed route. If two or more nets enter a pin-auction, the first net processed with a maximal bid has current control of the pin.

A net is considered to have a *complete detailed route* if the set of pin-auctions that it is currently winning comprise a path that would be a legal detailed route for the net. There are no "freeze points" where if a net is winning a pin, it can hold it. Doing so would violate the simultaneous nature of the router. As stated before, there is only one chip-auction for the entire run of the router, and that all nets must constantly have bids on the pins that they need to realize their route.

The chip-auction completes *successfully* on the first iteration in which all nets have a complete detailed route. The chip-auction completes *unsuccessfully* if at some iteration it is determined that it is not possible for a given net to realize a complete detailed route. It is possible for the chip-auction to continue indefinitely, with neither of these criteria being met. However, this situation does not seem to arise in practice.

Figure 1 is an example of the bidding process at work on a sample block. Observe that several nets have active bids on different pins of that block at the current time. It is responsibility of each net to ensure that it is bidding sufficiently high in its various pinauctions to have control of pins that realize a complete detailed route for itself.

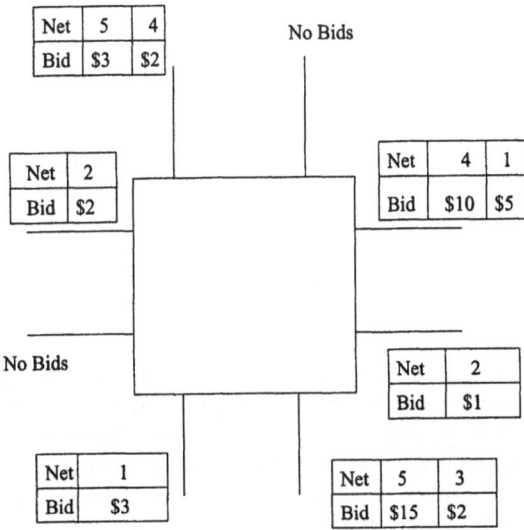

Fig. 1. Hypothetical bidding scenario.

3 Design Details

We now turn our attention to design details. Firstly, there is the issue of the bidding agents. In our model, each net is an active bidding agent, making decisions on which pin-auctions to participate, and once participating, on how much to bid. As a consequence, routing decisions can be made on a low level, which in turn means a priori that there is no general routing policy for deciding which nets gain which pins. Thus, each net individually determines how to best realize its route. Because the nets are simultaneously performing bidding operations, it makes sense to give each net its own view of the state of the various pin-auctions.

An important benefit of delegating these low-level decisions is that a routing tool is then free to allow different nets to use different routing algorithms for determining on which pins to bid. We call these local decision processes the net's *personality*.

While it is important to enable the nets to act individually, and simultaneously, it is also helpful to have a central authority, which we call the *overseer*, for policy decisions that require more information than can be found with regard to an individual net or pin.

3.1 Income and Priorities

When assigning a net an initial amount of money at the beginning of the chip-auction, the amount should be sufficiently large, to give the net some flexibility with regard to the bids of other nets. For example, suppose a net was given only $10 and was limited to bidding integral amounts. If the net needed ten pins to realize its route, it could only bid $1 per pin. It would have no excess money in a competition with other nets for its pins.

The initial income given to each net is a very natural way to implement a priority structure. In some designs, there is a need for a *critical net*—a net of high priority that must be given preference in finding its preferred path. In our design, a critical net can be given more initial funds. Heuristically, it is more likely in practice to win the pinauctions along its preferred path. If a net is absolutely critical and must be given its preferred path at the expense of the other nets, then it can be effectively given infinite funds to ensure that it gets that path.

3.2 Overseer

To ensure that the flow of the auction is simultaneous, it is important that there be a rapid flow of information regarding bids amongst the nets. The overseer must process bids quickly, and neither go into a failure state nor allow the router to end prematurely because nets have had an inadequate opportunity to bid against each other.

The overseer can determine whether a chip-auction has successfully completed by simply polling the nets whether they have won a complete detailed route. However, determining whether a chip-auction will be ultimately unsuccessful cannot be answered algorithmically (The problem is a variant of the Halting problem). Thus, the overseer must implement a heuristic to recognize this eventuality.

To decrease the likelihood of an unsuccessful routing, an overseer can be helpful in the case of a net lacking sufficient funds to complete any possible path to realize its detailed route. The routing analogy of eminent domain can be applied in such situations —the overseer steps in and in some manner assists the net in acquiring its path. Care needs to be taken so that the implementation of eminent domain does not effectively turn the router into a sequential one by simply assigning a path outright to the indigent net.

It is the overseer that also extracts the technology files; creates an internal representation of the FPGA that meets the specifications of the technology files specify; and extracts the global routing information for each net.

3.3 Pins

Our methodology gives individual pins the responsibility of processing the pin-auctions associated with them. As a result, the pin reports which net is currently winning its pin-auction. Furthermore, the pin ensures that only the current winner of the pinauction has the right to be routed on it.

3.4 Nets

Our net representation is also markedly different from similar representations found in traditional routers. For example, nets must have the ability to receive money and to bid this money in the pin-auctions. Most importantly, the nets must have the ability to bid on pin-auctions in a effective manner that realized a complete detailed route. This decision making is quite different from that found in previous routers. Routers traditionally have enforced routing decisions at a high-level and have required all nets to all perform homogeneously.

The placing of routing decisions at the net level has several benefits. By allowing different nets to have different personalities, we add great flexibility to the router's functionality. By changing the personality of an individual net, it is possible to optimize that net's routing path goal without impacting the routing decisions of the other nets. In addition, it is also possible to individually change the routing qualities that the nets are trying to optimize. For example, it is possible to route certain nets while trying to minimize delay, and to route other nets to minimize net length.

4 Personalities

The concept of a net personality is a novel one, and leads to many benefits in the routing process. As mentioned earlier, a net personality encapsulates a collection of routing decision processes. Thus allowing the router to give different nets different personalities based on their structure and importance. Our implementation provides three personalities.

The three personalities do have some commonality due to the nature of the routing task. Firstly in our environment, a net must have the ability to receive money from the overseer. One allocation activity always happens at the beginning of the chip-auction. An allocation can also occur as the result of an eminent domain request.

Secondly, a net must have the ability to make a constructive bidding decision in reaction to a request for bid(s) from the overseer. The algorithms that determine these bidding decisions are the central component of a net's personality. Although personalities will make different decisions as to where to bid, they will all share the goal of finding a complete detailed route. A net typically has three main bidding actions: increasing its bids when losing, decreasing its bids when winning, or deciding to bid on an alternate path if it deems its current path is too expensive. Besides being able to perform such actions, a net must be able to determine when they apply. For example, when a net is being outbid in a pin-auction and has sufficient unallocated funds to gain the pin, the net needs a process for determining what bid is sufficient. It will be an aspect of the net's personality to determine that new bid size.

Thirdly, each personality must have a mechanism for finding an alternate path if it deems the current path that it is attempting to acquire as too expensive. In implementing this mechanism, a net must be able to remove bids, modify existing bids, and place new bids.

Fourthly, a net needs a mechanism for determining when there are more affordable possible paths that it can construct. If an eminent domain request is viable in such a case, the request is signaled to the controller.

The personalities have been designed to optimize for the traditional FPGA routing problem of minimizing channel width and total net length. The first personality that we present is a very basic one. As such, we have named it the *baseline personality*. Because the baseline personality makes very simple decisions, it served as a framework for the succeeding personalities. The other two personalities—the *multiple personality* and the *focused personality*—extended the baseline personality in very different directions with regard to the management of resources. Both directions are heuristically promising in practice. The extensions were

- Allowing a net to bid on more than one possible path.
- Allowing a net to concentrate its resources in a small number of important locations.

We next discuss the three individual personalities in more detail.

4.1 Baseline Personality

The primary goal of a baseline personality net is to realize a detailed routing by trying to win the pin-auctions along its current global routing. (The global routing is maintained as the list of blocks that connect the various terminals of the net together.)

At the start of the chip-auction, the net is given the global route as determined by a global router in a previous step of the design process. Using the global route, an arbitrarily- determined legal detailed route is chosen. A minimal bid is placed upon each of these pins in the detailed route.

The personality favors pins that have no interest among the other nets. This bias helps the net to avoid congestion and to minimize overall resource expenditure. Once every pin in the detailed route has a bid of $1 on it, the net is ready to enter the main phase of the auction.

In the main phase, when a baseline personality net is signaled to react to the current state of the chip-action, the net first checks the status of each pin-auction along its current detailed route. If a net is winning a pin-auction by a substantial amount, the net heuristically concludes that some other net has lost interest in this pin-auction. The reason being that the net has minimally raised its bid in the past. As a result, the net reduces its bid so that it is only winning by only $1. The bidding decrease frees up funds to bid in other pin-auctions.

If instead a net is losing a pin-auction and has sufficient funds to gain the pin, the net bids the minimal amount to lead the pin-auction.

If the needed funds are not there, the net enters a re-route state. In this state, the net first tries to make changes that minimally alter the current detailed route. For example, the net might try to switch to a cheaper pin on the same block. However, switch blocks on most conventional FPGAs do not have great flexibility. Therefore, these attempts may not be successful. In such cases, the baseline personality follows a maze-routing strategy similar in nature to the UPSTART detailed router [McC02]. The UPSTART strategy considers a bounding region around the block that contains the expensive pin for the net. The width and length of the bounding region are specified by parameters. All of the net's bids on pins in the interior of this bounding region are removed. The locations where the detailed routing intersects the bounding box are marked as virtual terminals. An alternate detailed route is calculated that connects these virtual terminals together. If such a path can be found and if the net has sufficient funds to gain the associated pin auctions, then minimal winning bids are placed on these pins.

4.2 Multiple Personality

The baseline personality is clearly a very simple one and definitely has much room for improvement. One interesting way to improve the quality of the routing is to exploit some of the flexibility provided by the auction methodology. For example, we can have a net simultaneously show interest in multiple detailed routes. Sometime later in the course of the chip-auction, the net can commit to some particular routing. We have implemented such a personality and call it our *multiple personality*.

As a way of also demonstrating the fact that different personalities can be designed for different types of nets, we have designed the multiple personality to work on two-terminal nets whose terminals do not lie within the same channel. (If the terminals did lie so, there is a unique shortest path with regard to blocks.)

The key to understanding our multiple personality is to recognize that a two-terminal net with terminal blocks in differing channels has multiple minimum-length paths that are of the dogleg form. A dogleg is a connection in the form of a single stair step, where the step can be oriented either horizontally or vertically. The flat segment can be preceded and/or followed by riser segments of orthogonal orientation. To instantiate a dogleg, we only need to specify its orientation and the length of the risers. Our multiple personality starts by bidding all of its initial allocation as possible in equal amounts on each of these paths.

In subsequent bids, the personality first determines whether there are routes that it is currently bidding on which cannot be won with the available funds. If so, one such route is removed from consideration and the funds are reallocated to the other routes. In particular, the funds are first spent on routes that are most in danger of being lost and then on the other routes.

If the path that was out-bid was the last one being considered by the personality, then the net enters a re-route phase. In rerouting, the personality examines all possible dogleg paths. If there is a path that the net can afford, it is taken. The net then bids all of its funds along this path, in the expectation that the bids will be sufficient enough to retain control. If this action proves unsuccessful, later iterations will try a different dogleg path that the net can afford. The process continues until a net retains the path that was found, or the net realizes that all possible paths cost more money than the net has. In this case, an eminent domain request is considered.

4.3 Focussed Personality

Just as the multiple personality was designed to be used for two-terminal nets, so have we designed the focused personality to be used on particular types of nets. We expect that a "focus attention on a set of pins that are important to win" strategy would work well if we focus the bidding on certain pins of a multi-terminal net. We define a *branching point* of a multi-terminal net to be a point that has more than two edges incident on it In QUARK, we treat all blocks where the global route has greater than two edges incident on the block as a branching point. The personality assumes that the global router has intelligently placed these branching points. Therefore, it is important to win the bid on these pins.

The focused personality divides the initial money resources into two main parts. The amount to be used in bidding on branching points, *branching money*; and the amount to be used in bidding on the remainder of the routing, *connection money*. Clearly, the balance between these two amounts is important. If one component does not receive sufficient funds, then the net will be out-bid by some other net and thus have trouble completing a routing.

QUARK allocates enough connection money such that a bid of $2 can be made on each pin in the given global route. This allows the router to find connections that are longer than the length given by the global route. In doing so, it will be necessary to give some pins $1 bids. Another benefit of this behavior is that a net can bid more than $2 on certain pins by reducing the bid of other pins to $1. The different gives the router some flexibility in finding a connection path that it can actually keep.

At the beginning of the chip-auction, the division between branching money and connection money is made. The branching money is initially divided evenly among all pins on all branching points.

Once the branching points are bid, the net bids on the connections between branching points. It is important to observe that all effective routings segments leaving branching points either eventually lead to a single terminal, or a single pin on another branching point. Our implementation uses a quick maze-routing strategy to connect the two pins. The resulting routing can be of arbitrary length and go in arbitrary directions. The limitations are based only on the amount of connection money.

When the router signals a focused personality net during the main bidding phase, the net makes two separate checks: one for the connection paths, and one for the branching points. Each of the current connection paths are examined to make sure that the net is winning the pin-auctions along the entire path. If a net is winning, then an additional check is made to see if connection money can be freed. If a net is not winning and there is enough connection money remaining to make the bid higher, then that money is spent to increase the bid. If there still is not enough excess money, the net enters the reroute state.

Note that we do not use branching money to help these connection paths become routed. Doing so would slowly siphon funds away from the branching points into the connection paths. This action would violate the principles of this personality, because it would result in the branching points being more likely to be out-bid.

If a branching pin is being out-bid, then the net examines the bids on all of the other branching pins. The net searches for a branching pin that is winning by a amount that allows for some reduction. The reduction will free up branching money to help the out-bid branching pin to regain control of the pin-auction. The process continues until either the out-bid pin has enough excess money to control the pin-auction again, or it is the case that none of the other branching pins can lower their bid without losing their pin-auction. If this condition results, the net makes an eminent domain request.

The re-routing phase for the connection paths uses a maze-routing algorithm to find a different inexpensive connection path.

5 Experiments

We now analyze a series of experiments that we performed on the Quark router. The experiments were conducted using the standard set of benchmarks maintained by the University of Toronto. The benchmarks have been the basis for the testing of a significant number of tools. Different benchmarks have different numbers of nets, blocks, and block layouts. In our experiments, we use the QUARK tool as a detailed router. The placement and global routing were performed by the SPIFFY tool [SELF]. Because SPIFFY is nondeterministic with regard to its output, five separate runs of SPIFFY were generated for each benchmark. Besides producing a placement and global routing, each run also heuristically specified an expected channel width given complete switchboxes.

Our first set of experiments tested the functionality of the various personalities relative to each other. For these tests, we created four versions of QUARK. The versions differed in their use of various personalities. By using these tests, we are able to make a preliminary decision on which versions of QUARK are best in practice.

- QUARK-BASELINE: every net uses the baseline personality.
- QUARK-MULTIPLE: all nets that qualify for the multiple personality use that personality, and all other nets use the baseline personality.
- QUARK-FOCUSSED: all multi-terminal nets use the focussed personality, and all two-terminal nets use the baseline personality.
- QUARK-ALL: all multi-terminal nets use the focussed personality, all two-terminal nets that qualify for the multiple personality use that personality, and the remaining two-terminal nets use the baseline personality.

A summary of these results are presented in Table 1. This table shows the minimum channel width successfuly routed to by the various versions of Quark, along with the average smallest channel width of several runs.

We also note best results tend to be generated by the multiple personality and the baseline personality. Although the baseline personality is slightly better on average, it is significantly slower to run. The results suggest an interesting composite algorithm— run QUARK-MULTIPLE until it discovers a channel width that is too small for it to successfully route. The router then switches to using the slower but more effective QUARK-BASELINE. We call this version QUARK-PRIME. QUARK-PRIME required several minutes to produce its solution for the smaller instances. For the largest instances, it required slightly more than 200 minutes on a Solaris workstation.

Table 2 compares QUARK-PRIME to several state-of-the-art routers: Graph-based router [ALEX98], SEGA tool [LEMI93], CGE tool [BROW92], and VPR tool [BETZ97]. Because the SEGA and CGE tools are designed for different architectures, their results are combined. We note that only VPR betters QUARK-PRIME performance. However, VPR is not running the same instances as QUARK-PRIME. The VPR layout system uses a technology mapper to reduce the number of logic blocks that is unavailable to QUARK-PRIME. This table shows the minimal channel width routed to by each of the routers on these benchmarks.

Table 1. Channel widths successfully routed

Circuit	QUARK BASELINE		QUARK MULTIPLE		QUARK FOCUSSED		QUARK ALL	
	Avg	Best	Avg	Best	Avg	Best	Avg	Best
dfsm	8.8	8	10.4	9	11	10	10.6	10
9sym	8.0	8	7.4	7	7.6	7	8	8
term1	5.8	5	5.8	5	7	5	7	6
apex7	6.2	6	6.6	6	6.2	6	7.4	7
alu2	8.4	7	8.6	8	9.4	9	9.2	9
alu4	8.4	7	9.8	9	12	10	11.4	11
Total	45.6	41	48.6	44	53.2	47	53.6	51

Table 2. *Inter-router comparison.*

Circuit	QUARK PRIME	Graph-Based	SEGA/ CGE	VPR
dfsm	8	9	10	—
9symml	6	8	10	5
z03	9	11	13	—
term1	5	8	10	5
apex7	5	14	10	4
alu2	7	9	11	6
alu4	8	11	15	7
Total	48 (31)	70	79	(27)

References

[ALEX98] M. Alexander, J. P. Cohoon, J. L. Ganley, and G. Robins, Placement and Routing for High-Performance FPGA layout, *VLSI Design: International Journal of Custom-Chip Design, Simulation, and Testing*, 97-110, January 1998.

[BETZ97] V. Betz and J. Rose, VPR: a new packing placement and routing tool for FPGA research. International Workshop on Field Programmable Logic and Application, 1997.

[Brow92] S. D. Brown, J. S. Rose, and Z. G. Vranesic, A detailed router for field programmable gate arrays, *International Conference on Computer-Aided Design*, 382-385, 1990.

[CHAN00] P. K. Chan and M. D.F. Schlag, New parallelization and convergence results for NC: a negotiation-based FPGA route, *International Symposium on Field Programmable Gate Arrays*, 165-174, 2000.

[GAG95] R. A. Gagliano, M. D. Fraser, and M. E. Schaefer, Auction allocation of computing resources, *Communications of the ACM*, J88-102, June 1995.

[Lemi93] G. G. Lemieux and S. D. Brown, A detailed routing algorithm for allocating wire segments in field programmable gate arrays, *ACM-SIGDA Physical Design Workshop*, April 1993.

[McCu03] Sean T. McCulloch, Auction-based routing for FPGAs, University of Virginia, Doctoral Dissertation, 2002.

[McMu95] L. McMurchie and C. Ebeling, Pathfinder: A negotiation-based performance-driven router for FPGAs, *International Symposium on Field Programmable Gate Arrays*, 111-117, 1995.

[SHRA87] E. Shragowitz and S. Keel, A global router based on a multicommodity flow model, *INTEGRATION: the VLSI Journal*, 3-16, 1987.

Global Routing for Lookup-Table Based FPGAs Using Genetic Algorithms[1]

Jorge Barreiros[1,2] and Ernesto Costa[2]

[1] Departamento de Engenharia Informática e Sistemas,
Instituto Superior de Engenharia de Coimbra.
[2] Centro de Informática e Sistemas da Universidade de Coimbra.
jmsousa@isec.pt , ernesto@dei.uc.pt

Abstract. In this paper we present experiments concerning the feasibility of using genetic algorithms to efficiently build the global routing in lookup-table based FPGAs. The algorithm is divided in two steps: first, a set of viable routing alternatives is pre-computed for each net, and then the genetic algorithm selects the best routing for each one of the nets that offers the best overall global routing. Our results are comparable to other available global routers, so we conclude that genetic algorithms can be used to build competitive global routing tools.

Introduction

With the increasing growth of the complexity of electronic devices, the good performance of synthesis tools is critical for the success of design and manufacturing of non-trivial projects. Concerning the development of FPGA (*Field Programmable Gate Arrays*) systems, one of the fundamental tasks of these tools is to perform the routing of the circuit constrained to the limited resources available in the device. Although a lot of research has been made in this area [4,5,6,7,8,9,10,11,12,13], little has been done concerning the study of the feasibility of using genetic algorithms [2,3] to generate the required routing (see on [17] for a survey of genetic algorithms used for VLSI design). With this purpose, in this work we developed a global routing tool based on the application of a genetic algorithm for selection for viable routing paths on a LUT-based FPGA. Results are very close to those of other available tools, so we believe that, with further refinement, the genetic approach can became competitive with current techniques.

---

[1] This work was partially supported by the Portuguese Ministry of Science and High Education, under program POSI

P.Y.K. Cheung et al. (Eds.): FPL 2003, LNCS 2778, pp. 141–150, 2003.

FPGA Architecture

The FPGA architecture we used to test our algorithm is similar to the one described in [1] (see Figure 1). In this model, there are four distinct types of blocks, interconnected by several routing channels:

- **L-Blocks** – These blocks implement any logic function with N inputs. These inputs are equally distributed over all sides of the block, and a single output is provided on the right side of the block. We chose to select N=4 for all our experiments. These blocks can also be referred to as *logic cells* or *logic blocks*.
- **IO Blocks** – These blocks represent the input and output pins of the FPGA. Since the number of pins in real FPGAs is far greater than those that fit along the square formed by the L-Blocks, we consider that multiple I/O pins are contained in each I/O block. In this work, we consider that two I/O pins are available for each I/O block.
- **C Block** – These are the blocks through which the L and IO blocks are connected to routing channels. The number of alternative tracks to which a connection from a L-block or IO-block can be made, characterizes these blocks. In this work, this value was predefined to be equal to the width of the routing channel.
- **S-Block** – These blocks allow switching the tracks between different routing channels. Their *Flexibility* is defined as the number of different outputs to which a given input can be connected. We use S-Blocks with a *Flexibility* of 3 in this work (see Figure 2).

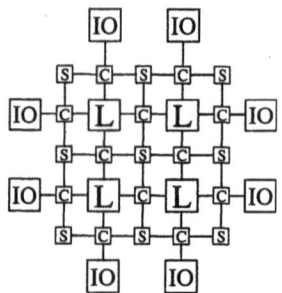

Fig. 1.- Architecture of the LUT-based FPGA

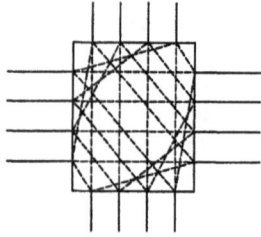

Fig. 2. S-Block configuration used in this work

Genetic Algorithms

Genetic algorithms (GA) are a group of stochastic optimization techniques [2,3]. These algorithms work with a set of candidate solutions to the problem (a population of individuals, using GA terminology) and seek to evolve them using concepts derived from genetics and natural selection. Each individual holds enough information (the *genes*) to describe a possible solution to the problem. They are evaluated regarding the quality of that solution (i.e. their *fitness* is computed), and a probabilistic selection method (based on the fitness of each individual) is used to find group of individuals (the *parents*) that will be used to create the next *generation* of the population. They individuals of the new generation are created by applying genetic-inspired transformations (*operators*) to the parents. Among these transformations we can find *mutations* and *crossover*. The mutation operator does random changes to the genes, while the crossover combines parts of the genes of two parents to create a single individual. After multiple iterations, the quality of the population will increase, and when a predetermined stopping condition is met, the solution for the problem will be found on the genes of the best individual of the last generation.

```
1.Randomly initialize population
2.While stopping condition is not met
    a)  Evaluate population
    b)  Select parents
    c)  Crossover
    d)  Mutation
    e)  Substitute old population
```

Fig. 3. Simple Genetic Algorithm

There are, of course, multiple variants of this simple framework.

Global Routing for FPGAs

The synthesis of circuits for FPGAs can be decomposed in several steps:
1. **Logic optimization** – The circuits are optimized, shrunk and redundant logic is eliminated.
2. **Technology Mapping** – The elements of the circuit description are assigned to specific classes of resources available in the actual FPGA. For example, logic expressions are broken into sub-expressions implemented in a single logic block.
3. **Placement** – When there are multiple components in the FPGA capable of implementing a specific aspect of the circuit description, a selection needs to be made about which one of them will be used (for example, what logic blocks will actually be used to implement the sub-expressions generated by the technology mapping?)
4. **Routing** – The communication resources available in the FPGA are used to connect adequately all the components of the circuit. Ideally, the routing should

use the smallest routing channel width and have minimum source-to-sink distance, to maximize circuit speed.

Some approaches combine some of these steps or further refine them into additional iterations. The last step is sometimes separated in global and detailed routing. The objective of the global routing step is to ensure a balanced occupation of available channels. The global router decides which routing channels will be used, without deciding about specific track usage inside those channels. Detailed routing will complete the process, by making the necessary track assignments within those channels. Frequently, it may be impossible to perform detailed routing with the same channel width found by the global routing because of restrictions on the flexibility of S-Blocks (the *routing anomaly*, see [1,4]), so the global routing channel width is actually a low bound for the width of the routing channels after detailed routing.

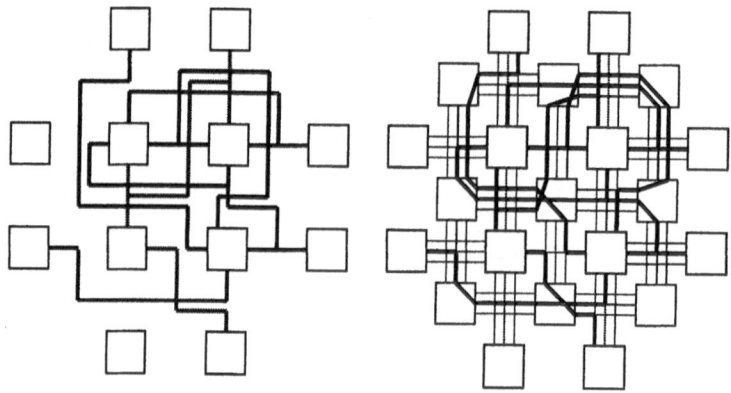

Fig. 4. Global vs. detailed routing

For further details about previous work on global and detailed routing, see [4,5,6,7,8,9,10,11,12,13]. For information regarding other steps of the synthesis process, please consult [14, 15, 16].

The Routing Algorithm

We used a two-step approach for building the global routing:

1. `For each net of the circuit, generate a set of alternative routing paths.`

2. `Find, for all nets, the combination of alternative routing paths that offers the best overall global routing`

This approach allows using different optimization techniques for each step. We developed three different heuristics for creating the alternate routing paths, and used a genetic algorithm for making the optimization described in step 2, according to the objectives of this work.

Generation of Alternate Routing Paths

The generation of the alternate paths is made with the following algorithm:

For each net,

1. Generate one routing path that connects all nodes in net

2. While the desired number of alternate paths isn't met, build new routing path by randomly selecting and mutating one of the previously built paths

This algorithm will first generate one model net. All other alternative paths are deviations from this model net. The rationale behind this option is that it should be more efficient to build alternate solutions by making minor changes to pre-existing ones than re-computing a new net from scratch.

Computing the Model Routing Path for Each Net

We developed three different algorithms for computing the first model routing. Two are graph-based search techniques and the other is based on a heuristic algorithm for computing rectilinear Steiner trees. The first graph-based technique is simply a greedy search from the source node to each sink node. This algorithm is fast but offers, as could be expected, poor quality solutions. The other graph algorithm is based on Dijkstra's algorithm [18] for finding the shortest path between two nodes. Every source/sink is connected sequentially using Dijkstra's algorithm, and the cost of previously taken segments is reduced. This algorithm offers reasonable performance, but is not as good as our heuristic for computing near optimal rectilinear Steiner trees (RST). A RST is the shortest rectilinear tree connecting a given set of points. Computing a RST instead of using graph-transversal algorithms eliminates the problem of having to route multipoint nets as groups of source/sink paths. The performance of those algorithms is very dependant on the order by which those paths are processed, and it isn't clear how to determine what the optimal order is, although some heuristic can be used (i.e. longest paths first). Our approach to computing the RST for a given set of points is based on a hill-climbing search algorithm that finds an optimal partition of those points into smaller RST, as explained bellow and illustrated in Figure 5, where a tree representation of the decomposition is presented.

The points are partitioned across horizontal and vertical axis that transverse de median point of each set, sharing one point to ensure a connected net is built. This partitioning is applied recursively until no more than a predetermined number of points (for illustration purposes, this number is 3 in Figure 5) are contained in every partition. The rectilinear tree at the nodes is built by constructing an axis along the median (vertical or horizontal) point, and connecting all other nodes to that is by transversal segments.

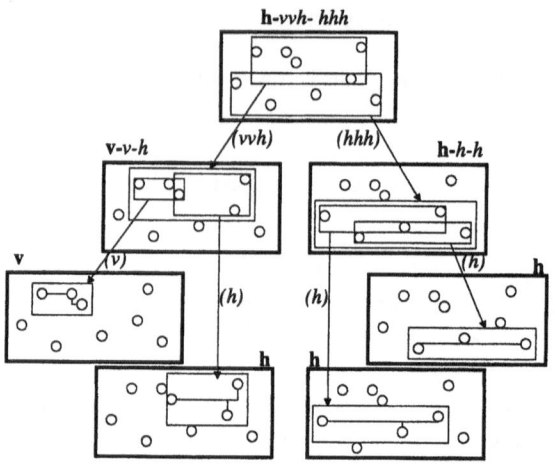

Fig. 5. Example of decomposition of point set into multiple RSTs

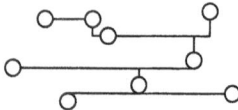

Fig. 6. (non-optimal) Rectilinear tree resulting from the decomposition illustrated in Figure 5

The actual decomposition is dependant on whether a vertical or horizontal axis is chosen on each node of the decomposition tree. This information can be represented linearly by a vector[2] where the first element represents the orientation of the axis on the root node, and the remaining information is split in half and interpreted accordingly by the left and right descendants of the root. For example, for the specific decomposition shown in Figure 5, this string is "hvvhhhh" for the root node.

The simple linear representation of the decomposition is important, because it enables us to use a simple hill-climbing optimization algorithm for finding the partition that offers the smallest length rectilinear tree. Figure 7 shows the rectilinear trees computed for a random point set with different values for maximum points per leaf. The dashed rectangles represent the points contained in a single leaf. (These trees are not optimal; they just represent a possible result of the algorithm.)

Generating Mutations from Model Path

Deviations from a model routing path are built by applying a mutation to previously built nets. The algorithm is based on the idea of generating an intersecting rectangle over the model network and re-routing the internal connections on the borders of that rectangle. For space considerations, some details are omitted (e.g. handling nodes internal to the rectangle, or multiple branches), but the general idea is illustrated in Figure 8.

---

[2] Although padding may be required for unbalanced decomposition trees.

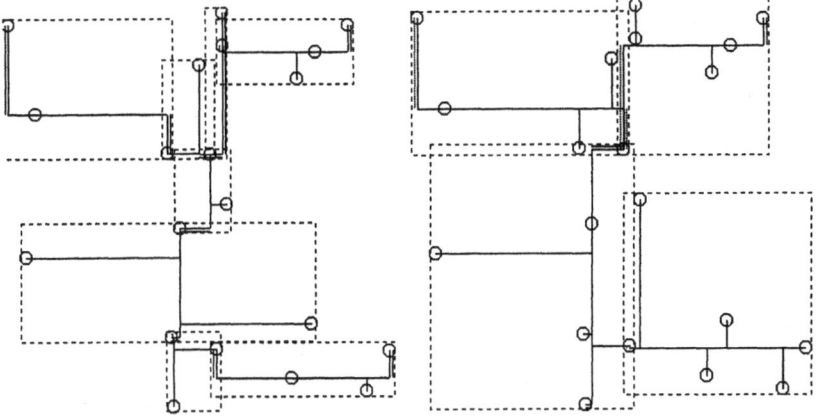

Fig. 7. Rectilinear trees for random set of points, with maximum of 4 points per leaf (left) and 8 points per leaf (right).

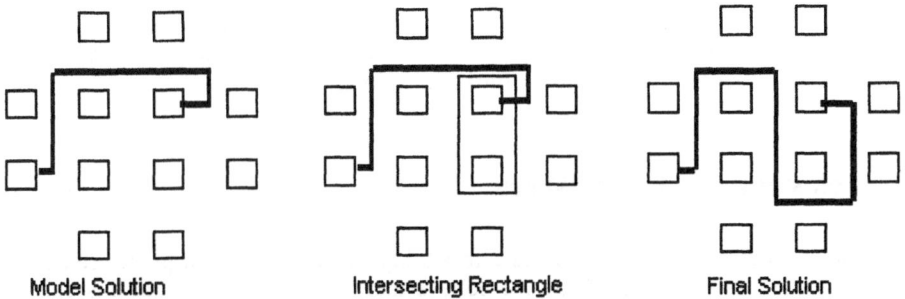

| Model Solution | Intersecting Rectangle | Final Solution |

Fig. 8. Generating alternative routing paths.

Genetic Algorithm

The genetic algorithm decides among the possible alternate routes which ones will be used in the global routing. Each individual (solution) is represented only by an integer vector[3]. Each position of this vector is associated to a net of the circuit, and in that position it is held an index that identifies what is the chosen routing alternative for that net.

After some tuning tests, the following configuration was found to offer the best results for the GA:

- Population size: 50
- Maximum number of generations: 2500
- Mutation rate: 8%
- Crossover rate: 80%

---

[3] Actually, some auxiliary data is also kept in each individual to help to speed up some computations. That detail isn't relevant to understand the main operation of the algorithm, so it is left out of the description.

- Elitism: 4 individuals
- Lamarckian learning[4]

Results

We used circuits from LGSynth93 test suite [19] to benchmark our algorithm. A set of 30 circuits of diverse complexity was chosen. The circuits were placed using VPR [10] and our tool was used to generate the global routing. For comparison purposes, VPR was also used to globally route the same circuits. A single run of both tools was used for all circuits. The test machine was a Pentium 3 processor at 866MHz with 384 MB RAMS. The results are presented in Table 1.

As we can see, although VPR offers slightly better results, both in terms of CPU time and track number. However, results are reasonably close, and in some cases the AG actually computed a better solution. We believe the small difference in performance can be further reduced if other heuristics for net generation are tried and more exhaustive tuning is made.

Additionally, the AG seems to have a very large performance advantage when compared with reported results for older global routers [11], which suggests that our approach seems valid and worthy of further development.

Conclusions and Future Work

Our main goal was to investigate the feasibility of constructing a global router for FPGAs using genetic algorithms. We believe that our results show that this class of algorithms is worthy of consideration when designing this kind of tools. Results seem to suggest that the AG offers performance that is comparable with that of other known algorithms, although in the current implementation it is on average slightly inferior. However, this difference in performance isn't very significant, and further improvement of the algorithm may offer better results. Although some tuning tests were made, a more exhaustive set of tests might reveal some better parameterization. Also, one improvement that might significantly improve efficiency would be to generate alternative routings dynamically, only when judged necessary, instead of pre-computing a fixed-size set of alternate paths. Further work could also be done by adapting the algorithm to use architectural features of current FPGAs, like routing segments of heterogeneous size.

---

[4] A quick, non-exhaustive local search is conducted for each individual, and if any improvement is found it the individual is changed accordingly.

Table 1. Results for LGSynth93 test suite benchmark

Circuit	Channel Width	Channel Width	CPU (s)	CPU (s)
	(VPR)	(AG)	(AG)	(VPR)
pdc	20	20	4299	8154
ex1010	11	13	2609	952
spla	16	17	1881	2654
s298	9	10	1247	814
seq	13	14	1158	278
alu4	11	11	807	184
misex3	12	12	661	210
ex5p	15	14	518	824
apex3	13	12	558	134
pair	8	9	476	59
C6288	6	6	89	11
i8	9	9	200	95
table5	11	12	109	33
cordic	10	10	99	68
C3540	8	9	40	27
i9	6	7	116	27
x3	6	6	273	12
vda	9	9	77	35
s1238	7	7	43	13
e64	9	8	158	45
planet	6	7	29	14
i7	4	5	12	93
mm9b	7	7	42	8
i6	5	5	13	61
alu2	7	7	33	8
too-large	7	8	8	8
C880	8	7	31	5
example2	6	6	48	3
term1	5	5	4	4
misex2	5	5	1	1
TOTAL	269	277	15639	14834

References

[1] "On two-step routing for FPGAs", Lemieux, G. ; Brown, S. ; Vranesic, D. , International Symposium on Physical Design, Abril 1997

[2] "Genetic Algorithms in Search, optimization and machine learning", Goldber, D. Addison Wesley, 1989

[3] "An introduction to Genetic Algorithms", Mitchel, M., MIT Press, 1996

[4] "New performance-driven FPGA routing algorithms", Alexander, M. ; Robins, G; Design Automation Conference, June 1995

[5] "A detailed router for Field-Programmable gate arrays", Brown, G. ; Rose, Z. ; Vranesic, G., IEEE Transactions on Computer-Aided Design, Vol. 11, No. 5, May 1992

[6] "Plane parallel A* maze router and it's application to FPGA's"; Palczewski, M. ; Proceedings of the Design Automation Conference, 1992.

[7] "New performance-driven FPGA routing algorithms", Alexander, M. ; Robins, G; Design Automation Conference, June 1995

[8] "Performance-oriented placement and routing for Field-Programmable gate arrays", Alexander, M; Cohoon , J. ; Ganley, J. ; Robins, G., European Design Automation Conference, Sept. 1995

[9] "New performance-driven FPGA routing algorithms"; Alexander, M ; Robins , G. ; IEEE Transactions on Computer Aided Design of Integrated Circuits and Systems, Vol. 15, N. 12, Dec. 1996.

[10] "Directional bias and Non-uniformity in FPGA global routing architectures", Betz, V.; Rose, J. ; IEEE/ACM International Conference on Computer Aided Design, 1996

[11] "LocusRoute: A parallel global router for standard cells"; Rose, J. ; Proceedings of the Design Automation Conference, 1988.

[12] "A detailed routing algorithm for allocating wire segments in field-programmable gate arrays"; Lemieux, G; Brown, S; Proceedings of the ACM Physical Design Workshop, 1993.

[13] "A performance and routability driven router for FPGAs considering path delays", Lee, Y. ; Wu, A.; IEEE Transactions on Computer Aided Design of Integrated Circuits and Systems, vol 16, n°2, Feb. 1997

[14] "Optimal FPGA mapping and retiming with efficient initial state computation", Cong ; J. ; Wu, C. Proceedings of the 35th Design Automation Conference, 1998.

[15] "Technology mapping for TLU FPGA's based on decomposition of binary decision diagrams", Chang, Shih-Chieh ; Marek-Sadowska, M. ; Hwang, T., IEEE Transactions on computer aided design of integrated circuits and systems, Vol. 15, N.10, 1996

[16] "Combining technology mapping and placement for delay-minimization in FPGA designs", Chen, C.-S. ; Tsay, Y.-W. ; Hwang, T.; Wu, A. ; Lin, Y.-L.; IEEE

[17] "Genetic Algorithms for VLSI design, layout & test automation", Mazumder, P., Rudnik, E., 1999, Prentice Hall, ISBN 0-13-011566-5

[18] "A note on two problems in connection with graphs", Dijkstra, E. W.; Numerische Mathematik, vol. 1, 1959

[19] CAD Benchmarking Laboratory, North Carolina State University, LGSynth93 suite, http://www.cbl.ncsu.edu/www/

Virtualizing Hardware
with Multi-context Reconfigurable Arrays

Rolf Enzler, Christian Plessl, and Marco Platzner

Swiss Federal Institute of Technology (ETH) Zurich*, Switzerland,
enzler@ife.ee.ethz.ch

Abstract. In contrast to processors, current reconfigurable devices totally lack programming models that would allow for device independent compilation and forward compatibility. The key to overcome this limitations is hardware virtualization. In this paper, we resort to a macro-pipelined execution model to achieve hardware virtualization for data streaming applications. As a hardware implementation we present a hybrid multi-context architecture that attaches a coarse-grained reconfigurable array to a host CPU. A co-simulation framework enables cycle-accurate simulation of the complete architecture. As a case study we map an FIR filter to our virtualized hardware model and evaluate different designs. We discuss the impact of the number of contexts and the feature of context state on the speedup and the CPU load.

1 Introduction

Reconfigurable computing fabrics have shown great potential in many high-performance applications that benefit from hardware customization while still relying on some amount of programmability. A major drawback of current reconfigurable devices, in particular field-programmable gate arrays (FPGAs), is the lack of programming models. Applications are compiled (synthesized) to given fixed-size hardware. The resulting configuration bitstream cannot be reused to program a device of different type or size. Thus, to leverage advances in VLSI technology, i.e. increased transistor count and higher clock rates, a complete recompilation is required.

The key to overcome this limitation is *hardware virtualization* [1–3]. In order to achieve hardware virtualization, we have to define a set of basic operators a hardware can execute. Together with a description of the data flow (communication paths between operators) and the control flow (sequencing of operators) a hardware programming model is defined that compilers can target. Processors use a well-established form of hardware virtualization and define an instruction set architecture that decouples the compiler from the actual hardware organization. Achieving virtualization of reconfigurable hardware is more complex. Reconfigurable hardware excels when computations are organized spatially. The

---

* This work is supported by ETH Zurich under the ZIPPY project and the Wearable Computing Polyproject.

P.Y.K. Cheung et al. (Eds.): FPL 2003, LNCS 2778, pp. 151–160, 2003.
© Springer-Verlag Berlin Heidelberg 2003

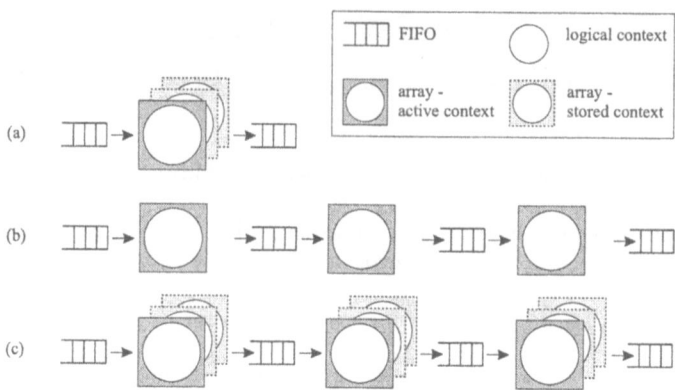

Fig. 1. Models for virtualized macro-pipelining

basic operators will thus have greater complexities than processor instructions and the number of possible operators is very large. Further, the reconfigurability allows to implement many basic operators with just one type of hardware block.

In this paper, we consider data streaming applications that map well to (macro-)pipelines, where one pipeline stage is implemented by one basic hardware block. Our basic hardware block is a 4 × 4 coarse-grained reconfigurable array. The inputs and outputs of the array connect to FIFO buffers to facilitate data streaming. One set of configuration data for the array is denoted as a *context*. Applications are organized by pipelining several *logical context* executions. Although this model is rather restrictive, it is amenable to true hardware virtualization and targets an important application domain.

Figure 1(a) shows our model with one physical array that is reconfigured to implement logical contexts as needed. To minimize or even hide the reconfiguration time the array stores multiple *physical contexts*. Figure 1(b) displays an alternative implementation with several single-context physical arrays arranged in a pipelined fashion. The arrays are still reconfigured to execute different logical contexts. However, as several contexts run in parallel the throughput increases. Both multiple contexts and physical pipelining can be combined which is shown in Fig. 1(c). All these architectures achieve virtualization as they provide the logical pipeline of array executions as programming model, but differ in their performance and hardware cost.

While there exists already a substantial body of work on coarse-grained arrays, macro-pipelining of stream computations and multi-context devices, a system-level evaluation of the performance and the various features of multi-context devices is missing. To this end, we form a reconfigurable hybrid system by coupling our multi-context array to a CPU. The CPU takes care of data I/O, context loading, and control of the multi-context array. We develop a system-wide, cycle-accurate architecture model and investigate the following issues by means of a co-simulation environment: First, we determine the performance gains for the hybrid over the CPU only, depending on the number of physical contexts.

Second, we try to identify whether and when the capability to resume the state of a previous context is advantageous. Third, we measure the CPU load for the different designs.

Section 2 summarizes related work. The hybrid architecture model and our co-simulation environment are discussed in Section 3. Section 4 presents an FIR filter case study, while Section 5 discusses the results. Finally, Section 6 summarizes our findings and points to further work.

2   Related Work

PipeRench [1, 4] is a reconfigurable architecture that supports hardware virtualization. The device is organized into a physical pipeline of stripes, which represent the minimal reconfigurable hardware blocks. A stripe's output is strictly pipelined and connects to the next stripe via an interconnection network. Thus, PipeRench is similar to the model in Fig. 1(b). Fast reconfiguration of stripes is supported by 256 contexts held on-chip. Each stripe comprises 16 processing elements, which implement addition/subtraction or a programmable logic function. Application kernels are mapped to virtual pipeline stages. During runtime, the virtual stages are configured to the physical stripes that are available on the device. The implementation described in [4] features 16 physical stripes.

Multi-context techniques for both fine-grained and coarse-grained reconfigurable devices have been investigated by several researchers. DeHon [5] demonstrated that adding multi-context support to FPGAs can increase computational density. Due to the moderate contribution of the configuration memory to the total chip area, a small number of contexts can be added with reasonable impact on cost. Trimberger et al. [2] introduce a multi-context extension of the Xilinx XC4000 architecture. The proposed device holds eight contexts on-chip. The flip-flops of the device are eight times replicated and each logic cell can write to any of these flip-flops. The authors propose to use the multi-context feature for emulation of arrays of arbitrary size. The fine-grained, memory-poor architecture prefers logic emulation rather than macro-pipelining.

PACT's XPP device [6] uses cells with a functionality similar to our model, but targets a different execution model. Data is transfered between the cells using a handshake protocol. This ensures that dataflow dependencies are met and makes the computation self-timed. Configurations are loaded on demand using a hierarchical configuration management. A configuration context is not necessarily activated for the whole device at the same point in time. For each cell, the new configuration is activated as soon as the current configuration is not used anymore. MorphoSys [7] integrates a CPU with a coarse-grained, multi-context ALU array. The device holds 32 contexts on-chip.

Our work targets a coarse-grained, multi-context reconfigurable hybrid. The main differences to related approaches are that we focus on macro-pipelining of contexts that execute for a longer time period, use FIFOs to transfer data between contexts, and couple the reconfigurable array with a CPU to form a hybrid device.

3   Architecture Model and Co-simulation

3.1   System Model

We investigate a hybrid reconfigurable device, which couples a coarse-grained reconfigurable unit closely to a CPU core. Figure 2 outlines the basic system model, which comprises the CPU core, instruction and data caches, and the reconfigurable unit (RU). The reconfigurable unit is attached to the CPU via a dedicated coprocessor interface and provides a number of coprocessor registers.

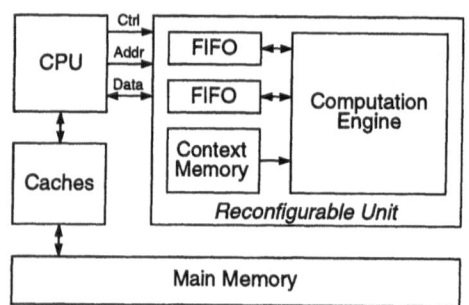

RU coproc. registers	CPU
RU reset	W
FIFO 1	R/W
FIFO 1 fill level	R
FIFO 2	R/W
FIFO 2 fill level	R
Cycle count register	R/W
Context memory $[1...n]$	W
Context select register	W
R: read access, W: write access	

Fig. 2. System model outline

We have developed a co-simulation framework that combines a cycle-accurate CPU model with an RU model specified in VHDL and allows for cycle-accurate simulation of the whole system. Details on the design and implementation of the co-simulation framework have been published in [8].

Currently the RU does not have its own memory access port, but all data communicated to and from the RU is passed via the CPU's register file. On the RU side, data transfers are performed via the FIFO buffers. Both FIFOs are readable and writable by the CPU as well as the RU.

The synchronization mechanism between CPU and RU is similar to the one proposed in the Garp processor [9]. The execution of the RU is started by writing the number of clock cycles the RU shall perform to the *cycle count register*. In every clock cycle, the cycle count register is decremented by one and stops the execution of the RU when reaching zero. By reading the cycle count register the remaining execution cycles can be determined.

3.2   CPU Model

For CPU simulation, we leverage on the SimpleScalar processor simulator [10]. SimpleScalar's CPU model is based on a 32-bit RISC processor architecture and has a MIPS-like instruction set. The CPU core's data and control path as well as the memory hierarchy are widely parameterizable. Thus, the CPU model can be configured to resemble a broad range of CPU architectures, from small embedded CPUs to powerful high-end CPUs.

In order to couple the RU to the CPU, we have extended SimpleScalar with a coprocessor interface. To this end, coprocessor read and write instructions have been added to the instruction set, which allow the CPU to access the coprocessor registers of the RU.

3.3 Model of the Reconfigurable Unit

The RU model comprises two FIFO buffers, the context memory, and the computation engine. Some RU characteristics are parameterizable: the data path width, the depth of the FIFO buffers, and the number of configurations the context memory holds. Another RU parameter determines whether the contexts contain state or not. An RU with context states replicates the registers in the data path in a way that each context is assigned a separate set of registers. An RU without context states provides only one register set that all contexts must share.

The context memory holds a set of configurations for the computation engine. The configuration data is written from the CPU to the RU via the configuration interface. The RU supports the download of full and partial configurations for any of the contexts. The CPU selects a context on the RU for execution by writing the number of the context to the *context selection register*. The context is immediately switched and the CPU can trigger the RU to run by writing the desired number of cycles to the *cycle count register*.

The computation engine is a 4×4 array of homogeneous, coarse-grained cells, which are connected by a 2-level network: direct interconnects between certain adjacent cells, Fig. 3(a), and horizontal buses between cell rows, Fig. 3(b). The computation engine has two input and two output ports, which are connected to the two FIFOs of the RU. Inside the computation engine, they are routed via the horizontal buses.

Figure 4 outlines the data path of a cell consisting of a fixed-point arithmetic logic unit (ALU), several multiplexers and registers. Figure 4(a) shows a cell without context state; all contexts have to share the same registers which are reset on context switches. Figure 4(b) displays a cell supporting context state. All the registers are replicated according to the number of physical contexts. This allows to preserve register values over several context switches. Alternatively, the register can also be reset on a context switch. The ALU implements the common arithmetic and logic operations (addition, subtraction, shift, OR, NOR, NOT, etc.) as well as multiplication. The control signals for the ALU and the multiplexers are part of the RU's configuration. The configuration contains also a constant operator, which can be routed to both ALU inputs.

The configuration of the computation engine is responsible for the functionality of the cells and the routing of the data path between the cells, from the input ports to the cells, and from the cells to the output ports. Since the configuration incorporates constant cell operators, the amount of required configuration bits depends on the datapath width. Given a datapath width of 16 bit, the configuration data results in 918 bits.

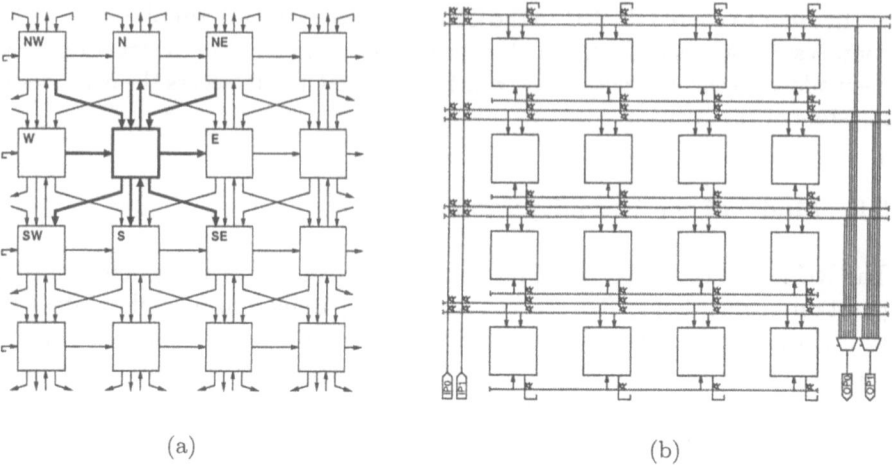

(a)   (b)

Fig. 3. 2-level interconnect scheme of the computation engine: (a) direct interconnects (*highlighted connections of one cell*), and (b) horizontal buses and I/O ports (*IPx, OPx*)

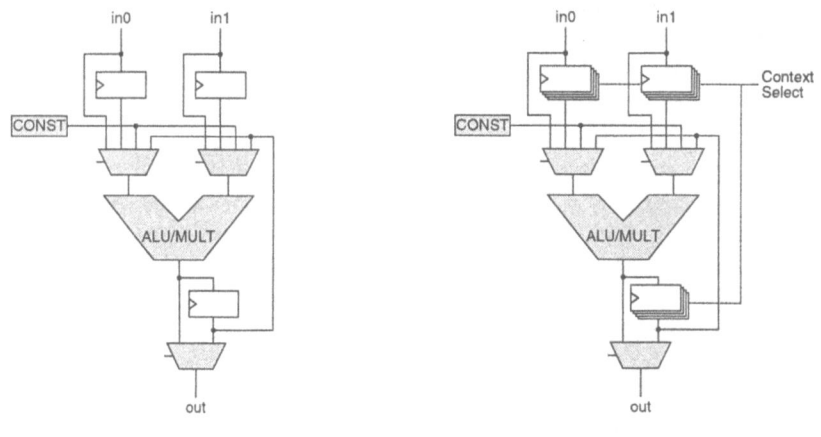

(a)   (b)

Fig. 4. Data path of a cell: (a) with *a single register set* for all contexts, and (b) with *a dedicated register set* for each context. The *shaded parts* are controlled by the configuration

4   Case Study and Experimental Setup

4.1   FIR Filter Partitioning and Mapping

As a case study, we have implemented a 56th-order FIR filter on our virtualized hardware. The filter is implemented as a cascade of eight subfilters of 7th-order. The input samples are processed in data blocks. An FIR filter implementation requires delay registers. These registers form the state of the context, which

Table 1. CPU model resembling an embedded CPU

Parameter Class	Setup
Computation units	1 int. ALU, 1 int. multiplier 1 FP ALU, 1 FP multiplier
Caches	32-way 16K L1 I-cache, 32-way 16K L1 D-cache, no L2 cache
Memory interface	32-bit memory bus, 1 memory port
Queue sizes[1]	instruction fetch: 1, register update unit: 4, load/store: 4
Bandwidths[1]	decode width: 1, issue width: 2, commit width: 2
Execution order	in-order
Branch prediction	always not-taken

[1] in number of instructions

must be saved between two executions of the same context. Depending on the capabilities of the reconfigurable array, there are two ways to achieve this:

- If all contexts of the RU share the same register set, the state must be explicitly saved and later on restored. For the filter implementation this is achieved by overlapping subsequent data blocks, which forms an execution overhead.
- If the RU provides dedicated register sets for each context the state is kept automatically. For the filter implementation, no extra cycles are needed for state handling if we can hold all logical contexts on the array.

4.2   System Model Setups

We have set up our system model to study the following cases: CPU only (no RU present), CPU with attached single-context RU, CPU with attached multi-context RU having 2, 4 and 8 contexts, and finally a CPU with attached 8-context RU incorporating a dedicated register set for each context.

The SimpleScalar CPU model is configured such that it resembles an embedded CPU. Table 1 lists the most important CPU parameters. In each experiment, 64K samples organized in data blocks are processed. The size of the data blocks depends on the FIFO depth available on the RU (cf. Fig. 2). We vary the depth of the FIFO buffers between 128 and 1k words.

For the coprocessor cases, a data block is written to the RU, processed sequentially by the eight FIR filter stages (the eight logical contexts), and finally read back. At the beginning, a controller task running on the CPU downloads as many contexts as fit onto the RU. If not all logical contexts fit, the contexts are loaded on demand. Each time a filter context is required that is not present, the controller performs the download by overriding always the same physical context. This is done to hold as many contexts as possible unchanged on the array with the goal to reduce the amount of reconfiguration data that has to be downloaded onto the RU.

5   Results and Discussion

Figure 5 illustrates the results of the experiments as functions of the device architecture (shown on the horizontal axis) and the FIFO buffer size. The execution
time of the filter for the CPU only is 110.65 million cycles. Figure 5(a) shows
the speedups relative to this computation time and Fig. 5(b) presents the CPU
load. We assume a real-time system that filters blocks of data samples at a given
rate. When the filter computation is moved from the CPU to the reconfigurable
array, the CPU is relieved from these operations and can use this capacity for
running other tasks. However, the CPU still has to transfer data to and from the
FIFOs, write contexts to the RU on demand, and control the context switches.
The load given in Fig. 5(b) determines the spent CPU cycles normalized to the
CPU only system. We point out the following observations:

- Hardware virtualization is an extremely useful concept. We were able to run
  the same filter implementation on reconfigurable array models with different
  features without resynthesizing the application.
- Using an RU we achieve significant speedups, ranging from a factor of 2.4 for
  a 128 word FIFO single-context device up to a factor of 8.9 for an 8-context
  RU that restores context state.
- The performance of the system in terms of speedup and CPU load depends
  on the length of the FIFO buffers. Enlarging the FIFOs increases the performance and at the same time the filter delay. Practical applications could
  limit these potential gains by imposing delay constraints. For instance, a
  2-context RU using a FIFO with 1k words instead of 128 words improves
  the speedup by a factor of 2.7, while increasing the latency by a factor of 8.
- Figure 5 shows that a multi-context array storing the context states greatly
  benefits our application if we can store *all* logical contexts. In this case,
  we can avoid the overlapping of data blocks. For an 8-context array with a
  128 word FIFO the speedup increases by factor of 2.0. In addition, as no
  reconfiguration is required, the speedup becomes almost independent of the
  FIFO size.
- Employing a reconfigurable coprocessor not only speeds up the computation
  but also lowers the CPU load significantly. As Figure 5(b) displays, for a
  single-context RU the CPU load drops from 100% to 28.4% for a 128 word
  FIFO and to 6.4% for a 1k word FIFO. Increasing the number of physical
  contexts the load approaches the asymptotic value of 4.8%, because the CPU
  task reduces to data transfer and context switches.

6   Summary and Future Work

In this paper, we have discussed the concept of hardware virtualization and
the use of multi-context architectures to achieve it. We have presented a co-
simulation framework based on a hybrid system model consisting of a reconfigurable unit attached to a CPU. As a case study, we have mapped an FIR

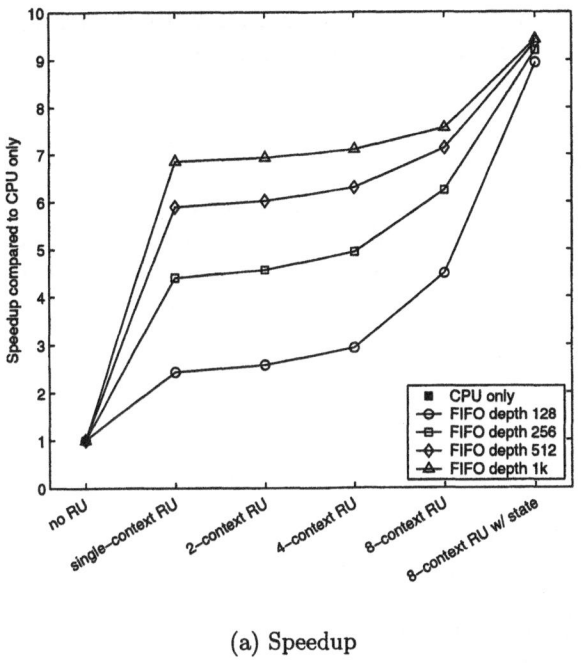

(a) Speedup

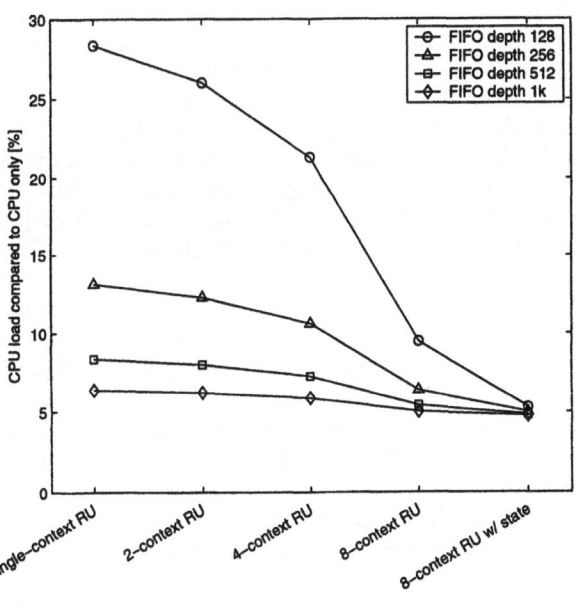

(b) CPU load

Fig. 5. Performance figures in comparison to the CPU only case

filter to our virtualized hardware and run it on various architectures by cycle-accurate simulation. The results show that hardware virtualization is a valuable concept and that multi-context features can be successfully employed. Further work includes:

- Implementation of application types that require more complex context sequences (control flow).
- Integration of a dedicated RU memory port.
- Investigation of context prediction and prefetching techniques.
- Development of an area model for the reconfigurable unit in order to quantify the hardware overhead introduced by the multi-context features.

References

1. Goldstein, S.C., Schmit, H., Budiu, M., Cadambi, S., Moe, M., Taylor, R.R.: PipeRench: A reconfigurable architecture and compiler. IEEE Computer **33** (2000) 70–77
2. Trimberger, S., Carberry, D., Johnson, A., Wong, J.: A time-multiplexed FPGA. In: Proc. 5th IEEE Symp. on Field-Programmable Custom Computing Machines (FCCM). (1997) 22–28
3. Caspi, E., Chu, M., Huang, R., Yeh, J., Wawrzynek, J., DeHon, A.: Stream computations organized for reconfigurable execution (SCORE). In: Field-Programmable Logic and Applications (Proc. FPL), LNCS 1896, Springer-Verlag (2000) 605–614
4. Schmit, H., Whelihan, D., Moe, M., Levine, B., Taylor, R.R.: PipeRench: A virtualized programmable datapath in 0.18 micron technology. In: Proc. 24th IEEE Custom Integrated Circuits Conf. (CICC). (2002) 63–66
5. DeHon, A.: DPGA utilization and application. In: Proc. 4th ACM Int. Symp. on Field-Programmable Gate Arrays (FPGA). (1996) 115–121
6. Baumgartne, V., May, F., Nückel, A., Vorbach, M., Weinhardt, M.: PACT XPP – a self-reconfigurable data processing architecture. In: Proc. 1st Int. Conf. on Engineering of Reconfigurable Systems and Algorithms (ERSA). (2001) 64–70
7. Singh, H., Lee, M.H., Lu, G., Kurdahi, F.J., Bagherzadeh, N., Chaves Filho, E.M.: MorphoSys: An integrated reconfigurable system for data-parallel and computation-intensive applications. IEEE Trans. on Computers **49** (2000) 465–481
8. Enzler, R., Plessl, C., Platzner, M.: Co-simulation of a hybrid multi-context architecture. In: Proc. 3rd Int. Conf. on Engineering of Reconfigurable Systems and Algorithms (ERSA). (2003)
9. Hauser, J.R., Wawrzynek, J.: Garp: A MIPS processor with a reconfigurable coprocessor. In: Proc. 5th IEEE Symp. on Field-Programmable Custom Computing Machines (FCCM). (1997) 12–21
10. Austin, T., Larson, E., Ernst, D.: SimpleScalar: An infrastructure for computer system modeling. IEEE Computer **35** (2002) 59–67

A Dynamically Adaptive Switching Fabric on a Multicontext Reconfigurable Device

Hideharu Amano[1], Akiya Jouraku[1], and Kenichiro Anjo[2]

[1] Dept. of ICS, Keio University,
[2] NEC Electronics,
drp@am.ics.keio.ac.jp

Abstract. A framework of dynamically adaptive hardware mechanism on multicontext reconfigurable devices is proposed, and as an example, an adaptive switching fabric is implemented on NEC's novel reconfigurable device DRP(Dynamically Reconfigurable Processor).

In this switch, contexts for the full crossbar and alternative hardware modules, which provide larger bandwidth but can treat only a limited pattern of packet inputs, are prepared. Using the quick context switching functionality, a context for the full crossbar is switched by alternative contexts according to the packet inputs pattern. Furthermore, if the traffic includes a lot of packets for specific destinations, a set of contexts frequently used in the traffic is gathered inside the chip like a working set stored in a cache.

4×4 mesh network connected with the proposed adaptive switches is simulated, and it appears that the latency between nodes is improved three times when the traffic between neighboring four nodes is dominant.

1 Introduction

Techniques on dynamically reconfigurable systems have been widely researched especially for mobile terminals whose hardware resources are strictly limited. Advanced dynamically reconfigurable systems or devices whose functions can be changed so as to adapt for surrounding conditions have been reported[3][4]. Such systems are also useful for quick delivery of an improved version of hardware which supports extended services. By further extension of such technologies, there becomes a possibility of a dynamically adaptive hardware, which modifies its structure automatically to fit the surrounding conditions.

Unfortunately, the configuration time and consuming power required for changing hardware structure of commodity reconfigurable devices are so large that the benefits of dynamically reconfiguration are lost in most cases. Thus, target applications of the dynamically adaptive hardware are mainly limited to specific functions of mobile devices.

However, recent advanced reconfigurable devices[1][2] which support quick and run-time reconfiguration can extend the target field of dynamically adaptive hardware drastically. For example, DRP(Dynamically Reconfigurable Processor)[1] developed by NEC in 2002 provides 16 hardware contexts inside the

P.Y.K. Cheung et al. (Eds.): FPL 2003, LNCS 2778, pp. 161–170, 2003.

chip, and can switch them with a clock cycle. The chip is partitioned into several regions where context switchings can be controlled independently. That is, a partial context switching is possible. Run-time reconfiguration from outside the chip to the context which is not currently used is also supported.

Here, by making the best use of such facilities, a novel framework of dynamically adaptive hardware is proposed. As a design example, a dynamically adaptive switching fabric is implemented on DRP, and the performance is evaluated.

2 Dynamically Adaptive Hardware

2.1 Concept of Dynamically Adaptive Hardware

Fig. 1A shows an execution unit of dynamically adaptive hardware proposed here. It consists of a fixed region and a multicontext reconfigurable region[7] which switches the context with a clock. The target circuit for adaptation is implemented on the multicontext reconfigurable region, while the context switching is managed with the scheduler on the fixed region. Here, a multicontext device which provides tens of contexts inside the chip is assumed. As described later, NEC's DRP provides sixteen contexts. The configuration data outside the chip can be loaded to currently unused context without disturbing the current available context like a virtual hardware mechanism[5].

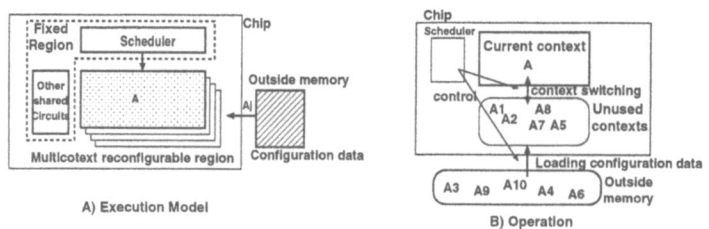

Fig. 1. Concept of dynamically adaptive hardware

Assuming that A is fully equipped hardware module which can execute every required function. $A_1, A_2, ...A_n$ are alternative hardware modules which can execute only a part of $A's$ functions with higher speed or lower power consumption. Generally, if required functions are limited, it is not difficult to design a circuit whose performance or power consumption is better than A.

When the system starts, A is executed on the multicontext region, and other specialized hardware modules A_j are held in unused contexts. Some of A_j which cannot be stored inside the chip are located in the memory outside the chip as shown in Fig. 1B. The scheduler in the fixed region inspects inputs and states of the executing logic. If the scheduler judges that another hardware module A_j

with better performance/power consumption can be used in the current situation, the context is replaced to that for A_j. If the situation changes and functions which A_j cannot treat are required, another context A_k which can treat the situation comes up. If there is no other candidates, fully equipped A is used again.

If the context corresponding to A_j is not inside the chip, A is used until the configuration data for A_j is loaded from outside the chip, and then the context is switched. The over-written context number by the configuration data from outside the chip is selected with the LRU (Least Recently Used) algorithm. The fully equipped A is assigned into the context 0, and never be over-written. If there are a lot of alternative hardware modules, only frequently used candidates are loaded inside the chip like a cache mechanism of common computers. That is, the system is adopted for the frequently occurring situations. Note that the system does not stop during loading configuration data for A_j, since fully equipped A is continuously working on the current context.

There are some open problems in this framework. First is the algorithm for selecting alternative hardware modules with limited functions $A_1, ... A_n$. If circuits for evaluating the execution performance or power consumption can be provided in the scheduler, learning algorithms developed in the artificial intelligence can be applied. However, such a complicated scheduler will increase the chip area and power consumption. Thus, for most applications, a simple learning mechanism is useful.

The next problem is the timing of the context switching and data communication between them. This is a common problem of multicontext reconfigurable devices, and a design methodology as a solution is proposed[7]. As described later, DRP provides distributed memory modules which are shared with all contexts. Communication between contexts can be done using such shared memory modules.

If the target system consists of several components, and the target device enables partial context switching, this framework can be applied to each component independently. In general, components are often designed by using IPs (intellectual Properties). For such adaptive systems, an IP which includes a set of configurations consisting of the fully equipped A, other candidates $A_1, .., A_n$, and corresponding conditions for application is desirable.

2.2 A Dynamically Adaptive Switch

As an example, a simple packet switching fabric with four inputs/outputs for system area network or parallel machines is designed. A 64bit width packet buffer which can store two full-size packets at maximum is provided to each input as shown in Fig. 2. Each link is 32bit width, and the transmission bit-late on the link is assumed to be twice as that of the clock frequency inside the chip. A flexible size packet with 256 flits at maximum is transferred in the asynchronous wormhole manner [1]. A header flit includes the destination and size of the packet.

---

[1] In a real switch, the link transfer rate becomes often eight times that of inside the switch[8]. Here, from the restriction of the DRP chip, the rate is set to be double.

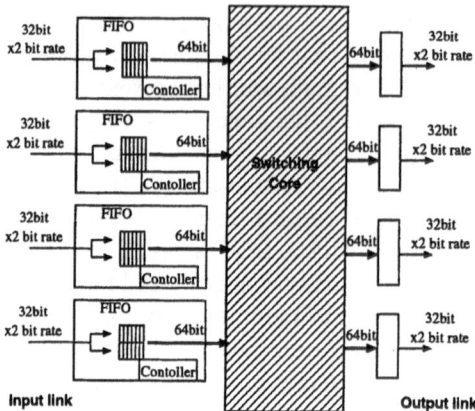

Fig. 2. Diagram of the switching fabric

The input port controller including packet buffer is assigned into the fixed region, while the multicontext region is used for the switching core. Here, fully equipped A for the switching core is full size (4×4) crossbar with 32bit bandwidth shown in Fig. 3A. However, if packets arrive at only an input port, the crossbar can be replaced by a configuration with only wires as shown in Fig. 3B. We call this "0-to-1 wire" configuration, and in general, "n-to-m wire" configuration is possible for ($n < 4, m < 4$). Compared with the full size crossbar, "0-to-1 wire" configuration requires only area for wires. That is, 64bit width wires, the double width of 4×4 crossbar can be used. That is, this configuration is possibly advantageous both in the performance and power consumption.

Similarly, a combination of wires (Fig. 3C) and a small size 2×2 crossbar shown in Fig. 3D can be used when packets are arrived at two input ports. Since the area required by these configurations is small compared with the full size crossbar, double bandwidth wires (64bits) also can be used. Thus, the design shown in Fig. 3B-D can be treated as alternative hardware modules with limited functions A_j.

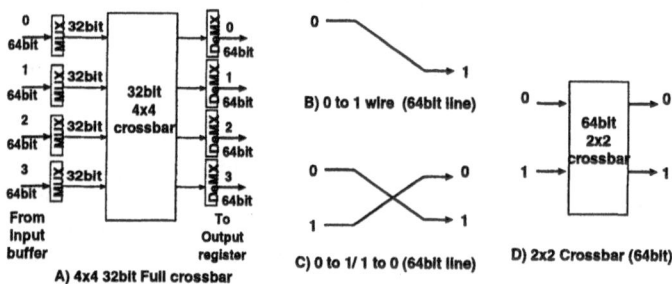

Fig. 3. Various Configurations for switching core

Since such alternative hardware modules can be formed for every combination of input/output, a lot of hardware modules A_j can be prepared for the switching core. If the context corresponding to requested alternative hardware module is not inside the chip, it is loaded from outside chip to currently unused context memory, then replaced with the full size crossbar. If the traffic includes a lot of packets for specific destinations, a set of contexts frequently used in the traffic is gathered inside the chip like a working set stored in a cache. That is, the switch can adapt to the traffic.

3   Implementation on DRP-1

3.1   DRP Overview

DRP is a coarse-grain reconfigurable processor core, which can be integrated into ASICs and SOCs. The primitive unit of DRP Core is called 'Tile', and DRP Core consists of arbitrary number of Tiles. The number of Tiles can be expandable, horizontally and vertically.

The primitive modules of Tile are processing elements(PEs), State Transition Controller(STC), 2-ported memories (VMEMs: Vertical MEMories), VMEM Controller(VMCtrl) and 1-ported memories (HMEMs: Horizontal MEMories). The structure of Tile is shown in Fig. 4.

There are 8x8 PEs located in one Tile. The architecture of PE is shown in Fig. 5. It has an 8-bit ALU, an 8-bit DMU, an 8-bit×16-word register file, and an 8-bit flip-flop. Those units are connected by programmable wires specified by instruction data. These bitwidths are ranging from 8B to 18B according to the location. PE has 16-depth instruction memories and supports multiple context operation. Its instruction pointer is delivered from STC.

STC is a programmable sequencer in which certain FSM (Finite State Machine) can be stored. STC has 64 states, and each state is associated with the

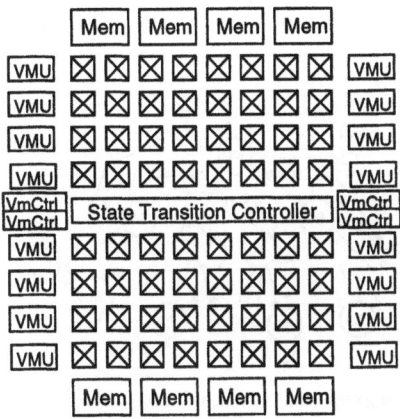

Fig. 4. Structure of a Tile

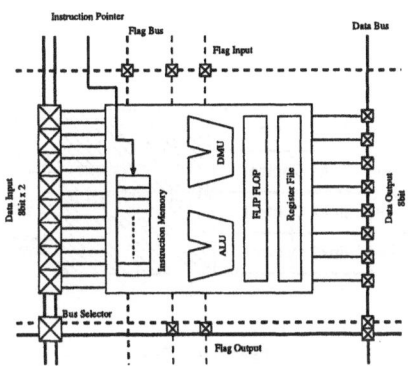

Fig. 5. Structure of a PE

instruction pointer. FSM of STC operates synchronized with the internal clock, and generates the instruction pointer for each clock cycle according to the state. Also, STC can receive event signals from PEs to branch conditionally. The maximum number of branch is four.

As for the memory units, Tile has eight 2-ported VMEMs on right and left sides, and four 1-ported HMEMs on upper and lower boundary. The capacity of a VMEM is 8-bit×256-word, and four VMEMs can be handled as a FIFO, using VMCtrl. HMEM is single-ported, thus has large capacity compared with VMEM. It has 8-bit×8K-word entries. Contents of these memories, flip-flops, register files of PE are shared with the datapath of all the contexts.

DRP Core, consisting of several Tiles, can change its contexts every cycle by instruction pointer distribution from STCs. Also, each STC can run independently, by programming different FSMs.

DRP-1 is the prototype chip, using DRP Core with 4×2 Tiles. It is fabricated with 0.15-um 8-metal layer CMOS processes. It consists of 8-Tile DRP Core, eight 32-bit multipliers, an external SRAM controller, a PCI interface, and 256-bit I/Os. The maximum operation frequency is 100-MHz.

3.2   Dynamically Adaptive Switch on DRP-1

A simple dynamically adaptive switching fabrics with 4-input/output shown in Fig. 3A-D is implemented on DRP-1.

In this implementation, a left four Tiles are assigned into the fixed region including input packet buffers and controllers as shown in Fig. 6a). Remaining four Tiles are used for the multicontext reconfigurable region where the various switching cores are placed. In this implementation, the logic overhead for controlling context switch is almost included in the arbiter, and the switching core is a combinatorial circuit except for the final buffer. Thus, there is almost no logic overhead for management of context switching.

ALUs and DMUs in the fixed region are almost fully used, and about a half of them are used for wiring. On the contrary, a lot of empty PEs are remained

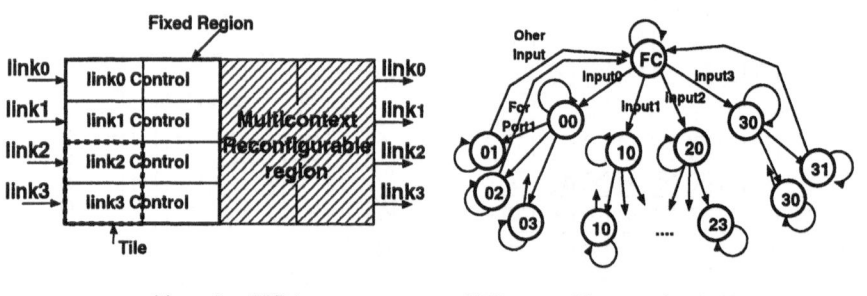

a) Layout on DRP-1   b) State transition associated with context switching

Fig. 6. Layout and State transition on DRP-1

in dynamically reconfigurable part especially for 1 to 1 wiring (Fig. 3B and 2 × 2 crossbars (Fig. 3D).

The evaluated maximum frequency is about 32MHz independent from the configuration in the switching core, since the packet controllers in the fixed region make critical paths in the total design. However, since the transfer bandwidth of Fig. 3B, Fig. 3C and Fig. 3D is enhanced compared with the fully equipped hardware module shown in Fig. 3A), the performance can be improved by replacing the configuration.

In DRP, STC controls the context switching by a state transition in which each state is associated into a context. Unfortunately, with the restriction in DRP-1, only four states can be designated as destinations from a state.

In this implementation, sixteen states each of which is corresponding to a context are used as shown in Fig. 6b). When the switch is initialized, the state "FC" corresponding to the full crossbar is used.

The control logic in the fixed region generates a signal (input0) when a packet arrives at input port 0. Detecting the signal as an event, STC moves its state into "00" corresponding to "0-to-0-wire" configuration. During a clock for the context switching, the arbiter in the fixed region checks the packet header, and decides the destination output port. If the destination output port of the packet is 1, 2 or 3, the state moves to "01" for "0-to-1-wire", "02" for "0-to-2-wire", or "03" for "0-to-3-wire" , respectively. Otherwise (that is output port 0 is selected), the state "00" is used without context switching. That is, the state transition for context switching is overlapped with the arbitration sequence, and no extra cycles are required.

In this implementation, sixteen contexts are assigned into full crossbar and "n-to-m-wire" configurations, where n and m is 0,1,2 or 3 except "3-to-3-wire" configuration which cannot be included by the limitation of the contexts. That is, if the arbiter detects the destination of the packet is output port 3 in "30" state, the state moves back to "FC" where the full crossbar is used.

A virtual hardware mechanism, which enables to use a full set of configurations including "2 × 2 crossbars", is under designing now, since it requires supporting hardware outside the chip. In this mechanism, state number is extended to hold contexts which cannot be held inside the chip. The fixed part checks whether the context is inside of the chip or not by referring the mapping table. If the context corresponding to the destination state is not inside the chip, the configuration data is loaded to unused context selected with the LRU algorithm.

4   Performance Evaluation

4.1   Evaluation Conditions

We evaluated the performance of the proposed dynamically adaptive switch with a flit level network simulator. The structure of the switching fabrics is almost the same as the implementation on the DRP shown in the previous section except with the size of switch. That is, the number of ports is extended to 5 × 5 for forming a two dimensional mesh structure in the simulation. A port is used for connecting with a host processor and the rest ports are connected to four neighboring switches.

A virtual hardware mechanism now under development is assumed to be available in this simulation, that is, runtime configuration is triggered when the required configuration is not inside the chip. Although several ways of configurations are supported in DRP-1, the most commonly used is configuration through the PCI interface. Configuration data corresponding to PEs in all Tiles, memory modules and other interfaces is mapped into a single logical address with 20bit address and 32bit data. It can be accessed as a simple byte-addressed 32bit memory from the host, and the configuration data is transferred through PCI bus usually in the burst mode. In the current implementation, the context switching can be done at any time, but requires a clock delay. After loading the configuration, an extra clock delay is assumed for re-writing the state transition table.

The size of configuration data for dynamically reconfigurable parts (4 Tiles) is dependent on the design. Since both designs of "n-to-m wire" and "2×2 crossbar" are simple, the runtime configuration requires about 200 clock cycles in total. Other simulation parameters are shown in Table 1.

Table 1. Simulation Parameters

Simulation time	3,000,000 clocks (ignore the first 50,000 clocks)
Topology	4 x 4 two dimensional mesh
Packet length	256 flits
Flow control	virtual cut-through
Routing	e-cube routing with a virtual channel
Traffic pattern	uniform/localized

4.2   Evaluation Results

Fig. 7 shows the throughput versus latency under the uniform traffic, that is, the desitination of packets are randomly distributed. The line marked "full crossbar" is corresponding to the case that only fully equipped configuration (5×5 crossbar in this case) is used.

"full crossbar + n-to-m wire" uses "n-to-m wire" configurations shown in Fig. 3B. The latency is slightly improved when the traffic is not severe, but becomes worse when the traffic becomes severe. This comes from the overhead for context switching and an extra overhead after the loading configuration.

"full crossbar + 2×2 crossbar" uses 2×2 crossbar configuration shown in Fig. 3D when arriving packets are stored in two input ports. In this case, the latency is reduced independent on the traffic load. Although the possible configurations of "2×2 crossbar" increases to 100, the number of traffic patterns

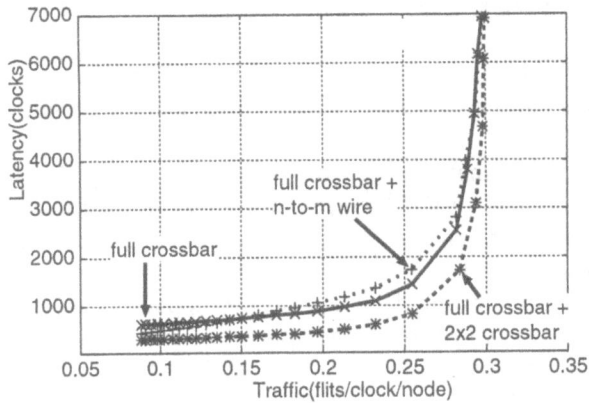

Fig. 7. Bandwidth versus Latency (Uniform traffic)

which a "2×2 crossbar" can treat is four times that of "n-to-m wire", and the performance is improved.

When a lot of scientific applications including Partial Differential Equations are executed on mesh-connected parallel machines, a large part of communication is done between neighboring four nodes. For such a condition, we used the localized traffic in which 90% of packets are to the neighboring nodes and the rest is randomly distributed. Fig. 8 shows the evaluation results under this localized traffic. In this case, both "n-to-m wire" and "2×2 crossbar" improves bandwidth as well as latency, since it does not require run-time loading of configuration. Especially, "2×2 crossbar" improves three times in latency and 9% in bandwidth.

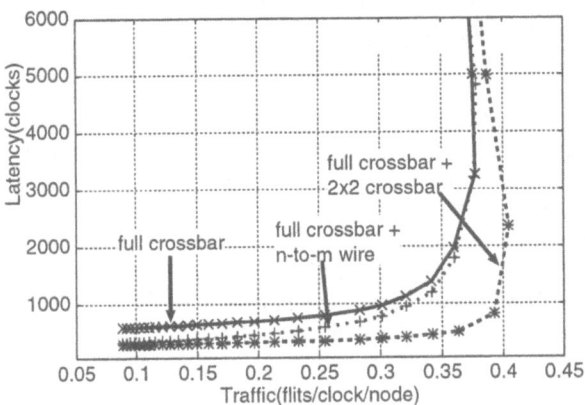

Fig. 8. Bandwidth versus Latency (Localized Traffic)

5   Related Work and Conclusion

A framework of dynamically adaptive hardware is proposed, and as an example, a dynamically adaptive switching fabric is implemented on NEC's multicontext device DRP.

The configuration management in the proposed mechanism is similar to virtual hardware mechanisms[5][9] with multicontext reconfigurable devices including our previous work. The cache effect of configuration data in reconfigurable devices is also discussed in [10]. However, the system never stop during loading of configuration data for A_j, since fully equipped A is continuously working on the current context. Although the approach in dynamic hardware plugins[4] is similar to our approach in some points, their approach focuses on the functional adaptation. Researches called "evolvable hardware"[11], which combines reconfigurable logics and hardware genetic algorithm engines have been exerted. Although genetic algorithm may be useful for selection of alternative logics, the functions supported in our mechanism do not change beyond the design.

References

1. M.Motomura:"A Dynamically Reconfigurable Processor Architecture," Microprocessor Forum, Oct. 2002.
2. P.Master: "The Age of Adaptive Computing Is Here," Proc. of FCCM, pp.1-3 (2002).
3. G.J.M.Smit, P.J.M.Havinga, L.T.Smit, P.M.Heysters: "Dynamic Reconfiguration in Mobile Systems," Proc. of FPL2002, pp.162-170, (2002).
4. E.L.Horta, J.W.Lockwood, D.Partour, "Dynamic Hardware Plugins in an FPGA with Partial Run-time Reconfiguration." Proc. of DAC2002, June (2002).
5. X.-P. Ling, H. Amano, "WASMII: A Data Driven Computer on a Virtual Hardware" Proc. FCCM, pp. 33–42, (1993).
6. S. Trimberger, D. Carberry, A. Johnson and J. Wong "A Time-Multiplexed FPGA" Proc. FCCM pp.22-28, (1997).
7. N.Kaneko, H.Amano, "A General Hardware Design Model for Multicontext FPGAs," Proc. of FPL, pp.1037–1047, (2002).
8. S.Nishimura, et al., "High-speed network switch RHiNET-2/SW and its implementation with optical interconnections", Hot Interconnects 8, pp.31–38, Aug. (2000).
9. G.Brebner, "The Swappable Logic Unit: a Paradigm for Virtual Hardware," Proc. of FCCM, pp.82–91, (1997).
10. Z.Li, K.Compton, S.Hauck, "Configuration Caching Management Techniques for Reconfigurable Computing," Proc. of FCCM, pp.22–36, (2000).
11. T.Higuchi and N.Kajihara, "Evolvable hardware chips for industrial applications," Commun. ACM, Vol.42, No.3, pp.60–66, (1999).

Reducing the Configuration Loading Time of a Coarse Grain Multicontext Reconfigurable Device

Toshiro Kitaoka[1], Hideharu Amano[1], and Kenichiro Anjo[2]

[1] Dept. of ICS, Keio University,
[2] NEC Electronics,
drp@am.ics.keio.ac.jp

Abstract. High speed and low cost configuration loading methods for a coarse grain multicontext reconfigurable device DRP(Dynamically Reconfigurable Processor) are proposed and implemented. In these methods, the configuration data is compressed on the host computer before loading, and decoded at the time of loading by circuits implemented on a part of logics. Unlike conventional reconfigurable device, the logic for decoder circuits is switched with application circuits immediately after loading in multicontext reconfigurable devices. Thus, the circuit does not use a real estate of the chip during the execution. Two compression methods LZSS-ARC and Selective coding are implemented and evaluated. LZSS-ARC achieves better compression ratio, while Selective coding can work at the same frequency of the data loading.

1 Introduction

A run-time reconfigurable system, which changes its structure dynamically, has been widely researched especially for mobile terminals whose hardware resources are strictly limited. Recent advanced reconfigurable devices[2][3] which support quick and run-time reconfiguration can extend target fields of dynamic reconfigurable systems drastically. A coarse grain cell is adopted as an element in such devices for efficient use of the streaming applications. Some of these chips include multi-context functionality which provides several configuration data sets inside the chip, and changes them quickly.

However, even using such advanced devices, the time for loading of the configuration data from outside the chip often bottlenecks the system performance for some dynamically reconfigurable applications. Reducing the amount of configuration data with compression techniques is one of hopeful approaches to improve the configuration speed. Efficient runtime decoding techniques have been researched for traditional FPGAs[9][10].

Unfortunately, such compression methods are not efficient for recent coarse grain reconfigurable devices because of the complicated cell architecture. In this paper, we propose and evaluate runtime decoding methods of the beforehand compressed configuration data for a recent multi-context reconfigurable device,

P.Y.K. Cheung et al. (Eds.): FPL 2003, LNCS 2778, pp. 171–180, 2003.

NEC's DRP(Dynamically Reconfigurable Processor)[2]. This method enables high-speed loading of configuration data with a small overhead of the decoder hardware by making the best use of the multi-context facility.

2   Configuration Code Compression and Runtime Decoding

2.1   Code Compression Techniques for Conventional FPGAs

For the SRAM type FPGAs/CPLDs, which are now popularly in use, the size of configuration data is sometimes more than 5 Mbit and they need an msec-order time for loading. From technical point of view, it is possible to make the configuration time almost equal to that of the access time of common static RAMs. Even using such a high speed configuration circuit which can load 64 bit data at every 50 MHz clock, it takes 1.6 msec to load 5 Mbit configuration data.

Since the configuration of common FPGAs/CPLDs includes a large amount of data corresponding to unused LUT(Look Up Table)s, wires and switches, common compression techniques are efficiently applied. Several techniques are proposed for compression and decoding configuration data of common FPGAs/ CPLDs[9][10]. In these methods, configuration data is prepared in a beforehand generated by certain software after place and routing. In such researches, an FPGA/CPLD assumed to provide hardware which decodes compressed configuration data and writes them into the configuration memory directly. This approach requires dedicated decoding hardware inside the FPGA chip as shown in Fig. 1, and requires a certain chip area. Thus, researches focus on the decoding circuits which enables runtime decoding with small hardware cost as possible.

Above compression algorithms for conventional FPGA with LUTs are not suitable for recent coarse grain reconfigurable devices. Table 1 shows the ratio of compressed data size to the original one when LZ77 and Huffman compression used in traditional compression methods are applied to the configuration data of two different reconfigurable devices. One is NEC's coarse grain reconfigurable device DRP, and another is its previous version called DRL[1] which uses LUTs. It appears that they can only reduce about 30% data size, while the configuration data of LUT based device can be compressed to less than a half size of original.

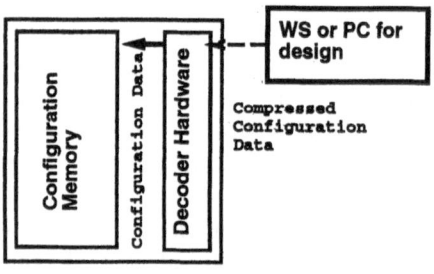

Fig. 1. FPGA with a decoder

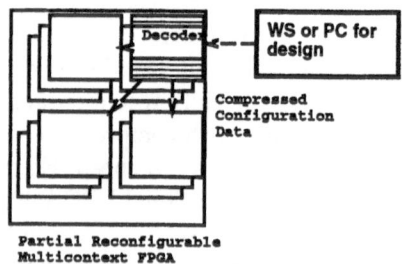

Fig. 2. partially multicontext reconfigurable device

Table 1. Ratio of the compressed data size (Neural Net)

	Coarse grain (DRP) (%)	LUT based device (DRL) (%)
LZ	72	27
Huffman	73	28

Although a coarse grain logic element requires less configuration data size than common FPGAs with LUTs, it does not include data with a long sequence of of '0's or '1's. This is the main reason why the traditional LZ or Huffman techniques are not so efficient.

2.2   Combination with a Multicontext Reconfigurable Device

On the other hand, some of recent reconfigurable devices are equipped with a multicontext facility which can be efficiently used for decoder of the compressed configuration.

Multicontext reconfigurable devices[5][11][1] can store multiple sets of configuration data. By switching the output of each configuration memory by the multiplexor, the configuration can be immediately switched. Configuration data on each configuration memory is called a context and switching them for reconfiguration by the multiplexer is called context switching. In many multicontext devices, context switching requires only a clock, the hardware structure can be changed quickly. Now, DRP, a multicontext device with sixteen contexts developed by NEC[2] is available, and Quicksilver's ACM[3] and IPFlex's DNA chip[4] also provide the similar facility.

For high-speed but cost effective configuration, we propose to implement decoding hardware for compressed configuration data in one or small number of the contexts. After configuration is done with the decoding hardware, the context is switched and the loaded circuit starts quickly.

Furthermore, in DRP, a partial context switching mechanism is available. In such a device, as shown in Fig. 2, it is possible to implement the decoding hardware in a partition and to rewrite the configuration data of unused contexts in other partitions. By scheduling context-rewriting in each partition, it is expected to cover up the time for configuration from outside the chip. This mechanism enables high speed run-time configuration for dynamically reconfigurable applications[6].

3   DRP and Its Configuration

3.1   DRP Overview

DRP is a coarse-grain reconfigurable processor core, which can be integrated into ASICs and SOCs. The primitive unit of DRP Core is called 'Tile', and DRP Core consists of arbitrary number of Tiles. The number of Tiles can be expandable, horizontally and vertically.

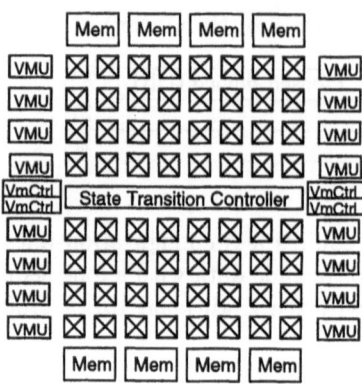

Fig. 3. Structure of a Tile

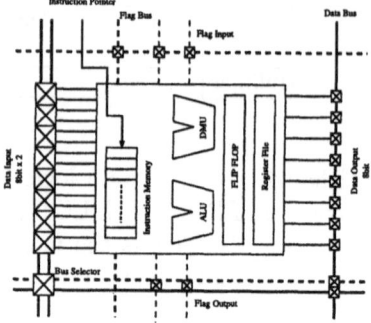

Fig. 4. Structure of a PE

The primitive modules of Tile are processing elements(PEs), State Transition Controller(STC), 2-ported memories (VMEMs: Vertical MEMories), VMEM Controller(VMCtrl) and 1-ported memories (HMEMs: Horizontal MEMories). The structure of Tile is shown in Fig. 3.

There are 8x8 PEs located in one Tile. The architecture of PE is shown in Fig. 4. It has an 8-bit ALU, an 8-bit DMU, an 8-bit×16-word register file, and an 8-bit flip-flop. Those units are connected by programmable wires specified by instruction data. These bitwidths are ranging from 8B to 18B according to the location. PE has 16-depth instruction memories and supports multiple context operation. Its instruction pointer is delivered from STC.

STC is a programmable sequencer in which certain FSM (Finite State Machine) can be stored. STC has 64 states, and each state is associated with the instruction pointer. FSM of STC operates synchronized with the internal clock, and generates the instruction pointer for each clock cycle according to the state. Also, STC can receive event signals from PEs to branch conditionally. The maximum number of branch is four.

As for the memory units, Tile has eight 2-ported VMEMs on right and left sides, and four 1-ported HMEMs on upper and lower boundary. The capacity of a VMEM is 8-bit×256-word, and four VMEMs can be handled as a FIFO, using VMCtrl. HMEM is single-ported, thus has large capacity compared with VMEM. It has 8-bit×8K-word entries. Contents of these memories, flip-flops, register files of PE are shared with the datapath of all the contexts.

DRP Core, consisting of several Tiles, can change its contexts every cycle by instruction pointer distribution from STCs. Also, each STC can run independently, by programming different FSMs.

DRP-1 is the prototype chip, using DRP Core with 4×2 Tiles. It is fabricated with 0.15-um 8-metal layer CMOS processes. It consists of 8-Tile DRP Core, eight 32-bit multipliers, an external SRAM controller, a PCI interface, and 256-bit I/Os. The maximum operation frequency is 100-MHz.

3.2 DRP Configuration

Although several ways of configurations are supported in DRP-1, the most commonly used is configuration through the PCI interface. Configuration data corresponding to PEs in all Tiles, switches, memory modules and other interfaces is mapped into a single logical address with 20bit address and 32bit data. It can be accessed as a simple byte-addressed 32bit memory from the host, and the configuration data is transferred through PCI bus usually in the burst mode. Although the address is not continuously used in all the 20-bit address space, uppermost 5bit of address is the same for a long fixed sequence. That is, address can be easily compressed into 15bit.

In the implementation shown here, a Tile is assigned for decoder circuits of compressed configuration data transferred from the PCI interface. Unfortunately, the configuration path and data path are completely separated in DRP-1, and the compressed data cannot transferred to the configuration memory. A simple bus from the output data of the Tile to the configuration path is assumed here.

4 Coding Methods and Decoder Implementation

Following properties are required for coding methods for compression of configuration data. (1) The decoding must be done quickly at the time of loading the configuration, while a complex encoding by the software is allowed. (2) The decoding must be done with small amount of hardware which can be implemented on a small part of a multicontext reconfigurable device. Methods which require large amount of buffers or logics cannot be adopted.

To satisfy the above conditions, we adopted LZSS[8] and modified it so as to fit the configuration data of DRP. This method is called LZSS with address run-length compression or LZSS-ARC. We also proposed another method which changes coding methods by switching decoding hardware based on the properties of the target context. This method is called Selective coding.

4.1 LZSS with Address Run-Length Compression

Coding Method. The LZ77 coding is a data-coding method which was proposed by J.Ziv and A.Lempel in 1977[7]. In this method, the dictionary for coding is not prepared beforehand but built dynamically while reading symbols for coding.

The sequence of symbols which is to be coded is replaced with special symbols such as (position, length), if the sequence appeared before. For example, in the sequence "ABCPQABCRS", ABC which appeared for the second time agrees with the 3-letter symbol, five symbols preceding, and this sequence is coded as ABCPQ(5,3)RS. LZSS is modified LZ method proposed by Storer and Szymanski in 1982[8]. In LZ77, an uncoded sequence and the tag (position, length) is alternatively appears, while LZSS uses flag bits to identify the tag or uncoded sequence. That is, the above example is coded to a sequence ABCPQ53RS and flag bits 000001100, whose bits corresponding to '53' are set.

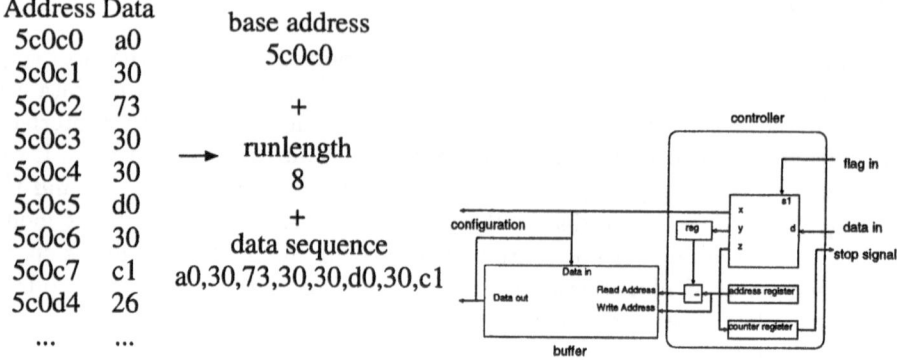

Address	Data
5c0c0	a0
5c0c1	30
5c0c2	73
5c0c3	30
5c0c4	30
5c0c5	d0
5c0c6	30
5c0c7	c1
5c0d4	26
...	...

base address
5c0c0

+

→ runlength
8

+

data sequence
a0,30,73,30,30,d0,30,c1

Fig. 5. Address run-length compression   **Fig. 6.** LZSS-ARC decoder circuit

However, direct use of LZSS is not efficient for the configuration data of DRP, since it consists of pairs of 20bit address and 32bit data. So, we propose to apply a pre-processing before LZSS compression, and call it LZSS with address run-length compression or LZSS-ARC. In this method, byte addressing data sequence is represented with a base address, run-length and a sequence of byte data as shown in Fig. 5. In DRP-1, a byte configuration data is used for each element (ALU, DMU, flip-flop and register files) of PE, and the same configuration is used for the same functions. Thus, the ratio of coded part in the data sequence is increased as well as the data size itself is reduced with this method.

Design of the LZSS-ARC Decoder. At the input stage of decoder, the address run-length compression is decoded, and the sequence of data and flag are transferred to the LZSS-ARC decoder circuit.

In the process of LZSS-ARC decoding, the sequence of symbols which is already decoded is stored in the buffer, and if a special symbol appears, the symbol-sequence of its length is output from the position in the buffer. As shown in Fig. 6, the LZSS-ARC decoder consists of a controller and a symbol buffer. In DRP, Vmem modules, which is 8-bit × 256 dual-port memory, is used as a symbol buffer. The controller provides a counter for counting output symbols and a register used as a pointer of the window.

1. The flag bit indicating whether the data is encoded or not is checked.
2. If input symbol is not encoded, it is directly sent to the output.
3. If a special symbol (position, length) comes, the decoder starts operation and input is suspended. First, the length is set into the counter. Symbols in the buffer are sent to the output and the counter is decremented. At the same time, the encoded symbol itself is also added to the buffer. When the counter becomes zero, the decoding is finished, and the next symbol starts being inputed.

4. The value of the address register increases every time when a new symbol is added to the buffer. When the buffer becomes full, the address register returns to the first position, and the content of the buffer is cleared.

The total circuits are divided into four blocks: Uncoded data input, Matching position input, Matching symbol length, and Output. Each block is enough small to be assigned into a Tile, and they are implemented in four contexts in the same Tile. Since input data must be suspended during decoding operation, an FIFO is required, and another Vmem is used. Thus, this implementation requires two Vmem modules (8bit × 512 in total) and four contexts assigned into the same Tile. DRP-1 provides 16 contexts with 4 × 2 Tiles and 80 Vmem modules, that is, the decode only requires 3% of PEs and 2.5% of Vmem modules. From the critical path analysis, the maximum frequency of the decoder is 200MHz. That is, configuration data loading with 33MHz clock can be done almost without suspending.

4.2 Selective Coding

Coding Method. The loading method with LZSS-ARC decoder requires an internal clock whose frequency which is a multiple of loading data rate. So, it can only accept slow loading clock frequency due to the maximum boundary of the internal clock frequency. We propose another method which can work with the same speed of the loading rate.

As mentioned before, configuration data for DRP includes a lot of similar data patterns. Thus, the following two compression techniques are useful. In both techniques, 20bit address is compressed into 15bit, for uppermost 5bit is the same for a long fixed sequence in DRP.

- Compression directory: A dictionary which enumerates frequent appearing 32bit data pattern referred by 8bit index is generated from statistics of several DRP design examples. As shown in Fig. 7a), 32bit data is compressed into 8bit index if the pattern is included in the directory.
- Data key sorting: When a single data appears many times in the target compress data, corresponding addresses are sorted as shown in Fig. 7b). This technique can be applied with the former technique as shown in Fig. 7c).

In this method, two compression methods are selectively used for the property of the compressed target. Data to which both techniques cannot be used is transferred as it is.

Design of the Decoder for Selective Coding. We designed three decoders for compressed data with Compression directory, Data key sorting, and the combined method. The decoder for Compression directory consists of a simple controller and directory implemented with five Vmem modules. The decoder for Data key sorting is also simple, since configuration data can be written into configuration memory of DRP in any order. Nine contexts are used in total, and

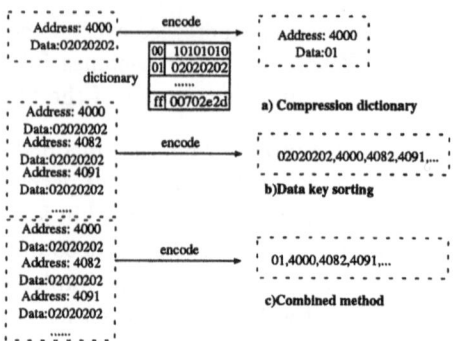

Fig. 7. Selective coding

assigned into the same Tile. Appropriate context corresponding to each technique is selectively applied depending on the coding method, and we call this method Selective coding. This implementation requires 7% of PEs and 6.25% Vmem modules of DRP-1 in total. From the critical path analysis, the maximum frequency of the decoder is 156MHz.

5   Evaluation

5.1   Compression Ratio

The compression data size of several DRP applications with LZSS-ARC and Selective coding are shown in Table 2.

The compression ratio of LZSS-ARC is slightly better than that of Selective coding especially for encryption algorithms in which various kind of functions of DMU/ALU are used. Van del Pol uses a lot of adders, and Wavelet uses a lot of data selector. This is the reason why Selective coding is effective for such designs.

Table 2. Compressed data size

Application	Config. data size	size with LZSS-ARC (ratio)	size with Selective coding (ratio)
Neural Net	78104 bit	36944 bit (47%)	41910 bit (53%)
Van del Pol	48080 bit	31888 bit (66%)	21544 bit (44%)
AES	69296 bit	39352 bit (56%)	48758 bit (70%)
DES	56920 bit	31752 bit (55%)	40629 bit (71%)
MD5	69296 bit	42776 bit (61%)	49663 bit (71%)
Wavelet	116720 bit	76440 bit (65%)	69349 bit (59%)

5.2   Loading Clock Cycles

Fig. 8 shows the required configuration time for loading configuration data, when the 33MHz PCI interface of DRP-1 is used. Note that, the loading of configuration for decoder hardware itself is not included in this figure, since it is only done at initializing the system. The required clock cycles are normalized to the case when the data is not compressed. LZSS-ARC is assumed to run at 198MHz (6 times of 33MHz), and in this case, the stall of the PCI rarely occurs. That is, LZSS-ARC achieves better performance. However, the Selective coding method allows to use the same clock frequency for both internal and loading, thus it can exploit the maximum throughput of the decoder inside. Thus, Selective coding becomes advantageous when loading frequency is increased in the future.

Here, the absolute clock cycles required for DRP configuration are compared with conventional FPGA with fine grain structure. The same circuits for DES encryption is also implemented on Xilinx's Virtex XC2V250[11], and its configuration data is compressed with LZSS. Ths size of data and clock cycles for loading are shown in Table 3.

Although the compression ratio of LZSS-ARC for DRP is worse, the absolute clock cycles are much less than that of LZSS for XC2V250. That is, the configuration speed is almost 80 times of the traditional FPGA with fine grain structure.

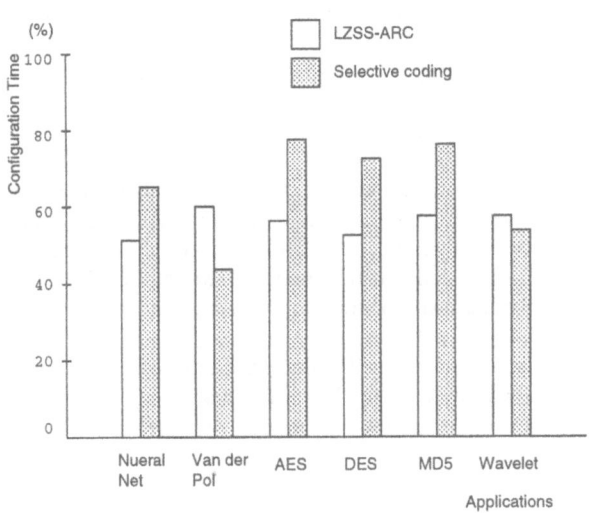

Fig. 8. Configuration time through PCI bus

Table 3. Compressed data size

Device	Config. data size	size with LZSS/LZSS-ARC (ratio)	Clock cycles for loading
DRP	56920 bit	31752 bit (55%)	992 cycles
XC2V250	1697736 bit	662112 bit (39%)	82746 cycles

In this evaluation, only decompression speed is considered, since the compression is done beforehand in the host PC. All compression methods described here require only a few seconds for compression on recent PCs with Pentium III, and it is negligible considering the time required for palce-and-routing stage in the FPGA design.

6   Summary

Two compression methods LZSS-ARC and Selective coding suited for a coarse grain device DRP are proposed and implemented. Using these methods, about 80 times high speed configuration compared with conventional fine grain FPGA is achieved. Applications implemented on DRP sometimes use multiple contexts with similar structure. By using the similarity between contexts, another configuration compression method can be deigned, and it is our future work.

References

1. T. Fujii et. al. "A Dynamically Reconfigurable Logic Engine with a Multi-Context/Multi-Mode Unified-Cell Architecture" Proc. International Solid State Circuit Conference 1999.
2. M.Motomura:"A Dynamically Reconfigurable Processor Architecture," Microprocessor Forum, Oct. 2002.
3. P.Master: "The Age of Adaptive Computing Is Here," Proc. of FCCM, pp.1-3 2002.
4. htpp://www.ipflex.com/.
5. X.-P. Ling, H. Amano, "WASMII: A Data Driven Computer on a Virtual Hardware" Proc. FCCM, pp. 33–42, 1993.
6. E.L.Horta, J.W.Lockwood, D.Partour, "Dynamic Hardware Plugins in an FPGA with Partial Run-time Reconfiguration." Proc. of DAC2002, June, 2002.
7. Jcob Ziv and Abraham Lempel. "A universal algorithm for sequential data compression" IEEE Transactions on Information Theory, IT-23(3):337-343, 1977.
8. J.A.Storer, T.G.Syzmanski, "Data compression via textual substitution," J.of ACM, 29 (4), pp.928-951, 1982.
9. D.William, S.Wilson, S.Hauck, "Runlength Compression Techniques for FPGA Configurations," Proc. FCCM, pp.276-277, 1999.
10. Z.Li, S.Hauck, "Configuration Compression for Virtex FPGA," Proc. FCCM, pp.143-154, 2001.
11. http://www.xilix.com/.

Design Strategies and Modified Descriptions to Optimize Cipher FPGA Implementations: Fast and Compact Results for DES and Triple-DES

Gaël Rouvroy, François-Xavier Standaert,
Jean-Jacques Quisquater, and Jean-Didier Legat

UCL Crypto Group, Laboratoire de Microélectronique,
Université catholique de Louvain,
Place du Levant, 3, B-1348 Louvain-La-Neuve, Belgium,
{rouvroy,standaert,quisquater,legat}@dice.ucl.ac.be

Abstract. In this paper, we propose a new mathematical DES description that allows us to achieve optimized implementations in term of ratio *Throughput/Area*. First, we get an unrolled DES implementation that works at data rates of 21.3 Gbps (333 MHz), using Virtex-II technology. In this design, the plaintext, the key and the mode (encryption/decrytion) can be changed on a cycle-by-cycle basis with no dead cycles. In addition, we also propose sequential DES and triple-DES designs that are currently the most efficient ones in term of resources used as well as in term of throughput. Based on our DES and triple-DES results, we also set up conclusions for optimized FPGA design choices and possible improvement of cipher implementations with a modified structure description.

Keywords: cryptography, DES, FPGA, efficient implementations, design methodology.

1 Introduction

The rapid growth of secure transmission is a critical point nowadays. We have to exchange data securely at very high data rates. Efficient solutions, with huge data rate constraints, have to be hardware implemented and flexible in order to evolve with the permanent changes in norms. FPGA (Field Programmable Gate Arrays) implementations of the triple-Data Encryption Standard (triple-DES) efficiently meet these constraints. Triple-DES is based on three consecutive DES[1]. As detailed in [6–8, 3], DES is very well suited for FPGA solution. Accurately, we have first to focus our attention to achieve very fast DES designs with limited hardware resources.

Some high-speed DES hardware implementations have been published in the literature. These designs unroll the 16 DES rounds and pipeline them. Patterson

---

[1] Without intermediate *IP* and *IP-1* permutations.

P.Y.K. Cheung et al. (Eds.): FPL 2003, LNCS 2778, pp. 181–193, 2003.

[7] made a key-dependent data path for encryption in an FPGA which produces a bitstream of about 12 Gbps. Nevertheless the latency to change keys is tenth of milliseconds. A DES implementation is also downloadable from FreeIP [6] and encrypts at 3.05 Gbps. Last known implementations were announced by Xilinx company in [3, 8]. They were FPGA implementations of a complete unrolled and pipelined DES encryptor/decryptor. The 16-stage and 48-stage pipelined cores could achieve data rates of, respectively, 8.4 Gbps and 12 Gbps[2]. The 48-stage pipelined version could also produce a throughput of 15.1 Gbps on Virtex-II. It also allowed us to change the plaintext, the key and the encryption/decryption mode on a cycle-by-cycle basis.

In this paper, based on our new mathematical DES description [17], we finally get also an optimized complete unrolled and pipelined DES design that encrypts with a data rate of 21.3 Gbps with 37 cycles of latency[3]. This design is the best currently known one in term of ratio *Throughput/Area*.

Concerning sequential designs, some DES and triple-DES FPGA implementations have been published in the literature. These designs encrypt or decrypt most of time every 16 cycles. In academic literature, we mainly found the paper of C. Paar [16] with a DES data rate of 402.7 Mbps. The most recent publication (October, 2001) comes from K. Gaj et al. [15]. They propose a tripe-DES design with a bitstream of 91 Mbps with less resources, using Virtex1000 component. For commercial Xilinx IP cores, different DES and triple-DES implementations are also available from Helion Technology, CAST, and InSilicon [10–14]. The best triple-DES one in term of ratio *Throughput/Area* comes from CAST and gives a throughput of 668 Mbps on Virtex-II FPGA, using 790 slices (January, 2002).

Concerning our DES and triple-DES sequential implementations, we get designs that can encrypt or decrypt every 18 cycles. Our triple-DES design uses 604 slices and produces a throughput of 917 Mbps. This solution is the best sequential one known nowadays in term of resources used as well as in term of throughput.

Based on our DES and triple-DES results, we also set up conclusions for accurate design choices. We also mention the importance of cipher modified structure descriptions.

The paper is organized as follows: section 2 refers the Data Encryption Standard; section 3 explains our previous published new mathematical DES description; section 4 describes our unrolled and pipelined DES implementations and proposes a comparison with the previous Xilinx implementations; section 5 details our sequential DES and triple-DES implementations and also shows comparisons; our design strategies to optimize cipher FPGA implementations are explained in section 6; finally, section 7 concludes this paper.

---

[2] These results were obtained with Virtex-E technology.

[3] Using Virtex-II technology.

2   The DES Algorithm

In 1977, the Data Encryption Standard (DES) algorithm was adopted as a Federal Information Processing Standard for unclassified government communication. It is still largely in use. DES encrypts 64-bit blocks with a 64-bit key, of which only 56 bits are used. The other 8 bits are parity bits for each byte. The algorithm counts 16 rounds. DES is very famous and well known. For detailed information about DES algorithm, refer to the standard [5] or our preceding work [17].

3   New Mathematical DES Description

The original DES description is not optimized for FPGA implementations regarding the speed performance and the number of LUTs used. It is not an efficient way to fit as well as possible into the Virtex CLBs. In order to achieve a powerful design, we explain a new mathematical description of DES previously detailed in our work [17]. The aim is to optimize its implementations on FPGA, reducing resources used as well as the critical path of one DES round.

An FPGA is based on slices composed by two 4-bit input LUTs and two 1-bit registers. Therefore, an optimal way to reduce the LUTs use is to regroup all the logical operations to obtain a minimum number of blocks that take 4-bit input and give 1-bit output. In addition, it is important to mention that all permutation and expansion operations (typically *P, E, IP, IP-1, PC-1* and *PC-2*) do not require additional LUTs, but only wire crossings and fanouts (pure routing).

First, we transform the round function of the enciphering computation. This transformation has no impact on the computed result of the round. Figure 1 shows a modified round representation, where we move the E box and the XOR operation.

This involves the definition of a new function denoted R (like reduction):

$$R = E^{-1},$$
$$\forall x, R(E(x)) = x,$$
$$\exists y \mid E(R(y)) \neq y. \tag{1}$$

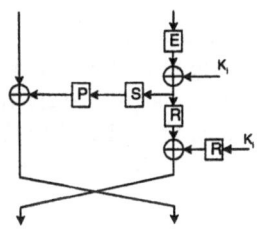

Fig. 1. Modified description of one DES-round.

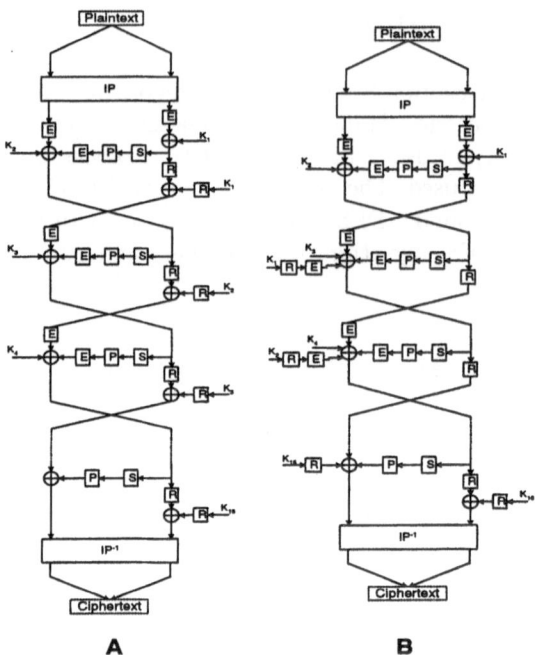

Fig. 2. Two modified descriptions of the DES algorithm.

Now, if we change all the enciphering parts of DES with this modified round function and if we combine the E and XOR block with the left XOR block of the previous round, we get the **A** architecture detailed in Figure 2. Another more efficient solution is to move the R and XOR of the right part of the round into the left XOR operator of the next round. As a result, we obtain the **B** architecture shown in Figure 2.

In the B arrangement, the first and last rounds are quite different from intermediate ones. Therefore, we obtain an irregular DES description. In addition, we increase the number of E and R blocks, which will not alter the number of LUTs consumed. We also keep exactly the same number of S-boxes that is the expensive part of the architecture. Concerning the modulo two sum operators, we really decrease their number. We spare 15×32 2-bit XORs comparing with the A architecture[4]. We can directly conclude that this design will consume less logic than traditional implementations.

Therefore, modifying the original DES description according to the B architecture of Figure 2, we are able to significantly reduce the number of resources used. In addition, we also reduce the critical path of one round function. These transformations allow us to obtain very efficient DES FPGA implementations as detailed in the next sections.

---

[4] The design exactly counts 14×48 4-bit XORs, 1×48 3-bit XORs, 1×48 2-bit XORs, 1×32 3-bit XORs and 1×32 2-bit XORs.

4   Our Unrolled and Pipelined DES Implementations

To be speed efficient, we propose designs that unroll the 16 DES rounds and pipeline them, based on the B mathematical description of Figure 2. In addition, we implemented solutions that allow us to change the plaintext, the key and the encryption/decryption mode on a cycle-by-cycle basis, with no dead cycle. We can achieve very high data rates of encryption/decryption.

The left part of Figure 3 illustrates how the critical path, in our solution, is hugely decreased. We only keep one S-boxes operator and one XOR function (= two LUTs + one $F5$ function + one $F6$ function)[5]. With this solution we obtain a 1-stage pipeline per round. Due to the irregular structure of our design, we have to add an additional stage in the first round. To be speed efficient for implementation constraints, we also put a 2-stage pipeline respectively in the input and in the output. This approach allows an additional degree of freedom for the place and route tool. As mentioned in the figure, first and last registers are packed into IOBs. Therefore, we obtain a 21-stage pipeline.

In the right part of Figure 3, we put an extra pipelined stage in each round in order to limit the critical path to only one S-box. As a consequence, we finally get a 37-stage pipeline. Table 1 shows our 21-stage and 37-stage pipelined results.

Comparing to the preceding efficient Xilinx implementations [3,8], our 21-stage gives better results in terms of speed and logical resources. Nevertheless, it is quite more registers consuming. For the 37-stage pipeline, we use less LUTs again. We also reduce the registers needed. This is due to the fact that we only have a 2-stage pipeline per round. In addition, this design uses shift registers for the key schedule calculation. In Virtex FPGAs, *SRL16* blocks can directly

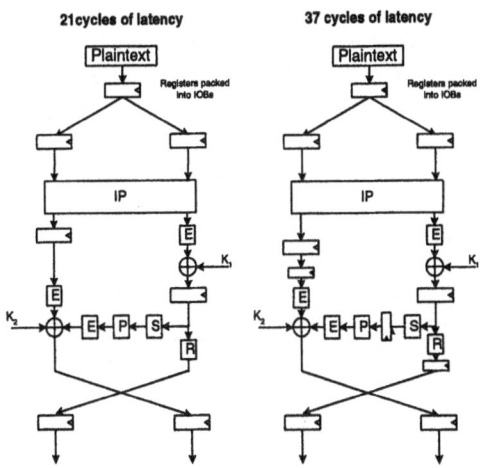

Fig. 3. Our two unrolled and pipelined DES designs.

---

[5] E and R operators do not increase the critical path.

Table 1. Final results of our pipelined implementations.

Pipeline	Xil. 16-stage	Xil. 48-stage	Our 21-stage	Our 37-stage
LUTs used	4216	4216	**3775**	**3775**
Registers used	1943	5573	**2904**	**4387**
Frequency in XCV300-6	100 MHz	158 MHz	**127 MHz**	**175 MHz**
Data rate in XCV300-6	6.4 Gbps	10.1 Gbps	**8.1 Gbps**	**11.2 Gbps**
Frequency in XCV300E-6	132 MHz	189 MHz	**176 MHz**	**258 MHz**
Data rate in XCV300E-6	8.4 Gbps	12.0 Gbps	**11.2 Gbps**	**16.5 Gbps**
Frequency in XC2V1000-5	/	237 MHz	**227 MHz**	**333 MHz**
Data rate in XC2V1000-5	/	15.1 Gbps	**14.5 Gbps**	**21.3 Gbps**

implement a 16-bit shift register into one LUT. This design uses 484 extra LUTs for the key schedule calculation[6].

The reason why we have a better speed result for the 37-stage pipeline is quite strange. Obviously, in their design, they do not put registers into IOBs and an additional pipelined stage before and after encryption. Without such registers, the critical path is in the input and output paths.

In addition, after checking and simulating their available source code on the web, we found two errors. First, they forgot to put a 1-stage pipeline after the XOR between key and R part. Actually, Xilinx implemented this 1-stage pipeline but they sent the XOR directly between key and R part into S-boxes, instead of the corresponding registered value. They also forgot to register the key just before the XOR function. Therefore, their critical path is quite longer. Finally, their solutions do not implement a correct DES that can encrypt every cycle.

To conclude, we propose two efficient and different solutions in terms of space and data rate. Depending on environment constraints, we really believe that one of the two designs should be well appropriate. Especially for high ratio *Throughput/Area*, the 37-stage pipelined solution is very efficient in terms of speed and slices used. Table 2 compares ratio *Throughput/Area* between 48-stage Xilinx implementation [3, 8] and our 37-stage implementation. We directly see the significative improvement.

Table 2. Comparisons with Xilinx implementation on Virtex-II-5

DES Version	Xil. 48-stage	Our 37-stage
Slices used	3900	**2965**
Data rate	15.1 Gbps	**21.3 Gbps**
Throughput/Area (Mbps/slices)	3.87	**7.18**

---

[6] Estimation of the corresponding Xilinx implementation gives the use of about 900 LUTs for shift registers. No accurate values are given in [3, 8].

5   Our Sequential DES and Triple-DES Implementations

To be space efficient, we propose one DES and triple-DES sequential design, based on the B mathematical description of Figure 2. Instead of keeping a critical path of one LUT (as the best one of our pipelined designs), we prefer to limit the critical path to two LUTs (+ one *F5* function + one *F6* function). Indeed, if we limit the critical path to one LUT, we will encounter some critical problems with the routing part of the control signals. For example, to select the input data of a round with MUX functions, we will need 64 identical control signals. An unique signal will be generated with the control part. Then, this signal will be extended to all MUX operators. The propagation delay, of this high fanout signal, will approximatively correspond to one LUT delay. Therefore, we will have an critical path of about two LUTs. So, our sequential DES and triple-DES are designed to have a critical path of two LUTs.

First, we obtain one DES design that encrypts/decrypts, with possible different keys, every 18 cycles with no dead cycles. Our solution proposes almost the same interface as commercial cores [10, 12, 14]. Indeed, they propose modules with 16 cycles for encryption/decryption. Nevertheless, we get better throughput because of higher work frequency.

Figure 4 explains the key scheduling part for the computation of K1 and K2, needed in the data path. The selection of the input key is carried out with XOR operations. To control the correct functionality, all registers have to be cleared in the right cycles. The *reset_regKIN* and *reset_K1* signals ensure this constraint.

As shown on the B scheme of Figure 2, due to the irregular new DES description, we also need additional XOR operators in the first and last round. To carry out these operations, we use the data path part where we clear the output regis-

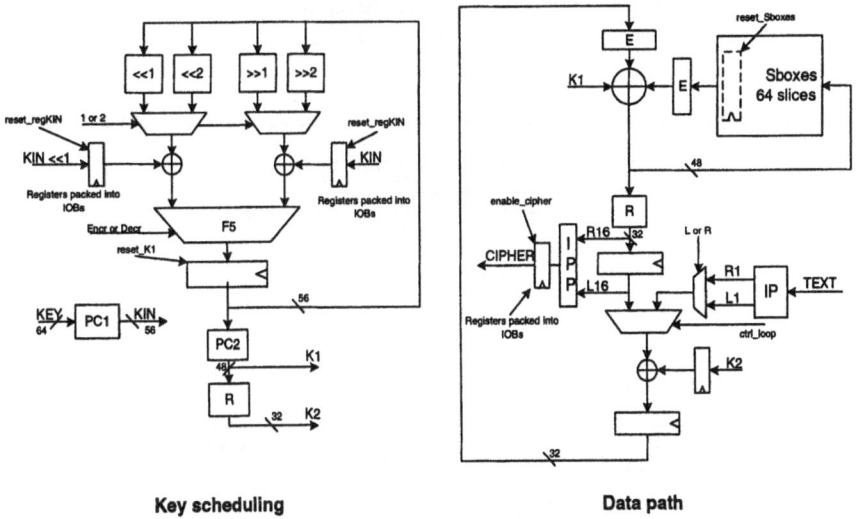

Key scheduling   **Data path**

Fig. 4. Our sequential DES encryptor/decryptor.

ters of S-boxes, in the first and 18th cycles. This is done using the *reset_Sboxes* signal. If we decide to spare these XOR operators, we loose 2 cycles for the encryption/decryption. This is why we get a 18-cycle encryptor/decryptor.

In addition, we mainly focus our attention to maximize the utilization of resources in a slice. For example, we prefer to register the S-boxes outputs instead of other signals because it does not consume additional slices. If we do not this, using these slice registers for other signals will not allowed. Therefore, we will waste these registers. To manage all the signals, we develop a control unit that produces registered signals.

On the other hand, we also obtain one triple-DES design that encrypts/decrypts, with possible different keys, every 18 cycles with no dead cycles. This circuit is obtained using three of our DES sequential implementations. Our solution proposes almost the same interface as commercial cores [10, 11, 13]. Indeed, they mainly propose modules with 16 cycles for encryption/decryption. Nevertheless, we get better throughput because of higher work frequency.

Table 3 summarizes our result for DES and triple-DES sequential designs. In table 4 and 5, we compare our sequential DES and triple-DES with other existing implementations, in term of ratio *Throughput/Area*.

Table 3. Final results of our sequential DES and triple-DES implementations.

Sequential	DES	triple-DES
LUTs used	365	1103
Registers used	202	672
Slices used	189	604
Latency (cycles)	20	58
Output every (cycles)	1/18	1/18
Freq. in XCV300E-6	176 MHz	165 MHz
Freq. in XC2V1000-5	274 MHz	258 MHz

Table 4. Comparisons with other sequential DES implementations on Virtex-II -5.

Sequential DES	CAST	Helion	Ours
Slices used	238	$\simeq 450$	189
Throughput (Mbps)	816	640	974
Throughput/Area (Mbps/slices)	3.43	1.42	5.15

Table 5. Comparisons with other sequential triple-DES implementations on Virtex-II-5.

Sequential 3-DES	CAST	Gaj et al.	Ours
Slices used	790	614	604
Throughput (Mbps)	668	91 (XCV1000)	917
Throughput/Area (Mbps/slices)	0.85	0.15	1.51

6 Design Strategies to Optimize Cipher FPGA Implementations

FPGAs allow computing in parallel with high work frequencies. Nevertheless, wrong uses of their CLB structure can dramatically increase the speed efficiency as well as the resources used. Accurate slice mapping will permit to get fast and compact designs.

In this section, we set up conclusions for accurate cipher FPGA[7] designs. Our conclusions are directly based on our DES and triple-DES implementations. We also mention the possible improvement of cipher implementations with a modified structure description. We propose systematic design rules to achieve very fast and compact cipher FPGA implementations, depending on the environment constraints. The following methodology is proposed:

1. **Analyze and Modify the Mathematical Cipher Description:**
   Most block ciphers have a regular structure with a number of identical repeated round functions, depending on different subkeys. Subkeys are also generated using a number of identical repeated keyround functions. This regular and repeated round structure allows achieving efficient FPGA designs. Regular cipher structures permit very suitable sequential designs. Nevertheless, original cipher descriptions are not always optimized regarding the speed performance and number of slices used. An FPGA is based on slices composed by two 4-bit input LUTs and two 1-bit registers. Therefore, an optimal way to reduce the LUTs used is to regroup all the logical operations to obtain a minimum number of blocks that take 4-bit input[8] and give 1-bit output.
   Therefore, the aim of this step is to reduce the number of LUTs used as much as possible, modifying the structure description without changing the mathematical functionality. The following methodology has to be applied:
   (a) In round and keyround descriptions, try to move and/or inverse and/or merge blocks as well as possible.
   (b) Ignore round and keyround descriptions, and try to regroup blocks from different rounds. Try to keep a regular structure as much as possible for sequential designs.
   If no improvements are possible, go to the next point.

2. **Choose Your Design Based on Data Rate Constraints:**
   Depending on the data rate constraints, adopt one of the following designs:
   (a) For more than ± 10 Gbps constraints, choose a complete unrolled and pipelined implementation.
   (b) For less than ± 1 Gbps constraints, choose a sequential design.
   (c) For other data rate constraints, choose an uncomplete unrolled and pipelined design or multiple sequential designs.

---

[7] For Virtex technologies.

[8] Sometimes 5-bit input.

3. **Fit as Well as Possible into CLB Slices:**
   (a) **For a complete unrolled and pipelined design,** it is important to achieve huge data rates. Therefore, we will adopt designs where we limit the critical path inside one slice (one LUT + one $F5$ function + one $F6$ function at most). Left part of Figure 6 shows some typical slice uses for one slice critical path designs.
   (b) **For sequential design,** it is important to achieve compact design. Due to control signals, it is better to limit the critical path to two LUTs (+ one $F5$ function + one $F6$ function). Indeed, if we limit the critical path to one LUT, we will encounter some critical problems with the routing part of the control signals. For example, to select the input data of a round with MUX functions, we will need β^9 identical control signals. A unique signal will be generated with the control part. Then, this signal will be extended to all MUX operators. The propagation delay, of this high fanout signal, will approximatively correspond to one or more LUT delay.

   Therefore, we recommend to limit the control signals to one LUT and routing. The routing part generates additional delays. With an average 60-fanout signal, the delay corresponds to more and less one LUT delay. For the data path part, if we do not use fanout more than three, we can limit the critical path to two LUTs. Figure 5 summarizes our advices.

   Figure 6 shows some typical slice uses for two slice critical path designs. We show how to use slice flip flops to register input signals. Figure 6 does not illustrate all the slice configuration possibilities. It is again important to mention that we need to describe the VHDL code at a slice level in order to avoid bad slice mapping and extra slice uses.
   (c) **For an uncomplete unrolled and pipelined design or multiple sequential designs,** previous advices are always valid depending on the pipelined or sequential choice.

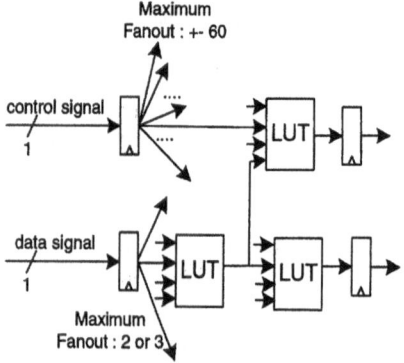

Fig. 5. Recommended architecture for sequential designs.

---

[9] Typically β equals 64, 128 or 256.

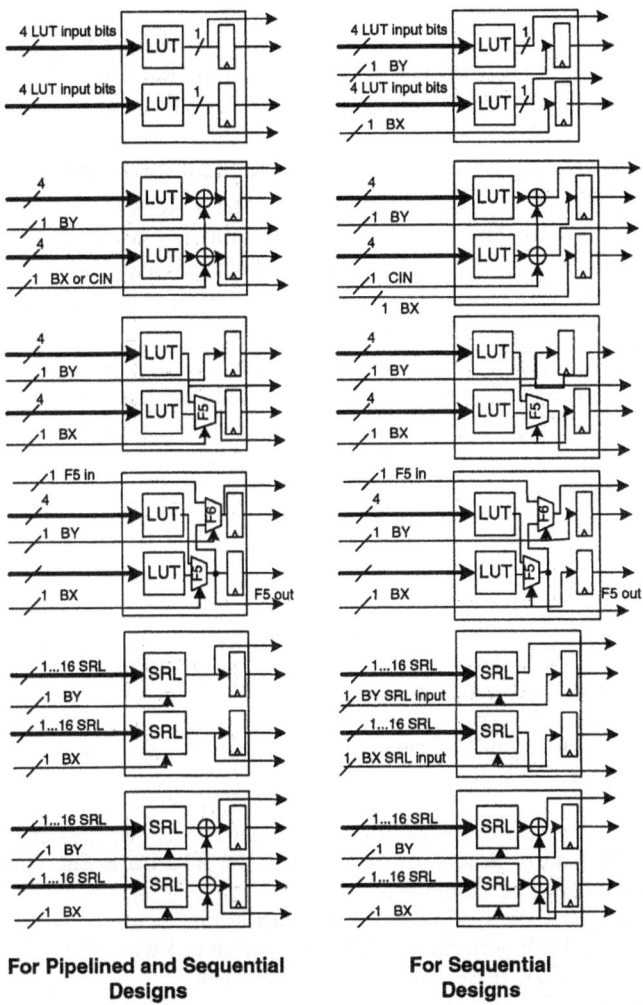

Fig. 6. Slice uses for different designs.

4. Deal with Input and Output Buffers:

To achieve good frequency results, input and output registers have to be packed into IOBs. In addition, we also recommend to put one more register stage in the input and output paths. This approach allows an additional degree of freedom for the place and route tool.

The previous explained rules set up a methodology to implement cipher algorithms. We do not affirm that this methodology is the most optimized systematic method. Nevertheless, it gives efficient results for DES and triple-DES algorithms.

7 Conclusion

We propose a new mathematical DES description that allows us to achieve optimized FPGA implementations. We get one complete unrolled and pipelined DES, and sequential DES and tripe-DES implementations. Finally, we get the best DES and triple-DES FPGA implementations known nowadays in term of ratio *Throughput/Area*. The fastest DES gives a data rate of 21.3 Gbps (333 MHz). In this design, the plaintext, the key and the mode (encryption or decrytion) can be changed on a cycle-by-cycle basis with no dead cycles. Our sequential triple-DES computes ciphertexts every $\frac{1}{18}$ cycle and produces a data rate of 917 Mbps, using 604 slices.

Based on our DES and triple-DES results, we also set up conclusions for optimized cipher FPGA designs. We recommend to modify the mathematical cipher description. Depending on our data rate constraints, we advice to choose unrolled and pipelined designs or sequential designs. We also show how to fit in slices as well as possible and how to deal efficiently with input and output buffers.

References

1. J.M. Rabaey. *Digital Integrated Circuits*. Prentice Hall, 1996.
2. Xilinx. Virtex 2.5V field programmable gate arrays data sheet. available from http://www.xilinx.com.
3. Xilinx, V. Pasham and S. Trimberger. High-Speed DES and Triple DES Encryptor/Decryptor. available from http://www.xilinx.com/xapp/xapp270.pdf, Aug 2001.
4. B. Schneier. *Applied Cryptography*. John Wiley & Sons, Inc., second edition, 1996.
5. National Bureau of Standards. *FIPS PUB 46*, The Data Encryption Standard. U.S. Departement of Commerce, Jan 1977.
6. FreeIP. http://www.free-ip.com/DES/index.html.
7. C. Patterson. High performance DES encryption in Virtex FPGAs using Jbits. In *Proc. of FCCM'01*, IEEE Computer Society, 2000.
8. S. Trimberger, R. Pang and A. Singh. A 12 Gbps DES encryptor/decryptor core in an FPGA. In *Proc. of CHES'00*, LNCS, pages 156–163. Springer, 2000.
9. M. Davio, Y. Desmedt, M. Fosséprez, R. Govaerts, J. Hulsbosch, P. Neutjens,P. Piret, J.J. Quisquater, J. Vandewalle and P. Wouters. Analytical Characteristics of the DES. In David Chaum, editor, *Advances in Cryptology - Crypto '83*, pages 171–202, Berlin, 1983. Springer-Verlag.
10. Helion Technology. High Performance DES and Triple-DES Core for XILINX FPGA. available from http://www.heliontech.com.
11. CAST, Inc. Triple DES Encryption Core. available from http://www.cast-inc.com.
12. CAST, Inc. DES Encryption Core. available from http://www.cast-inc.com.
13. inSilicon. X_3 DES Triple DES Cryptoprocessor. available from http://www.insilicon.com.
14. inSilicon. X_DES Cryptoprocessor. available from http://www.insilicon.com.
15. P.Chodowiec, K. Gaj, P. Bellows and B. Schott. Experimental Testing of the Gigabit IPSec-Compliant Implementations of RIJNDAEL and Triple DES Using SLAAC-1V FPGA Accelerator Board. In *Proc. of ISC 2001: Information Security Workshop*, LNCS 2200, pp.220-234, Springer-Verlag.

16. J.P. Kaps and C. Paar. Fast DES Implementations for FPGAs and Its Application to a Universal Key-Search Machine. In *Proc. of SAC'98 : Selected Areas in Cryptography*,LNCS 1556, pp. 234-247, Springer-Verlag.
17. G. Rouvroy, FX. Standaert, JJ. Quisquater, JD. Legat. Efficient Uses of FPGA's for Implementations of DES and its Experimental Linear Cryptanalysis. Accepted for publication on April 2003 in IEEE Transactions on Computers, Special CHES Edition.

Using Partial Reconfiguration in Cryptographic Applications: An Implementation of the IDEA Algorithm

Ivan Gonzalez, Sergio Lopez-Buedo, Francisco J. Gomez, and Javier Martinez

Escuela Politecnica Superior, Universidad Autonoma de Madrid,
Cantoblanco E-28049
Madrid, Spain
{Ivan.Gonzalez, Sergio.Lopez-Buedo, Francisco.Gomez,
Javier.Martinez}@uam.es

Abstract. This paper shows that partial reconfiguration can notably improve the area and throughput of symmetric cryptographic algorithms implemented in FPGAs. In most applications the keys are fixed during a cipher session, so that several blocks, like module adders or multipliers, can be substituted for their constant-operand equivalents. These counterparts not only are faster, but also use significantly less resources. In this approach, the changes in the key are performed through a partial reconfiguration that modifies the constants. The International Data Encryption Algorithm (IDEA) has been selected as a case-study, and JBits has been chosen as the tool for performing the partial reconfiguration. The implementation occupies an 87% of a Virtex XCV600 and achieves a throughput of 8.3 GBits/sec.

1 Introduction

Most operations in the IDEA algorithm [1-4] involve data coming from both the plaintext and the key. This is not a peculiarity of IDEA: most symmetric algorithms work in the same way, like DES [5,6] and Rijndael [7]. Thus, if the key were fixed, then a major portion of the algorithm operations will be by constant. This looks very appealing, as it will imply a benefit in both area and performance. It is widely known that the operations by constant not only are faster, but also occupy less area [8,9].

Unfortunately, a ciphering hardware that has a fixed key does not make too much sense, because it would offer a poor security. However, the key remains constant during the ciphering sessions, which are relatively long periods. For example, in a Virtual Private Network connection, once the encryption key has been negotiated, it is maintained during all the session. Then, in an FPGA environment, one could think of using a fixed-key hardware that would be reconfigured each time the key changes. This will be the optimal solution, because it will take advantage of the benefits of a constant-key hardware, but anyway the key can be changed by the user. This is right as long as the reconfiguration is fast enough not to stop the ciphering for an unacceptable amount of time. Therefore, the hardware should be designed in such a way that

P.Y.K. Cheung et al. (Eds.): FPL 2003, LNCS 2778, pp. 194–203, 2003.

the key can be modified just by altering the contents of a few LUTs, so the changes can be done fast and easily. This approach has been followed in this paper, using JBits for performing the reconfiguration. As it is detailed in the results, it provides a significant improvement in both area and performance when compared to a conventional (variable key) solution.

2 The IDEA Algorithm

IDEA is a symmetric algorithm [1-4], that is, the same key is used for both encryption and decryption. IDEA encrypts 64-bit blocks with a 128-bit key, using three simple operations: or-exclusive, module 2^{16} addition and module $2^{16}+1$ multiplication.

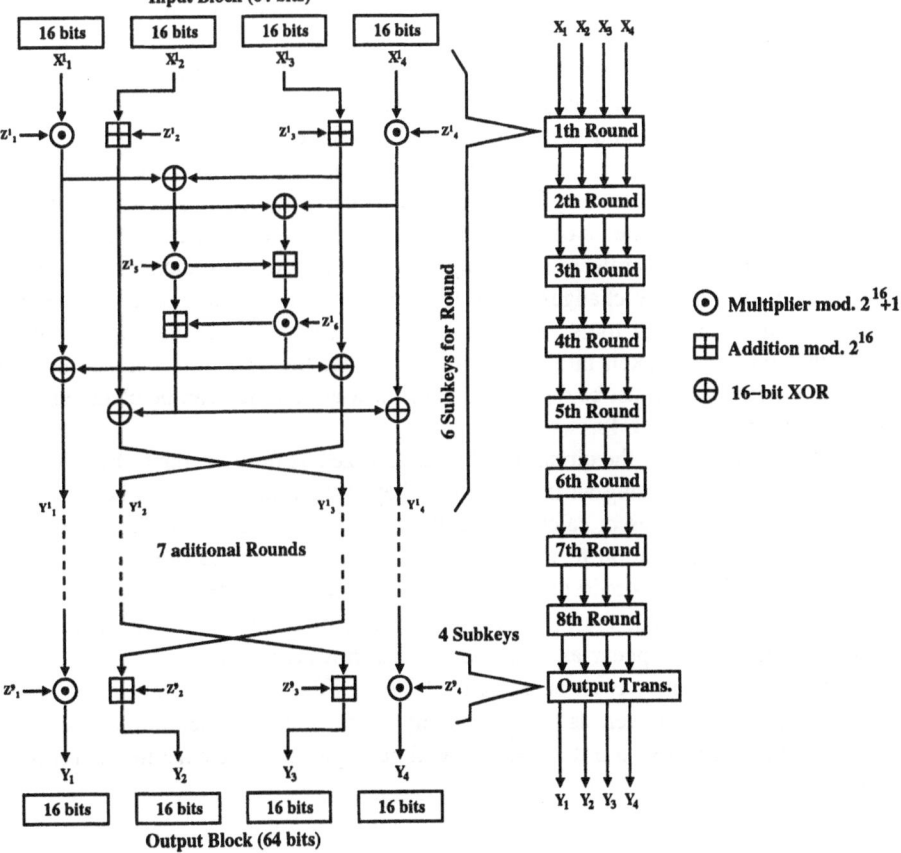

Fig. 1. The IDEA Algorithm.

As it can be seen in Fig.1, the input data passes through 8 similar stages, called rounds, and there is a last one, called output transformation. The only difference in the rounds is the 16-bit subkeys, noted in the figure as Z_j^i. The first 8 subkeys are obtained directly from the original 128-bit key. Then, this key is rotated 25 times to the left to obtain the second set of 8 subkeys, and so on, till all the 52 subkeys are created. That is, six subkeys per stage plus four subkeys for the last transformation. This process needs to be done only when the key is changed. Finally, the only difference between encryption and decryption is the subkey generation; the same datapath depicted in Fig. 1 is used for both actions.

3   Design Methodology

The approach taken in this paper for implementing IDEA is replacing all the operational units involving the key with its constant-operand equivalents; as Fig. 1 shows, these units are the module 2^{16} addition and the module $2^{16}+1$ multiplication. The benefits from this methodology come from two sides. First, as it was stated in the introduction, arithmetic or logic operators by constant are faster and occupy less area, and this is especially true in FPGAs. And second, the subkey generation hardware is eliminated [4,10], because this operation is done in an associated microprocessor which calculates the new constants and reconfigures the FPGA.

For this methodology to be useful, the modifications to be done in the circuit when the key is changed must be easy to perform. This certainly implies that the changes will involve only the contents of the LUTs that build up the operators. Changing the routing or adding new hardware is a complex task, and few tools support it. Another advantage of limiting the changes to the LUT contents is that the circuit can be created using the traditional tools; in this case, it was created using VHDL. The only aspect that has to be taken into consideration is that the constant-operand module adders and multipliers must be designed in such a way that they can be changed only by reconfiguring a few LUTs.

Thus, the operation of the circuit can be summarized as follows. First, the FPGA is loaded with the IDEA using an initial session key. When the key has to be changed, an associated microprocessor calculates all the subkeys and the new LUT values, and sends the new configuration to the FPGA. Here, this process has been done using JBits, but any other tool could have been used, or even the bitstream could have directly been altered, using the information provided by Xilinx [11]. The need of a microprocessor might appear as a drawback. But nowadays, nearly all systems include at least one. It may even be embedded on the FPGA, so no additional devices would be needed. Moreover, as the complexity of the subkey generation and the reconfiguration is very low, a little overhead will be imposed to the existing microprocessor.

3.1   JBits

Basically, JBits [12,13] is a Java API (application program interface) that allows the designer to describe digital circuits using Java. Currently, it is available only for Virtex FPGAs. The main advantage of JBits is that the designs can be parameterised at run-time. That is, the circuit is generated by running a Java program, which at run-time takes the decisions in order to adapt the generated circuit to the current circumstances: area available, device present... This methodology is called run-time parameterizable (RTP) cores [14].

Another advantage of JBits, closely related to the previous one, is its support for partial reconfiguration [15]. The interface for handling it is straightforward; only a few lines of code are necessary to load a new bitstream and reconfigure the FPGA. JBits automatically infers the differences between them current bitstream and the new one, and only reconfigures the frames that have changed. That is the main reason why JBits has been selected in this paper.

3.2 Addition Module 2^{16}

In this case there is not a significant improvement regarding the non-constant operand version. Only the register needed for storing the subkey is saved. The interest of the proposed solution is to give an example of how to avoid undesirable optimizations that could be carried out by the synthesis tool.

The basic block of a generic n-bit adder is a two bits adder with 1 bit carry. A structural description in VHDL using the primitives of the Virtex slice was developed. In this implementation the XOR operation to add the two input bits is mapped into a 2 input LUT.

The version with a constant-operand is obtained by transforming the previous LUT into an 1 input LUT. The LUT operation inverts the input when the bit of the constant is 1 or it transfers the input to the output when the bit of the constant is zero. As the value of the constant may be modified at execution time by partial reconfiguration, this LUT should be configured by default like an inverter, avoiding undesirable optimization, and later proceed to change its functionality by means of JBits, like it is shown in the following code:

```
int[] f_zero = Expr.F_LUT("~F1");
int[] f_one = Expr.F_LUT("F1");

if(((constant >> bit)&0x1) == 0)
    jbits.set(row,column,LUT.SLICE0_F,f_zero);
else
    jbits.set(row,column,LUT.SLICE0_F,f_one);
```

3.3 Multiplier Module $2^{16}+1$

A constant (k) coefficient multiplier (KCM) [8,9] has been considered because its size and layout are independent of the constant encoded within it, and the mechanism for implementing the multiplication is contained in look-up tables. The block diagram of a 16-bit KCM multiplier is shown in figure 2.

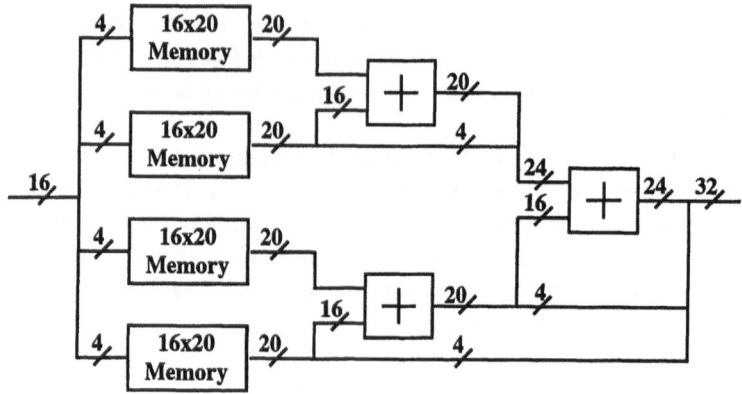

Fig. 2. A 16-bit KCM Multiplier.

The 16-bit version of a KCM holds 4 copies of the constant (k) multiplication table, each of which contains 16 entries. In a KCM implemented in Virtex, when multiplying an n-bit operand by a constant, n/4 copies of the multiplication table are required (because the LUTs have 4 inputs). Each group of n/4 bits of the operand is used to address a table, and the partial products thus obtained are added to get the product. This KCM multiplier is between 3 and 3.8 [8] times smaller than a generic one.

It can be seen in figure 2 that the overall structure of the KCM is independent of the constant encoded within it. Only the contents of the look-up tables vary, and therefore they have to be reconfigured when the constant is changed. A first approach would be implementing these memories as LUT ROMs. However, there is a potential problem: the place and route tool might swap the address lines to simplify the routing. This can be solved by using the SRL16 primitive [9] instead of a ROM. This primitive corresponds to a shift register, not to a memory. But if the write enable is tied low, and the contents of the register are initialized correctly, it works in the same way as memory. But, as it is not strictly a memory, the routing tool can not swap the input lines.

The calculation of the module $2^{16}+1$ is based in the Low-High algorithm:

```
uint16 modmult(uint16 operand) {
    word32 p;
    uint16 c, d;

    if(k) {
        if(operand) {
            p = (word32)k * operand;
            c = low16(p);
            d = p >> 16;
            return c - d + (c < d);
        }
        else return 1 - k;
    } else return 1 - operand;
}
```

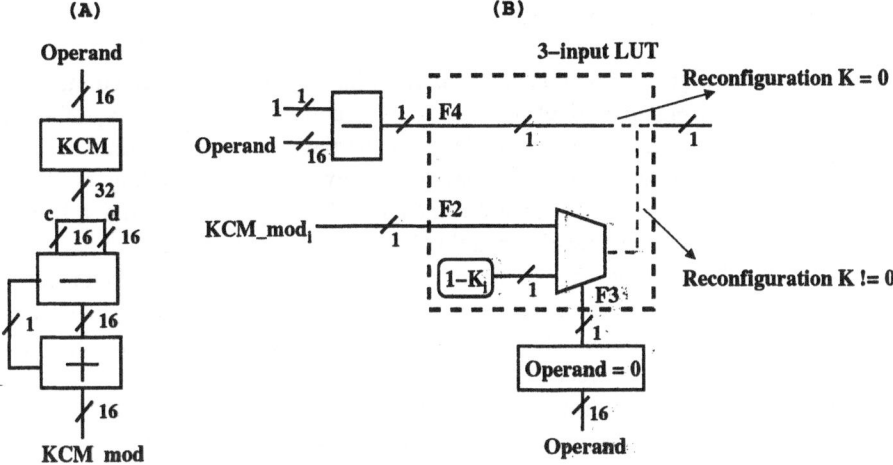

Fig. 3. Module $2^{16}+1$ multiplier. (A) KCM module $2^{16}+1$ with non zero operands. (B) Output selector.

An adaptation of this algorithm for reconfigurable hardware is shown in figure 3. As it can be seen in Fig. 3A, the output carry of the 'c-d' subtraction is used to perform the 'c<d' comparison.

An output selector (Fig. 3B) using LUT reconfiguration has implemented the multiple-choice selection in the Low-High algorithm. At configuration time, there are two possibilities to be considered according to the constant value. When the constant is zero the LUT works as a buffer that translate their input F4 to the output. In this case the LUT input F4 is connected to the result of the (1-Operand)$_i$ subtraction.

Otherwise, with a non-zero constant, the (1-K)$_i$ value are hardwired into the LUT and also, the LUT is configured as a multiplexer. When the operand is zero the output is (1-K)$_i$, otherwise the value of the KCM_mod block connected to the input LUT (F2).

This design also presents the problem that the LUT inputs might be swapped by the routing tool, and it should be implemented using SRL16 components.

The JBits code for this reconfigurable output selector is the following:

```
int aux = (1-constant)&0xFFFF;
int[] f_operand = Expr.F_LUT("~F4");
int[] f_function1 = Expr.F_LUT("~(~F3 | (F3 & F2))");
int[] f_function0 = Expr.F_LUT("~(F3 & F2)");

if(constant == 0)
    jbits.set(row,column,LUT.SLICE0_F,f_operand);
else if(((aux >> bit)&0x1) == 0)
    jbits.set(row,column,LUT.SLICE0_F,f_function0);
else
    jbits.set(row,column,LUT.SLICE0_F,f_function1);
```

3.4 Relative Location Constrains

In order to reconfigure a component, JBits must know where it is located inside the FPGA. Using placement attributes in the VHDL code, and particularly, LOC and RLOC, this problem is resolved completely or partially. LOC establishes the fixed position of each component, so it is possible to know a-priori where they are found, but it is a hard constraint for the place&route tool. RLOC is a soft constraint, because it permits the user to give a relative position between the basic elements of each component. Then, the tool can choose the component position. Generally, better results are obtained because RLOC is more flexible than LOC attribute. The drawback is that, after the implementation, the reports have to be examined to find the position of the different components to reconfigure.

These relative location attributes are a slice level constrains. If the two LUTs of the slice are used, the BEL attribute indicates which element goes in the F LUT and which one in the G LUT.

All reconfigurable components have been relatively placed in columns using RLOC. This location constrain decreases the number of frames that are necessary to reconfigure during the process, therefore minimizing the reconfiguration time.

4 Pipelined Algorithm

Pipelining is necessary to increase the throughput. In the IDEA case it is straightforward, because if Electronic CodeBook (ECB) chaining mode of operation is employed, the datapath has no feedbacks (Fig. 1). In a first approach, the segmentation registers will be added after every basic operation. Then, the clock period will be limited by the delay of the most complex operator, the module $2^{16}+1$ multiplier. But the throughput attainable using this degree of pipelining is fairly poor, because of the complexity of the multiplier. Thus, in the final design the multiplier has been pipelined again.

The only drawback of pipelining ECB-mode IDEA is that it causes a large area increment because has many nets which cross the segmentation boundaries. In FPGAs, pipelining after a logic operator does not generally cause an area increment, because there are as many LUTs as flip-flops. But if the pipelining crosses a net, the flip-flops have to be obtained from new CLBs, which probably will have their LUTs unused.

4.1 Pipelined Module $2^{16}+1$ Multiplier

The module multiplier is the most complex component of IDEA, having a significantly greater delay than the other operations. In order to reduce this delay, it has been segmented in 4 stages. The first three stages correspond to the pipelining of the KCM multiplier, adding registers after the 16x20 memories and the two adder stages (Fig. 4A). This segmentation affects directly to the rest of the module $2^{16}+1$ multiplier, so it is necessary to spread it to the other blocks (Fig. 4B). A fourth stage is

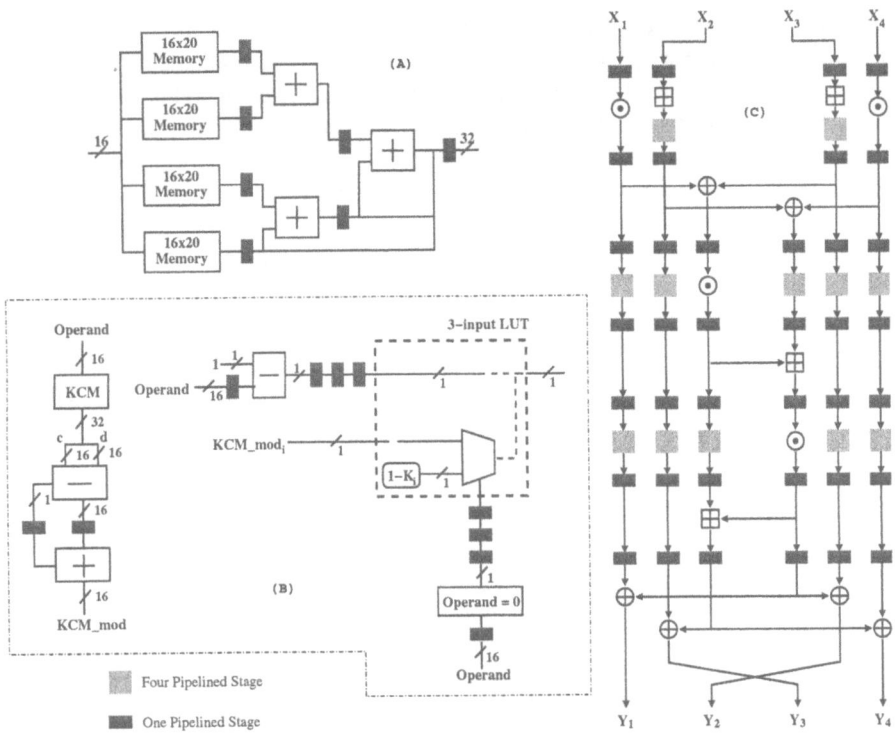

Fig. 4. (A) Pipelined KCM. (B) Pipelined Module Multiplier (C). Pipelined IDEA Round.

added before the adder generating KCM_mod. Finally, another segmentation stage is added after the multiplier (Fig. 4C).

4.2 Final Design for the Pipelined IDEA

The final design of the IDEA algorithm uses 19 pipelined stages per round and 6 pipelined stages for the output transformation. It implies a total of $(8\times19)+6 = 158$ pipelined stages. So 158 clock cycles are necessary for a 64-bit input plaintext block to arrive at the output. This latency is very high for applications that encrypt/decrypt small quantities of data, but usually the cryptographic algorithms work on a stream of bytes, so it represents no problem.

5 Results

The fully pipelined IDEA algorithm uses only an 87% of a XCV600 and works at 131 MHz, obtaining a remarkable throughput of 8.3 GBits/sec. This result is compared in table 1 with several implementations. By comparing with the alternative pipelined at module multiplier level [16], this implementation shows an area reduction of 18%, while the throughput is five times better. The best FPGA implementation of IDEA that has been referenced [2] has a throughput of 6.8 GBits/sec. This throughput has been improved in the results obtained in this paper, and the area used is smaller, nearly 33% less. The implementation presented in [4] also uses reconfiguration, but it is not comparable with the solution offered in this paper, because it is not a fully unrolled implementation of IDEA.

Table 1. Results.

Implementation	Slices	Device	Frequency	Throughput	Latency
This paper	6078	87% XCV600-6	131.1 MHz	8.3 GBits/sec	1.20 us
Deeply Pipelined [2]	9052	73% XCV1000-6	105.9 MHz	6.8 GBits/sec	1.24 us
Cheung et al. [4] 2 round cores	2444	79% XCV300-6	82 MHz	1.2 GBits/sec	2,13 us
Pipelined [16]	7410	60% XCV1000-6	24.5 MHz	1.5 GBits/sec	2.37 us
Combinational [16]	6863	55% XCV1000-6	1.4 MHz	0.8 GBits/sec	0.71 us

6 Conclusions

The results show that using key replacement through partial reconfiguration can significantly improve the throughput and area of FPGA implementations of cryptographic algorithms. The selected case-study, fully unrolled IDEA implementation, using other design methodologies had to be implemented on a XCV1000 [2,16]. But if this approach is employed, it fits in a XCV600 because it saves a 33% in area. This method not only provides area savings: the throughput is improved by 23% in comparison to the deeply pipelined solution. The area comparison is not so favourable in less pipelined implementations (only 18% less), but in this case the throughput is much higher, more than 500%. Although the key can only be changed using reconfiguration, this can not considered a problem, because the keys remain constant during the ciphering sessions, and the partial reconfiguration can be performed very fast, in less than 4 ms for the example used in this paper. Finally, the design was made with the standard design tools and VHDL. By employing a set of techniques which have been described in this paper, the encryption key can be changed by altering the contents of only a few LUTs. This has successfully been done using JBits.

Acknowledgement

This work has been partially supported by the Spanish Government under project number TIC2000-0464.

References

1. J. L. Beuchat, J.O. Haenni, H.F. Restrepo, C. Teuscher, F. J. Gomez, and E. Sanchez, "Approches matérielles et logicielles de l'algorithme de chiffrement IDEA", *Technique et Science Informatiques (TSI)*, vol. 1, pp. 203-224, 2001 (in French).
2. Antti Hämäläin, Matti Tommiska, and Jorma Skyttä, "6.78 Gigabits per Second Implementation of the IDEA Cryptographic Algorithm", *Proc. 12th International Conference FPL 2002*, pp. 760-769, Montpellier, France, September 2002. Berlin: Springer Verlag, 2002.
3. M. P. Leong, O. Y. H. Cheung, K. H. Tsoi, and P. H.W. Leong, "A Bit-Serial Implementation of the International Data Encryption Algorithm IDEA", *Proc. 2000 IEEE Symposium on Field-Programmable Custom Computing Machines*, pp. 122-131, Napa, California, 2000.
4. O.Y.H. Cheung, K.H.T. Soi, P.H.W. Leong, and M.P. Leong, "Tradeoffs in Parallel and Serial Implementations of the International Data Encryption Algorithm IDEA", *Proceedings of CHES*, pp. 333-347, Paris, 2001
5. Cameron Patterson, "High Perfomance DES Encryption in Virtex FPGAs using JBits", *Proc. 2000 IEEE Symposium on Field-Programmable Custom Computing Machines*, pp. 113-121, Napa, California, 2000.
6. J. Leonard and W. Magione-Smith, "A Case Study of Partially Evaluated Hardware Circuits: Key-Specific DES", *Proc. Seventh International Workshop on Field-Programmable Logic and Applications FPL '97*, London, UK, 1997. Berlin: Springer Verlag, 1997.
7. J. Daemen and V. Rijmen, *"AES Proposal: Rijndael, NIST AES Proposal"*, 1998.
8. Xilinx XAPP 054 (*http://www.xilinx.com/xapp/xapp054.pdf*)
9. Philip James-Roxby and Brandon J. Blodget, "A Study of high-perfomance reconfigurable constant coefficient multiplier implementations", Xilinx Inc., Tech Notes Archive Chipcenter (*http://www.chipcenter.com/pld/images/pldf085.pdf*)
10. P.H.W. Leong and K.H. Leung, "A Microcoded Elliptic Curve Processor using FPGA Technology", *IEEE Transactions on VLSI Systems*, accepted for publication, 2002
11. Xilinx XAPP 290 (*http://www.xilinx.com/xapp/xapp290.pdf*)
12. S. A. Guccione and D. Levi, *"JBits: A Java-based Interface to FPGA Hardware"*, San Jose, California: Xilinx Inc., 1998.
13. Steven A. Guccione, Delon Levi, and Prasanna Sundararajan, "JBits: A Java-based Interface for Reconfigurable Computing", *Proc. 2nd Annual Military and Aerospace Applications of Programmable Devices and Technologies Conference (MAPLD)*.
14. S. A. Guccione and D. Levi, *"Run-Time Parameterizable Cores"*, San Jose, California: Xilinx Inc., 1999.
15. S. McMillian and S. A. Guccione, *"Partial Run-Time Reconfiguration using JRTR"*, San Jose, California: Xilinx Inc., 2000.
16. I. Gonzalez, "Codiseño en Sistemas Reconfigurables basado en Java", *Internal Technical Report*, UAM, Madrid, 2002 (in Spanish).

An Implementation Comparison
of an IDEA Encryption Cryptosystem
on Two General-Purpose Reconfigurable Computers

Allen Michalski[1], Kris Gaj[1], and Tarek El-Ghazawi[2]

[1] ECE Department, George Mason University
4400 University Drive, Fairfax, VA 22030, U.S.A.
{emichals, kgaj}@gmu.edu
[2] ECE Department, The George Washington University
801 22nd Street NW, Washington DC 20052, U.S.A.
tarek@seas.gwu.edu

Abstract. The combination of traditional microprocessors and Field Programmable Gate Arrays (FPGAs) is developing as a future platform for intensive computational computing, combining the best aspects of traditional microprocessor front-end development with the reconfigurability of FPGAs for computation-intensive problems. Several prototype PC-FPGA machines have demonstrated significant speedups compared to standalone PC workstations for computationally intensive problems. Cryptographic applications are a clear candidate for this type of platform, due to their computational intensity and long operand lengths. In this paper, we demonstrate an efficient implementation of IDEA encryption, using two of the leading reconfigurable computers available, SRC Computers' SRC-6E and Star Bridge Systems' HC-36. We compare the hardware architecture and programming model of these reconfigurable computers, and the implementation of a common IDEA encryption architecture in both platforms. Detailed analyses of FPGA resource utilization for both systems, data transfer and reconfiguration overheads for the SRC system, and a comparison between SRC and a public domain software implementation are given in the paper.

1 Introduction

The need for faster computational processing methods has grown along with the desire to process large amounts of data in shorter periods of time. Approaches to this problem have typically involved microprocessor-based solution, utilizing concurrent processing with multiple processors within a single machine or in a cluster of workstations over a network. Since their introduction, FPGAs have emerged as a low cost complement to traditional microprocessor software and hardware solutions for computationally intensive applications, due to their hardware reconfigurability [1].

Several workstation-based FPGA systems are available today for commercial applications. These "reconfigurable computers" make use of workstation microprocessor(s) for front-end processing, and provide one or more FPGAs for computations less suited for a microprocessor. The choice of how a design can be divided between a microprocessor and an FPGA depends on the development

P.Y.K. Cheung et al. (Eds.): FPL 2003, LNCS 2778, pp. 204–219, 2003.

environment available for the system. In this paper, we explore the efficient implementation of pipelined IDEA encryption within two such available reconfigurable computers: the SRC-6E from SRC Computers, and the HC-36 from Star Bridge Systems. This paper discusses the hardware architectures and development tools of both systems in detail, and provides FPGA timing and utilization results for both systems. A summary including the state of development of both systems is also presented.

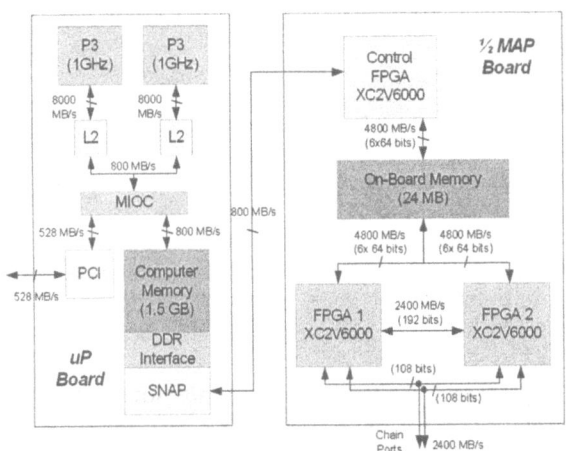

Fig. 1. SRC-6E Hardware Architecture

2 FPGAs and Reconfigurable Computing

An FPGA is a regular structure of basic modules called Configurable Logic Blocks (CLBs), which can be interconnected to provide hardware implementations of algorithms required by a designer. FPGA interconnects between modules are under the designer's complete control [2]. The FPGAs used in both the SRC and Star Bridge systems are the Xilinx Virtex II 6000 series, which have a capacity of six million system gates, and provides dedicated logic for fast carry propagation for addition operations, multipliers that handle operand sizes up to 18 bits, and block RAMs for local memory access. FPGAs are reconfigurable, meaning that the FPGA device can be configured to carry out a specific function, and can be reconfigured to carry out a different function at a later time.

The term "reconfigurable computing" is used to describe a combination of microprocessor systems with FPGAs to provide a reconfigurable hardware environment. Both the SRC and Star Bridge systems are examples of reconfigurable computing platforms. Reconfigurable computers offer benefits over microprocessor-based hardware solutions because FPGAs can more easily exploit computational precision and operand sizes required by the design, and can implement operation pipelining and parallelism specific to the needs of the application being developed. Microprocessor instruction sets have fixed operand lengths that may not match operand sizes specific to the design, and the implementation of algorithms typically involves multiple-instruction executions for one algorithmic operation. In addition, much of the microprocessor's capability is unutilized, since a general-purpose microprocessor is designed to implement a wide variety of operations specific to a workstation computing environment, whereas most design requirements for cryptographic systems only use a subset of the full capabilities of a microprocessor. All of the resources of an FPGA can be dedicated to the needs of a design, which provides a more efficient implementation versus a single or multi-processor-based design solution.

3   The SRC-6E

3.1   SRC-6E Hardware Overview

The SRC-6E system architecture frontend consists of two dual Intel processor motherboards. Each motherboard contains two Intel P3 Xeon processors and 1.5 GB of memory. Each system hosts a multi-processor version of the Linux operating system, and provides two distinct Linux-based microprocessor-FPGA reconfigurable computers.

An SRC MAP® processor is attached to each Intel motherboard, as shown in Fig. 1. Each MAP processor consists of two Xilinx Virtex II 6000 FPGA chips available for user logic, and a control processor, which is also a Xilinx Virtex II 6000 FPGA, all running at a clock rate of 100 MHz. The control processor implements fixed control logic, and is responsible for direct memory access (DMA) transfers between Intel system memory and the onboard memory of the MAP processor. The user logic and control FPGAs have access to six banks of dual port 512k x 64 bit static RAM providing a total of 24MB of memory external to the FPGAs. The MAP control processor communicates with the Intel processors through a SNAP interconnect. The SNAP interconnect is a high speed, low latency interface which functions as a Double Data Rate (DDR) memory interface, and plugs into a DDR SDRAM slot on the motherboard. SNAP provides higher data throughput between the Intel processor and the MAP processor versus component interfacing using the PCI-X bus.

3.2   SRC-6E Programming Environment

SRC has created a development environment that uses traditional programming paradigms in addition to hardware description language (HDL) development for implementing a design with the MAP FPGAs, as shown in Fig. 2. The SRC environment provides the ability to implement FPGA user logic using either C or FORTRAN sourcecode alone, or in combination with HDL sourcecode. GNU compilers are provided for C or FORTRAN sources that target the Intel processors, and SRC provides its own C or FORTRAN compilers that target the MAP processors. The MAP processor compilers produce Verilog code which is then synthesized using Synplify Pro, and Xilinx tools perform map, place and route. The GNU compilers allow the use of ANSI-compliant C or FORTRAN code and libraries, which allows for integration with other existing UNIX applications.

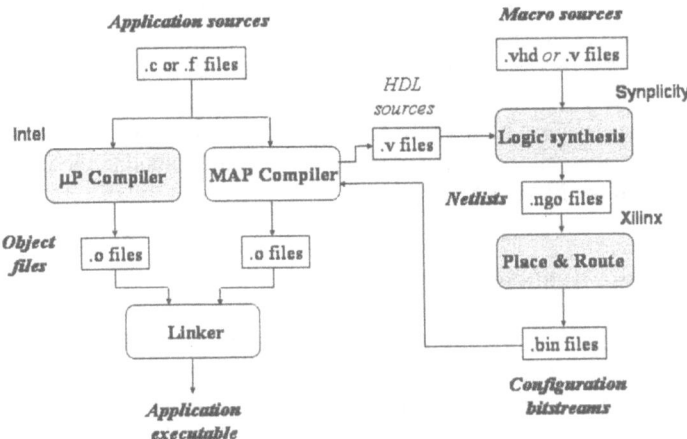

Fig. 2. Overview of the SRC Compilation Process

The design environment is hosted in each Linux platform within the SRC-6E. The compilation process consists of compilation of user logic HDL files, if present, that will execute on the MAP processor, compilation of C or FORTRAN code that will also execute on the MAP processor, and compilation of the C or FORTRAN code that will execute on the Intel processors. The compilation process places wrapper code around logic that resides within the FPGAs to facilitate data transfer and control synchronization between the microprocessor and MAP processor. The binary files produced by each compile process are combined into one single executable for the host Intel platform, which is responsible for loading the Intel and MAP processors with the code compiled for each.

3.3 Designing within the SRC-6E

Designs in SRC can have required operations performed on either the Intel or MAP processors. Within the MAP processor, FPGA designs can be implemented using a C or FORTRAN source alone, a C or FORTRAN source in combination with HDL, or completely in HDL, using a C or FORTRAN source to provide data transfer services. A simple API is available in the high-level language (HLL) to provide control functionality and data transfer functions between the Intel and MAP processors, and the passing of data to and from included user HDL designs. MAP C sourcecode data types are limited to 32 and 64 bit types.

HDL source that is targeted for the MAP processor is called a macro. SRC supports either VHDL or Verilog sources within macros. A macro's primary characteristics are defined as functional, stateful, or external. Additional characteristics define latency of the design and whether the design is pipelined. An "info" file is used to describe these characteristics. A functional macro is one that carries no state information, and therefore doesn't require the reset of a state machine. A stateful macro is one that caries state information and may need to be reset during a data cycle. An external macro is one that needs direct hardware access to memory, versus using HLL calls to read memory and supply data to the design. Latency defines how many clock cycles are needed before a result is available, and the pipeline attribute defines whether the macro can take data on each clock cycle. The combination of all these attributes defines the HDL macro supplied by the user.

If a C or FORTRAN source is used alone or in combination with HDL source to implement the design within the MAP processor, the compiler attempts to extract the maximum parallelism from the code and generate pipelined hardware logic for instantiation in the MAP FPGAs [3].

3.4 Debugging within the SRC-6E

Debugging within SRC can be attained by traditional functional testing of HDL outside of SRC and traditional HLL debugging techniques, or using SRCs MAP debugger. The MAP debugger provides hardware debugging of code compiled for the MAP processor, and emulation debugging of emulation or simulation code. Emulation is based on specially formatted "data flow graph" C routines, provided by the compiler from existing HLL source if that is used to implement logic, or by the user in an additional file if HDL is also used, that emulates the functionality of the design being implemented in the MAP processor. Simulation mode actually simulates the HDL produced by the compiler, which requires a license for a Verilog simulator not included in the development environment.

4 The Star Bridge HC-36

4.1 HC-36 Hardware Overview

The Star Bridge HC-36a consists of a Tyan S2720 dual Intel P4 Xeon processor motherboard with 4 GB of memory. Attached to the motherboard through the PCI-X bus is a PCI card containing two Virtex II 4000 FPGAs for dedicated PCI and hardware control, one Virtex II 6000 FPGA allocated for user-designed FPGA control functions, and 4 Virtex II 6000 FPGAs available for user logic, as shown in Fig. 3. FPGA clocks are set to the PCI-X clock speed, with the FPGA digital clock manager used to provide clock speeds other than the PCI-X clock speed. Star Bridge uses the name "HyperComputer®" to refer to this combination of FPGAs, an Intel-based frontend, and Star Bridge's VIVA development environment [4]. The environment is hosted in Microsoft's Windows 2000 Server operating system.

FPGAs available for user logic are referred to as PE1 to PE4. These four PEs (Processing Elements) have dedicated 50-bit connections to each other. Each PE, along with the XPoint FPGA mentioned below, has four banks of memory, each bank having 512 MB of RAM with a 64-bit PE interface. This gives a total of 10 GB of memory with 20 64-bit independent memory channels within the user FPGAs.

Control FPGAs available to the user consist of an FPGA known as XPoint and a router FPGA. XPoint connects to the four user FPGAs using dedicated 32-bit data connections, and is allocated to provide user-defined control interfacing to the user FPGAs. The router FPGA has dedicated 94-bit connectivity to the user FPGAs, and provides additional user FPGA connectivity.

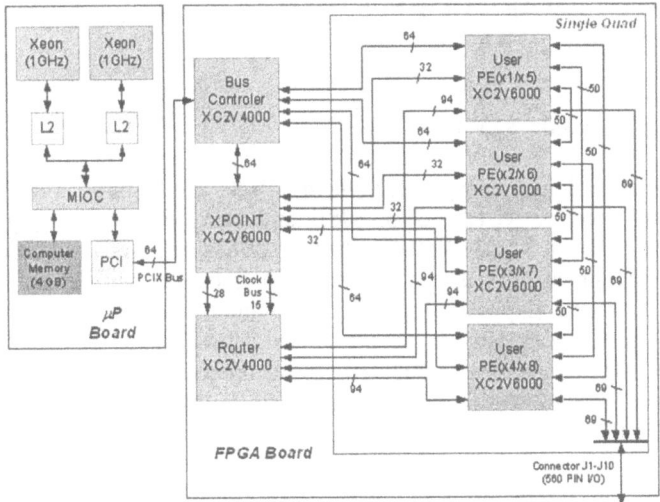

Fig. 3. Star Bridge HyperComputer FPGA Configuration

The base grouping of four PEs is called a Quad Element. Only one Quad Element is available in the HC-36a, however each XPoint control FPGA can be used to control two Quad Elements. This structure of one XPoint control FPGA along with two Quad Elements can be replicated to create larger FPGA arrays, and is used to create larger Star Bridge platforms [4].

4.2 The Star Bridge Programming Environment

VIVA® is Star Bridge's proprietary development environment. It provides a graphical user interface based on "drag and drop" design principles. VIVA's design language, "Implementation Independent Algorithm Description Language" or IIADL, is object-based and uses object attributes to specify different options within the design. Fig. 4 shows the VIVA Integrated Development environment (IDE).

Designs within VIVA can be allocated to execute on either the Intel platform or the FPGA array. Objects that execute on the Intel processor are implemented using pre-built VIVA libraries within Windows. VIVA is able to use Windows COM (Component Object Model) to communicate with other Windows applications, allowing Windows applications that make use of COM methods to be integrated in with VIVA FPGA designs.

VIVA provides its own proprietary synthesis tool for designs that will run on user FPGAs. The target of the synthesis tool is defined by a system description. VIVA includes system descriptions for the Pentium x86, which allows simulation of the design in an x86, and for specific PEs in the FPGA array. VIVA's system descriptions allow VIVA to abstract out the target hardware description from the design to be implemented, allowing VIVA to target hardware other than its own.

If the design is targeted for a PE, the synthesis EDIF output is fed into Xilinx place and route tools to produce a final binary, which VIVA then loads onto the Intel processors and the FPGA array. After initial execution, the VIVA binary can be saved in a file format that can be loaded using VIVA command-line tools or within the VIVA development environment.

4.3 Designing within VIVA

VIVA defines basic data types, from which objects can be built. The basic types are Bit, Variant, Vector, NULL, LSB, MSB and BIN. Bit is the only non-abstract fundamental data set. Within VIVA, data sets are basically recursive combinations of Bits. Each type therefore has two basic objects associated with it: an Exposer and a Collector. A data set Exposer takes the data set and outputs its two children data sets. A collector does the opposite: it takes two child data sets and combines them into one parent data set. For instance, a BIN016 (16-bit binary) data set, when exposed, will produce two BIN008 (8-bit binary) data sets. A MSB016, when Exposed, will produce a MSB001 and a MSB015. This is equivalent to splitting out the most-significant bit of a std_logic_vector(15 downto 0), and returning a std_logic and a std_logic_vector(14 downto 0).

The above definition of types allows for recursion of large data sets into small data sets, which is useful when a large operation is comprised of smaller operations of the same type. For instance, a 16-bit AND gate can be built using a tree of 2-bit AND gates, which can be built recursively. In order for recursion of an operation to be properly implemented, two objects for the operation must be created: a base object of the smallest acceptable operand size, and a recursive version of the object, which tells VIVA how to recurse down to the base operand object. Operands can be overloaded by simply copying the operand object design and changing the input data types, effectively overloading the operand type to handle different data sets. Objects can incorporate other types of objects, which allows for the hierarchical building of basic objects.

VIVA provides basic operands such as AND, OR, and INVERT, which are standard to VIVA. These basic objects can be implemented in multiple system descriptions. VIVA objects, in general, can be designed to be implemented in multiple system descriptions or in a specific system description only. VIVA provides a library of additional operands, and data types built from the standard data types. This library, Corelib, has objects ranging from registers through state machines used for design control to file I/O objects for data transfer from and to a Windows file (see Fig. 4). This library is meant to be the basis for designs built in VIVA.

4.4 Debugging within VIVA

Debugging a design within VIVA is performed by loading the x86 system description into a design simulate design components within the x86 environment. There is no in-place hardware debugger, although one is planned for future release. The simplest method of input and output testing is displaying inputs and outputs using widgets, which can be displayed using scrollbars, textboxes, graphics, and a number of other selections, as in Fig. 4.

The clock can be manually single-stepped or continuously run. Windows COM objects can be integrated in to provide advanced debugging of the data output of a design.

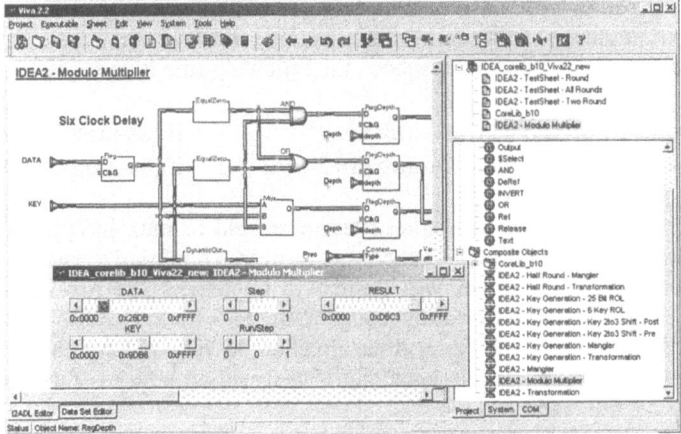

Fig. 4. VIVA Synthesis Output using the X86 System

5 IDEA

5.1 The IDEA Algorithm

The International Data Encryption Algorithm (IDEA) is a symmetric block cipher developed by Xuejia Lai and James Massey of the Swiss Federal Institute of Technology and published in 1990[5]. At that time it was suggested as a candidate to replace DES, however its widest adoption has been in PGP, which has insured widespread use of the algorithm.

IDEA uses a 128-bit key to encrypt data blocks of 64 bits. IDEA consists of eight rounds follow by a Transformation half-round that provides a 64-bit encrypted output. A round consists of a Transformation half-round and a Mangler half-round, of which an excellent description is provided by William Stallings in the book "Cryptography and Network Security: Principles and Practice" [6]. IDEA makes use of three basic operations to carry out encryption and decryption: a 16-bit XOR operation, a 16-bit modulo addition operation (mod 2^{16}), and a 16-bit modulo multiplication operation (mod $2^{16} + 1$). In addition, an all-zero operand input to a modulo multiplier within IDEA equals 2^{16} for internal calculations.

Each round requires six keys: four for the Transformation half-round and two for the Mangler half-round. For the total 8.5 rounds required for IDEA encryption, 52 16-bit keys are required. The first eight keys are provided by the input key. Each additional set of eight keys is generated by performing a circular left shift of 25 bits of the previous eight key set.

5.2 A Common Design Choice for IDEA Implementation within SRC and Star Bridge

The XOR and modulo addition operations of IDEA can be easily implemented. The most difficult operation, multiplication mod ($2^{16} + 1$) with the input of $0 = 2^{16}$, can be broken down into three cases: multiplication of two nonzero inputs, multiplication where one input is zero and multiplication where both inputs are zero. For multiplication of two nonzero numbers, the following rule is used [6]:

$$ab \bmod(2^n + 1) = (ab \bmod 2^n) - (ab \operatorname{div} 2^n) \qquad \text{if } (ab \bmod 2^n) \geq (ab \operatorname{div} 2^n)$$
$$ab \bmod(2^n + 1) = (ab \bmod 2^n) - (ab \operatorname{div} 2^n) + 2^n + 1 \quad \text{if } (ab \bmod 2^n) \leq (ab \operatorname{div} 2^n)$$

For the remaining two cases, zero tests are used to mux in appropriate nonzero results and constants to provide the correct answer. To minimize latency, the use of a Virtex II hardware multiplier was desired to implement the two equations above. Since a 16x16 multiply operation using a single Virtex II 6000 block multiplier requires over 10 ns, a two-level three pipeline "divide-and-conquer" strategy was used to meet the 10 ns timing constraint. This method requires four 8x8 multiplies along with column additions [7].

Design choices for IDEA centered on making each half-round modular to create a repetitive instantiation, therefore key scheduling is broken into a unit that can be implemented in a modular round. The solution for key scheduling requires the generation of six keys for each round. After eight keys have been consumed, though, a 25-bit rotate left operation is required. To accommodate the above two constraints, a unique constant is input to each round's key scheduler, which selects a mux that determines where the key rotate operation occurs within that round, if required.

The design was pipelined in order to achieve high throughput. Both data and key scheduling were pipelined, which allows this core to be used in IDEA breaking or encryption since different data blocks can be introduced with either the same or a different key at each clock cycle. Pipeline placement was chosen based on synthesized VHDL and Xilinx place and route results for a target Xilinx II 6000 FPGA and a timing constraint of 10 ns (100 MHz). The final design has a pipeline latency of 116 clocks: each Transformation half-round requires four clock cycles and each Mangler half-round requires ten clock cycles. Fig. 5 shows a block diagram of the design.

6   IDEA within SRC

6.1   Design Implementation within SRC

To implement an algorithm within SRC, at least two source files are required. *main.c* is responsible for reading in data and calling a user-defined function that loads a user-logic bitstream into the MAP processor. The second source *IDEA_test.mc* implements the MAP function that is called from *main.c*, and describes a bitstream to be loaded into the MAP processor. This MAP function calls SRC functions that control the data transfer between the Intel platform and the MAP processor, and uses a C for-loop structure for passing data to C commands that implement data processing within the MAP processor and to user HDL macros (see Table 2) that are also loaded within the MAP. In addition, an "info" file is required to specify attributes of the IDEA VHDL instantiation for the MAP compiler.

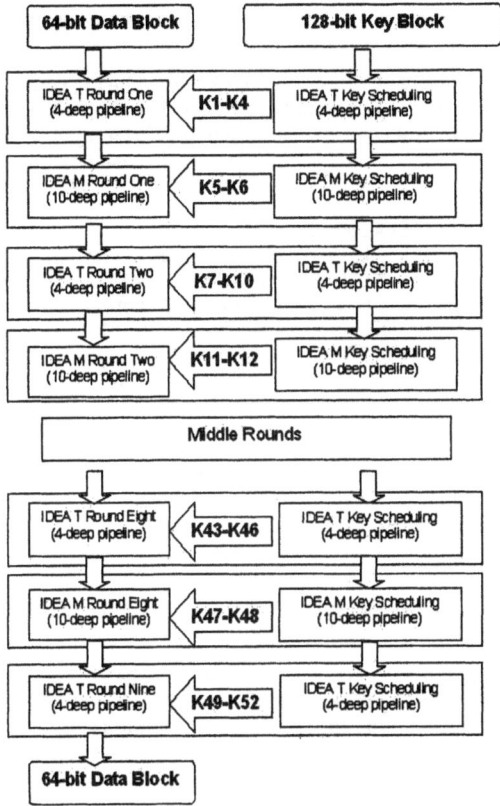

Fig. 5. IDEA Block Diagram

Two test cases were implemented within SRC, as shown in Tables 1 and 2. The first test case instantiates the whole IDEA algorithm within VHDL, and uses the MAP C function to pass data. The SRC info file for case one defines the VHDL user macro as a functional pipelined macro with a 116-clock latency. The second case instantiates half-rounds within VHDL, and uses the MAP C function to instantiate the 8.5 rounds required by IDEA. The SRC info file for case two describes two VHDL macros, both being defined as functional pipelined macros with a 4 or 10-clock latency. These test cases are representative of SRC design options, and allow for an FPGA resource and timing comparison between the two methods.

To test for latency results, a Unix high-resolution time structure was used to time of the HLL function call within *main.c*, while SRC timer calls were used within *IDEA_test.mc* to time the input and output DMA data transfers and the time to process the data. HLL time reads are in *us*, while timer reads within SRC provide the current clock tick. A conversion to ms was performed to arrive at the final timing calculations.

Table 1. Portions of IDEA Rounds Instantiated using VHDL

```
-- Round 8.
signal outkeyhigh_half_round8,    outkeylow_half_round8,
       outblockdata_half_round8,  outkeyhigh_round8,
       outkeylow_round8,          outblockdata_round8:
                                  std_logic_vector(63 downto 0);
begin
  -- Round 1.
  round1_T: component IDEAroundT
    port map (clk => clk, inkeyhigh => inkeyhigh, inkeylow => inkeylow,
              mux_sel => X"00000000", inblockdata => inblockdata,
              outkeyhigh => outkeyhigh_half_round1,
              outkeylow => outkeylow_half_round1,
              outhalfblockdata => outblockdata_half_round1);

  round1_M: component IDEAroundM
    port map (clk => clk, inkeyhigh => outkeyhigh_half_round1,
              inkeylow => outkeylow_half_round1, mux_sel => X"00000000",
              inhalfblockdata => outblockdata_half_round1,
              outkeyhigh => outkeyhigh_round1,
              outkeylow => outkeylow_round1,
              outblockdata => outblockdata_round1);
```

Table 2. Portions of IDEA Rounds Instantiated using C

```
/* Round 8 */
uint64_t outkeyhigh_half_round8, outkeylow_half_round8,
         outblockdata_half_round8, outkeyhigh_round8,
         outkeylow_round8, outblockdata_round8;
......
int i;
......
cm2obm_c(c, data_in, size*8);
wait_server_c();
 read_timer(start_loop);

for (i=0; i<size; i++) {
  keyhigh = a[i];
  keylow  = b[i];
  data    = c[i];

  /* Round 1 */
  IDEAroundT(keyhigh, keylow, (uint32_t)0x0, data,
             &outkeyhigh_half_round1, &outkeylow_half_round1,
             &outblockdata_half_round1);
  IDEAroundM(outkeyhigh_half_round1, outkeylow_half_round1,
             (uint32_t)0x0, outblockdata_half_round1,
             &outkeyhigh_round1, &outkeylow_round1,
             &outblockdata_round1);
```

6.2 Results of SRC Timing

Results of the SRC implementation of IDEA are summarized in Tables 3 and 4. The timing results were independent of which of the two SRC implementations of IDEA described in Section 6.2 was used for the measurements.

Table 3. FPGA Processing and Configuration Times

Data Size	PC End-to-End Time (ms)	MAP FPGA Conf Time (ms)	MAP Total Processing Time (ms)	MAP Data Transfer In (ms)	MAP FPGA Processing (ms)	MAP Data Transfer Out (ms)
5 MB	185.6	102.0	83.6	53.0	6.6	24.0
10 MB	269.7	102.0	167.7	104.4	13.1	50.1
15 MB	350.8	102.2	248.7	153.5	19.7	75.4
20 MB	432.6	101.8	330.8	203.0	26.2	101.5

Table 4. FPGA Throughput Results

Data Size	MAP Total Processing Throughput (MB/s)	MAP Transfer In Throughput (MB/s)	MAP Transfer Out Throughput (MB/s)	MAP FPGA Processing Throughput (MB/s)
5 MB	59.8	282.9	207.9	761.1
10 MB	29.8	287.3	199.4	762.0
15 MB	20.1	293.1	198.8	762.3
20 MB	15.1	295.6	196.9	762.5

In Table 3, the execution times are provided for four input data sizes ranging from 5 MB to 20 MB. In each case, the encryption was accomplished by ten calls to the MAP function with a proportional amount of input data to the total data processed. An end-to-end time includes a single MAP configuration time in the range of 102 ms, and the total MAP processing time, which is proportional to the amount of data being encrypted. Since the reconfiguration of the MAP FPGA can be performed before any input data becomes available for processing, this configuration may be treated as a part of a one-time set up routine.

It is worth noting that the time spent for processing data inside of the MAP FPGA (MAP FPGA Processing Time) constitutes less than 8% of the Total MAP Processing time. Instead, the majority of the time is spent for transferring data to and from the MAP using the DMA transfer between the microprocessor's Computer Memory and the MAP's On-Board Memory. The MAP Transfer In Time is greater than the MAP Transfer Out Time because in our implementation of IDEA, each input consists of both a data block (64 bits) and the corresponding key (128 bits), while an output includes only an encrypted data block (64 bits).

In Table 4, the corresponding MAP Processing and Data Throughputs are calculated. All throughputs are for the most part independent of the amount of data being processed. The MAP FPGA Processing Throughput approaches the theoretical maximum of 64 bits per clock cycle = 800 MB/s. The only reasons for a slightly smaller value of this throughput are the latency of the pipelined architecture of IDEA (116 clock cycles) and a control overhead associated with implementing a loop structure within the MAP function, measured to be equal to 47 clock cycles [13]. The MAP FPGA Processing time expressed in the number of clock cycles is equal to:

$$Encryption\ Unit\ Latency + (Number\ of\ Data\ Blocks\ Processed - 1)$$
$$+ Loop\ Control\ Overhead \qquad (1)$$

The MAP Transfer In Throughput is greater than the MAP Transfer Out Throughput because of the larger number of 64-bit words being processed without changing the OBM address. Finally, the Total MAP Processing Throughput (not including reconfiguration) is in the range of 60 MB/s, which is only 7.5% of the theoretical maximum of 800 MB/s. The limited data transfer bandwidth and a lack of overlapping between data transfers and data computations inside of the MAP FPGA contribute to this considerable slow down.

Table 5. SRC PAR Results

Test Case	Slices	Slice FFs	LUTs	Mult 18x18	Clock Period
VHDL Only	9006 (26%)	13221 (19%)	10111 (15%)	136 (94%)	11.579 ns
VHDL-C	11442 (33%)	18460 (27%)	10435 (15%)	136 (94%)	11.379 ns

6.3 Results of SRC FPGA Timing and Resource Utilization

SRC v1.4 compilers were used for MAP compilation. Synopsys v7.2 was used for systhesis, and Xilinx v5.2 was used for map, place and route. For the VHDL-only case, synthesis times averaged around two minutes, map times averaged around three minutes, while place and route (PAR) times averaged approximately 66 minutes. For the VHDL-C case, synthesis times averaged around two minutes, map times averaged around four minutes, while place and route (PAR) times averaged approximately 94 minutes. Table 5 shows Xilinx PAR results for both the SRC VHDL and VHDL-C instantiation cases. Using a VHDL-C combination to instantiate the higher level hierarchy of IDEA shows an 8% increase in the number of FPGA slice flip-flops used. This is due to SRC control logic that is inserted in its instantiation of the IDEA rounds. Clock timing differences were negligible.

7 IDEA within VIVA

VIVA was used to build up a design that matches the VHDL implementation. Fig. 6 shows portions of the Mangler half-round, and Fig. 7 shows portions of an IDEA test sheet. The VIVA PE system descriptions are not complete; a limitation of this is that the full I/O capabilities of the FPGA were not available. As a result, a limitation in this design is that the key inputs are constant, as the minimum number of inputs required is 192 bits (64 data + 128 key), while only 128 bits are available for inputs in the system description. Functional verification was performed using the same IDEA data file as in SRC.

Measuring of timing results of data transfer or data processing on the Intel side requires using a COM wrapper around a Windows timer, as a standard VIVA timer object is not available within the current library. A COM object was developed for timing; however, the CoreLib library's beta state of development prevented its integration with CoreLib's I/O objects.

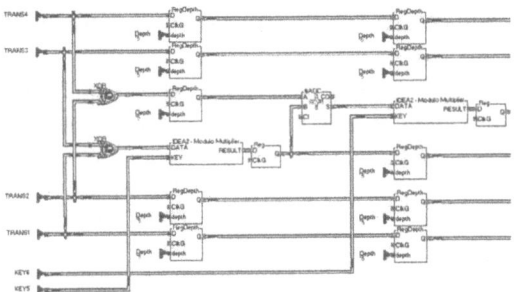

Fig. 6. VIVA Half Round Mangler Design

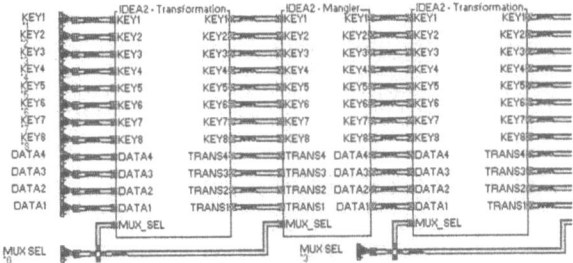

Fig. 7. IDEA in VIVA

Table 6. Star Bridge PAR Results

Test Case	Slices	Slice FFs	LUTs	Mult 18x18	Clock Period
VIVA	17292 (51%)	16118 (23%)	7639 (11%)	136 (94%)	12.191 ns

7.1 Results of Star Bridge FPGA Timing and Resource Utilization

VIVA v2.2 was used for synthesis, and Xilinx v4.2 was used for map and PAR. VIVA synthesis and PAR times scaled exponentially depending on the number of rounds synthesized: one round took 25 minutes, five rounds took approximately 7 hours, and the full 8.5 rounds took approximately 36 hours, due to the alpha state of their synthesis algorithms. The Xilinx map time for the full round implementation was 14 minutes, and PAR for the full round implementation was 85 minutes.

Table 6 shows Xilinx PAR results. Compared to the SRC VHDL-only design, the number of Slices used was 24% greater, while the number of Slice Flip Flops used was only 3% greater. An explanation for this discrepancy is that VIVA may be using only one LUT per CLB slice. The total number of LUTs used was 4% smaller. Clock timing was slightly longer. This comparison is tentative since the key inputs are constant versus the SRC design.

8 Crypto++ Software Implementation versus SRC

Crypto++ v5 was used to implement IDEA encryption within software on SRC's Linux-based workstation, to normalize the microprocessor capability between the software and the software-FPGA implementations. 20 MB of data was processed using ten calls to the Crypto library's IDEA encryption algorithm, using the same input data as SRC and VIVA. Timing measurements for IDEA encryption processing of the data were taken using a UNIX high resolution timer.

The time required to process 20 MB of data was approximately 8 seconds, which corresponds to a throughput of 2.5 MB/s. When compared to SRC's MAP Processing Throughput for IDEA, the SRC design is 24 times faster than the software implementation. When compared to the time including the FPGA configuration time, the SRC design was 18.5 times faster. Given continuous processing of large data blocks, the throughput advantage of SRC will approach the 24-times speed increase maximum.

9 SRC and Star Bridge Limitations

Both systems have limitations, which should be noted before conclusions can be understood.

SRC's compiler can only handle certain HLL constructs within a MAP processor subroutine. It is not a limitation in this design, and SRC has been adding significant functionality with each release.

Star Bridge's synthesis technology is in an alpha state. Significant improvements have been seen with the last few product releases; however the product is still in its early development stages. Star Bridge's CoreLib library is in a beta state, and design elements within the library are not final.

Both systems plan on implementing multiple FPGA use within their next software releases. SRC and Star Bridge currently cannot make use of more than one FPGA.

10 Conclusion

Reconfigurable computing is a platform that can provide larger and more efficient computing resources for implementing computationally intensive design solutions Both SRC and VIVA have unique reconfigurable computer environments that provide complimentary ways achieving this goal. This paper shows how the designer can use SRC's development environment to implement design objectives using different combinations of traditional HLL development with HDL development, and how Star Bridge's VIVA environment provides a new paradigm for FPGA design entry and synthesis. An IDEA encryption algorithm is implemented in both environments for comparison.

Both SRC and Star Bridge take a different approach in their design environment. The SRC system is based on traditional programming languages, builds upon known development paradigms, and can incorporate existing HLL source code for designs targeted at FPGAs. Star Bridge's VIVA provides a new development paradigm that has the potential to provide an easier method of design entry, and can target several different architectures in addition to its own.

The combination of a PC with FPGAs provides a high-speed design alternative to designs implemented solely in PC software. This paper shows that the FPGA's maximum throughput can be significantly reduced due to time required to configure the FPGA and data transfer between PC memory and memory available to the FPGA. While the effect of reconfiguration can be reduced with large data processing, the effect of data transfer is a consistent bottleneck to maximum throughput if large amounts of data need to be transferred.

References

1. Singleterry, R., Sobieszczanski-Sobieski, J. Brown, S., "Field-Programmable Gate Array Computer in Structural Analysis: An Initial Exploration", available at http://www.starbridgesystems.com
2. Parnell, K., Mehta, N., "Programmable Logic Design Quick Start Handbook", available at http://www.xilinx.com
3. SRC Inc. Web Page, http://www.srccomp.com
4. Star Bridge Systems Web Page, http://www.starbridgesystems.com
5. Lai, X., Massey, J., "A Proposal for a New Block Encryption Standard", Proceedings, EUROCRYPT '90, 1990.
6. Stallings, W., "Cryptography and Network Security: Principles and Practice", 2nd Edition, pgs. 102-109, 128, 1999.
7. Parhami, B., "Computer Arithmetic Algorithms and Hardware Designs", pgs 191-192, 2000.
8. Dai, W., Crypto++ v. 5, http://www.cryptopp.com, Sep. 2002.
9. VIVA User's Guide, Version 2.2
10. The SRC-6E C Programming Environment Guide, v1.2, 2003, available from SRC Computers.
11. The SRC-6E MAP Hardware Guide, 2003, available from SRC Computers.
12. 'Reconfigurable, Inherently Parallel "Hypercomputing" ', PowerPoint presentation, 2002, available from Star Bridge Systems.
13. Fidanci, O.D., Diab, H., El-Ghazawi, T., Gaj, K., Alexandridis, N., "Implementation trade-offs of Triple DES in the SRC-6e Reconfigurable Computing Environment," 2002 MAPLD International Conference, Sep. 2002.
14. Fidanci, O.D., Poznanovic, D., Gaj, K., El-Ghazawi, T., Alexandridis, N.,"Performance and Overhead in a Hybrid Reconfigurable Computer," Reconfigurable Architecture Workshop, RAW 2003, Apr. 2003.

Data Processing System with Self-reconfigurable Architecture, for Low Cost, Low Power Applications

Michael G. Lorenz, Luis Mengibar, Luis Entrena, and Raul Sánchez-Reillo

Electronic Technology Department
Universidad Carlos III de Madrid. Spain.
{lorenz, mengibar, entrena, rsreillo}@ing.uc3m.es

Abstract. In this paper, a low cost self-reconfigurable data processing system with a USB interface is presented. A single FPGA performs all processing and controls the multiple configurations without any additional elements, such as microprocessor, host computer or additional FPGAs. This architecture allows high performances at very low power consumption. In addition, a hierarchical reconfiguration system is used to support a large number of different processing tasks without the penalty in power consumption of a big local configuration memory. Due to its simplicity and low power, this data processing system is specially suitable for portable applications.

1 Introduction

There are several examples of architectures for data processing based on reconfigurable platforms. Usually, a PCI interface is found [1], or a custom system like in SPLASH-2 [2], DISC [3]. However, the cost and the complexity of this kind of systems are high because they are intended for investigation or high-end applications. More examples can be seen in [4],[5].

Most of these systems require plenty of resources, one or more FPGAs with several hundred (or millions) of gates each, a high-end personal computer or a workstation in order to implement the dynamic reconfiguration. In this scenario, it is difficult to think in a low power consumption product, since only the quiescent power dissipation in these systems can be quite important [6].

This paper proposes a completely different system, very simple, with only a commercial FPGA, an ATMEL AT40K family FPGA [7]. The control of the data processing system and the reconfiguration flow is performed by the FPGA itself; this approach eliminates the need for additional hardware, another FPGA [8],[9] CPLD ,[10],[11], or microprocessor [12],[13], to perform this task, which would increase the cost and the power dissipation of the system. On the other hand, this approach can be implemented by exploiting the partial reconfiguration capability of the FPGA [14]. It results in a powerful system for signal processing and at a very low cost, less than 50€, instead of thousand of Euros.

Besides the FPGA, the system contains only a few more elements: a non-volatile reconfiguration memory and a block of data memory. The link to the external world is made with a USB interface, easy to find in many systems. Most of the data is stored locally in the board. This makes unnecessary to send a huge amount of data through

P.Y.K. Cheung et al. (Eds.): FPL 2003, LNCS 2778, pp. 220–229, 2003.

the connection link, so that the USB interface does not slow down the data processing, because all the data processing is made inside this board. The signal processing application chosen to test the system is a simple pattern recognition system.

In the following sections, the architecture of the system, two levels of hierarchy of the reconfiguration mechanism, the simple operating system, and the example chosen to evaluate the data processing system, are described in detail. Finally, we summarize the results, in performance and power dissipation, and present the conclusions of this work.

2   Reconfigurable Architecture

In the following paragraphs the hardware of the system, the reconfigurable data memory and the operating system designed are described in detail.

2.1   Hardware Architecture

Figure 1 shows the system's architecture diagram. The system is built around a dynamic partial reconfigurable FPGA (AT40K40 from ATMEL). The rest of the system consists of a reconfiguration cache memory, a data memory, a USB interface and a programmable expansion port. The reconfiguration cache memory is a FLASH memory devoted to store FPGA configuration data. The size of the reconfiguration cache memory allows to store up to eight different configurations for the FPGA, which correspond to up to eight different data processing tasks.

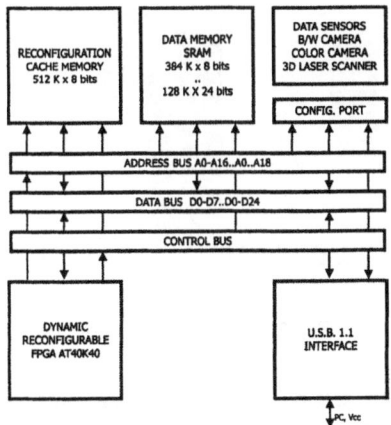

Fig. 1. Architecture of the Data Processing System with reconfigurable data memory

The data memory is a fast SRAM memory provided to allocate temporal data. It can be reconfigurable; both in data bus size and address bus width. The USB interface is intended to send data or commands to and from external equipment. In particular, it allows exchanging process data, and configuration data, with a host computer attached externally, which can be a PC or a simple microcontroller with USB interface.

A configurable expansion port for data acquisition purposes is provided. Several different sensors can be attached to this port, such as infrared sensors, ultrasound sensors, 3D laser scanner sensors, linear cameras, digital cameras, etc...

2.1.1   Reconfiguration Mechanisms

The reconfiguration process is built in two levels of hierarchy, in a similar way to a computer cache memory system. With this approach a versatile system can be developed, with enough reconfiguration speed for critical tasks, but also with the ability to support a large number of different tasks. The first level has been named *Reconfiguration Cache* and the second level has been named *Reconfiguration Library*. The reconfiguration cache is implemented in the board, near the FPGA. The reconfiguration library is stored in the attached host computer, if necessary.

2.1.2   Reconfiguration Cache Memory

The reconfiguration data is stored in a non-volatile FLASH memory, which can allocate up to eight different tasks [15]. The first one, cache page zero, manages the reconfiguration flow, the reconfiguration cache control, and controls the data processing and the task downloads from the reconfiguration library, from the host computer. This module remains resident in the FPGA during all the time the data processing system is active. The rest of the tasks present in the reconfiguration cache, pages one to seven, are intended for different data processing purposes.

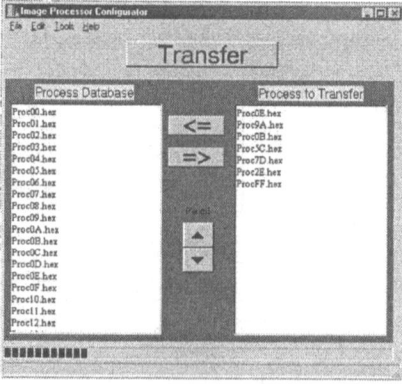

Fig. 2.. Configuration Program

In order to setup the self-reconfigurable data processor for a particular data processing application, it is necessary to download the data processing tasks from the reconfiguration library For this purpose, a simple download program (Figure 2) has been developed.

These tasks are loaded into the FPGA, in a sequential way, using a partial dynamical reconfiguration scheme. The execution order is previously determined, during the setup of the system, and implemented in the resident control task to obtain the desired data processing flow. Only if the seven processes are not enough, the resident control task can recall new tasks from the reconfiguration library, in the same way as in a second level cache memory. These new tasks are downloaded through the USB interface into the reconfiguration cache. Only in the case the system requires more than seven different data processing tasks, it reverts to a non-autonomous system.

2.1.3   Reconfiguration Library

The reconfiguration library is stored in a database located in the host computer. The system's functionality can be increased with this database, as follows. If the seven different tasks are not enough, or it is necessary to adjust the processing sequence, the control task, permanently resident in the FPGA, requests to the PC the configuration bit-stream corresponding to the task needed, from up to 256 available. This command request is sent to the host computer through the USB interface. The control process task erases one of the seven local tasks stored in the reconfiguration cache memory, replacing it with the new process task received. Finally, the control task generates the signal to start a partial dynamical reconfiguration process for the new task. Both the data memory and control process are not altered during this reconfiguration process. The new task will be kept in the reconfiguration cache for future use and can be eventually removed to allocate other tasks. The number of processes stored in the reconfiguration library is currently limited to 256. This number is enough to store any task needed in the image processor system and it is easy to implement with an eight-bit wide command.

2.1.4   Reconfigurable Data Memory Map

To improve the performance of the system, it is possible to reconfigure the data memory map to adapt its data width to the more advantageous size in each case. Due to the dynamical reconfiguration possibilities of the FPGA, it is possible to do this *on the fly*, without disturbing the normal process of the system.

In some applications the size is implemented with only 8 bits wide. This is the case of image capture with a black and white camera. The data captured is of the same size, and the capture is easily implemented, without wasting any portions of the data memory.

In processes like convolution, correlation, filtering etc... it is necessary to get a large amount of data from the memory. This part of the process can be improved if the system is able to get more than a single data value each clock cycle. In this case the system reconfigure itself to a 24 bits data width. At this data width, it is possible to get up to three single data values at the same time, making the system almost three times faster.

In table 1 the different memory configurations provided in the systems are shown. They depend on the width of the data collected from the sensor port, (e.g. 16 bits for a 3D laser scanner or 8 bits for each primary color in a color camera), and also on the active process. For example in a convolver the data memory is arranged in 24 bits data mode. This approach permits a wider bandwidth in the data processing stage, and a better management of the data memory during the acquisition and communications stages. This allows a lower memory size, minimizing the overall power dissipation of the data processing system.

Table 1. Different Data Memory Configurations

	Data Bus	Address Bus
B/W Camera	8 bits	A0..A18
Colour Camera RGB	24 bits	A0..A16
Stereoscopic Camera	2 x 8 bits	A0..A17
3D Laser Scanner	16 bits	A0..A17
Ultrasonic Sensor	8 bits	A0..A18
Send Data to Host Comp.	8 bits	A0..A18
Convolver (Sobel 3x3)	24 bits	A0..A16

2.2 Operating System

In order to control the system a simple operating system with a small instruction set has been defined. The complete set of instructions can be seen in Table 2.

Table 2. Set of Commands defined in the operating system.

Command Definition			
Host to Processing System			
Configuration memory commands (Level 1 cache)			
Codification	Designation	Operand	More
00000001.. 00000111	Record Config Memory Block	Block Nr. 1.. Block Nr. 7	Data
00010001.. 00010111	Read Config Memory Block	Block Nr. 1.. Block Nr. 7	
00100001.. 00100111	Erase Config Memory Block	Block Nr. 1.. Block Nr. 7	
Secuence control Commands			
Codification	Designation	Operand	More
0100XXXX	Stop Process		
0101XXXX	Start Process		
0110XXXX	Next Process		
0111XXXX	Last Process		
1000XXXX	Goto first Process		
10010000.. 10010111	Goto Process Nr. (oper)	Process Nr. 0.. Process Nr. 7	
1010XXXX	Read Process		
1011XXXX	RESERVED		
Processing System to Host			
Codification	Designation	Operand	More
1100XXXX	Request Process		(0..255)
11010000.. 11011111	Send Data	Data Tipe 0.. Data Type 15	Data
1110XXXX	RESERVED		
Headers			
Codification	Designation	Operand	More
1111XXXX	Data		
0011XXXX	Configurations		

The first group of instructions is only intended for the setup of the system. It is intended to provide a simple way to properly configure the data processing system. With these instructions it is possible to easily change all data processing tasks, and to accomplish the different data processing operations. The configuration program shown in figure 2 uses these instructions.

The second group of instructions is intended for use during the normal operation of the data processing system. These instructions are designed to accomplish interchange of data between the system and the host computer. These data can be process data or reconfiguration data, in order to perform more than the eight different data processing tasks stored in the reconfiguration cache memory.

3   Example of the Data Processing System Operation

To test the system, the system was configured to implement a complete pattern recognition process, having several image processing steps consecutively. The first step is usually the image capture. Then, a sequence of image processing steps is performed by partially reconfiguring the FPGA for each step. Typical image processing steps include a median filter to eliminate the image's noise, image convolution with an edge detection kernel (Sobel 3x3 mask size), and image correlation (18x18 size mask) [16]. Finally, in the last step the system sends the image processing results, reduced to a 1 bit data per pixel, to the host computer. Additional post-processing or display can be performed in the host computer, if needed.

The image capture is done with a programmable digital camera, based on a CMOS sensor chip from Omnivision [17]. The image size is programmable, up to a maximum of 384 x 288 pixels, and has a B/W digital output with 256 gray levels. In this case, in order to improve the data memory usage, a lower resolution 256 x 192 pixels has been chosen. The camera has low power consumption with only 100 milliwatts during the capture stage and 500 microwatts in stand-by mode. The captured images are stored in the data memory of the system.

4   Experimental Results

Figure 4 shows a picture of the system prototype. The configurable sensor port is shown in front. Several types of data sensors can be attached to this port, by only properly configuring this expansion port. The system is able to implement a median filter or a 3x3 convolution, with the image of 256x192 pixels, at a rate of nearly seven mega pixels per second. Pattern recognition with an 18 x 18 pixel mask is made in about 135 milliseconds, including the time used to send the results through the USB interface to the host computer. All these results have been obtained with a system clock of just 10 MHz. By using SRAM with 45ns instead of the installed memory of 90 ns, it is possible to double the system clock, and the performance of the system. The image processing system is used for navigation of an autonomous robot.

Table 3. Performance of the System.(System Clock 10 MHz).

	Configuration Time ms	Processing Time ms	Total Time ms	Images per Second
Image Capture (256x192 Pixels)	0.84	3.80		
Filter (Median)	0.90	19.42		
Convolver (Sobel 3x3)	1.08	19.42		
Correlation (18x18 Mask)	2.85	84.82		
Image Send USB (1 bit Data)	0.79	4.90		
	6.46	132.36	138.82	7.20
Time Percent	4.65%	95.35%	100.00%	

The time required to completely configure the FPGA is about 6 milliseconds, with a system clock of 10 MHz. This time includes configuration of all the 2300 FPGA's cells. However, if a partial reconfiguration is made, as in this case, the time required is proportional to the number of cells to be reconfigured, at a rate of three μsecs per cell approximately. In conclusion, reconfiguration time is just a small fraction of the total processing time. These configuration times, for each task in the sample, can be seen in table 3 and table 4. The data shown in table 3, correspond to the system arranged without data memory reconfiguration. In each clock cycle, the system can get only a pixel from memory (data bus eight bits wide).

Table 4. Performance of the System with Dynamical Memory Data Reconfiguration.(System Clock 10 MHz).

	Configuration Time ms	Processing Time ms	Total Time ms	Images per Second
Image Capture (256x192 Pixels)	0.84	3.80		
Filter (Median)	1.18	7.17		
Convolver (Sobel 3x3)	1.21	7.17		
Correlation (18x18 Mask)	3.84	33.90		
Send Image USB (1bit Data)	0.79	4.90		
	7.86	56.94	64.80	15.43
Time Percent	12.13%	87.87%	100.00%	

The data shown in table 4, correspond to the system arranged with data memory reconfiguration. In each clock cycle, the system can get up to three pixels from memory (data bus twenty four bits wide). The system performance was increased by a figure of 2.7 times, but with a higher overhead in total power dissipation, design size, and reconfiguration time. However the overall power dissipation per frame processed is lowered due to a better usage of memory access; 28 mW per frame without any reconfiguration of the data memory and 22 mW per frame processed with data memory reconfiguration.

Figure 3 shows the supply current to the self-reconfigurable data processor system during the dynamic reconfiguration of the FPGA. Note that the current scale is set in a negative range. Measures that are more accurate have been made using the procedure discussed in [18]. With this method, it is possible to measure power consumption even in each clock cycle.

The actual mechanism of FPGA reconfiguration is as follows. The process begins with the configuration of the look up tables, multiplexers and the flip-flops inside each individual cell. In a second stage all the interconnections paths are configured. From Figure 3 we can see that most of the energy is wasted in this second stage.

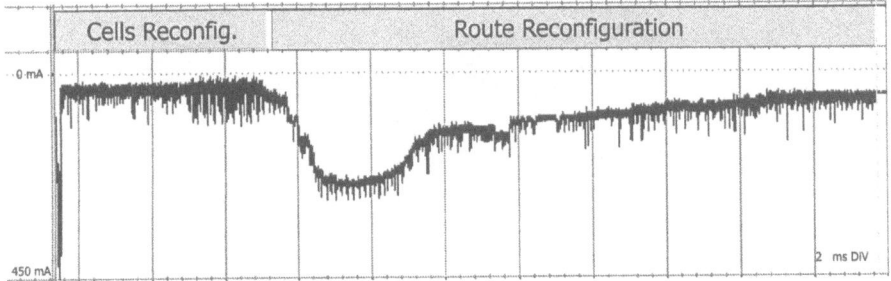

Fig. 3. Power consumption during dynamic partial reconfiguration (Atmel AT40K40 FPGA)

In Table 5 and 6 the measurements of system's power dissipation are provided. The data collected in Table 5 correspond to the system without any data memory reconfiguration, and a data memory width of eight bits. The last column represents the average power dissipation of the self-reconfigurable processing system.

Table 5. Power Consumption of the System.

	Reconfiguration Power µWs	Operation Power µWs	Average Power mW
Image Capture	33	53	
Filter (Median)	75	1457	
Convolver (Sobel 3x3)	111	1942	
Correlation (18x18 Mask)	972	23326	
Send Image USB (1bit Data)	36	59	
	1226	26836	202
Percent Consumption	4.37%	95.63%	

The data shown in Table 6 correspond to the system endowed with data memory reconfiguration. An increment in power dissipation during the configuration phase can be observed, that is due to the necessary pipelining of the data.

From this point of view, the system can be used for many data processing applications, despite its simplicity. It is remarkable that the system due to its low power consumption can be powered through the USB interface, making unnecessary any additional power supply.

Table 6. Power Consumption of the System with Data Memory Reconfiguration.

	Consumption Reconfiguration µWs	Consumption Operation µWs	Average Power mW
Image Capture	33	53	
Filter (Median)	98	1040	
Convolver (Sobel 3x3)	125	1470	
Correlation (18x18 Mask)	1309	18137	
Send Image USB (1bit Data)	36	59	
	1600	20758	345
Percent Consumption	7.16%	92.84%	

Fig. 4. Prototype Board.

5   Conclusions

In this paper we have described a low power self-reconfigurable data processing system that is able to make high speed data processing using dynamic reconfigurations to improve the system's versatility. The system is built with only one FPGA, some memory modules and a USB interface. Despite of the simplicity of the system, a large variety of processing tasks can be supported thanks to the use of a hierarchical reconfiguration scheme. The reconfiguration memory is used as a cache of the reconfiguration library, allowing the FPGA to be reconfigured for a large number of tasks while keeping the average reconfiguration time to a small fraction of the processing time.

The control of the data processing system and the reconfiguration flow is performed by the FPGA itself. This approach eliminates the need for additional hardware, which would increase the cost and the power requirements of the system. The system can be equipped with larger memories, enough for the 256 tasks supported, but at the expense of a higher cost and higher power consumption. The system has been evaluated with a classical pattern recognition but it can be used for many other digital signal processing applications, by simply reassigning the configurable expansion port to other types of sensors or interfaces.

References

[1]. F.Lisa. F.Cuadrado. D. Rexachs. J. Carrabina, "A Reconfigurable Coprocessor for a PCI-Based Real Time Computer Vision System", Field-Programmable Logic and Applications, 1997, 392-399.

[2]. D. Buell, J. Arnold et al. "SPLASH-2" Proceedings of the 4th Annual ACM Symposium on parallel Algorithms and architectures, June 92, 316-324.

[3]. M.J.Wirthling, B.L. Hutchings. "A Dynamic Instruction Set Computer". IEEE Workshop on FPGAs for Custom Computing Machines. Napa, CA, April 95. 99-107.

[4]. Hartenstein, R.; "A decade of reconfigurable computing: a visionary retrospective" Design, Automation and Test in Europe, 2001. 642 -649

[5]. K. Compton, S. Hauck, "Reconfigurable Computing A Survey of System and Software", ACM computing surveys, Vol 34, No 2, June 2002. 171-210

[6]. L. Benini, G. de Micheli, E. Macii, "Designing Low-Power Circuits: Practical Recipes", IEEE Circuits and Systems Magazine, Vol 1(1) First Quarter 2001, 6-25.

[7]. "Configurable Logic:Design and Aplication Book" Atmel, San Jose, CA. 1998

[8]. Haenni J.O., Beuchat J.L. and E Sanchez. "RENCO: A reconfigurable Network Computer". Sixth Annual IEEE Symposium on Field-Programmable Custom Computing Machines (FCCM'98) April 21-23, 1998 Napa California

[9]. M.A.Aguirre, J. Tombs. A. Torralba and L.G. Franquelo, "Experience on VLSI for Digital Signal Processing Using Adavnced FPGAs: The Unshades Framework", 4th European Workshop on Microelectronics Education, Vigo May 2002, 193-196.

[10]. C. Carmichael."Configuring Virtex FPGAs from Parallel EPROMs with a CPLD" Xilinx Application Note 137, 1999.

[11]. J. Khan, J. Handa and R. Vemuri, "iPACE-V1 A portable adaptative computing engine for real time applications", Field-Programmable Logic and Applications , 2002, 69-78..

[12]. "AT94K Field programmable system level Integrated circuit FPSLIC Data sheet" Atmel, San Jose, CA. 2002

[13]. J. Faura, J.M. Moreno et al, "Multicontext Dynamic reconfiguration and real time probing on a novel mixed signal programmable device with on chip microprocessor", Proceedings 7th International Workshop FPL97 London Sept 97. 1-10

[14]. Lysaght P., Dunlop J.; "Dynamic reconfiguration of FPGAs" European FPL'93 Oxford, 1993. 82-94

[15]. Huesung Kim; Somani, A.K.; Tyagi, A. "A reconfigurable multifunction computing cache architecture" Very Large Scale Integration (VLSI) Systems, IEEE Transactions on , Volume: 9 Issue: 4, Aug. 2001, 509-523

[16]. Derbyshire A. And Luk W.;Combining Serialisation and Reconfiguration for Convolver designs Eighth Annual IEEE Symposium on Field-Programmable Custom Computing Machines (FCCM'00) April 17-19, 2000 Napa, California

[17]. "OV5510 Data sheet". Omnivision Technologies Inc. July 1998

[18]. L. Mengibar, M.G. Lorenz. et al, "Experiments in FPGA characterization for low power design", XIV Design of Circuits and Integrated Systems Conference, Mallorca Nov 1999. 385-390.

Low Power Coarse-Grained Reconfigurable Instruction Set Processor

Francisco Barat[1], Murali Jayapala[1], Tom Vander Aa[1], Rudy Lauwereins[13], Geert Deconinck[1], and Henk Corporaal[2]

[1] ESAT K.U.Leuven, Kasteelpark Arenberg 10, B–3001 Leuven–Heverlee, Belgium,
firstname.lastname@esat.kuleuven.ac.be
[2] TUEindhoven, Den Dolech 2, 5612 AZ Eindhoven, The Netherlands
[3] Imec, Kapeldreef 75, B–3001 Leuven, Belgium

Abstract. Current embedded multimedia applications have stringent time and power constraints. Coarse-grained reconfigurable processors have been shown to achieve the required performance. However, there is not much research regarding the power consumption of such processors. In this paper, we present a novel coarse-grained reconfigurable processor and study its power consumption using a power model derived from Wattch. Several processor configurations are evaluated using a set of multimedia applications. Results show that the presented coarse-grained processor can achieve on average 2.5x the performance of a RISC processor with an 18% increase in energy consumption.

1 Introduction

Current and future multimedia applications such as 3D rendering, video compression or object recognition are characterized by computationally intensive algorithms with deeply nested loop structures and hard real time constraints. Implementing this type of applications in embedded systems (e.g. multimedia terminals, digital assistants or cellular phones) leads to a power-optimizing problem with time and area constraints. By adequate use of the high parallelism available in the inner loops of these applications, it has been shown that reconfigurable instruction set processors, composed of a standard instruction set processor tightly coupled to some sort of reconfigurable logic, have the required computational power to fulfill the time constraints [7, 6, 4]. What has not been shown is whether some of these reconfigurable processors can also meet the power consumption requirements of these applications.

This paper presents CRISP, a coarse-grained reconfigurable instruction set processor designed for multimedia applications that can accelerate multimedia applications in a power efficient manner. The power of this architecture lies in the reconfigurable logic, which is composed of complex blocks such as ALUs or multipliers, that operate on the data sizes typically found in multimedia applications (8 to 32 bits), and is divided in independently enabled slices in order to reduce overall energy consumption and reconfiguration times. Also important is the tight coupling to the main microprocessor (the reconfigurable logic is seen

P.Y.K. Cheung et al. (Eds.): FPL 2003, LNCS 2778, pp. 230–239, 2003.
© Springer-Verlag Berlin Heidelberg 2003

as an extra functional unit) that allows quick control and data communication between the processor and the reconfigurable logic.

The proposed processor architecture has been evaluated using a variation of the Wattch power estimation framework [3]. Several CRISP processors with different amounts of reconfigurable hardware are compared to a RISC processor. Results on a set of multimedia applications show that the reconfigurable processor is able to achieve on average 2.5 times the performance of a RISC processor with just an average of 18% energy increase.

This paper is organized as follows. The proposed CRISP architecture and the associated compilation/synthesis techniques are discussed in sections 2 and 3. Experimental setup and results are presented and discussed in sections 4 and 5, respectively. Related work is discussed in section 6 and the paper is closed with conclusions and future work.

2   A Low Power Reconfigurable Architecture

CRISP (Coarse-grained Reconfigurable Instruction Set Processor) is an instruction set processor composed of a main processor core tightly coupled to some coarse-grained reconfigurable logic. The coarse-grained reconfigurable logic is placed in a reconfigurable functional unit (RFU), and just like any other functional unit, an operation can be issued to it every clock cycle. The RFU reads/ writes data from/to the main register file. The main processor can be any type of processor, though in this paper we will assume the processor is a simple RISC (Reduced Instruction Set Computer) processor.

Figure 1 presents the overall architecture of the complete processor. The main processor core reads its instructions from the level 1 instruction cache and obtains data via the level 1 data cache. Both caches are connected to a unified level 2 cache, which is in turn connected to an external memory. The reconfigurable fabric, in the center of the figure and directly controlled by the main processor core, contains configuration memory that is loaded via the unified level 2 cache (this allows reuse of configurations loaded from external memory and reduces reconfiguration times). The reconfigurable fabric can directly access the data cache via several data ports. Additionally, the reconfigurable logic communicates with the main processor core via a functional unit interface.

As shown in figure 1, the reconfigurable functional unit is divided in reconfigurable slices, one of which is shown with more detail on figure 2. Each slice contains several coarse-grained processing elements (PEs), a register file, interconnect, and a small configuration memory. Each processing element is either an ALU, shifter, multiplier or memory unit. Such complex processing elements are better suited than the traditional logic blocks based on look up tables (LUTs) for the execution of the operations typically found in multimedia applications, which are word-oriented and not bit-oriented. These complex PEs allow the reconfigurable logic to operate at higher frequencies with lower power consumption when compared to traditional FPGAs.

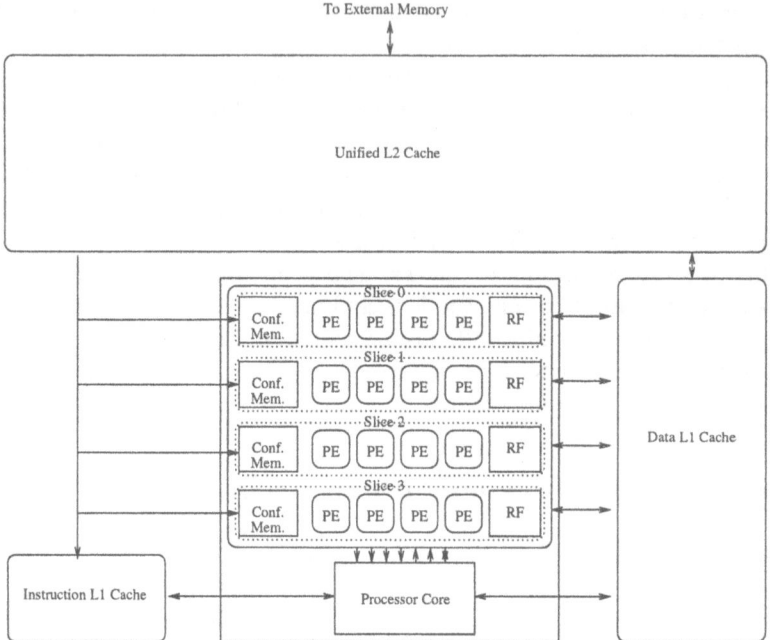

Fig. 1. Example CRISP instance (RFU: Reconfigurable Functional Unit, PE: Processing Element, FU: Functional Unit)

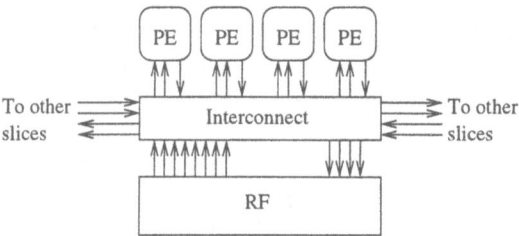

Fig. 2. Internals of a reconfigurable slice

The processing elements inside a slice are connected together through programmable interconnect. This interconnect is a full crossbar that operates on words and has the same complexity as the bypass network typically found in current VLIW (Very Long Instruction Word) microprocessors. This crossbar can connect the output of any processing element to the input of any other processing element. It also connects the processing elements to the register file and to the other slices. In most cases, each processing element writes its output to the register file of the slice, but this behavior can be optionally bypassed, just like in traditional FPGAs, and the result routed to a different processing element. By combining this optional register write and the interconnect crossbar, it is possible to perform spatial computation such that elements in a data flow chain are connected together through the crossbar. The processing element at the end of the chain is connected to the register file.

The number of processing elements has to be kept small in order to reduce the complexity of the interconnect and register files. From the results of our simulations, four processing elements represent a good tradeoff between power consumption and performance (though for space reasons these results are not presented here).

Each reconfigurable slice also contains a configuration memory. This configuration memory stores the configuration for the slice's datapath components. Since the typical loop requires several configurations to be quickly alternated (as will be discussed in section 3), the configuration memory must be multi-contexted, (i.e. it must be able to store several configurations). Switching from one context (or configuration) to another takes just one clock cycle and is equivalent to reading from a shallow and wide memory. In the case of a slice with four processing elements, the width of this configuration memory is around 128 bits, much less than the bits required for a slice of a typical FPGA. The number of configurations in the configuration memory typically ranges between 8 and 32 contexts. Ideally, the number of contexts should be kept as small as possible to reduce the energy consumption of the configuration memory.

The reconfigurable functional unit is activated via a special reconfigurable instruction as shown in Figure 3. This reconfigurable instruction contains two main pieces of information. First, it contains a reconfigurable instruction identifier (RID) that specifies which of the many available configurations must be used. This identifier can select among a larger number of configurations than the number of contexts available in the configuration memory. If the required configuration is not currently loaded in the configuration memory, which behaves like a small cache indexed by this RID, the system is halted and the adequate configuration is loaded from the unified level 2 cache.

Aside from the RID, the reconfigurable instruction includes several fields of one bit length that specify which slices are going to be activated. Figure 3 shows these slice enable fields (named ENx in the figure). For those parts of the application that require a small number of processing elements, only the first slice will be activated. For those parts with higher parallelism requirements, more slices will be used. This mechanism, which can be considered as a form of partial reconfiguration, reduces the size of the configuration stream that must be loaded in the case of a configuration miss. Additionally, the slices of the reconfigurable datapath that are not required can be switched off, thus providing an effective form of energy consumption control.

3   Compilation Techniques

Code generation for any reconfigurable instruction set processor involves main two steps: synthesis of the different configurations for the reconfigurable array and generation of the code for the main processor (not mapped to the recon-

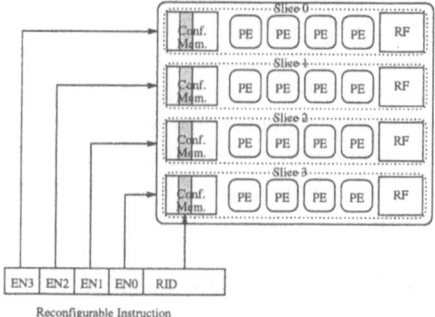

Fig. 3. Fields of a reconfigurable instruction. EN: slice enable bit, RID: reconfigurable instruction identifier

figurable array). In the case of CRISP, with processing elements of complexity similar to standard functional units, existing VLIW techniques have been reused.

On our research compiler (based on Trimaran [1]), code generation for loops is based on software pipelining. In software pipelining, iterations are initiated at regular intervals and execute simultaneously but in different stages of the computation. This allows mapping the available parallelism onto the number of resources of CRISP. With this technique [2], the code generated for a loop will contain as many configurations as the initiation interval of the loop (the number of cycles of the loop kernel). It is therefore important to check that an iteration does not last more than the number of available contexts in the configuration cache. If this was not the case, the generated code would need constant reconfiguration and performance would fall down.

Software pipelining can also be modified to exploit the ability to perform spatial computation by chaining operations [2]. This allows a reduction of the critical path length of inner loops, with the corresponding decrease in execution time. The process of code generation with spatial computation requires a proper model of the timing delay of the processing elements and the interconnect, since the process is similar to the place and route stage in FPGAs.

Additionally, our compiler studies for each loop the required number of slices. Only the necessary number of slices are used, in order to reduce both reconfiguration times and energy consumption.

4   Experimental Setup

In order to evaluate the performance and power consumption of the proposed architecture, we ran a set of multimedia applications on a simulated processor. The list of benchmarks can be seen on table 1. The applications were compiled using a modified version of the Trimaran [1] compiler that includes the compilation techniques described in the previous section. This compiler is able to exploit the coarse-grained reconfigurable unit present in the reconfigurable processor. Spa-

Table 1. Benchmarks

Benchmark	Application type
ADPCM decode	Audio
ADPCM encode	Audio
AES	Encryption
JPEG decode	Image
MPEG2 decode	Video

tial computation is not currently implemented in our prototype compiler and therefore was not used.

The compiled aplications were simulated in a cycle accurate simulator that models the processor core, the reconfigurable array and the memory hierarchy, including all cache effects. This simulator provided a series of trace files that were later used to compute the power consumption of each application.

In order to calculate the power consumption of the reconfigurable processor, we have used Wattch [3] as a starting point for our power calculations. Wattch is a power consumption model for superscalar processors that can be customized to different architecture models. The power contribution of the different components of the processor is calculated using analytical models (for array structures like memories or register files), through empirical models (for the functional units) or via a combination of the two (for the clock distribution network).

Power consumption in a coarse-grained reconfigurable instruction set processor comes from two main sources: i) the main processor core, and ii) the reconfigurable unit. Since our main processor core is a RISC processor, it is straightforward to calculate the power consumption of the processor core using the Wattch building blocks. The power of the coarse-grained reconfigurable logic is modeled using these same Wattch blocks as shown in table 2.

Our reference processor is a RISC processor whose parameters can be seen on table 3. This RISC processor was extended with a reconfigurable functional unit with varying number of reconfigurable slices. The contents of each slice are shown in table 4. The number of slices was changed from 1 to 4.

The multibanked data memory was modeled as several independent SRAM memories. The performance and power consumption effect of the interconnect required in such a multibanked memory were not modeled.

Table 2. Mapping from reconfigurable logic components to Wattch power models

Reconfigurable logic component	Wattch power model
Register file	Register file
Processing element	Functional unit
Configuration memory	SRAM memory
Interconnect	Result bus
Clock	Clock

Table 3. RISC processor parameters

Parameter	Value
Instruction Level 1 Cache	2Kbytes, block size 16, 128 sets, direct mapped
Unified Level 2 Cache	64Kbytes, block size 32, 512 sets, 4 way set associative
External memory	SDRAM, 18 cycles latency
Registers in main processor core	32 32-bit registers
Data memory	Multi-banked SRAM 8Kbytes
Frequency	600MHz
Technology	0.35um

Table 4. Slice parameters

Parameter	Value
Processing elements	1 Multiplier, 1 ALU/shifter and 2 ALUs
Register file	32 32-bit registers
Configuration memory	32 configurations of 128 bits

5   Results

Figure 4 left shows the normalized execution time of the set of benchmarks in several configurations. RISC represents the baseline RISC processor and the other entries represent the RISC processor with the number of reconfigurable slices ranging from 1 to 4. From this graph we can see that as the number of slices is increased, the execution time drops until the curve saturates. After this saturation point adding extra slices does not improve the performance. The saturation point depends on characteristics of the benchmark. ADPCM encode saturates with just one slice, while others like ADPCM decode or mpeg2dec profit with 3 or more slices. From this figure we see that the average performance increase is 2.5 and the maximum is 5.

Figure 4 right shows the normalized energy consumption for the benchmark set. In general, the total energy consumption increases as more resources are added to the processor. The fact that the increase in energy consumption is not as fast as one might expect is derived from a better utilization of the processing elements of the processor as more slices are added. After the performance saturation point, the energy consumption is only increased by the extra load in the clock network (better clock gating would reduce this effect). The extra slices that are not using are basically shut off and hence their contribution to the energy consumption is negligible.

Figure 5 left shows the normalized energy delay product (calculated as application execution time in cycles times the application energy consumption). It can be observed from there that all reconfigurable processors have a better energy delay product than the baseline RISC. The reason for this is that the reconfigurable processors are able to exploit the parallelism in the application reducing the number of instructions that need to executed.

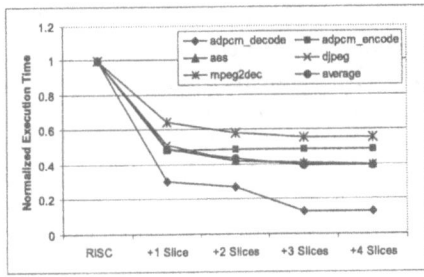 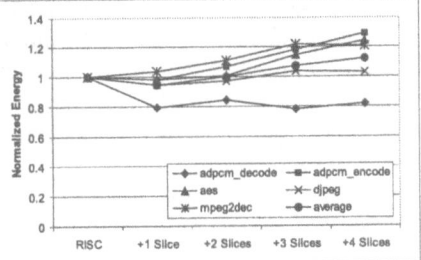

Fig. 4. Normalized execution time versus number of reconfigurable slices (left) and normalized energy versus number of reconfigurable slices (right)

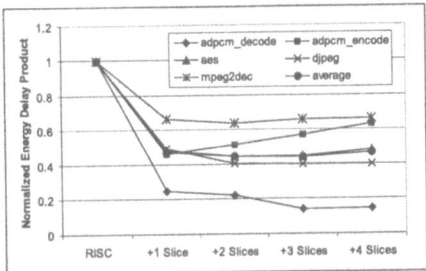 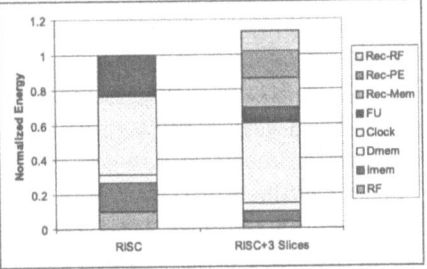

Fig. 5. Energy-delay product versus number of reconfigurable slices (left), and, normalized energy distribution for a RISC processor and a RISC processor with 3 slices for benchmark AES (right)

Figure 5 right shows the different components to the energy consumption of the RISC processor and a RISC processor with 3 slices for the benchmark AES. The benchmark AES runs at 2.5 the speed of the RISC processor and consumes only around 15% more energy consumption. We can see that a significant part of the energy consumption of the RISC processor is transfered to the reconfigurable components in the reconfigurable processor. Energy in the instruction memory decreases at an increase in the energy of the configuration memory. Energy from the functional units decreases and is transfered to the processing elements (resulting in almost the same energy consumption). In the case of the register file energy, the energy in all register files (processor core plus reconfigurable logic) is increased due to the usage of more power consuming units in the reconfigurable logic (multiported register files). Finally, the clock power is also increased, and from what can be seen from the figure, represents a significant amount of the energy consumption. Clock power reduction techniques would certainly reduce the power of both the RISC and reconfigurable processor.

As mentioned in the introduction, the proposed architecture is low power. The reasons for this are several:

- Coarse-grained datapath elements: they are much more energy efficient than bit level elements due to the small control overhead they require.
- Small and sliced configuration memory: Even though that constant switching in the configuration memory is required, the total power contribution of the configuration memory is marginal. Additionally, the fact that the configuration memory slices can be selectively enabled contributes to an even smaller energy consumption.
- Sliced datapath that limits the complexity and energy consumption of the different elements.
- Intelligent compiler that is energy aware and tries to minimize the number of active slices without compromising performance

6   Related Work

Coarse-grained reconfigurable processors like [4, 7, 6] have been shown to be extremely efficient for executing multimedia and DSP applications. [6, 7] present processors with a reconfigurable fabric similar in complexity to ours. One of the main differences are the way registers are spread over the array and the way the configuration memories are controlled. None of them provide for a mechanism to selectively load a configuration to only a subset of the reconfigurable array. [4] has a concept similar to the reconfigurable slices but is applied for a different reason, namely increasing the virtual size of the hardware. Another important difference of our work with previous others is that none have studied the power implications of their architectures.

Regarding the topic of power estimation for reconfigurable architectures, there has been work on modeling FPGA power consumption, like [8]. We are currently unaware of power models for coarse-grained processors, but fortunately, models for superscalar and VLIW processors can be reused. Some of them are [3, 9, 5].

7   Conclusions

This paper has presented a coarse-grained architecture designed for low power multimedia applications. The reconfigurable logic is divided in slices that can be independently activated to reduce the power consumption of the processor. The processor achieves an average 2.5 performance increase over a standard RISC processor with just an 18% energy overhead.

Future work will study in more detail the power consumption of the processor. Also, we will study the effects of spatial computation on the performance and power consumption of the processor.

Acknowledgements

This project is partially supported by the IWT through MEDEA+ project A502 MESA and by the Fund for Scientific Research -Flanders (FWO) through the postdoctoral fellowship of G.Deconinck.

References

1. Trimaran: An infrastructure for instruction level parallelism. http://www.trimaran.org, 1999.
2. Francisco Barat, Murali Jayapala, Pieter Op de Beeck, and Geert Deconinck. Software pipelining for coarse-grained reconfigurable instruction set processors. In *Proc. ASP-DAC*, pages 338–344, January 2002.
3. David Brooks, Vivek Tiwari, and Margaret Martonosi. Wattch: a framework for architectural-level power analysis and optimizations. In *Proc. 27th Int'l Symp. Computer Architecture (ISCA 2000)*, pages 83–94, June 2000.
4. Seth Copen Goldstein, Herman Schmit, Matthew Moe, Mihai Budiu, Srihari Cadambi, R. Reed Taylor, and Ronald Laufer. PipeRench: a coprocessor for streaming multimedia acceleration. In *Proc. 26th Int'l Symp. Computer Architecture (ISCA 1999)*, pages 28–39, May 1999.
5. Weiping Liao and Lei He. *Compilers and Operating Systems for Low Power*, chapter Power Modeling and Reduction of VLIW Processors. Kluwer Academic Publishers, 2002.
6. Guangming Lu, Hartej Singh, Ming-Hau Lee, Nader Bagherzadeh, Fadi J. Kurdahi, and Eliseu M. Chaves Filho. The morphosys parallel reconfigurable system. In *European Conference on Parallel Processing*, pages 727–734, 1999.
7. Takashi Miyamori and Kunle Olukotun. REMARC: Reconfigurable multimedia array coprocessor. In *Proc. 6th Int'l Symp. Field-Programmable Gate Arrays (FPGA98)*, February 1998.
8. Kara K. W. Poon, Andy Yan, and J. E. Steven Wilton. A flexible power model for fpgas. In *Proc. 12th Int'l Workshop Field-Programmable Logic and Applications (FPL 2002)*, September 2002.
9. W. Ye, Narayanan Vijaykrishnan, Mahmut T. Kandemir, and Mary Jane Irwin. The design and use of simplepower: a cycle-accurate energy estimation tool. In *Proc. DAC*, pages 340–345, June 2000.

Encoded-Low Swing Technique
for Ultra Low Power Interconnect

Rohini Krishnan, Jose Pineda de Gyvez, and Harry J.M. Veendrick

Philips Research Laboratories,
Prof. Holstlaan 4, 5656AA Eindhoven, The Netherlands,
rohini.krishnan@philips.com

Abstract. We present a novel encoded-low swing technique for ultra low power interconnect. Using this technique and an efficient circuit implementation, we achieve an average of 45.7% improvement in the power-delay product over the schemes utilizing low swing techniques alone, for random bit streams. Also, we obtain an average of 75.8% improvement over the schemes using low power bus encoding alone. We present extensive simulation results, including the driver and receiver circuitry, over a range of capacitive loads, for a general test interconnect circuit and also for a FPGA test interconnect circuit. Analysis of the results prove that as the capacitive load over the interconnect increases, the power-delay product for the proposed technique outperforms the techniques based on either low swing or bus encoding. We also present the signal to noise ratio (SNR) analysis using this technique for a CMOS $0.13\mu m$ process and prove that there is a 8.8% improvement in the worst case SNR compared to low swing techniques. This is a consequence of the reduction in the signal switching over the interconnect which leads to lower power supply noise.

1 Introduction

As process geometries continue to shrink and we enter the nanometer era, the interconnects and the drivers and receivers associated with them belong to the major energy consumers on an integrated circuit. As more complex circuits are integrated in a single chip, with global busses, clock lines and timing circuits running across the chip, the fraction of energy consumed by the interconnect is ever increasing. The fraction of energy dissipated over interconnect and clock lines was found to be 40% for gate array based designs, 50% for cell-library based designs and 90% for traditional FPGA devices [7, 8].

Methods to reduce the amount of energy consumed by the interconnect have been extensively researched in the literature. In the past, encoding techniques like work zone encoding[9], bus invert coding[3] etc. have been proposed for inter and intra chip interconnects while low swing techniques [1] have been used for intra-chip interconnects. Encoding has been mainly applied only for I/O circuits due to the presence of huge external and parasitic capacitances at I/O pads. But with the advent of reconfigurable FPGAs, even on chip interconnect (with pass

P.Y.K. Cheung et al. (Eds.): FPL 2003, LNCS 2778, pp. 240–251, 2003.

transistors in connection boxes and switch boxes), have a rather large capacitive load, warranting the use of encoding techniques on-chip. By using bus-invert encoding as proposed in [3], the average power on the bus cannot be reduced by more than 25%. Other techniques have to be used to obtain larger energy reductions. Reducing the voltage swing of the signal on the wire has been the most efficient technique for reducing the power quadratically and power-delay product linearly. However low swing techniques suffer from lower noise immunity and reduced signal to noise ratios. Encoding technique for noise reduction has been proposed in [2] but it is not energy optimal.

In this paper we present a novel encoded-low swing technique that combines the advantages of encoding and low swing. We present exhaustive simulation results over a benchmark test architecture (for general and for FPGA interconnect) which includes the driver and receiver circuitry. The analysis is done for a range of capacitive loads. We compare the proposed technique with existing techniques namely, "no encoding no low-swing" (full-swing CMOS) approach, "low swing" approach, and the "encoding" approach.

We examine the results for a random set of data to be transmitted over the interconnect keeping in mind that the best case energy savings for encoding is when the number of transitions over the interconnect is reduced to zero (ideal case when the data values are allowed to flip every clock cycle) and the worst case for encoding is when N/2 values over a N bit wide interconnect flip every cycle. We do our study for a particular low swing technique, namely Static Driver with Voltage Sense Translator, abbreviated as SDVST-II[6] and a particular encoding technique, bus-invert coding[3]. We chose to compare against these two techniques since we use the SDVST-II and bus-invert coding for generating the encoded-low swing signals as well (the same transmitters and receivers are used for generating the low-swing, encoded and encoded-low swing signals thus ensuring a fair comparison). The results are general and hold for other low swing and encoding techniques also.

The paper is organized as follows. First, the benchmark circuits (general and FPGAs) used for all our simulations and comparisons are presented. The encoded-low swing technique and an efficient circuit implementation of the same is then presented. This is followed by a comparison with existing techniques. The effect of crosstalk and process variations over the schemes is also analysed. This is followed by a signal to noise ratio analysis. Finally the summary and the conclusions drawn are presented. Throughout this paper, we refer to the "no encoding no low-swing" approach by the acronym NENL, "encoded-low swing" approach by the acronym EL, "low swing" approach by the acronym L, the "encoding" approach by the acronym E.

2   Test Architecture

We use the test architecture of Fig. 1[1] for the purpose of evaluating the effectiveness of our approach for general interconnects. Fig. 1 shows the schematic of the benchmark interconnect circuit. An inverter prior to the driver and after

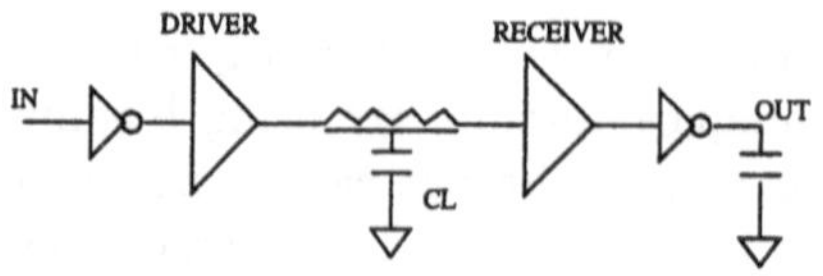

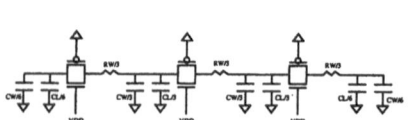

Fig. 1. *Benchmark test architecture*

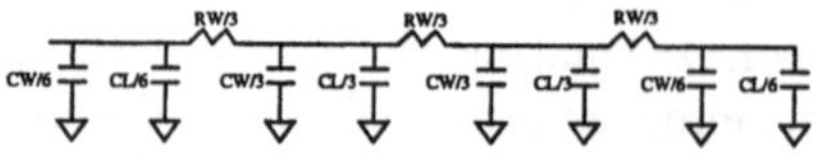

Fig. 2. *General interconnect model*

the receiver is added with 20fF capacitive load. Both the inverters are sized with Wp=1.2um and Wn=0.6um. Fig. 2 shows the general interconnect line which is a metal-3 layer wire with a length of 1 mm modeled by a $\pi 3$ distributed RC model with an extra capacitive load, CL, distributed along the wire for fanout. We distribute CL since the exact location of the fanout load is not known, and so we cannot use a lumped capacitance. RW and CW stand for the wire resistance and wire capacitance respectively.

Fig. 3. *FPGA interconnect model for short length*

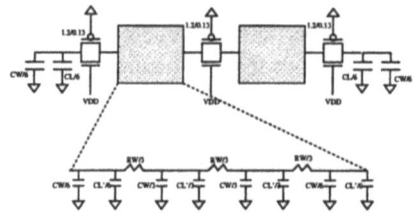

Fig. 4. *FPGA interconnect model for long length*

We have developed the test architecture in Fig. 3 and Fig. 4 for FPGA interconnects. In the FPGA interconnect model, we incorporate pass transistor switches to replicate the path over which signals travel between logic blocks[5]. Note that we actually use complementary pass gates instead of NMOS pass gates to get rid of the threshold drop and simplify our analysis. Usage of complementary pass gates instead of NMOS pass gates increases the delay of the signal. But since we use the same test architecture for comparing all the techniques, the delay is equally degraded for all the schemes and our comparison remains fair. Fig. 3 shows the RC interconnect model which is used when the wire connecting the complementary pass transistor switches is short. When the wires are long, the RC model in Fig. 3 is replaced by a distributed RC model as shown in Fig. 4. In Fig. 4, CL' represents the capacitive load due to fanout between two complementary switches.

As already mentioned, we compare four cases, the NENL, EL, L and E approaches. In the NENL approach, the driver and the receiver are just buffers. In the EL approach, the driver converts the input signal into an encoded-low swing signal, which is converted back to its original value and level by the receiver. In the L approach, the driver is a low swing converter and the receiver is a level restorer. In the E approach, the driver is an encoder and the receiver is a decoder. All circuit comparisons are based on CMOS $0.13\mu m$ process parameters and spice models. The minimum drawn channel length for this process is set to $0.13\mu m$.

3 Encoded Low Swing Technique

In this section, a brief overview of the proposed technique is presented. In the proposed encoded-low swing scheme, the current values to be transmitted on the bus are compared with the previous state of the bus. When the number of bits flipping is greater than $\frac{N}{2}$ where N is the width of the bus, the decision to transmit the inverted signal values is made. In addition, an "invert" signal is also sent to the receiver to indicate whether the bus values are inverted or not. These encoded values are then converted into their low swing equivalents and transmitted. In this way, we ensure that the energy consumed over the interconnect is minimum. Our two pronged strategy not only reduces the probability of transitions over the interconnect but also transmits only low swing values to achieve tremendous energy reductions. This energy saving can only be achieved if we have an efficient driver and receiver circuit, which does not consume more energy than is saved over the interconnect. For that, an efficient circuit implementation has been developed.

We first theoretically estimate the energy savings that are possible using the proposed technique. We can estimate the average number of transitions using probabilistic analysis for a N bit wide bus. The dynamic switching energy of the bus is given by Eqn.1[1].

$$E_{dyn} = C_{average} V_{ref}^2 T \tag{1}$$

In Eqn. 1, T is the total number of transitions over the wire. Without encoding, the transitions, T_{NE}, for an average case for a N bit wide bus is

$$T_{NE} = \sum_{M=1}^{N} P(M).M \tag{2}$$

where T_{NE} denotes the number of transitions without encoding, P(M) denotes the probability that M bits flip in a N bit wide bus and is given by

$$P(M) = \frac{1}{2^N} C \binom{N}{M} = \frac{1}{2^N} \frac{N!}{(N-M)!M!}. \tag{3}$$

By using the bus-invert coding method , we compute the transitions for an average case for a N bit wide bus. We differentiate between the cases when N is odd and N is even. This is shown next.

Case a: When N is Odd. Using bus invert coding, the number of transitions is given by Eqn. 4. T_E indicates the number of transitions over the bus in the presence of encoding. Here, when the number of bit flips exceeds $\frac{N+1}{2} - 1$, the decision to invert the data bits is made. Counting the extra transition due to the invert signal, the number of transitions over the bus, when $\frac{N+1}{2}$ data bits flip, is $N - \frac{N+1}{2} + 1 = \frac{N+1}{2}$.

$$
\begin{aligned}
T_E = \frac{1}{2^N} [& 1C\binom{N}{1} + 2C\binom{N}{2} + \cdots + \\
& + \left(\frac{N+1}{2}\right) C\binom{N}{\frac{N+1}{2}} + \left(\frac{N+1}{2} - 1\right) C\binom{N}{\frac{N+1}{2}+1} + \\
& + \left(\frac{N+1}{2} - 2\right) C\binom{N}{\frac{N+1}{2}+2} + \cdots + 1C\binom{N}{N}] .
\end{aligned}
\tag{4}
$$

Case b: When N is Even. Here, when the number of bit flips is exactly N/2, there is no advantage in encoding. We then take the decision of inverting the values on the bus if it does not cause a transition over the "invert" signal itself. This means that when N is even, an extra state flip flop for storing the state of the "invert" signal is needed which is not the case when N is odd.

$$
\begin{aligned}
T_E = \frac{1}{2^N} [& 1C\binom{N}{1} + 2C\binom{N}{2} + \cdots + \left(\frac{N}{2} + 1\right) C\binom{N}{\frac{N}{2}} + \\
& + \frac{N}{2} C\binom{N}{\frac{N}{2}+1} + \left(\frac{N}{2} - 1\right) C\binom{N}{\frac{N}{2}+2} + \cdots + 1C\binom{N}{N}]
\end{aligned}
\tag{5}
$$

Substituting values in (1),(2),(4),(5) the expected average energy per unit capacitance curves (NENL, EL_average, L, E_average) for different values of N varying from 1 to 16 for the different techniques, is shown in Fig. 5. For the NENL technique and E technique, we use $V_{ref} = V_{DD} = 1.2V$. For the L and EL technique we use $V_{ref} = 0.8V$. The best case energy savings for the bus-invert coding technique, is when all the bits over a N bit wide bus flip.

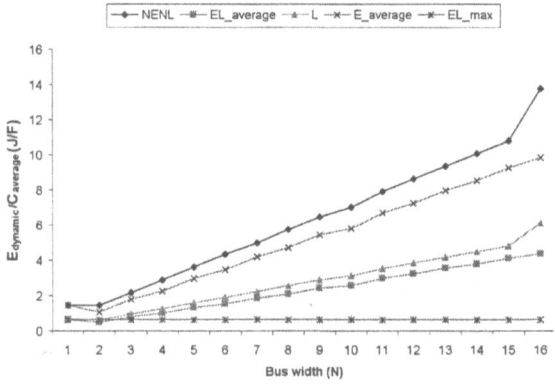

Fig. 5. *Theoretical energy for different schemes*

In this case, for the techniques which do not employ encoding, the number of transitions would be N, and for the techniques which employ encoding, the number of transitions would be 1 (due to the invert signal). So the lower bound for the energy/capacitance value is $V_{ref}^2 * T_E = V_{ref}^2 * 1$, this is indicated by the EL_max curve in Fig. 5. The curves in Fig. 5 do not show the energy consumed by the driver and receiver circuits. They only show the theoretical estimate of the amount of energy per unit capacitance that is consumed over the interconnect. The actual energy consumed, the delay and the power-delay curve inclusive of the contributions from the driver and receiver circuit are shown in the section on simulation results.

4   Circuit Implementation

We developed an efficient implementation of the driver for an 8 bit wide bus using an analog majority voter circuit as shown in Fig. 6. The receiver circuit is shown in Fig. 7. The current state of the bus (D0T, D1T, ... , D7T, INV) is compared with the new values to be transmitted. If majority of the bits have flipped, the analog majority voter sets the INVB signal (shown in Fig. 6) to high. The advantage of using the analog majority voter circuit is that it is easily scalable to larger bus widths with very little extra area overhead. The encoded signal values are then converted into a low swing value using the NMOS only push-pull driver[1,6]. The driver and receiver circuits consume very little power as is illustrated in our simulation results as well. In the driver, in the analog majority voter circuit, by using the clock as the gate signal for the PMOS transistors in the latch and for the NMOS transistor (at the bottom) acting as a current source, we ensure that there is never a path from the power supply to ground except during the clock transitions. In the receiver, since we use cascode circuitry and differential circuits, the short circuit current is reduced.

The receiver consists of a low-swing restorer and a decoder as shown in Fig. 7. The decoder consists simply of XOR gates, which uses the "invert" signal to either invert or not-invert the received values depending on whether the "invert" signal is 1 or 0.

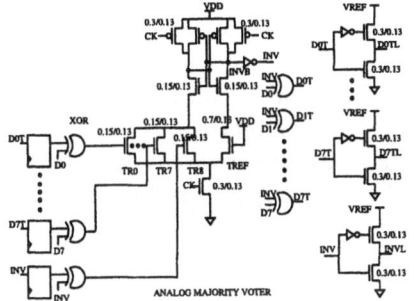

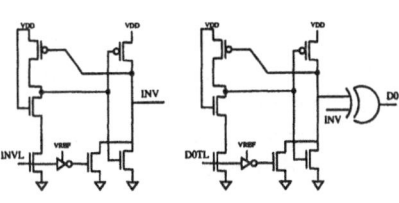

Fig. 7. *Encoded low swing receiver*

Fig. 6. *Encoded low swing transmitter for 8 bits*

5   Simulation Results and Comparison

The set up that we use for comparing the four techniques, the NENL, EL, L and E was illustrated in Fig. 1. It has to be noted that the power-delay product that we obtain for the E scheme and the EL scheme are data dependent. The best case energy savings for E and EL is when all the bits flip, and the worst case is when N/2 bits flip. We take care to ensure that the data sequence does not consist of bits which flip every cycle, since this represents the best case for bus-invert coding and this would result in an unfair comparison. The simulation results that we obtain are averaged over a random sequence of data bits. All the results(power, delay and power-delay product) are inclusive of the transmitter and receiver circuits.

5.1   FPGA Interconnect for 8 Bits

We perform the simulations over the benchmark FPGA interconnect circuit shown in Fig. 3 and Fig. 4. The proposed EL technique is particularly interesting here due to the presence of large capacitive loads over FPGA interconnects. Even though each wire in a FPGA channel can have different sources and sinks depending on the configuration that is loaded, the proposed technique can be applied to FPGAs when the logic block is doing datapath operations where normally the granularity is higher like a 4 bit addition, 8 bit multiplication etc. The energy and delay over FPGA interconnects are larger than over general interconnects due to the presence of series complementary pass transistor switches which increase the resistance and capacitance over the path of the signal. This is confirmed in our simulation results as well. As the capacitive load begins to increase, the encoded-low swing technique consumes the lowest energy. By reducing the number of transitions, and subsequently the number of times the capacitance has to be charged and discharged, over the interconnect, the schemes employing encoding (E and EL) have lower delays than the L scheme, but higher than the NENL scheme. Fig. 8 shows the plot of power-delay product against capacitive load. For low capacitive loads (CL≤100fF), the L scheme has the best power-delay product. But, as CL increases, the EL scheme outperforms the rest and has the lowest power-delay product.

The above simulations were performed for a fixed length of interconnect (L=1mm) and the capacitive load due to fanout (CL) was varied from 0fF to 2pF. We also performed simulations over the FPGA test interconnect circuit keeping CL constant at an average value of 400fF and varied the length of the interconnect from 0.1mm to 1mm. The results are summarised in Fig. 9. Fig. 9 essentially shows how the power-delay product varies with wire resistance, RW, and wire capacitance, CW. It can be seen that the proposed technique has the best power delay product.

5.2   General Interconnect for 8 Bits

Analysis of Fig. 10 illustrates that power values increase almost linearly against capacitive load (CL) but with different slopes for different schemes. As the ca-

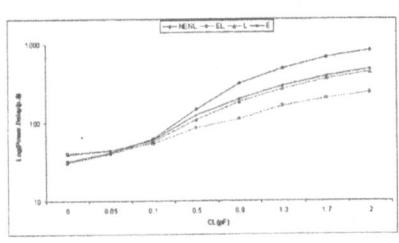

Fig. 8. *Power-Delay vs CL for L=1mm for FPGA interconnect*

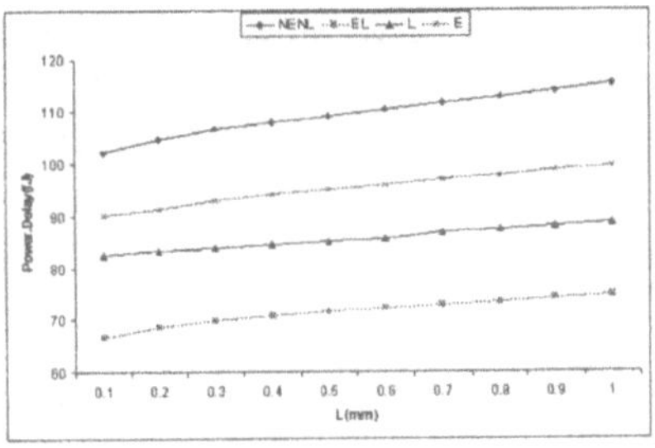

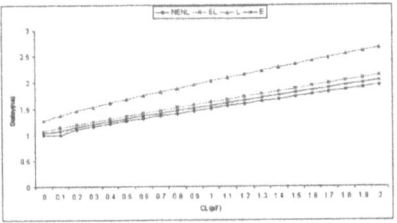

Fig. 9. *Power-Delay vs Length for CL=400fF for FPGA interconnect*

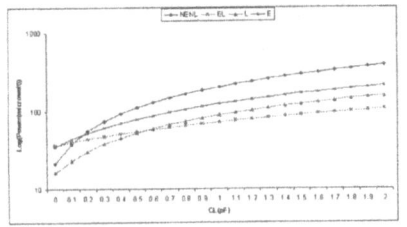

Fig. 10. *Power vs CL for L=1mm for general interconnect*

Fig. 11. *Delay vs CL for L=1mm for general interconnect*

pacitive load begins to increase, the encoded-low swing technique outperforms the other techniques. It can be seen from Fig. 11 that the signal delay over the interconnect is the lowest for the NENL(full-swing CMOS) scheme and is the highest for the L scheme, as expected. By reducing the number of transitions, and subsequently the number of times the capacitance has to be charged and discharged, over the interconnect, the schemes employing encoding (E and EL) have lower delays than the L scheme. Fig. 12 shows the plot of power-delay product against capacitive load. For low capacitive loads (CL≤200fF), the L scheme

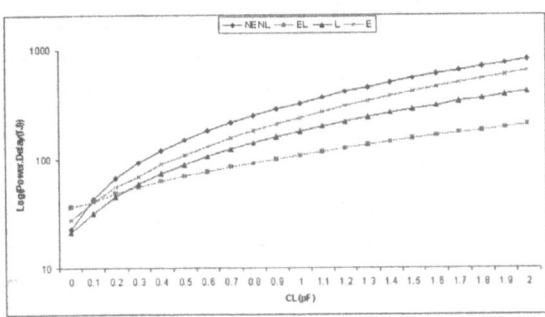

Fig. 12. *Power-Delay vs CL for L=1mm for general interconnect*

has the best power-delay product since the complexity of the driver and receiver circuit is less than that of the driver and receiver of the EL scheme. But, as CL begins to increase, the savings from driver and the wire begin to dominate and the energy overhead of the receiver remains almost constant. Then the proposed EL technique has the best power-delay product.

6   Effect of Crosstalk and Process Variations

We simulated the effect of crosstalk and process variations for the four techniques that we are comparing. It is summarized in Table 1. To simulate the effect of crosstalk we assume that the crosscoupling capacitance value is as specified in the CMOS 0.13μm process for a metal 3 wire at minimum spacing. We assume a victim wire surrounded by two aggressors(3 bit wide interconnect). We assume the worst case scenario, in which the aggressors switch in the same direction and induce a noise over the victim wire, which is switching in the opposite direction. In this way, we are able to measure the effect on signal delay, of crosstalk induced slow down as well. We simulate this condition over the slow, fast and typical process corners. We denote this for a L=1mm interconnect line with a capacitive load for fanout (CL) assumed to be 100fF. It can be seen from the table that in the presence of crosstalk noise, the energy consumed and the delay over the interconnect increases for all the schemes. But, the techniques employing encoding are, on an average, less affected by the presence of crosstalk noise because of the reduction in the number of transitions. In the presence of crosstalk noise, the best power-delay product is of the proposed encoded low swing technique.

Table 1. Effect of crosstalk and process variations

Scheme	Process	Power(P)(μW)	Delay(D)(ns)	PD(fJ)
NENL	slow	51.81	1.051	54.45
	typical	51.86	1.016	52.68
	fast	52.25	0.946	49.428
Proposed EL	slow	22.425	1.257	28.188
	typical	23.55	1.264	29.767
	fast	24.805	1.209	29.989
L	slow	25.15	1.434	36.065
	typical	26.285	1.405	36.93
	fast	29.51	1.376	40.605
E	slow	24.195	1.209	29.25
	typical	25	1.208	30.2
	fast	25.965	1.141	29.626

7   Signal to Noise Ratio Analysis

We use the worst case analysis method presented in [4] to measure the signal integrity of each circuit. The noise sources are classified into two categories: the proportional noise sources and the independent noise sources

$$V_N = K_N V_S + V_{IN} \tag{6}$$

$K_N V_S$ represents those noise sources that are proportional to the magnitude of the signal swing(V_S) such as crosstalk and signal induced power supply noise. V_{IN} includes those noise sources that are independent of V_S such as receiver input offset(due to process variation), receiver sensitivity, and signal-unrelated power supply noise.

Table 2 summarises the noise sources and their contributions. The crosstalk coupling coefficient is defined as $K_C = \frac{C_C}{C_B + C_C}$ where C_C represents the coupling capacitance and C_B represents the bottom capacitance. In the CMOS 0.13μm process, the value of C_C is 109fF/μm and C_B is 113fF/μm. The crosstalk noise attenuation factor given by Atn_C[1] is defined to be 0.2 for a static driver and 1 for a dynamic driver. The signal induced power supply noise is estimated to

Table 2. Parameters in signal to noise ratio analysis

K_C	Crosstalk coupling coefficient=0.49 for 1mm wires with 3fF load and 0.18μm spacing
Atn_C	Crosstalk noise attenuation=0.2 for static driver
K_{PS}	Power supply noise due to signal switching=0.05 for single ended signalling
Rx_O	Receiver input offset = 30mV for inverter
Rx_S	Receiver sensitivity = 30mV for inverter
PS	Unrelated power supply noise = 5% of V_{DD} 0.05*1.2 = 0.06
Atn_{PS}	Power supply noise attenuation= $\Delta V_{TH}/\Delta V_{DD}$=0.5
Tx_O	Transmitter offset=30mV

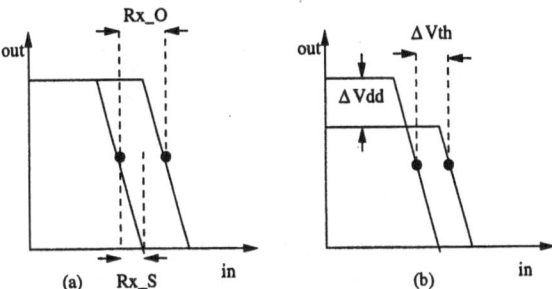

Fig. 13. *Voltage transform curves*

Table 3. Signal to noise ratio calculations

Scheme	K_{PS}	K_N	V_S	V_N	SNR
Proposed EL	0.025	0.123	0.8	0.2184	1.83
NENL	0.05	0.148	1.2	0.2976	2.016
L	0.05	0.148	0.8	0.2384	1.67
E	0.025	0.123	1.2	0.2676	2.242

be 5% of the signal-swing for single-ended signaling. Process variations such as device size mismatch, threshold voltage variation etc., will induce receiver input offset noise which is denoted by Rx_O. Rx_S indicates the receiver sensitivity. These are indicated in Fig. 13(a). Through simulations for the CMOS 0.13μm process, we find that Rx_O and Rx_S are 30 mV for an inverter.

The power supply noise attenuation Atn_{PS} is defined as $\frac{\Delta V th}{\Delta V dd}$ and is indicated in Fig. 13(b). Through simulations for the CMOS 0.13μm process we find that ΔVth is 30mV, and so we calculate Atn_{PS} to be $\frac{0.03}{0.05*1.2} = 0.5$. K_N and V_{IN} in Eqn.6 are defined by Eqn.7.

$$K_N = Atn_C K_C + K_{PS}, V_{IN} = Rx\_O + Rx\_S + Atn_{PS} PS + Tx\_O \qquad (7)$$

The noise margin is defined as $SNR = \frac{0.5 V_S}{V_N}$ where V_S stands for the signal swing and V_N is the total noise induced voltage which is defined by Eqn.6. The power supply noise is maximum when all the signals switch. The techniques employing encoding (namely EL and E) prevents all the signals from switching simultaneously, infact only 50% of the signals are allowed to switch. So, in the worst case, the number of transitions is 50% compared to unencoded case. So the probability of the worst power supply noise is correspondingly reduced. This leads to a reduction in the factor K_{PS}. Using the values in Table 2, we compute the SNRs for the four schemes that we are comparing. This is shown in Table 3. The techniques employing full swing, namely NENL and E, have the best signal-to-noise ratios. This is due to the fact that the signal swing indicated by V_S in the SNR equation is higher (1.2V) than the signal swings (0.8V) of the techniques employing low-swing, namely EL and L. But it has to be noted that the power-delay product of the NENL and E techniques are much higher than the EL and L techniques. For comparable power-delay products, it can be seen from Table 3 that the worst-case SNR of the proposed EL technique is better than the L technique by 8.8%. This can be attributed to a reduction in the power supply noise due to signal switching. The worst case flipping of all bits is prevented by the encoding, leading to a direct reduction in power supply noise.

8   Conclusion

We introduced a novel encoded-low swing technique and an efficient circuit implementation of the same. This achieves the best power-delay product over the existing schemes when the capacitive load over the interconnect begins to increase above 200fF. Analyses of the simulation results show that the average power-delay product of the proposed technique is superior by 45.7% with respect to techniques using only low swing, and by 75.8% with respect to techniques using only encoding for random data streams. In the presence of crosstalk noise, we show that the proposed technique has the best power-delay product even for small capacitive loads (CL$\leq$200fF). The signal to noise ratio of the proposed technique is superior to existing low swing techniques by 8.8%. We perform the case study for both general and FPGA interconnects and prove the feasibility of the proposed technique for both. Since the power-delay product over FPGA interconnects is larger than over general interconnects the proposed technique is ideally suitable for FPGA interconnects and the energy savings are significant with increased loads. A disadvantage in the current circuit implementation is the need to have a reference low voltage power supply, but this can be easily overcome by using some of the techniques that exist in literature for generating a low swing signal without a reference low voltage power supply. In the future, the possibility of reducing the number of buffered switches on a FPGA as a consequence of the proposed technique needs to be investigated and quantified.

References

1. H. Zhang et al, "Low-Swing On-Chip Signaling Techniques: Effectiveness and Robustness", *IEEE Transactions on VLSI systems*, Vol.8, No.3, pp.264-272, June 2000.
2. K. Nakamura et. al. "A 50% Noise Reduction Interface Using Low-Weight Coding", *1996 Symposium on VLSI Circuits Digest of Technical Papers*, Pg.144-145, 1996.
3. Mircea R. Stan "Bus-Invert Coding for Low-Power I/O", *IEEE Transactions on VLSI systems*, Vol.3, No.1, pp.49-58, March 1995.
4. W. Dally and J. Poulton, *Digital Systems Engineering, Cambridge, U.K., Cambridge Univ. Press, 1998.*
5. Jonathan Rose, Vaughn Betz *Architecture and CAD for deep-submicron FPGAs,*
6. Varghese George et. al., "Design of a low energy FPGA", *International Symposium On Low Power Electronics and Design*, pp.188-193, 1999.
7. D.Liu et. al., "Power consumption estimation in CMOS VLSI chips", *IEEE Journal Solid-State Circuits*, Vol.29, pp. 663-670, June 1994.
8. E.Kusse, "Analysis and circuit design for low power programmable logic modules", *M.S. thesis, Univ. Calif., Berkeley, 1997.*
9. Musoll, E. et. al., "Working-zone encoding for reducing the energy in microprocessor address buses" *IEEE Transactions on VLSI Systems*, Vol. 6, Issue: 4, pp.568-572, Dec 1998.

Building Run-Time Reconfigurable Systems from Tiles

Gareth Lee and George Milne*

Department of Computer Science & Software Engineering,
The University of Western Australia.
[gel,george]@csse.uwa.edu.au

Abstract. This paper describes a component-based methodology tailored to the design of reconfigurable systems. Systems are constructed from *tiles*: localised, self contained blocks of reconfigurable logic which adhere to a specified interface. We present a state-based model for managing a hierarchical structure of tiles in a reconfigurable system and show how our approach allows automatic garbage collection techniques to be applied for reclaiming unused FPGA resources.

1   Introduction

Despite rapid growth in the commercial applications of field-programmable logic, few current systems utilise run-time reconfiguration. We believe this reflects the fact that designers of reconfigurable systems lack methodologies and tools to guide the design process. In a previous paper [1] we advocated the benefits of applying software engineering ideas to the design of reconfigurable computing devices. We propose a methodology based on composition of *tiles*: localised, self-contained blocks of logic, adhering to a specified interfaces [4] (see Section 2).

The tiles approach draws from established ideas in object oriented programming and software components and builds on previous work on reconfigurable cores [2]. Since each tile extends an *abstract tile* (which provides a formal abstract data type, much as an interface in java or an abstract class in C++) this opens the possibility of top-down design. But, in common with reconfigurable cores, our approach also allows tiles to be defined as a composition of simpler tiles, allowing the traditional bottom-up approach.

This paper focuses on how tiles can be used to construct systems which are both partially and dynamically reconfigurable. To avoid confusion we take *partial reconfiguration* to denote systems which allow adaptation of a subsection of the overall circuit, while the remaining circuit elements remain unchanged. Similarly we consider *dynamic reconfiguration* to be a case where a circuit is modified over time in a series of discrete steps. (For the synchronous logic considered here, this requires that the global clock be temporarily halted while reconfiguration occurs, but the methodology also applies to asynchronous logic.) Thus a system

---

* This research was funded by a grant from the Australian Research Council. We would also like to thank Xilinx Inc., San Jose, for the donation of devices and software.

P.Y.K. Cheung et al. (Eds.): FPL 2003, LNCS 2778, pp. 252–261, 2003.

may be dynamically reconfigurable but not partially reconfigurable, implying that the entire device must be reconfigured at each step. However a system which is partially reconfigurable must also be dynamically reconfigurable.

The reconfiguration benefits offered by our approach, stem from the strict adherence to an abstract tile (or interface type) for every tile. During dynamic reconfiguration we do not allow the interface type of a tile to change, only the internal realisation. This ensures that the new concrete tile shares a common interface with the old tile and will therefore be *plug-compatible*. It also ensures the side-effects of reconfiguration are limited, since reconfiguration can be localised to the region occupied by the tile (described in Section 3.1). The tiles methodology also enforces a formal life-cycle for each tile (Section 3.2).

The approach promises widespread applicability since many reconfigurable computing applications are constructed from repetitive sub-components: for example, digital filter pipelines, systolic arrays and cellular automata.

We demonstrate our ideas in the form of a reconfigurable regular expression matcher (Section 4). The choice of application stems from our previous work, however we see the scope of this approach as being much wider than this particular type of example. The application allows many patterns to be concurrently searched for, within a continuous stream of text. It allows patterns to be added or removed at any time during operation, resulting in partial reconfiguration of the FPGA. The system manages its internal resources by applying a copying garbage collection algorithm [3] to reclaim discarded space (see Section 5).

This paper makes a contribution to the field in its application of a fixed-interface tile approach for design and realisation of run-time reconfigurable systems. The state-model we present and message interchange approach is novel in this application, as is the use of automated garbage collection techniques for managing FPGA resources.

2   The Tiles Methodology

This section provides an overview of the salient points of our approach. We provide a more detailed treatment of tiles and a comparison with other work in this area in a previous publication [1].

2.1   Abstract Tiles

The basis for our methodology is the *abstract tile*, which is a rectangular region of logic with a defined interface to its neighbours. The abstract tile consists of a set of signal ports, declared much as in other HDLs [4], but each signal is also attributed to a specific edge of the rectangle: north, south, east or west. Thus the interface also contains spatial information about the directions in which signals flow in and out of a tile, which draws on previous work on structural descriptions of systems [5]. In addition to the interface information the abstract tile also has a specification which states what function the tile performs. The abstract tile does not provide a circuit which implements this function (hence the term abstract).

Multiple *concrete tiles* can be implemented which adhere to a single abstract tile. Each concrete tile is a system of logic elements which adheres to the interface defined by a single abstract tile and functions within the scope of the abstract tile's specification. The set of concrete implementations of a particular abstract tile form an equivalence class since they all perform the same function and all have the same I/O signals and the same spatial characteristics.

In our example (Section 4) we create an abstract tile which corresponds to a finite state machine for detecting regular expressions. We implement multiple concrete tiles which contain the circuits for various regular expression operators: for instance, one to match sequences such as "abc" and one to match one or more occurrences of a pattern "a+". Each of these tiles has the same set of input and output signals: we know that signal 'x' enters through the eastern edge.

Abstract tiles do not have the size of the circuit as an attribute since they do not presuppose an implementation. The circuit for detecting patterns of the form "a+" may be physically larger than that for detecting "abc", for example.

2.2   Primitive and Composite Tiles

Concrete tiles may be formed in two ways. They may be designed at the circuit level using existing HDLs, but in our case we build these *primitive tiles* out of established run-time parameterisable cores [2], hereafter referred to simply as *cores*. In previous work [1] we referred to primitive tiles as 'component tiles', but feel the new name better describes their role.

Cores provide a way of localising and controlling the layout of the design elements constituting static systems, but the designer still has to resort to *ad hoc* techniques when reconfiguring the elements; our approach provides a more methodical strategy for programming reconfiguration (described in Section 3).

Alternatively, *composite tiles* may be formed by composing one or more primitive tiles together. This establishes a hierarchical relationship between tiles and sub-tiles which we utilise heavily. Thus an entire system consists of a tree of tiles with each composite tile having one or more offspring and primitive tiles only occurring as leaf nodes (for example, see Figure 1).

The controlling mechanisms provided by this hierarchy lead directly to benefits in FPGA resource management. Each tile has a state variable attached indicating whether it is currently realised on the FPGA substrate and whether a tile is in use or has been declared redundant. We use these attributes to explore the benefits of automated garbage collection for managing FPGAs by periodically reclaiming CLBs occupied by discarded tiles (see Section 5).

2.3   Controlling Objects

In common with JBits cores [2] and approaches such as JHDL [6] each tile has two distinct aspects: a physical circuit configured into a small rectangular region of the FPGA and a *controlling object* existing within the memory of a supervising computer. This computer re-programs the FPGA, modifying its internal state each time reconfiguration is required. Our approach differs from others since the

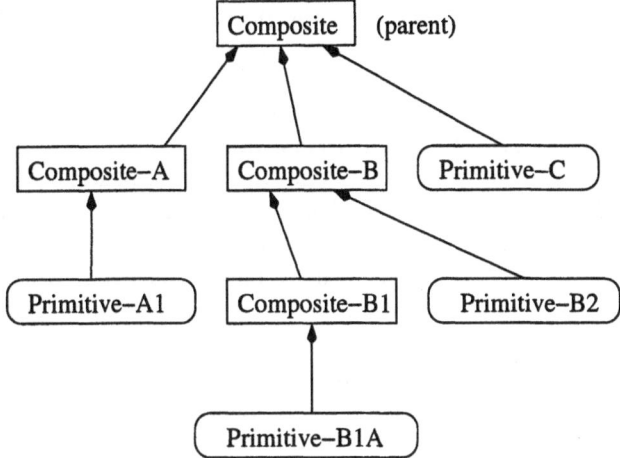

Fig. 1. An example of a tile hierarchy. Composite tiles are shown in rectangles and primitive tiles in boxes with rounded corners. We construct primitive tiles from JBits' ULPrimitive cores, but this has been omitted for clarity.

controlling object not only acts as an *on-the-fly* generator for the physical circuit, but persists for the lifetime of the circuit, coordinating future reconfiguration. We have designed these controlling objects as a class library in Java.

Centralised control also allows us to ensure that the overall system configuration remains valid at all times, guaranteeing correct behaviour and avoiding deadlocks. In general, such problems can be solved by applying formal modelling techniques, during design, to the reconfiguration state-space.

3   Reconfiguration Approach

As mentioned in the previous section, a system of tiles can be thought of as a tree with composite tiles forming the branches and primitive tiles as the leaves. Each tile within the tree has a separate map of configuration attributes (and values) associated with it. Both types of tile can be reconfigured.

During reconfiguration tiles may be thought of as *active objects*, rather than passive circuits on the FPGA substrate, since each has a controlling object associated with it in the memory of the supervising processor. Therefore, in the following discussion the term *tile* refers both to the circuit which is being reconfigured and its controlling object.

3.1   Reconfiguring Tiles

Primitive tiles can be be reconfigured when their configuration attributes are modified. For instance a tile might be created, as part of a digital filter, which multiplies a variable input by a constant value C. When the configuration attribute C changes value it results in the primitive tile reconfiguring. This reconfiguration may result in minor changes to the internal connections within the tile (such as *constant folding*, when C changes from 6 to 7) or more radical reorganisation. For instance, when C changes to 0 the tile can discard all its internal circuitry and set its output to constant zero.

Each *composite tile* has a set of one or more sub-tiles, which along with any additional cores, define the tile. The sub-tiles and sub-cores are embedded within the composite tile's area on the FPGA substrate. Each of the offspring tiles has an immutable type (specified by an abstract tile) which is set when the composite tile is first defined. However the implementation of each sub-tile cannot be resolved until run-time since it may depend on external events.

Several distinct concrete tiles may be created adhering to the same abstract tile, forming an equivalence class since its members are interchangeable. The members of this set will share a common structure (or interface to their neighbours) but each will have a different behaviour, albeit within the scope of the specification of their abstract tile. A parent composite tile may only be reconfigured when one member of the class is swapped for another. Assuming sufficient space is allocated by the parent composite tile for the physically largest member of the equivalence class, the effect of implementation will be limited to the region of the sub-tile (*i.e.* CLBs within the sub-tile will need to be reconfigured to implement the new circuit). However the effect of swapping a tile may *spill over* slightly, since connections to neighbouring tiles must also be re-routed.

3.2   Tile State Model

We allow tiles to be dynamically created, modified and destroyed, much as an object in an object oriented program. Tiles must therefore follow a well defined life-cycle so that they can allocate and deallocate their resources in an ordered fashion. This is achieved by associating a state variable with each tile, which is held within its controlling object. The state variable can take one of six states as show in Figure 2, which also shows the valid state transitions tiles can follow.

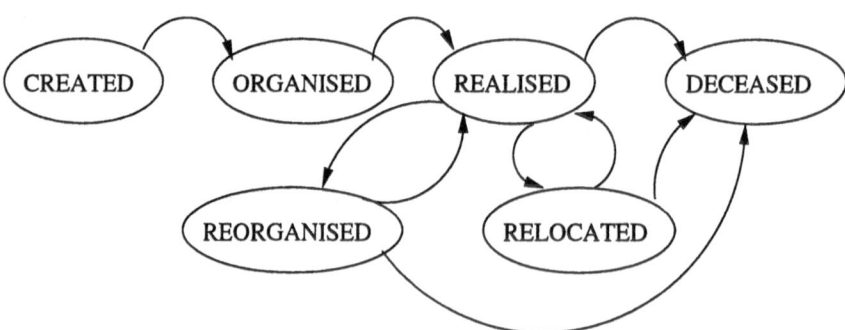

Fig. 2. The tile states and valid transitions. All tiles originate in the CREATED state and progress through configuration and optional reconfiguration to the DECEASED state, at which point their resources can be recycled.

Tiles coordinate their reconfiguration by exchanging messages between controlling objects (Section 3.3) instructing each to change state and by changing the configuration maps associated with their offspring (Section 3). To design a concrete tile using out approach the user has to program call-back handler functions in Java for some of the state transitions in Figure 2.

Tiles start in the CREATED state and must be configured before use. When a tile is instructed to move to the ORGANISED state by its parent, a tile reads its configuration parameters (if any) and creates its constituent parts. Composite tiles will organise themselves by creating and organising sub-tiles, whereas primitive tiles just create and configure their constituent cores. We refer to this organisational phase as *structural generation* [1].

Once all the tiles within the hierarchy have been organised a further parent message triggers transitions to the REALISED state. This instructs each controlling object to generate the appropriate circuit within the configuration bit-stream, which we refer to as *physical generation* [1].

Run-time reconfigurable systems will utilise three further states. Firstly, a tile can be instructed to move to the REORGANISED state when its configuration parameters are changed by its parent. A composite tile will reorganise its constituent parts, creating and deleting sub-tiles as appropriate. This may result in the circuit being regenerated and will move back to state REALISED once the bit-stream has been updated.

Secondly, a transition can occur to the RELOCATED state, when a parent tile instructs a sub-tile to physically relocate on the FPGA substrate. This can be used to create extra space for siblings when a parent tile is reorganising (or in garbage collection, Section 5.1). The state of relocating tiles must persist, therefore the controlling object detects all the memory elements within the tile and reads-back the values stored in these latches. Once the circuit corresponding to the tile is rebuilt at a new location on the FPGA substrate the latches will be preset to the stored values. (This poses a technical problem with Virtex FPGAs which have no direct I/O command to 'write-in' state to selected latches.)

Finally, superfluous tiles may be instructed to move to the DECEASED state so that they can disconnect all external connections. The region on the substrate corresponding to such a tile lays dormant. When a tile dies its controlling object can be garbage collected within the Java virtual machine. We describe a similar approach for reclaiming space on the FPGA substrate in Section 5.

3.3   Inter-tile Messaging

When tiles must be reconfigured their controlling objects coordinate by exchanging messages. Typically messages are sent from parent tiles to their offspring. Each message is an instruction to a sub-tile to change state, along with an optional map of configuration parameters. For instance, a parent tile might send a message to a REALISED sub-tile instructing it to enter the REORGANISED state, along with the changes to its configuration set that warrant the change. Alternatively, a message instructing a sub-tile to move to the DECEASED state does not require parameters (and is equivalent to the C++ delete operator).

Our current implementation treats message dispatch synchronously; the sender is blocked until the recipient has acted on the message. In future work we hope to investigate asynchronous dispatch of messages since this would allow the controlling objects to be executed on multiple supervisory processors. This would allow supervision to originate from multiple embedded processors such as the four PowerPC 405 processors included within Xilinx's VirtexPro FPGAs [7].

4   Regular Expression Application

A regular expression (regex) matching application was used as a vehicle to test this tile-based approach. Our interest in this application stems from the fact that it is representative of a wide class of reconfigurable systems [8]. Here, we extend an application presented previously [1], [9], to make it fully reconfigurable.

We assume the system is deployed in an environment where it receives a continuous stream of text, in which it is to detect a set of regular expression patterns. It generates an interrupt in real time when one or more patterns match and indicates which pattern(s) caused the interrupt. At any point the user may add additional patterns to the system or remove existing patterns. Any such command results in the clock being temporarily halted and the FPGA partially reconfigured. We therefore assume the data stream is buffered and that the FPGA, once restarted, can clear the backlog.

We previously presented a design process for finite state machines which match arbitrary regular expressions and how they can be implemented with tiles [1]. This system groups together a number of these regex sub-tiles into a 'Slice' composite tile which occupies a column within the FPGA. A Slice serves two purposes: (i) it groups the regex tiles together into a chunk, akin to a page in a virtual memory system, to facilitate garbage collection (described in Section 5) and; (ii) it combines the match outputs of an arbitrary number of regex tile via a chain of OR gates to detect whether any pattern is matched.

The top level composite tile, or 'Slice Manager', composes a user-specified number of Slices to build the system, as shown in Figure 3. Similarly, it ORs together the outputs of each Slice to generate a single aggregate match signal which acts as the interrupt source. When a match occurs this temporarily stops the data source sending more text and tests to see which Slice(s) are reporting a match. The state-persistence mechanism described previously in Section 3.2 provides a useful short-cut for reading-back the internal state of the regex tiles within the matching Slices to see which regular expressions are matching.

We have implemented this design on a Celoxica RC1000 board with a Virtex XCV1000 FPGA. The host computer acts as a supervisor, running our tiles library which uses Xilinx's JBits 2.8 API [2]. The application has not been optimised for performance but can be clocked up to 50 MHz. But since it uses four-phase handshaking to transfer text data through an 8-bit control port into the FPGA, this limits throughput to about 3000 chars/sec. However the regex tiles are fully pipelined and by using DMA to transfer data onto the board, much greater throughput should be possible.

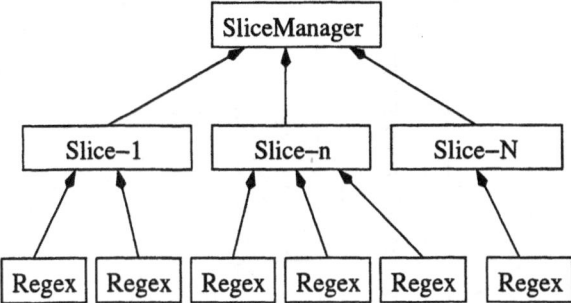

Fig. 3. The hierarchical structure of the regular expression matcher. The physical structure is shown in Figure 4. The Regex composite tiles are composed of, still simpler, primitive tiles.

Our current implementation does not consider how reconfiguration affects combinational logic delays. However the approach allows propagation delays to be precomputed for each tile and keeps inter-tile connections short, and therefore unlikely to impact on the global clock rate. This is a topic for further research.

5   Resource Management

The assumption underlying our approach is that a future system of field programmable logic will need to be able to adapt in complex ways to the environment in which it operates. This means that the system will need to be able to react to external events, such as someone plugging in a peripheral device, by dynamically creating internal circuit sub-systems to act as *drivers* [7].

One way this could be achieved would be with some form of *virtualisation* technology, providing virtual CLBs, similar to the virtual memory used by modern operating systems. Such a system would realise all the possible logic within its 'virtual' matrix of CLBs and then *page-in* the circuits on demand. There has been little progress in this direction; applying this concept to FPGAs, without diminishing the benefits of massively concurrent computation and communication, is a formidable technical problem.

Our approach assumes that exponential increases in FPGA size over time will ensure there is sufficient space available to implement all the sub-systems needed to perform a particular task. However, we assume that an application will need a mechanism for recycling its resources: old circuits will become redundant and must be deallocated, to allow CLBs to be recycled to create new sub-systems. In this section we explore whether tiles offer an appropriate level of granularity to support garbage collection.

5.1   Garbage Collection

As part of the regular expression matching system we have implemented a form of *incremental incrementally-compacting garbage collection* (GC) [3]. The im-

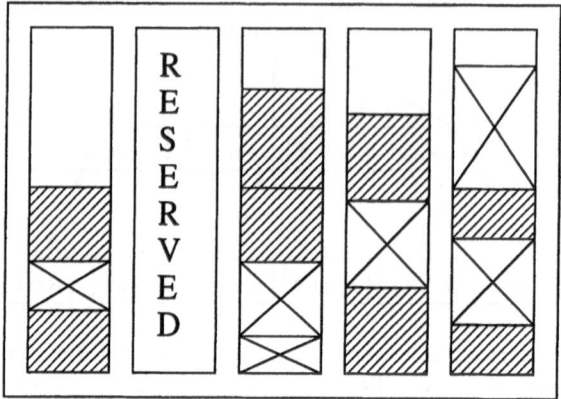

Fig. 4. The arrangement of tiles to permit GC. The outer rectangle corresponds to the SliceManager, whereas vertical columns represent Slices and each is filled with crosshatched regex tiles from the bottom. The white rectangle at the top of each column represents space which has not yet been used. Some tiles have been discarded and are marked with an X. Note the reserved Slice which is kept empty to permit GC.

plementation was simplified by the fact that each of the regex tiles is a single column wide and a variable number of rows high, depending of the complexity of the pattern. (This allows us to exploit the partial reconfiguration of individual columns provided by Virtex FPGAs.) Our design stacks up the expression tiles within columns of the FPGA as each new pattern is added. We store patterns within a user-specified number of columns but reserve a single column for GC. The arrangement of tiles within the FPGA is shown in Figure 4.

The tiles which correspond to discarded patterns are instructed by their parent Slice to enter state DECEASED. In response they disconnect themselves from their neighbours, but continue to occupy space within their column. When there is no longer space to add a new pattern the tile hierarchy is traversed to ascertain which column has the most 'dead' space. The active tiles within this column are copied across to the free column and the old Slice tile is instructed to enter the DECEASED state. The column is then overlaid by a new Slice tile, which is instructed to become ORGANISED then REALISED creating an empty column. This column then becomes the new reserved column to act as the destination for the next cycle of GC.

All GC techniques need to be able to traverse the available resources, usually the memory heap, following pointers to detect the memory regions that are used and those which, being unreachable, can be discarded. Our tile hierarchy, along with the state variable in each controlling object provides an ideal mechanism for following the same approach with FPGA resources.

Extending GC to arbitrary two dimensional arrangements of tiles would complicate the copying operation somewhat but doesn't fundamentally change the approach. We intend to investigate this in future work.

6   Conclusions

This paper has argued the need for design of fine-grained reconfigurable systems using a more methodical approach than current reconfigurable cores. We believe that the tiles methodology sensibly builds on existing reconfigurable cores and bridges the gap between APIs, such as JBits, and system design methodologies such as object orientation. It also provides a useful starting point for investigating the automation of reconfiguration and resource management, allowing design and implementation of reconfigurable systems in a more abstract way. Construction of a real-time regular expression matcher has both proved the efficacy of this approach and provided an initial implementation of a garbage collector designed to manage FPGA resources.

References

1. Lee, G., Milne, G.: A methodology for design of run-time reconfigurable systems. In: Proceedings IEEE International Conference on Field-Programmable Technology, CUHK, Hong Kong, IEEE Computer Society Press (2002) 60–67
2. Guccione, S.A., Levi, D.: Run-time parameterizable cores. In Lysaght, P., Irvine, J., Hartenstein, R.W., eds.: Field-Programmable Logic and Applications, Springer-Verlag, Berlin (1999) 215–222 Proceedings of the 9th International Workshop on Field-Programmable Logic and Applications, FPL 1999. Lecture Notes in Computer Science 1673.
3. Jones, R., Lins, R.: Garbage Collection – Algorithms for Automatic Dynamic Memory Management. John Wiley & Sons (1996)
4. Milne, G.J.: CIRCAL and the representation of communication, concurrency, and time. ACM Transactions on Programming Languages and Systems 7 (1985) 270–298
5. Bjesse, P., Claessen, K., Sheeran, M., Singh, S.: Lava: Hardware design in Haskell. ACM SIGPLAN Notices 34 (1999) 174–184
6. Bellows, P., Hutchings, B.: JHDL – an HDL for reconfigurable systems. In: IEEE Symposium on FPGAs for Custom Computing Machines, IEEE Computer Society Press (1998) 175–184
7. Lysaght, P.: FPGAs as meta-platforms for embedded systems. In: Proceedings IEEE International Conference on Field-Programmable Technology, CUHK, Hong Kong, IEEE Computer Society Press (2002) 7–12
8. Lee, G.: Expression of information processing systems using diverse programming systems. Technical Report 2002/1, Reconfigurable Computing Research Group, The University of Western Australia, http://www.csse.uwa.edu.au/~gel/reports/ (2002)
9. Gunther, B., Milne, G., Narasimhan, L.: Assessing document relevance with run-time reconfigurable machines. In Arnold, J., Pocek, K.L., eds.: Proceedings of IEEE Workshop on FPGAs for Custom Computing Machines, Napa, CA (1996) 10–17

Exploiting Redundancy
to Speedup Reconfiguration of an FPGA

Irwin Kennedy

Division of Informatics, University of Edinburgh,
Mayfield Road, Edinburgh EH9 3JZ, United Kingdom,
iok@dcs.ed.ac.uk

Abstract. Reconfigurable logic promises a flexible computing fabric
well suited to the low cost, low power, high performance and fast time
to market demanded of today's computing devices. This paper presents
an analysis of what exactly occurs when a fine grain FPGA, specifically
the Xilinx Virtex, is reconfigured, and proposes a tailorable approach
to configuration architecture design trading off silicon area with reconfiguration time. It is shown that less than 3% of the bits contained in
a typical Virtex reconfiguration bitstream are different to those already
in the configuration memory, and a highly parallelisable compression
technique is presented which achieves highly competitive results - 80%
compression and better.

1 Introduction

Reconfigurable Field Programmable Logic (FPL) based Custom Computing Machines (CCM) provide the ability to alter their function by writing to the contents of their configuration memory at run-time. This paper explores ways of
speeding up reconfiguration by analysing the changes that occur and proposing
architectures and algorithms to leverage the observations.

One of the central aims of dynamic reconfiguration is to perform useful work
by making the most efficient use of the silicon resources available. In applications
where many thousands of specialisations per second are necessary to have the
optimal algorithm implementation, the time to reconfigure the fabric would need
to be much less than a milli-second, otherwise the benefit of specialisation is lost.

This paper examines the specific problem of speeding up partial dynamic
reconfiguration of a fine grain FPGA. Section 2 introduces dynamic reconfiguration, gives relevant details of the Xilinx Virtex configuration architecture and
presents an overview of previous work done in the area of speeding up reconfiguration. Section 3 analyses the resource redundancy in a circuit implemented on
the Virtex. Section 4 presents an advanced configuration technique leveraging
the redundancy in an FPGA to speed up partial reconfiguration. Section 5 describes three partial bitstream compression algorithms together with an example
of a configuration architecture for using them.

P.Y.K. Cheung et al. (Eds.): FPL 2003, LNCS 2778, pp. 262–271, 2003.
© Springer-Verlag Berlin Heidelberg 2003

2   Background

The Xilinx Virtex device [4] is chosen as the FPGA fabric for experimentation in this paper since it is a commercial device with a large established market. Another advantage of choosing the Virtex is the availability of the Java JBits API, enabling low level access and manipulation of bitstreams produced by the Xilinx tool flow.

The Virtex is composed of Configurable Logic Blocks (CLBs), Input Output Blocks (IOBs), block RAMs (BRAM), clock resources, routing resources and configuration circuitry. All of these resources are configured by a configuration bitstream that is read and written through an 8-pin configuration port.

The configuration memory [5] is arranged into a series of columns, each of which can be visualised as stretching from the top to the bottom of the device. A column is divided into 48 frames, and the frame forms the atomic unit of configuration, meaning it is the smallest piece of memory that can be read or written to. There are several different types of column in the Virtex device: a central clock column, two IOB columns and multiple CLB and Block RAM columns (depending on the part). For each frame in a CLB column the first 18 bits contribute to the control of the two IOBs at the top of the column, then there are 18 bits for each CLB in the column and finally another 18 bits for controlling the two IOBs at the bottom of the column.

The content of a frame is a seemingly unrelated subset of the configuration for the inter-connect, IOBs and CLBs. An example of the problems created by this for fast reconfiguration is that changing the configuration of the two IOBs at either end of a column of CLBs can require most of the frames in the column to be written to the device.

Configuration is done through a shift register called the Frame Data Register (FDR) into which the configuration data is loaded before being transferred in parallel to a configuration memory frame.

2.1   Existing Techniques for Speeding Up Reconfiguration

The time taken to perform reconfiguration depends on a number of factors: the number of resources to be configured, off-chip configuration bandwidth, granularity (or atomic unit) of the configuration memory and the configuration memory organisation.

The importance of the first three factors to configuration time is obvious. The organisation of the configuration memory is important, since it can adversely affect the (naively) expected linear relationship between the number of resources being configured and the amount of data that must be loaded into the device. If configuration bits controlling unrelated resources are contained in the same memory locations, then there is a high likelihood that, with a small change to one area of the fabric, a disproportionately large number of memory locations will need to be written in order to bring about the change.

Configuration compression [1] exploits the similarities between frames but requires a large hardwired decompression unit which means the technique does not scale well. The multi-context FPGA [2, 3] has multiple memory bits per configuration bit forming configuration planes. Although this provides extremely fast switching between preloaded contexts, the additional memory planes can require significant area and since a plane's content is likely to change often, the off-chip configuration bandwidth is still a major bottleneck. Further, small changes are extremely wasteful since they require an entire context plane to implement.

3   Reconfiguration Bitstream Analysis

It is well known that due to the highly flexible nature of an FPGA's interconnect, only a small fraction of the configuration bits loaded into a device for a particular circuit are important. This section of the paper explores the configuration changes necessary to switch between a pair of circuits.

The largest member of the Virtex device family has a configuration bit stream of over 1 million bits. Research topics of interest include finding how many of the configuration bits of a large device such as the Virtex are essential for the configuration of a circuit, and of those bits, how they break down across configuring routing resources, lookup-table (LUT) contents and the multiplexors providing I/O to the LUTs.

3.1   Technique

Analysis. The Xilinx JBits API provides a low level method of creating and manipulating Virtex bitstreams. It provides several layers of abstraction — from the provision of high-level utilities such as routers and tracers (JRoute and Route-Tracer), to the connection of individual wires and the setting of multiplexors (JBits.set()). It is possible to take circuits produced by an HDL synthesis flow and manipulate them using JBits.

To provide a means for analysing and observing reconfiguration in detail, a piece of software was written using the JBits API. The software takes as input two bitstreams generated by the standard Xilinx toolflow that have some CLB usage in common. It may be that one circuit uses all the CLBs used by the other circuit, or simply uses some of them, so the portion of the fabric where both circuits use the same CLBs is also specified as an input to the program. As output, the program produces a detailed list of the minimum number of changes necessary to reconfigure the fabric from implementing one circuit to the other.

The minimal set of changes necessary to reconfigure the fabric is produced by writing special wrapper functions for the low-level JBits calls and then using these wrappers to perform the reconfiguration. The wrappers record the setting of the resource being reconfigured in addition to modifying the bitstream. This means that when all necessary changes have been made, the final setting of each resource can be compared to its setting before any changes were made, producing the minimal set of changes.

The algorithm for implementing the reconfiguration is composed of three stages. The first stage involves reading the LUT and internal CLB multiplexors configuration settings from the resultant circuit and writing these settings to the same CLBs in the starting circuit bitstream. The starting circuit bitstream is now a mix between the settings for CLB internals of the resultant circuit and the interconnect configuration of the starting circuit. The second stage consists of tracing the route of every possible source in the resultant circuit and extracting the settings of each resource in its path. The resource settings are then written to the starting circuit bitstream. After stages one and two, the starting circuit bitstream now contains the complete implementation of the resultant circuit with any non-overlapping resources for the first circuit left intact.

Unfortunately this bitstream will not necessarily configure the fabric to produce a working second circuit, this is because extraneous nets and partial nets belonging to the original circuit may overlap and interfere with the resultant circuit. For example, it is possible that a net will have more than one source. To remove this problem, all the sources in the new bitstream are traced and any sinks present which should not be are removed. The resulting netlists are now functionally correct, i.e., they connect a source to a list of sinks, but there maybe additional 'antenna' wires hanging off the netlist not performing any function. Although not altering the functional correctness of the netlist these antennae do potentially affect timing. For the purposes of this implementation, a simple timing analysis of the resultant circuit is performed and compared with the lean version. Those nets in the new circuit which are larger than the critical net in the lean version are trimmed appropriately.

The analysis tool enables a large number of different investigations to be carried out on the details of Virtex reconfiguration. The analyses carried out for this paper focus on revealing ways to reduce configuration time by providing a detailed view of what exactly occurs when reconfiguring.

Metric for Analysis. To put the number of bits that flip during reconfiguration in perspective, the actual number of bits controlling the configuration of a CLB is used as a metric. It is equal to the 18 bits in a frame allocated to a CLB multiplied by the 48 frames in a column, $18 \times 48 = 864$ bits.

3.2 Results

Two fundamental analyses of reconfiguration are presented in this section. The first is the percentage of bits in the configuration memory that flip during reconfiguration. The second is a breakdown of the percentage of bits that flip in each of the different resource types.

The experiments were conducted using a selection of circuits typically implemented on FPGAs: FIR and IIR filters, a DES encryption core, an FFT and a CORDIC. The circuits were placed and routed automatically, with area minimised and they ranged in size from 528 to 2320 slices.

It was found that the number of bits that change during reconfiguration is consistently between 8% and 10% of the total size of the configuration memory controlling that area of the fabric. It was also found that when the number of bits that flip during reconfiguration is compared to the size of the bitstream loaded to perform the change, the percentage change is smaller again - less than 3%. The percentage number of bits that actually flip is consistently between 8% and 10% across the circuit pairs showing the generallity of the result despite the large variation in circuit sizes.

Figure 1 shows that the majority of changes during reconfiguration occur in the multiplexors feeding the LUT inputs and the LUT contents themselves.

Sections 4 and 5 give two different approaches to leveraging the knowledge gained by the reconfiguration bitstream analysis. Section 4 gives details of a method called the overlay technique, which identifies and loads those changes that can be made before the fabric is taken off-line for the residual reconfiguration. Section 5 proposes a new configuration architecture design space, and explores some complementary reconfiguration compression algorithms that operate on the configuration changes required.

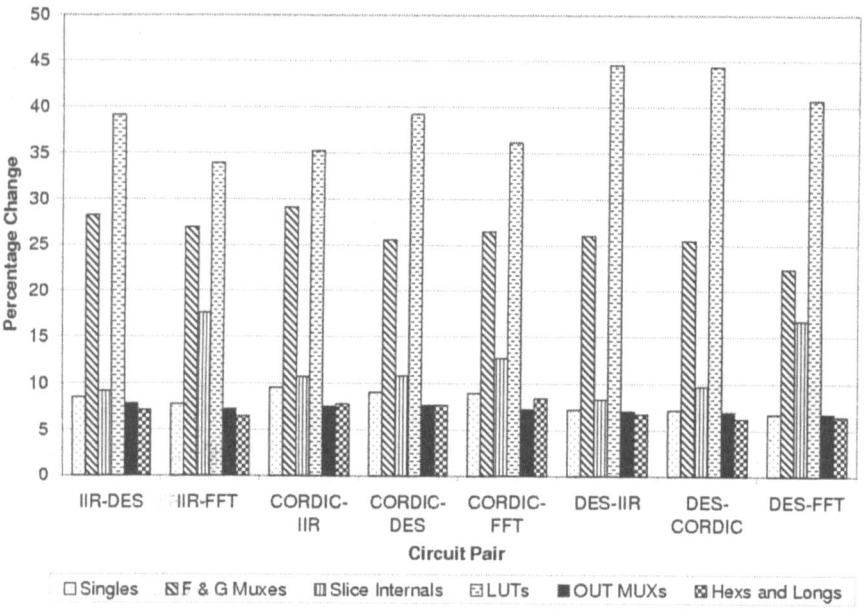

Fig. 1. The number of changed bits broken down by resource type and expressed as a percentage of the total number of change bits required to configure the area of fabric occupied.

4   Overlay Technique

From the analysis of the number of bits required for each resource type, it can be seen that a fair proportion of the bits that require changing are Programmable Interconnect Points (PIPs) controlling single connections. These are of interest because many of them could be set in advance of reconfiguration without affecting the operation of the existing circuit. This section presents and evaluates a technique for exploiting the potential of this redundancy.

A software tool was written, which takes in the bit streams describing two circuits, A and B, that time-share a section of the FPGA fabric and produces two new bit streams, called the advance bits and the residual bits. The advance bits describe a circuit equivalent to circuit A, except that in addition any resources used by B that can safely be set in advance are configured. The residual bit stream contains the remaining configuration data necessary to minimally switch from the advance bit configuration to a equivalent configuration for circuit B.

This overlay technique is suited to an embedded application where the computing requirements are easily predicted. In an application where circuit B is not necessarily required after circuit A, it may be desirable to separate the advance change bits from the definition of circuit A and only load the appropriate circuit's advance bits when the system's requirements can be better predicted. This is a simple extension of the algorithm implemented, producing three bitstreams instead of two - circuit A, advance bits for circuit B and the residual bits of circuit B.

When two circuits of different sizes are paired together, the area of the smaller circuit is chosen for the overlay experiment. Only PIPs are considered for overlaying in the experiments. Only CLB configuration data is manipulated, the circuits are not connected to IOBs, and any clocking or BRAM configuration is ignored. This is acceptable since the goal of this technique is the minimisation of the data required to reconfigure an area of logic, and logic is controlled by the CLB configuration data. It is expected that the technique can be extended to include IOB and clock configuration data.

4.1   Results

The overlay technique results in the identification that 10% of the change bits can be overlayed in advance of the fabric being taken offline for the second residual stage of reconfiguration. It is likely that with less densely packed circuits, and a more exhaustive set of techniques to identify resources that can be overlayed, e.g., LUT contents, CLB MUXes and CLB internal MUXes, the fraction of bits than can be overlayed at the advance stage could be increased.

5   Configuration Architecture Design Space

5.1   Overview

The reconfiguration bitstream analysis in Section 3 suggests that simply loading the changes required instead of the complete bitstream may improve reconfiguration time. This Section investigates this within a new configuration architecture design space. A specific point within the new design space is selected and described before a number of algorithms are proposed to compress configuration changes. This approach can be used independently or enhanced through a combination with the overlay technique given in Section 4.

5.2   Architecture Description

RAM can be added to the configuration sub-system to the point where there are two (or more) RAM cells for every configuration bit (the multi-context configuration architecture), resulting in instantaneous whole-chip reconfiguration time in the order of a single clock cycle. This section of the paper aims to provide an illustrative example of how the analysis of what exactly occurs during reconfiguration presented in Section 3, identifies a new area of configuration architectures spanning the single context device and the multi-context device. The aim is not the definition of a new architecture, but more an exploratory feasibility study of a large family of architectures.

The specific architecture point chosen for study in this paper has a small RAM at the top of each column of configuration frames (or alternatively makes use of existing block RAM) and an associated configuration controller, as shown in Figure 2. The idea is to have the small configuration memory contain the changes that must be applied to the column, and a configuration controller applies the changes to the existing contents of the configuration memory. The configuration frames of each column can be read and written to by its configuration

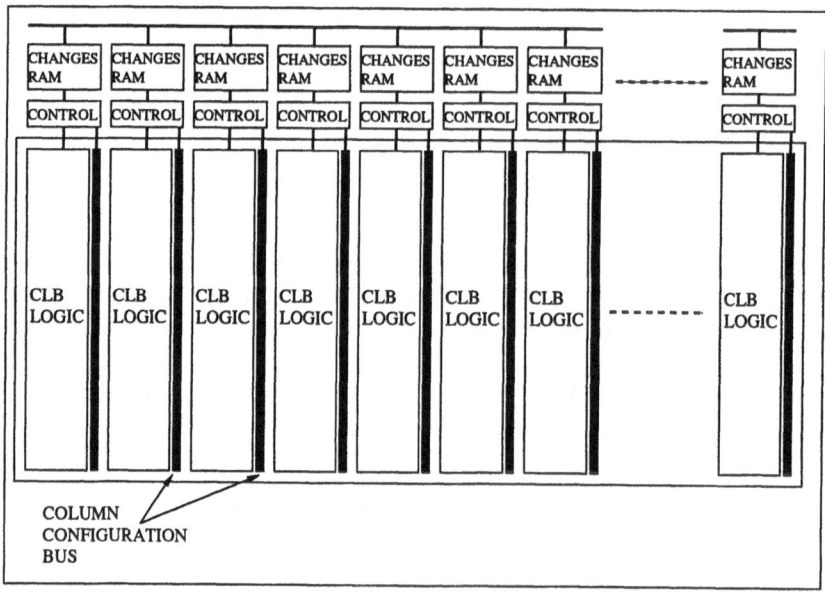

Fig. 2. Example of embedding extra RAM into the configuration subsystem.

controller, parallelising reconfiguration across columns. This architecture enables the sparse nature of the configuration bitstream changes to be compressed and stored on chip efficiently in advance of the reconfiguration stage. At the reconfiguration stage, the controller circuitry makes the changes by reading the existing configuration memory contents and applying the difference by inverting certain bits within the read back frame.

The configuration controller consists of an address generator unit, for reading from the small RAM, a frame address generator unit, two of registers for applying the difference function and the necessary control circuitry.

5.3 Configuration Change Compression Algorithms

This section proposes three possible compression techniques that leverage the observations from the earlier analysis and examines their effectiveness and cost in terms of silicon area.

Vanilla Compression Technique. The Vanilla compression scheme exhaustively searches and finds the optimum constant data chunk size for representing the changes and the corresponding optimum relative addressing scheme.

Banded Compression Technique. Change bits tend to be clustered into bands within the column because the change bits for every frame in the column are concentrated around the rows that are being changed. The banded technique expresses only the banded region of each frame. This reduces the space that must be covered by the addressing scheme of the change dataset.

Partitioned Compression Technique. Figure 1 reveals that expressing the LUT's value explicity instead of attempting to compress it may be a better approach. The partitioned compression technique separates the LUT contents from the other configuration data, states its content explicitly and uses the banded compression algorithm on the remaining configuration bits.

Compression Results. Figure 3 shows the results for the three techniques presented as a percentage of the actual data configuring the portion of the FPGA. The partitioned technique provides the best results, reducing the amount of data that needs to be loaded into the proposed architecture to 45% of the bits configuring that part of the fabric. However, since the performance difference between the banded and partitioned techniques is small, the banded technique seems to be the best solution, as its implementation in terms of logic and memory resources is significantly simpler and is hence likely to require less silicon area despite its slightly bigger storage requirements.

When the banded techniques results are expressed as a percentage of the Xilinx bitstream that is loaded to configure the fabric the compression is around 80%, although this figure is highly dependent on the number of CLBs occupied in each column. This result competes well with the best reported configuration compression method for a complete bitstream [1].

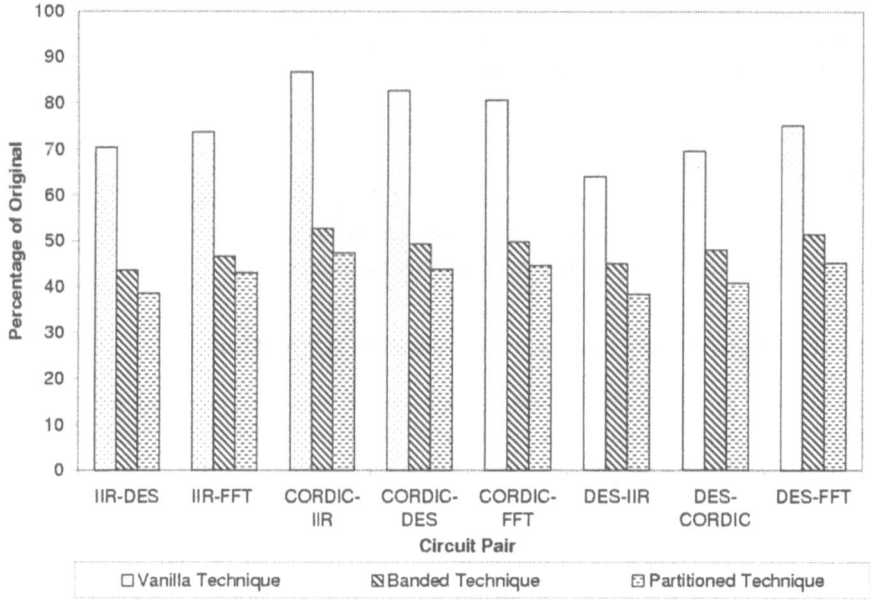

Fig. 3. The compressed changes dataset size expressed as a percentage of the number of bits required to configure the area of fabric occupied.

5.4   Architecture Evaluation and Discussion

A CLB has 864 configuration bits, so the configuration changes RAM is required to have 432 RAM cells per CLB to contain a complete configuration change for the column. At 5 transistors per cell, the changes RAM needs 2160 transistors which is about 1/3 of the estimated 6000 transistors in a CLB [6]. The control circuitry is negligible in size compared to the changes RAM. This is particularly true if the architecture is such that a frame may be read in a series of small chunks and hence doesn't require a large register to store its contents. Considering that IOBs, BRAM and other existing parts of the architecture mean that CLBs do not use all the silicon area it is estimated that the changes RAM would increase die size by less than 15%. This estimation is particularly conservative since it does not take into account the smaller area required by a single large RAM block per column compared to that of the highly distributed configuration RAM.

The time taken to reconfigure is proportional to the maximum number of CLBs requiring to be changed in any column. It is very dependent on the design of the control circuitry and the width of the changes RAM, but, assuming a column's control logic is capable of processing 1 change per clock cycle, then on average (864/10=87) cycles are required to reconfigure a CLB. So, as a concrete example, the XCV1000 part which has 72 CLBs per column would require a total of 72×87 (approx. 6300) cycles to entirely reconfigure. This is two orders of magnitude less than the time to reconfigure using the existing architecture.

The example architecture given is only one point in the design space of this configuration architecture domain and can be tailored to suit the application. For example, the number of change RAMs used could be reduced if parallel reconfiguraion of the entire device is never necessary; say a design's requirements demonstrate no more than half the fabric's columns need to be reconfigured then the reconfiguration architecture would increase die size by less than 7.5%.

6 Conclusion

This paper's analysis of the actual changes made during reconfiguration shows the extent of redundancy present in a modern FPGAs bitstream to be 90%. Leveraging this insight, a configuration architecture design space was proposed that spans the area between the two main existing configuration architectures, and three complementary algorithms for compressing the reconfiguration changes were presented and tested. The compression technique proposed produces a 50% compression of the minimum bitstream required to configure a specific area of fabric, and an average of 80% compression of the bitstream used to configure the fabric in present Virtex devices. This result compares favourably with the existing algorithms for full bitstream compression, and brings the advantages of being less expensive to decompress and is highly parallelisable.

Acknowledgements

This work was sponsored by Lucent Bell Labs and the UK Engineering and Physical Sciences Research Council.

References

1. **Zhiyuan L. and Hauck S.**, Configuration Compression for Virtex FPGAs, *IEEE Symposium on FPGAs for Custom Computing Machines*, April, 2001.
2. **Motomura M., Aimoto yr., Shibayama A., Yabe Y. and Yamashina M.**, An embedded DRAM-FPGA chip with instantaneous logic reconfiguration, *Symposium on VLSI Circuits Digest of Technical Papers* pp. 55-56, June 1997.
3. **Trimberger S., Carberry D., Johnson A. and Wong J.**, A time multiplexed FPGA, *IEEE Symposium on Field-Programmable Custom Computing Machines*, April 1997.
4. **Xilinx Corporation**, Virtex 2.5V Field-Programmable Gate Arrays, *DS003-1 (v.25)* April 2, 2001.
5. **Xilinx Corporation**, Virtex Series Configuration Architecture User Guide, *XAPP151 (v1.5)* September 27, 2000.
6. **Franklin N.**, Re: Silicon Area for Xilinx FPGAs, *comp.arch.fpga* 15:31:59 PST, 2 August 2002.

Run-Time Exchange of Mechatronic Controllers Using Partial Hardware Reconfiguration

Klaus Danne, Christophe Bobda, and Heiko Kalte

Heinz Nixdorf Institute, University of Paderborn,
Fürstenallee 11, 33102 Paderborn, Germany,
danne@upb.de, bobda@upb.de, kalte@hni.upb.de

Abstract. We present an efficient technique to implement multi-control-ler systems using partial reconfigurable hardware (FPGA). The control algorithm is implemented as a dedicated circuit. Partial runtime recon-figuration is used to increase the resource efficiency by keeping just the currently active controller modules on the FPGA while inactive controller modules are stored in an external memory. [1]

1 Introduction

Control systems can be implemented in reconfigurable hardware as an efficient and high-performance alternative to control algorithms executed by processors [4, 7, 10]. The large design space offered by reconfigurable hardware allows an exploration of different area/time trade-offs and to customize the data-word width to the problem.

Complex mechatronic systems in a changing environment require adaptive control to perform well. The concepts vary from adaptive parameter control to the *multiple model approach* [15, 12, 8]. In the latter the plant is modeled as a process operating in a limited set of operating regimes. The control system consists of a set of controller modules, each optimized for a different operating regime. It automatically switches to the corresponding controller module during runtime. When using just one static FPGA configuration to implement the *multi-controller architecture*, most of the system resources would stay inactive. We use partial runtime reconfiguration to overcome this drawback and to increase the efficiency of the system. The following section introduces linear controller systems and our reconfigurable hardware solution for these controllers. In section 3 we introduce the enhanced class of multi-controller systems. Section 4 presents our prototyping system that allows the implementation of multi-controller systems based on reconfigurable hardware. We close with a conclusion and future work.

---

[1] This work was partly developed in the course of the Graduate College 776 -Automatic Configuration in Open Systems- and the Collaborative Research Center 614 - Self-Optimizing Concepts and Structures in Mechanical Engineering - University of Paderborn, and was published on its behalf and funded by the Deutsche Forschungs-gemeinschaft.

P.Y.K. Cheung et al. (Eds.): FPL 2003, LNCS 2778, pp. 272–281, 2003.

2   Digital Linear Controller

The task of a controller is to influence the dynamic behavior of a system referred as *plant*. If the input values for the plant are calculated on basis of the plant's outputs, we refer to a control feedback (fig. 1).

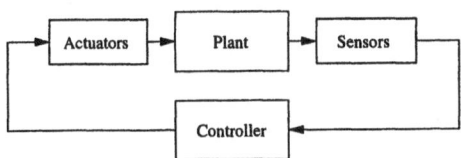

Fig. 1. Control Feedback Loop

A common basic approach is to model the plant as a linear time-invariant system. Based on this model and the requirements of the desired system behavior, a linear controller is systematically derived using formal design methods. The controller as a result of the synthesis considered above, is described as a linear time-invariant system and a time discretization is performed which results in eq. 1. The input vector of the controller is represented by $\mathbf{u}$ (measurements from sensors of the plant), $\mathbf{y}$ is the output vector of the controller (regulating variable to actuators of the plant) and $\mathbf{x}$ is the inner state vector of the controller. The matrices $\mathbf{A}, \mathbf{B}, \mathbf{C}$ and $\mathbf{D}$ are used for the calculation of the outputs based on the inputs.

$$\begin{aligned} \mathbf{x}_{k+1} &= \mathbf{A}\mathbf{x}_k + \mathbf{B}\mathbf{u}_k \\ \mathbf{y}_k &= \mathbf{C}\mathbf{x}_k + \mathbf{D}\mathbf{u}_k \end{aligned} \tag{1}$$

$$p = dim(\mathbf{u}), \qquad n = dim(\mathbf{x}), \qquad q = dim(\mathbf{y})$$

The task of the digital system is to calculate eq. 1 during one sampling interval. That includes determining the new state $\mathbf{x}_{k+1}$ and the output $\mathbf{y}_k$ before the next sampling point $k + 1$.

2.1   Hardware Architecture of a Linear Control System

In previous work [4, 7] we developed a framework for the implementatione of conventional digital linear controllers on reconfigurable hardware. It consists of transformation tools and a generic HDL (hardware description language) description of the controller architecture. Input to our design flow is the controller in form of eq. 1. Our framework generates fixed point arithmetic hardware which uses a fixed range of $[-1, +1[$. Thus a scaling of the equation system is necessary. The parameters of the matrices, the values of the input and output vectors, as well as the values of the state vector must not exceed this range during any time of operation. To perform the scaling transformation our tool requires the original range of the inputs and outputs determined by the physical sensors and actuators and the original range of the state variables determined by a simulation or analytical methods [4, 7]. The scaled equation system is used as an input to our generic HDL controller architecture. The generated hardware structure is described below.

Both parts of eq. 1 share the same form and are independent of each other. So a generic *MEC* block *(Matrix Equation Calculator)* which processes eq. 2 is instantiated twice to process both parts of eq. 1 in parallel.

$$c = Ma + Nb \tag{2}$$

Inside a *MEC* block the two terms Ma and Nb are also computed in parallel. For this purpose two *SMUL* blocks *(scalar-multiplier)* and an adder are instantiated. The first *SMUL* block multiplies one row of M with vector a at a time (the second *SMUL* block does the same for Nb respectively). Therefore the rows of M (and N) are processed sequentially. A *SMUL* block multiplying two vectors of dimension n (a matrix row and a vector) consists of n multiplier instances and an adder tree to sum the results. The multipliers themselves are implemented as Booth multipliers which have a cycle delay that is equal to the word width of the operands. The parameters of the matrices are directly synthesized into the design. Some additional control logic is instantiated to control the sequential computing of the scalar products and to periodically start the computation at each sampling point.

The area/time trade-off of this architecture leads to $f_1(p, n)$ parallel instantiated multipliers and $f_2(n, q)$ sequential performed multiplications (eq. 3). $f_3(n, q, w)$ is the number of overall clock cycles per sampling point where w is the word width of the input values of u. (see eq. 1 for p, n, q)

$$\begin{aligned} f_1 &= 2(p + n) \\ f_2 &= max(n, q) \\ f_3 &= (w + 2) \times max(n, q) + 2 \end{aligned} \tag{3}$$

In our implementation, the quadratic complexity of the problem (eq. 1) results into linear space complexity $O(p + n)$ (parallel computing) and linear time complexity $O(max(n, q))$ (sequential computing). However, other mappings with different area/time trade-offs are possible. Moreover, the data width can be chosen at bit granularity according to the accuracy required by the application. Therefore, by using reconfigurable hardware the system can be implemented efficiently, i.e. using minimal hardware resources while meeting all constraints on computation time and accuracy.

3   Multi-controller Architecture

In the *multiple-model approach* the plant, e.g. a complex mechatronic system in an changing environment, is modeled as a physical process that is operating in a limited set of operating regimes. From time to time the plant changes the operating regime. With conventional methods it might be possible to design one robust controller that controls the plant in all operating regimes, but it will not work optimal for the current operating regime. Parameter adaptive controllers can be used, but they may respond too slow to abrupt changes of the plant's dynamic behavior [13].

In the *multiple-model approach*, the plant is controlled by the architecture in fig. 2 [12]. It is composed of a set of *controller modules (CM)*, each optimized for a special operating regime of the plant. The *supervisor* is able to switch between the controller modules to determine the active module. The decision to switch from one *CM* to the next is made on basis of measurements of physical values of the plant. The strategy of the *supervisor* can vary from simple functions of the measurements to agent-based techniques [15].

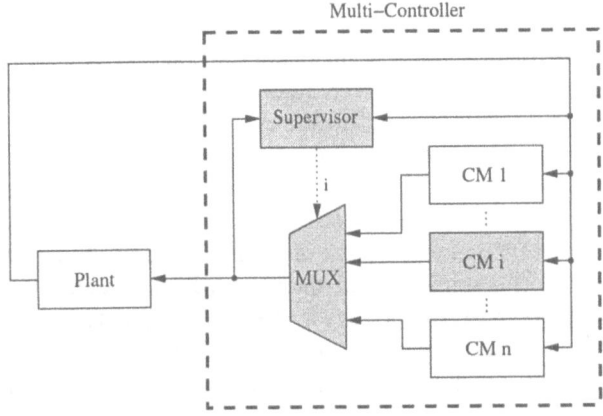

Fig. 2. Multi-Controller Architecture (The gray parts of the multi-controller show the active elements. The active controller module changes from time to time.)

4   Multi-controller on Reconfigurable Hardware

We assume that every *controller module* CM_i of fig. 2 can be represented by a linear controller. Therefore, we can use our framework to implement each *controller module* as a dedicated circuit. By using the standard approach, all modules would be instantiated in parallel. This leads to a design with a high area and power consumption, proportional to the number of controller modules. This number can be quite large, making this approach impossible due to the physical bounds of the FPGAs. Moreover, this design is very inefficient since only one controller module is active at a time (fig. 2). Considering just the resources for the controller modules, $n - 1$ of n resources stay inactive in contrast to efficient designs where almost all resources are active all the time. Our system proposed in the next section uses partial reconfiguration to overcome this drawback.

4.1   Architecture

To map the multi-controller architecture (fig. 2) to our prototyping system we divide the FPGA into three parts (fig. 3). In the middle the static module *(SM)* can be seen, which is not reconfigured and operates all the time. It contains the *supervisor*, a multiplexer and the communication interface to the plant. The left and the right parts of the FPGA are slots for reconfigurable modules *(RM-Slot-A, RM-Slot-B)*. In these areas a controller module can be implemented. Like in the original architecture (fig. 2) the supervisor and the controller modules get their input from the output of the plant. The calculated outputs of the currently active controller module are propagated back to the inputs of the plant. The active controller module is set by the supervisor. The initial controller is configured into the left slot *(RM-Slot-A)* of the FPGA and the supervisor activates this controller by setting the MUX to A. If the supervisor decides to switch to a new controller module CM_k, it request the configuration manager to reconfigure the right slot *(RM-Slot-B)*. After the reconfiguration process is finished the MUX is set to B. Every time a switch event occurs the supervisor will toggle to the other target slot. An example sequence is illustrated in the timing diagramm in fig. 4.

In our prototyping environment the task of the host is to simulate the plant and to store the partial bit-streams of the controller modules (fig. 3). In a final product, the control system will consist of the FPGA, a flash memory to store the partial bit-streams, the configuration manager which is a simple state machine and the I/O converters to connect the system to the plant.

The reason the system has two reconfigurable slots (instead of just one) is the relatively slow reconfiguration time of FPGAs which is in the range of milliseconds. If the sampling frequency of the controller is in the range of kHz, it might not be possible to reconfigured a controller module within one sampling interval. With our method, the new CM can be reconfigured and synchronized

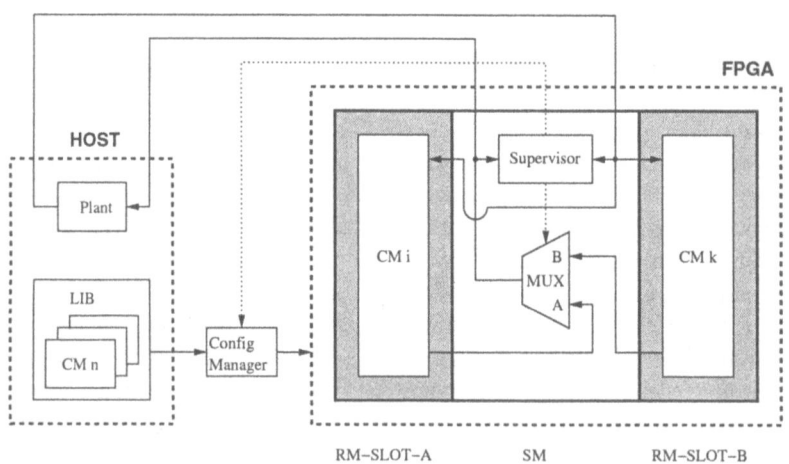

Fig. 3. Multi-Controller based on Reconfigurable Hardware

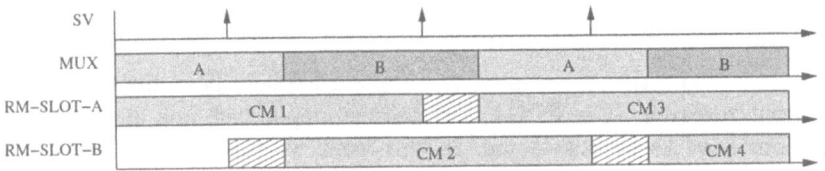

Fig. 4. State of the multi-controller system over time (The first row shows the reconfiguration requests generated by the *supervisor* and the second row shows the state of the MUX. The content of the two reconfigurable slots is shown in row three and four, where the fasciated area refers to the reconfiguration phase.)

while the other one controls the plant. This enables the switching between the two controllers within two sampling points, additionaly smooth fading between the controller outputs becomes possible.

The FPGA area of our system is two times the area of a controller module A_{RM} plus the area of the static module A_{SM} (eq. 4). As a boundary condition, the computation time t_c of a CM must be less than the sampling period. In an optimized implementation of a CM the FPGA-area/computation-time trade-off is chosen in such a way, that the area is minimized while the computation time is close to the sampling period. Approximatly, we can assume a linear trade-off between the area and the computation time of a controller module since we can highly parallelize the control algorithm (eq. 5).[2]

$$A_{twoslotsystem} = 2A_{RM} + A_{SM} \tag{4}$$

$$A_{RM} \approx c_1 \frac{1}{t_c}, \qquad t_c \leq T \tag{5}$$

A system which uses only one RM slot and the staic module SM can be an alternative in some cases. Since the controller module slots require the mass portion of the FPGA area, using only one slot seems to be a great reduction of the overall system resource cost. However, if the application requires a new controller output **y** within every sampling interval T, the new CM must be configured and compute the new output during one period. The reconfiguration time t_r plus the computing time t_c must be less than the period. The FPGA area of such a system is the area of one reconfigurable module slot A_{RM} plus the area of the static module A_{SM} (eq. 6). The reconfiguration time t_r increases approximatly linear with the module size (eq. 7). [3]

$$A_{oneslotsystem} = A_{RM} + A_{SM} \tag{6}$$

$$A_{RM} \approx c_1 \frac{1}{t_c}, \qquad t_r + t_c \leq T, \qquad t_r \approx c_2 A_{RM} \tag{7}$$

Finaly, it depends on the reconfiguration speed and the sampling period, which system should be preferred. When t_r exceeds T, or when fading between the controller modules is required, the one slot solution cannot be used.

---

[2] c_1 is an application specific constant
[3] c_2 is an FPGA device specific constant

4.2 Partial Reconfiguration Design Flow

Partial reconfiguration is not part of a standard design flow, but is the subject of current research. It is FPGA architecture dependent and many boundary conditions have to be considered. In this work we target the XILINX Virtex FPGAs [2]. These devices can be partially reconfigured column wise. During a complete reconfiguration a bit-stream with configuration data for all logic resources of the gate array is send to the device. In the partial reconfiguration mode the bit-stream contains just the configuration data for one or more columns of the gate array. The other columns keep their old configuration and their logic blocks stay active during the reconfiguration.

The design of a system using partial reconfiguration starts with the partitioning of the system into modules that should be reconfigured at runtime. Much theoretical research in the area of reconfigurable computing focuses on the systematic partitioning of the system and the optimal scheduling and placing of the reconfigurable modules, e.g. [5, 6, 14, 3]. Due to the complexity of the design flow, we started with a manual partitioning as described in section 4.1.

To implement the system the bit-streams for all modules have to be created during the design time. In the standard design flow the design entry is HDL and the synthesis tools are employed to generate the bit-stream. To generate partial bit-streams two methods exist. The first omits a synthesis by the standard tools and uses tools that allow access to the FPGA resources on a low abstraction level [1, 16]. The second uses the standard synthesis tools in a modified, restricted design flow [11]. The latter, which we used in our work, is described in the following.

The basic idea of the modified design flow for partial configuration is the constraint of the place and route process in that way, that all logic of a module is placed inside a restricted area. For this purpose, a net-list for every module is synthesized. For each net-list the place and route process is invoked, with the constraint to use only FPGA resources in a defined area. Some restrictions that have to be met to generate a reconfigurable module for XILINX Virtex FPGAs are listed below:

- The area of two modules that should work simultaneously must not overlap.
- The height of the module area is always the full height of the FPGA. (This is due to the column wise reconfiguration of Virtex FPGA)
- The position to place a module is fixed and determined during design time. Relative placement during runtime is not supported. (e.g. currently, a module bit-stream for the left FPGA half cannot be used for the right half)
- An active module occupies all resources in its area. That is, other modules cannot use logic blocks, routing resources, block ram, I/O pins or any other resource of the active module.
- Communication between adjacent modules is done via fixed well defined connections at the modules boarders.
- Communication between not adjacent modules can be realized by feed through wires of the between module.

– The modules must not share any other connection (except the clock nets, which are global resources). That means, that the modules must not share one reset signal or constant nets like VCC or GND.

4.3   Implementation and Results

For the implementation of the system we used our self developed flexible and modular rapid prototyping environment RAPTOR2000 [9]. In this work, the RAPTOR2000 motherboard hosts a module that carries a XILINX Virtex 800 FPGA. A PCI-bridge implements the connection to a host computer. A configuration manager implemented in a CPLD supports the fully and partially configuration of the FPGA. The bit-streams are taken from the host-memory and fed to the SelectMap interface of the FPGA allowing a reconfiguration rate of 50 MByte/s. To communicate between the host computer and the FPGA a graphical as well as an C-library interface exist. They allow to access the internal local bus which is connected to the FPGA. In the FPGA a module called *local-bus access* implements the bus-protocol and allows the host to access internal registers of the FPGA design. This logic as well as the *supervisor* are part of the static module.

Our first implementation differs from the proposed system in section 4.1. We use the discussed solution with one RM-slot instead of two RM-slots. While we still propose the three slot solution, many challanges and problems of partial reconfiguration can be explored using this first prototype. Our example controller (implementing the control of an inverse pedulum) is a module with three inputs of 16 bit width, and two inner states and three outputs of 32 bit width. The communication between the SM and the RM is done via special bus-macros. It consists of a 32-bit data bus form SM to RM, a 32-bit data bus from RM to SM and an 32-bit address bus from SM to RM. The module uses 1066 FPGA slices which is approximately 10 percent of the Virtex 800 FPGA logic resources. From this it follows that either smaller FPGAs can be used or more complex controllers can be implemented. The worst net delay is about 13.5 ns which results in a maximum clock frequency of more than 70 MHz. Since the computation time of the module is 56 clock cycles, it can operate with an sampling frequency above 1 MHz. There are just few applications requiering such high sampling rates. Nevertheless, this is the performance archived by our CM-architecture of section 2.1 for the dimensions mentioned above. Other area/time trade-offs can be implemented which reduce the perfomance to the timing requirements of the application and minimize the area of the CMs. The area we reserved for the RM covers about the half of the FPGA. The resulting bit-stream has a size of 274 kB which leads to a reconfiguration time of 5.5 ms using the XILINX SelectMAP interface at 50 MHz. However, the slice count assumes a much smaller area which might lead to an reconfiguration time of about one millisecond. The seperated modules RM and SM as well as the bus-macros can be seen in the screenshot of the routed FPGA design in fig. 5.

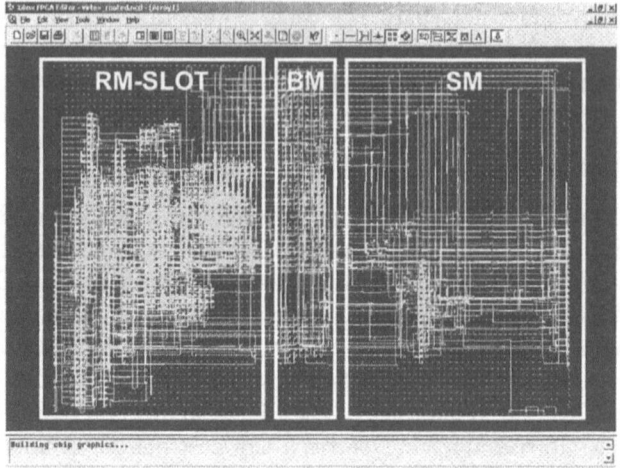

Fig. 5. Screenshot of Routed FPGA Design (left: reconfigurable module slot, right: static module, middle: bus-macros for communication)

5   Conclusion

In this paper we presented an architecture for hardware based multi-controller systems. We motivated the implementation of controllers using reconfigurable logic as a good alternative to control algorithms executed on processors. After introducing the class of digital linear controllers we presented our framework to create them in reconfigurable logic. Later, we introduced control applications that follow the multiple-model approach. They target systems whose dynamic behavior is modeled as a process working in a limited set of operating regimes. While controllers derived by formal design methods result in low performance, a set of controllers optimized for different operating regimes is more suitable. In these multi-controller architecture, only one controller module is active at a time. In section 4 we enhanced our framework for linear controllers to multi-model controller systems. We showed, how partial dynamic reconfiguration of the FPGA led to an efficient design. In the architecture of our prototype, the host simulates the plant while the FPGA implements the controller. Therefore the FPGA is divided into three parts. The static part realizes communication between the host and the controller as well as the administration of the reconfiguration process. The other parts are reserved for controller modules and can be reconfigured. This architecture can guaranty that always one controller module is active, while the other slot is being reconfigured. The results of the implementation proved our concept and showed that reconfigurable hardware can be used to implement quite complex controller system with high performance. Also this is an example for close to reality employment of partial dynamic reconfiguration, increasing the efficiency of the system.

In our future work we will enhance the implementation to the two slot version and extend the supervisor allowing more intelligent self reconfiguration. In addition we will enhance our framework for the controller modules to allow to chose different FPGA-area/computation-time trade-offs. Also our research will target the hardware implementation of non-linear controllers and floating point solutions.

References

1. *XILINX ISE 5 Software Manuals - FPGA Editor*, 2002.
2. *XILINX Virtex Data Sheet, DS003(1-4)*, 2002.
3. K. Bazargan, R. Kastner, and M. Sarrafzadeh. Fast template placement for reconfigurable computing systems. *IEEE Design and Test of Computers*, Mar. 2000.
4. M. Bednara, K. Danne, M. Deppe, O. Oberschelp, F. Slomka, and J. Teich. Design and implementation of digital linear control systems on reconfigurable hardware. *EURASIP Journal on Applied Signal Processing*, 2003.
5. C. Bobda. IP based synthesis of reconfigurables systems. In *Tenth ACM International Symposium on Field Programmable Gate Arrays(FPGA 02)*, page 248, Monterey, California, 2002. ACM/SIGDA.
6. C. Bobda. Temporal partitioning and sequencing of dataflow graphs on reconfigurable systems. In *International IFIP TC10 Stream on Distributed and Parallel Embedded Systems (DIPES 2002)*, pages 185–194, Montreal, Canada, 2002. IFIP.
7. K. Danne. Implementierung digitaler regelungen in hardware. Bachelor's thesis, University of Paderborn, 2000.
8. Johanson and Murray-Smith. The operating regime approach to nonlinear modelling and control. *Multiple Model Approaches to Modelling and Control*, 42, 1997.
9. H. Kalte, M. Porrmann, and U. Rückert. A prototyping platform for dynamically reconfigurable system on chip designs. In *Proceedings of the IEEE Workshop Heterogeneous reconfigurable Systems on Chip (SoC)*. Hindawi, 2002.
10. R. Kasper and T. Reinemann. Gate level implementation of high speed controllers and filters for mechatronic systems. *Mechatronic Workshop*, 2000.
11. D. Lim and M. Peattie. *Xilinx Application Note XAPP290: Two Flows for Partial Reconfiguration: Module Based or Small Bit Manipulations*, May 2002.
12. A. Morse. Control using logic-based switching. *Trends in Control, Springer, London*, 1995.
13. Narendra and Balakrishnan. Adaptive control using multiple models: Switching and tuning. *Yale Workshop on Adaptive and Learning Systems*, 1994.
14. J. Teich, S. Fekete, and J. Schepers. Optimization of dynamic hardware reconfigurations. *The J. of Supercomputing*, 19(1):57–75, May 2000.
15. A. van Breemen and T. de Vries. An agent-based framework for designing multi-controller systems. *Proc. of the Fifth International Conference on The Practical Applications of Intelligent Agents and Multi-Agent Technology, pp. 219-235, Manchester, U.K*, Apr. 2000.
16. www.xilinx.com/products/software/jbits.

Efficient Modular-Pipelined AES Implementation in Counter Mode on ALTERA FPGA

François Charot[1], Eslam Yahya[2], and Charles Wagner[1]

[1] IRISA/INRIA
Campus de Beaulieu
35042 Rennes Cedex, France
charot@irisa.fr, wagner@irisa.fr
[2] Information Technology Institute-ITI
Benha High Institute of Technology-BHIT
241-El Haram Street-Cairo/El Estad Street-Benha-Kaliobia, Egypt
esyahya@iti-idsc.net.eg

Abstract. This paper describes a high performance single-chip FPGA implementation of the new Advanced Encryption Standard (AES) algorithm dealing with 128-bit data/key blocks and operating in Counter (CTR) mode. Counter mode has a proven-tight security and it enables the simultaneous processing of multiple blocks without losing the feedback mode advantages. It also gives the advantage of allowing the use of similar hardware for both encryption and decryption parts. The proposed architecture is modular. The architecture basic module implements a single round of the algorithm with the required expansion hardware and control signals. It gives very high flexibility in choosing the degree of pipelining according to the throughput requirements and hardware limitations and this gives the ability to achieve the best compromised design due to these aspects. The FPGA implementation presented is that of a pipelined single chip Rijndael design which runs at a rate of 10.8 Gbits/sec for full pipelining on an ALTERA APEX-EP20KE platform.

1 Introduction

With the very large growth of network applications, the issue of security is becoming one of the most important aspects in network design. High security applications running on the Internet require hardware implementation of fast and secure protocols. In a protocol like IPsec, developed to secure the Internet traffic, there is a need for underlying ciphering algorithms. One of the best suitable is the Rijndael algorithm selected by NIST as the Advanced Encryption Standard (AES) in October 2000 and approved in the summer of 2001 as the Federal Information Processing Standard (FIPS) [5]. The suitability of this algorithm comes from its symmetry and high key agility as well.

This paper presents a fully-modular partially-pipelined AES architecture, dealing with 128-bit data/key blocks and operating in Counter (CTR) mode. CTR mode of operation is fully parallelizable, and has a proven-tight security. It also gives the advantage of allowing the use of similar hardware for both encryption and decryption. This architecture is suitable for secure network applications, especially when high

P.Y.K. Cheung et al. (Eds.): FPL 2003, LNCS 2778, pp. 282–291, 2003.
© Springer-Verlag Berlin Heidelberg 2003

data rates are required. The proposed architecture exploits pipelining. Pipelining gives the ability to achieve higher throughput by processing multiple input blocks simultaneously.

The proposed architecture is modular. The architecture basic module implements a single round of the algorithm with the required expansion hardware and control signals. It gives very high flexibility in choosing the degree of pipelining according to the throughput requirements and hardware limitations, that gives the ability to achieve the best compromised design due to these two aspects. The key generation part of such an architecture is of major importance. Key agility, that is the possibility of calculating keys on the fly, is a necessity in most practical applications like IPsec, encrypted routers and secure ATM networks. In case of partial pipelining, key agility is a complicated part, especially from the synchronization point of view due to calculating multiple keys on the fly at the same time.

AES implementation needs storage RAMs, look-up table ROMs, registers, shift registers and simple boolean operations, which make the fine grain structure of Field Programmable Gate Arrays (FPGAs) quite suitable. Moreover FPGAs allow inherent parallelism and flexibility of the algorithm to be easily exploited with a fast development time. The proposed modular architecture has been implemented on an ALTERA APEX FPGA [1]. The first experiments, based on a full pipelining of the algorithm, allow a rate of 10.8 Gbits/sec to be achieved. Section 2 of this paper describes the Rijndael algorithm with a focus on the CTR mode of operation. The design of the pipelined Rijndael implementation is outlined in section 3. Performance results are given in section 4. Finally, concluding remarks are provided in section 5.

2 Rijndael Algorithm Overview

2.1 Basic Algorithm Features

The Rijndael algorithm is a block cipher using 128, 192 and 256-bit input/output blocks and keys [2]. The size of blocks and keys can be chosen independently. The encryption is done in a certain number of rounds, which may vary between 10, 12, and 14, and it depends on the block length and key length chosen.

In a 128-bit block encryption, plain text and cipher text are processed in blocks of 128 bits. The transformation considers the input data block as a four column rectangular array of 4-byte vectors called *State*. The Rijndael cipher algorithm consists of four basic operations:

- ByteSub: non-linear substitution based on Galois Field applied to each *State* byte.
- ShiftRow: cyclically shifting the last three rows of the *State* by different offsets.
- MixColumn: each *State* column is processed based on Galois Field polynomial multiplication.
- AddRoundKey: each round key byte is added (Xored) to the corresponding *State* byte.

The single round structure is repeated 10 times. An additional initial round is applied in which the original key is added to the input data. The first 9 rounds perform all the four transformations described above whereas the final round omits the MixColumn operation. The key entering the cipher is expanded so that a different sub-key (round key) is created for each round of the algorithm. This round key generation is a process consisting of S-boxes, XORs and word rotation operations.

2.2 Modes of Operation

As described in [3], different modes of operation may be used in conjunction with any symmetric key block cipher algorithms. Modes of operation can be divided into two main categories: non-feedback modes, in which the main advantage is from architecture point of view, since any kind of parallelism is enabled and feedback modes, in which the main advantage is from security point of view but slow down the hardware implementation due to loop carried dependencies.

Among the different block cipher modes of operation proposed in the NIST recommendation detailed in [3], Counter (CTR) mode has a proven-tight security and it enables the simultaneous processing of multiple blocks without losing the feedback mode advantages. It also gives the advantage of allowing the use of similar hardware for encryption and decryption parts.

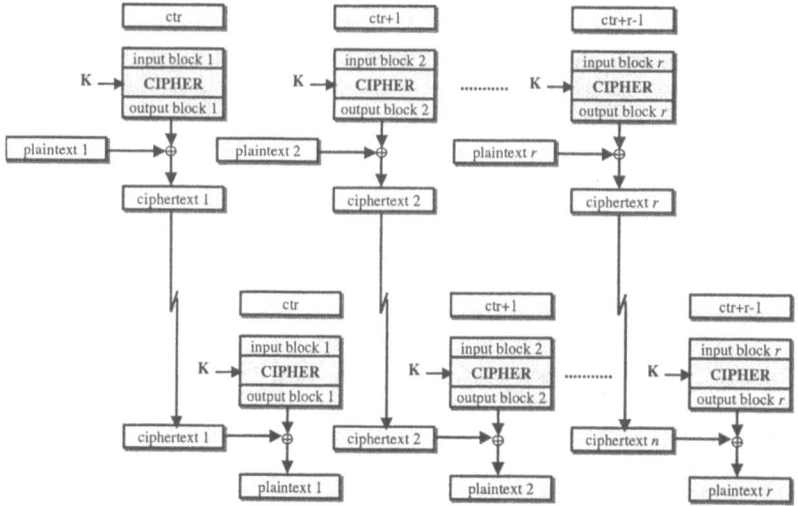

Fig. 1. Encryption and decryption process in CTR mode

In CTR mode, as illustrated in Figure 1, the cipher is applied to a set of input blocks, called counters, to produce a sequence of output blocks that are exclusive-ORed with the plaintext to produce the ciphertext. The sequence of counters must have the property that every block in the sequence is different from every other block.

3 Hardware Architecture

A block cipher consists generally of a single round module, which is applied to a data block multiple times, with a different sub-key applied in each round. The optimum hardware implementation for Rijndael is a single round with agile sub-key generation. Key agile implementation is preferable to stored key implementation due to the flexibility it allows and also to the reasonable size of sub-key generation. When high throughput is required, the single round and associated sub-key generation have to be pipelined. The pipelining greatly improves throughput. The throughput is proportional to the number of duplicated rounds at a price of an increase of silicon area.

The AES hardware architecture presented in this paper implements a 128-bit data/key Rijndael algorithm. In order to implement full encryption, 10 rounds of transformation have to be performed [2], [5]. Since high security applications and fast protocols implementations are aimed, the architecture design has been guided by the following decisions:

- The use of pipelining which allows the throughput constraints to be adapted according to the application requirements and which is mandatory when high throughput is required.
- Key agility, which is a demand for all practical applications.
- The use of CTR mode, which is one of the best known modes since it has significant efficiency advantages over the standard encryption modes without weakening the security.
- Modularity of the architecture which gives high flexibility and reconfigurability.

Through the next subsections, every aspect introduced above is discussed in details.

3.1 Pipelining

The simplest definition for pipelining is using multiple processing units running in parallel on multiple input blocks. A number of different architectures can be considered when designing encryption algorithms [4]. They are generally described as follows. Iterative looping, where only one round is designed. Loop unrolling involves the unrolling of multiple rounds. Pipelining consists in replicating the round and placing registers between each round to control the flow of data. When the round is complex, sub-pipelining can be used. Sub-pipelining decreases the pipeline delay between stages, but it increases the number of clock cycles required to perform the encryption.

Two pipelining approaches are considered here:

- Partial pipelining which consists in a partial replication of the round module, with a registering of the intermediate data between rounds and the need to iterate a given number of times on the pipeline structure in order to calculate ten rounds according to the number of replications.
- Full pipelining, a specific form of a partial pipeline, which consists in replicating the round module ten times, a ten-pipeline stage structure implements all rounds of the algorithm.

The higher is the degree of pipelining, the higher is the silicon area and the power consumption. So, very careful compromising between the required throughput and the chosen degree of pipelining has to be done. For efficiency reasons (throughput versus hardware utilisation), a 5-stage partial pipelined module has been chosen. This pipelined processor requires two iterations for producing an output cipher text.

3.2 Key Agility

The Rijndael processor is fed by a 128-bit input key but every round of the algorithm needs its own round key. That means that the original key has to be expanded. There are two methods to do that:

- Calculation and storage, the key is expanded once and stored in a buffer, then used for all coming data blocks until a reset operation occurs.
- On the fly calculation, the key is expanded from step to step for every new coming data block, producing 10x128-bit key rounds.

The first method is common and it is used in most of the AES implementations proposed so far. It is simpler and faster but not practical. It supposes that all coming data will be encrypted using the same key. The second method allows a new key to be used with every new coming data block. This is required with most applications like for instance encrypted routers. But this method is slower and more hardware consuming since the total delay of the round is the sum of the round transformation delay and of the key expansion delay.

As reported in most papers, like in [6], the key expansion delay is much higher than the round transformation delay itself. That means that key agility has a price from the speed point of view, but it is a necessity in all practical applications.

3.3 Modes of Operation

The different standardized modes of operations are summarized in table 1. They are divided into two main categories: non-feedback modes and feedback modes.

Table 1. Modes of operation

	Non-feedback mode	**Feedback mode**
Security point of view	Same key + same data = Same ciphered output	Same key + same data = Different ciphered output
Architecture point of view	Pipelining is enabled	Pipelining is disabled

Table 1 shows that fast and secure implementation of AES is hard to achieve, but CTR mode solves the conflict. Its main advantages are the following [3], [7], [10]:

- Hardware efficiency since any kind of parallelism is enabled;
- Pre-processing since encryption can be done even before the input plaintext is known;
- Provable security since analysis proved that CTR mode has a tight security;

- Simplicity, important advantage of this mode with AES is that, typical hardware for encryption and decryption can be used, which prevent implementing different hardware for both, especially that the encryption is more complex.

3.4 Modularity

With a deep look to the Rijndael algorithm, it may be noticed that it is a modular one. The full algorithm can be implemented as repeated identical modules and the advantages of that way of thinking are:
- Efforts are done in designing a small module;
- Highly reconfigurable design where any modification in the system specifications can be achieved by only modifying the basic module;
- Flexible degree of pipelining, where it is very easy to instantiate modules exactly as necessary to achieve the required throughput.

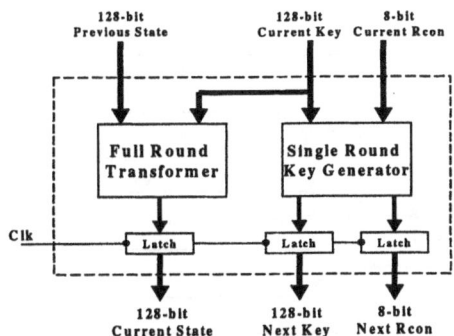

Fig. 2. The basic module block diagram

The implemented module, as illustrated in Figure 2, has only one disadvantage. Some unnecessary hardware will be added –for example the round constant *Rcon* [2] is computed and not used as a constant– which means more hardware is required but the gain is very valuable. Round transformation operates in parallel with key generation, and during a given iteration the round key required by the next iteration is computed.

3.5 System Architecture

Two architectures are introduced. The first one corresponds to the initial design based on the choices previously explained, the second is characterized by some enhancements which greatly affect the final system throughput. In the next subsections we will discuss these two architectures, their differences and their performance results.

3.5.1 Initial System Architecture

Figure 3 introduces the full block diagram of the proposed AES processor. The main component in the system is the 128-bit pipelined Rijndael module. This module is composed of 5 instances of the basic module described in Figure 2. It implements the ten rounds. An additional Xor module is introduced before the 5-stage module instance to perform the initial key addition operation. The basic module implements the four transformations described in 2.1. The S-box transformation is implemented as a 256-entry 8-bit wide lookup table. This transformation is mapped into the dedicated ALTERA APEX Embedded System Block (ESB) resources. The other transformations can be reduced to a series of lookup table operations and bitwise XOR operations, which are easily implemented in ALTERA APEX logic elements (LE).

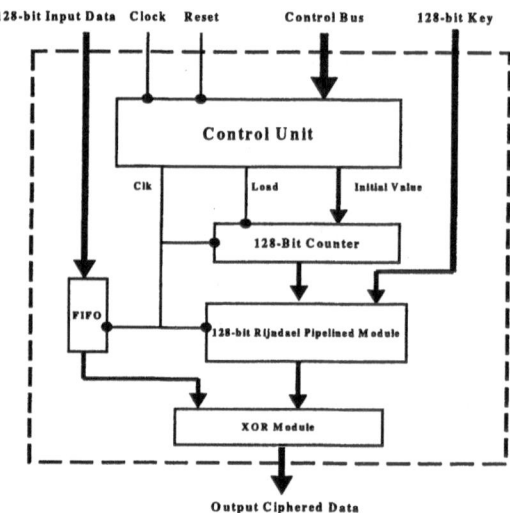

Fig. 3. AES system architecture

The control unit defines the initial value of the counter, *Reset*, *load* and *Clock* control signals. The Rijndael pipelined module is a 5-stage pipelined Rijndael processor. Registers and FIFO allow the synchronization requirements to be fully satisfied.

3.5.2 Optimised System Architecture

By analysing the architecture, some possible enhancements were studied. The floor plan analysis shows that all critical paths are located between I/O pins and internal logic. This has the effect of affecting the internal logic propagation delay since the later includes the input set-up time.

The solution consisting in inserting register set directly after/before I/O allows the critical path to be broken. Even if it introduces an additional pipelining stage, it greatly reduces the critical path delay.

Considering the concept above and by reviewing synthesis results of table 3, another enhancement can be achieved. Two clock speeds can be used, a first one for the I/O and a second one for the core. In case of a 5-stage partial pipelining, good results can be achieved by forcing the core to use a clock with a twice frequency of the I/O clock. It is clear from Figure 4 that the core will be triggered two times every one I/O operation.

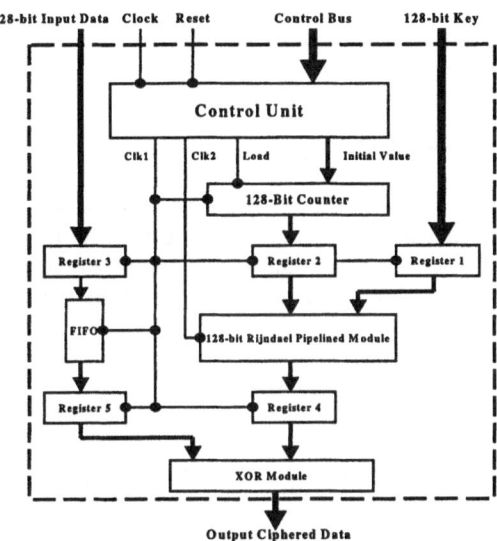

Fig. 4. Enhanced architecture

4 Performance Results

The Rijndael processor proposed in this paper is implemented using Leonardo Spectrum as the synthesis tool and Quartus II as the Placement and Routing tool, targeting ALTERA APEX FPGA [1]. The design is expressed in VHDL language. As illustrated in figure 2, the module is composed of two main components: the full round transformer and the round key generator. Both components run in parallel and feed the next stage. So the slowest clock of both components corresponds to the main system clock. In the next subsections, performance results for both initial and optimised architectures are presented.

4.1 Initial Architecture

Table 2 summarizes the synthesis results for the initial architecture. Since the critical path is as long as 25 ns, the system could operate under a clock speed of 40 MHz. When the cipher and the key are 128-bit, the throughput is 5.12 Gbits/sec for full pipelining. A rate of 2.56 Gbits/sec can be achieved in case of half pipelining.

Table 2. Initial architecture synthesis results

	Hardware	Critical Path Delay
Full Round Transformer	16 ESB, 356 LE, 128 FF	19.687 ns 50 MHz
Round Key Generator	4 ESB, 131 LE, 141 FF	25 ns 40 MHz

4.2 Optimised System Architecture

Table 3 summarizes the synthesis results for the optimised architecture. For a 5-stage partial pipelining, the achieved throughput is 7.1 Gbits/sec. It is even higher than the throughput achieved with full pipelining of the initial architecture. In this case the core clock frequency is 111 MHz and the I/O clock will be half of that frequency.

Table 3. Optimised architecture synthesis results

	Hardware	Critical Path Delay	
		Core	**I/O**
Full Round Transformer	16 ESP, 740 LE, 512 FF	8.934 ns 111.93 MHz	10.549 ns 94.79 MHz
Round Key Generator	4 ESP, 272 LE, 272 FF	7.225 ns 138.4 MHz	11.8 ns 84 .7 MHz

For full pipelining, the throughput is 10.8 Gbits/sec. This throughput is determined by the path delay of the I/O in the key generator part.

4.3 Performance Comparison

In order to evaluate our work we did a comparison with the fastest reported designs. In this comparison we divided the known designs into two categories key agile and non-key agile designs, because of the importance of the key agility property and its great effect on the throughput.

The ASIC chip proposed in [6] is the only one we know that reports a key agile design. It achieves a throughput of 1.82 Gbits/sec in case of full pipelining. The fastest known designs are reported in [8] (7 Gbits/sec), and in [9] (7.68 Gbits/sec), but none of them offers the key agility property.

From the point of view of silicon area, the proposed design is more memory block consuming than the one presented in [8]. The latter uses 82 Xilinx BRAMS, which corresponds to 164 ALTERA ESB whereas we used 200 ESB (for full pipelining). The difference comes from the part devoted to the key generation part of our design (key agility).

5 Conclusion

This paper describes a high performance FPGA implementation of the Rijndael algorithm. This 128-bit data/key encryption design performs at a data rate of 10.8 Gbits/sec for full pipelining and it can be considered as one of the fastest key agile AES design. It is 6 times faster than the fastest reported key agile chip. In comparison with the fastest reported AES chip, a rate of 7.1 Gbits/sec is achieved in case of half pipelining. The achieved throughput is nearly the same as in [8], [9] with half pipelining while they used full pipelining, and the design is 1.5 times faster for full pipelining. Modularity gives very high flexibility and great compromising between degree of pipelining and required throughput without any core modifications. Moreover, this design can be considered as the first one reporting AES chip running in CTR mode.

For future work, more detailed analysis and estimations especially on the basic module implementation should be done. Sub-pipelining can be used to enhance the core performance and hand review for the floor plan can reduce the I/O path delay. Combinational implementation of the S-box transformation is also under study.

Acknowledgments

This work has been partially funded by the French ministry of Defence under research contract of French procurement agency (DGA).

References

1. Altera APEX20K, APEX 20K Devices: System-on-a-Programmable-Chip Solutions, http:/www.altera.com.
2. J. Daemen, V. Rijmen, The Rijndael Block Cipher: AES Proposal, First AES Candidate Conference (AES1), August 1998.
3. M. Dworkin, Recommendation for Block Cipher Modes of Operation, NIST special Publication 800-38A, December 2001.
4. A.J. Elbirt, W. Yip, B. Chetwynd, C. Paar, An FPGA Implementation and Performance Evaluation of the AES Block Cipher Candidate Algorithm Finalists, The Third AES Candidate Conference (AES3), April 2000.
5. FIPS Publication 197, Specification for the Advanced Encryption Standard (AES), Federal Information Processing Standards Publication 197, U.S. DoC/NIST, November 2001.
6. H. Kuo, I. Verbauwhe, Architectural Optimisation for a 1.82Gbits/sec VLSI Implementation of the AES Rijndael Algorithm, CHES 2001, LNCS 2162, 2001.
7. H. Lipmaa, P. Rogaway, D. Wagner, Comments to NIST concerning AES Modes of Operations CTR-Mode Encryption, Symmetric Key Block Cipher Modes of Operation Workshop, October 2000.
8. M. McLoone, J.V. McCanny, High Performance Single-Chip FPGA Rijndael Algorithm Implementations, CHES 2001, LNCS 2162.
9. M. Alam, W. Badawy, G. Jullien A Novel Pipelined Threads Architecture for AES Encryption Algorithm, ASAP 2002, July 2002.
10. Report on the Symmetric Key Block Cipher Modes of Operations Workshop. (NIST), October 2000.

An FPGA-Based Performance Analysis of the Unrolling, Tiling, and Pipelining of the AES Algorithm

G.P. Saggese[1], A. Mazzeo[1], N. Mazzocca[2], and A.G.M. Strollo[1]

[1] University of Naples Federico II, Via Claudio 21, 80125 Napoli, Italy
[2] Second University of Naples, Via Roma 29, 81031 Aversa, Italy
{saggese,mazzeo,n.mazzocca,astrollo}@unina.it

Abstract. In October 2000 the National Institute of Standards and Technology chose Rijndael algorithm as the new Advanced Encryption Standard (AES). AES finds wide deployment in a huge variety of products making efficient implementations a significant priority. In this paper we address the design and the FPGA implementation of a fully key agile AES encryption core with 128-bit keys. We discuss the effectiveness of several design techniques, such as accurate floorplanning, the unrolling, tiling and pipelining transformations (also in the case of feedback modes of operation) to explore the design space. Using these techniques, four architectures with different level of parallelism, trading off area for performance, are described and their implementations on a Virtex-E FPGA part are presented. The proposed implementations of AES achieve better performance as compared to other blocks in the literature and commercial IP core on the same device.

1 Introduction

Symmetric-key block ciphers are important in many cryptographic systems. Individually they provide confidentiality which means keeping information secret from all but those who are authorized to see it. As a basic building block, they allow construction of pseudo-random number generators, stream ciphers, message authentication code, and hash functions. These primitives have a crucial role in many real-world applications. In October 2000 the National Institute of Standards and Technology (NIST) chose Rijndael algorithm [1] as the new Advanced Encryption Standard (AES) [2]. This standard is become effective on May 2002. The use of AES and the AES replacement of DES and triple DES, is encouraged to provide the desired security of electronic data in commercial and private organizations.

Even if most new algorithm designs cite efficiency in software as a design objective, the software implementations of symmetric algorithms are often computationally expensive. Therefore in many applications the use of hardware-based solutions has become unavoidable in order to meet high performance requirements. Reconfigurable devices, like Field-Programmable Gate Arrays (FPGA),

P.Y.K. Cheung et al. (Eds.): FPL 2003, LNCS 2778, pp. 292–302, 2003.

are a promising alternative for the implementation of block ciphers, since they include many advantages of software implementations (like algorithm agility, algorithm modification capability [9], and cost efficiency) with physically security and high throughput of an Application-Specific Integrated Circuits (ASIC). In order to achieve maximum performance on FPGA, designs making use of architecture-specific features are required, which means that circuit implementations are not portable between different FPGA technologies.

In this paper we address the design and the FPGA implementation of a fully key agile AES encryption core with 128-bit (128b) keys. We discuss the effectiveness of several techniques, such as accurate floorplanning, the unrolling, tiling and pipelining transformations (also in the case of feedback modes of operation) to explore the design space, allowing to tradeoff area for performance. The proposed AES blocks achieve better performance (with respects to area and throughput metrics) than other implementations on the same Xilinx device, available in the technical literature and as commercial IP cores. The application of the above-mentioned techniques allow an effective design space exploration, since our least area (iterative) design requires 446 Virtex-E slices and 10 Select BlockRAM delivering 1 gigabit per second (Gbps) encryption rate, and our fastest AES block (fully unrolled and deeply pipelined) reaches a throughput of 20,3 Gbps.

The rest of the paper is organized as follows. In Section 2 we describe the AES algorithm and the block cipher modes of operation. Section 3 addresses main issues related to the design and the implementation of AES and discusses our solutions. Section 4 describes four architectures differing in unrolling and pipelining level and presents their floorplan aware implementations. Section 5 reports performance results and analyzes area vs throughput trade-offs, with respect to other academical and commercial implementations. Finally, Section 6 concludes the paper with some final remarks.

2   AES Encryption Algorithm and Modes of Operating

The AES algorithm selected by NIST as the successor of the DES encryption algorithm is Rijndael [1]. In AES standard [2] the length of the data block is always equal to 128b and so the parameter Nb, which represents the number of 32b words composing a data block, is 4. As far as different security requirements are concerned, the key length for AES can be chosen as either 128b, 192b, or 256b long, and the corresponding parameter Nk is 4, 6, or 8 respectively. The pseudo-algorithm for the AES encipherment and for the KeyExpansion procedure with a 128b or 192b key are reported in the left and right boxes of Fig. 1, respectively. The main computation of AES encryption is a loop repeated Nr times, where Nr is equal to either 10, 12, or 14, based on the key size. The cipher algorithm has the goal of obscuring (in a reversible manner) the plain text with a set of $Nb(Nr + 1)$ 32b words w (named *sub-keys*) derived by the expansion of the key. AES interprets the incoming data and the intermediate result (state) as a 4×4 array of 8b words (byte). The processing of a 128b plain text starts

```
Cipher (byte in[4*Nb],
   word w[Nb*(Nr+1)], byte out[4*Nb])
begin
   byte state[4,Nb] = in
   AddRoundKey(state, w[0, Nb-1])     } Initial Round
   for round = 1 step 1 to Nr-1
      SubBytes(state)
      ShiftRows(state)
      MixColumns(state)                 } Round
      AddRoundKey(state,
      w[round*Nb, round*Nb+Nb-1])
   end for
   SubBytes(state)
   ShiftRows(state)
   AddRoundKey(state,                    } Final Round
   w[Nr*Nb,Nr*(Nb+1)-1])
   out = state
end
```

```
KeyExpansion (byte key[4*Nb],
   word w[Nb*(Nr+1)])
begin
   word temp, i = 0
   while (i < Nk)
      w[i] = (key[4*i], key[4*i+1],
         key[4*i+2],key[4*i+3])
      i = i+1
   end while
   while (i < Nb*(Nr+1))
      temp = w[i-1]
      if (i mod Nk = 0)
         temp = SubWord(RotWord(temp))
            xor Rcon[i/Nk]
      end if
      w[i] = w[i-Nk] xor temp
      i = i+1
   end while
end
```

Fig. 1. The AES Cipher algorithm in the case $Nb = 4$ and the KeyExpansion procedure particularized for $Nk = 4$ and 6, in the left and right box, respectively.

copying it into the state. The state is mixed with the first part of the sub-keys by means of an AddRoundKey step (*InitialRound* in Fig. 1). Then the Round transformation, involving the four basic steps of Fig. 1, is applied $Nr - 1$ times. At last the *FinalRound* block, which differs slightly from the *Round* since it does not include MixColumns, is executed. The basic operations which compose AES are: 1) SubBytes – each byte of the state is transformed using an s-box function which is composed of two transformations over $GF(2^8)$, namely an inversion and an affine transformation (i.e. a matrix per vector multiplication); 2) ShiftRows – each row of the matrix state is circularly shifted with a row-dependent amount of positions; 3) MixColumns – is a multiplication over $GF(2^8)$ of each column for a constant, giving the corresponding column of the updated state; 4) AddRoundKey – is a bitwise modulo-2 addition, i.e. an addition over $GF(2^8)$, of a sub-key to the state. The AES algorithm takes the cipher key and performs a KeyExpansion procedure to generate the linear array of $Nb(Nr + 1)$ sub-keys, which are used in the $Nr + 1$ steps of the cipher algorithm. In the KeyExpansion procedure of Fig. 1 RotWord is a circular rotation of the bytes in a word and SubWord applies an s-box transformation to each byte of a $32b$ word. The constant Rcon[h] is $[\{02\}^{h-1}, 00, 00, 00]$ where the power is considered over $GF(2^8)$. Several modes of operation for use with an underlying symmetric key block cipher algorithm are recommended by NIST [4] and are widely employed. The main goal of these modes of operation is twofold [3]: 1) improving the security of block ciphers avoiding repeated inputs from being encrypted to the same value. To achieve this goal, the output of the block cipher is fed back, introducing a dependence of current output upon previous inputs. 2) Providing different security features (such as hashing, integrity check, etc.) using the block cipher as a building block. The main modes of operation are: Electronic Codebook (ECB), Cipher Block Chaining (CBC), Cipher Feedback (CFB), Output Feedback (OFB), and Counter (CTR). Specific information about the security

	ECB	CBC	CBC-MAC	CFB	OFB	OCB	CTR
Sender	Enc	Enc	Enc	Enc	Enc	Enc	Enc
Receiver	Dec	Dec	Enc	Enc	Enc	Dec	Enc
Feedback	No	Yes	Yes	Yes	Yes	No	No

Fig. 2. Block cipher modes of operation.

properties of these modes of operation, with respect to errors, manipulations, and substitutions, and the associated schematics, can be found in [3] [4]. The Fig. 2 reports, for each mode of operation, whether encrypter (Enc) or decrypter (Dec) is needed in transmitting and receiving phase, and the feedback loop in the processing flow of a data stream is present.

3   Main AES Design Choices

Unrolling, tiling and pipelining AES. – Many cipher algorithms are based on a basic looping structure (Feistel or other substitution-permutation network) whereby data are iteratively passed through a transformation, namely the "round" function. Three techniques are known in the technical literature to exploit the intrinsic parallelism of this kind of algorithm at different levels providing architecture with different performance: the unrolling, the tiling and the pipelining transformation. The *unrolling* [9] consists in allocating the body of different instances of a loop some number of times (called the unrolling factor U) and iterates the loop by step U. Referring to U, the number of instanced loop bodies, it is possible to distinguish an iterative (serial) architecture, a partially unrolled or a completely unrolled one. The *iterative* architecture consists in the instantiation of a single body of the loop. This block is fed back through a mux, in order to realize successive iterations of the loop. This approach can minimize the hardware requirement giving low throughput. In an *unrolled* architecture the algorithm is implemented by allocating U rounds as a single combinatorial logic block. This approach requires more area and does not necessarily increase the throughput but enables pipelining, as it is shown in the following of the paper. The *tiling* [5] transformation replicates the instances of a block in order to increase the functional parallelism. Finally, the *pipelining* increases the number of blocks of data that are being simultaneously operated upon, inserting registers to store the partial results.

In the recent past, some authors [5] asserted that, in order to increase the throughput, it is better tiling a serial AES block than unrolling it. They justified this remark noting that: 1) an unrolled architecture is faster than the iterative one in the encryption time of a block, but this difference is often negligible; 2) the time saving comes with a greatly increase of the area. We agree with this analysis (that can also be supported by the quantitative measurements presented in [9]), but we would emphasize that tiling has different flaws with respect to unrolling as a throughput increasing technique. First of all, the unrolling enables further increase of pipelining levels improving the throughput, and this can justify the extra area. Second, when the design is tiled, additional logic (with corresponding area and time penalty) is usually involved: demultiplexing logic to feed each one of the blocks, and a block collecting the results. An unrolled and pipelined architecture does not require any other additional logic than pipeline registers. In this paper we propose (see Section 4) several AES implementations, differing in the number of rounds actually instanced and in the level of pipelining. These demonstrate that the throughput per area of an iterative AES block is lower than the one of an unrolled and pipelined version of AES, even without considering the extra logic to route the data in a tiled architecture.

Pipelining and Feedback Loops – As we already mentioned, in many real-world applications, cipher modes of operation are employed. The pipelining can be effectively used for applications requiring modes of operation not relying on feedback (such as ECB, OCB, and CTR) to enhance the performance. Some of these modes imply feedback (see last row of Fig. 2). The feedback introduces a loop carried dependence, i.e. the processing of a block can start only if the previous block has been processed, and this prohibits a pipelined implementation. For this reason, some cipher modes, such as like interleaved CFB or CFB-k [4], [3], have been proposed to allow security and high encryption rate when implemented in hardware. In the applications requiring feedback modes (such as CBC, CFB, and OFB), a method of increasing the throughput is to pipeline the cipher even in presence of feedback, and to interleave multiple data streams keeping full the pipeline This technique [6] is called *innerroundpipelining* (or c-slow technique). Please note that it is possible to resort to the same technique also in an iterative architecture, with or without feedback cipher mode. As general rule in both cases, if the pipeline is composed by N stages, N different data flows should be available. The availability of many independent data flows is a common scenario in encrypted Internet router whereas multiple routes are active and have to be encrypted or authenticated. Since an (iterative or feedback) AES block can be pipelined, provided that a sufficient number of data flows is available, in the following of this paper we focus on the application of pipelining on the AES algorithm.

Decryption and Key-Length – In this paper we focus on the design and the implementation of an AES encrypter for two main reasons: 1) in many communication systems only the encipherment can suffice (see first and second row of Fig. 2); 2) the AES decryption process is roughly the same as the encryption one. In fact, the same loop structure of the encryption and similar steps (InvByteSubs, InvShiftRows, etc.) are used in the decryption and thus the same solutions to architectural and implementation issues we propose can be extended to an AES decryption core.

As far as the key-length is concerned, we consider the design of an AES encryption block for 128b keys. However in the same way, it is possible to design AES encryption cores with 192b and 256b keys, since they differ in the number of rounds and in a slightly different KeyExpansion algorithm.

On/Off-Line Subkey Generation – The sub-keys for encipherment/decipherment can be either stored in a local memory (*stored-key* approach) or generated concurrently with the encryption/decryption process (*key-agile* approach). The stored-key approach requires a preprocessing phase every time the key is changed, and it needs a (quite large) memory block for the sub-keys. The key-agile approach allows the block cipher to work at full speed, even if the key is changed, and saves the extra memory for the sub-keys. It is worth noting that in an unrolled and pipelined version of a key-agile cipher, also the KeyExpansion has to be unrolled and further pipeline registers are needed. However in the stored-key approach, when multiple flows are processed in a pipelined AES block, the KeyExpansion and the sub-keys buffer have to be duplicated. The stored key approach is used by [7] and [9], while other AES implementations rely on key-agile approach [5], [8], [10]. In order to guarantee high-performance and maximum cipher flexibility we decided to resort to the key-agile approach.

Data-Path Placement – *Data-path placement* involves using the high-level structure of the computation and the data flow to better place the design. This allows many benefits, including shorter wires, more physically compact layout, and faster and easier place-&-route step. The shortening of the wires and their associated delays enables improved performance on FPGA, since a considerable contribution to the critical path is due to the net delay. Unfortunately a post synthesis hand layout on the FPGA of a given block demands for a high effort and time. Instead, we designed and implemented each basic block of AES cipher using appropriate VHDL attributes to address synthesis and placement, then we placed the overall architecture in accordance with the data flow. This approach is much more handy than a hand layout with a florplan editor. Some details about the implementation and the resulting placed data-paths are reported in Section 4 and in Fig. 4, respectively.

Target Device, Design Flow and Tools – The FPGA part we chose is a Xilinx Virtex-E 2000-8bg560. The Virtex device family appears to be a good representative for a modern FPGA, and is not fundamentally different from devices from other vendors. The Xilinx Virtex series of FPGAs, consists of Configurable Logic Blocks (CLBs) and interconnection circuitry tiled to form a chip. Each CLB consists of two slices, and each slice contains two 4 inputs Look-Up Tables (LUT), 2 flip-flops (FF), and associated carry chain logic. Each LUT can either be used as a 16x1 bit RAM, or as a 1-16 cycle delay shift register. The Xilinx Virtex-E 2000 presents 19200 slices and 19520 tristate buffers and incorporates also fully synchronous dual-ported $4096b$ block memories named Block SelectRAM. Block SelectRAM memories (BRAM) are organized in columns and each BRAM is four CLBs high. As far as design tools are concerned, we used Aldec Active VHDL 4.2 for simulation, Synplicity Synplify Pro 7.1, Synopsys FPGA Express, Xilinx XST for synthesizing VHDL, and Xilinx ISE 4.1 for place-&-route and timing analysis. In our AES design experience Synplify appeared to achieve better synthesis results with respect to Xilinx XST and FPGA Express.

4   AES Blocks Design and Implementation

Iterative Architecture – The schematic of the proposed iterative AES block, the *Round* and the *KeyGen* blocks are reported in Fig. 3. The *Round* block can be configured to implement both the `Round` and the `FinalRound` functions of Fig. 1, while the `InitialRound` is instanced as a single block. The AES block (see Fig. 3) presents a latency equal to $N_R + 1$ clock ticks, but the processing of a new plain text can start and a cipher-text is output every N_R ticks. This is allowed by the overlapping of the first serial step of the processing (*InitialRound*) of a plain text block with the last *Round* of the previous plain text processing. We implemented the *InitialRound* (that is a 128*b* xor gate) and the following multiplexer together in the same set of 128 LUTs, since each LUT realizes any 4 inputs function. In other 128 LUTs the 2-to-1 multiplexer and the *AddRoundKey* can be accommodated. The matrix multiplication involved in the *MixColumns* block can be realized as suggested in [2] using the `xtime` function. In fact the *MixColumns* can be realized as a net of xor gate, whereas the maximum fan-in of a xor is 5 and this gate requires 2 LUTs. For implementing the *SubBytes* we used the Xilinx Virtex-E BRAMs. Each BRAM is configured as a dual-port synchronous 256 × 8*b* words RAM, and it implements 2 s-boxes accessing two locations concurrently. Hence 8 BRAMs suffice for each *SubBytes* blocks. Since the BRAM are synchronous, they can also realize the *Register* before *SubBytes*. Finally the *ShiftRows* can be simply realized by hardwiring. As far as the *KeyGen* block is concerned, the *SubWord* is made up of 4 s-boxes and requires 2 BRAMs which implement also the 32*b* *Register* of Fig. 3. Four 32*b* registers balance the latency due to the 32*b* *Register*. The *Controller* has 4 main states and uses a counter to store the number of round functions applied to the current plain text. The *Controller* runs concurrently in pipelining with the data-path, in a such way that elaboration of data and sequencing of operations can be overlapped. The control signals are buffered into flip-flops that also break up the propagation of

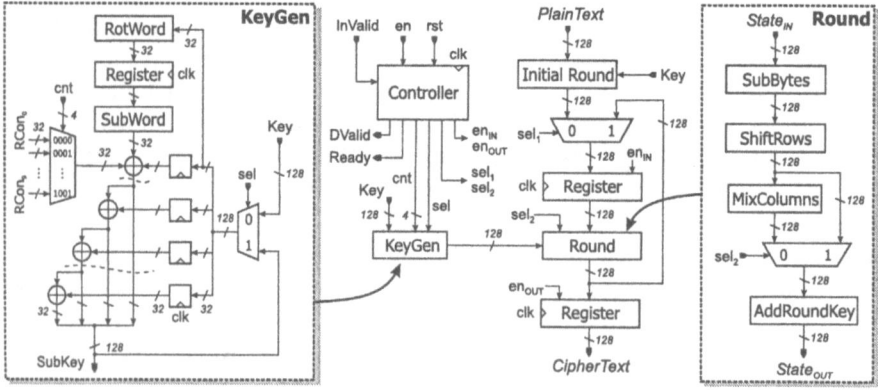

Fig. 3. The schematic of the iterative AES architecture.

these signals to the data-path, greatly limiting the contribution of the net delay on the critical path.

As far as the placement of the blocks is concerned, we decided to enclose the data-path of the iterative AES block in a box 4 CLBs high. Actually a *SubBytes* block requires 8 BRAMs and each BRAM is 4 CLBs high. 32 CLBs can accommodate 128 slices and 128 FF-slices and the data-path parallelism is exactly 128b. Therefore each block of Fig. 3 is placed in order to occupy a different number of 32 CLB columns, and the blocks are ordered in the same way as they process the data. The *Round* function requires a single column. The *MixColumns* is accomodated in 3 columns, since the xtime function is allocated in the first one but it does not fill all the available LUTs, while the remaining 2 columns are completely used for 5-inputs xor-s. The *KeyGen* employees 2 BRAMs and so it is enclosed in a box 8 CLBs high. For the *Controller* we did not impose any placement constraints. The resulting floorplan of the iterative AES block is reported in Fig. 4a. The implementation results are reported in Fig. 5. The same AES block without any optimization and placement constraints requires 1736 slices and presents a T_{CK} of 19,9 ns, and so our optimizations achieve great area saving (-74%) and throughput improvement (+55%).

5-Pipelined Iterative Architecture – Interleaving encryption flows allows to use the pipelining technique also for the iterative architecture. We analyzed the delay of each block in the AES core, in order to insert pipeline registers balancing the delay of the stages. We introduced in the schematic of Fig. 3 one 128b register between *SubBytes* and the wiring implementing *ShiftRows*, and two registers before the inputs of *AddRoundKey*, to obtain a 5 pipeline stages version of the iterative architecture presented in the previous paragraph. The introduction of the pipeline registers does not remarkably impact on the slices count, since many slices are partially occupied by the LUTs of the iterative non-pipelined AES block. As far as the sub-keys generator is concerned, several

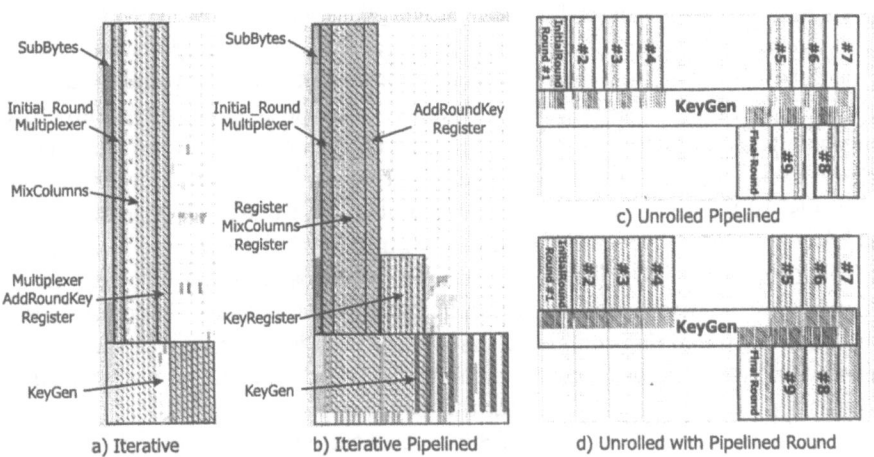

Fig. 4. The floorplans of all the proposed implementations of AES.

Design	Device	Pipeline [stages]	Clock [Mhz]	Th. [Gbps]	Slices (BRAM)	Mbps/ Slices
Weaver et al. [5]	XVE600-8	1	60	0,69	460 (10)	1,5 (0,40)
Labbé et al. [7]	XCV1000-4	1	31	0,39	2151 (4)	0,18 (0,15)
Amphion CS5220 [10]	XVE-8	1	101	0,29	421 (4)	0,69 (0,31)
Amphion CS5230 [10]	XVE-8	1	91	1,06	573 (10)	1,9 (0,57)
Iterative	XVE2000-8	1	79	**1**	446 (10)	**2,3 (0,58)**
Weaver et al. [5]	XVE600-8	5	158	1,75	770 (10)	2,3 (0,85)
Labbé et al. [7]	XCV1000-4	5	30	1,91	8767 (4)	0,22 (0,21)
5-Pipe Iterative	XVE2000-8	5	142	**1,82**	648 (10)	**2,8 (0,94)**
Elbirt et al. [9]	XCV1000-4	4	40	0,98	5449 (0)	0,18 (0,18)
Elbirt et al. [9]	XCV1000-4	10	31	1,88	10923 (0)	0,17 (0,17)
1-Pipe Fully Unr.	XVE2000-8	10	70	**8,9**	2778 (100)	**3,2 (0,57)**
5-Pipe Fully Unr.	XVE2000-8	50	158	**20,3**	5810 (100)	**3,5 (1,1)**

Fig. 5. Performance results for the proposed, commercial and academical implementations of AES.

changes to the schematic of Fig. 3 are needed to realize a 5-stages pipelined key agile *KeyRegister*. Registers are inserted in the place of the dotted lines in Fig. 3. The registers after the multiplexer become shift-registers to keep aligned the data: the first 32*b* register on the top remains unchanged, the second and the third are delayed of 2 clock ticks, while the last of 3 clock ticks. Also the '1' input of the feedback mux is delayed of other 2 clock ticks. A register (indicated as *KeyRegister* in Fig. 4b) on the output *SubKey* to split the propagation of the sub-keys towards the *Round*. We used LUTs configured as shift registers to realize efficiently these flip-flop chains. Finally the *Controller* has to be slightly modified in order to keep track of the current blocks being processed.

1-Pipelined and 5-Pipelined Fully Unrolled Architecture – The 1-pipelined Architecture and the 5-pipelined Architecture implement the completely unrolled AES loop and differ in the level of pipelining. The *1-pipelined Fully Unrolled Architecture*, which uses one pipeline stage per round, is obtained simply juxtaposing ten replica of the *Round* and *KeyGen* depicted in Fig. 3, and using a pipeline register for each round. The first round is an *InitialRound*, while the last round is a *FinalRound* block, and both are simpler than the intermediate *Round*. The *5-pipelined Fully Unrolled Architecture* makes use of 5 levels of pipelining per round, and is obtained from the 1-pipelined Architecture, increasing the pipelining level per round up to 5, following the same remarks described in the design of the 1-pipelined Iterative Architecture. In order to place 10 rounds, 10 BRAM columns are needed, but in a Virtex-E 2000 there are only 8 BRAM columns. Since each BRAM column has 20 BRAMs and each *Round* and *KeyGen* occupy 10 BRAMs, we decided to dispose the first 6 rounds on the upper half of the device, the round 7 split into two parts on the upper and lower half, while the remaining rounds on the lower part. The net delays in the round 7 are relevant, since the basic blocks are scattered on the device, so we inserted an additional pipe register to avoid that this could impact on the critical path. The resulting floorplans of these two AES implementations are reported in Fig. 4c,d and the performance results are given in the last two rows of the Fig. 5.

5   Performance Results and Comparison with Related Work

We adopted three different parameters to evaluate proposed implementations of AES: 1) the throughput T considered as the maximum encryption rate; 2) the cost A in terms of area as number of Virtex-E slices and BRAMs; 3) the throughput per area T/A, which measures the contribution of each slice to the throughput and hence the efficiency of the implementation. Performance results for each AES block of Section 4, and for other academical and commercial implementations are summarized in Fig. 5. The results are expressed in terms of the pipelining levels, the maximum clock frequency, the throughput, the area (as slices and as BRAMs) and finally T/A. It is worth noting that a dual-port $256 \times 8b$ BRAM can be replaced by a distributed memory composed of 256 LUTs (128 slices). Therefore T/A is also reported (in brackets) whereas A is given by the total number of slices using distributed memories and no BRAMs. The reported implementations lie only on 2 types of devices, namely XCV-4 and XVE-8, with differently available resources. These devices are based on the same building block, i.e. a slice containing 2 LUTs and 2 FFs. Our iterative architectures with 1 and 5 pipelining levels have the best T (ranging from 1 to 1,82 Gbps) and T/A among the other iterative implementations. Our fully unrolled architecture with 50 pipeline stages provides the highest encryption throughput (20,3 Gbps), the highest clock frequency (158 MHz) and the best T/A, among the reported implementations. These results demonstrate that the use of techniques such as unrolling and pipelining allow to explore the design space tailoring the performance and area requirements. These techniques are also able to deliver very high performance. Furthermore the accurate data-path placement is effective in reducing the area and improving the clock period of the AES blocks. The placed and routed AES blocks with the placement constraints present a clock period improvement ranging between the 36% and the 55%, with respect to the corresponding design without constraints. In [8] an iterative architecture implemented in a chip using a 0,18 micron CMOS technology with core supply at 1,8V is reported. It can deliver 1,6 gigabit per second encryption rate with $128b$ keys, while our best FPGA iterative implementation reaches 1 gigabit per second. So the floorplan aware design, the use of the architectural features of an FPGA family, and the deep pipelining enable a low-cost technology such as FPGA to reach performance comparable to an ASIC implementation, at least for algorithms which can be mapped in a highly regular structure.

6   Conclusions

This paper has addressed the design and the FPGA implementation of a fully key agile AES encryption core with 128-bit keys. We discussed the effectiveness of several techniques, such as accurate floorplanning, the unrolling, tiling and pipelining transformations to explore the design space. We dealt with the issues concerning the use of the pipelining in presence of feedback loops and analyzing several solutions. Based on these solutions, four architectures with different level of parallelism have been designed and implemented on a Virtex-E FPGA part. Performance results show that the presented implementations achieve better performance with respect to other academical and commercial AES block.

References

1. J. Daemen and V. Rijmen, "AES Proposal: Rijndael, AES Algorithm Submission", September 1999, available at http://www.nist.gov/CryptoToolkit
2. National Institute of Standards and Technology (NIST), FIPS Publication 197, "Advanced Encryption Standard (AES)", November 2001.
3. A. Menezes, P. van Oorschot, and S. Vanstone, "Handbook of Applied Cryptography", CRC Press, 1996.
4. National Institute of Standards and Technology (NIST), Special Publication 800-38A, "Recommendation for Block Cipher Modes of Operation", December 2001.
5. N. Weaver, J. Wawrzynek, "Very High Performance, Compact AES Implementation in Xilinx FPGAs", September 2002, available at http://www.cs.berkeley.edu/~nweaver/sfra/rijndael.pdf
6. C. Leiserson, F. Rose, and J. Saxe, "Optimizing synchronous circuitry by retiming", Third Caltech Conference On VLSI, March 1993.
7. A. Labbé, A. Pérez, "AES Implementations on FPGA: Time-Flexibility Tradeoff", Proceedings of FPL02, pp. 836-844.
8. P.R. Schaumont, H. Kuo, and I.M. Verbauwhede, "Unlocking the Design Secrets of a 2.29 Gb/s Rijndael Processor", Proceedings of the International Design Automation Conference 2002 (DAC02), pp. 634-639.
9. A.J. Elbirt, W. Yip, B. Chetwynd, and C. Paar, "An FPGA-Based Performance Evaluation of the AES Block Cipher Candidate Algorithm Finalists", IEEE Trans. on VLSI Systems, Vol.9, No.4, August 2001, pp. 545-557.
10. Amphion Semiconductor, "CS5210-40: High Performance AES Encryption Cores", Amphion(TM), Febrauary 2003, available at http://www.amphion.com/cs5210.html

Two Approaches for a Single-Chip FPGA Implementation of an Encryptor/Decryptor AES Core

Nazar A. Saqib, Francisco Rodríguez-Henríquez, and Arturo Díaz-Pérez

Computer Science Section, Electrical Engineering Department
Centro de Investigación y de Estudios Avanzados del IPN
Av. Instituto Politécnico Nacional No. 2508, México D.F.
nabbas@computacion.cs.cinvestav.mx
{francisco, adiaz}@cs.cinvestav.mx

Abstract. In this paper we present a single-chip FPGA full encryptor/decryptor core design of the AES algorithm. Our design performs all of them, encryption, decryption and key scheduling processes. High performance timing figures are obtained through the use of a pipelined architecture. Moreover, several modifications to the conventional AES algorithm's formulations have been introduced, thus allowing us to obtain a significant reduction in the total number of computations and the path delay associated to them. Particularly, for the implementation of the most costly step of AES, multiplicative inverse in $GF(2^8)$, two approaches were considered. The first approach uses pre-computed values stored in a lookup table giving fast execution times of the algorithm at the price of memory requirements. Our second approach computes multiplicative inverse by using composite field techniques, yielding a reduction in the memory requirements at the cost of an increment in the execution time. The obtained results indicate that both designs are competitive with the fastest complete AES single-chip FGPA core implementations reported to date. Our first approach requires up to 11.8% less CLB slices, 21.5% less BRAMs and yields up to 18.5% higher throughput than the fastest comparable implementation reported in literature.

1 Introduction

Recently, Rijndael block cipher algorithm was chosen by NIST as the new Advanced Encryption Standard (AES) [2, 13]. Rijndael is a block cipher that can process blocks of 128, 192 and 256 bits and keys of same lengths, but for official AES version, the only legal block length is 128 bits.

FPGA AES implementations are attractive since costs of VLSI design and fabrication can be reduced. However, AES hardware implementation poses a challenge since encryption and decryption processes are not completely symmetrical [2, 7, 13]. Designing separated architectures for encryption and decryption processes would imply the allocation of a large amount of FPGA resources. Designs reported in [3, 4, 5] have considered only the encryption part of AES. A single-chip FPGA implementa-

P.Y.K. Cheung et al. (Eds.): FPL 2003, LNCS 2778, pp. 303–312, 2003.

tion of a full encryptor/decryptor AES core has been reported in [9]. Performance results for these designs are broadly variable; they range from 300 Mbit/s to 3.2 Gbit/s, approximately.

In this paper, we describe a fully pipelined AES implementation core for an FPGA device. It is a complete encryptor/decryptor core for which encryption, decryption and key scheduling work efficiently. We propose two approaches to implement multiplicative inverse for $GF(2^8)$ which is the most costly operation of AES. The first design uses pre-computed values through a lookup table requiring fast memory access to obtain a good throughput. The second approach eliminates memory requirements at the cost of more FPGA resources. Obtained results show that both designs are competitive with the previous full AES core reported in [9].

In Section 2, AES algorithm is briefly described. Several modifications to AES algorithm's expressions to gain performance are explained in Section 3. In Section 4, we discuss a three-stage computation to calculate multiplicative inverse in finite field $GF(2^8)$. In Section 5, a fully pipelined AES FPGA implementation is presented and performance results are provided. Finally, concluding remarks are included in section 6.

2   The AES algorithm

The AES cipher treats the input 128-bit block as a group of 16 bytes organized in a 4×4 matrix called *State* matrix. Fig. 1 depicts the AES cipher algorithm flow for encrypting one block of input data.

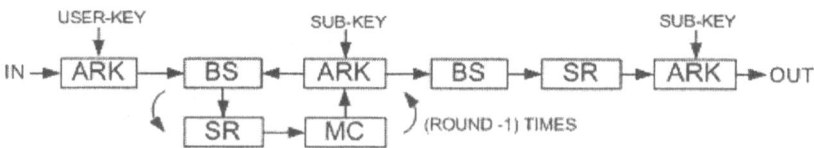

Fig. 1. Basic AES encryption flow.

The algorithm consists of an initial transformation, followed by a main loop where nine iterations called *rounds* are executed. Each round is composed of a sequence of four transformations: *Byte Substitution* (BS), *ShiftRows* (SR), *MixColumns* (MC) and *AddRoundKey* (ARK). For each round of the main loop, a round key is derived from the original key through a process called *Key Scheduling*. Finally, a last round consisting of three transformations, BS, SR and ARK, is executed. The AES decryption algorithm operates similarly by applying the inverse of all the transformations described above in reverse order. In the rest of this section we shall briefly describe the four AES round transformations BS, SR, MC and ARK.

In *ByteSubstitution* (BS), each input byte of the *State* matrix is independently replaced by another byte from a look-up table called *S-box*. The AES *S-box* is a 256-entry table composed of two transformations: First each input byte is replaced with its multiplicative inverse in $GF(2^8)$ with the element {00} being mapped onto itself;

followed by an affine transformation over $GF(2)$ [2, 13]. For decryption, inverse S-box is obtained by applying inverse affine transformation followed by multiplicative inversion in $GF(2^8)$ [13]. *ShiftRows* (SR) is a cyclic shift operation where each row is rotated cyclically to the left using 0,1,2 and 3-byte offset for encryption while for decryption, rotation is applied to the right. In *MixColums*(MC) transformation, each column of the *State* matrix is multiplied by a constant fixed matrix as follows,

$$
\begin{bmatrix} c'_{0,i} \\ c'_{1,i} \\ c'_{2,i} \\ c'_{3,i} \end{bmatrix} = \begin{bmatrix} 02 & 03 & 01 & 01 \\ 01 & 02 & 03 & 01 \\ 01 & 01 & 02 & 03 \\ 03 & 01 & 01 & 02 \end{bmatrix} \begin{bmatrix} c_{0,i} \\ c_{1,i} \\ c_{2,i} \\ c_{3,i} \end{bmatrix} \qquad i = 0,1,2,3 \tag{1}
$$

Similarly, for decryption, we compute Inverse MixColumns, by multiplying each column of the *State* matrix by a constant fixed matrix as shown below

$$
\begin{bmatrix} c'_{0,i} \\ c'_{1,i} \\ c'_{2,i} \\ c'_{3,i} \end{bmatrix} = \begin{bmatrix} 0E & 0B & 0D & 09 \\ 09 & 0E & 0B & 0D \\ 0D & 09 & 0E & 0B \\ 0B & 0D & 09 & 0E \end{bmatrix} \begin{bmatrix} c_{0,i} \\ c_{1,i} \\ c_{2,i} \\ c_{3,i} \end{bmatrix} \qquad i = 0,1,2,3 \tag{2}
$$

Finally, in *AddRoundKey* (ARK), output of MC is XOR-ed with the corresponding round sub-key derived from the user key. The ARK step is essentially same for AES encryption and decryption processes.

3   Novel Computational Expressions for Implementing AES on FPGA Devices

In this section, we introduce some novel techniques for implementing AES algorithm on FPGA devices. The three main AES algorithms, key schedule, encryption and decryption, are considered for optimization. Our optimization criteria are based on three main factors: To maximize path delay reductions, reutilization of pre-computed blocks and to exploit full resources of the target device. In addition, we have tried to use 4-input logic gates wherever is possible, since they can be efficiently implemented using FPGA CLBs.

3.1   AES Algorithm Optimizations

S-box and inverse S-box are required for computing the BS step of AES. Both may be computed by implementing affine (AF) and inverse affine (IAF) transformations together with a look-up table for Multiplicative Inverse (MI). In this way, the combination MI + AF provides S-box for encryption, while IAF + MI computes the Inverse S-box needed for decryption. To use only one MI module, a multiplexer is used to switch the data path for encryption/decryption, as shown in Fig. 2.

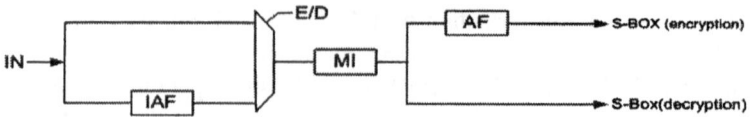

Fig. 2. S-box and inverse S-box implementation

SR and ISR Implementations do not require FPGA resources as they can be implemented by rewiring. For the sake of symmetry, ISR step is embedded into IAF while SR and AF steps are joined together. *MC and IMC transformations* are reviewed in deep. Encryption's MC can be efficiently computed by using only 3 steps [1]: a sum step, a doubling step and a final sum step. Further optimization consists on embedding ARK step to fully exploit 4-input FPGA slice resources. Let the elements of State matrix's column one be $a[0]$, $a[1]$, $a[2]$, $a[3]$, and let $k[0]$, $k[1]$, $k[2]$ and $k[3]$ represent first four bytes of a key block, then transformed matrix (MC + ARK) column $a'[0]$, $a'[1]$, $a'[2]$ and $a'[3]$ can be efficently obtained as shown in Table 1

Table 1. The modified MC expression with ARK.

Step 1	Step 2	Step 3
$v = a[1] \oplus a[2] \oplus a[3]$	$xt_0 = xtime(a[0])$	$a'[0] = k[0] \oplus v \oplus xt_o \oplus xt_1$
$v = a[0] \oplus a[2] \oplus a[3]$	$xt_1 = xtime(a[1])$	$a'[1] = k[1] \oplus v \oplus xt_1 \oplus xt_2$
$v = a[0] \oplus a[1] \oplus a[3]$	$xt_2 = xtime(a[2])$	$a'[2] = k[2] \oplus v \oplus xt_2 \oplus xt_3$
$v = a[0] \oplus a[1] \oplus a[2]$	$xt_3 = xtime(a[3])$	$a'[3] = k[3] \oplus v \oplus xt_3 \oplus xt_0$

Here $xtime(v)$ represents finite field multiplication of $02 \times v$, where 02 stands for constant polynomial x in $GF(2^8)$. Note that all the computations shown in the same column of Table 1 can be performed in parallel.

Table 2. The modified IMC expression with IARK.

Step 1	Step 2	Step 3	Step 4	Step 5
$t = a[0] \oplus a[1] \oplus a[2] \oplus a[3]$			$u = xtime(u)$	$t' = t \oplus u$
		$u = s_0' \oplus s_1' \oplus s_2' \oplus s_3'$	Step 6	
$s_0 = xtime(a[0])$	$s_0' = xtime(s_0)$	$v = s_0 \oplus s_1 \oplus s_0' \oplus s_2'$	$a'[0] = a[0] \oplus t' \oplus v \oplus k[0]$	
$s_1 = xtime(a[1])$	$s_1' = xtime(s_1)$	$v = s_1 \oplus s_2 \oplus s_1' \oplus s_3'$	$a'[1] = a[1] \oplus t' \oplus v \oplus k[1]$	
$s_2 = xtime(a[2])$	$s_2' = xtime(s_2)$	$v = s_2 \oplus s_3 \oplus s_0' \oplus s_2'$	$a'[2] = a[2] \oplus t' \oplus v \oplus k[2]$	
$s_3 = xtime(a[3])$	$s_3' = xtime(s_3)$	$v = s_3 \oplus s_0 \oplus s_1' \oplus s_3'$	$a'[3] = a[3] \oplus t' \oplus v \oplus k[3]$	

The same strategy applied for MC would yield up to seven steps to compute IMC (four sum steps and three doubling steps). The difference is due to the fact that coefficients in equation (2) have a higher Hamming weight than the ones in equation (1). To overcome this drawback we use the strategy depicted in Table 2 where IMC manipulation is re-structured and seven steps are cut to five steps.

As explained above, for *ARK Implementation*, ARK step is embedded into MC step. For final round (Round 10), MC and IMC steps are not executed, therefore, a separate implementation of ARK is made.

The ideas discussed in this section can be implemented as shown in Fig. 3, where the block-diagram represents proposed architecture for implementing AES encryption/decryption processes on FPGA devices. Hence the critical path for encryption is MI•AF•SR•MC•ARK, while ISR•IAF•MI•IMC•IARK is the critical path for decryption as shown below.

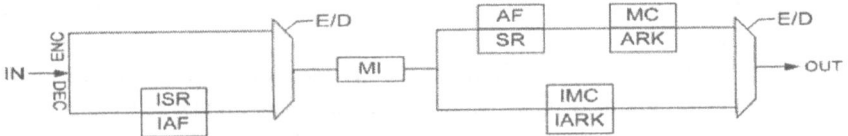

Fig. 3. AES algorithm encryptor/decryptor implementation.

3.2 Key Schedule Optimization

The original user key consists of 128 bits arranged as a 4 x 4 matrix of bytes. Let $w[0]$, $w[1]$, $w[2]$, and $w[3]$ be the four columns of the original key. Then, those four columns can be recursively expanded to obtain 40 more columns, as follows: Let the columns up to $w[i-1]$ have been defined then,

$$w[i] = \begin{cases} w[i-4] \oplus w[i-1] & \text{if } i \bmod 4 \neq 0 \\ w[i-4] \oplus T(w[i-1]) & \textit{otherwise} \end{cases} \tag{3}$$

Where $T(w[i-1])$ is a non-linear transformation based on the application of the *S-box* to the four bytes of the column, an additional cyclic rotation of the bytes within the column and the addition of a round constant (*rcon*) for symmetry elimination [2]. Let $[k_0,k_4,k_8,k_{12}]$, $[k_1,k_5,k_9,k_{13}]$, $[k_2,k_6,k_{10},k_{14}]$ and $[k_3,k_7,k_{11},k_{15}]$ be the first four columns of the key. By combining and parallelizing expressions, first column of the new key $[k'_0,k'_4,k'_8,k'_{12}]$ can be calculated as shown in Table 3.

Table 3. Modified expressions for key schedule.

Step 1		
$k'_0 = k_0 \oplus Sbox(k_{13}) \oplus rcon$		
Step 2		
$k'_4 = k_4 \oplus k'_0$,	$k'_8 = k_4 \oplus k_8 \oplus k'_0$,	$k'_{12} = k_{4.} \oplus k_8 \oplus k_{12} \oplus k'_0$

$Sbox(k_{13})$ refers to the data obtain by substituting k_{13}. The same process is used for calculating other 3 columns of the new key. In the same manner, next nine keys are obtained.

4   AES S-Box Design Based on Composite Field Techniques

The most costly operation in the BS transformation described in section 2.1, is the computation of the inverse multiplicative of a byte in the finite field $GF(2^8)$ defined by the AES cipher. In an effort to reduce the costs associated to this operation, several authors have designed AES *S-Box* based on the composite field technique reported first in [10, 11, 12]. That technique uses a three-stage strategy: Map the element $A \in GF(2^8)$ to a composite field F using a isomorphism function δ. Compute the multiplicative inverse over the field F and finally map the computation results back to the original field.

In [6] an efficient method to compute the inverse multiplicative based on Fermat's little theorem was outlined. That method is useful because it allows us to compute the multiplicative inverse over a composite field $GF((2^m)^n)$ as a combination of operations over the ground field $GF(2^m)$. It is based on the following theorem:

Theorem [7,12]: The multiplicative inverse of an element A of the composite field $GF((2^n)^m)$, $A \neq 0$, can be computed by

$$A^{-1} = \left(A^r\right)^{-1} A^{r-1} \bmod P(x), \text{ Where } A' \in GF\left(2^n\right) \text{and } r = \frac{2^{nm} - 1}{2^n - 1} \qquad (4)$$

An important observation of the above theorem is that the element A' belongs to the ground field $GF(2^n)$. This remarkable characteristic can be exploited to obtain an efficient implementation of the inverse multiplicative over the composite field. By selecting $m = 4$ and $n = 2$ in the above theorem we obtain $r = 17$ and,

$$A^{-1} = \left(A^r\right)^{-1} \cdot A^{r-1} = \left(A^{17}\right)^{-1} \cdot A^{16} \qquad (5)$$

In case of AES, it is possible to construct a suitable composite field F, by using two degree-two extensions based on the following irreducible polynomials [10]:

$$\begin{aligned}
F_1 &= GF\left(2^2\right): & P_0(x) &= x^2 + x + 1 \\
F_2 &= GF\left(\left(2^2\right)^2\right): & P_1(y) &= y^2 + y + \phi \\
F_3 &= GF\left(\left(\left(2^2\right)^2\right)^2\right): & P_2(z) &= z^2 + z + \lambda
\end{aligned} \qquad (6)$$

where $\phi = \{10\}_2$, $\lambda = \{1100\}_2$.

The inverse multiplicative over the composite field F_2 defined in the equation (6), can be found as follows. Let $A \in F_2 = GF((2^2)^2)$ be defined in polynomial basis as $A = A_H y + A_L$, and let the Galois field F_1, F_2, and F_3 be defined as shown in equation (5). Then it can be shown that

$$\begin{aligned}
A^{16} &= A_H y + \left(A_H + A_L\right); \\
A^{17} &= A^{16} \cdot A = 0 \cdot y + \left(\lambda A_H^{16} A_H + A_L^{16} A_L\right) = \lambda A_H^2 + A_L^{16} A_L
\end{aligned} \qquad (7)$$

Fig. 4 depicts the corresponding block-diagram of the three-stage inverse multiplier represented by equations (5) and (7). The circuits shown in Fig. 5 and Fig. 6 present a gate-level implementation of the aforementioned strategy.

As we explained above, in order to obtain the multiplicative inverse of the element $A \in F = GF(2^8)$, we first map A to its equivalent representation (A_H, A_L) in the isomorphic field $F_2 = GF((2^4)^2)$ using the isomorphism δ (and its corresponding inverse δ^1).

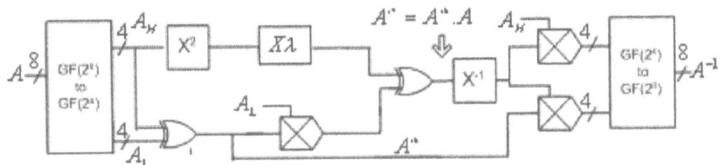

Fig. 4. Three-stage strategy to compute multiplicative inverse in composite fields.

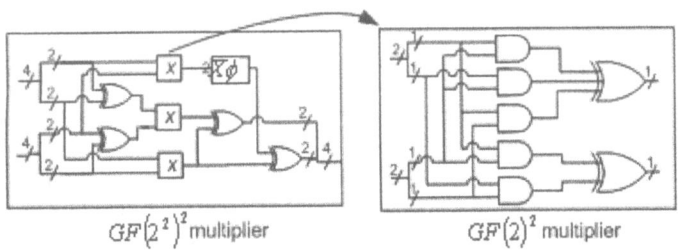

Fig. 5. $GF((2^2)^2)$ and $GF(2^2)$ multipliers

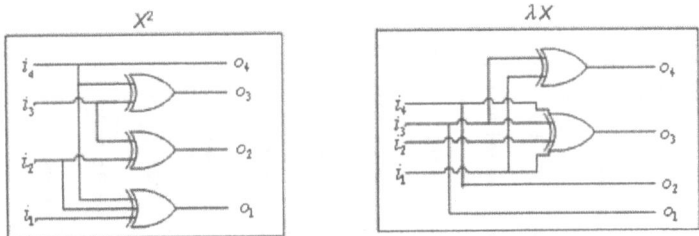

Fig. 6. Gate-level implementation for X^2 and λX

In order to map a given element A from the finite field F to its isomorphic composite field F_2 and vice versa, we only need to compute the matrix multiplication of the said element A, by the isomorphic functions shown in equation (8) given by [10]: Also by taking advantage of the fact that A^{17} is an element of F_2, the final operation $\left(A^{17}\right)^{-1} \cdot A^{16}$ of equation (7), can be easily computed with further gate reduction. Last stage of the algorithm consists of mapping computed value in the composite field, back to the field $F = GF(2^8)$.

$$
\delta = \begin{bmatrix}
1 & 0 & 1 & 0 & 0 & 0 & 0 & 0 \\
1 & 1 & 0 & 1 & 1 & 1 & 1 & 0 \\
1 & 0 & 1 & 0 & 1 & 1 & 0 & 0 \\
1 & 0 & 1 & 0 & 1 & 1 & 1 & 0 \\
1 & 1 & 0 & 0 & 0 & 1 & 1 & 0 \\
1 & 0 & 0 & 1 & 1 & 1 & 1 & 0 \\
0 & 1 & 0 & 1 & 0 & 0 & 1 & 0 \\
0 & 1 & 0 & 0 & 0 & 0 & 1 & 1
\end{bmatrix} ; \delta^{-1} = \begin{bmatrix}
1 & 1 & 1 & 0 & 0 & 0 & 1 & 0 \\
0 & 1 & 0 & 0 & 0 & 1 & 0 & 0 \\
0 & 1 & 1 & 0 & 0 & 0 & 1 & 0 \\
0 & 1 & 1 & 1 & 0 & 1 & 1 & 0 \\
0 & 0 & 1 & 1 & 1 & 1 & 1 & 0 \\
1 & 0 & 0 & 1 & 1 & 1 & 1 & 0 \\
0 & 0 & 1 & 1 & 0 & 0 & 0 & 0 \\
0 & 1 & 1 & 1 & 0 & 1 & 0 & 1
\end{bmatrix} \tag{8}
$$

5   Performance Results

To achieve high throughput, we have designed a pipelined architecture for AES as shown in Fig. 7. Eleven AES rounds have been unrolled to develop a pipelined design. The design is symmetric in the sense that the same steps used for encryption are re-used for decryption. The corresponding round-keys for encryption or decryption are generated from the input key accordingly eleven stages of the pipeline. Each stage is clock triggered and data is transferred to next stage at rising-edge of the clock. The data blocks are accepted at each clock cycle and then after 11 cycles, output encrypted/decrypted blocks appear at the output at consecutive clock cycles.

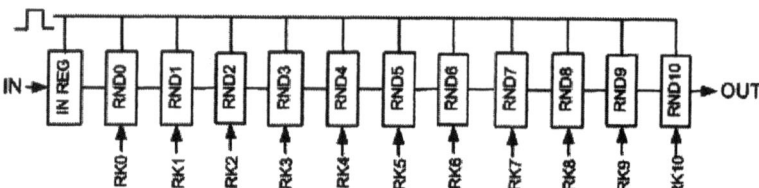

Fig. 7. Pipeline approach for 128 bit AES Encryption/decryption

AES memory requirements could be too high, especially for a pipeline design. To overcome this requirement, the design is implemented on Xilinx Virtex-E XCV2600. The Virtex and Virtex-E family of devices contains more than 280 BRAMs which are well suited for fast memory access [14]. A dual port BRAM can be configured into two single port BRAMs with independent data access. Xilinx Foundation Series F4.1i tool is used for design entry, verification and synthesis. Both approaches to implement MI, previously discused, follow the same pipeline architecture. Table 4 details the total area required for each block as shown in Fig. 3.

The first design uses 80 BRAMs (43%) needed for MI implementation and occupies 386 I/O Blocks (48%) and 5677 slices (22.3%). The system runs at 30 MHz and data is processed at 3840 Mbits/s. The second design, using a three-stage MI computation, occupies 13416 CLB slices (52%) and 386 I/O Blocks(48%); no BRAMs are required here. The allowed system frequency is 24.5 MHz and the design achieves a throughput of 3136 Mbits/s.

Table 4. Summary of features for AES encryptor/decryptor cores.

BLOCK	E/D $GF(2^8)$ CLB SLICES	E/D $GF(2^4)$ CLB SLICES
KeyBlock	1278	1278
MI (10 rounds)	(80 BRAMS)	6676
SR	------	------
AF (10 rounds)	800	800
IAF (10 rounds)	640	640
MC + ARK (combined) (9 rounds)	1368	1368
IMC + IARK (combined) (9 rounds)	2088	2088
ARK (1st + last round)	128	128
Misc. (Timing + I/O registers etc.)	374	438
TOTAL	6676 + (80 BRAMs)	13416 (NO BRAMs)
Throughput (Mbits/s)	3840	3136
Throughput/Area (Mbits/s/Slices)	0.58	0.24

Some FPGA implementations [3, 4, 5, 9, 15] for AES do exist. But some of them deals only with encryption [3, 4, 5] while others like [15] show better results but can not be categorized as a single-chip implementation. A fair comparison of our design can be made with the design reported in [9] which is an encryptor/decryptor core realization. It was implemented on Virtex-E XCV3200 FPGA device, using 7576 CLB slices, 102 BRAMs and achieving a throughput of 3239 Mbits/s. That design uses the same set of BRAMs for encryption/decryption and consumes 256 cycles to prepare BRAMs for encryption or decryption. That design also occupies 20 BRAMs for key scheduling. Our first design compared with [9] has reduced the area require-ments up to 11.8% while expected throughput has been increased more than 18.5%. For our second design, area requirements are high but throughput is competitive with [9]. An advantage of this design is the portability factor since it does not use BRAMs.

6   Conclusions

We have presented two AES encryptor/decryptor core designs following a pipelined architecture. Both designs take advantage of using only one S-box (to compute multi-plicative inverse). The difference lies in the implementation of MI. In the first design, 80 BRAMs are utilized for MI pre-computed values. In the second design, a three-stage architecture has been adopted for MI where calculations are made in $GF(2^4)^2$ and $GF(2^2)^2)^2$ instead of $GF(2^8)$. The goal was to reduce memory requirements as much as possible, and then this design does not require BRAMs. The encryp-tor/decryptor starts reporting results only after 11 cycles needed to latch the round keys. We have re-organized MC and IMC computations to reduce data-path and to take advantage of four-input/one-output organization of CLBs. The ARK step has been embedded with MC and IMC step to enhance the timing performance.

Our designs have been implemented on Virtex-E family of devices (XCV2600) using Xilinx Foundation Series F4.1i. Our results have shown competetive behavior compared to the existing AES FPGA encryptor/decryptor known cores. In the first design, we have outperformed design reported in [9]. The second design can be implemented on any FPGA family of devices. Future work includes extending the design for variable key and block lengths for AES algorithm.

References

1. Guido Bertoni et al: Efficient Software Implementation of AES on 32-bits Platforms: CHES2002, LNCS 2523, Springer-Verlag, 2002.
2. Joan Daemen, Vincent Rijmen: The Design of Rijndael, AES-The Advanced Encryption Standard: Springer-Verlag Berlin Heidelberg, New York, 2002.
3. A. Dandalis, V.K. Prasanna, J.D.P. Rolim: A Comparitive Study of Performance of AES Candidates Using FPGAs: The Third Advanced Encryption Standard (AES3) Candidate Conference, 13-14 April 2000, New York, USA.
4. J. Elbirt, W. Yip, B. Chetwynd and C. Paar: A FPGA implementation and Performance Evaluation of the AES Block Cipher Candidate Algorithm Finalists: The Third AES3 Candidate Conference, 13-14 April 2000, New York.
5. K. Gaj, P. Chodowiec: Comparison of the Hardware Performance of the AES Candidates using Reconfigurable Hardware: The Third Advanced Encryption Standard (AES3) Candidate Conference, 13-14 April 2000, New York, USA.
6. Brian Gladman: The AES Algorithm (AES) in C and C++: URL: http://fp.gladman.plus.com/cryptography_technology/rijndael/index.htm, April 2001.
7. Guajardo, C. Paar: Efficient Algorithms for Elliptic Curve Cryptosytems: CRYPTO '97, August 17-21, Santa Barbara, CA, USA.
8. T. Ichikawa, T. Kasuya, M. Matsui: Hardware Evaluation of the AES Finalists: The Third Advanced Encryption Standard (AES3) Candidate Conference, 13-14 April 2000, New York, USA.
9. Maire McLoone and J.V McCanny: High Performance FPGA Rijndael Algorithm Implementations: C. Koc, D. Naccache, and C.paar(Eds): CHES2001, LNCS 2162, pp. 65-76, Springer-Verlag, 2001.
10. S. Morioka and A. Satoh: An Optimized S-Box Circuit Architecture for Low Power AES Design: CHES2002, LNCS 2523, Springer-Verlag, 2002.
11. C. Paar: Efficient VLSI Architectures for Bit Parallel Computation in Galois Fields: PhD thesis: Universitat GH Essen, VDI Verlag, 1994.
12. A. Rudra et al: Efficient Rijndael Encryption Implementation with Composed Field Arithmetic: CHES2001, LNCS 2162, pp. 171-184, Springer-Verlag, 2001.
13. W. Trappe and L. C. Washington: Introduction to Cryptography with Coding Theory: Prentice-Hall, Upper Saddle River, 2002.
14. Xilinx Virtex TM-E 1.8V Field Programmable Gate Arrays, URL: www.xilinx.com, November 2000.
15. URL: http://ece.gmu.edu/crypto/rijndael.htm

Performance and Area Modeling
of Complete FPGA Designs
in the Presence of Loop Transformations

K.R. Shesha Shayee, Joonseok Park, and Pedro C. Diniz

University of Southern California / Information Sciences Institute
4676 Admiralty Way, Suite 1001
Marina del Rey,California 90292, USA,
{shesha,joonseok,pedro}@isi.edu

Abstract. Selecting which program transformations to apply when mapping computations to FPGA-based architectures leads to prohibitively long design exploration cycles. An alternative is to develop fast, yet accurate, performance and area models to understand the impact and interaction of the transformations. In this paper we present a combined analytical performance and area modeling for complete FPGA designs in the presence of loop transformations. Our approach takes into account the impact of input/output memory bandwidth and memory interface resources, often the limiting factor in the effective implementation of these computations. Our preliminary results reveal that our modeling is very accurate allowing a compiler tool to quickly explore a very large design space resulting in the selection of a feasible high-performance design.

1 Introduction

The application of loop-level transformations, important to expose vast amounts of fine-grain parallelism and to promote data reuse, substantially increases the complexity of mapping computations to FPGA-based architectures. Figure 1 illustrates a typical behavior for a design mapped to an FPGA exploring loop unrolling. At first as the unrolling factor increases the execution time decreases and the amount of consumed space resources increases. However, with additional unrolling, there is the need to exploit more memory resources and memory parallelism. Before the FPGA capacity limitation is reached (in this case for an unrolling of 16) another limit is reached for an unrolling factor of 8. At this point, the design requires more memory channels than the implementation can provide. As such, these resources must be time-multiplexed leading to a non-trivial increase in the execution time (illustrated by the heavy-shaded bar).

Understanding, and mitigating the impact of hardware limitations is important in deriving the parameters of the loop transformations in the quest for the best possible design. In the example illustrated below if a compilation tool is not aware of the fact that memory channels resources are limited, it incorrectly selects the design with an unrolling factor of 8. This selection is only better

P.Y.K. Cheung et al. (Eds.): FPL 2003, LNCS 2778, pp. 313–323, 2003.
© Springer-Verlag Berlin Heidelberg 2003

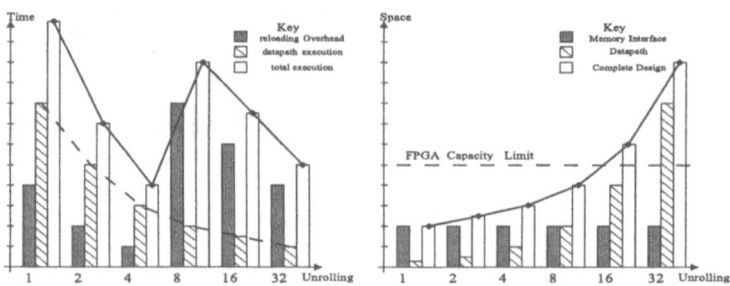

Fig. 1. Qualitative Execution and Space Plots in the Presence of Limited I/O Resources

than the design without any unrolling and far from the actual best design which corresponds to an unrolling amount of 4.

As this example illustrates, ignoring the real memory interface resource limitation can lead extremely poor design choices. Given the extremely large number of possible loop transformations and their interaction, a solution to the problem of finding the best performing (and feasible in terms of space) design is to develop performance and area estimation for the designs resulting from the application of a sequence of loop transformations. To this extend we focus on the development of a set of analytical models combined with behavioral estimation techniques for the performance and estimation of complete FPGA designs. In this work we model the application of a set of important loop transformations, *unrolling*, *tiling*, and *interchange*. We explicitly take into account the impact of the transformations on the limited resources of the design's memory interfaces.

This paper makes the following specific contributions:

- It presents an area and performance modeling approach that combines analytical and estimation techniques for complete FPGA designs.
- It describes the application of the proposed modeling to the mapping of computations to FPGAs in the presence of loop *unrolling*, *tiling*, *interchange* and *fission*.
- It validates the proposed modeling for a sample case study application on a real Xilinx Virtex™ FPGA device. This experience reveals our model to be very accurate even when dealing with the vagaries of synthesis tools.

Overall this paper argues that performance and area modeling for complete designs, either for FPGA or not, is an instrumental technique in handling the complexity of finding effective solutions in the current reconfigurable as well as future architectures.

This paper is organized as follows. In section 2 we describe the analysis and modeling approach in detail. In section 3 we present experimental results for a sample image processing kernel and – binary image correlation. In section 4 we describe related work and conclude in section 5.

2   Modeling

We now describe our analytical execution time and area modeling for complete designs as implemented in our target FPGA system.

2.1   Target Design Architecture and Execution Model

The designs considered in our modeling are composed of two primary components as illustrated in Figure 2(a). The first component is the *core datapath* that implements the core of the computation and is generated by the synthesis tools from a high-level description such as VHDL. This datapath is connected to an external memory via an *memory interface* component. This memory interface (see [1]) is responsible for generating the physical memory addresses of various data items as well as the electrical signaling with the external memory.

In many computations, such as image processing kernels, there is a substantial opportunity to reuse data in registers across iterations of a given loop. This is the case, for example, in window-based algorithm in which a computation is performed over shifted windows of the same image. the overlap between windows allows for a subset of the data to be reused in registers, therefore avoiding to load all of the data from memory. For computations with these features, our datapath implementations include an internal *tapped-delay* line.

Our model exploits the pipelined execution *mode* of memory accesses and the core datapath. To adequately support this execution mode for the class of regular image processing computations, our memory interface supports the concept of *streamed data channels* or simply *streams*. Associated with each of these *stream* the memory interface include resources to generate the corresponding sequence of memory addresses. Setting up and resetting these resources whenever the datapath requires data from channels not currently set-up (or set up at a different base address in memory) incurs an overhead. The important parameters for the peformance modeling of the pipelined execution mode of the datapath are its *initiation interval* and *latency*. As to the memory interface, its important parameters are the *reloading overhead*, the *read* and *write* latencies and the *setup* for reading and writing in pipelined memory access mode.

Figure 2(b) illustrates the basic parameters of the execution model, which will be the basic parameters for the performance modeling described in the next section. We show a 2-channel implementation with 2 *buffers* in the datapath,

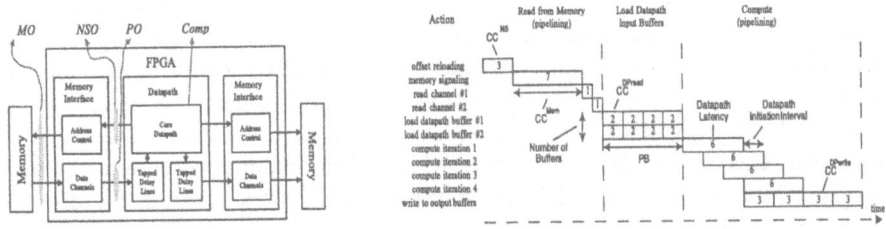

Fig. 2. (a) FPGA Design Architecture (b) Execution Mode

each with a depth of 4 elements and corresponding to 2 datapath input ports. The computation executes 4 loop iterations per block of data read from memory.

2.2   Performance Modeling

As with other compile-time modeling approaches we assume known compile-time loop bounds. In addition we assume that in the presence of control-flow the expressions must capture the longer execution path and that the datapath implementation will preemptively fetch all of the data items required in each of the possible control paths. While in general these assumptions would lead to excessively inflated performance results, for the target set of digital image applications, and due to their amenability to be modeled into perfectly loop nests without severe control flow, this potential phenomenon is rare, if at all observed. Under these assumptions we split the overall execution into 5 components summarized below and whose analytical expressions are presented below.

$$
\begin{aligned}
Comp &= LoopCnt * initInter \\
NSO &= \sum_{0 \leq i \leq m} LoopCnt_i^{NS} * CC^{NS} \\
PO &= \prod_{0 \leq i \leq m} LoopCnt_i^{P} * max(BD_j^{B}) * CC^{DPread} \\
MO &= \sum_{0 \leq i \leq m} LoopCnt_i^{Mem} * CC^{Mem} \\
Exec &= (Comp + NSO + MO + PO) * Clk
\end{aligned}
$$

Computation Time (Comp) models the aggregate time the datapath spends actively computing results. We define it as the product of the overall number of iterations times the datapath initiation interval.

Non-Streaming Access Overhead (NSO) models the aggregate overhead in the reloading of the base and offset for non-streaming memory accesses and is defined by the NSO expression. In this expression $LoopCnt_i^{NS}$ captures the number of iterations the datapath performs a non-streaming memory accesses; and CC^{NS} is the individual cost of such accesses. The value m is the total number of array variables in the particular implementation with non-streaming accesses. For streaming memory access this metric is defined as 0.

Prologue Overhead (PO) models the overhead of filling buffers associated with a data stream before the datapath is ready to start its pipelined execution. The component PO defined above captures this cost of buffer preloading where $LoopCnt_i^{P}$ corresponds to the iteration count of the transformed loop nest that requires prologue loading. BD_i^{B} is the buffer depth of the B buffer (in the data-path) for variable i. The factor $max(BD_i^{B})$ defines the maximum length of time required to fill the longest buffer.

Memory access overhead (MO) models the latency on the memory interface for retrieving data from the memory. In the MO expression LC_i^{Mem} is the ratio of the number of memory accesses (or loop count accounting the memory access) to that of the granularity, for a variable i. Granularity is defined as the ratio of the data width of memory access to that of the data consumed in the datapath. Granularity is 1 if the data width of the memory access is the same as that of the data consumed in the datapath. CC^{Mem} is the number of cycles required for a memory data access.

Clock rate (Clk) We include this metric in the evaluation of the performance of a design as different designs will have can exhibit a wide disparity of clock rates due to radically different internal hardware implementations. Conducting relative comparison based on the clock cycles alone could lead to the wrong decision in selecting the best performing design.

Execution Time (Exec) simply defines the aggregate execution time and is simply the product of the number of execution cycles with the actual values of Clk as derived from the synthesis tools.

2.3 Area Modeling

This modeling is essential so that a compiler can use the area estimates derived from the synthesis tool and combine the predicted area for the memory interface when judging the area of a complete FPGA design.

To achieve this goal, we have derived a linear model using linear regression for a set of data points for various choices of number of memory interface channels. For 32-bit wide data channels the overall FPGA (Xilinx Virtex™ 1K device) area (in slices) for the memory interface can be approximated by the expression $Area_{32} = 137 * NumberChannels + 1514$. A similar empirical approach can be used for other channels widths and FPGA devices.

As to the modeling of the area for the datapath we use estimation results provided by synthesis tools such as the Mentor Graphics' Monet™ tool. The overall area estimation therefore combines the empirical model for the memory interface with the synthesis estimates.

2.4 Deriving the Model Parameters

In the current implementation we manually apply the various loop transformations, and derive the parameters of our modeling effort are derived using a crude emulation of the actual execution by actually running the computation in software and accumulating the number of occurrences of each metric. We are in the process of automating the application of the transformations and developing more sophisticated, and automated, approach to extract the values of the modeling parameters.

The modeling parameters that depend on the target FPGA device such as the maximum clock rate or the actual implementation of the memory interfaces (*e.g.*, the latency or the number of cycles for non-pipelined operations) are architecture dependent and known at compile time. The remaining model parameters are either derived from analysis of the source code, as is the case of the classification of which array data accesses are non-sequential, where the compiler can use data dependence analysis as described in [2].

3 Case Study: Binary Image Correlation

We now validate the proposed performance and area modeling for one image processing kernel — a binary image correlation (BIC) computation.

3.1   Original Computation and Role of Loop Transformations

Figure 3 depicts the pseudo-code for the example under study. This computation consists of a 4-loop nest with known loop bounds and implements binary image correlation using a mask variable (2D mask array) over an input image (2D image array). The computation scans the input image by sliding an $e \times e$ window over the input and accumulates the values of the corresponding image window for non-zero mask values in th (another 2D array variable).

The input image(image[m+i][n+j]) is accessed in a row-wise fashion. Unrolling the j-loop fully, $e - 1$ of the e values in a single row, can be reused in consecutive iterations of the n-loop if the e values are stored in a *tapped-delay line*. And, as the data access is row-wise, unrolling of the i-loop enhances the performance by increasing the reuse to 2 dimensions. As a consequence, after the first iteration(where $e \times e$ data are loaded), only e data values corresponding e data streams needs to be loaded in the sub-sequent iterations of the n-loop.

```
for m = 0 to m < (t-e)
 for n = 0 to n < (t-e)
  for i = 0 to i < e
   for j = 0 to j < e
    if(mask[i][j] != 0)
     th[m][n] += image[m+i][n+j];

     (a) Original code.


for m = 0 to m < t
 for q = 0 to q < t by e
  for p = 0 to p < s by f
   for n = q to e-1
    for i = p to p+f-1 // unrolled loop
     if(mask[i][0] != 0)
      th[m][n] += image[m+i][n];
     ...
     if(mask[i][s-1] != 0)
      th[m][n] += image[m+i][n+s-1];

     (b) Using e × f tiling.
```

```
for p = 0 to s/k
 for m = 0 to t
  for n = 0 to t
   for i = p*(s/k) to p*(s/k)+k // fully unrolled
    for j = 0 to s // fully unrolled
     if(mask[i][j] != 0)
      temp[p][m][n] += image[m+i][n+j];
// loop-fission and loop interchange
for m = 0 to t
 for n = 0 to t
  for p = 0 to s/k // fully unrolled
   th[m][n] += temp[p][m][n];

(c) Loop fission after interchange (loops m and p)
```

Fig. 3. BIC codes after applying loop transformations.

Full unrolling of the two inner most loops, as suggested above, may not always be possible as unrolling leads to the replication of the loop body operators generating very large designs. As such we can apply tiling of the innermost loops as illustrated in Figure 3(b). By tiling the i and j-loops by a factor of e and f respectively, we reduce the amount of hardware resources devoted to the datapath. The trade-off, however, is in terms of reduced reuse. The tapped-delay lines that, in case of unrolling, held the data of a row of image, now holds data of length e. But, of greater significance is the fact that the exectution time increases as the memory interface resources, associated with address generation, requires reloading of address (and as a consequence increased data access) for the data streams associated with each tile. The performance of the overall design is thus substantially reduced.

In the presence of associative operations it is possible to use loop interchange with loop tiling to avoid the need of reloading the data in a tile. The implementation saves partial results of the computation in temporary datapath buffers thereby avoiding the need to reload data from external memory. This strategy comes at the cost of a, potentially large, multi-dimensional array for storing the partial results in the datapath. The size of this temporary array is in the order of the input image size, making it an infeasible solution for most cases.

To mitigate this issue of buffer space, we use yet another set of transformation: array privatization (to privatize temporal variable) and loop fission. The privatized temporal value is now stored in the external memory rather internal buffers (we pay the price for memory access). A second loop generates final results by adding up temporal values in the correct order.

The application of these 3 loop transformations to this case study illustrates the point of this paper. As these transformations affect the data reuse patterns and, in turn, the way the data needs to be streamed in and out of the corresponding datapath implementation, the overall performance in FPGA-based implementations is not only affected by the number of iterations executed but also, and infact more significantly, by the number of times the memory interface resources need to reset.

3.2   Experimental Results

We have manually derived behavioral specification for the BIC code applying the loop transformations described above for a variety of parameter choices. We then use Mentor Graphics' Monet high-level synthesis tool to derive the structural VHDL specification and obtain the estimated area and initiation interval and latency for their pipelined implementation. To derive a complete design we merge the structural VHDL with the structural code of the memory channel interfaces. Given a complete VHDL design we used Synplicity Synplify Pro 6.2 and Xilinx ISE 4.1i tool sets for logic synthesis and Place-and-Route (P&R) [3] targeting a Xilinx Virtex™ XCV 1000 BG560 device.

Performance Model Validation. We validate the performance modeling describe in section 2 for the VHDL code resulting from the application of the various loop transformations described above. In this modeling we use the numerical parameters as, $CC^{DPRead} = 2$, $CC^{Mem} = 7$ and $CC^{NS} = 3$ and omit the symbolic loop expressions here for space considerations.

Figure 4 plots the measured vs. predicted performance for the various, loop unrolled (for the inner loop); tiled (the i and j loops and tiled-with-interchange. The vertical axis corresponds to the simulated clock cycle counts. We limit our experimental set to tiling versions of $1 \times f$ because other transformations do not deliver additional performance.

These results indicate that our performance modeling tracks the overall performance for the various implementations very well. The gap between the predicted performance and simulation results is due to our channel controller behavior. The channel controller used in our memory interface uses a round-robin

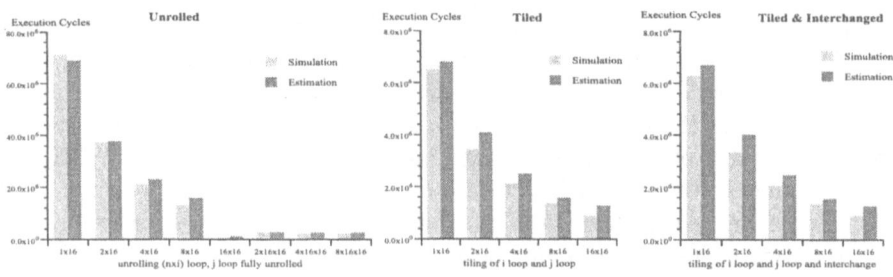

Fig. 4. Performance Estimation .vs. Simulation Results for BIC.

scheduling strategy which introduces a small number of additional cycles, as we verified through simulation.

Area and Implementation Results. We now validate the overall area estimation approach using the empirical model for the memory interface and the area estimation extracted from behavioral synthesis. In this validation we compare the results a compiler would obtain off-line (*i.e.*, without actually synthesizing any designs) with the real results synthesizing the actual complete designs.

Table 1 presents the synthesis results and complete design results for the various tilled and tiled&interchange designs. For the tiled&interchange design we present here only the design corresponding to the first loop that carries out the bulk of the computation. Area estimates produced by Monet behavioral synthesis tool are not compatible, in terms of units, with those of the P&R tools. However, these estimates to a large extent do capture the 'trend' of the results(area) produced by the synthesis and P&R tools as is evident from table 1. Table 1 compares the estimation numbers with those of the synthesis and P&R numbers for the tiled implementations. Estimation numbers for datapath and interfaces

Table 1. Synthesis results for tiling (top) and tiling with interchanging and fission (bottom).

Tile	Datapath Only		Interface	Full Design			
	Estimates (Monet)	P&R (Slices)	Estimates (Monet)	Estimates (Monet)	P&R		
					Slices(%)	Clk (MHz)	
(1x1)	14621	4214	1788	16409	4558 (37)	35.3	
(1x2)	15804	3808	2602	18406	6661 (54)	29.9	
(1x4)	18191	5163	2610	20801	7070 (57)	21.7	
(1x8)	22868	6369	3706	26574	9734 (79)	28.3	
(1x16)	19704	5053	5898	25602	10417 (84)	24.2	

Tile	Datapath Only		Interface	Full Design			
	Estimates (Monet)	P&R (Slices)	Estimates (Monet)	Estimates (Monet)	P&R		
					Slices(%)	Clk (MHz)	
(1x1)	1411.1	408	1788	3199.1	2326 (19)	43.4	
(1x2)	2599.1	725	2602	5201.1	2984 (24)	36.1	
(1x4)	4963.1	1333	2610	7573.1	4038 (33)	26.6	
(1x8)	9694.9	2620	3706	13400.9	6119 (50)	16.7	
(1x16)	19704.0	5053	5898	25602.0	10417 (84)	24.2	

are provided in columns 2 and 4 while columns 3 and 6 are the synthesis and P&R results for datapath only and full design (datapath+interface) respectively. Column 5 is the estimation number for the full design (col. 2 + col. 4).

As can be seen the combined estimation results for the complete designs track very well the real area results even for the very large designs that occupy more than 50% of the FPGA area. This is important as the application of loop transformations that expose vast amount of instruction level parallelism require the replication of functional operators and invariably lead to large designs. For some of the implementations the internal tapped-delay lines used to reduce the number of memory accesses is an important factor on the area as well.

We investigated the fact that the (1×8) tiled-version maps to more slices than the (1×16) version (see table 1 (top) col. 3). We concluded that although the computational for the (1×8) version are approximately half of those for the (1×16) version, the area consumed by the registers used to store intermediate results across the m-loop were responsible for this area anomaly.

3.3   Discussion

Experimental results reveal that our performance and area modeling correlates well with our simulation/synthesis results for various implementations of BIC. Our performance model is capable of identifying effects of various loop transformations, which ultimately lead to the best combination of loop transformations. In the BIC case study our model accurately captures the effect of tiling with different tiling factors, and we can find the best tiling shape/s. The results also reveal that our modeling successfully captures the performance behavior of the more sophisticated combination of loop interchange, fission and privatization.

Overall, we believe the modeling approach described in this paper to be applicable to a wider range of reconfigurable architectures as our model does not exploit any feature of the underlying architecture. In fact, our case study reveals that the major source of discrepancies was either due to characteristics of FPGA synthesis (*e.g.* the large impact of temporary buffers) or due to FPGA implementation execution features (*e.g.*, scheduling of memory accesses).

4   Related Work

Other researchers have also recognized the need for fast area and performance estimation to guide the application of compiler high-level loop transformations for deriving alternative designs. Derrien and S. Rajoupadyhe[4] describe a processor array partitioning that takes into account the memory hierarchy and I/O bandwidth and apply tiling to maximize the performance of the mapping of a loop nest onto FPGA-based architectures. In this context they use an analytical performance model to determine the best tile size. So *et. al.* [5] have expanded the loop transformations to include unrolling and use behavioral synthesis estimates directly from commercially available synthesis tools in an integrated compilation and synthesis. The PICO project [6] takes functions defined as C loop nests to a synchronous array of customizable VLIW processors. PICO uses estimates from its scheduling of the iterations of the nest to determine which loop transformation(s) lead to shorter scheduling time and therefore minimal completion time. The MILAN project [7] provides design space exploration and simulation environments for System-on-Chip(SoC) architecture. MILAN evaluates several possible partitions of the computation among the various system components (processor, memory, special purposed accelerators) using simulation techniques to derive estimates used in the evaluation of a given application mapping.

The work presented in this paper differs from these approaches in several respects. While other projects have focused on more general computation we have focused on the domain of image processing algorithms specified as tight loop nests. Second, rather than using profiling or simulation based estimates we have analytically modeled the performance and relied on the accuracy of behavioral synthesis estimation commercial tools for area modeling. In terms of loop transformations we have focused not only on loop unrolling and tiling but also on loop interchange in the presence of associative operators. Loop interchanging introduces the complication of having to save intermediate results for subsequent computations. Designs with large set of registers for temporary values lead to an explosion of area and substantial degradation of clock estimates.

5   Conclusion

In this paper we presented a performance and area modeling approach using analytical and empirical techniques for complete FPGA designs. Our modeling is geared towards computations expressed as loop nests in high-level programming languages such as C. Using the proposed modeling, compiler tools can evaluate the impact of multiple loop transformations on FPGA resources as well as that of the memory interface resources, often the limiting factor in the effective implementation of these computations. The preliminary results reveal that our approach delivers area and performance estimations that correlate very well with the corresponding metrics from the actual implementation. This experience suggests the proposed modeling approach to be an effective technique that allow compilers to quickly explore a wider range of loop transformations for selecting feasible and high-performance FPGA designs.

References

1. Park, J., Diniz, P.: Synthesis of Memory Access Controller for Streamed Data Applications for FPGA-based Computing Engines. In: Proc. of the 14th Intl. Symp. on System Synthesis (ISSS'2001), IEEE Computer Society Press (2001)
2. Diniz, P., Park, J.: Automatic synthesis of data storage and contol structures for FPGA-based computing machines. In: In Proc. IEEE Symp. on FPGAs for Custom Computing Machines (FCCM'00), IEEE Computer Society Press (2000)
3. (Virtex 2.5v FPGA product specification. ds003(v2.4)) Xilinx, Inc., 2000.
4. Derrien, S., Rajoupadyhe, S.: Loop tiling for reconfigurable accelerators. In: Proc. of the Eleventh Int. Symp. on Field-Programmable Logic (FPL'2001). (2001)

5. So, B., Hall, M., Diniz, P.: A compiler approach to fast hardware design space exploration for fpga systems. In: Proc. of the 2001 ACM Conference on Programming Language Design and Implementation (PLDI'00), ACM Press (2001)
6. Kathail, V., Aditya, S., Schreiber, R., Rau, B., Cronquist, D., Sivaraman, M.: PICO: Automatically designing custom computers. In: IEEE Computer. (2002)
7. Bakshi, A., Prasanna, V., Ledeczi, A.: Milan: A model based integrated simulation framework for design of embedded systems. In: Proc. of the ACM Workshop on Languages, Compilers, and Tools for Embedded Systems (LCTES 2001). (2001)

Branch Optimisation Techniques
for Hardware Compilation

Henry Styles and Wayne Luk

Department of Computing, Imperial College, 180 Queen's Gate, London, England

Abstract. This paper explores using information about program branch probabilities to optimise reconfigurable designs. The basic premise is to promote utilization by dedicating more resources to branches which execute more frequently. A hardware compilation system has been developed for producing designs which are optimised for different branch probabilities. We propose an analytical queueing network performance model to determine the best design from observed branch probability information. The branch optimisation space is characterized in an experimental study for Xilinx Virtex FPGAs of two complex applications: video feature extraction and progressive refinement radiosity. For designs of equal performance, branch-optimised designs require 24% and 27.5% less area. For designs of equal area, branch optimised designs run upto 3 times faster. Our analytical performance model is shown to be highly accurate with relative error between 0.12 and 1.1×10^{-4}.

1 Introduction

For most computer programs, the execution frequency of each basic block is controlled by the runtime behavior of conditional branches. Optimal resource allocation between basic blocks requires that execution frequencies be known. Software profilers collect execution frequencies for a representative dataset to support static resource allocation. Microprocessors demonstrate that branch probability information can be used at runtime to aid dynamic resource allocation. In this paper we explore the analogous use of branch probability information to optimize resource allocation in hardware compilation. In particular, the novel aspects of our work include:

- a compiler that maps programs written in a subset of C to a set of hardware designs that are optimised for different branching probabilities;
- analytical methods, including a queueing network model, for elucidating the properties of the proposed compilation procedure;
- evaluation of our approach based on both analytical and experimental methods for two large applications: video feature extraction and radiosity.

The rest of the paper is organised as follows. Section 2 describes the two basic compilation phases: dependency analysis and circuit synthesis. Section 3 then presents the branch-optimised compilation path. Section 4 deals with the models for studying the analytical properties of this compilation procedure, and is followed by Section 5 which evaluates both analytically and experimentally the effectiveness of the proposed approach. Finally, Section 7 summarises our current and future research.

P.Y.K. Cheung et al. (Eds.): FPL 2003, LNCS 2778, pp. 324–333, 2003.

2   Compilation

Our compilation procedure consists of two phases: dependency analysis and circuit synthesis. The input language for the compiler is a streaming subset of the C language in which arbitrary pointers and loop carry dependencies are not supported. Each input program specifies the body of a single loop, with flow control specified by an *if..then..else* branch construct. These restrictions preclude certain types of program such as the Fibonacci generator, however an extensive set of applications can be automatically transformed [10] into this form. A simple example program, shown on the left of Fig. 1, will be used to illustrate our compilation procedure in the following sections.

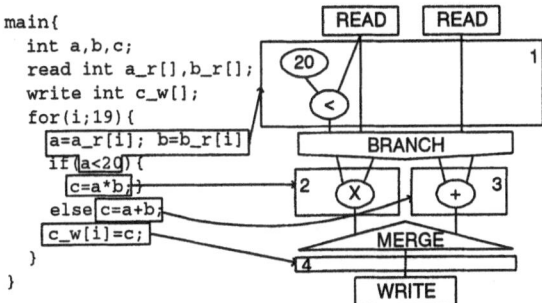

Fig. 1. A simple example program and its two-level data flow graph. The four basic blocks map to four numbered DAG subgraphs. The *if..then..else* maps to BRANCH and MERGE nodes. Array accesses map to READ and WRITE nodes.

The dependency analysis phase constructs a two-level data flow graph from the input program. The data flow graph for our simple example program is shown on the right of Fig. 1. It includes a numbered direct acyclic graph (DAG) for each basic block. Flow control between DAGs is represented by BRANCH and MERGE nodes with firing rule semantics as described in data flow computing literature [9]. Reads and writes to vector variables at the start and end of the data flow graph are mapped to READ and WRITE nodes.

The circuit synthesis phase transforms the dependency graph into a unidirectional pipeline captured in structural VHDL. It consists of module selection, scheduling, binding and instantiation of appropriate flow control circuits. The *initiation interval* of a library block is the number of cycles between each output. An XML library block database specifies the initiation interval, latency in cycles, and area of available library blocks. A static pipelined list scheduler [2] is provided for basic block scheduling.

Circuit synthesis is specialized to form two compilation paths: control study compilation path and branch-optimised compilation path. The control study compilation path is inspired by the StReAm [4] compiler. It creates pipelines which perform equally well under all branching conditions. Designs are parameterised with a global initiation interval parameter $b_{pipeline}$. The control study compilation path circuit with $b_{pipeline} = 1$ for our simple example program is shown on the left of Fig. 2.

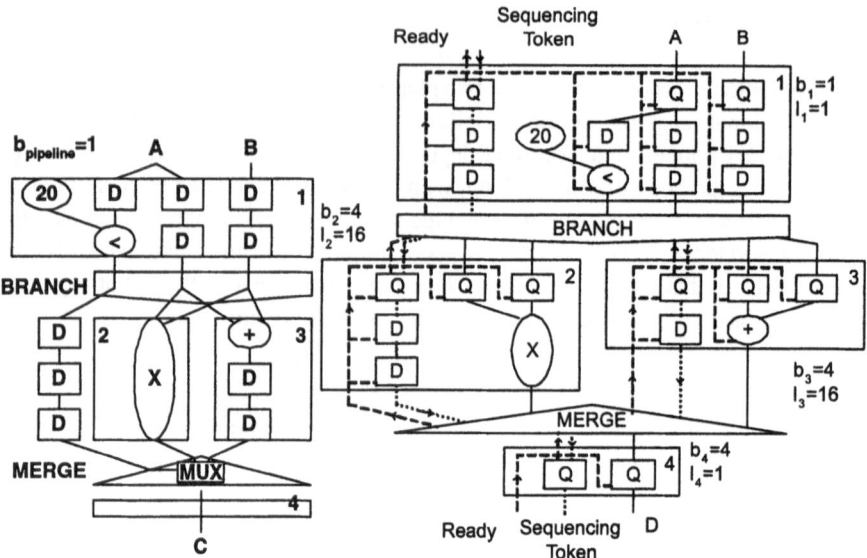

Fig. 2. Uniform (left) and multiple (right) rate circuits for the simple program of Fig. 1. For the control study compilation path, additional flip-flops (D) are inserted to synchronise data flow. For the branch-optimised compilation path, rate-smoothing queues (Q) and sequencing token (dotted line) are inserted to synchronise data flow. The add and multiply modules are pipelined and the flow can be halted by the *ready* control signal (dashed line) shown as an additional input to these components.

3   Branch-Optimised Compilation Path

In this section we introduce a new compilation scheme which promotes efficiency in the presence of branch probability information. A branch-optimised circuit for the simple example program is shown on the right of Fig. 2. The branch-optimised compilation scheme transforms the data flow graph into a set of hardware configurations in which different basic blocks run at different initiation intervals. From this set, a configuration can be chosen in which the resources assigned to different branches match the observed computational load. The branch-optimised compilation scheme creates designs with the following characteristics.

1. Basic blocks can run at independent rates, with different degrees of sequentialization. The rate of basic block i is controlled by the initiation interval parameter b_i.
2. Each basic block propagates a sequencing token downwards, shown as a dotted line in Fig. 2. Different paths through the pipeline run at different rates, and so computations may retire out of order. The sequencing token identifies the loop index associated with a set of results at the pipeline outputs, enabling the original ordering to be recovered. The width of the sequencing token is determined from the upper loop bound and maximum length path through the compiled design.
3. Basic blocks have rate smoothing FIFOs, labeled Q in Fig. 2, for the sequencing token and data inputs. The FIFO length for basic block i is given by parameter l_i.

4. Each basic block propagates a ready signal back up through the pipeline, shown as a dashed line in Fig. 2. The ready signal allows basic blocks to stall incoming computation when input queues are full. For each basic block, the incoming ready signal fans out to the clock-enable input of all registers in the datapath. In our current implementation we adopt a fully synchronous design style. However, a globally asynchronous locally synchronous (GALS) design style could potentially be adopted, in which each basic block operates in a separate clock domain and the ready signal is replaced with true asynchronous handshaking.

5. The BRANCH node routes sequencing token and data to the branch target specified by the branch condition. It receives ready signals from the two branch targets, and blocks computation if the branch target set by the branch condition is not ready.

6. The MERGE node forwards data and sequencing tokens from true and false branch targets. If sequencing tokens arrive from both branch targets simultaneously, the MERGE node blocks the branch targets alternately in a round robin fashion.

4   Analytical Modeling

In this section we describe analytical models of the area-throughput design space for the control study and branch-optimised compilation paths. These models are used to determine the best compilation path and parameterization from observed branch probability information. In the experimental study presented in Section 5, branch probability information is collected at compile time by profiling. In a future system, branch probability information could be collected and acted upon at runtime. Analytical techniques are of increasing importance, as severe time constraints on the optimisation process would almost certainly preclude more complex modelling.

We model the cycle count throughput of branch-optimised designs using a queueing network model. Branch-optimised designs introduce finite queue lengths, blocking, and the possibility of correlated arrival rates. Queuing networks which model these properties are generally solved by simulation [7]. We adopt a simple analytical model based on a $M/M/1/\infty/FCFS$ queueing network with saturating external arrivals to node one [5]. Given information about steady state branch probabilities, known variables in the model are:

1. The node initiation intervals vector $b \in \Re^N$. In the model, element b_i is the exponentially distributed mean initiation interval of node i and b_i is set as the basic block initiation interval for block i.

2. The "routing matrix" $Q \in \Re^{N \times N}$. In the model, element Q_{ij} is the steady state probability that a job, completing node i, routes to node j. A BRANCH after node x to select between branch targets y and z is modeled with $Q_{xy} + Q_{xz} = 1$. The summation of probabilities $q_i = \sum_{j=1}^{N} Q_{ij} \leq 1$ where $i = 1, 2, .., N$. If $q_i < 1$, then a job, on completing node i, exits the queuing network with probability $1 - q_i$. Q is filled with the known branch probability information.

To estimate performance, we determine the maximum sustainable external arrival rate to node one. In the model, external arrival rates are captured in $\gamma \in \Re^N$ where element γ_i is the Poisson process mean external arrival rate for node i. For compiled designs, γ_i is of the form $\gamma = [\gamma_1, 0, 0, ..]$. The procedure to determine the maximum sustainable value of γ_1 is as follows.

1. Solve the traffic equations (eq.1) to determine the net arrival rate at each node in terms of the external arrival rate at node one. The mean net arrival at each node is an element in $\lambda \in \Re^N$. An equation is formed for each element λ_i in terms of γ_1.

$$\lambda(I - Q) = \gamma \tag{1}$$

2. Determine the maximum arrival rate at node one given that the utilization of each node is less than or equal to one. In the model, the utilization of each node is an element in $\rho \in \Re^N$. We maximize γ_1 subject to the utilization constraint (eq.3).

$$\rho_i = \lambda_i b_i \tag{2}$$
$$\rho_i \le 1 \qquad i = 1, 2, .., N \tag{3}$$

Any design with $\rho_i = 1$, $i \ne 1$ for maximum λ_1 will exhibit steady state blocking.

The control study compilation path is parameterised with the global initiation interval $b_{pipeline}$. Designs sustain throughput $1/b_{pipeline}$ for all branching probabilities.

5   Case Studies

In this section we compare the performance of both compilation paths and evaluate the accuracy of analytical models for two case study applications. The input programs and their corresponding top-level data flow graphs for the case study applications are shown in Fig. 4 and Fig. 5. The test scenes are shown in Fig. 3.

Fig. 3. Left and center panes show Video Quality Expert Group test sequence 10 (VQEG10) before and after video feature extraction. Right pane shows the radiosity test scene.

Video Feature Extraction. The algorithm [11] consists of edge detection, thresholding and 3x3 sum-squared difference. There are four basic blocks and one branch.

Progressive Refinement Radiosity. Radiosity algorithms [8] simulate radiation of energy between surfaces. There are ten basic blocks guarded by three branches.

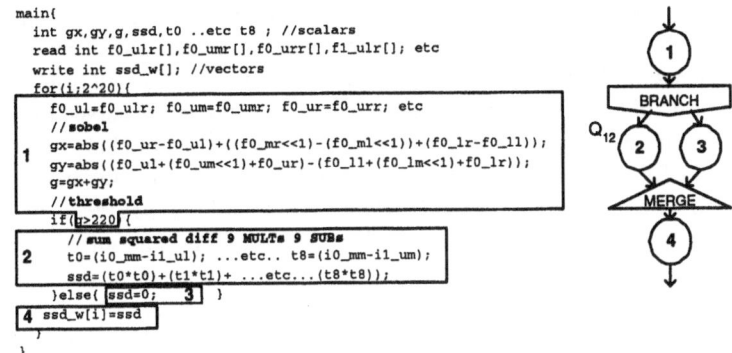

```
main{
    int gx,gy,g,ssd,t0 ..etc t8 ; //scalars
    read int f0_ulr[],f0_umr[],f0_urr[],f1_ulr[]; etc
    write int ssd_w[]; //vectors
    for(i;2^20){
        f0_ul=f0_ulr; f0_um=f0_umr; f0_ur=f0_urr; etc
        //sobel
        gx=abs((f0_ur-f0_ul)+((f0_mr<<1)-(f0_ml<<1))+(f0_lr-f0_ll));
        gy=abs((f0_ul+(f0_um<<1)+f0_ur)-(f0_ll+(f0_lm<<1)+f0_lr));
        g=gx+gy;
        //threshold
        if(g>220){
            //sum squared diff 9 MULTs 9 SUBs
            t0=(i0_mm-i1_ul); ...etc.. t8=(i0_mm-i1_um);
            ssd=(t0*t0)+(t1*t1)+ ...etc...(t8*t8));
        }else{ ssd=0;    }
        ssd_w[i]=ssd
    }
}
```

Fig. 4. Input program and the corresponding top-level data flow graph for video feature extraction.

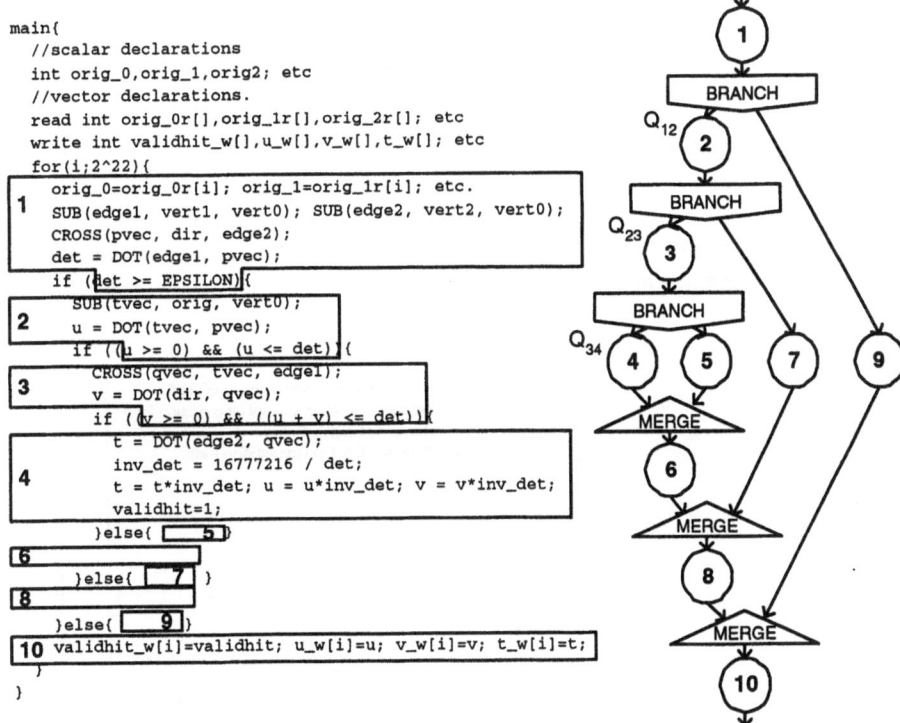

```
main{
    //scalar declarations
    int orig_0,orig_1,orig2; etc
    //vector declarations.
    read int orig_0r[],orig_1r[],orig_2r[]; etc
    write int validhit_w[],u_w[],v_w[],t_w[]; etc
    for(i;2^22){
        orig_0=orig_0r[i]; orig_1=orig_1r[i]; etc.
        SUB(edge1, vert1, vert0); SUB(edge2, vert2, vert0);
        CROSS(pvec, dir, edge2);
        det = DOT(edge1, pvec);
        if (det >= EPSILON){
            SUB(tvec, orig, vert0);
            u = DOT(tvec, pvec);
            if ((u >= 0) && (u <= det)){
                CROSS(qvec, tvec, edge1);
                v = DOT(dir, qvec);
                if ( (v >= 0) && ((u + v) <= det)){
                    t = DOT(edge2, qvec);
                    inv_det = 16777216 / det;
                    t = t*inv_det; u = u*inv_det; v = v*inv_det;
                    validhit=1;
                }else{    5 }
            }else{    7 }
        }else{    9 }
        validhit_w[i]=validhit; u_w[i]=u; v_w[i]=v; t_w[i]=t;
    }
}
```

Fig. 5. Input program and the corresponding top-level data flow graph for radiosity Moller-Trumbore ray-triangle intersection.

6   Results

For the purposes of the experiments, all designs have a uniform word length of 32 bits. All results use the Xilinx XCV3200E-8 device. Arithmetic library blocks are generated using

Xilinx Core GENERATOR 5.1.02i, with b_i =1, 2, 4, or 8 for multipliers and dividers. Other library blocks do not scale for initiation interval. VHDL output by the compiler is synthesised with Synplify Pro 7.1. Area and clock rate are collected from Xilinx 5.1i. Run-time basic block utilization and queue length behavior are observed by simulation using ModelSim SE Plus. A wrapper was constructed in Handel-C [1] to demonstrate designs on the RC1000-PP FPGA platform. The relative accuracy of the analytical model is calculated with the formula $(analytical \ \gamma_1 - observed \ \gamma_1)/observed \ \gamma_1$.

Branch probability information was collected by profiling. The software implementation of the video feature extraction case study was profiled with the test sequence VQEG for 100 frames. Routing table entries relating to Fig. 4 are $Q_{12} = 0.0891$. The software implementation of the radiosity case study was profiled for 4 refinements, involving approximately 800K ray-triangle intersection tests. Routing table entries relating to Fig. 5 are $Q_{12} = 0.520$, $Q_{23} = 0.164$ and $Q_{34} = 0.367$.

The analytical and experimental results for both compilation paths and case studies are shown in Fig. 6 and Tables 1, 2, 3 and 4. Fig. 7 illustrates the effects of different probabilities on the performance of both compilation paths for the video feature extraction case study. The key results of the experimental study are as follows.

1. The branch-optimised compilation path automatically identifies the basic blocks that can benefit from branch probability information and produces designs with different parameterizations of b, the initiation interval vector. For the video feature extraction application, the compiler identifies basic block 2 and produces 10 different designs; for the progressive refinement radiosity application, the compiler identifies basic blocks 1, 2, 3 and 4 and produces 35 designs.

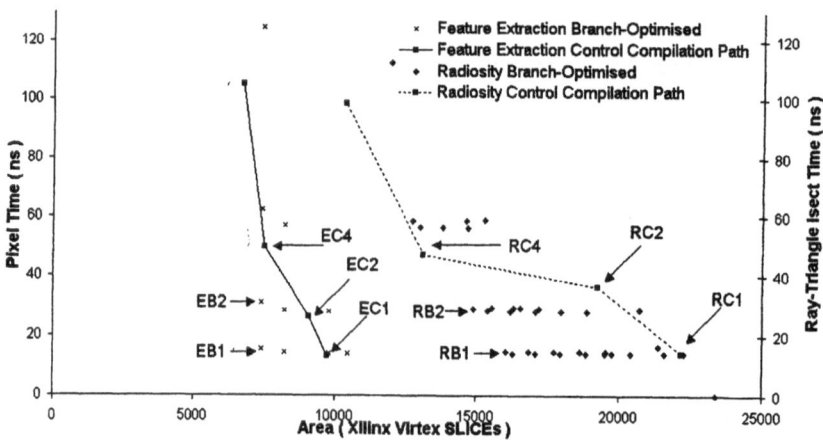

Fig. 6. Combined area-performance design space for video feature extraction (left) and radiosity (right) case study applications. The lines on the graph represents the control study compilation path designs, with $b_{pipeline}$ varying from 1 to 8. The clusters of points are different branch-optimised designs with different parameterizations of b. EC1, EC2, EC4, EB1, EB2, RC1, RC2, RC4, RB1 and RB2 correspond to the optimal designs for each compilation path under different performance constraints as shown in Tables 1, 2, 3 and 4.

Table 1. Complete area-throughput design space with control study compilation path for video feature extraction case study, with input scenes shown in Fig. 3. Designs EC1, EC2 and EC4 are the smallest control study compilation path designs which meet performance constraints 64Mpixel/set, 32Mpixel/sec and 16Mpixel/sec. These designs are labeled in Fig. 6.

	Area			Clk	Pixel Time	
$b_{pipeline}$	(slice)	(FFs)	(LUTs)	(ns)	(ns/pixel)	(Mpixels/sec)
1	9753	11137	14623	13.32	13.316	75.10 EC1
2	9097	10006	11392	13.23	26.46	37.79 EC2
4	7514	7003	8764	12.47	49.86	20.06 EC4
8	6733	5405	7469	13.13	105.05	9.52

Table 2. Selected area-throughput results for branch-optimised compilation path in the video feature extraction case study with input scene shown in Fig. 3. 10 designs are automatically generated. Designs EB1 and EB2 are the smallest branch-optimised designs which meet performance constraint 64Mpixel/sec and 32Mpixel/sec. Clock period for both designs is 13.96ns. ρ_1 can be calculated as $\gamma_1 \times b_1$. EB1 and EB2 are labeled in Fig. 6.

b		Area			Experimental Performance				Analytical Performance			
					Utilization		Pixel Time		Utilization		Time	Relative
b_1	b_2	slice	FF	LUT	ρ_2	γ_1	(ns/pix)	(Mpix/s)	ρ_2	γ_1	(ns/pix)	Error
1	8	7411	5562	8322	.635	.894	15.61	64.04	.7130	1	13.96	0.12 EB1
2	8	7411	5562	8322	.635	.447	31.22	32.03	.7130	0.5	27.92	0.12 EB2

Table 3. Complete area-throughput design space with control study compilation paths for radiosity case study with input scene shown in Fig. 3. Designs RC1, RC2 and RC4 are the smallest branch-optimised designs which meet performance constraint 70Mray-triangle intersections/sec, 33Mray-triangle intersections/sec and 17.5 Mray-triangle intersections/sec. RC1, RC2 and RC4 are labeled in Fig. 6.

	Area			Clk	Intersection Time	
$b_{pipeline}$	(slice)	(FFs)	(LUTs)	(ns)	(ns/Isect)	(MIsects/sec)
1	22182	34338	34,823	14.12	14.12	70.84 RC1
2	19239	28638	25,238	18.27	36.55	27.36 RC2
4	13116	18852	16,070	11.78	47.13	21.22 RC4
8	10340	13657	11,809	12.32	98.53	10.15

2. For a given area, branch-optimised designs can often run significantly faster than non-branch-optimised designs. In Fig. 6, for instance, EB1 (7411 slices) is slightly smaller than EC4 (7514 slices), and at 15.61 ns/pixel is more than 3.2 times faster than EC4 at 49.86 ns/pixel. Similarly while RB1 and RB2 are respectively 22.6% and 13.5% larger than RC4, they run 322% and 162% faster than RC4.

3. For a given performance, branch-optimised designs often require smaller areas than non-branch-optimised designs. In Fig. 6, for instance, at 64 Mpixels/sec EB1 is 24% smaller than EC1 and at 32 Mpixels/sec EB2 is 18% smaller than EC2. Similarly at 70 Mray-triangle intersections per second, RB1 is 27.5% smaller than RC1 while at 35 Mray-triangle intersections per second, RB2 is 27.5% smaller than RC2.

Table 4. Selected area-throughput results for branch-optimised compilation path, radiosity case study with input scene in Fig. 3. 35 designs are automatically generated. Designs RB1 and RB2 are the smallest branch-optimised designs with approximate performance 70Mray-triangle intersections/sec and 35Mray-triangle intersection/sec. ρ_1 can be calculated as $\gamma_1 \times b_1$. Processing rate can be calculated as $1/\text{Itime(ns)}$. RB1 and RB2 are labeled in Fig. 6.

b				Area	Clk	Experimental						Analytical						
						Utilization				Itime		Utilization					Itime	Relative	
b_1	b_2	b_3	b_4	slice	ns	ρ_2	ρ_3	ρ_4	γ_1	ns		ρ_2	ρ_3	ρ_4	γ_1	ns	Error	
1	2	8	8	16074	14.09	.99	.657	.241	.96	14.65	1	.654	.240	.962	14.64	5.3×10^{-4}	**RB1**
2	4	8	8	14888	14.03	1	.328	.121	.48	29.16	1	.327	.120	.481	29.16	8.1×10^{-5}	**RB2**
4	8	8	8	12723	14.08	1	.164	.060	.24	58.51	1	.164	.060	.241	58.52	8.1×10^{-5}	
8	8	8	8	11924	14.08	.52	.085	.031	.13	112.7	.52	.085	.031	.125	112.7	1.0×10^{-6}		

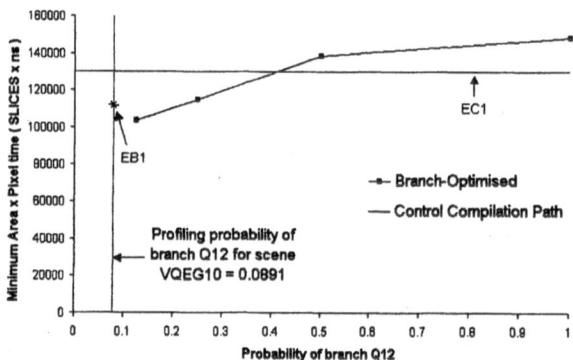

Fig. 7. Minimal area-time versus probability of branch Q_{12} for branch-optimised and control study compilation paths. Video feature extraction case study application is shown with performance constraint of 64Mpixels/sec. The observed probability for Q_{12} of 0.0891 is indicated with a vertical line through the graph. EC1 and EB1 correspond to the optimal designs for each compilation path as shown in Table 1 and Table 2. The trend line for branch-optimised compilation path with different probabilities is produced using our analytical model. The intersection of trend lines for branch-optimised compilation path and control study compilation path shows that branch-optimised compilation is favourable when $Q_{12} < 0.41$. As the probability Q_{12} decreases, branch-optimised compilation becomes increasingly attractive. EB1 performs worse than the analytical model trend line due to intermittent blocking.

4. The analytical performance model is shown to be accurate. For video feature extraction, the relative error varies between 0.12 and 2.4×10^{-5}; for progressive refinement radiosity, the worst case relative error is smaller than 1.1×10^{-4}.

5. As the probability of a branch tends towards zero or one the branch becomes more biased and branch-optimised compilation becomes more attractive. Fig. 7 shows that for video feature extraction, branch-optimised compilation is favourable if branch probability Q_{12} is below a threshold of $Q_{12} < 0.41$. As Q_{12} tends towards zero, the performance gap between branch-optimised and non-branch-optimised designs increases.

7 Conclusion

This paper explores using branch probability information to optimise hardware compilation. We demonstrate that this technique can result in significant improvements in area and performance. Future work will focus on extending the analytical model and compilation system. In the long term we intend to develop a dynamically reconfigurable system in which branch optimisation techniques are applied at runtime.

References

1. Celoxica Limited, *Handel-C Language Reference Manual*, version 3.1, document number RM-1003-3.0, 2002.
2. G. De Micheli, *Synthesis and Optimization of Digital Circuits*, McGraw-Hill, 1994.
3. T. Harriss, R. Walke, B. Kienhuis and E. Deprettere, "Compilation from Matlab to process networks", *Design Automation for Embedded Systems*, Vol. 7, pp. 385–403, 2002.
4. O. Mencer, H Huebert, M. Morf and M.J. Flynn, "StReAm: Object-oriented programming of stream architectures using PAM-Blox", in *Field-Programmable Logic: the Roadmap to Reconfigurable Systems*, LNCS 1896, Springer, 2000, pp. 595–604.
5. I. Mitrani, *Probabalistic Modelling*, Cambridge University Press, 1998.
6. T. Moller and B. Trumbore, "Fast, minimum storage ray-triangle intersection", *Journal of Graphics Tools*, 2(1), pp. 21–28, 1997.
7. R.O. Onvural, "Survey of closed queueing networks with blocking", *ACM Computing Surveys*, 22(2), pp. 83–121, June 1990.
8. H. Styles and W. Luk, "Accelerating radiosity calculations using reconfigurable platforms", in *Proc. IEEE Symp. on Field-Programmable Custom Computing Machines*, 2002, pp. 279–281.
9. A.H. Veen, "Dataflow machine architecture", *ACM Computing Surveys*, 18(4), pp. 365–396, 1986.
10. M. Weinhardt and W. Luk, "Pipeline vectorisation", *IEEE Trans. on Comput.-Aided Design*, 20(2), pp. 234–248, February 2001.
11. H. Ziegler, B. So, M. Hall and P.C. Diniz, "Coarse-grain pipelining on multiple FPGA architectures", in *Proc. IEEE Symp. on Field-Programmable Custom Computing Machines*, 2002, pp. 77–86.

A Model for Hardware Realization of Kernel Loops

Jirong Liao, Weng-Fai Wong, and Tulika Mitra

Department of Computer Science, School of Computing,
National University of Singapore, 3 Science Drive 2, Singapore 117543,
{liaojiro,wongwf,tulika}@comp.nus.edu.sg

Abstract. Hardware realization of kernel loops holds the promise of accelerating the overall application performance and is therefore an important part of the synthesis process. In this paper, we consider two important loop optimization techniques, namely loop unrolling and software pipelining that can impact the performance and cost of the synthesized hardware. We propose a novel model that accounts for various characteristics of a loop, including dependencies, parallelism and resource requirement, as well as certain high level constraints of the implementation platform. Using this model, we are able to deduce the optimal unroll factor and technique for achieving the best performance given a fixed resource budget. The model was verified using a compiler-based FPGA synthesis framework on a number of kernel loops. We believe that our model is general and applicable to other synthesis frameworks, and will help reduce the time for design space exploration.

1 Introduction

A standard practice in synthesis of application specific hardware is to focus attention at kernel loops. In many applications, they account for the bulk of the execution time and are thus natural candidates for hardware acceleration. A key difficulty in synthesizing hardware for kernel loops is that there are many loop optimizations available and the complex interactions among these optimizations make it difficult to predict the cost-benefit of applying each. In particular, one cannot tell how much more or less resources a particular optimization will take or what its impact will be on performance. This means that one has to either settle for sub-optimal results or go through a costly process of trial-and-error in order to arrive at the correct combination of loop optimizations that fits the need of the user. Having a model of how a particular loop optimization will impact resource and performance is therefore necessary.

Two important loop optimizations applicable to kernel loops are *loop unrolling* and *software pipelining*. Loop unrolling is a technique to expand the loop such that a new iteration consists of 2 or more of the original iterations. This is performed by a compiler to expose more instruction level parallelism and reduce the overhead of updating index variables. The number of times the loop is expanded is called the *unroll factor*. If the loop iteration count is not a multiple

P.Y.K. Cheung et al. (Eds.): FPL 2003, LNCS 2778, pp. 334–344, 2003.
© Springer-Verlag Berlin Heidelberg 2003

of unroll factor, then the remainder of the loop iterations needs to be executed at the end as it is.

Software pipelining [1] tries to achieve higher level of instruction level parallelism by moving operations across iteration boundaries. This optimization achieves overlap among the iterations by pipelining the execution of the iterations. The loop body is scheduled such that (a) all iterations have identical schedule and (b) each iteration is scheduled to start some fixed number of cycles later than the previous iteration. The delay between the start cycles of two successive iterations is called the *Initiation Interval (II)*. The modulo scheduling algorithm attempts to achieve the smallest value of II such that no intra- or inter-iteration dependencies and resource constraints are violated.

As multiple iterations are executed in parallel, both loop unrolling and software pipelining increase register pressure and resource requirement but in different ways. Furthermore, it is possible to use them in combination, i.e. it is possible to software pipeline unrolled loops. The complex interaction between the two optimizations makes it difficult to decide how they should be deployed in optimizing a loop given a particular resource constraint. Often the only way to tell is to exhaustively try various combinations of these two optimizations to obtain the optimal one.

In this paper, we propose a model for the performance and resource requirement for the hardware realization of unrolled and software pipelined loops. The novelty of our model lies in the use of the compiler to extract certain key parameters of the loop in question that characterize the code including the data dependences present for a given hardware. For example, the platform we use allows at most four parallel reads to memory and only if they do not hit the same memory bank. Such characteristics are hard to model. So instead we rely on the instruction scheduler of a compiler to capture these. From these parameters reported by the compiler, the model will inform the user if given a certain resource constraint, unrolling alone, or software pipelining used in combination with loop unrolling would deliver the better performance. It will also output the optimal unrolling factor that should be used. The contribution of this model is that without exhaustively trying a large number of possibilities, it can very quickly recommend a solution that we believe is optimal or very near it.

2   Related Work

Hardware realization of kernel loops has been actively studied by many research groups. However, the focus has been mainly on automatic synthesis of kernel loops from high level language constructs. The exploitation of compiler optimizations such as loop unrolling and modulo scheduling has largely remained unexplored. Even a few commercial synthesis tools that apply these compiler optimizations depend on user feedback to choose unroll factor or decide between unrolling and modulo scheduling. Our work bridges this gap in automatic hardware realization of kernel loops.

There are two main approaches towards hardware synthesis from high level constructs. One approach is to design new languages for hardware design which are at much higher level than traditional hardware description languages such as Verilog and VHDL. The claim is that the productivity gap will be reduced as software programmers can easily learn these new languages. An example is Handel-C [2] programming language which has C-like syntax with support for explicit hardware parallelism, communication, and hardware structures such as memory, bus etc.

The other approach attempts to map a subset of commonly used software programming languages such as C to hardware automatically. These efforts include SA-C [3], PipeRench C Compiler [4], Garp C compiler [5], work by Weinhardt et. al. [6] [7], Babb et. al. [8] and Snider et. al. [9]. The PACT project [10] at Northwestern University performs C to hardware synthesis by taking power/performance trade off into account. The PICO project [11, 12] performs static timing analysis to identify chain of operators to minimize number of cycles while maintaining cycle time constraints.

The only existing tool that allows application of high level compiler optimizations in hardware synthesis is Monet [13]. However, it requires user feedback in deciding unroll factor for example. Among research projects, Derien et. al [14] have developed an analytical model to choose a tiling strategy that will minimize loop execution time. The closest to our work is So et. al. [15]. They perform fast and automatic design space exploration to choose the right loop unrolling factor that satisfies the area constraints and maximizes performance. However, they do not use other compiler optimizations such as software pipeline which can potentially improve the performance significantly.

3 Our Model

In this section, we will present our proposed model. The novelty of the model lies in the use of key parameters supplied by the compiler in characterizing aspects of the kernel loop as well as the machine that are hard to model correctly.

3.1 Model for Performance

For the discussion below, we will assume a loop L that is executed N times. Let S_1 be the *schedule length* of the loop. In our model, S_1 is a quantity reported by the compiler as it performs instruction scheduling. As we are realizing the loop in hardware, we assumed infinite registers by skipping the traditional register allocation phase. In the quantity S_1, various complex issues such as the machine's configuration, instruction type distribution, data dependencies etc. are encapsulated. The user, for example, can choose to use the machine configuration to constraint the amount of parallelism or number and types of functional units to be realized in hardware. We will also generalize S_1 to S_u which is the schedule length of the kernel when it is unrolled u times. The following formula gives the total number of cycles the unrolled kernel will take to execute N iterations.

$$C_{\text{unrolled}}(u) \approx \left\lfloor \frac{N}{u} \right\rfloor \times S_u + \left(N - \left\lfloor \frac{N}{u} \right\rfloor \times u \right) \times S_1 \tag{1}$$

After unrolling, the loop size is $\lfloor N/u \rfloor$ and the schedule length is S_u. Therefore the first term in Eq. 1 accounts for the total number of cycles executed by the unrolled loop. However, if N is not divisible by u, a compensation loop of size $N - \lfloor N/u \rfloor \times u$ and a schedule length of S_1 will be generated. In practice, we would not want to have to get all S_u's from the compiler as that requires multiple runs. Rather, we estimate S_u given S_1. In particular, we assumed that

$$S_u = S_{u-1} + c_S \tag{2}$$

where c_S is a constant. From the experience gained from our experimentation, we chose

$$c_S = \frac{(S_3 - S_1)}{2}$$

This is because we found that there may be a case where it so happens that empty resource slots available at the end of the instruction schedule can be filled up by a new instance of the loop.

To model software pipelining, we assumed the technique of *iterative modulo scheduling* given by Rau [1] that uses *predicated execution* and *rotating registers* [16]. It is characterized by two important parameters also obtained from the compiler, the initiation interval, II, and the epilog counter e. The initiation interval is the gap (in machine cycles) between two successive software pipelined iterations. In effect, after a successful modulo scheduling, each iteration of the software pipelined kernel loop takes exactly II cycles. The epilog count is the number of iterations in the epilog of the software pipelined loop. Again, in II and e, the complexity of machine configuration, resource requirements, and data dependencies are hidden away. Since we would like to combine software pipelining with unrolling, we will introduce II_u and e_u which are the II and e for a software pipelined loop that has been unrolled u times. We have the following formula for the total number of cycles a software pipelined loop that has been previous unrolled u times will take:

$$C_{\text{swp}}(u) \approx \left(\left\lfloor \frac{N}{u} \right\rfloor + e_u \right) \times (II_u + 1) + 3 + \left(N - \left\lfloor \frac{N}{u} \right\rfloor \times u \right) \times S_1 \tag{3}$$

A constant of 1 is added to II_u because at the end of each iteration, it is necessary to perform a shift of the content of the rotating registers so as to prepare for the next iteration. These shifts can be done in parallel in hardware and thus cost one cycle. The constant of 3 is needed because in our scheme, we needed one clock cycle at the beginning of the loop to set up the rotating registers, another clock cycle to initiate the loop and epilog counters, and one more at the end of the loop to copy out the content of the rotating registers.

S_u is obtained from Eq. 2. As is the case for S_u, we do not redo modulo scheduling over all possible u's for II_u and e_u. Given a machine configuration, M, and a loop, L, the following holds:

$$II_u = II_{u-1} + c_{II} \tag{4}$$

$$e_u = \left\lceil \frac{S_u}{II_u} \right\rceil - 1 \qquad (5)$$

where c_{II} is dependent on M and L. However, we also found that the simple recurrent relation for II_u do not necessarily end with the unroll size of 1. In particular, for software pipelining, if there is sufficient resources, then $II_i = II_{i-1}$ and the recurrent relations are not established until resource over-subscription comes into play. In our experiments, we used a machine that has only four memory port but otherwise has unlimited resources. The former condition is to reflect the limitation of the FPGA board that we are using. We used the following strategy: we perform software pipelining with $II_1, II_2, ...$ until $II_i \neq II_{i-1}$.

e_u can be derived from S_u and II_u through Eq. 5. This relationship is apparent once we see the idealized diagram for software pipelining shown in Fig. 1. In this example, $S_u = 4$, and $II_u = 1$, giving $e_u = 3$. Since $S_u > II_u$, $e_u \geq 1$.

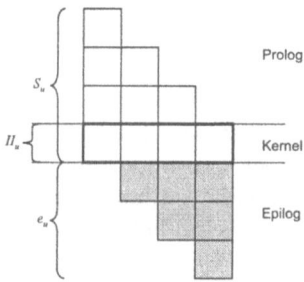

Fig. 1. Relationship between S_u, II_u and e_u.

Estimating FPGA Frequencies. The total running time of an implementation of a loop in a FPGA is given by the product of the number of cycles it takes to execute the code and the frequency of the FPGA which permits the safe operation of the realized design. It turns out that it is difficult to use static compiler information to obtain an accurate model of the final realizable frequency. In order to overcome this problem, we use the following strategy. We run place and route for three instances of the loop, namely the loop unrolled two, three, and four times. These three runs are also used in our resource estimation process described in the next section. Let the actual frequencies obtained from the three runs be $f_l(2)$, $f_l(3)$ and $f_l(4)$, respectively where l is either 'unrolled' or 'swp'. We set the predicted frequency as follows:

$$F_l(u) = \begin{cases} \max(f_l(2), f_l(3), f_l(4)) & \text{if } u = 1 \\ f_l(u) & \text{if } u = 2, 3, \text{ or } 4 \\ \min(f_l(2), f_l(3), f_l(4)) & \text{if } u > 4 \end{cases} \qquad (6)$$

Using these equations, we can finally approximate the time taken to execute the realized design to be

$$T_{\text{unrolled}}(u_1) = C_{\text{unrolled}}(u_1) \times F_{\text{unrolled}}(u_1) \quad \text{and} \quad T_{\text{swp}}(u_2) = C_{\text{swp}}(u_2) \times F_{\text{swp}}(u_2)$$

3.2   Model for Resource Usage

While we can easily count the various operators emitted by the compiler, optimizations further down the synthesis chain, in particular, the place and route pass, introduce non-trivial relationships between the high level hardware description our compiler output and the final resource usage. From experimental results, we found this to be especially true for the case of software pipelined loops. From the same three place and route runs used to obtain the frequencies, we also obtained the resource consumption information by means of linear regression. In particular, for a machine M and loop L, we model resource usage as:

$$R_{\text{unrolled}}(u) = m_{\text{unrolled}} \times u + c_{\text{unrolled}} \qquad (7)$$

$$R_{\text{swp}}(u) = m_{\text{swp}} \times u + c_{\text{swp}} \qquad (8)$$

where m_{unrolled}, c_{unrolled}, m_{swp}, and c_{swp} are constants obtained from the linear regression.

3.3   Putting It Together

The model is used as follows. The user will decide on a certain amount of resource, R_{user}, that he would like to use for realizing the loop in hardware. Using Equations 7 and 8, we obtained two maximal unroll factors u_1 and u_2 such that

$$R_{\text{unrolled}}(u_1) \leq R_{\text{user}} \qquad \text{and} \qquad R_{\text{swp}}(u_2) \leq R_{\text{user}}$$

Next we examine all unroll factors less than u_1 and u_2 to look for a $u_1' \leq u_1$, and a $u_2' \leq u_2$ such that $T_{\text{unrolled}}(u_1')$ and $T_{\text{swp}}(u_2')$ are the respective minimum. If $T_{\text{unrolled}}(u_1') > T_{\text{swp}}(u_2')$ then we will get better performance by using software pipelining with the loop unrolled u_2' times and vice versa.

4   Compilation Framework

We used the Trimaran [17] compiler infrastructure to experiment with the model. The compiler targets for a parameterized Explicitly Parallel Instruction Computing (EPIC) architecture called HPL-PD [16]. We modified the compilation framework as follows:

- An EPIC machine with infinite resources except for four memory ports was defined. The four memory port was a constraint of the FPGA board which we used in our experiments. It has four banks of memory that can be simultaneously accessed with only one access to a bank at any time. Consequently, we also had to modify the instruction and modulo schedulers of Trimaran. We assumed that an entire array is stored in a single bank. Thus any two access to the same array has to be performed in different machine cycles.
- Trimaran uses some heuristics to guide unrolling. Furthermore, it does not always emit compensation loops during unrolling as these can be folded into the unrolled loop using predicated execution. For our purpose, we forced unrolling to be performed as per our requirements.

- Finally, we added a phase to generate Handel-C [18] code for Trimaran's Elcor intermediate representation. Handel-C is a C-like behavioral hardware description language. The Handel-C compiler compiles our output into a EDIF [19] file for the FPGA vendor's synthesis tools to process.

In the resultant design flow, we are able to utilize the advanced features used by Trimaran including predicated execution and rotating registers and translate them into Handel-C. From Handel-C's EDIF output, we synthesis the bitmap for a Xilinx XCV1000 FPGA and execute it on a Celoxica RC1000 board.

5   Results

We used six kernel loops to verify our model:

- **Edge detection.** A 32×32 mask is computed over 128×128 image to detect edges.
- **Matrix multiplication.** Integer multiplication of 160×320 and 320×40 matrix.
- **Finite impulse response filter.** A 128-tap FIR filter on 256 integer data values.
- **Livermore Loop 1.** Hydro fragment loop of size 1001.
- **Jacobi.** 4-point stencil averaging computation over an array with loop size of 100.
- **Histogram.** Mapping from the old to the new grey levels with loop size of 1024.

The accuracy of our performance model is given in Table 1. The first set of columns present the result for loop unrolling and the second set of columns present the result for unrolling and software pipelining. "Est." is the predicted execution time, i.e. $T_{\text{unrolled}}(u)$ and $T_{\text{swp}}(u)$. "Act." is the actual execution time taken to execute the loop. This is obtained from multiplying the actual frequency obtained after place and route with the actual number of cycles executed. "Diff_T" represents the percentage difference between "Est." and "Act." while "Diff_C" represents the percentage difference in estimating $C_{\text{unrolled}}(u)$ and $C_{\text{swp}}(u)$. The average value for "Diff_C" for loop unrolling and loop unrolling with modulo scheduling are 2.84% and 2.19%, respectively. In addition, the values for S_u^p, II_u^p and e_u^p in Table 1 were computed using Equations 2, 4 and 5 while S_u^a, II_u^a and e_u^a were obtained from the actual compilation. The average relative error for "Diff_T" are 3.6% and 8.4% respectively for loop unrolling alone and software pipelining with unrolling. Given that the average relative difference between the actual execution time of the two strategies is 36%, we conclude that our performance estimation model is within the necessary margin and is accurate.

Fig. 2 shows the accuracy of our resource model. Due to space limitation, we will show the results for two benchmarks: Edge and LM1. The results for other benchmarks are similar. "Unroll" and "SWP" show the actual resource usage

Table 1. Accuracy of Performance Model.

Bench-mark	(u)	Unrolling Only						Unrolling + SWP							
		S_u^p	S_u^a	Est. msec	Act. msec	Diff_T (%)	Diff_C (%)	II_u^p	II_u^a	e_u^p	e_u^a	Est. msec	Act. msec	Diff_T (%)	Diff_C (%)
Edge	1	4	4	0.421	0.391	7.46	-2.38	1	1	3	3	0.221	0.223	-1.09	-4.11
	2	5	4	0.282	0.296	-4.90	-4.90	2	2	2	2	0.177	0.189	-6.75	-6.75
	3	6	6	0.227	0.244	-7.01	-7.01	3	3	1	1	0.172	0.188	-8.52	-8.52
	4	7	7	0.184	0.201	-8.38	-8.38	4	4	1	1	0.145	0.161	-9.65	-9.65
	5	8	8	0.197	0.207	-5.00	-9.86	5	5	1	1	0.166	0.178	-7.05	-8.82
	6	9	9	0.187	0.201	-7.25	-10.36	6	6	1	1	0.166	0.198	-16.46	-7.22
	7	10	10	0.197	0.199	-0.77	-9.86	7	7	1	1	0.184	0.218	-15.37	-5.03
	8	11	11	0.155	0.158	-1.68	-12.22	8	8	1	1	0.15	0.226	-33.49	-4.24
MM	1	4	4	731.5	743.4	-1.60	-0.16	1	1	3	3	411.0	382.2	7.53	-0.47
	2	5	5	483.4	485.2	-0.38	-0.34	2	2	2	2	318.8	322.1	-1.03	-1.03
	3	6	6	368.0	370.3	-0.62	-0.63	3	3	1	1	291.2	295.2	-1.36	-1.36
	4	7	7	361.2	363.8	-0.72	-0.72	4	4	1	1	258.4	262.2	-1.47	-1.47
	5	8	8	330.3	308.6	7.00	-0.78	5	5	1	1	260.7	281.8	-7.50	-1.52
	6	9	9	312.8	289.8	7.97	-1.23	6	6	1	1	258.0	298.5	-13.54	-1.54
	7	10	10	303.2	275.0	10.24	-1.27	7	7	1	1	253.4	290.3	-12.70	-1.57
	8	11	11	283.8	283.5	0.12	-1.36	8	8	1	1	243.4	310.9	-21.70	-1.63
FIR	1	3	3	4.431	4.499	-1.51	-1.29	1	1	2	2	3.236	3.115	3.89	-1.89
	2	4	4	3.039	3.098	-1.93	-1.94	2	2	1	1	2.496	2.567	-2.75	-2.74
	3	5	5	2.492	2.559	-2.59	-2.58	3	3	1	1	2.391	2.476	-3.45	-3.45
	4	6	6	2.420	2.503	-3.30	-3.30	4	4	1	1	2.067	2.133	-3.09	-3.09
	5	7	7	2.319	2.390	-2.96	-3.41	5	5	1	1	2.219	2.330	-4.77	-3.09
	6	8	8	2.193	2.250	-2.51	-4.09	6	6	1	1	2.153	2.340	-7.99	-2.58
	7	9	9	2.118	2.283	-7.23	-4.21	7	7	1	1	2.127	2.312	-8.01	-2.00
	8	10	10	2.017	2.044	-1.36	-4.37	8	8	1	1	2.061	2.108	-2.26	-1.42
Lm1	1	8	8	0.754	0.779	-3.22	-0.06	2	2	3	3	0.303	0.286	6.05	-0.13
	2	9	9	0.441	0.441	-0.29	-0.29	3	3	2	2	0.209	0.209	-0.1	-0.70
	3	10	10	0.33	0.332	-0.6	-0.60	4	4	2	2	0.178	0.177	0.36	-1.07
	4	11	11	0.26	0.26	-0.22	-0.22	5	5	2	2	0.153	0.153	-0.2	-0.60
	5	12	12	0.239	0.229	4.17	-0.25	6	6	1	1	0.149	0.163	-8.33	-0.42
	6	13	13	0.218	0.208	4.52	-1.86	7	7	1	1	0.145	0.152	-4.74	-1.04
	7	14	14	0.198	0.19	4.52	-0.30	8	8	1	1	0.137	0.163	-15.97	-0.38
	8	15	15	0.187	0.19	-1.75	-0.32	9	9	1	1	0.134	0.164	-18.6	-0.32
Jacobi	1	10	10	5.712	5.367	6.42	-0.30	8	8	1	1	6.371	5.411	17.75	0.44
	2	13	13	3.713	3.736	-0.61	-3.21	8	8	1	2	3.29	3.249	1.29	1.29
	3	17	17	3.359	3.389	-0.87	-0.87	12	12	1	1	3.85	3.884	-0.89	-0.89
	4	20.5	21	3.095	3.125	-0.95	-3.31	16	16	1	1	4.695	4.642	1.13	1.13
	5	24	25	2.948	3.186	-7.47	-4.95	20	20	1	1	4.684	4.286	9.28	2.06
	6	27.5	29	2.972	3.139	-5.34	-5.70	24	24	1	1	4.821	4.931	-2.23	0.43
	7	31	33	2.842	3.137	-9.42	-6.78	28	28	1	1	4.758	5.405	-11.97	2.26
	8	34.5	37	2.854	3.177	-10.17	-7.16	32	32	1	1	4.863	5.912	-17.74	2.21
Histogram	1	5	5	0.313	0.312	0.36	-0.04	1	1	4	4	0.122	0.126	-3.29	0.44
	2	6	6	0.188	0.188	-0.1	-0.10	2	2	2	2	0.092	0.092	-0.19	1.29
	3	7	7	0.151	0.151	-0.17	-0.17	3	3	2	2	0.091	0.092	-0.36	-0.89
	4	8	8	0.126	0.126	-0.15	-0.15	4	4	1	1	0.102	0.102	-0.23	1.13
	5	9	9	0.116	0.121	-4.43	-1.23	5	5	1	1	0.1	0.111	-10.34	-1.24
	6	10	10	0.107	0.113	-5.09	-1.56	6	6	1	1	0.097	0.109	-11.18	-1.56
	7	11	11	0.102	0.105	-2.86	-0.31	7	7	1	1	0.095	0.135	-29.84	-0.31
	8	12	12	0.097	0.105	-7.57	-0.20	8	8	1	1	0.092	0.12	-23.07	2.21

due to unroll and unroll with software pipeline respectively. These points are obtained from the reports of the FPGA synthesis tool. The "Linear of Unroll" and "Linear of SWP" show the estimated resource usage using linear regression of $u = 2, 3$ and 4. As can be seen from the figures, the estimated resource usage closely follows the actual resource usage.

It seem that in most cases, unrolling alone yields better performance under the same resource constraints. However, if we set $R_{user} = 100,000$, then for the

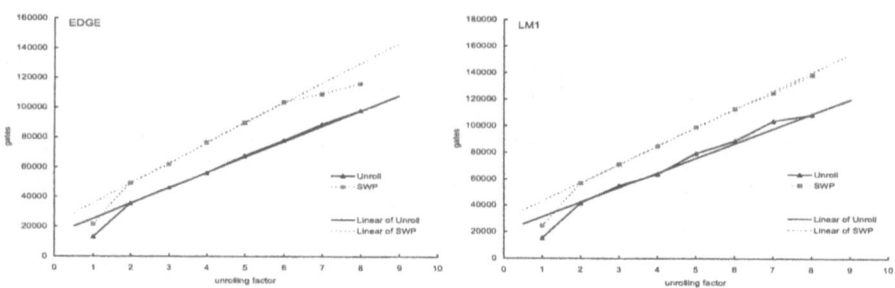

Fig. 2. Resource requirement for the benchmarks.

Lml benchmark, the unroll factor to be used for unrolling and software pipelining are 7 and 5, respectively. Using these unroll factors, our model predicts that we should use software pipelining instead of unrolling. The actual execution time given in Table 1 confirms that our prediction is correct.

Table 2 shows the various constants of Equations 7 and 8 obtained in our model. The results show that our model is fairly accurate and can significantly cut down the design space exploration time.

Table 2. Accuracy of Linear Regression.

Benchmark	Unrolling				Unrolling + SWP			
	$m_{unrolled}$	$c_{unrolled}$	Max Err.	Min Err.	m_{swp}	c_{swp}	Max Err.	Min Err.
Edge	10,371	14,826	1.67%	0.53%	13,471	22,104	11.67%	0.92%
MM (large)	10,468	15,333	4.55%	1.52%	13,365	24,686	20.86%	0.71%
FIR	9,496	13,115	4.18%	3.91%	11,701	20,457	9.87%	0.32%
Lm1	11,112	19,995	5.78%	0.44%	13,926	28,955	1.42%	0.08%
Jacobi	6,604	12,157	2.02%	0.63%	9,039	25,908	4.08%	0.37%
Histogram	3,973	5,676	13.50%	2.25%	3,714	1,6505	2.51%	0.01%

6   Conclusion

In this paper, we proposed a model that projects the data obtained from a small number of compilation and synthesis runs to obtain a global picture of the tradeoffs the designer faces in selecting between two loop optimizations, namely loop unrolling and software pipelining. The novelty of our approach is in the use of key parameters reported by the compiler to capture information about the machine configuration, data dependencies, and resource requirement patterns. This allowed us to obtain a very accurate model of the the cycle counts of the loops' execution. In the worst case, we are less than 5% off the actual cycle counts for larger loops.

The big challenge has been in modeling the two key parameters obtainable only after place and route, namely the circuit's realizable frequency and the resource consumption. For resource usage, we found good linear relations in the growth of resource consumption as unrolling increases especially within the realistic unroll factors that we studied.

Our approach is not very satisfying in modeling the frequency of software pipelined loops. In the worse case for software pipelined loop with high unroll numbers, we are can be off by 30%. Nonetheless, taken together as a whole, the average relative error in estimating $T_{\text{swp}}(u)$ is 8.4%. We would certainly like to improve this in future works.

Combining the resource model and the performance model, we have a methodology for deciding the optimal unroll factor as well as predict whether software pipelining will be beneficial given a certain resource constraint given by the user. We believe our model will reduce the time for design space exploration.

References

1. Rau, B.R.: Iterative Modulo Scheduling. The International Journal of Parallel Processing **24** (1996)
2. Page, I., Luk, W.: Compiling OCCAM into FPGAs. In: Proceedings of the International Symposium on Field Programmable Logic (FPL). (1991)
3. Rinker, R., et al.: An Automated Process for Compiling Dataflow Graphs into Reconfigurable Hardware. IEEE Transactions on VLSI Systems **9** (2001)
4. Goldstein, S.C., et al.: Piperench: A Reconfigurable Architecture and Compiler. IEEE Computer (2000)
5. Callahan, T., Hauser, J.R., Wawrzynek, J.: The Garp Architecture and C Compiler. IEEE Computer (2000)
6. Weinhardt, M.: Compilation and Pipeline Synthesis for Reconfigurable Architectures. In: Proceedings of the Reconfigurable Architecture Workshop (RAW). (1997)
7. Weinhardt, M., Luk, W.: Pipeline vectorization for reconfigurable systems. In: Proceedings of the IEEE Symposium on Field Programmable Custom Computing Machines (FCCM). (1999)
8. Babb, J., et al.: Parallelizing Applications into Silicon. In: Proceedings of the IEEE Symposium on Field Programmable Custom Computing Machines (FCCM). (1999)
9. Snider, G., Shackleford, B., Carter, R.J.: Attacking the Semantic Gap between Application Programming Languages and Configurable Hardware. In: Proceedings of ACM FPGA. (2001)
10. Jones, A., et al.: PACT HDL: A C Compiler Targeting ASICs and FPGAs with Power and Performance Optimizations. In: Proceedings of International Conference on Compilers, Architecture. and Synthesis for Embedded Systems (CASES). (2002)
11. Schreiber, R.: High-Level Synthesis of Nonprogrammable Hardware Accelerators. In: Proceedings of the IEEE International Conference on Application Specific Systems, Architectures, and Processors (ASAP). (2000)
12. Sivaraman, M., Aditya, S.: Cycle-time Aware Architecture Synthesis of Custom Hardware Accelerator. In: Proceedings of International Conference on Compilers, Architecture. and Synthesis for Embedded Systems (CASES). (2002)

13. Mentor Graphics Inc.: Mentor Graphics Monet User's Manual (release r42). (1999)
14. Derrien, S., Rajopadhye, S.: Loop Tiling for Reconfigurable Accelerators. In: Proceedings of the International Symposium on Field Programmable Logic (FPL). (2001)
15. So, B., Hall, M.W., Diniz, P.C.: A Compiler Approach to Fast Hardware Design Space Exploration in FPGA-based Systems. In: Proceedings of the International Conference on Programming Language Design and Implementation (PLDI). (2002)
16. Kathail, V., Schlansker, M., Rau, B.: Hpl-pd architectural specifications: Version 1.1. Technical Report Technical Report HPL-93-80(R.1), Hewlett-Packard Laboratories (Revised 2000)
17. Trimaran Consortium: TRIMARAN: An Infrastructure for Research in Instruction Level Parallelism. (http://www.trimaran.org)
18. Celoxica Inc.: Handel-C. (http://www.celoxica.com/tech/handel-c/)
19. Electronic Industries Alliance: Electronic Design Interface Format. (http://www.edif.org)

Programmable Asynchronous Pipeline Arrays

John Teifel and Rajit Manohar

Computer Systems Laboratory, Electrical and Computer Engineering,
Cornell University, Ithaca, NY 14853, U.S.A.,
{teifel,rajit}@csl.cornell.edu, http://vlsi.cornell.edu

Abstract. We discuss high-performance programmable asynchronous pipeline arrays (PAPAs). These pipeline arrays are coarse-grain field programmable gate arrays (FPGAs) that realize high data throughput with fine-grain pipelined asynchronous circuits. We show how the PAPA architecture maintains most of the speed and energy benefits of a custom asynchronous design, while also providing post-fabrication logic reconfigurability. We report results for a prototype PAPA design in a $0.25\mu m$ CMOS process that has a peak pipeline throughput of 395MHz for asynchronous logic.

1 Introduction

We present programmable asynchronous pipeline arrays (PAPAs) as a high-performance FPGA architecture for implementing asynchronous circuits. Asynchronous design methodologies seek to address the design complexity, energy consumption, and timing issues affecting modern VLSI design [10]. Since most experimental high-performance asynchronous designs (e.g., [1, 13]) have been designed with labor-intensive custom layout, we propose the PAPA architecture as an alternative method for prototyping these asynchronous systems.

Previously proposed asynchronous FPGAs have shown that it is possible to port a clocked FGPA architecture to an asynchronous circuit implementation (e.g., [2, 14]). However, in an asynchronous system, logic computations are not artificially synchronized to a global clock signal and hence we can explore a larger programmable design space. In this paper we present one such exploration into the design of high-performance pipelines suitable for programmable asynchronous systems.

The PAPA architecture is inspired by high-performance, full-custom asynchronous designs [1, 13] that use very fine-grain pipelines. Each pipeline stage contains only a small amount of logic (e.g., a 1-bit full-adder) and combines computation with data latching, such that explicit output latches are absent from the pipeline. This pipeline style achieves high data throughput and can also be used to design energy-efficient systems [15]. As a result, we use fine-grain asynchronous pipelines as the basis for our high-performance FPGA architecture.

Existing work in programmable asynchronous circuits has concentrated on three design approaches: (1) mapping asynchronous logic to clocked FPGAs (e.g., [3, 5]), (2) asynchronous FPGA architectures for clocked logic [4, 16], and

P.Y.K. Cheung et al. (Eds.): FPL 2003, LNCS 2778, pp. 345–354, 2003.

(3) asynchronous FPGA architectures for asynchronous logic [2, 6, 8, 14]. The first approach suffers from an inherent performance penalty because of the circuit overhead in making a hazard-prone clocked FPGA operate in a hazard-free (the absence of glitches on wires) manner, which is necessary for correct asynchronous logic operation. Likewise, the second approach is not ideal because clocked logic does not behave like asynchronous logic and need not efficiently map to asynchronous circuits. The third approach runs asynchronous logic natively on asynchronous FPGA architectures. The work in this area has largely been modeled from existing clocked FPGA architectures, with the most recent running at an unencouraging 20MHz in $0.35\mu m$ CMOS [6].

In this paper we introduce the PAPA architecture as a new asynchronous FPGA that is designed to run asynchronous logic, yet differs from existing work because it is based on high-performance custom asynchronous circuits and is not a port of an existing clocked FPGA. The result is a programmable asynchronous architecture that is an order-of-magnitude improvement over [6]. Section 2 describes the asynchronous pipelines that our FPGA targets. In Section 3 we present the programmable asynchronous pipeline array architecture and in Section 4 describe its circuit implementation. Section 5 analyzes the performance of the PAPA architecture and Section 6 discusses logic synthesis results.

2   Asynchronous Pipelines

We design the logic that runs on PAPAs and other asynchronous systems as a collection of concurrent hardware processes that communicate with each other through message-passing channels [11]. Asynchronous pipelines can be constructed using such processes by connecting their channels in a FIFO configuration, where each pipeline stage consists of a single process. We refer to data items in a pipeline as *tokens* (i.e., messages passed on channels).

Since there is no clock in an asynchronous design, processes use handshake protocols to send and receive tokens on channels. All PAPA channels use three wires, two data wires and one acknowledge wire, to implement a four-phase handshake protocol. The data wires encode bits using a dual-rail code, such that setting "wire-0" transmits a "logic-0" and setting "wire-1" transmits a "logic-1". The four-phase protocol operates as follows: the sender sets one of the data wires, the receiver latches the data and raises the acknowledge wire, the sender lowers both data wires, and finally the receiver lowers the acknowledge wire. The *cycle time* of a pipeline stage is the time required to complete one four-phase handshake.

In PAPA logic designs we enforce the following constraints on channels and processes: (1) no shared variables, (2) no shared channels, (3) no arbiters, and (4) the ability to add an arbitrary number of pipeline stages on a channel without changing the logical correctness of the original system. These system restrictions are reasonable for many high-performance asynchronous systems, including entire microprocessors [13], and in the rest of this paper we restrict our attention to asynchronous pipelines and circuits satisfying them.

A system that satisfies the aforementioned constraints is an example of a *slack-elastic* system [9] and has the nice property that a designer can locally add pipelining anywhere in the system without having to adjust the global pipeline structure. This property allows PAPA logic cells to be implemented with a variable number of pipeline stages and enables channels with long routes to be pipelined to improve performance. Any non-trivial clocked design will *not* be slack elastic, since changing local pipeline depths in a clocked system may require global retiming of the entire system. Adding high-speed retiming hardware support to a clocked FPGA incurs a significant register overhead [17], which the PAPA architecture can avoid because its logic cells are inherently pipelined and its channels are slack elastic.

Asynchronous (fine-grain) pipeline stages perform one or more of the following dataflow operations: (1) compute arbitrary logical functions, (2) store state, (3) conditionally receive tokens on input channels, (4) conditionally send tokens on output channels, and (5) copy tokens to multiple output channels. While strategies for implementing these pipeline operations in custom circuitry have been described in [7], the goal of the PAPA architecture is to implement these operations in a programmable manner.

Techniques for implementing operations 1 and 2 are well-known in both the clocked and asynchronous FPGA circuit literature (e.g., [2, 14]). PAPAs have a *Function* unit to compute arbitrary functions and use feedback loops to store state. However, because operations 3, 4, and 5 involve tokens they are inherently asynchronous pipeline structures. The PAPA architecture provides a *Merge* unit to conditionally receive tokens, a *Split* unit to conditionally send tokens, and an *Output-Copy* unit to copy tokens. Since a clocked FPGA circuit has no concept of a token, it uses multiplexers, demultiplexers, and wire fanout to implement structures similar to operations 3, 4, and 5, respectively. The main difference is that these clocked circuits are destructive (i.e., wire values not used are ignored and overwritten on the next cycle), whereas an asynchronous circuit is non-destructive (i.e., tokens remain on channels until they are used).

3   The PAPA Architecture

The PAPA architecture is a RAM-based, coarse-grain FPGA design and consists of *Logic Cells* surrounded by *Channel Routers*. Figure 1a shows the basic PAPA logic cell and channel router configuration that is used in this paper. Logic cells communicate through 1-bit wide, dual-rail encoded channels that have programmable connections configured by the channel routers.

Logic Cell. The pipeline structure of a PAPA logic cell is shown in Figure 1b. The *Input-Router* routes channels from the physical input ports (Nin, Ein, Sin, Win) to the three internal logical input channels (A, B, C). This router is implemented as a switch matrix and is unpipelined. If an internal input channel is not driven from a physical input port, a token with a "logic-1" value is internally sourced on the channel (not shown in the figure). The internal input channels are shared between four logical units, of which only one unit can be enabled.

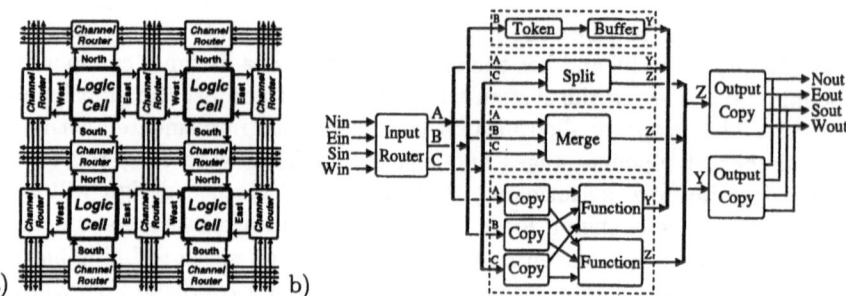

Fig. 1. PAPA architecture: (a) logic cell and channel router configuration, (b) pipeline structure of logic cell.

The logical units are as follows:

- *Function* Unit (2 pipeline stages): Two arbitrary functions of three variables. Receives tokens on channels (A, B, C) and sends function results on output channels (Y, Z). (e.g., this unit efficiently implements a 1-bit full-adder).
- *Merge* Unit (1 pipeline stage): Two-way controlled merge.
  Receives a control token on channel C. If the control token equals "logic-0" it reads a data token from channel A, otherwise it reads a data token from channel B. Finally, the data token is sent on channel Z.
- *Split* Unit (1 pipeline stage): Two-way controlled split.
  Receives a control token on channel C and a data token on channel A. If the control token equals "logic-0" it sends the data token on channel Y, otherwise it sends the data token on channel Z.
- *Token* Unit (2 pipeline stages): Initializes with a token on its output.
  Upon system reset a token (with a programmable value) is sent on channel Y. Afterwards the unit acts as a normal pipeline (i.e., it receives a token on channel B and sends it on channel Y). Unit is used for state initialization.

The *Output-Copy* pipeline stage copies result tokens from channels Y and Z to one or more of the physical output ports $(Nout, Eout, Sout, Wout)$ or sinks the result tokens before they reach any output port.

A PAPA logic cell uses 44 configuration bits to program its logic. The configuration bits are distributed as follows: 15 bits for the *Input-Router*, 4 bits for the logical unit enables, 16 bits for the *Function* unit, 1 bit for the *Token* unit, and 8 bits for the *Output-Copy* stages.

Unlike most existing FPGA architectures, PAPA logic cells do not have internal state feedback. Instead, state feedback logic is synthesized with an external feedback loop through an additional logic cell that is configured as a *Token* unit. This ensures that the state feedback loop is pipelined and operates at close to full throughput without adding additional area overhead to the logic cell to support an internal feedback path [7].

Channel Router. A PAPA channel router is an unpipelined switch matrix that *statically* routes channels between logic cells. PAPA channel routers route all channels on point-to-point pathways and all routes are three wires wide (necessary to support the dual-rail channel protocol). Each channel router has 12 channel ports (6 input and 6 output) that can route up to six channels. Four of the ports are reserved for connecting channels to adjacent logic cells and the remaining ports are used to route channels to other channel routers. To keep the configuration overhead manageable, a PAPA channel router does not allow "backward" routes (i.e., changing a channel's route direction by 180 degrees) and requires 26 configuration bits.

By examining numerous pipelined asynchronous logic examples, we empirically determined the PAPA logic cell and channel router interconnect topology (Fig.1a) as a good tradeoff between performance, routing capability, and cell area. We make no claims that it is the most optimal for this style of programmable asynchronous circuits and in fact it has several limitations. For example, it is not possible to directly route a channel diagonally on a 3x3 or larger PAPA grid using *only* channel routers (routing through one logic cell is required, which will improve performance for long routes). However, since most asynchronous logic processes communicate across short local channels we have not found this long-diagonal route limitation to be overly restrictive. More complicated channel routing configurations (such as those used in clocked FPGAs) could be adapted for the PAPA architecture, with the added cost of more configuration bits and cell area.

4   Pipelined Asynchronous Circuits

The asynchronous circuits we use are quasi-delay-insensitive (QDI). While they operate under the most conservative delay model that assumes gates and most wires have arbitrary delays [12], we believe QDI circuits to be the best asynchronous circuit style in terms of performance, energy, robustness, and area.

Although high-throughput, fine-grain QDI pipelined circuits have been used previously in several full-custom asynchronous designs [1,13], the PAPA architecture is the first to adapt these circuits for programmable asynchronous logic. A detailed description on the design and behavior of this style of pipelined asynchronous circuits is in [7]. What follows is a summary of their salient features.

- **High throughput** – Minimum pipeline cycle times of ∼10-16 FO4 (fanout-of-4) delays (competitive with clocked domino logic).
- **Low forward latency** – Delay of a token through a pipeline stage is ∼2 FO4 delays (superior to clocked domino logic).
- **Data-dependent pipeline throughput** – Operating frequency depends on arrival rate of input tokens (varies from idle to full throughput).
- **Energy efficient** – Power savings from no extra output latch, no clock tree, and no dynamic power dissipation when the pipeline stage is idle.

Figure 2 shows the two pipeline circuit templates used in the PAPA architecture. L_0 and L_1 are the dual-rail inputs to the pipeline stage and R_0 and R_1 are the dual-rail outputs. We use inverted-sense acknowledge signals ($L_{\overline{ACK}}$, $R_{\overline{ACK}}$)

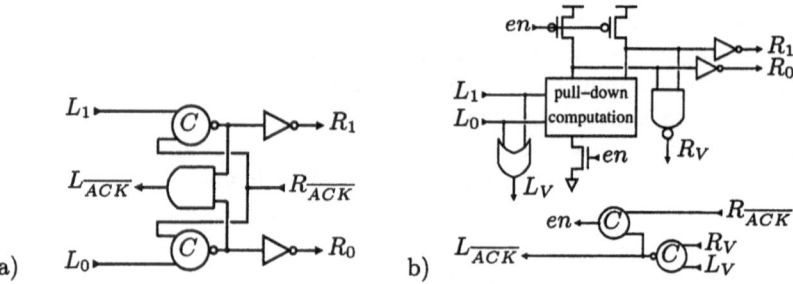

Fig. 2. Fine-grain pipelined asynchronous circuit templates: (a) weak-condition (dual-rail) pipeline stage, (b) precharge (dual-rail) pipeline stage.

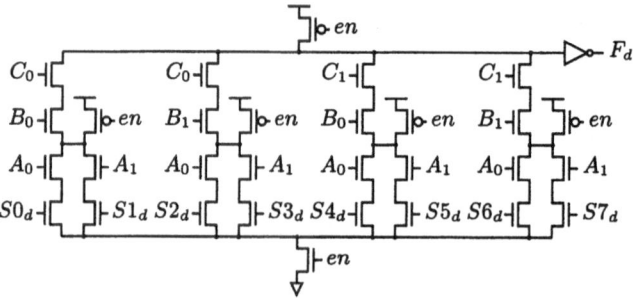

Fig. 3. One rail of a dual-rail precharge computation block for a 3-input function unit.

for circuit efficiency. The weak-condition pipeline stage (Fig.2a) is most useful for token buffering and token copying, while the precharge pipeline stage (Fig.2b) is optimized for performing logic computations (similar to dual-rail clocked domino circuits). Since the weak-condition and precharge pipeline stages both use the dual-rail handshake protocol, they can be freely mixed together in the same pipeline. Weak-condition pipeline stages are used in the *Token* unit, *Output-Copy*, and in the copy processes of the *Function* unit. The *Split* unit, *Merge* unit, and the evaluation part of the *Function* unit use precharge pipeline stages.

A partial circuit used in the evaluation part of the *Function* unit is shown in Figure 3. A, B, and C are the input channels and $S0_d \ldots S7_d$ are the configurations bits that program the function result F_d, where d specifies the logic rail (e.g., $d{=}0$ computes F_0). As noted in [2], a function computation block of this style will suffer from charge sharing problems, which we solved using aggressive transistor folding and internal-node precharging techniques.

Physical Design. A prototype PAPA device has been designed and preliminarily layed out in TSMC's $0.25\mu m$ CMOS process (FO4 delay$\approx$120ps) available via MOSIS. An arrayable PAPA cell that includes one logic cell and two channel routers is 144 x 204 μm^2 (1200 x 1700 λ^2) in area, which is 50-100% larger than a conventional clocked FPGA cell but 50-33% the size of the pipelined clock FPGA in [17]. To minimize cell area and simplify programming, configuration bits are programmed using JTAG clocked circuitry. The area breakdown for the architecture components is: function unit (14.4%), merge unit (2.5%), split unit (2.9%), token unit (2.6%), output copies (12.5%), configuration bits (37.7%), channel/input routers (18.2%), and miscellaneous (9.1%).

We have simulated our layout in SPICE (except for inter-cell wiring parasitics) and found the maximum inter-cell operating frequency for PAPA logic to be 395MHz. Internally the logical units can operate much faster, but are slowed by the channel routers. To observe this we configured the logical units to internally source "logic-1" tokens on their inputs and configured the *Output-Copy* stages to sink all result tokens (bypassing all routers). The results are: *Function* unit (498MHz, 26pJ/cycle), *Merge* unit (543MHz, 11pJ/cycle), *Split* unit (484MHz, 12pJ/cycle), and *Token* unit (887MHz, 7pJ/cycle). These measurements compare favorably to the pipelined clock FPGA in [17] that operates at 250MHz and consumes 15pJ/cycle of energy per logic cell. Our current work focuses on intelligently pipelining the channel routers to match the internal cycle times of the logical units and using improved circuit techniques to reduce the energy consumption of the PAPA logic cells.

5   Performance Analysis

The pipeline dynamics of asynchronous pipelines, due to their interdependent handshaking channels, are quite different from the dynamics of clocked pipelines. To operate at full throughput, a token in an asynchronous pipeline must be physically spaced across multiple pipeline stages, whereas in a clocked pipeline the optimum results when there is one token per stage [18]. The optimal number of pipeline stages, n_0, per token in an asynchronous pipeline is attained when $n_0 = \tau_0/l_0$, where τ_0 is the cycle-time of a pipeline stage and l_0 is its forward latency. For circuits used in the PAPA design, n_0 ranges from 5 to 8 pipeline stages per token (for pipelines without switches). If a pipeline has fewer stages per token than n_0, it will operate at a slower than maximal frequency but consume less energy [15]. On the other hand, if the pipeline has more stages per token than n_0, it will both operate slower and consume more energy than the optimal case.

To observe the pipeline dynamics when there are programmable switches between pipeline stages, we modeled a PAPA pipeline with a linear pipeline of n weak-condition pipeline stages that contain a variable number of routing switches between each pipeline stage. This model uses layout from the *Token* unit, has n_0=5, and measures all results from full SPICE simulations (including inter-cell wiring parasitics). This model gives an upper bound on the performance of PAPA pipelines and shows the behavioral trends of inserting switches between fine-grain asynchronous pipeline stages.

Figure 4a shows the maximum operating frequency curves for our model pipeline when there are K routing switches between every pipeline stage (K=0 is the "custom" case when there are no switches between stages). We observe that as K increases, n_0 decreases from 5 stages to 4 stages and the frequency curves shift downward because the switches uniformly increase the cycle time of

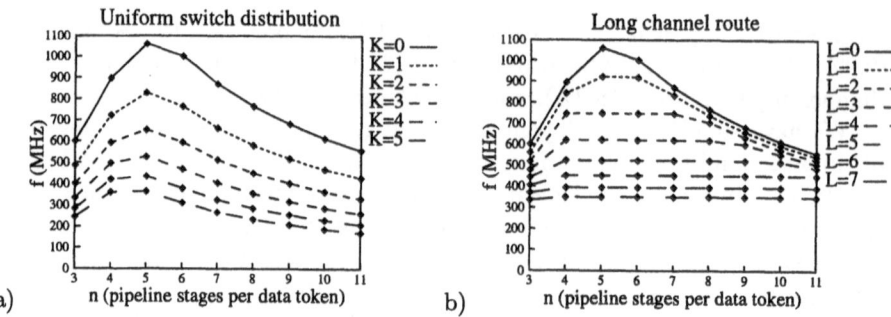

Fig. 4. Maximum operating frequency curves for one token in a linear pipeline of n weak-condition pipeline stages, when (a) there are K routing switches between every pipeline stage and (b) one pipeline stage has a long route through L switches.

every pipeline stage. Figure 4b shows the effect of one pipeline stage having a long route through L switches (when the other pipeline stages have no switches). In this case, the frequency curves flatten as L increases because the cycle time of the pipeline is mainly determined by the cycle time of the stage containing the long route (i.e., the long route behaves as a pipeline bottleneck).

In addition to decreasing their operating frequency, the energy consumption of asynchronous pipelined circuits also increases when routing switches are added between pipeline stages. To observe the energy effect of adding switches to asynchronous pipelines we use the $E\tau^2$ energy-time metric [13, 15]. E is the energy consumed in the pipeline per cycle and τ is the cycle time ($1/f$). Since E is proportional to V^2 and τ is proportional to $1/V$, to first order this metric is independent of voltage and provides an energy-efficiency measure to compare both low-power designs (low voltage) and high-performance designs (normal voltage). Figure 5 shows energy-efficiency curves for our model pipeline under the two switch scenarios examined earlier (lower values imply more energy efficiency).

The maximum operating frequency and energy-efficiency curves for a PAPA pipeline will look like a mixture of the two switch scenarios we investigated, since

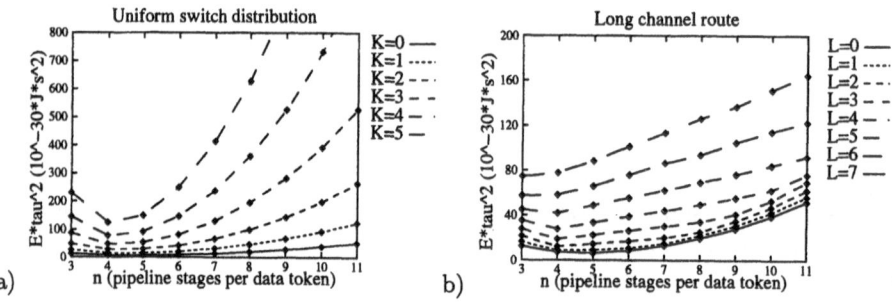

Fig. 5. Energy-efficiency curves for one token in a linear pipeline of n weak-condition pipeline stages, when (a) there are K routing switches between every pipeline stage and (b) one pipeline stage has a long route through L switches.

some pipeline stages will have no switches between them (channels inside of the logic cell) and some will have two or more (channels going through the input and channel routers). We have found that in synthesized PAPA logic there is at most six switches between logic cells, and on average two to four (including input routers). While the plots in this section show that (as expected) adding routing switches to full-custom, high-throughput pipelined circuits decreases both their speed and energy efficiency, they also show that there is still much performance remaining ($\approx 50\%$) to make them attractive for high-speed programmable asynchronous logic.

6   Logic Synthesis Results

High-level logic synthesis for PAPA designs borrows heavily from the formal synthesis methods we use to design full-custom asynchronous circuits [10]. We begin with a sequential description of the logic that is written in the CHP (Communicating Hardware Processes) hardware description language and apply (already existing) semantics-preserving program transformations to get a set of fine-grain concurrent CHP processes. Each of the resulting processes can be implemented in a single PAPA logic cell. The processes are then physically mapped onto PAPA logic cells. While this procedure is currently only semi-automated, it is not as tedious a task as for gate-level FPGAs. Finally, channels connecting logic cells are automatically routed and a configuration file generated.

We report SPICE simulations for several synthesized logic examples:

- N-bit ripple-carry adder (N logic cells) – Throughput of 292MHz, with a data input-to-output latency of 1.91ns, and a carry input-to-output propagation latency of 1.04ns per bit (the router was directed to minimize carry latency).
- Pipelined Booth encoded multiplier 1-bit cell (12 logic cells) – Throughput of 222MHz (original full-custom version ran at 190MHz in $0.8\mu m$ [1]).
- Register bit (5 logic cells) – Throughput of 272MHz, can read and/or write on same cycle.

7   Summary

We introduced a new high-performance asynchronous FPGA architecture. The architecture uses fine-grain asynchronous pipelines to implement a coarse-grain FPGA and is suitable for prototyping pipelined asynchronous logic. Our preliminary circuit simulations demonstrate that PAPA logic systems are a promising alternative to full-custom asynchronous designs.

Acknowledgments

This research was supported in part by the Multidisciplinary University Research Initiative (MURI) under the Office of Naval Research Contract N00014-00-1-0564, and in part by an NSF CAREER award under contract CCR 9984299. John Teifel was supported in part by an NSF Graduate Research Fellowship.

References

1. Cummings, U.V., Lines, A.M., Martin, A.J.: An Asynchronous Pipeline Lattice-structure Filter. *Proc. Int'l Symp. on Asynchronous Circuits and Systems* (1994)
2. Hauck, S., Burns, S., Borriello, G., Ebeling, C.: An FPGA for implementing asynchronous circuits. *IEEE Design & Test of Computers* **11**(3) (1994) 60-69
3. Ho, Q.T., et al.: Implementing asynchronous circuits on LUT based FPGAs. *Proc. 12th Int'l Conf. on Field Programmable Logic and Applications* (2002)
4. How, D.L.: A Self Clocked FPGA for General Purpose Logic Emulation. *Proc. of the IEEE 1996 Custom Integrated Circuits Conf.* (1996)
5. Keller, E.: Building Asynchronous Circuits with JBits. *Proc. 11th Int'l Conf. on Field Programmable Logic and Applications* (2001)
6. Konishi, R., et al.: PCA-1: A fully asynchronous self-reconfigurable LSI. *Proc. 7th Int'l Symp. on Asynchronous Circuits and Systems* (2001)
7. Lines, A.M.: *Pipelined Asynchronous Circuits*. M.S. Thesis, California Institute of Technology (1996)
8. Maheswaran, K.: *Implementing Self-Timed Circuits in Field Programmable Gate Arrays*. M.S. Thesis, U.C. Davis (1995)
9. Manohar, R., Martin, A.J.: Slack Elasticity in Concurrent Computing. *Proc. of the 4th Int'l Conf. on the Mathematics of Program Construction* (1998)
10. Manohar, R.: A Case for Asynchronous Computer Architecture. *Proc. of the ISCA Workshop on Complexity-Effective Design* (2000).
11. Martin, A.J.: Compiling Communicating Processes into Delay-insensitive VLSI circuits. *Distributed Computing,* **1**(4) (1986)
12. Martin, A.J.: The Limitations to Delay-Insensitivity in Asynchronous Circuits. *Sixth MIT Conf. on Advanced Research in VLSI* (1990)
13. Martin, A.J., Lines, A., Manohar, R., et al.: The Design of an Asynchronous MIPS R3000. *Proc. of the 17th Conf. on Advanced Research in VLSI* (1997)
14. Payne, R.: Asynchronous FPGA architectures. *IEE Computers and Digital Techniques* **143**(5) (1996) 282-286
15. Teifel, J., Fang, D., Biermann, D., Kelly, C., Manohar, R.: Energy-Efficient Pipelines. *Proc. 8th Int'l Symp. on Asynchronous Circuits and Systems* (2002)
16. Traver, C., Reese, R.B., Thornton, M.A.: Cell Designs for Self-timed FPGAs. *Proc. of the 2001 ASIC/SOC Conf.* (2001)
17. Tsu, W., et al.: HSRA: High-Speed, Hierarchical Synchronous Reconfigurable Array. *Proc. 7th Int'l Symp. on Field-Programmable Gate Arrays* (1999)
18. Williams, T.E.: *Self-Timed Rings and their Application to Division*. Ph.D. thesis, Stanford University (1991)

Globally Asynchronous Locally Synchronous FPGA Architectures

Andrew Royal and Peter Y.K. Cheung

Department of Electrical & Electronic Engineering, Imperial College, London, UK,
{a.royal,p.cheung}@imperial.ac.uk

Abstract. Globally Asynchronous Locally Synchronous (GALS) Systems have provoked renewed interest over recent years as they have the potential to combine the benefits of asynchronous and synchronous design paradigms. It has been applied to ASICs, but not yet applied to FPGAs. In this paper we propose applying GALS techniques to FPGAs in order to overcome the limitation on timing imposed by slow routing.

1 Introduction

Most Field Programmable Gate Arrays (FPGAs) are designed with one or more global clocks. FPGAs with multiple clock domains must provide some mechanism for synchronising data passing between them, which will increase latency and be prone to metastability.

In addition a signal routed over a great distance on an FPGA must pass through many long wires, transistor switches and buffers. Consequently these long connections usually prove to exhibit the longest delays in the whole device. If transmission is to occur over one clock cycle or even a fraction of a clock cycle, this routing delay will limit the clock frequency.

One solution is to pipeline the routing, as used in [1]. A potential problem of this is that the number of clock cycles allocated to the routing must be determined at the routing stage and may impact upon any cycle allocation assumed at the circuit design stage. Also, concurrent data travelling along different routing paths need to contain exactly the same number of pipeline stages or data will arrive on different cycles.

We could also use asynchronous circuits. Rather than assume that data takes a clock period to pass through a pipeline stage, asynchronous circuits use handshake protocols to indicate when each stage has data to pass to the next stage. This allows each stage to be independently timed. Were we to use asynchronous routing for FPGAs, the speed of the rest of the circuit would no longer be limited the speed of the routing. As they have no clocks, there is no need for multiple clock domains and no problem with synchronisation.

Asynchronous FPGAs have already been proposed for prototyping asynchronous circuits. However, the drawbacks of asynchronous circuits deter designers from using them in preference to synchronous arrays. A compromise is to use a Globally Asynchronous Locally Synchronous (GALS) system. In such a

P.Y.K. Cheung et al. (Eds.): FPL 2003, LNCS 2778, pp. 355–364, 2003.

system synchronous modules with locally generated clocks are used, with asynchronous connections between them. Hence we retain the advantages of synchronous circuits, but can also exploit the advantages of asynchronous routing. This technique has not previously been applied to FPGAs.

In this paper we propose adding asynchronous routing to a synchronous FPGA. In section 2 a brief overview of asynchronous communication, asynchronous FPGAs and Globally Asynchronous Locally Synchronous Systems is given. We go on to describe an architecture for applying this work to synchronous FPGAs in section 3 and assess its viability in section 4. Finally, in section 5 we describe our future work into improving this architecture.

2   Background and Related Work

2.1   Asynchronous Systems

In this paper, we use the term "Asynchronous" to refer to circuits designed without clocks, also known as Self-Timed circuits, where the clock is replaced by handshaking signals. In a synchronous system, all blocks are assumed to have finished computation when a clock edge arrives. The blocks in Self-timed systems independently indicate completion by sending out a request and only proceed when that request has been acknowledged. Blocks only operate as needed, there is little redundant processing. Also, a self-timed system is very composable as blocks can be individually optimised and timing of one block does not affect another. We wish to exploit this feature for our FPGA architecture.

Data is transmitted using a bundled data protocol [2]. This means that data signals are grouped into a bundle and the validity of the whole bundle is indicated by a single request/acknowledge handshake pair. The bundling constraint states that a request must arrive after the corresponding data, so data can be latched safely, so the delay of request lines often needs tuning.

Asynchronous designs require some circuit components which are rarely used in synchronous design. Ebergen [3] showed many of the circuit elements required to build delay insensitive circuits. We require some of these components for building our architecture. A C-element is a component which fires a transition on its output when each of its inputs have made transitions in the same direction, it is effectively an AND for events. An isochronic fork is simply a signal which branches out to two destinations, where the delay on the wires is negligible compared to the delays of the gates they are driving and so can be assumed to arrive at the same instant. A mutual exclusion element (often ME element or MUTEX) is used to arbitrate between requests. ME elements may go metastable if the requests arrive close to each other, but the Seitz arbiter [4] is a mutual exclusion element designed such that any metastability is internal and not propagated to its outputs. Finally, Micropipelines [2] are an asynchronous equivalent of synchronous pipelines, where the registers are controlled by request/acknowledge signals rather than a clock.

2.2   Asynchronous FPGAs

There have been several attempts to implement asynchronous circuits on FPGAs designed for synchronous circuits [5], [6]. The main problems with this are that synchronous FPGAs are not designed to be hazard free and do not provide many of the components commonly used in asynchronous design.

There have also been FPGAs designed specifically for implementing asynchronous circuits. MONTAGE [7] is an extension of the synchronous TRIPTYCH architecture [8]. Fast feedback in functional units allows asynchronous-specific components to be built and specific arbiter blocks are also provided. Routing is organised to allow isochronic forks. PGA-STC [6] is similar to MONTAGE, but also includes a reconfigurable delay line which can be used in the implementation of a bundled data protocol. The delay uses a ring coupled oscillator which is unfortunately very large and power consuming. STACC [9], [10] is loosely based on Sutherland's micropipeline design [2]. The data array can be like that of any synchronous FPGA, but the clock is replaced by control signals from a timing array which consists of a micropipeline-like structure.

Asynchronous FPGAs are not widely used. They are fraught with problems with hazards, critical races and metastability. Asynchronous circuits are hard to design and tools have only recently begun to reach maturity. Asynchronous buses are difficult to construct [11]. We also find that the additional completion detection circuitry required takes considerable area and power and slows the circuit down. Hence we propose a Globally Asynchronous Locally Synchronous FPGA as a compromise between synchronous and asynchronous styles.

2.3   Globally Asynchronous Locally Synchronous (GALS) Systems

Globally Asynchronous Locally Synchronous (GALS) Systems combine the benefits of synchronous and asynchronous systems. Modules can be designed like modules in a globally synchronous design, using the same tools and methodologies. Each block is independently clocked, which helps to alleviate clock skew. Connections between the synchronous blocks are asynchronous.

Early work on GALS systems ([12] and [13]) introduced clock stretching or pausing. When data enters a synchronous system from an asynchronous environment, registers at the input are prone to metastability. To avoid this, the arrival of data is indicated by an asynchronous handshaking protocol. When data arrives, the locally generated clock is paused: in practice the rising edge of the clock is delayed. Once data has safely arrived, the clock can be released so data is latched with zero probability of metastability on the datapath. [14] used ME elements to arbitrate between the clock and incoming requests, which helped to eliminate metastability. [15] introduced asynchronous wrappers, standard components which can be placed around synchronous modules to provide the handshake signals and make them GALS modules.

The local clock generator is constructed from and inverter and a delay line, similar to an inverter ring oscillator. The problem with using inverters alone as a delay line is that it is difficult to accurately tune the clock period as process variations and temperature affect the delay. Hence accurate delay lines have been developed which are capable of maintaining a stable clock frequency [16], [17]. These use a global reference clock for calibration. The former can use either standard cells or full custom blocks for the tunable delay and was shown to exhibit less than 1% jitter around the chosen frequency.

To make the clock pausable, an ME element is added to the ring as shown in figure 1(a). This arbitrates between the rising edge of the clock and an incoming request. Hence the clock is prevented from rising as the input registers are being enabled by the request and metastability is prevented. For each bundle of data a port controller, request and ME element is required. Only when all of the ME elements have been locked out by the clock is the rising clock edge permitted to occur.

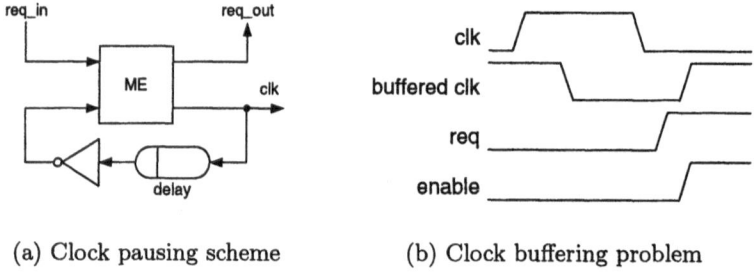

(a) Clock pausing scheme   (b) Clock buffering problem

Fig. 1. Pausable clock

Port controllers are required to generate and accept handshaking signals at the inputs and outputs of modules. These port controllers are asynchronous state machines, which are similar to on inputs rather than a clock.

A problem with clock pausing is that the clock is delayed as it is distributed across the clock domain, but the clock must be paused at its source. When the clock releases the ME elements there may still be a clock edge in the buffer tree. Hence it is possible that registers will be enabled as the clock rises, as shown in figure 1(b). But while the source clock is high, ME elements will remain locked so for this phase of the cycle no requests are permitted. For this reason, we must ensure that the delay of the clock buffer is shorter than the duration of the high phase of the clock. Limiting this delay limits the size of the clock tree, hence defining the size of GALS blocks.

3   System Architecture

We propose converting a conventional, synchronous FPGA into a GALS system. To do this we partition the FPGA into smaller blocks of FPGA cells. Within one of these blocks, the local connections are synchronous to a local clock for that block and hence the block resembles the original FPGA. However, longer communication channels between blocks become asynchronous.

Figure 2 shows the proposed architecture in place around a block of FPGA cells. Note in particular the dividing line between the synchronous and asynchronous domains. All of the FPGA cells are in an isolated block above the line in the synchronous domain. Internally, the FPGA block could resemble any synchronous FPGA as it is hidden from the rest of the system. Below the line there is an asynchronous wrapper: this interfaces between the synchronous and asynchronous domains. Outside the asynchronous wrapper blocks are connected together using asynchronous routing. All of these blocks are explained in detail in the following section.

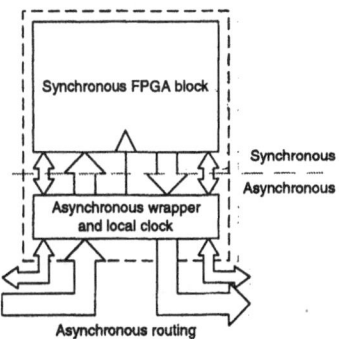

Fig. 2. FPGA block with asynchronous routing

3.1   Additional Synchronous FPGA Fabric

Figure 3 shows the boundary of a synchronous FPGA block. The FPGA block could contain any type of FPGA cell and its associated routing. There could be a number of different levels of routing within the block, for example nearest neighbour routing or fast interconnect spanning the block, as there are within a conventional FPGA. At the boundary, we can use the same routing schemes to connect to the asynchronous interface.

In this instance only one input port and one output port is shown for clarity, though it would be possible to design a system in which each FPGA block has several input and output ports to allow communication with a number of different FPGA blocks. As well as the data itself, the asynchronous ports needs to communicate with the synchronous to indicate when data is valid. To accomplish this, a wire for each port spans the routing leaving the block. For nearest neighbour or other unidirectional connections, one of the inputs to each block is allowed to connect to the data valid wire for each input port. Similarly, one of the output wires from each block may connect to the data valid wire of the output port. All other inputs and outputs from the FPGA cells can be used for data transfer.

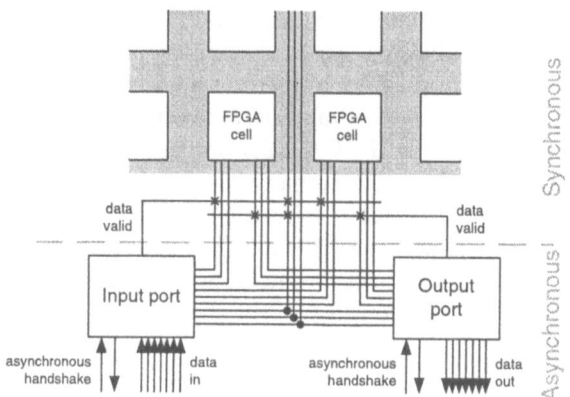

Fig. 3. Interconnect between FPGA cells and GALS ports

We work under the assumption that should we use part of an FPGA cell to create a data valid signal, that signal could be mapped to any of the inputs or any output as appropriate. In the case of a look-up table (LUT) based FPGA this can easily be done. Hence by only allowing one wire per cell to map to an input data valid signal and one for the corresponding output signal, we save on switching with little impact on flexibility.

Similarly, we also allow longer routing channels to be connected to the data valid signals. Again, in order to cut down on the number of switches used we only allow a fraction of the wires in the channel to be connected to the data valid signals. However, as long routing wires can usually be configured with data flow in either direction, we allow the chosen wires to be connected to both the input and output data valid signals. Here, we can expect some loss in flexibility as each routing wire could be connected to a completely different area of the FPGA block. But as each port can only have a single data valid signal, we should not need to allow many connections to the routing. In many cases it will be necessary to route to an FPGA cell and perform some logic function to merge several data valid signals into a single signal, which could be done in a boundary FPGA cell with a nearest-neighbour connection to the port's signals.

In figure 3, data is shown to enter the ports. This is merely a grouping of input wires and output wires, there is no need for any processing or even latching as that is handled in the synchronous part of the FPGA block. However, handshake signals accompany each "bundle" of data and so inside the synchronous FPGA block the data valid signal corresponding to the handshake must control the same data.

To complete this part of the system, we need to place a few requirements on any circuit mapped to the FPGA block. As discussed above, data valid signals need to be generated or processed. Data must leave the module with corresponding data valid signal and be latched when incoming data is valid. Any data signals also need to be routed to the boundary where they will leave the FPGA block accompanied by the handshake. For our implementation of the output port, we require that the data valid signal be "differential", i.e. it indicates valid data by changing its value and that no valid data is present by remaining at the same value.

3.2 Asynchronous Wrapper

The asynchronous wrapper shown in figure 2 is formed of 2 components: a local clock generator and port controllers. These components were described in more detail in section 2.3. For our implementation we use a four phase bundled data protocol. The interface between the synchronous and asynchronous domains is facilitated by making the synchronous signals differential, so an event is created whenever a signal changes. Our port controllers have been designed under the assumption that the synchronous block produces these differential signals and so they are a requirement of the circuit mapped to the FPGA block.

Note that we require a separate clock tree for each locally synchronous block. Clearly using the global trees featured in current FPGAs would be wasteful, hence it is preferable to use a dedicated local clock buffer. As mentioned in section 2.3, to prevent the size and delay of the clock buffers from becoming too large a limit of the size of the FPGA blocks within each wrapper is imposed.

3.3 Asynchronous Routing

The four phase bundled data transmission protocol continues into our routing. Not only must the data be routed, but also the corresponding handshakes. Furthermore, the bundling constraint must by maintained by delaying the request lines sufficiently that they always arrive after the corresponding data. When data arrives at a register, that register must be disabled so it will not go metastable if the rising edge of the clock arrives at the same instant. Once the register has won control of the ME element ahead of the clock, the register can be enabled and the ME element released.

Transfer of data is facilitated by inserting micropipelines into the routing. We exclusively use unidirectional wires to make point-to-point connections rather than using buses. The configuration in the routing is greatly simplified and in particular if micropipeline stages are used they need only operate in a single direction. The overall routing scheme is shown in figure 4. We have no long lines as long, slow lines are what we are trying to avoid. Instead, all long connections are made through a series of block-to-block connections. Each wire entering the FPGA block can either be routed to an input port or bypass the block completely. A connection between any two modules must pass through at least one micropipeline, which helps reduce the time each module remains paused and eliminate deadlock. Connections between rows must pass through at least two micropipeline stages, which adds a little latency.

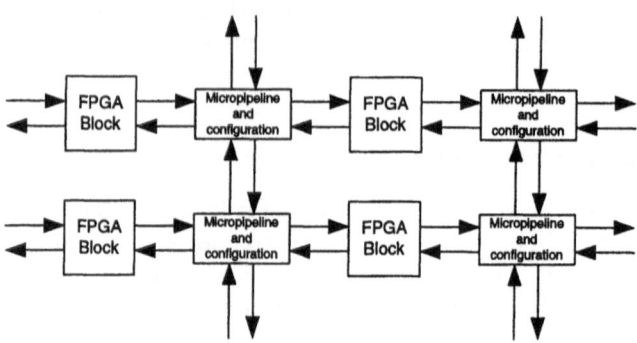

Fig. 4. Routing Scheme

4   Analysis of Architecture

This system confers several advantages. Our aims of metastability free independent clock domains and synchronous blocks whose timings are independent of the routing are met. The asynchrony of the routing allows data tokens to take an arbitrary number of clock cycles to reach their destinations. Indeed, as the source and destination are not in the same clock domain the number of clock cycles which pass in each domain as the data is transmitted can be expected to be different. Additionally, the FPGA system becomes more composable, i.e. the asynchrony of blocks makes it easy to design the component blocks independently without too much concern on how they will operate together as a system. It is no longer timing but the handshaking protocol which ensures data integrity. Independent design also allows the blocks to be independently optimised. As well as preventing metastability, clock pausing can be used to force a module to wait for data to arrive before proceeding, without wasting energy on redundant clocking.

However, the scheme does have several disadvantages. Firstly, we have now forced a separation of local routing signals, which are contained within the synchronous FPGA block, and global routing signals. This reduces the flexibility of the FPGA as a router is no longer able to use long routing tracks for local routing or join shorter routing tracks to make longer connections.

Secondly, we force additional constraints on placement. Any circuit which is too big to fit into a single FPGA block will necessarily need to be partitioned across several different blocks which will be in different clock domains. In some cases this may be inappropriate. For example feedback loops should ideally be contained on a single block due to the latency incurred. Due to the differing data rates, partitioning which results in data being transferred between blocks every cycle should be avoided.

To interface with the port controllers the synchronous block may need some additional circuitry. Some form of synchronous handshake is required to indicate to the output controller when data is valid and to accept data from the input port when incoming valid data is present. In our implementation we require differential data valid signals. If the design is contained within a single block, the external ports may already include these signals. However, this need not necessarily be the case and if a design needs to be partitioned the required signals will almost certainly not be present at the block boundary and therefore need to be inserted. This may prove to be problematic as the timing of such signals requires not only the circuit information but also some knowledge of the pattern of the data.

Finally, as the size and complexity of the system implemented grows, it may become necessary to partition across many blocks. In this case, data being sent from a number of blocks to a single destination may be required to arrive at the same time. To make allowances for this, rendezvous elements are needed at each input port. These elements must be fully configurable to allow several possible combinations of data channels or to allow a bypass when rendezvous is not required. This will however move the system away from one in which the only configurable components are within the synchronous blocks.

5   Conclusion and Future Work

We have presented an extension to existing FPGA architecture with the potential to prevent long routing delays from dominating FPGA performance. However, the solution is not without its drawbacks.

To address some of these problems, some alternative architecture may be required. It has already been mentioned that configurable rendezvous elements are required at the input ports which adds configuration required in the routing. It may also be possible to join local clock trees to effectively combine two smaller blocks into a larger block under a single clock domain, though great care must be taken to maintain the balance of the tree. If this is possible, it may also be possible to join all clock trees and allow a globally synchronous mode to be retained alongside the GALS mode. Request lines need some tuning to force requests to arrive after the data and meet the bundling constraint, but alternatively we can use a dual rail protocol to provide delay insensitive routing. Though we currently use a four phase protocol, a two phase protocol may have some advantages.

We have yet to fully verify the scheme and prove its effectiveness. The difficulty lies in simulation as we need to concurrently simulate synchronous, asynchronous and FPGA components. Furthermore we have yet to extensively investigate how circuits may map to the system.

References

1. Mirsky, E., DeHon, A.: MATRIX: A Reconfigurable Computing Architecture with Configurable Instruction Distribution and Deployable Resources. In: Proceedings of the IEEE Symposium in Field-Programmable Custom Computing Machines. (1996) 157–166

2. Sutherland, I.E.: Micropipelines. Communications of the ACM (1987) 720–738
3. Ebergen, C.: A Formal Approach to Designing Delay-Insensitive Circuits. Distributed Computing 5 (1988) 107–119
4. Seitz, C.L.: System timing. In Mead, C.A., Conway, L.A., eds.: Introduction to VLSI Systems. Addison-Wesley (1980)
5. Brunvand, E.: Using FPGAs to Implement Self-Timed Systems. J. VLSI Signal Process. 6 (1993) 173–190
6. Maheswaran, K.: Implementing Self-Timed Circuits in Field Programmable Gate Arrays. Master's thesis, University Of California Davis (1995)
7. Hauck, S., Burns, S., Borriello, G., Ebeling, C.: An (fpga) for Implementing Asynchronous Circuits. In: IEEE Design & Test of Computers. Volume 11. (1994) 60–69
8. Borriello, G., Ebeling, C., Hauck, S., Burns, S.: The Triptych FPGA Architecture. IEEE Transactions on Very Large Scale Integration (VLSI) Systems 3 (1995) 491–501
9. Payne, R.E.: Self-Timed FPGA Systems. In: Proceedings of the 5th International Workshop on Field Programmable Logic and Applications. (1995)
10. Payne, R.: Self-Timed Field Programmable Gate Array Architectures. PhD thesis, University of Edinburgh (1997)
11. Molina, P.: The Design of a Delay-Insensitive Bus Architecture using Handshake Circuits. PhD thesis, Imperial College (1997)
12. Chapiro, D.M.: Globally Asynchronous Locally Synchronous Systems. PhD thesis, Stanford University (1984)
13. Pěchouček, M.: Anomalous response times of input sychronisers. IEEE Transactions on Computers C-25 (1976) 133–139
14. Yun, K.Y., Donohue, R.P.: Pausible clocking: A first step toward heterogeneous systems. In: Proceedings of the International Conference on VLSI in Computers and Processors. (1996) 118–123
15. Bormann, D.S., Cheung, P.Y.K.: Asynchronous wrapper for heterogeneous systems. In: Proceedings of the International Conference on Computer Design (ICCD). (1997) 307–314
16. Moore, S.W., Taylor, G.S., Cunningham, P.A., Mullins, R.D., Robinson, P.: Self-calibrating clocks for globally asynchronous locally synchronous systems. In: Proceedings of the International Conference on Computer Design (ICCD). (2000) 37–78
17. Olsson, T., Nilsson, P., Meincke, T., Hemam, A., Tokelson, M.: A digitally controlled low-power clock multiplier for globally asynchronous locally synchronous designs. In: The IEEE International Symposium on Circuits and Systems (ISCAS). Volume 3. (2000) 13–16

Case Study
of a Functional Genomics Application
for an FPGA-Based Coprocessor*

Tom Van Court[1], Martin C. Herbordt[1], and Richard J. Barton[2]

[1] Department of Electrical and Computer Engineering,
Boston University, Boston, MA 02215,
herbordt|tvancour@bu.edu
[2] Department of Electrical and Computer Engineering,
University of Houston, Houston, TX 77204,
rbarton@uh.edu

Abstract. Although microarrays are already having a tremendous impact on biomedical science, they still present great computational challenges. We examine a particular problem involving the computation of linear regressions on a large number of vector combinations in a high-dimensional parameter space, a problem that was found to be virtually intractable on a PC cluster. We observe that characteristics of this problem map particularly well to FPGAs and confirm this with an implementation that results in a 1000-fold speed-up over a serial implementation. Other contributions involve the data routing structure, the analysis of bit-width allocation, and the handling of missing data. Since this problem is representative of many in functional genomics, part of the overall significance of this work is that it points to a potential new area of applicability for FPGA coprocessors.

1 Introduction

Microarrays measure simultaneously the expression products of thousands of genes in a tissue sample and so are being used to investigate a number of critical biology questions. Among these are (paraphrasing from pages 19-20 of [4]): Given the effect of 5000 drugs on various cancer cell lines, which gene is most predictive of the responsiveness of the cell line to a particular chemotherapeutic agent? or Is there a group of genes that can serve to distinguish the outcomes of patients with disease ijk who are otherwise indistinguishable? or Which of all known genes have a pattern of expression similar to those genes regulated by factor xyz?

As exciting as this usage is, microarray analysis is extremely challenging. One issue is that the data are noisy and the noise is often difficult to characterize. Another issue is that the number of measured quantities is invariably much larger than the number of samples. This results in an underconstrained system

---

* This work was supported in part by the National Science Foundation through CAREER award #9702483 and by a grant from the Compaq Computer Corporation.

not amenable to traditional statistical analysis such as finding correlations. As a result of these and other difficulties, techniques are used (often derived from machine learning) that provide a focus of attention, or a visualization, from which biological significance can be inferred. Among these are various forms of clustering, inference nets, and decision trees. Their computational complexity ranges from the trivial to the intractable. However, given the cost of obtaining microarray data, the fact that further biological interpretation is usually still required, and the value of many of the most rudimentary computations, most analysis applications are tailored to run fairly quickly on ordinary PCs.

It is undoubtedly the case, however, that biologists would like to ask far more complex questions and that increased computational capability would help to answer them. With applications such as those listed above, however, even a slightly harder question can result in an increase by orders of magnitude in computation. This is true of the problem we investigate here.

Kim [3] would like to find a set of genes whose expression can be used to determine whether liver tissue samples are metastatic or non-metastatic. For biological reasons, it is likely that three genes is an appropriate number to make this determination. Kim further proposes that use of linear regression would be appropriate to evaluate the gene subsets over the available samples. Since there are thousands of potential genes, 10^{10} to 10^{12} data subsets need to be processed. Although simple to implement, he reported that this computation was intractable even on his small cluster of PCs.

We decided to investigate the prospects of supporting this computation on an FPGA-based coprocessor. There are many reasons:

- It is an excellent match computationally. The data-set size, the amount of computation per datum, the nature of the individual computations, and the data-type size all are favorable to FPGAs.
- This computation is representative of a large number of similar computations in microarray analysis as well as in broader bioinformatics and computational biology (BCB). Therefore, demonstrating the efficacy for this problem would have much wider significance. Note that Yamaguchi et al. [11] have shown similar success (to what we show here) for a different application in bioinformatics, but one that has substantially different computational characteristics.
- There is a large number—perhaps thousands—of potential users. Therefore a low-cost distributed solution is much more attractive than a centralized resource such as a large-scale cluster. This is especially true since (as we will show) it would take a very large cluster to match performance.
- The state of algorithmics in microarray analysis (and some other areas of bioinformatics) is one of flux. Therefore a solution based on a generic PC and coprocessor may be more attractive than hardwired alternatives such as ASICs. The same would be true with respect to turn-key software/hardware systems, such as those provided by a number of vendors, assuming that they provided solutions to these problems at all. Also, both of these "hardwired" alternatives are extremely expensive.

What we have found is that we can obtain a speed-up of a factor of more than 1500 over an optimized serial version running on a 1.7GHz Pentium IV PC. These results have so far been achieved in simulation using post place-and-route timing for the Xilinx XC2VP100-7.

Our most basic contribution is the speed-up for this particular problem: a set of 10,000 genes can be examined in 10 minutes instead of 19 days. Other contributions have to do with the actual implementation, including a novel data routing structure, the analysis of the bit-width allocation, and our handling of missing data. Finally, as this problem has similar characteristics to many others in microarray analysis, we show that there is potential for broader applicability of FPGA coprocessors in this very important domain.

The rest of this paper is organized as follows. In the next section we present the problem formally and describe the serial implementation. There follows a description of the FPGA design and implementation including an analysis of data path widths. We conclude with a discussion.

2 Application Detail and Serial Implementation

The data to be analyzed are derived from n microarrays; each consists of a binary diagnosis and an expression value for each of the genes being analyzed. Expression values are tabulated in a matrix with row vectors corresponding to microarrays and column vectors corresponding to particular genes. The outcomes are tabulated in a column vector Y. The technique used is to compute the linear regressions for all 3-way combinations of genes. Pearson's R^2 defines the goodness of fit for each regression. Examining a standard statistics reference [8], we find that the estimators $\hat{\beta}_0, \ldots, \hat{\beta}_n$ comprising the column vector $\hat{\beta}$ can be computed as follows

$$\hat{\beta} = (X^T X)^{-1} X^T Y$$

The first column of X consists of n 1s and each remaining column consists of the n expression values for one gene of the subset being considered in this particular combination. However, since much of computational complexity results from the inversion, it is important to reduce its rank. This is done by centering the data, which results in a $\hat{\beta} = \hat{\beta}_1, \ldots, \hat{\beta}_n$ of

$$\hat{\beta} = [(X^+)^T X^+]^{-1} (X^+)^T Y^+$$

and

$$R^2 = \frac{\hat{\beta}(X^+)^T Y^+}{(Y^+)^T Y^+}$$

where X^+ is the $n \times 3$ matrix containing column vectors with elements $X_{ij} - \overline{X_i}$ and Y^+ is the column vector containing the values of $Y_j - \overline{Y}$. Note that in centered mode we do not need to compute $\hat{\beta}_0$ and thus do not need the column of ones in X. We therefore operate on 3×3 matrices rather than 4×4.

The covariance matrix

$$c_{ij} = (\mathbf{X_i} - \overline{\mathbf{X_i}})^{\mathbf{T}}(\mathbf{X_j} - \overline{\mathbf{X_j}}),$$

can also be written as

$$nc_{ij} = n\Sigma_k(X_{ik}X_{jk}) - (\Sigma_k X_{ik})(\Sigma_k X_{jk}).$$

The factor of n cancels in later operations, and eliminates the need to perform the division implied by $\overline{X_i}$. The entire computation for each gene combination thus partitions into two parts: the first consists of the 9 dot products and 4 summations while the second consists of computing the covariance matrix, inverting the matrix, and using those results to obtain $\hat{\beta}$ and R^2.

In our tests we used data derived from a human breast tumor study by Perou, et al. [7]. The data are typical of those generated in microarray studies and consist of expressions of 9218 genes (including controls) from 84 microarray samples. Raw expression data are ratios. As is standard practice, we took the log, normalized, and rounded the data to integer values in the range -4 to +4. Using four bits is on the high end of information per sample; often only two bits are used. The result vector is binary: 0/1 for diseased/healthy.

An important consideration in microarray analysis is dealing with missing data. That is, the microarray value for a gene/sample expression is sometimes unreadable; in that case no value at all is reported. In the Perou data set, 46% of genes contain missing data. If not handled properly, missing data can dominate the regressions and render results meaningless. We take a simple approach: for a given combination of three genes, for each sample with missing data, we eliminate that sample from consideration for all genes. The combination is rejected completely if too many healthy samples are dropped.

Statistical literature [5] refers to this as a form of complete-case analysis within any one combination of genes, and available-case analysis in selecting combinations of genes. A χ^2 test of Perou's data shows that the frequency of missing values is not decisively different among healthy samples than among diseased samples. Thus, the missing data matches the Missing At Random (MAR) assumption and complete-case analysis appears justified.

When implementing the algorithm, one notes that each dot-product is used a large number of times. It seems to makes sense to precompute *all* of the $n \times n$ dot products, then use them in the $\binom{n}{3}$ inversions later. This eliminates roughly a factor of n dot products. Unfortunately, this does not account for missing data: each dot-product can have drop-outs not just from data missing from the two vectors being multiplied, but also from the third vector of the set. Since this third vector changes for every set, it follows that all dot products must be recomputed for every iteration.

We created two versions of the serial code. One was a mirror of the FPGA implementation and was used to verify results, especially with respect to maintaining precision. The other was used to generate timing and so was highly optimized for serial execution on a modern processor.

The R^2 for each set of genes was computed in 10.1us on a 1.7GHz Pentium IV PC. Because of the tremendous data locality, there was no drop-off in performance with respect to the number of gene combinations evaluated. This means that evaluating 3-way combinations of 1,000 genes takes about half an hour, while 3-way combinations of 10,000 genes takes more than 19 days, and 20,000 genes takes more than five months. Clearly L1 cache and available ILP are being used to a very high degree, the latter not surprisingly due to the numerous multiply-accumulate (MAC) computations and the hardwired invert.

Still, these results confirm our initial assumption: that this computation, while perhaps not "heroic" in the grand-challenge sense, is still outside the realm of usage in exploratory data analysis. For this and similar computations to be readily usable as part of an analysis toolkit, days need to be reduced to minutes.

3   FPGA Methods, Implementation, and Results

3.1   Description

Hardware is assumed to be an FPGA on a commercial PCI board plugged into a PC. As the amount of reuse per datum is very high, details about the particular board and interface do not have a significant impact on our results: only a KB/second input rate needs to be supported.

The rest of this discussion describes simulations, synthesis, and place-and-route in the Xilinx ISE 5.2.02i environment [10] for Virtex-II Pro XC2VP100 gate array [9] and anticipates implementation on a generic coprocessor board when one becomes available later this year. The Virtex-II Pro product family is especially interesting here because of its large gate count, large on-chip memory, and dedicated multipliers. Implementations on other devices in this family follow from the one described here using analogous optimizations.

The circuitry consists of three parts: (i) vector storage and distribution (VSD) and the two computational segments described in Section 2: (ii) dot-products and summations (DPS), and (iii) covariance matrix, inverse, and regression (CIR). We now describe these, starting with the computations. In the following subsections we talk about optimizations and safety checks and then about integration and speed-matching. At that point the responsibilities of the vector storage and distribution unit are clear and it is then described.

Each DPS accepts four data vectors, three X values and one Y. The Y represents the diagnosis, 0 or 1 for cancerous or healthy samples respectively. The X values represent expression levels, encoded as four bit values. The encoding represents a symmetric range from $-N$ to $+N$. Earlier, we noted that the application works well when expression data is quantized to a range of -4 to 4 (encoded as 0 to 8). A special value (15, binary 1111) is a Not-a-Number (NaN) code that represents missing or invalid input data. The DPS unit, illustrated in Figure 1, consists of: counters to tally valid data sets and Y values, accumulators to total the X vectors and XY dot products, and MAC sections to total the $X_i X_j$ dot products and X_i^2.

Note the handling of missing data. In Figure 1, the boxes labeled = NaN? detect missing data and propagate that fact to the MAC units where the accumulation of the invalid summands is blocked. If, for example X_{1i} is a missing

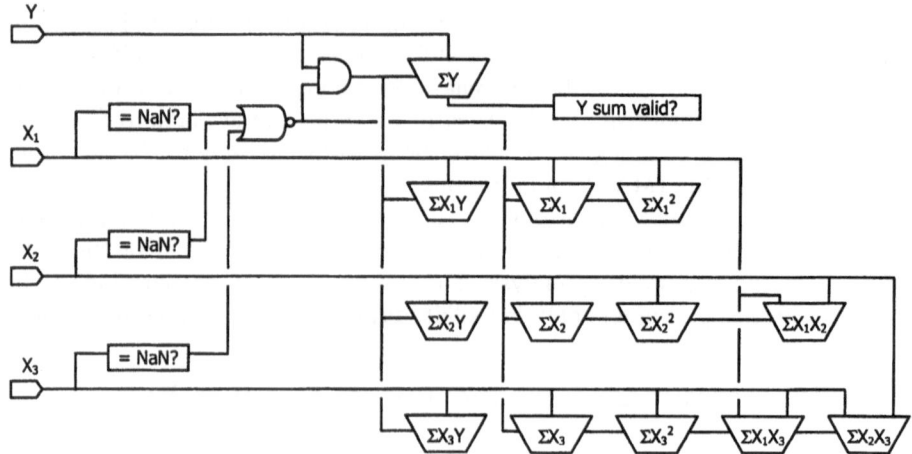

Fig. 1. Dot products and summations.

value, then the summands X_{1i}^2, $X_{1i}X_{2i}$, $X_{1i}X_{3i}$ and $X_{1i}Y_i$ are obviously meaningless. Following [5], X_{2i}, X_{2i}^2, X_{3i}, X_{3i}^2, $X_{2i}X_{3i}$, $X_{2i}Y_i$, $X_{3i}Y_i$, and Y_i are also omitted from their respective sums and from the vector length count.

Once the vector is processed, DPS results are latched for input to the CIR. DPS results consist of outputs from all accumulators in Figure 1, a valid data counter, and the validity indicator (overflow and minimum-Y tests—details below). The result is presented broadside to the CIR. That section consists entirely of unclocked combinational logic, including adders, subtracters, and multipliers. It first computes the 3×3 covariance matrix from the dot products. That feeds the closed form inversion of the covariance matrix.

3.2   Optimizations and Safety Checks

One fundamental optimization is to minimize the datapath width at all points. Worst-case calculations of the minimum width are simple and safe, but result in a far more conservative implementation than necessary. We address this by: (i) *a priori* determining the maximum datapath widths and (ii) confirming that overflow has not occurred. There are two aspects to datapath width determination: the bits that can be dropped at the high end (because of worst cases that never occur in practice) and bits that can be dropped at the low end (because the loss of precision is insignificant).

The *a priori* path width determination is done in two ways: empirical and theoretical. On the theoretical side, the need for less significant bits is estimated using interval arithmetic [1]. Precision is verified empirically by sampling test data off-line. Also on the theoretical side, worst-case analysis sets an upper bound on the number of high-order bits required. It is interesting that most authors (e.g. [6]) perform worst-case bit allocation in whole bit increments, even when their bitwidth allocation is otherwise sensitive to application data values.

For example, consider the accumulator needed to add up 84 terms in a dot-product of vectors with 9 element values, 0 to 8, as described above. Naively, that would require $7 + 4 + 4 = 15$ bits. In fact, the accumulator need only hold $\lceil \log_2(84 \times 8 \times 8) \rceil = 13$ bits to handle the worst-case for this application.

Empirical estimates of high order bit requirements also come from measurements of calculations on sample data. The calculation is performed off-line, for a meaningful subset of the application cases, and the number of bits used at each step is measured. This gives statistics of actual bit usage. High-order bits are allocated to cover the samples observed, plus some margin based on observed standard deviations.

The implementation, then, handles all data values that can reasonably be expected. To be thorough, however, the FPGA logic checks for overflow at each step. The DPS and CIR stages both present validity indicators, stating whether an erroneous calculation was detected at any point.

As noted earlier, some regressions may be invalid because missing data values leave too few $Y = 1$ (healthy) samples. Figure 1 illustrates the *"Y sum valid?"* check, a configurable test requiring some minimum number of healthy samples in a regression. If inadequate data with $Y = 1$ appears in a regression, that also marks the whole calculation as invalid. A similar test of the number of $Y = 0$ samples is not necessary for this data set.

In the CIR, one optimization is the use of closed form inversion. Although this method has many problems when applied to large matrices, it works well with these matrix, vector, and data sizes. Besides the obvious speed advantage, hardwiring allows us to keep cofactors and determinant separate for independent use in computing correlation coefficients and regression.

Another optimization is that all multiplications in the CIR use the FPGA's block multipliers. The DPS, with smaller operands, uses multipliers built from logic. Further optimization may be possible in the CIR by exchanging some block multipliers for multipliers built from logic for sufficiently small operands.

Pipelining the CIR (not currently done) may also be advantageous. The CIR unit is implemented as three combinational logic elements in series: the correlation matrix, its inverse, and the regression results. These represent natural boundaries for pipelining.

It may also be possible to speed up the DPS by processing vectors in parallel instead of serially as is currently done. However, timing and resource allocation tradeoffs make it unlikely that this will improve performance significantly.

3.3   Timing and Speed-Matching Optimization

Timing is computed statically, using post place-and-route (PAR) results. PAR is completely automatic, with no manual guidance, timing constraints, or floor-planning. Timing estimates are based on synthesis of 9 CIR units, each supplied by 10 DPS units (90 DPSs total; see Figure 2 for one CIR/10 DPS unit) and VSD as described below. The target is a XC2VP100 gate array with speed grade -7. The entire design, minus 'glue' needed to attach the application logic to its host environment, occupies 73% of the logic slices and 94% of the block multipliers.

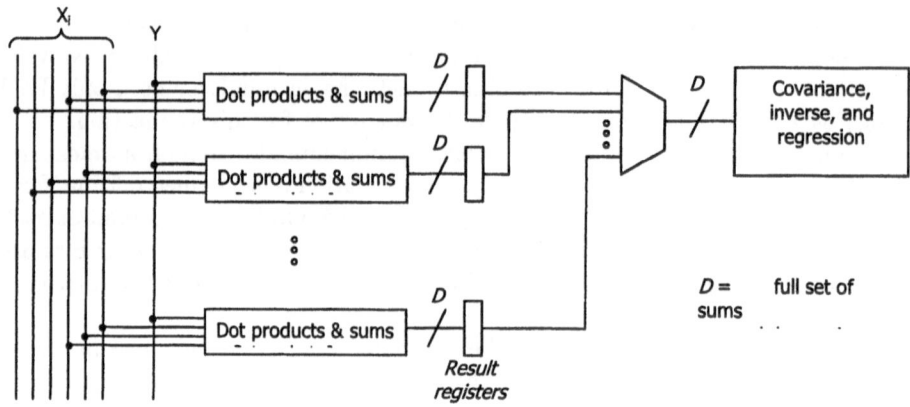

Fig. 2. Computation balancing.

Post-PAR timing analysis indicates a DPS clock of 5.35ns while propagation delay through the CIR to its result register is 47.36ns. Using conservative approximations, this implies a 5.5ns DPS clock rate and very roughly a 10:1 ratio of CIR delay to DPS rate. The system clock is set to the DPS rate. A separate counter divides the system rate down so that CIR results are latched and new results presented to the CIR at 1/10 the DPS clock rate.

Given data vectors of length ~ 80, the DPS requires about 500ns to compute a result for one set of vectors. The CIR versus DPS imbalance, about 50ns vs. 500ns, is conspicuous. The other conspicuous imbalance is in the resources used by each section. Each CIR requires 48 of the target FPGA's 444 block multipliers. The DPS, however, uses only 4×4-bit products, which can be built more effectively from ordinary logic. A DPS uses smaller, less specialized logic resources, under 1% of the chip total. Considering both execution time and logic resources used, to achieve maximum logic activity, each CIR serves 10 DPS units. Figure 2 shows how this is done.

The DPSs all start and end their computations at the same time, latch their results, and begin processing the next set of vectors. After latching the DPS results, separate logic presents results from each DPS to its CIR sequentially. A counter (not shown in Figure 2) holds input stable for the CIR propagation delay, then latches the CIR result and reads the next saved DPS result.

These design values (vector length, CIR propagation delay, etc.) and available resources are all parameters specific to the problem and technology at hand. Different parameters would yield different specific results, but the analysis would follow the same general pattern over a wide range.

3.4 Vector Store and Distribution

Recall that each DPS unit processes 3 vectors and that there are 90 DPS units. Therefore, the VSD must simultaneously distribute 270 vector streams. A general (but impractical) solution would be to store all $v = 10,000$ vectors on chip in a

270-ported memory. Size itself is not the problem: the entire data set fits in 1MB. Our solution leverages the v^2 reuse of each vector in a hierarchical structure.

We observe that 9 vectors form $\binom{9}{3} = 84$ combinations and so are nearly sufficient to keep the 90 DPS units busy. This leaves six DPSs idle, the price paid for efficient memory access. A simple network (shown on the left in Figure 2 and the right in Figure 3) distributes vector data to the DPSs.

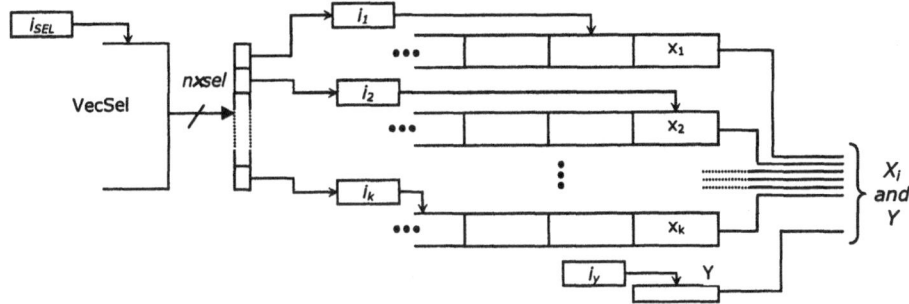

Fig. 3. Vector store and distribution to the DPS units.

Each vector store unit (labeled $x_1 \ldots x_k$ in Figure 3) feeds one of the distribution network's nine X inputs and is indexed by an independent address register (labeled $i_1 \ldots i_k$). The sequence of vectors is stored in VecSel. The chip's 16Kb RAM blocks can each hold 50 vectors, so vector storage uses only a small amount of the available RAM. Vector store units and the VecSel RAM are loaded from off-chip. Double buffering ensures that loading does not interfere with data access for computation. Note that many more values are read from the vector stores than are loaded into them, since each vector is reused in many size-9 combinations, so the DPS should never be idle while the VSD is reloaded.

Operation is as follows. A set of 9 vectors is chosen at the beginning of a vector computation, one from each vector store unit, as indicated by the address registers. Successive vector elements are read by incrementing all of the data address registers in unison, through the length of the vector. At the end of a vector, the address registers are reloaded from the VecSel and the next set of data vectors is ready for transfer.

This VSD design supports the solution of the following combinatorial problem: choosing sets of size 9 (or k) from a set of 9218 (or v) elements such that every size-3 subset of those v appears in one size-k set. This is exactly the Steiner System $S(3, k, v)$. Generating algorithms exist for these size-k sets [2], but are not amenable to FPGA implementation. Off-line computation generates that set of size-k sets, and loads the vector stores and VecSel accordingly. The VecSel RAM and vector stores must be reloaded $10^5 - 10^6$ times to cover all size-3 sets; however, this is a trivial requirement for a run of several minutes.

3.5 Throughput

As described, the basic component of our design consists of a pipeline with a single CIR ($\sim$ 50 ns) and single DPS ($\sim$ 500 ns for a length-80 vector). This gives a speed-up of a factor of 20 over the serial PC version. The chosen Virtex II Pro easily fits 9 of these units (the critical resource being the multipliers) increasing the speed-up to 180 $\times$. Each CIR serves 10 DPSs reducing the cycle time of the original unit to 50 ns and increasing the total speed-up to a factor of 1800 $\times$. Combinatoric inefficiency in the VSD reduces this factor to 1600 $\times$.

4   Discussion and Extensions

The 1000-fold+ speed-up derived from our FPGA implementation achieves our goal of reducing the duration of this computation from days to minutes. This is done while keeping the system cost (hardware and IT support) low. It is therefore quite plausible that this and related techniques could indeed become part of a computational toolbox for functional genomics, broadening the types of inferences possible from microarray data.

A necessary extension to this work is achieving some generality in implementation: it is certainly expensive to buy a large cluster and hire a parallel applications programmer; it is perhaps even more problematic to find good FPGA designers. We are currently working in this direction.

References

1. Hansen, E., et al.: Topics in Interval Analysis. Clarendon Press, Oxford, U.K. (1969)
2. Hartman, A.: The fundamental construction for 3-designs. Discrete Mathematics **124** 107-132 (1994)
3. Kim, S.: Finding Genes for Cancer Classification: Many Genes and Small Number of Samples. 2nd Ann. Houston Forum on Cancer Genomics and Informatics (2001)
4. Kohane, I.S., Kho, A.T., Butte, A.J.: Microarrays for an Integrative Genomics. MIT Press, Cambridge, MA (2003)
5. Little, R.J.A., Rubin, D.B.: Statistical Analysis with Missing Data John Wiley and Sons, Hoboken, NJ (2002)
6. Mahlke, S., Ravindran, R., Schlansker, M., Schreiber, R., Sherwood, T.: Bitwidth Cognizant Architecture Synthesis of Custom Hardware Accelerators. IEEE Trans. on CAD of Integrated Circuits and Systems **20** (2001) 1355-1370
7. Perou, C.M., et al.: Molecular Portraits of Human Breast Tumors. Nature **406** (2000) 747-752
8. Ryan, T.P.: Modern Regression Methods. John Wiley and Sons, Inc., New York (1997)
9. Xilinx, Inc.: Virtex-II Pro Platform FPGA User Guide. (2002)
10. Xilinx, Inc.: Integrated Software Environment. (2002)
11. Yamaguchi, Y., Miyajima, Y., Maruyama, T., Konagaya, A.: High Speed Homology Search Using Run-Time Reconfiguration. In Proceedings of the 12th International Conference on Field Programmable Logic and Applications (2002) 281-291

A Smith-Waterman Systolic Cell

C.W. Yu, K.H. Kwong, K.H. Lee, and P.H.W. Leong

Department of Computer Science and Engineering,
The Chinese University of Hong Kong,
Shatin, New Territories, Hong Kong SAR,
{cwyu1,khkwong,khlee,phwl}@cse.cuhk.edu.hk

Abstract. With an aim to understand the information encoded by DNA sequences, databases containing large amount of DNA sequence information are frequently compared and searched for matching or near-matching patterns. This kind of similarity calculation is known as sequence alignment. To date, the most popular algorithms for this operation are heuristic approaches such as BLAST and FASTA which give high speed but low sensitivity, i.e. significant matches may be missed by the searches. Another algorithm, the Smith-Waterman algorithm, is a more computationally expensive algorithm but achieves higher sensitivity. In this paper, an improved systolic processing element cell for implementing the Smith-Waterman on a Xilinx Virtex FPGA is presented.

1 Introduction

Bioinformatics is becoming an increasingly important field of research. With the ability to rapidly sequence DNA information, biologists have the tools to, among other things, study the structure and function of DNA; study evolutionary trends; and correlate DNA information with disease. For example, two genes were identified to be involved in the origins of breast cancer in 1994 [1]. Such research is only possible through the help of high speed sequence comparison.

All the cells of an organism consist of some kind of genetic information. They are carried by a chemical known as the deoxyribonucleic acid (DNA) in the nucleus of the cell. DNA is a very large molecule and nucleotide is the basic unit of this type of molecule. There are 4 kinds of nucleotides and each have different bases, namely adenine, cytosine, guanine and thymine. Their abbreviated forms are "A", "C", "G" and "T" respectively. In this paper, the sequence is referred to as a string, and the bases form the alphabet for the string.

It is possible to deduce the original sequencing in DNA which codes for a particular amino acid. By finding the similarity between a number of "amino-acid producing" DNA sequences and a genuine DNA sequence of an individual, one can identify the protein encoded by the DNA sequence of the individual. In addition, if biologists succeed in finding the similarity between DNA sequences of two different species, they can understand the evolutionary trend between them. Another important usage is that the relation between disease and inheritance can also be studied. This is done by aligning specific DNA sequences of individuals

P.Y.K. Cheung et al. (Eds.): FPL 2003, LNCS 2778, pp. 375–384, 2003.

with disease to those of normal people. If correlations can be found which can be used to identify those susceptable to certain diseases, new drugs may be made or better techniques invented to treat the disease. There are many other applications of bioinformatics and this field is expanding at an extremely fast rate.

A human genome contains approximately 3 billion DNA base pairs. In order to discover which amino acids are produced by each part of a DNA sequence, it is necessary to find the similarity between two sequences. This is done by finding the minimum string edit distance between the two sequences and the process is known as sequence alignment.

There are many algorithms for doing sequence alignment. The most commonly used ones are FASTA [2] and BLAST [3]. BLAST and FASTA are fast algorithms which prune the search involved in a sequence alignment using heuristic methods. The Smith-Waterman algorithm [4] is an optimal method for homology searches and sequence alignment in genetic databases and makes all pairwise comparisons between the two strings. It achieves high sensitivity as all the matched and near-matched pairs are detected, however, the computation time required strongly limits its use.

Sencel Bioinformatics [5] compared the sensitivity and selectivity of various searching methods. The sensitivity was measured by the coverage, which is the fraction of correctly identified homologues (true positives). The coverage indicates what fraction of structurally similar proteins one may expect to identify based on sequence alone. Their experiments show that for a coverage around 0.18, the errors per query of BLAST and FASTA are about two times that of the Smith-Waterman Algorithm.

Many previous ASIC and FPGA implementations of the Smith-Waterman algorithm have been proposed and some are reviewed in Section 4. To date, the highest performance chip [6] and system level [7] performance figures have been achieved using a runtime reconfigurable implementation which directly writes one of the strings into the FPGA's bitstream.

In this work, an FPGA-based implementation of the Smith-Waterman algorithm is presented. The main contribution of this work is a new 3 Xilinx Virtex slice Smith-Waterman cell which is able to achieve the same density and performance as an earlier reported cell [6], without the need to perform runtime reconfiguration. This has advantages in that the design is less FPGA device specific and thus can be used for non-Xilinx FPGA devices as well as ASICs. Whereas the runtime reconfigurable design requires JBits, a Xilinx specific API for runtime reconfiguration, the design presented in this paper was written in standard VHDL. Moreover, in the proposed design, both strings being compared can be changed rapidly as compared to a runtime reconfigurable system in which the bitstream must be generated and downloaded, which is typically a very slow process since a large bitstream must be manipulated and downloaded via a slow interface. This reconfiguration process may become a bottleneck, particularly for small databases. Furthermore, other applications may require both strings to change quickly. The design was implemented and verified using Pilchard [8], a memory-slot based reconfigurable computing environment.

2   The Smith-Waterman Algorithm

The Smith-Waterman Algorithm is a dynamic programming technique which utilizes a 2D table. As an example of its application, suppose that one wishes to compare sequence S ("ACG") with sequence T ("ATC"). The intermediate values a, b and c (shown in Fig. 1(b)) are then used to compute d according to the following forumla:

$$d = min \begin{cases} a & if \ S_i = T_j \\ a + sub & if \ S_i \neq T_j \\ b + ins \\ c + del \end{cases} \qquad (1)$$

If the strings being compared are the same, the value a is used for d. Otherwise, the minimum of a plus some substitution penalty sub, b plus some insertion penalty ins and c plus some deletion penalty del is used for d. Data dependencies mean that entries d in the table can only be calculated if the corresponding a, b, c values are already known and so the computation of the table spreads out from the origin as illustrated in Fig. 2. As an example, the first entry that can be computed is that for "AA" in Fig. 1(a). Since $S_i = T - i = \text{`A'}$, according to Equation 1, $d = a$ and so the entry is set to 0. In order to complete the table, the template of Fig. 1(b) is moved around the table constrained by the dependencies indicated by Fig. 2.

The substitution, insertion and deletion penalties can be adjusted for different comparison requirements. If the presence of redundant characters is relatively less acceptable than just a difference in characters, the insertion and deletion penalties can be set to a higher value than the substitution penalty. In the

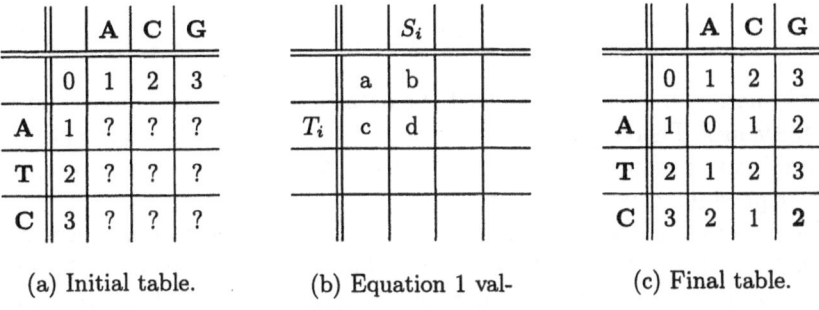

		A	C	G
	0	1	2	3
A	1	?	?	?
T	2	?	?	?
C	3	?	?	?

(a) Initial table.

		S_i		
T_i	a	b		
	c	d		

(b) Equation 1 values.

		A	C	G
	0	1	2	3
A	1	0	1	2
T	2	1	2	3
C	3	2	1	2

(c) Final table.

Fig. 1. Figure showing the progress of the Smith-Waterman algorithm, when the string "ACG" is compared with "ATC".

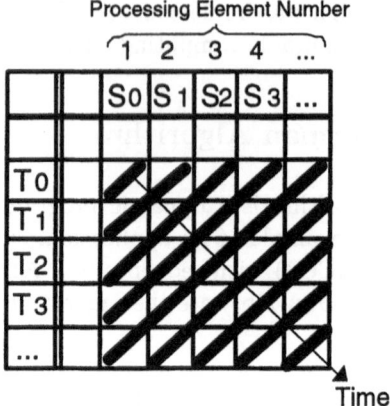

Fig. 2. Data dependencies in the alignment table. Thick lines show entries which can be computed in parallel and the time axis is arranged diagonally.

alignment system presented, the insertion and deletion penalties were fixed at 1 and the substitution penalty set to 2, as is typical in many applications.

If S and T are m and n in length respectively, then the time complexity of a serial implementation of the Smith-Waterman algorithm is $O(mn)$. After all the squares have been processed, the result of Fig. 1(c) is obtained. In a parallel implementation, the positive slope diagonal entries of Fig. 2 can be computed simultaneously. The final edit distance between the two strings appears in the bottom right table entry.

3   FPGA Implementation

In 1985, Lipton and Lopresti observed that the values of b and c in Equation 1 are restricted to $a \pm 1$ and the equation can be simplified to obtain [9]:

$$d = \begin{cases} a & if \ ((b \ or \ c) \ = \ a-1) \ or \ (S_i = T_j) \\ a+2 & if \ ((b \ and \ c) \ = \ a+1) \ and \ (S_i \neq T_j) \end{cases} \tag{2}$$

Using Equation 2, it can be seen that the data elements b, c and d only have two possible values. Therefore, the number of data bits used for the representation of b, c and d can be reduced to 1 bit. Furthermore, two bits can be used to represent the four possible values of the alphabet.

The processing element (PE) shown in Fig. 3 was used to implement Equation 2. A number of PEs are then connected in a linear systolic array to process diagonal elements in the table in parallel. As shown in Fig. 2, PEs are arranged horizontally and are responsible for its corresponding column. In the description that follows, the sequence that changes infrequently is S and the sequences from the database are T. In each PE, two latches are used to store a character S_i.

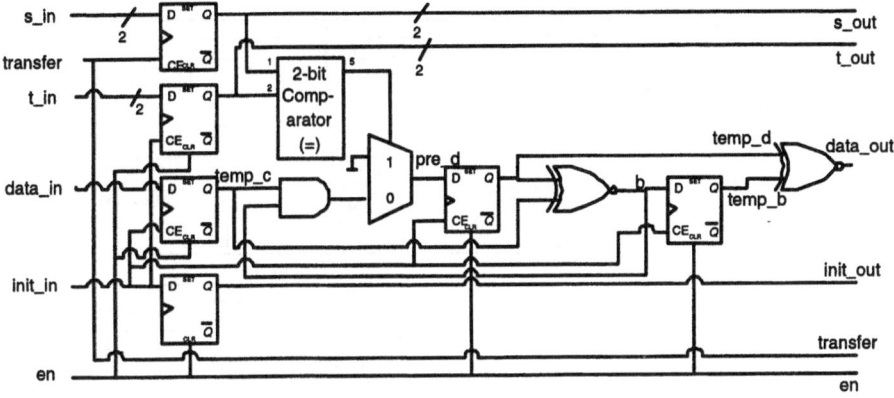

Fig. 3. The Smith-Waterman processing element (PE). Boxes represent D-type flip-flops.

These characters are shifted and distributed to every processing element before the actual comparison process beings. The base pairs of T are passed through the array during the comparison process, during which the d of Equation 2 is also computed.

In order to calculate d, inputs a, b and c should be available. In the actual implementation, the new values b, c and d are calculated during the comparison of the characters as follows:

1. data_in is the new value of c and it is stored in a flip-flop. At the same time, this new c value and the previous d value (from a flip-flop) determines the new b value ($b = temp\_c$ XNOR $temp\_d$)
2. The new b value is stored in a flip-flop. At the same time, the output of a 2-to-1 MUX is then selected depending on whether $S_i = T_i$. The output of the MUX (a '0' value or (b AND $temp\_c$)) becomes the new d value. This new d value is stored in a flip-flop.
3. Values of b and d determine the data output of the PE ($data\_out = temp\_b$ XNOR $temp\_d$). The data output from this PE is connected to the next PE as its data input (its new c value)

When the *transfer* signal is high, the sequence S is shifted through the PEs. When the *en* signal is high, all the flip-flops (except the two which store the string S) are reset to their initial values. The *init* signal is high when new signals from the preceeding PE are input and the new value of d calculated. When the *init* signal is low, the data in all the flip-flops are unchanged.

Each PE used 8 flip-flops as storage elements and 4 LUTs to implement the combinational logic. Thus the design occupied 4 Xilinx Virtex slices. Guccione and Keller [6] used runtime reconfiguration to write one of the sequences into the bitstream, saving 2 flip-flops and implementing a PE in 3 slices. In the proposed approach, two otherwise unused LUTs were configured as shift register elements

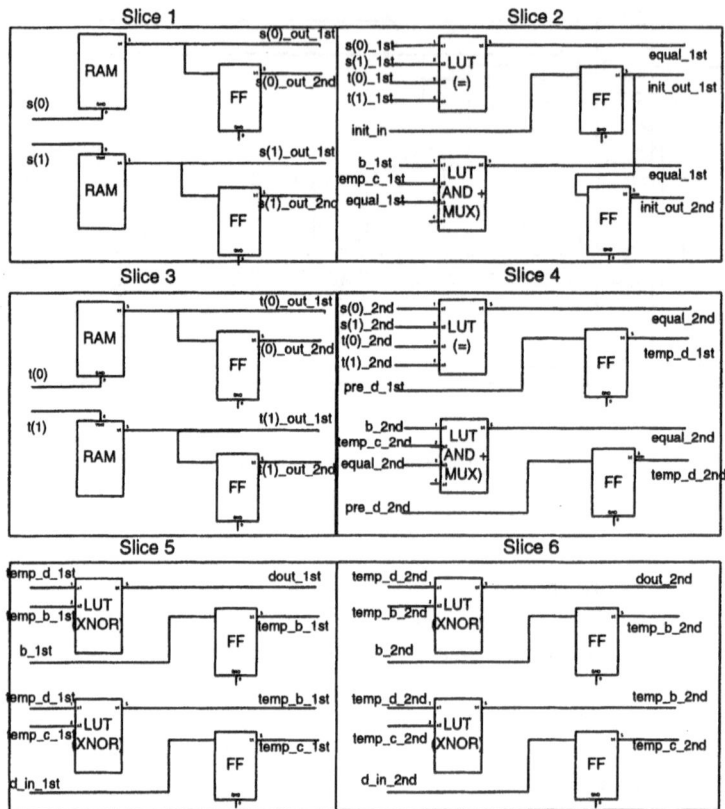

Fig. 4. Two processing elements mapped to 6 Virtex slices.

using the Xilinx distributed RAM feature [10]. Thus the design occupies 3 Xilinx Virtex slices per PE, without requiring runtime reconfiguration to change S. In the actual implemention, 2 PEs were implemented in 6 slices since sharing of resources between adjacent PEs was necessary in the actual implementation.

Fig. 4 shows the mapping of the PEs to slices. All the signals ending with "_1st" were used in PE Number 1, and signals ending with "_2nd" were used for PE2. The purpose of each signal can be understood by referring back to Fig. 3. It was necessary to connect the output of the RAM-based flip-flops directly to a flip-flop (FF in the diagram) in the same logic cell (LC) since internal LC connections do not permit them to be used independently (i.e. it was not possible to avoid connecting the output of the RAM and the input of the FF). Thus, Slice 1 was configured as a 2 stage shift register for consecutive values of S_i and Slice 3 was used for two consecutive values of T_i.

Fig. 5 shows the overall system data path. Since the T sequences are shifted continuously, the system used a FIFO constructed from Block RAM to buffer the sequence data supplied by a host computer. This improves throughput of the system since a large number of string comparisons can be completed before

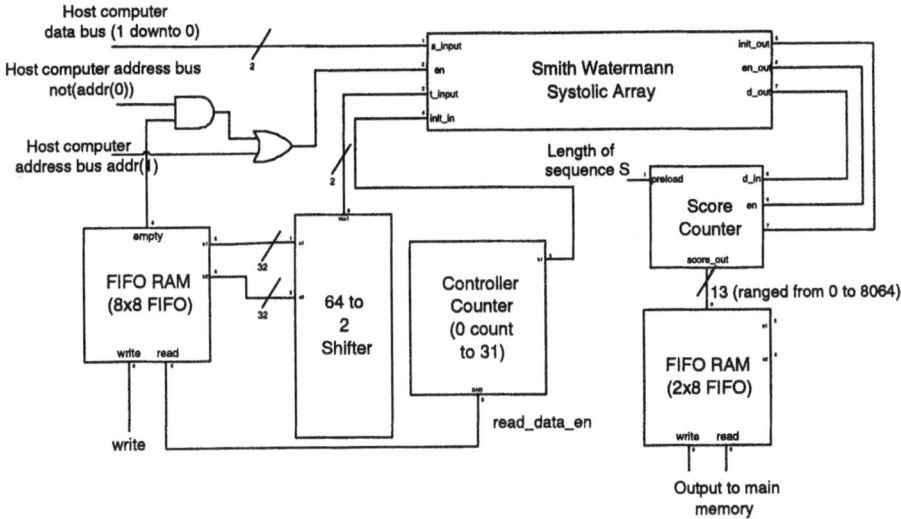

Fig. 5. System data path.

all of their scores are read from the controller, reducing the amount of idle time in the systolic array. The input and output data width of the FIFO RAM were both 64 bits. The wide input data width helped to improve IO bandwidth from the host computer to the FIFO RAM. A 64-to-2 shifter and a controller counter were used for reducing the output data width of the FIFO RAM from 64 bits to 2 bits, so as to allow data to be fed into the systolic array.

The Score Counter computes the edit distance by accumulating results calculated in the last PE of the systolic array. The output of the last PE is actually the d value in the squares of the rightmost column of the matrix, and differences in values of consecutive squares in the rightmost column must be 1. The $data_{out}$ of the last PE is '0' when $d = b$ - 1 , and the output '1' when $d = b +$ 1. Therefore, a Shift Counter was initialized to the length of the sequence S. It was decremented if the output value is '0', otherwise it was incremented. After the entire string T is passed through the systolic array, the counter contains the final string comparison score.

4   Results

The design was synthesized from VHDL using the Xilinx Foundation 5.1i software tools and implemented on Pilchard, a reconfigurable computing platform [8] (Fig. 6). The Pilchard platform uses a Xilinx Virtex XCV1000E-6 FPGA (which has 12288 slices) and uses a SDRAM memory bus interface instead of the conventional PCI bus to reduce latency.

Fig. 6. Photograph of the Pilchard board.

A total of 4,032 PEs were places on an XCV1000E-6 device (this number was chosen for floorplanning reasons). As reported by the Xilinx timing analyzer, the maximum frequency was 202 MHz.

A number of commercial and research implementations of the Smith Waterman algorithm have been reported and their performance are summarized in Table 1. Examples are Splash [11], Splash 2 [12], SAMBA [13], Paracel [14], Celera [15], JBits from Xilinx [6], and the HokieGene Bioinformatics Project [7]. The performance measure of cell updates per second (CUPS) is widely used in the literature and hence adopted for our results.

Splash contains 746 PEs in a Xilinx XC3090 FPGA performing the Smith-Waterman Algorithm. Splash 2's hardware was different from Splash, which used XC4010 FPGAs with a total of 248 PEs. SAMBA [13] incorporated 16 Xilinx XC3090 FPGAs with 128 PEs altogether dedicated to the comparison of biological sequences.

ASIC and software implementations have also been reported. Paracel, Inc. used a custom ASIC approach to do the sequence alignment. Their system used 144 identical custom ASIC devices, each containing approximately 192 processing elements. Celera Genomics Inc. reported a software based system using an 800 node Compaq Alpha cluster.

Both the JBits and the HokieGene Bioinformatics Project were the latest reported sequence alignment systems using the Smith-Waterman Algorithm and use the same PE design. JBits reported performance for two different FPGA chips, the XCV1000-6 and the XC2V6000-5. The HokieGene Bioinformatics Project used an XCV6000-4. As can be seen from the table, the performance of the proposed design is similar to the JBits design on the same size FPGA (a XCV1000-6), and the JBits and HokieGene implementations on an XCV6000 gain performance by fitting more PEs on a chip, and our performance on the same chip would be similar.

The implementation was successfully verified using the Pilchard platform whcih provides a 133 MHz, 64-bit wide memory mapped bus to the FPGA. The processing elements and all other logic of the implementation operate from the same 133 MHz clock. The interface logic occupied 3% of the Virtex device. The working design was used mainly for verification performance and had a disappointing performance of approximately 136 B CUPS, limited by the simple polling based host interface used. A high speed interface which performs more buffering and is able to cause the memory system to perform block transfers between the host and Pilchard is under development.

Table 1. Performance and hardware size comparison of previous implementations (processor core not including system overheads). Device performance is measured in cell updates per second (CUPS).

System	Number of Chips	PEs per chip	System Performance (CUPS)	Device Performance (CUPS)	Run-time reconfiguration requirement
Splash(XC3090)	32	8	370 M	11 M	No
Splash 2(XC4010)	16	14	43 B	2,687 M	No
SAMBA(XC3090)	32	4	1,280 M	80 M	No
Paracel(ASIC)	144	192	276 B	1,900 M	N/A
Celera (software implementation)	800	1	250 B	312 M	N/A
JBits (XCV1000-6)	1	4,000	757 B	757 B	Yes
JBits (XC2V6000-5)	1	11,000	3,225 B	3,225 B	Yes
HokieGene (XC2V6000-4)	1	7000	1,260 B	1,260 B	Yes
This implementation (XCV1000-6)	1	4,032	742 B	742 B	No
This implementation (XCV1000E-6)	1	4,032	814 B	814 B	No

5   Conclusion

A technique, commonly used in VLSI layout, in which two processing elements are merged into a compact cell was used to develop a Smith-Waterman systolic processing element design which computes the edit distance between two strings. This cell occupies 3 Xilinx Virtex slices and allows both strings to be loaded into the system without runtime reconfiguration. Using this cell, 4032 PEs can fit on a Xilinx XCV1000E-6, operate at 202 MHz and achieve a device performance of 814 B CUPS.

References

1. Y. Miki, et. al. A Strong Candidate for the Breast and Ovarian Cancer Susceptibility Gene, BRCA1. *Science*, 266:66–71, 1994.
2. European Bioinformatics Institute Home Page, FASTA searching program, 2003. http://www.ebi.ac.uk/fasta33/.
3. National Center for Biotechnology Information. NCBI BLAST home page, 2003. http://www.ncbi.nlm.nih.gov/blast.

4. T. F. Smith and M. S. Watermann. Identification of common molecular subsequence. *Journal of Molecular Biology*, 147:196–197, 1981.
5. Sencel's search software, 2003. http://www.sencel.com.
6. Steven A. Guccione and Eric Keller.   Gene matching using JBits, 2002. http://www.ccm.ece.vt.edu/hokiegene/papers/GeneMatching.pdf.
7. K. Puttegowda, W. Worek, N. Pappas, A. Danapani and P. Athanas. A run-time reconfigurable system for gene-sequence searching. In *Proceedings of the International VLSI Design Conference*, page (to appear), Jan 2003.
8. P. Leong , M. Leong , O. Cheung , T. Tung , C. Kwok , M. Wong and K. H. Lee. Pilchard - a reconfigurable computing platform with memory slot interface. In *Proceedings of the IEEE Symposium on Field-Programmable Custom Computing Machines*, page (to appear), April 2001.
9. Richard J.Lipton and Daniel Lopresti. A systolic array for rapid string comparison. In *Proceedings of the Chapel Hill Conference on VLSI*, pages 363–376, 1985.
10. Xilinx. *The programmable logic data book*, 2003.
11. D. T. Hoang. A systolic array for the sequence alignment problem. *Brown University, Providence, RI, Technical Report*, pages CS–92–22, 1992.
12. D. T. Hoang. Searching genetic databases on splash 2. In *Proceedings 1993 IEEE Workshop on Field-Programmable Custom Computing Machines*, pages 185–192, 1993.
13. Dominique Lavenier. SAMBA: Systolic Accelerators for Molecular Biological Applications, March 1996.
14. Paracel, inc, 2003. http://www.paracel.com.
15. Celera genomics, inc, 2003. http://www.celera.com.

Software Decelerators

Eric Keller, Gordon Brebner, and Phil James-Roxby

Xilinx Research Labs, Xilinx Inc., U.S.A.,
{Eric.Keller,Gordon.Brebner,Phil.James-Roxby}@xilinx.com

Abstract. This paper introduces the notion of a *software decelerator*, to be used in logic-centric system architectures. Functions are offloaded from logic to a processor, accepting a speed penalty in order to derive overall system benefits in terms of improved resource use (e.g. reduced area or lower power consumption) and/or a more efficient design process. The background rationale for such a strategy is the increasing availability of embedded processors 'for free' in Platform FPGAs. A detailed case study of the concept is presented, involving the provision of a high-level technology-independent design methodology based upon a finite state machine model. This illustrates easier design and saving of logic resource, with timing performance still meeting necessary requirements.

1 Introduction

The research literature in field-programmable logic contains many examples of 'hardware accelerators'. In short, these concern a processor-centric system model: algorithms are executed on a processor, with certain key functions being performed by an associated programmable logic array, the intention being to achieve greater overall performance. The exact arrangements may vary from the tightly-integrated case of a processor with an augmented instruction set (e.g. [3]) to the more loosely-coupled case of a processor interacting with a co-processing logic device (e.g. [6]).

This paper is concerned with a *logic-centric* system model, of the sort described for example in earlier papers by Brebner [2]. Here, the main computational focus is on logic circuitry, with other components — in particular processors — viewed as additional system components rather than central system components. Aside from computational differences, there are implied architectural differences, notably concerning serial data access and shared buses, features both tailored to processor behavior. The logic-centric model is well-suited to systems that react to, and are driven by, inputs received over time. Thus, in contrast to the usual processor-centric model, the environment, not the computer, is the controlling force. We feel that this view of will be of increasing relevance to real-life systems in the future.

In the logic-centric model, it is fairly natural to invert accepted wisdom about system organization. In this paper, we consider the benefits of 'software decelerators', inverting the notion of hardware accelerators. The basic idea is that algorithms are executed in programmable logic, with certain functions being

P.Y.K. Cheung et al. (Eds.): FPL 2003, LNCS 2778, pp. 385–395, 2003.

performed by an associated processor. In general, this direction of migration is not likely to lead to speed increases — indeed quite the opposite — which is why we use the word 'decelerator'; however, overall system speed requirements will still be met. It is intended that there are other motivations that lead to an overall increase in the quality of the design process and/or the resulting systems. We discuss various possible motivations, and then describe one detailed case study experiment where the main motivation was to provide a high-level, technology-independent design methodology based upon finite state machines.

In this case study, the software decelerator technique concerns finite state machines being implemented on the embedded processor of a Platform FPGA. With this approach, the cost of implementing a machine in terms of logic resource usage is greatly reduced, since the processor is always present on the Platform FPGA, whether it is used or not. The emphasis differs from that of some conventional state machine design methodologies, such as Esterel Studio, the principal aims being to consume as few logic resources as possible, optimize the interfacing between logic and processor, and run code directly from the processor's built-in cache.

The paper is organized as follows. Section 2 considers the technological background that motivates software decelerators as a viable concept in system design. Then, Section 3 describes the case study, and Section 4 discusses initial experimental results. Finally, Section 5 contains some conclusions and directions for future work.

2   Technological Background

2.1   Emergence of Platform FPGAs

Early Field Programmable Gate Array (FPGA) architectures consisted of an array of similar programmable logic elements interfaced to interconnection elements. With the emergence of the Platform FPGA, architectures have evolved to a pre-defined mix of very different elements — which in time will largely supersede monolithic logic arrays. Today, for example, the Xilinx Virtex -II Pro Platform FPGA contains configurable logic blocks, I/O blocks supporting many different I/O standards, distributed Block RAM memories, internal access to the configuration memory, digital clock managers, gigabit transceivers, dedicated multiplier blocks and embedded PowerPC 405 processors. Platform FPGAs present both unique design challenges and unique design opportunities. Importantly, the mix of resources is pre-determined by the FPGA vendor rather than by the designer, and so a particular set of resources will be present whether or not the designer makes use of them, a situation unlike that in ASIC design.

Therefore, in designs that aim to maximize the use of the resources of a certain device size, the designer may often make decisions which on the surface appear non-intuitive. For example, in designs which use little BlockRAM memory but use many configurable logic blocks (CLBs), a designer may choose to implement certain logic functions by lookup tables in BlockRAM rather than in CLBs. This is despite the fact that this approach wastes much of the Block-RAM data width, and does not take advantage of their dual-port nature. Such an approach would be unthinkable in ASIC design, but makes a lot of sense in Platform FPGA design. This use of pre-existing resources in unusual ways is likely to become more and more prevalent as the mixture of resources becomes richer and richer. The overall message is that one has to be very fluid in terms of how and where computation, storage and communication are carried out in systems.

The logic-centric system paradigm is one approach to assist in managing the potential design space. This focuses on inputs, outputs and programmable logic circuitry as the core system architecture, with all other components (memory, processors, etc.) being seen as *assists* to the circuitry. Of course, it is also possible to view the system in other ways. For example, in a more conventional processor-centric manner, the Virtex-II Pro can also be seen as a 'motherboard on a chip', and indeed it has been demonstrated as such, with the Linux operating system running on the PowerPC processor.

2.2   Motivation for Software Decelerators

The concept of using software decelerators derives directly from the discussion of using Platform FPGA resources in unusual ways, and focuses on processors in particular for non-standard treatment. In fact, what is sought here is not necessarily a completely different and unusual harnessing of processors, rather some balance between the long and rich legacies of processor development and software engineering, and the new context of the logic-centric system.

The more general backdrop to this reconsideration of the role of processors is an examination of the role of any kind of *universal machine* within a logic-centric system. Other examples, simpler than a traditional microprocessor, would be programmable state machines or microcontrollers. In all such cases, there is a basic trade-off between speed — one normally expects a universal machine implementation of a function to be slower than a bespoke implementation — and other issues as diverse as chip area requirements and ease and speed of implementing new functions.

So far, in the development of Platform FPGAs that include processors, there has largely been a drive to maximize processor clock rates, reflecting a view either that the processor is the central system component or perhaps that the processor plays a key time-critical supporting role in the system. This drive parallels the continuing race to increase clock rates of microprocessor chips (although, very recently, there has been acknowledgement that raw speed is not everything, power consumption being of importance in an increasing proportion of systems). It is our belief that, as far as Platform FPGAs are concerned, the clock rate of the processor may not be the dominant concern for many future systems. This follows from a prediction that the processor may either be called upon relatively infrequently to carry out work in the logic-centric system, or may be called upon to perform work of a non time-critical nature. One recent example of such a system is the high throughput, low latency mixed-version IP router developed for the Virtex-II Pro [2].

As soon as the clock rate requirement is relaxed, it becomes possible to focus the treatment of a processor on other goals, like saving logic resource by employing the processor plus its supporting cache memory. In this paper, the goal is to combine this particular trade-off with the provision of a particular high-level design flow that hides the nature of the implementation from the user. A more straightforward and obvious application of software deceleration is just to make the overall system implementation process easier by allowing a designer access to the wide range of software tools (and human expertise) to implement functions on the processor, rather than implementing logic circuitry.

To ensure that software decelerators provide the anticipated benefits, and do not impose resource demands on an overall logic-centric system design, nor impose unnatural design methodologies on the user, there are some particular attributes that are desirable:

- The overall area consumed by a software decelerator implementation should not be greater than its logic circuitry counterpart unless there are other strong benefits.
- The interfacing between logic circuitry and processor should consume minimal programmable logic resources, and should be designed to shield the processor from the logic and vice-versa.
- The method of capturing designer intent should be independent of the actual implementation mechanism chosen for the Platform FPGA unless there is a particular benefit in permitting the use of certain familiar tools.
- The designer should be able to get accurate timing information, and resource usage information in general, for the overall logic-centric system, taking into account the behavior of the various non-logic system components that are being harnessed.

The case study that is presented in the subsequent sections illustrates that it is feasible to meet these goals, in the framework of software decelerator use.

3   Case Study: FSM-Based Design Methodology

Finite state machines (FSMs) are an important component of many digital systems, and can be implemented well on FPGAs. Particular FSMs may contain a large number of states, and may involve much computation to determine the next state and the state outputs based on varying inputs. However, they may actually have relatively relaxed timing constraints compared to the rest of a system, an attribute that points to possible software decelerator implementation.

The case study problem tackled was to implement FSMs as an example of a software decelerator. Referring back to three desirable attributes defined in Section 2.2, the interfacing hardware should consume minimal resources and act as a shield between the processor and the rest of the system — the rest of the system should not know it is working with a processor. Capture should be implementation independent, that is, the designer should not be aware that an embedded processor is being targeted. Finally, accurate timing and resource usage information should be obtainable so that the designer is able to get hardware-like metrics. The net effect of using a software decelerator in a system where the processor would otherwise be idle is that logic resources are freed, without penalizing the designer in terms of design style or quality of timing information.

There has been a fairly substantial body of prior work on implementing finite state machines in software. Perhaps the most notable effort is the Berekeley POLIS system [1]. POLIS is a complete co-design solution, which uses the codesign finite state machine (CFSM) as the central representation of the required system behavior. One significant difference between a CFSM and a classical FSM is that a CFSM allows that the reaction time to events will be non-zero, unlike the FSM which assumes a synchronous communication model.

POLIS allows the designer to implement CFSMs in either software or hardware, and since this is a co-design solution — a single CFSM can be partitioned into multiple CFSM sub-networks, and have different target implementations. The hardware CFSM sub-networks are constructed using standard logic synthesis techniques, and in this case a CFSM can execute a transition in a single clock cycle.

A CFSM sub-network chosen for software implementation is transformed to a program and a simple custom real time operating system (RTOS). The program is generated from a control/data flow graph, and is coded in C. The designer can use a timing estimator to find quickly the speed of the software, or instead produce an instrumented version of the code and run this on an actual processor. The instrumented version counts the actual cycles used, giving a more accurate way of extracting timing information. The custom RTOS consists of a scheduler for organizing the execution of the procedures, using policies such as rate-monotonic or deadline-monotonic scheduling. The RTOS also includes I/O drivers. The case study here is rather different in nature from POLIS, as it has very different aims, including minimizing logic-processor interfacing and code size.

3.1 System Description

The case study involved producing a tool which takes a textual representation of a finite state machine and produces: a hardware platform that can be interfaced to existing logic circuitry; software to run on the embedded processor; and a timing report. The tool flow is shown in Figure 1.

3.2 Design Entry

A number of methods are available for capturing FSMs. Tools such as Esterel Studio or Xilinx's StateCAD allow a designer to capture graphically the FSM as a state transition diagram. Designers annotate the state transition diagram with conditions for taking branches, and define the calculations for the state outputs.

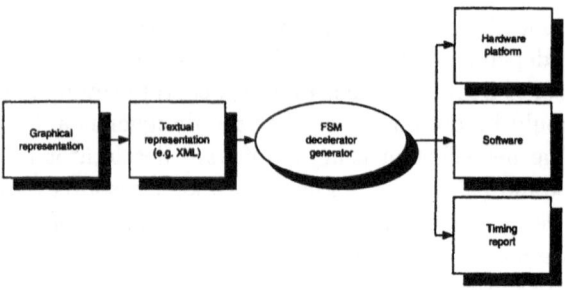

Fig. 1. Tool flow

Alternatively, FSMs can be described in a conventional HDL. In order to support the maximum number of possible design methods, an XML grammar was defined to capture the functionality of an FSM. In a file containing a description following this grammar, the interface of the machine is specified first. For example:

```
⟨variables⟩
  ⟨variable name="rst" dir="in" width="1" registered="true"/⟩
  ⟨variable name="clk" dir="in" width="1" registered="true" /⟩
  ⟨variable name="phy_ad" dir="in" width="5" registered="true" /⟩
  ⟨variable name="mdio_tristate" dir="out" width="1" registered="true" /⟩
⟨/variables⟩
```

After this, the description specifies the states. An initial tag specifies the global conditions for the state machine, such as reset input, clock input, reset state, and synchronous or asynchronous reset. A description of the individual states follows this. A state has equations and transitions associated with it. Each equation assigns some value to an output. Input, constants and basic operators (add, sub, and, or, etc.) are used to form the right-hand side of the equation. Transitions include the next state and the condition when the transition occurs. Equations can also be associated with transitions. The time when the equation is executed depends on where the equation is located — in the state or in the transition.

```
⟨state name="stateADD"⟩
  ⟨equations⟩ ⟨equation lhs="out0" rhs="in1 + in2" /⟩ ⟨/equations⟩
  ⟨transitions⟩
    ⟨transition condition="else" next="state1"⟩
      ⟨equations⟩ ⟨equation lhs="ready" rhs="1" /⟩ ⟨/equations⟩
    ⟨/transition⟩
  ⟨/transitions⟩
⟨/state⟩
```

3.3   Logic-Processor Interfacing

In a conventional SoC design containing processors, the processor is normally connected to the rest of the logic via a system bus or other on-chip network [5].

The processor is a master on the network, initiating and responding to transfers. Since multiple masters may be present in a bus-based system, an arbiter is needed, and masters must first request the bus from it. In the case of an SoC implementation using Virtex-II Pro Platform FPGAs, the logic required to generate the arbiter, the bus itself and the bus interfaces on each of the slaves is implemented in the fabric of the FPGA.

Examining the interface of the PowerPC to the logic fabric in the Virtex-II Pro, there is some flexibility in choosing a method of transferring information between the processor and the logic, and vice-versa. In essence though, for all methods, communication is done over a bus at the processor/logic boundary. Three data buses are natively interfaced to the PowerPC: the On-Chip Memory (OCM) bus, and the Device Control Register (DCR) and Processor Local Bus (PLB) buses from the CoreConnect family of SoC interconnect.

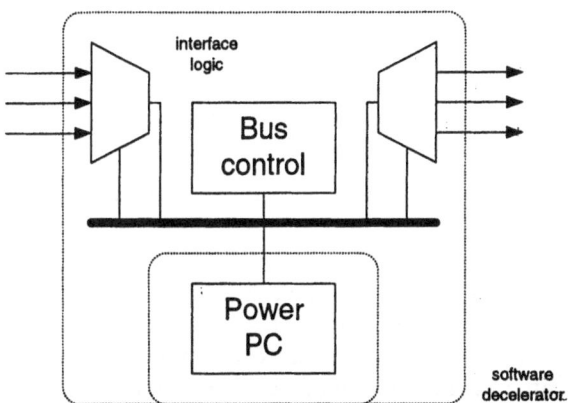

Fig. 2. Software decelerator architecture

The logic design task in the tool development was to take the XML description of the FSM, and produce the interfacing harness as illustrated in Figure 2. The role of this harness is to shield the software decelerator, so in effect the FSM code running on the processor looks like an FSM implemented in logic — that is, the rest of the system should not see any processor specific signals. Moreover, the philosophy is to allow logic to communicate with software using the minimum amount of interfacing.

To interface with the embedded PowerPC processor, it is necessary to use one of the three native buses mentioned above. Simplifications to the interface logic can be made by relying on the fact that only one master (the processor) is present and that it only talks to one slave (the logic circuitry). In each case, the slave can assume it is being addressed all the time, and can simply write outputs and read inputs directly to and from the data bus. For the DCR and the PLB buses, the master interface in the processor assumes the existence of an acknowledge signal, but the logic to do this is very simple.

State machines are normally driven by a clock that dictates when the machine should move between states. For software decelerators in general, there will not be a relationship between the clocks of the rest of the system and the processor. For the state machine and general decelerators, there are two methods for dealing with clocks.

The first method is to use the clock for the state machine directly, as if it were a normal input. Since the processor operates at a much higher frequency than the rest of the system, it is possible for the processor to poll the clock input, and begin processing when it detects a rising edge. The limitation with this method for finite state machines is that the worse case state execution dictates whether the system will meet the timing requirements or not. Figure 3 shows a simplified timing diagram using this method.

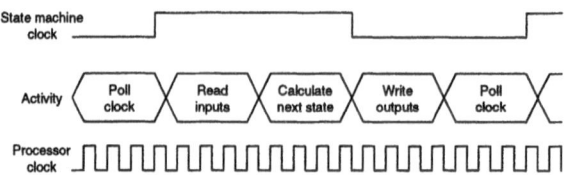

Fig. 3. Simplified timing diagram

A second method is to generate a clock pulse from the processor itself using a memory-mapped one-shot circuit. In this case, states would be allowed to take a different number of processor cycles to complete — the clock pulses would simply appear after a different number of processor cycles, but the external circuits would know when their inputs and outputs have been clocked.

3.4   Timing Generation

The ability to get accurate timing information is crucial to the success of the software decelerator technique. The designer needs to be confident that the overall system, including the decelerator, meets overall timing constraints. To do this, a measure of the delay through the software decelerator is required, in exactly the same way as in hardware design, where the delay through logical elements is required.

Li and Malik present a good discussion of the state of the art of determining the worst-case execution time for software [4]. Similar techniques would be needed for general software decelerator use, where arbitrary code structures are permissible. However, in the case study, the structure of the generated software is strictly under the control of the tool. Therefore, given that the execution times for each instruction type are documented, the tool can count the actual number of processor cycles. Since the software runs out of cache and the time to perform bus transfers to the rest of the system is known, this cycle count is very accurate. This is different to the approach used by POLIS, which generates execution characteristics by instrumenting and running code.

3.5   Software Design

Section 2.2 stated that interfacing should consume minimal resources, to make the software decelerator a value proposition to the designer. Similarly, other support for the processor (e.g. memory and clock control) should consume minimal resources. In the case of the PowerPC, it is possible to reduce external memory requirement to zero by using the instruction and data caches as main memory. This means that the whole executable needs to fit inside the 16Kb instruction cache. In the F SM case under consideration, where no lavish software support is required, extremely complex state machines would still fit inside the 16Kb limit.

It was decided to use assembly code directly for the software implementation. This had two specific advantages. The first advantage is that using assembler simplifies the problem of extracting timing information as described in Section 3.4. The other is that the PowerPC instructions *mfdcr* and *mtdcr*, which move data from and to the DCR bus respectively, can be used if the DCR bus is chosen as the interface. These instructions move data between a specified general-purpose register and the DCR, and thus are difficult to deal with from compiled high-level languages.

Every state uses the same template to create output assembly code. The most complex task is the translation to assembly code of the equations to determine the next state and the state outputs. Inputs are only read if they are used in these equations. The conditions for translations use the same method to calculate the value of the condition. Special care was taken to maximize the usage of registers and only use the cache memory if needed. This can lead to more efficient code — a useful feature since the state machine is already operating at a much lower frequency than the FPGA fabric.

4   Experimental Results

The tool was used on three state machines from the networking domain, with very different performance and I/O requirements. In each case, the clock is supplied by the system's environment. The first state machine (rs232echo) was an RS232 protocol handling machine, which echoed received inputs onto outputs, and the second one (miim) handled the Media Independent Interface (MII) of an Ethernet MAC which runs at 2.5MHz. The third state machine (tx_host_io) handles the host interface to a 10G Ethernet MAC. This machine clearly has a need for high performance, and is included here as an illustration of a case where a software decelerator is not likely to be chosen as the implementation technique. The first table shows the results for a direct logic circuit implementation from synthesized VHDL.

Machine	Input width	Output width	Number of states	Registers	LUTs	Required frequency
rs232echo	3	1	12	92	111	115 kHz
miim	33	20	33	26	61	2.5 MHz
tx_host_io	90	94	5	142	320	156 MHz

To determine the possible real estate savings through using a software decelerator, the new tool was run targeting each of the three available buses. The resource usage and the relative savings in terms of LUTs used compared with the direct logic implementation for each of the buses is shown in the next table.

Machine	OCM			DCR			PLB		
	Registers	LUTs	Ratio	Registers	LUTs	Ratio	Registers	LUTs	Ratio
rs232echo	1	4	3.6%	2	6	5.4%	4	8	7.2%
miim	20	38	62.3%	21	40	65.6%	23	42	68.9%
tx_host_io	94	75	23.4%	95	77	24.1%	97	79	24.7%

The OCM-based implementations are the smallest of the three for each example. Also, the OCM is as fast as the PLB bus for processor/logic interaction — in both cases they take four processor cycles, as opposed to the DCR which takes nine processor cycles. These figures assume the system bus operates at half the frequency of the PowerPC core, that is the PowerPC operates at 350MHz, and the bus clock runs at 175 MHz. In order to determine timing figures, each of these machines was implemented using the new tool, targeting the OCM bus. The final table shows the results for each of the example machines.

Machine	Worst-case performance (cycles)	Worst-case performance (MHz)	% of time in I/O	Code size (kbytes)	Code size as % of cache
rs232echo	40	8.75	30.95%	1416	8.6%
miim	74	4.730	25.22%	2968	18.1%
tx_host_io	135	2.593	33.99%	1952	11.9%

The rs232echo would work at all required baud rates, and the performance of the miim state machine is well inside the required 2.5MHz limit. These examples illustrate a software decelerator delivering the required performance, rather than an unnecessarily high performance. The tx_host_io state machine however, and as expected, does not operate at anything like the required frequency, and is an example of a case where high performance is very much the key focus. In each case, the code only occupies a fraction of the cache. Thus, it would be possible to run multiple state machines, that is multiple software decelerators, on the same processor, as long as any cumulative timing requirements are still be met.

It can be seen that I/O occupies a large proportion of time in each machine, and is clearly a limiting factor in the machine speeds that can be achieved. In the case of state machines, it may be possible to exploit the rich routing resources of the FPGA, and introduce parallel transfers by packing the inputs required for each state into a single word. This would require adjustments to the input multiplexor shown in Figure 2. The parallelized input signals could then be unpacked inside the processor by shift and masks or alternatively, instructions could be applied directly to the packed signals.

5 Conclusions and Future Work

This research has introduced software decelerators as a mechanism for harnessing the resources of an embedded processor in a Platform FPGA, making the point that maximizing raw processor speed is not likely to be an issue in many logic-centric systems. An application of this philosophy has been demonstrated through the FSM-based design methodology. This experiment shows encouraging initial results in terms of overall resource usage and ease of design. The authors are now evaluating the methodology on larger Xilinx customer designs containing multiple state machines. Future work foci will include: further study of the implications of adopting a logic-centric system model; automatic selection and synthesis of apt logic-processor interfaces; characteristics of soft and hard embedded processors; FSM-based architectural components; and the provision of domain-specific high-level design entry and tools.

References

1. F. Baloron, P. Giusto, A. Jurecska, C. Passerone, E. Sentovich, M. Chiodo, H. Hsieh, L. Lavagno, A. L. Sangiovanni-Vincentelli, and K. Suzuki. *Hardware-Software co-design of embedded systems: the POLIS approach.* Kluwer, 1997.
2. Gordon Brebner. Single-chip Gigabit mixed-version IP router on Virtex-II Pro. *IEEE Symposium on FPGAs for Custom Computing Machines (FCCM02)*, pp.35–44, April 2002.
3. M. Dales. The Proteus processor - a conventional CPU with reconfigurable functionality. *9th Int. Workshop on Field Programmable Logic (FPL99)*, pp.431–437, September 1999.
4. Y. S. Li and S. Malik. Performance analysis of embedded software using implicit path enumeration. *ACM/IEEE Design Automation Conference (DAC95)*, pp.456–461, June 1995.
5. M. Sgroi, M. Sheets, A. Mihal, K. Keutzer, S. Malik, J. Rabaey, and A. Sangiovanni-Vincentelli. Addressing the system-on-a-chip interconnect woes through communication-based design. *ACM/IEEE Design Automation Conference (DAC01)*, pp.667–672, June 2001.
6. S. Singh and R. Slous. Accelerating Adobe Photoshop with reconfigurable logic. *IEEE Symposium on FPGAs for Custom Computing Machines (FCCM98)*, pp.15–17, April 1998.

A Unified Codesign Run-Time Environment for the UltraSONIC Reconfigurable Computer

Theerayod Wiangtong[1], Peter Y.K. Cheung[1], and Wayne Luk[2]

[1] Department of Electrical & Electronic Engineering, Imperial College, London, UK,
{tw1, p.cheung}@imperial.ac.uk
[2] Department of Computing,Imperial College, London, UK,
wl@imperial.ac.uk

Abstract. This paper presents a codesign environment for the Ultra-SONIC reconfigurable computing platform which is designed specifically for real-time video applications. A codesign environment with automatic partitioning and scheduling between a host microprocessor and a number of reconfigurable processors is described. A unified runtime environment for both hardware and software tasks under the control of a task manager is proposed. The practicality of our system is demonstrated with an FFT application.

1 Introduction

Reconfigurable hardware has received increasing attention from the research community in the last decade. FPGA-based designs become popular because of their reconfigurable capability and short design-time which the old design style, like ASICs, cannot offer. Instead of using FPGAs simply as ASICs replacements, combining reconfigurable hardware with conventional microprocessors in a codesign system provides an even more flexible and powerful approach for implementing computation intensive applications, and this type of codesign system is our attention in this paper.

The major concerns in the design process for such codesign system are the synchronization and the integration of the hardware and the software design [1]. Examples are the partitioning between hardware and software, the scheduling of tasks and the communication between hardware and software tasks in order to achieve the shortest overall runtime. Decision on partitioning plays a major role in the overall system performance and cost. Scheduling is important to complete all the tasks in a real-time environment. These decisions are based heavily on design experience and are difficult to automate. Many design tools leave this burden to the designer by providing semi-auto interactive design environments. Fully automatic design approach for codesign system is generally considered to be impossible at present. In addition, traditional implementation employs a hardware model that is very different from that used in software. These distinctive views of hardware and software tasks can cause problem in the design process. For example, swapping tasks between hardware and software can result in a totally new structure in the control circuit.

P.Y.K. Cheung et al. (Eds.): FPL 2003, LNCS 2778, pp. 396–405, 2003.

This paper reports a new method of constructing and handling tasks in a codesign system. We structure both hardware and software tasks in an interchangeable way without sacrificing the benefit of concurrency found in conventional hardware implementations. At the same time, the hardware task model exploits the advantages of modularity, scalability, cohesion and structured approach offered by software tasks. We further present a codesign flow containing the partitioning and scheduling algorithms to automate the decision process of where and when tasks are implemented and run. Our codesign system using this unified hardware/software task model is applied to a reconfigurable computer system known as UltraSONIC [14]. Finally we demonstrate the practicality of our system by implementing an FFT computational engine. The novel contributions of this paper are: 1) a unified way of structuring and modelling hardware and software tasks in a codesign system; 2) proposing a codesign flow for a system with mixed programmable hardware and microprocessor resource; 3) propose a run-time task manager design to exploit the unified model of tasks; 4) the demonstration of implementing a real application by using our model in the UltraSONIC reconfigurable platform.

The rest of this paper is organized as follows. Section 2 presents some of work related to codesign development systems. In section 3, we explain how the hardware and software tasks are modelled in our codesign system. Section 4 describes the codesign flow and environments of the UltraSONIC system used as the realistic target. A case study of implementing the FFT algorithm on the UltraSONIC system is discussed in section 5. Section 6 concludes this paper.

2   Related Work

There are many approaches to hardware/software codesign. Most focus on some particular stages in the design process such as system specification and modelling, partitioning and scheduling, compilation, system co-verification, cosimulation, and code generator for hardware and software interfacing. For example, Ptolemy [2] concentrates on hardware-software co-simulation and system modelling. Chinook [5] focuses on the synthesis of hardware and software interface. MUSIC [6], a multi-language design tools between SDL and MATLAB, is applied to mechatronic system. For embedded systems, CASTLE [7] and COSYMA [8] design environments are developed.

Codesign system specifically targeted for reconfigurable systems are PAM-Blox [4] and DEFACTO [3]. PAM-Blox is a design environment focusing on hardware synthesis to support PCI Pamette board that consists of five XC4000 Xilinx FPGAs. This design framework does not provide a complete codesign process. DEFACTO is an end-to-end design environment for developing applications mapped to configurable platform consists of FPGAs and a general-purpose microprocessor. DEFACTO concentrates on raising the level of abstraction to a higher level, and develop the parallelizing compiler technique to achieve optimizations on loop transformations and memory accesses. Although both PAM-Blox and DEFACTO are developed specifically for reconfigurable platforms, they take no account of the existence of tasks in microprocessor. Consequentially, neither system is suitable for hardware/software codesign.

In this work, tasks can interchangeably be implemented in software or reconfigurable hardware resources. We suggest a novel way for task modelling and task management which will be described in details in the next section. We also present a novel idea of building infrastructure for dataflow-based applications implementing on this type of codesign system consisting of a single software resource (in the form of a microprocessor) and multiple reconfigurable hardware. Our approach is inherently modular and is suitable for implementing runtime reconfigurable designs.

3   System Architecture

Fig. 1 shows the hardware/software system model adopted in this work. We assume the use of a single processor (software) resource SW capable of multitasking, and a number of concurrent hardware processing elements PE0 to PEn, which are implemented on FPGAs. We employ a loosely coupled model with each processing element (PE) having its own single local memory. All system constraints such as shared resource conflicts, reconfiguration times (of the FPGAs) and communication times are all taken into account.

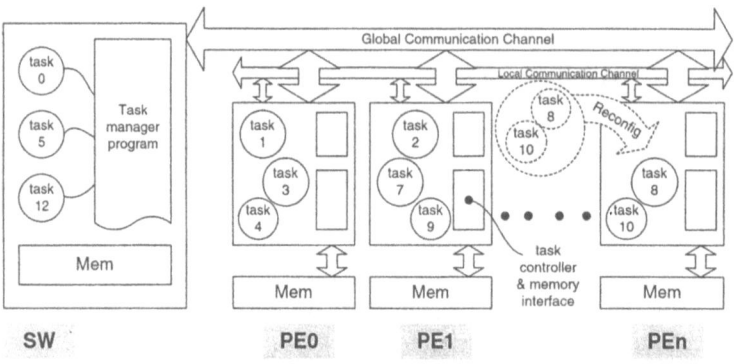

Fig. 1. Codesign System Architecture

The assumptions used in our model are: 1) tasks implemented in each hardware PE are coarse grain tasks which may consist of one or more functional tasks (blocks, loops). 2) Each PE has one local memory, only one task can access the local memory at any given time. Therefore multiple tasks residing in a given PE must execute sequentially; however, tasks residing across different PEs can execute concurrently. 3) Tasks for a PE may be dynamically swapped in and out through dynamic reconfiguration. 4) A global communication channel is available for the processor and the PEs to communicate with each other. 5) Local communication channels are available for neighboring PEs to communicate with each other in a pipeline ring.

Because of the reconfigurable capability of the hardware, we can build the hardware tasks very much like software tasks. In this way, the management of task scheduling, task swapping and task allocation can be done in a unified manner, no matter whether the task in question is implemented in hardware or in software. Concurrency is not affected as long as we map concurrent tasks onto separate PEs. Although conceptually different PEs are separate from each other, multiple PEs may be implemented on a single FPGA device.

3.1 Hardware Task Model

UltraSONIC is a reconfigurable computer system designed specifically for real-time video processing applications. In such an application domain, it is reasonable to assume that applications are dominated by dataflow behavior with few control flow constructs[12]. Algorithms can be broken down into tasks in coarse (or functional) granularity and are represented as a directed acyclic graph (DAG). Nodes in the graph represent tasks and edges represent data dependency between tasks. Each task is characterized by its execution time, resource occupied, and its communication cost with other tasks.

The tasks we implement on our system are assumed to conform the following restrictions:

- Tasks in the DAG are processed (once for each task) from top to bottom according to their precedence levels and priorities.
- Communication between tasks is always through local single-port memory.
- Task execution is done in three consecutive steps: read input data, process the data, and write the results. This is done repeatedly until input data stored in memory are all processed. Thus the communication time between memory and task while executing is considered to be a part of the task execution time [11].
- Exactly one task in a given PE is active at any one time. This is a direct consequence of the single port memory restriction. However, multiple tasks may run concurrently provided that they are mapped to different PEs. This is an improvement over the model proposed by others in [9].
- A task starts executing as soon as all the necessary incoming data from its sources have arrived. It starts writing outgoing data to destinations immediately after processing is completed [10].

4 The Design Environment

Fig. 2 depicts the codesign environment in our system. The system to be implemented is assumed to be described in some suitable high level language, which is then mapped to a DAG. Tasks are represented by nodes and communications by edges. The nodes are then partitioned into hardware or software tasks, and are

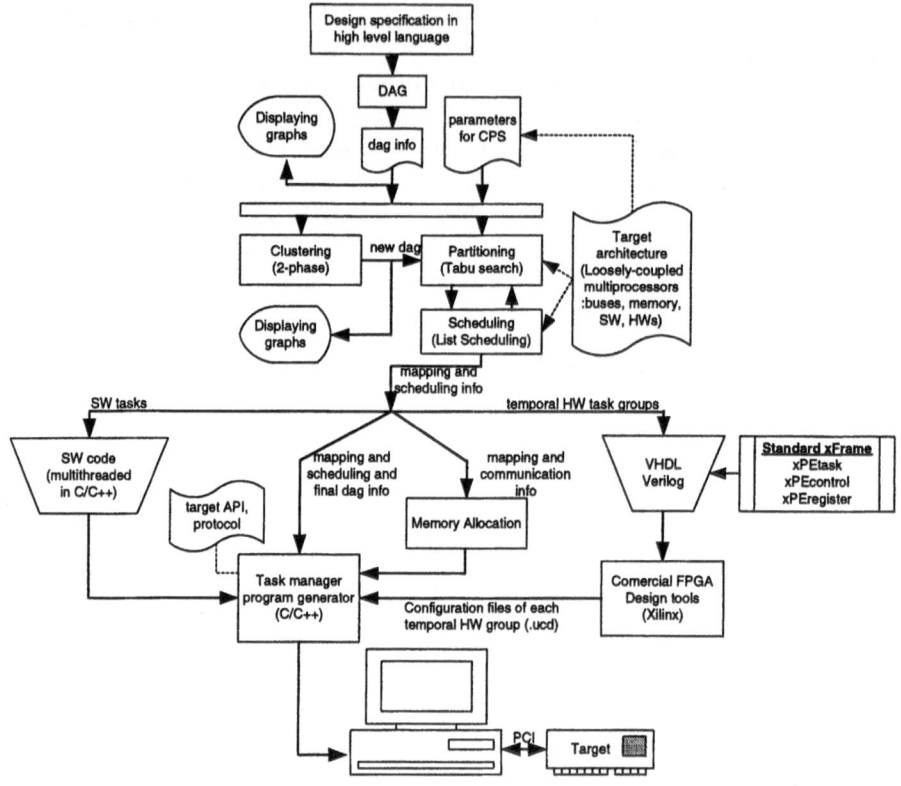

Fig. 2. Codesign environment

scheduled to execute in order to obtain the minimum makespan (total processing time). The partitioning and scheduling software used is adapted from tabu
search and list scheduling algorithms reported earlier by the authors [13]. The algorithms are, however, modified to be compatible with this architectural model.
A two-phase clustering algorithm is used as a pre-processing step to modify the
granularity of the tasks in the DAG in order to improve the critical path delay,
enable more task parallelism and provide results the achieve 13%-17% shorter
makespan [15].

   This group of algorithms, containing clustering, partitioning and scheduling,
is collectively called the *CPS* algorithm for short. The *CPS* algorithm reads
textual input files including DAG information and control information for clustering, partitioning and scheduling process. During this input stage, the user can
optionally specify the type of tasks as *software-only* task, *hardware-only* task,
or *dummy* task. A software-only task is a task that the user intentionally implements in software without exception, and similarly for a hardware-only task.
Dummy tasks are either source or sink for inputting and outputting data respectively, and are not involved in any computation. In our system, we assume that
input and output data are initially provided and written to the microprocessor

memory. Unspecified tasks are then free to be partitioned, scheduled and bound to either hardware or software resources.

The result of the partitioning and scheduling process are the physical and temporal binding for each task. Note that in case of hardware tasks, they may be divided into many temporal groups that can either be statically mapped to the hardware resource, or dynamically configured during runtime.

We currently assume that software tasks are manually written in C/C++, while hardware tasks are designed manually in a HDL (such as Verilog) using a library-based approach. Once all the hardware tasks for a given PE are available, they are wrapped in a standard frame with a pre-designed circuit (xPEtask, xPEregister and xPEcontrol) which is task independent. Commercially available synthesis and place-and-route tools are then used to produce the final configuration files for each hardware. Each task in this implementation method requires some hardware overhead to implement the task frame wrapper circuit. Therefore our system favours partitioning algorithms that generate coarse grain tasks.

The results from the partitioning and scheduling process, the memory allocator, the task control protocol, the API functions, the configuration files of hardware tasks, are used to automatically generate the codes for a *task manager program* that controls all operations in this system, such as dynamic configuration, task execution and data transfer. The resulting task manager is inherently multi-threaded to ensure that tasks are run concurrently.

4.1   The Task Manager Program

The center of our implementation is the task manager (TM) program running on the microprocessor (software) resource to manage both hardware and software tasks. This program controls the sequencing of all the tasks, the transfer of data and the synchronization between them, and the dynamic reconfiguration of the FPGA in the PEs when required. Fig. 3 shows the conceptual control view of the TM and its operation. The TM communicates with a local task controller on each PE in order to assert control. A message board is used in each PE to receive commands from the TM or to flag finishing status to the TM.

In order to properly synchronize the execution of the tasks and the communication between tasks, our task manager employs a message-passing, event-triggered protocol when running a process. However, unlike a reactive codesign system [16], we do not use external real-time events as triggers. Instead, we use the termination of each task or data transfer as event-triggers, and signaling of such events is done through dedicated registers. For example, in Fig. 3, messages indicating execution completion from tasks 1 are posted to registers inside PE0. The task manager program polls these registers, finds the message, then proceeds to the next scheduled task, in this case task 3. By using this method, tasks on each PE is run independently because the program operates asynchronously at the system level.

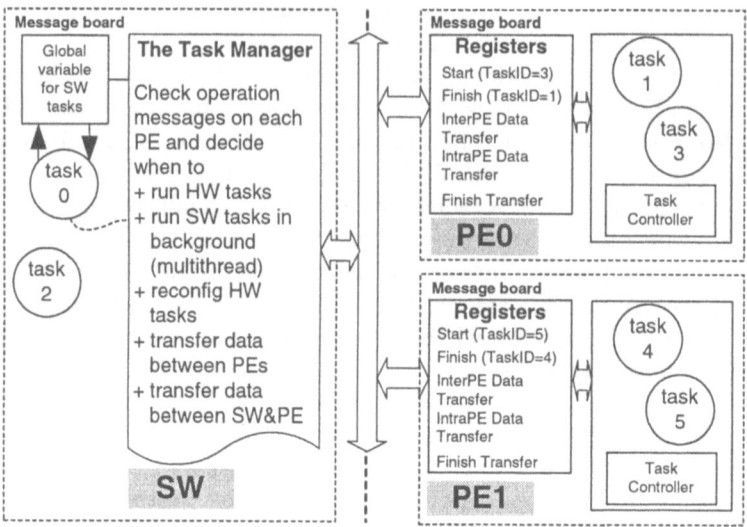

Fig. 3. The Task Manager control view

4.2   The UltraSONIC Reconfigurable Platform

The codesign described above is targetted for the UltraSONIC System [14]. UltraSONIC (see Fig. 4(a)) is a reconfigurable computing system designed to cope with the computational power and the high data throughput demanded by real-time video applications. The system consists of Plug-In Processing Elements (PIPEs) interconnected by local and global buses. The architecture exploits the spatial and the temporal parallelism in video processing algorithms. It also facilitates design reuse and supports the software plug-in methodology.

Fig. 4(b) shows how our codesign model is implemented in the UltraSONIC PIPE. The xPEcontrol implements the message passing protocol to control the operation of all the hardware tasks (xPEtask) resident in this PIPE. The message board is implemented in xPEregister. The total hardware overhead of using the Task Manager is modest. It consumes around 10% of the reconfigurable resource on each PIPE (which is implemented on Xilinx's XCV1000E).

5   An Example: FFT Implementation

In order to demonstrate the working of our system, we chose to implement the well known FFT algorithm. Although we can implement the algorithm for an arbitrary data length, we use an 8-point FFT implementation to illustrate the results of our codesign system. The DAG of an 8-point FFT can be straightforwardly extracted as shown in Fig. 5(a). Nodes 0, 1, 2 are used for arranging inputs data and are implemented as software-only tasks. Tasks 3 to 14 are butterfly computation nodes. The number shown inside the parenthesis (the second

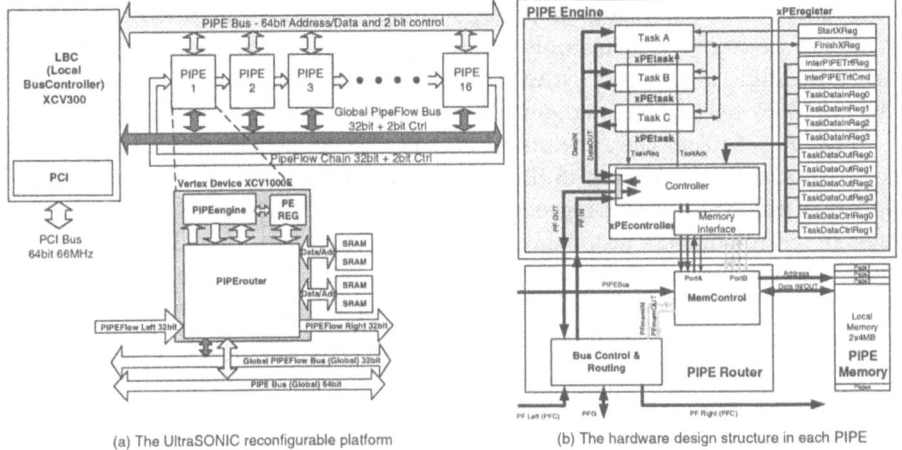

(a) The UltraSONIC reconfigurable platform   (b) The hardware design structure in each PIPE

Fig. 4. The hardware implementation in UltraSONIC architecture

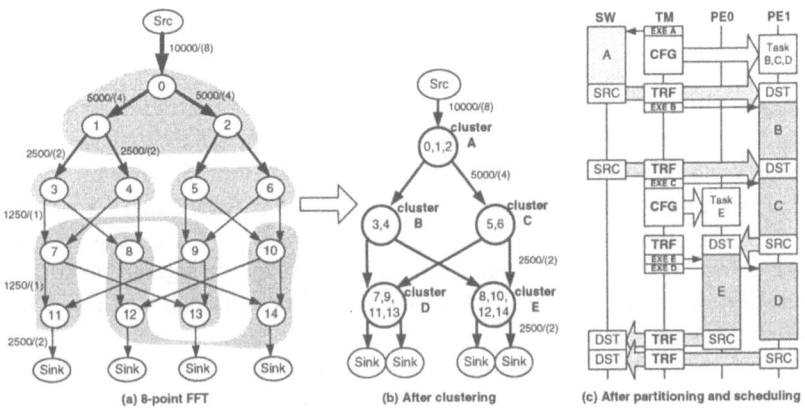

(a) 8-point FFT   (b) After clustering   (c) After partitioning and scheduling

Fig. 5. The DAG of 8-point FFT algorithm

number) on each edge is the number of data values needed for each iteration of the task. Each task is executed repeatedly until all data (shown as the first number on the edge) are processed. Initially in this DAG, the software execution times are obtained by profiling tasks on PC, while the hardware times and areas are obtained by using Xilinx development tools.

This DAG information is supplied to the CPS algorithm in our design environment (see Fig. 2) to perform automatic clustering, partitioning and scheduling. Parameters such as reconfiguration time, bus speed, FPGA size, are all based on the UltraSONIC system. The clustering algorithm produces a new DAG that contains tasks in higher granularity as shown in Fig. 5(b). These new tasks are then iteratively partitioned and scheduled. The computational part of each of the hardware tasks are designed manually in Verilog and the software tasks are written in C. Our tools then combine the results from the partitioning and scheduling algorithms, wrap the hardware task designs automatically with the standard task frame, and generate the task manager program.

Fig. 5(c) depicts the run-time profile of this implementation. Each column represents activities on the available resources which are software tasks (SW), and two hardware processing elements (PE0 and PE1). TM is the task manager program which is also running on the software resource. The execution of this algorithm proceeds from top to bottom. It shows all the runtime activities including configuration (CFG), transferring data (TRF), events that trigger task executions (EXE), the source of data (SRC), receiving data (DST) and the executions of the tasks (A to E).

The FFT algorithm for different data window sizes are also tested on the UltraSONIC system and are shown to work correctly. The method, although requires some manual design steps, is very quick. Implementing the FFT algorithm only took a few hours from specification to completion.

6   Conclusions

This paper presents a semi-automatic codesign environment for a system consisting of single software and multiple reconfigurable hardware. It proposes the use of a task manager to combine the runtime support for hardware and software in order to improve modularity and scalability of the design. Partitioning and scheduling are done automatically. Codes for software tasks are run in software concurrently (using multi-threaded programming) with the task manager program which is based on message-passing and event-triggered protocol. Implementation of the FFT algorithm on UltraSONIC demonstrates the practicality of our approach.

Future work includes testing our codesign system with more complex applications, tools to map behavioral or structural descriptions to DAG automatically and to improve the task management environment so that external asynchronous real-time events can also be handled.

Acknowledgement

The authors would like to acknowledge the continuing support of John Stone, Simon Haynes, Henry Epsom and the rest of the UltraSONIC team at Sony Broadcast Professional Research Laboratory in UK.

References

1. Ernst, R., "Codesign of embedded systems: status and trends", *IEEE Design & Test of Computers*, 1998.
2. Manikutty, G.; Hanson, H., "Hardware/Software Partitioning of Synchronous Dataflow Graphs in the ACS domain of Ptolemy", University of Texas, Literature Survey, Final Report May 12 1999.

3. Hall, M.; Diniz, P.; Bondalapati, K.; Ziegler, H.; et al., "DEFACTO:A Design Environment for Adaptive Computing Technology", *Proceedings of the 6th Reconfigurable Architectures Workshop*, 1999.
4. Mencer, O.; Morf, M.; Flynn, M.J., "PAM-Blox: high performance FPGA design for adaptive computing", *FPGAs for Custom Computing Machines*, 1998.
5. Chou, P.H.; Ortega, R.B.; Borriello, G., "The Chinook hardware/software co-synthesis system", *System Synthesis*, 1995.
6. Coste, P.; Hessel, F.; Le Marrec, P.; Sugar, Z.; et al., "Multilanguage design of heterogeneous systems", *Hardware/Software Codesign*, 1999.
7. Wilberg, J.; Kuth, A.; Camposano, R.; Rosenstiel, W.; et al., "Design Exploration in CASTLE", *Workshop on High Level Synthesis Algorithms Tools and Design (HILES)*, 1995.
8. Ernst, R., "Hardware/Software Co-Design of Embedded Systems", *Asia Pacific Conference on Computer Hardware Description Languages*, 1997.
9. Srinivasan, V.; Govindarajan, S.; Vemuri, R., "Fine-grained and coarse-grained behavioral partitioning with effective utilization of memory and design space exploration for multi-FPGA architectures", *IEEE Transactions on Very Large Scale Integration (VLSI) Systems*, vol. 9, pp. 140 -158, 2001.
10. Hou, J.; Wolf, W., "Process partitioning for distributed embedded systems", *Hardware/Software Co-Design*, 1996.
11. Pop, T.; Eles, P.; Peng, Z., "Holistic scheduling and analysis of mixed time/event-triggered distributed embedded systems", *Hardware/Softwarw Codesign*, 2002.
12. Chatha, K.S.; Vemuri, R., "Hardware-software partitioning and pipelined scheduling of transformative applications", *IEEE Transactions on Very Large Scale Integration (VLSI) Systems*, vol. 10, pp. 193-208, 2002.
13. Wiangtong, T.; Cheung, P.Y.K.; Luk, W., "Comparing Three Heuristic Search Methods for Functional Partitioning in HW-SW Codesign", *International Journal on Design Automation for Embedded Systems*, vol. 6, pp. 425-449, July 2002.
14. Haynes, S.D.; others, a., "UltraSONIC: A Reconfigurable Architecture for Video Image Processing", *Field-Programmable Logic and Applications (FPL)*, 2002.
15. Wiangtong, T.; Cheung, P.Y.K.; Luk, W., "Cluster-Driven Hardware/Software Partitioning and Scheduling Approach For a Reconfigurable Computer System", submit to *Field-Programmable Logic and Applications (FPL)*, 2003.
16. De Micheli, G., "Computer-aided hardware-software codesign", *IEEE Micro*, Vol 14, pp. 10-16, 1994.

Extra-dimensional Island-Style FPGAs

Herman Schmit

Dept. of Electrical and Computer Engineering, Carnegie Mellon University,
Pittsburgh, PA, 15213 USA,
herman@ece.cmu.edu

Abstract. This paper proposes modifications to standard island-style FPGAs that provide interconnect capable of scaling at the same rate as typical netlists, unlike traditionally tiled FPGAs. The proposal uses a logical third and fourth dimensions to create increasing wire density for increasing logic capacity. The additional dimensions are mapped to standard two-dimensional silicon. This innovation will increase the longevity of a given cell architecture, and reduce the cost of hardware, CAD tool and Intellectual Property (IP) redesign. In addition, extra-dimensional FPGA architectures provide a conceptual unification of standard FPGAs and time-multiplexed FPGAs.

1 Introduction

Island-style FPGAs consist of mesh interconnection of logic blocks, connection blocks and switchboxes. Commercial FPGAs from vendors such as Xilinx are implemented in an island-style. One advantage of island-style FPGAs is that they are completely tileable. A single optimized hardware tile design can be used in multiple products all with different capacity, pin count, package, etc., allowing FPGA vendors to build a whole product line out of a single relatively small design. The tile also simplifies the CAD tool development effort. The phases of technology mapping, placement, routing and configuration generation are all simplified by the regularity and homogeneity of the island-style FPGA. Finally, these tiled architectures are ammenable to a re-use based design methodology. A design created for embedding in larger designs, often called a core, can be placed to any location within the array, while preserving routing and timing characteristics.

The inherent problem with tiled architectures is that they do not scale to provide the interconnect typical of digital circuits. The relationship of design interconnect and design size is called Rent's rule. This relationship is described in [12], where it is shown that a partition containing n elements has Cn^p pins either entering or leaving the partition. The quantity p, which is known as the Rent exponent, has been empirically determined to be in the range from 0.57 to 0.75 in a variety of logic designs. In a tiled architecture, the number of logic elements in any square bounding box is proportional to the number of tiles contained in the square, and the number of wires crossing the square is proportional to the perimeter of the square. Therefore, in a traditional tiled architecture, the interconnect is a function of the square root of the logic capacity, and the "supplied"

P.Y.K. Cheung et al. (Eds.): FPL 2003, LNCS 2778, pp. 406–415, 2003.

Rent exponent is 1/2. Supply therefore can never keep up with interconnect demand, and at some point a tiled architecture will fail to have enough interconnect.

Commercial vendors have recognized this phenomenon. Succeeding generations of FPGAs always have more interconnect per logic block. This has the effect of keeping up with the Rent exponent by increasing the constant interconnect associated with each tile, i.e. increasing the C term in Rent's formula. There are several problems with this solution. First, it requires that the tile must be periodically redesigned, and all the CAD tools developed for that tile must be updated, tested and optimized. Second, all the cores designed for an early architecture cannot be trivially mapped to the new architecture.

What would be preferable would be to somehow scale the amount of interconnect between cells, while still providing the benefits of a tiled architecture, such as reduced hardware redesign cost and CAD tool development. This is impossible in conventional island-style FPGAs because they are two-dimensional, which means that the Rent exponent of supplied interconnect is fixed at one half. This paper proposes a new island-style architecture, based on three- and four-dimensional tiling of blocks, that can support Rent exponents greater than 0.5 across an entire family of FPGAs.

Three-dimensional FPGAs have been proposed frequently in the literature [14, 13, 1, 2, 8], but commercial three-dimensional fabrication techniques, where the transistors are present in multiple planes, do not exist. An optoelectrical coupling for three-dimensional FPGAs has also been proposed in [6], but there are no instances of such couplings in large commercial designs as yet. Four-dimensional FPGAs face even greater challenges to actual implementation, at least in this universe. Therefore, this paper proposes ways to implement three- and four-dimensional FPGAs in planar silicon technology. Surprisingly, these two-dimensional interconnect structures resemble double and quad interconnect lines provided in commercial island-style FPGAs, albeit with different scaling behavior.

2   Architecture

Extra-dimensional FPGAs can provide interconnect that matches the interconnect required by commercial designs and that scales with design size. This section describes how these extra dimensions provide scalable interconnect. Possible architectures for multi-dimensional FPGAs are discussed in the context of implementation on two-dimensional silicon. Three- and four-dimensional FPGA architectures are proposed, which will be the subject of placement and routing experiments performed in subsequent sections.

As discussed previously, a 2-D tiled FPGA provides a Rent exponent of 1/2 because of the relationship of the area and perimeter of a square. In a three-dimensional FPGA, assuming that all dimensions grow at an equal rate, the relationship between volume and surface area is governed by an exponent of 2/3. Therefore in a perfect three-dimensional FPGA, the supplied interconnect has a

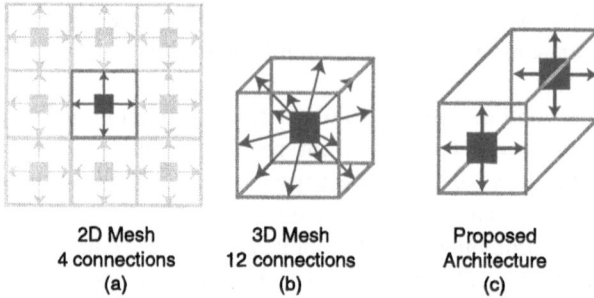

2D Mesh	3D Mesh	Proposed
4 connections	12 connections	Architecture
(a)	(b)	(c)

Fig. 1. Two and three dimensional interconnect points: as shown in (a) a CLB in a 2D mesh must connect to 4 wires. A CLB in a 3D mesh must (b) connect to twelve points. The proposed architecture has 2D CLBs interconnected to the squares in a 3D mesh.

Rent exponent of 2/3. By extension, the Rent exponent for a four-dimensional architecture is 3/4.

There are many ways to construct an extra-dimensional FPGA. A conventional island-style FPGA can be envisioned as an array of CLBs, each of which is surrounded by a square of interconnect resources. The CLB connects to each of the edges of the square that surrounds it, as shown in Figure 1(a). By extension, a 3-D FPGA would have a three dimensional interconnection mesh, with CLBs located in the center of the cubes that make up this mesh. Each CLB would have to connect to all twelve edges of the cube, as illustrated in Figure 1(b). This entails a three-fold increase in the number of places that CLB IOs connect, which either means greater delay, greater area, and perhaps worse placement and routing. In addition, this "cube" interconnect is very difficult to lay out in two-dimensional silicon. In a four-dimensional FPGA, the CLB would have to connect to thirty-two of the wires interconnecting a hypercube in the mesh, and is even harder to lay out in two dimensions.

An alternative proposal is to allow CLBs to exist only on planes formed by the x and y dimensions. These CLBs only connect to wires on the same xy plane. A multi-dimensional FPGA is logically constructed by interconnecting planes of two-dimensional FPGAs, as illustrated in Figure 1(c). This keeps the the number of IOs per CLB equal to conventional FPGAs, while still providing the benefits of an extra-dimensional FPGA.

In our proposed architecture, two adjacent xy planes of CLBs are interconnected by the corresponding switch boxes. For example, in a logically three-dimensional FPGA, the switch box at (x, y) in plane z is connected to the switch box at (x, y) in plane $z + 1$ and plane $z - 1$. Such a three-dimensional switchbox architecture was used in [1].

A problem with these extra-dimensional switch boxes is the increased number of transistors necessary to interconnect any wire coming from three or four dimensions. When four wires meet within a switch box, coming from all four directions in a two-dimensional plane, six transistors are necessary to provide any interconnection. With three dimensions, six wires come from every direction, requiring 15 transistors to provide the same flexibility of interconnect. With four dimensions, 28 transistors are necessary at each of these switch points. We will assume that extra-dimensional switch boxes include these extra number of transistors. We will later show that this overhead is mitigated by the reduction in channel width due to extra-dimensional interconnect. It is worth noting that there are only 58 possible configurations of switch point in a 3D switch box,[1] and 248 possible configurations of a 4D switch point. Therefore the amount of configuration for these switch points could potentially be significantly reduced to six or eight bits. It is also possible that all 15 or 28 transistors are not necessary to provide adequate interconnect flexibility. Future work will explore optimizations of the switch box architectures for extra-dimensional FPGAs.

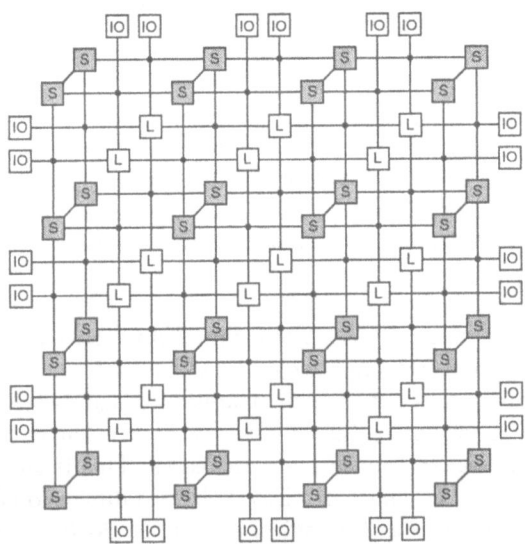

Fig. 2. Three dimensional FPGA in two dimensions: This particular FPGA is a 3 x 3 x 2 array of CLBs. There are IOs around the periphery. Switch Blocks are labelled "S" and CLBs are labelled "L". Connections between xy planes take place in the diagonal channels between switch boxes. All lines in this figure are channels containing multiple wires.

Figure 2 and Figure 3 show feasible layouts for three- and four-dimensional FPGAs in two-dimensional silicon. The four-dimensional layout is particularly interesting, because it allows for close interconnections between neighboring CLBs, and because it is symmetric in both x and y dimensions. The most interesting aspect of the four-dimensional FPGA is how the interconnect resembles the interconnect structure of commercial island-style FPGAs. Particularly, the interconnections between different xy planes resemble the double and quad lines

---

[1] This is determined by $1 + \binom{6}{2} + \binom{6}{3} + \binom{6}{4} + \binom{6}{5} + \binom{6}{6} = 58$

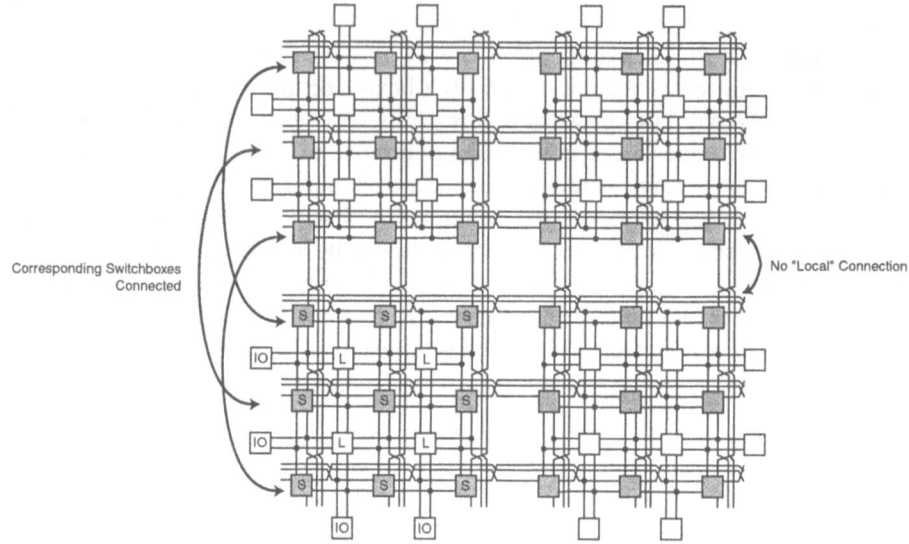

Fig. 3. Four dimensional FPGA: This FPGA has dimensions of 2 x 2 x 2 x 2 CLBs. The connections between dimensions take place on the long lines that go from one switch box to the corresponding box in another dimension.

in Xilinx 4000 and Virtex FPGAs. These lines skip across a number of CLBs, in order to provide faster interconnect over long distance.

The primary difference between the traditional island-style with double and quad lines and the proposed 4D FPGA is that the length of the long wires increases when the size of the xy plane increases. This is the key to providing scalable interconnect. The length of wires in the two-dimensional implementation of the 4D FPGA increases as the logical size of the device increases. This does mean that the delay across those long wires is a function of the device size, which presents an added complication to timing-oriented design tools. Fortunately, interconnection delay in commercial FPGAs is still dominated by the number of switches in any path and not the length of the net. [2]

The assumption that the physical channel width is proportional to the number of wires passing through that channel leads to the conclusion that the four-dimensional FPGA in Figure 3 would be much larger than the two-dimensional FPGA with the same number of CLBs. In real FPGA layouts however, the largest contributor to channel width is the number of programmable interconnect points in the channel, and the width of the switch box. Because these points require one to six transistors, they are much larger than the minimum metal pitch. Most wires in the channel of the four-dimensional FPGA do not contact any interconnect points however. These wires can be tightly routed at a higher level of metal, and may only minimally contribute to total channel width.

---

[2] The increasing "logical length" of these lines may also be counter-acted to some extent by the scaling of technology.

A second substantial difference between the proposed 4D FPGA and the traditional island-style FPGA is that in the commercial FPGA, the local connections are never interrupted. In the 4D FPGA, there is no local interconnection from the boundary of one xy plane to the corresponding boundary of the adjacent xy plane. The reason for this is to provide for tiling of IP blocks. Suppose we want to use a four-dimensional core in a four-dimensional FPGA. To guarantee that this core will fit in this FPGA, all four dimensions of the core must be smaller than the four dimensions of the FPGA. If there are local connections between xy planes, that cannot be guaranteed. A core that was built on a device with $x = 3$, and which used local connections between two xy planes would only fit onto other devices with $x = 3$.

The next section will describe an experiment that compares the scaling of interconnect in two- and four-dimensional FPGA. First, Rent's rule is demonstrated by measuring how the channel width of a two-dimensional FPGA increases as the size of the design increases. When the same suite of netlists is placed and routed on an 4D FPGA, the channel width remains nearly constant.

3   Experimental Evaluation

In order to demonstrate the effectiveness of extra-dimensional FPGAs, a large set of netlists have been placed and routed on both two- and four-dimensional FPGAs. The minimum channel width required in order to route these designs is compared. In the two-dimensional FPGA, the channel width grows with respect to the logical size of the design. The four-dimensional FPGA scales to provide exactly the required amount of interconnect, and channel width remains nearly constant regardless of the design size.

A significant obstacle to performing this experiment was to acquire a sufficiently large set of netlists in order to make conclusions based on measured data. Publicly available circuit benchmark suites are small, both in the number of gates in the designs, and the number of designs in the suites. In this paper we use synthetic netlists generated by the CIRC and GEN tools [10, 11]. CIRC measures graph characteristics, and has been run on large sets of commercial circuits. GEN constructs random netlists that have the same characteristics as the measured designs. We used the characteristics measured by CIRC on finite state machines to generate a suite of 820 netlists ranging from 20 to 585 CLBs.

The placement and routing software we will use is an extended version of the VPR tool [3]. The modifications to this software were relatively simple. The placement algorithm is extended to use a multi-dimensional placement cost. The cost function is a straight-forward four-dimensional extension of the two-dimensional cost, which is based on the RISA cost function [4]. The cost of routing in the two additional dimensions is scaled by a small factor to encourage nets to be routed in a single xy plane.

The logic block architecture used for this experiment is the standard architecture used in VPR. This architecture consist of a four-input lookup table (LUT) and a bypassable register. The switch box architecture is the Xilinx-type (also

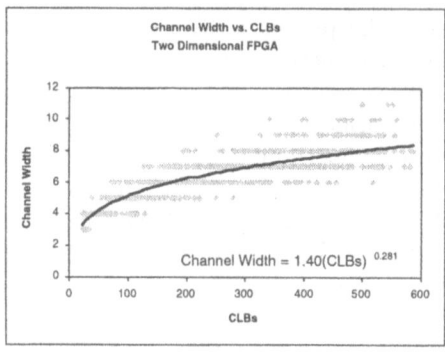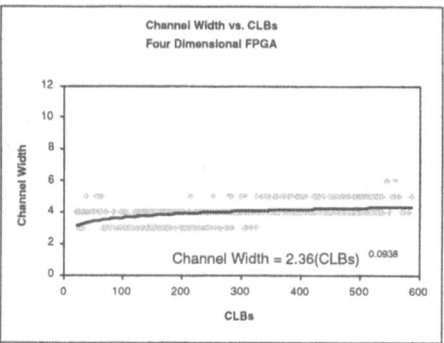

Fig. 4. Two- and Four-Dimensional FPGA Channel Width

known as subset or disjoint), where the n-th channel from one direction connects to the n-th channel in every other direction.

VPR works by first placing the design, and then attempting to find the minimal channel width that will allow routing of the design. It performs a binary search in order to find this minimal channel width. The results of routing the netlist suite on the two-dimensional and four-dimensional architecture are shown in Figure 4, which plots the maximum channel width versus the netlist size in CLBs. The best-fit power graph has been fit to this data, showing an exponent of 0.28 for the two-dimensional FPGA. As shown in [7] and [9], this indicates the Rent exponent of these designs is approximately $0.28 + 0.5 = 0.78$, which is large, but possibly reasonable for the design of an FPGA architecture.

Even a four dimensional FPGA cannot provide interconnect with a Rent exponent of 0.78. In our experiment, the extra dimensions of the FPGA scale at one-eight the rate of the x and y dimensions. If the x and y dimensions are less than eight, there is no other extra dimensions to the FPGA (the FPGA is simply two-dimensional). With x or y in the range of 8 to 16, there are two planes of CLBs in each of the extra dimensions.

The channel width for the four-dimensional FPGA, as shown in Figure 4 is comparitively flat across the suite of netlists. Figure 5 demonstrates the effectiveness of the technique by showing total wire length, in channel segments, for each design. When fit to a power curve, the two dimensional FPGA exhibits a much larger growth rate than the four dimensional FPGA. Superlinear growth of wire length was predicted by [7] as a result of Rent's rule. The four dimensional FPGA evidences a nearly linear relationship between wire length and CLBs, indicating that the supply of interconnect is keeping pace with the demand.

As mentioned in previously, the four dimensional swith box might have more than four times more bits per channel compared to the two dimensional switch box. Table 1 shows the number of configuration bits necessary to program the interconnect for a two- and four-dimensional FPGA with 576 CLBs. Using our data from the previous experiments, we have assumed a channel width of twelve for the two-dimensional FPGA, and six for the four-dimensional FPGA. Con-

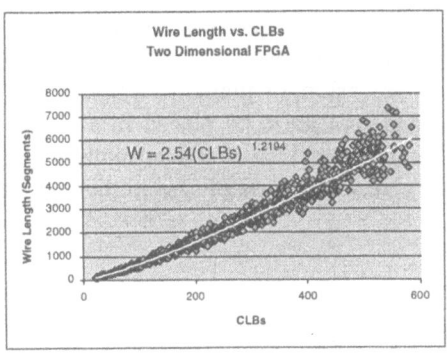 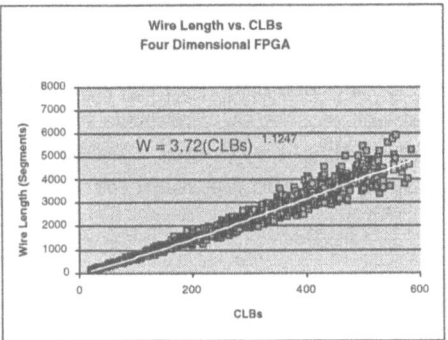

Fig. 5. Wirelengths for Two- and Four-dimensional FPGAs

Table 1. Computation of configuration bits required for large two- and four- dimensional FPGAs. The larger number of bits per switch point the four-dimensional FPGA is countered by the reduced channel width. If the switch point configuration is encoded with sixteen bits, there is no configuration overhead.

FPGA Dimensions	Channel Width	Switch Point Bits	Total Switchbox Bits	Total Channel Bits	Total Config Bits
24 x 24	12	6	41,472	27,648	69,120
2 x 12 x 12 x 2	6	28	96,768	13,824	110,592
2 x 12 x 12 x 2	6	16	55,296	13,824	69,120

sidering both switch box and connection box bits, the four-dimensional FPGA requires 60% more configuration bits. If we can control the four-dimensional switch point with just sixteen bits, then the reduction in channel width completely compensates for the more complex switch point.

Four-dimensional FPGAs provide scalable interconnect across a wide suite of netlists. Required channel width can remain constant across a large family of devices, allowing the hardware and CAD tool development efforts to be amortized over many devices, and reducing the waste of silicon in smaller devices.

4   Time Multiplexing and Forward-Compatiblity

Since their proposal [17, 5, 15], time-multiplexed FPGAs have been thought of as fundamentally different creatures from standard FPGAs. The Xilinx and Sanders FPGAs operate by separating every logical cycle into multiple micro-cycles. Results from a micro-cycle are passed to subsequent micro-cycles through registers.

Time-multiplexed FPGAs could also be constructed using a three-dimensional FPGA architecture like that shown in Figure 2. Micro-registers would be inserted on the inter-plane connections that exist between switchboxes. The number of registers between planes would correspond to the channel width, which allows it to be designed to provide scalable interconnect.

By viewing time as a third dimension, the scheduling task can be accomplished within the scope of a three dimensional placement algorithm. All combinational paths must go from a previous configuration, or plane, to a future one, and not vice-versa. Therefore real logical registers must be placed in a configuration that is evaluated after its entire fanin cone. This is conceptually much simpler than the separated scheduling and place-and-route phases implemented in [16].

5   Conclusions

This paper discussed the benefits of three- and four-dimensional FPGAs, and their implementation in conventional two-dimensional silicon. Primary among the benefits is the ability to provide interconnect that scales at the same rate as typical netlists. Four dimensional FPGAs resemble the double and quad lines in commercial FPGAs, although the key to providing scalable interconnect is to increase the length of those lines as the device grows.

References

1. M. J. Alexander, J. P. Cohoon, J. L. Colflesh, J. Karro, and G. Robins. Three-dimensional field-programmable gate arrays. In *Proceedings of the IEEE International ASIC Conference*, pages 253–256, September 1995.
2. M. J. Alexander, J. P. Cohoon, J. Karro J. L. Colflesh, E. L. Peters, and G. Robins. Placement and routing for three-dimensional FPGAs. In *Fourth Canadian Workshop on Field-Programmable Devices*, pages 11–18, Toronto, Canada, May 1996.
3. V. Betz and J. Rose. Effect of the prefabricated routing track distribution on FPGA area efficiency. *IEEE Transactions on VLSI Systems*, 6(3):445–456, September 1998.
4. C. E. Cheng. RISA: Accurate and efficient placement routability modeling. In *Proceedings of IEEE/ACM International Conference on CAD (ICCAD)*, pages 690–695, November 1996.
5. A. DeHon. DPGA-coupled microprocessors: Commodity ICs for the early 21st century. In D. A. Buell and K. L. Pocek, editors, *Proceedings of IEEE Workshop on FPGAs for Custom Computing Machines*, pages 31–39, Napa, CA, April 1994.
6. J. Depreitere, H. Neefs, H. Van Marck, J. Van Campenhout, R. Baets, B. Dhoedt, H. Thienpont, and I. Veretennicoff. An optoelectronic 3-D field programmable gate array. In R. Hartenstein and M. Z. Servit, editors, *Field-Programmable Logic: Architectures, Synthesis and Applications. 4th International Workshop on Field-Programmable Logic and Applications*, pages 352–360, Prague, Czech Republic, September 1994. Springer-Verlag.
7. W. E. Donath. Placement and average interconnection lengths of computer logic. *IEEE Transactions on Circuits and Systems*, pages 272–277, April 1979.
8. Hongbing Fan, Jiping Liu, and Yu-Liang Wu. General models for optimum arbitrary-dimension fpga switch box designs. In *Proceedings of IEEE/ACM International Conference on CAD (ICCAD)*, pages 93–98, November 2000.
9. A. El Gamal. Two-dimensional stochastic model for interconnections in master slice integrated circuits. *IEEE Transactions on Circuits and Systems*, 28(2):127–138, February 1981.

10. M. Hutton, J.P. Grossman, J. Rose, and D. Corneil. Characterization and parameterized random generation of digital circuits. In *Proceedings of the 33rd ACM/SIGDA Design Automation Conference (DAC)*, pages 94–99, Las Vegas, NV, June 1996.

11. M. Hutton, J. Rose, and D. Corneil. Generation of synthetic sequential benchmark circuits. In *5th ACM/SIGDA International Symposium on Field Programmable Gate Arrays, (FPGA 97)*, February 1997.

12. B. S. Landman and R. L. Russo. On pin versus block relationship for partions of logic circuits. *IEEE Transactions on Computers*, C-20:1469–1479, 1971.

13. M. Leeser, W. M. Meleis, M. M. Vai, W. Xu S. Chiricescu, and P. M. Zavracky. Rothko: A three-dimensional FPGA. *IEEE Design and Test of Computers*, 15(1):16–23, January 1998.

14. W. M. Meleis, M. Leeser, P. Zavracky, and M. M. Vai. Architectural design of a three dimensional FPGA. In *Proceedings of the 17th Conference on Advanced Research in VLSI (ARVLSI)*, pages 256–268, September 1997.

15. S. Scalera and J. R. Vazquez. The design and implementation of a context switching FPGA. In D. A. Buell and K. L. Pocek, editors, *Proceedings of IEEE Workshop on FPGAs for Custom Computing Machines*, pages 78–85, Napa, CA, April 1998.

16. S. Trimberger. Scheduling designs into a time-multiplexed FPGA. In *6th ACM/SIGDA International Symposium on Field Programmable Gate Arrays, (FPGA 98)*, pages 153–160, February 1998.

17. S. Trimberger, D. Carberry, A. Johnson, and J. Wong. A time-multiplexed FPGA. In J. Arnold and K. L. Pocek, editors, *Proceedings of IEEE Workshop on FPGAs for Custom Computing Machines*, pages 22–28, Napa, CA, April 1997.

Using Multiplexers for Control and Data in D-Fabrix

Tony Stansfield

Elixent Limited, Castlemead, Lower Castle Street, Bristol, United Kingdom
tony.stansfield@elixent.com

Abstract. This paper describes the use of dynamically controlled multiplexers in the Elixent D-Fabrix Reconfigurable Algorithm Processor (RAP) for both datapath functions and to implement simple logic functions for control circuits.

1   Introduction

Reconfigurable computing applications can be viewed as consisting of a mixture of control and datapath blocks. These two types of blocks have different characteristics, in particular the datapath blocks process word-based data and the control blocks manipulate bits. Applications vary in the relative amounts of control and datapath that they contain. At one extreme are DSP functions such as the discrete cosine transform (DCT) that can be expressed as almost 100% datapath, and at the other are control intensive tasks such as searching, sorting, and boolean satisfiability.

The majority of implementation approaches for reconfigurable computing use the same hardware for both control and datapath. Typically, FPGAs provide the hardware, and their bit-based lookup tables (LUTs) are used to construct both the bit-oriented control logic and the word-based datapaths. The use of LUTs to construct a datapath is an inefficient use of the underlying silicon resources, as it fails to exploit the repetitive nature of the datapath – all bits in the word are processed in the same way and could therefore share the same control state.

An alternative implementation approach is to use a Reconfigurable Computing device (such as Elixents D-Fabrix Reconfigurable Algorithm Processor) that consists of an array of word-based processing elements – in the case of D-Fabrix all the processing units are designed to handle 4-bit quantities and can be easily combined for larger words. Here there is the opposite problem to the FPGA case – the basic units are designed for the implementation of datapaths and lose efficiency when used to process 1-bit control signals – and if this efficiency loss becomes too great then the overall solution would be no better than the FPGA-based approach.

This paper describes the dedicated multiplexers in D-Fabrix, and how they can be used to assist with the implementation of control structures. The next section describes the kinds of control structures that may be encountered in applications and ways of implementing them as gates. Later sections then describe the D-Fabrix array, and ways to implement these structures on it. This is followed by an assessment of the ability of software to efficiently target the available hardware resources, and a comparison of the current work with previous work on heterogeneous architectures.

P.Y.K. Cheung et al. (Eds.): FPL 2003, LNCS 2778, pp. 416–425, 2003.

2   Types of Control Structures

Control can be divided into three parts:

- Control of the flow of data, as expressed in the C/Java conditional assignment operator: `a = b ? c : d;`
- Control of the flow of program control, as typified by programming language constructs such as `if(expr)` and `while(expr)`, and:
- Calculation of the expressions with Boolean results that are inputs to these control circuits (the `b` and `expr` in the above examples)

The control of the flow of data can be implemented with multiplexers, choosing between 2 (or more) word-wide data inputs.

Page and Luk [1] have described circuits that can be used to implement the flow of program control. Every program statement that affects control flow results in the creation of a simple sequential circuit that waits for a "start" token from a preceding statement and passes the token to an appropriate successor. For instance, in the case of an if...then...else... statement the token is passed to one branch or the other depending on the result of evaluating the condition. The overall circuit created by combining the individual circuits from the individual statements is effectively a distributed state machine with a one-hot encoding. The token is represented as a single bit, propagated from stage to stage on a single wire. The logic gates that control the path of the token are all either AND gates, OR gates or inverters, and (with the exception of the wide OR gates used to merge tokens after branches and PAR statements) are all 1 or 2 input gates.

This distributed state machine is one source of the control inputs of the data multiplexers. The other possible source is an expression that produces a single bit result – such as a comparison (A == B, A > B etc.).

3   D-Fabrix Architecture

The D-Fabrix architecture is an evolution of the Chess architecture [2] developed by Hewlett-Packard Laboratories. It shares with Chess the following features:

- The basic processing elements are 4-bit ALUs, each with:
  - Two 4-bit data inputs, A and B
  - One 4-bit data output, F
  - 1-bit Cin (input) and Cout (output) terminals to create carry chains linking ALUs to process wider words
  - A 4-bit instruction input, I. The ALU operation can be set either statically (via an internal configuration register) or dynamically (through the instruction input).
  - An optional output register on the 4-bit data output.
- The routing network propagates data as 4-bit quantities:
  - All routing buses are 4-bit buses
  - All routing switches connect two 4-bit buses
  - 1-bit data (carry and control) is carried in the LSB of a 4-bit word.

        o   There are direct 1-bit connections between Cin and Cout of neighboring ALUs, separate from the main routing network.
- ALUs and switchboxes are arranged in a checkerboard pattern.
- ALUs and Switchboxes are paired, and share some common signals within the pair
- Switchboxes contain additional 4-bit registers that can be used to pipeline applications.

The operations performed by the ALU are of four basic types:
- Arithmetic operations (using Cin and Cout as a carry chain):

     ADD: $F = A + B + Cin$ **SUB**: $F = A - B + Cin$

- Bitwise logical operations, such as:

     AND: $F = A \ \& \ B$ **OR**: $F = A \mid B$

     XOR: $F = A \wedge B$ **NOT**: $F = \sim A$

- Multiplexer operations (with Cout = Cin in all cases):

     MUX: $F = Cin \ ? \ A : \ B$ **INVMUX**: $F = Cin \ ? \sim\!A : \ \sim\!B$

- Comparison and test operations:

     NEQ: $Cout = (A == B) \ ? \ Cin : 1$

     When Cin = 0, Cout is 0 if A is equal to B and 1 otherwise.

     ORAND: $Cout = (A\&B == 0)? \ Cin : 1$

     When Cin = 1, Cout is 1 only if the AND of A and B is not 0. This is equivalent to an AND of A and B, followed by an OR reduction.

Less-than and Greater-than functions are available via the carry output of an appropriate subtraction.

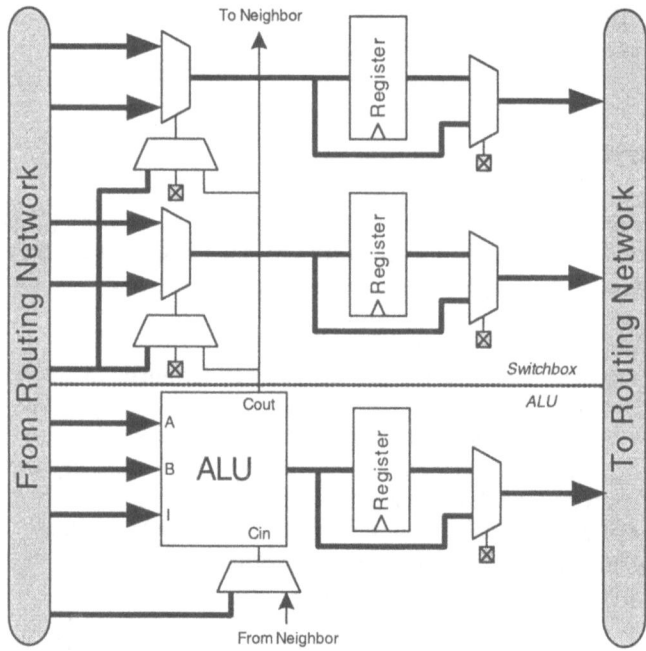

Fig. 1. Simplified diagram of logic in D-Fabrix ALU and Switchbox pair

There is however one significant difference between D-Fabrix and Chess, which is illustrated in figure 1. The multiplexers that select signals from the routing network to pass to the pipeline registers were statically configured in Chess, but D-Fabrix allows them to be dynamically controlled, using either the ALU Cout or a signal from the routing network as the multiplexer control signal. The sources of dynamic control for these multiplexers (normally referred to as "Switchbox Multiplexers" due to their being located in the switchbox) are broadly equivalent to those of the ALU Cin – either Cout from a nearby ALU or a signal from the global routing network.

These dynamic switchbox multiplexers are in addition to the MUX instruction of the ALU. It is possible to use both the switchbox multiplexers and the ALU as a multiplexer simultaneously, all controlled by the same signal, so that a single ALU and switchbox pair can implement a multiplexer for up to 12-bit data inputs.

This change to dynamic switchbox multiplexers has resulted in a slight overall increase in the area of the combined ALU and switchbox. This increase is due to the addition of the circuits that select the control inputs to the switchbox multiplexers and their configuration memories – the switchbox multiplexers themselves were already present, but with only static control. The area increase due to this change is no more than 10%. Compared to an ALU a switchbox multiplexer is one-third of the size, has half the propagation delay and 40% of the power dissipation. (Area comparisons are based on the Chess and D-Fabrix layouts, speed and power comparisons on spice simulation of the D-Fabrix circuit). Consequently, there are significant gains in area, speed and power dissipation to be made if an application can use these dedicated multiplexers instead of ALUs.

4   Multiplexer-Based Control

Multiplexers can be used to implement simple logic functions of two 1-bit inputs. Table 1 lists all 16 Boolean functions of up to 2 inputs, and possible implementations constructed from a multiplexer controlled by one of the function inputs, and with the multiplexer data inputs being connected to either a constant or the other function input.

Table 1. Boolean Functions of up to 2 Inputs

Function	Implementation	Function	Implementation
0	A ? 0 : 0	1	A ? 1 : 1
A & B	A ? B : 0	A | B	A ? 1 : B
A & !B	B ? 0 : A	A | !B	B ? A : 1
!A & B	A ? 0 : B	!A | B	A ? B : 1
!A & !B	(NOR)	!A | !B	(NAND)
A	A ? 1 : 0	B	B ? 1 : 0
!A	A ? 0 : 1	!B	B ? 0 : 1
A^B	(XOR)	!A^B	(NXOR)

Of these 16 functions, four are trivial (0, 1, A and B), and can be implemented without a multiplexer, eight have multiplexer-based implementations and four do not.

These four are NAND, NOR, XOR and NXOR. A simple logic optimization pass can commonly turn NAND and NOR into other AND and OR-type functions, so it is only XOR and NXOR that are difficult to implement in a compact form with multiplexers.

Historically, Actels antifuse-based FPGA families [9] have used multiplexers as the basis of their logic elements. Actel used multiplexers with 1-bit data and control inputs, but it is also possible to use larger multiplexers in the same way, with some of the available data bits being unused if necessary. The D-Fabrix switchbox multiplexers can therefore be used to construct 2-input logic gates, with one of the inputs connecting to the control input of the multiplexer and the other to one of the data inputs. Only one of the four available data bits is used, but since multiplexers are much smaller than ALUs it is more area efficient to use the multiplexers for this purpose than to use an ALU.

The control flow circuits referred to in section 2 are all constructed from AND and OR gates, some with inverters on their inputs. It can be seen from table 1 that such functions all have multiplexer-based implementations. Thus it appears that both the data flow and the program control flow aspects of control can be converted into arrangements of multiplexers controlled by the outputs of expressions with Boolean results. Figure 2 illustrates the use of the switchbox multiplexers for controlling the flow of (a) data and (b) application control. Figure2(b) is an implementation of the circuit shown in figure 4 of [1]. Both 2(a) and 2(b) consist of two parts – a group of ALUs that compute a Boolean function of their word-wide inputs, and a group of multiplexers whose control input is driven by Cout from one of the ALUs. All test operations computed by the ALUs produce their Boolean result on Cout, and (as shown in figure 1) D-Fabrix has direct support for Cout to multiplexer control connections.

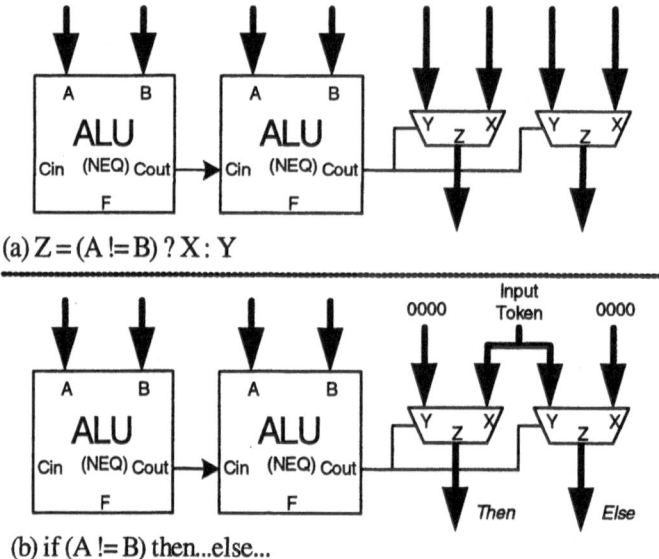

Fig. 2. Use of multiplexers to route (a) data and (b) application control

Furthermore, many of the control expressions themselves contain subexpressions that can be implemented with 2-input gates. For example, consider the expression:

$$Z = (A1 \, != \, B1) \, \& \, (A2! \, = \, B2)$$

The two comparisons are both functions of word-based data, but the AND that combines them is a function of single-bit inputs, and can therefore be implemented with a multiplexer. Figure 3 shows such an implementation of this expression.

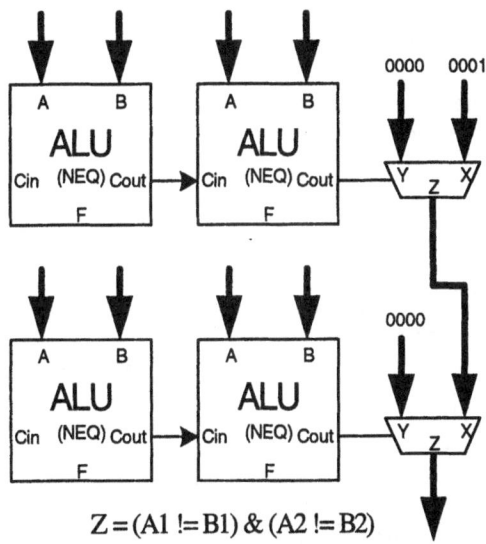

$$Z = (A1 \, != B1) \, \& \, (A2 \, != B2)$$

Fig. 3. Use of Multiplexers to combine results of Boolean expressions

5   Other Uses of Multiplexers

The switchbox multiplexers have other uses beyond the ones mentioned above. Some of these uses are described below.

5.1   1-Bit to 4-Bit Gateway

Figure 1 shows direct connections from the 4-bit routing network to the multiplexers that select among 1-bit signals to provide both the ALU Cin and the switchbox multiplexer control signals. These direct connections allow the least-significant bit of a 4-bit bus to be used as the source of these signals. The switchbox multiplexers can be used to provide the opposite 1-bit (i.e. Cout) to 4-bit connection. If the multiplexer has as its data inputs the binary constants 0000 and 0001 then its function is:

$$Z = Cout \, ? \, 0001 \, : \, 0000 \, = \, 000<Cout>$$

i.e. the output is a 4-bit version of the multiplexer control input. Such a use of the multiplexer is illustrated in the upper half of figure 3. It is possible to produce other functions of Cout simply by using different input constants for the multiplexers. For example:

```
Z = Cout ? 0000 : 0001 = 000<~Cout>
Z = Cout ? 1111 : 0000 = <Cout><Cout><Cout><Cout>
```
i.e. a 1-bit inverse of Cout, or a 4-bit version.

5.2   Register Control

The switchbox multiplexers are located at the inputs of the switchbox registers. This makes them easy to use to add control options to the registers, such as reset and enable, as illustrated in figure 4.

This usage is more flexible than having a dedicated reset connection to the register, as it is possible to change the reset value of the register by changing the constant supplied to the multiplexer input.

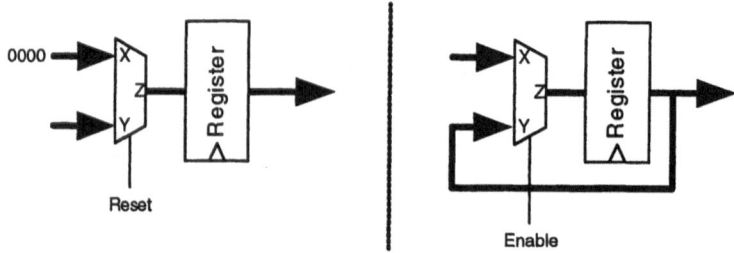

Fig. 4. Use of Multiplexers for register reset and enable

6   Software Support

Preceding sections have described the D-Fabrix hardware architecture. In this section we consider the ability of software to make use of the hardware.

Table 2 shows the relative proportions of the different types of ALU operations generated by the Elixent verilog compiler on 5 different stages of a JPEG encoder. JPEG is used as an example here because:

a) It is a widely used (and understood) example, and

b) It contains some sections that are almost all datapath (such as the DCT), and others that have a significant amount of control logic.

It therefore provides a reasonable test of an architectures abilty to handle both datapath-intensive and control-intensive designs.

Table 2 shows that the compiler consistently generates a high proportion of multiplex operations. Between 40% and 60% of operations are multiplexers. The variation in the proportion of multiplexers is lower than that for the other types of instruction – the maximum percentage is 1.5 times the minimum for multiplexers, more than twice the minimum for arithmetic operators and greater still for test and bitwise logic operations.

Table 2. 4-bit Operation Usage in JPEG Encoder

Block	Arithmetic	Multiplex	Test	Bitwise Logic
Fmerge	35.7%	42.9%	14.3%	7.1%
FDCT	35.1%	59.3%	4.0%	1.5%
Quantize	51.1%	42.1%	6.8%	0%
Zigzag	19.0%	42.9%	31.0%	7.1%
Ent&Pack	21.9%	57.2%	8.2%	12.7%

This result is not specific to JPEG. When measured across a larger set of benchmarks the fraction of operators that can be implemented with a multiplexer typically lies in the 40% to 70% range. The extremes of this range are represented at the high end by applications with a large amount of bit manipulation (such as DES, and an arithmetic codec for JPEG2000), and at the low end by multiplication intensive pure datapath functions (such as a CDMA matched filter). This is in accordance with expectation based on the description of multiplexer usage given above.

Given that multiplexers account for 40% – 70% of an application, it is clear that the switchbox multiplexers have a net positive impact on array density. If all the multiplex operations had to be implemented with ALUs then an array would need between 1.6 and 3 times the number of ALUs for the same application. Compared with this the 10% area increase required to make the multiplexers dynamic rather than static is a significantly smaller cost. There are also gains in speed and power dissipation as a result of the use of multiplexers instead of ALUs for large parts of an application

The array is constructed with 2 multiplexers per ALU, corresponding to 66% of the processing resources in the array. This ratio means that for almost all applications the number of ALUs is the limiting factor, not the number of multiplexers. If the ratio was increased to 3 multiplexers per ALU the majority of the extra multiplexers would be unused in almost all applications.

7   Comparison to Other Work

The D-Fabrix architecture as described above can be categorized as a "Spatially Uniform Heterogeneous Array" – heterogeneous in that it contains two distinct types of processing element (ALUs and multiplexers), but spatially uniform in that the array is a regular arrangement of these two types of resource (it does not have distinct large groups of ALUs, and of multiplexers). Previously published work on heterogeneous arrays falls into two main classes:

- Studies of the impact of using two or more different sizes of LUT [3][4]. The use of embedded RAMs as supplementary logic when not used as memory [5] can be regarded as a special case of this category.
- Studies of arrays containing both LUTs and product term arrays [6][7][8].

Neither of these categories is directly relevant to the current work:

- D-Fabrix has two types of logic element, but they differ in the functionality that they provide, not in the number of inputs.
- Product-term arrays generate multiple functions of the same inputs, whereas both the ALUs and Multiplexers in D-Fabrix generate a single function of independent inputs.

The key difference between D-Fabrix and these other examples is that the two types of processing element are chosen so that the functionality provided by one (the multiplexer) is a subset of that of the other (the ALU). This greatly simplifies the main problem identified in the earlier work – the increased complexity of synthesis and/or technology mapping due to the need to decide which parts of the application to map to the different types of processing element. Since the multiplexers provide a subset of the ALU functionality, the synthesis process need only target the ALUs, and placement can then determine whether a given multiplexer in the netlist maps to an ALU or a switchbox multiplexer. The only change made to the synthesis process is a modification of some logic optimization rules so that simple gates are converted to their multiplexer-based equivalents.

It is worth noting that although the D-Fabrix architecture described in this paper is significantly different to the arrays containing LUTs of two different sizes considered by He and Rose [3], the basic conclusion is the same – a heterogeneous array offers better density than a comparable homogeneous array.

The reasoning behind these similar conclusions is also the same. In [3] an array of mostly 4-input LUTs (4-LUTs) with some 2-LUTs is shown to be more efficient than a homogeneous array of 4-LUTs. The explanation given for this is that applications often contain functions with fanout greater than 1 but only a small number of inputs, and these are the functions that map well onto the 2-LUTs. In effect there is a significant fraction of the application for which the 4-LUT provides more logic than necessary, and the removal of this overhead results in a more compact solution. In the current case, an array of multiplexers and ALUs is denser than an array of ALUs alone, again because there is a significant fraction of the target applications for which the ALU provides more functionality than required, and so a simplified functional unit targeted at the needs of this fraction results in an increase in density.

8   Summary

This paper has described how the addition of dynamic multiplexers to an ALU array provides an overall improvement in performance and density. These multiplexers are used for both data manipulation and to implement simple 2-input gates to process control signals, and account for around half of the total logic required. Since multiplexers are smaller and faster than ALUs there is a net reduction in both area and delay, and software is easily able to make good use of this mixture of ALUs and multiplexers.

References

[1] I. Page and W. Luk, Compiling occam into FPGAs, in FPGAs, W. Moore and W. Luk (editors), Abingdon EE&CS Books, 1991, pp. 271-283.

[2] Marshall et. al., A Reconfigurable Arithmetic Array for Multimedia Applications. In Proc. of the International Symposium on Field Programmable Gate Arrays (FPGA 99), pp 135-143, February 1999.

[3] J. He and J. Rose, Advantages of Heterogeneous Logic Block Architectures, CICC 1993, pp7.4.1 – 7.4.5

[4] J. Cong and S. Xu, Delay-Optimal Technology Mapping for FPGAs with Heterogeneous LUTs, Proc. of 35th Design Automation Conf., San Francisco, California, pp. 704-707, June 1998

[5] J. Cong and S. Xu, Delay-Oriented Technology Mapping for Heterogeneous FPGAs with Bounded Resources, Proc. ACM/IEEE International Conference on Computer Aided Design, San Jose, California, pp. 40-45, November 1998.

[6] A. Kaviani and S. Brown, Hybrid FPGA architecture, In Proc. of the International Symposium on Field Programmable Gate Arrays (FPGA 96), pp 3-9, February 1996.

[7] F. Heile and A. Leaver, Hybrid Product Term and LUT Based Architectures Using Embedded Memory Blocks, In Proc. of the International Symposium on Field Programmable Gate Arrays (FPGA 99), pp 13-16, February 1999.

[8] E. Lin and S. Wilton, Macrocell Architectures for Product Term Embedded Memory Arrays, in Field-Programmable Logic and Applications (FPL 2001), pp 48-58

[9] Actel ACT1 datasheet at http://www.actel.com/docs/datasheets/db97s01d07.pdf

Heterogeneous Logic Block Architectures for Via-Patterned Programmable Fabrics

Aneesh Koorapaty[1], Lawrence Pileggi[1], and Herman Schmit[1]

Carnegie Mellon University, Pittsburgh PA 15213, USA,
aneeshk@ece.cmu.edu, pileggi@ece.cmu.edu, herman@ece.cmu.edu

Abstract. ASIC designs are becoming increasingly unaffordable due to rapidly increasing mask costs, greater manufacturing complexity, and the need for several re-spins to meet design constraints. Although FPGAs solve the NRE cost problem, they often fail to achieve the required performance and density. A Via-Patterned Gate Array (VPGA) that combines the regularity and design cost amortization benefits of FPGAs with silicon area and power consumption comparable to ASICs, was presented in [1]. The VPGA fabric consists of a regular interconnect architecture laid on top of an array of patternable logic blocks (PLBs). Customization of the logic and interconnect is done by the placement or removal of vias at a subset of the potential via locations. In this paper, we propose four heterogeneous PLBs for via-patterned fabrics and explore their performance, density and fabric utilization characteristics across several applications. Although this analysis is done in the context of the VPGA fabric, the proposed heterogeneous PLBs and the experimental methodology can be employed for any embedded programmable fabric.

1 Introduction

Application Specific Integrated Circuits (ASICs) consist of pre-designed logic cells and up to seven layers of metal wiring. The high degree of flexibility in placement and routing of the cells necessitates unique, customized masks for all fabrication layers. With shrinking feature sizes and increasing design complexity, mask costs and design costs for re-spins are becoming prohibitively expensive. As a result, programmable fabrics like Field Programmable Gate Arrays (FPGAs) are becoming increasingly attractive. Unlike ASICs, FPGAs amortize design costs across several applications and enhance manufacturability via greater layout regularity. However, FPGAs are at least three times slower and require at least ten times as much die area as ASICs. The area overhead is due to the island style topology in which the interconnect is placed adjacent to the logic array. The performance penalty is due to the RC delays of the SRAM controlled pass transistors in the interconnect switchboxes, and the SRAM based LUTs in the PLBs.

A Via-Patterned Gate Array [1] is a novel, regular fabric that bridges the cost-performance gap between ASICs and FPGAs. Like an FPGA, a VPGA

P.Y.K. Cheung et al. (Eds.): FPL 2003, LNCS 2778, pp. 426–436, 2003.

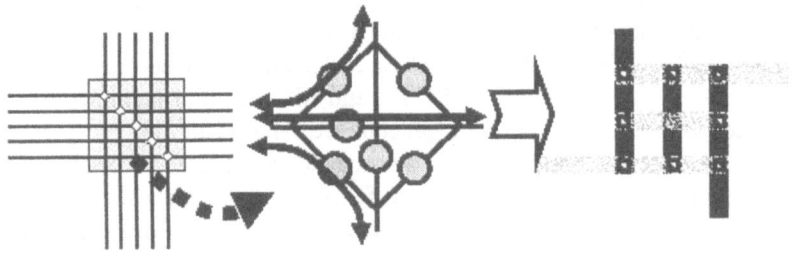

Fig. 1. VPGA Switchpoint Design

consists of an array of PLBs, and a fixed interconnect architecture. However, there are two key differences: First, the routing architecture is laid on top of, instead of adjacent to the PLB array, resulting in a significant reduction in die area. Second, the customization of the logic and interconnect is done by the placement or removal of vias at the potential via locations, as opposed to configuring SRAM bits.

Figure 1 illustrates a via-patterned switchpoint which replaces between 36 to 120 transistors for an FPGA switchpoint with 8 potential mask configurable vias. Grouping many of these via-patterned switchpoints results in a via-patterned switchbox. This switchbox is similar to an FPGA switchbox in that each track requires a switchpoint to connect to the same track in a neighboring switchbox. The difference, however, is that there are no pass transistors to define the connectivity of the switchpoints. The switchbox shown in Figure 2 is a 3x3 arrangement of metal lines with vias inserted at the possible connectivity points.

The key objective of the VPGA fabric is to combine the cost amortization benefits of FPGAs with a level of performance close to ASICs, for a *domain of ap-*

Fig. 2. VPGA Switchbox Layout

plications. The choice of PLB architecture is crucial for achieving this objective. In [4] the authors investigated the optimal LUT size for homogeneous LUT-based PLBs for the VPGA fabric. However, [2] and [3] showed that heterogeneous PLBs with a combination of logic gates, LUTs, and MUXes offer significantly better performance and density than homogeneous LUT-based PLBs.

In this paper, we investigate four heterogeneous PLB architectures for via-patterned programmable fabrics. Using a fabric-specific synthesis engine [2], we map a set of standard benchmarks to each of the PLBs, and compare their performance, density and fabric utilization characteristics across applications. Based on this analysis, we select a high performance heterogeneous logic block architecture for the VPGA fabric. Although this analysis is conducted in the context of the VPGA architecture, the proposed heterogeneous PLBs and the experimental methodology can be employed for any embedded programmable fabric.

The remainder of this paper is organized as follows. Section 2 illustrates the proposed heterogeneous architectures, followed by a description of the experimental methodology in Section 3. Section 4 compares the performance, density and fabric utilization characteristics of each of the PLBs across a host of applications. Based on these results, we select a PLB architecture for the VPGA fabric. Section 5 concludes the paper.

2   Heterogeneous Programmable Logic Blocks

There are two key design considerations for heterogeneous PLBs. The first is to determine which logic gates are suitable and which combinations of LUT sizes to employ. The second is to explore various configurations of these logic elements.

2.1   Logic Gates for Heterogeneous PLBs

Nand Gates with Programmable Inversion. It can be shown that a 2-input Nand gate with programmable inversion on the inputs and outputs (ND2WI) can implement all 2-input functions except exclusive or (XOR) and exclusive nor (XNOR). Similarly, it can be shown that a 3-input Nand gate with programmable inversion (ND3WI) can implement 46 of the 256 3-input functions. In [3], twenty MCNC benchmarks were mapped to 3-LUTs and 4-LUTs with Flowmap. For most of the benchmarks, between 40-80% of the functions in the 3-LUT mapped netlists could be implemented by a ND3WI gate. The ND3WI gate also achieved significant coverage of functions in the 4-LUT mapped netlists. These results show that ND3WI gates are good candidates for heterogeneous PLBs.

3-input Semi-LUTs. [3] proposed a logic structure called an S3 gate (3-input Semi-LUT) consisting of two ND2WI gates driving a 2:1 MUX. Since each of the ND2WI gates can implement 14 of the 16 2-input functions, this structure can implement at least 196 of the 256 3-input functions. For the same set of benchmarks as in the ND3WI analysis, [3] showed that the S3 gate could implement over 90% of the functions in the 3-LUT mapped netlists for well over half the benchmarks. Furthermore, the S3 gate consistently achieved a coverage between 20% to 40% for 4-LUT mapped netlists as well. Hence, we also consider the S3 gate for our heterogeneous PLBs.

2.2 Proposed Heterogeneous PLB Architectures

In previous work, [2] showed that a heterogeneous PLB consisting of two 3-LUTs and one 2-LUT is 30% more area-efficient than a homogeneous 4-LUT based PLB. Also, [5] showed that a combination of 4 and 2-input LUTs result in significant density benefits. Based on these results and the above analysis of logic gates, we propose four heterogeneous PLBs for the VPGA fabric. These PLBs are illustrated in Figure 3. Although not shown in Figure 3 all of these PLBs also contain I/O buffers and a D flip flop. The I/O buffers ensure that all primary inputs and outputs are available in any polarity.

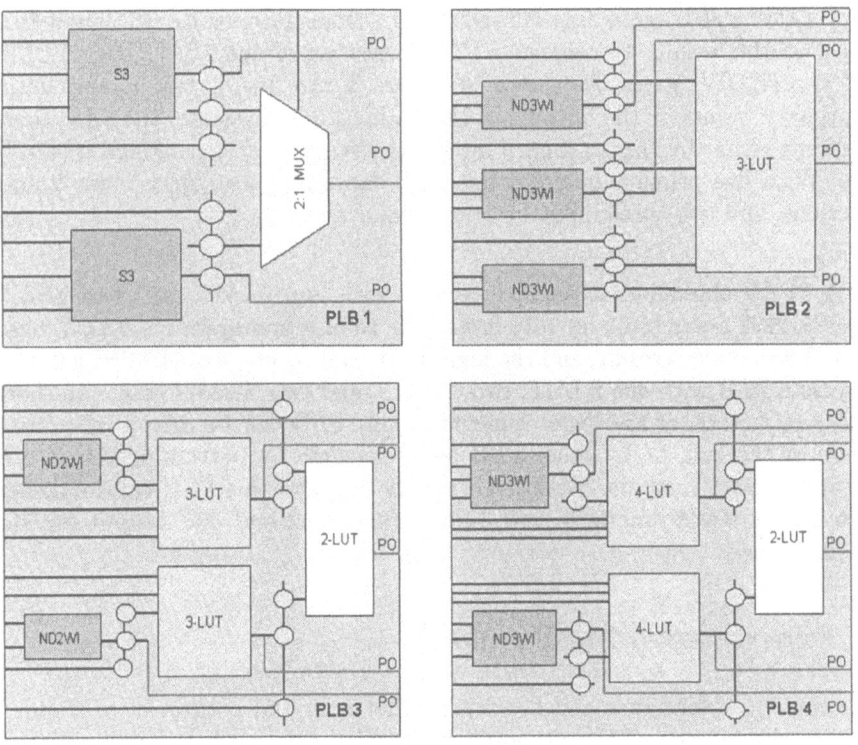

Fig. 3. Proposed Heterogeneous Logic Block Architectures

PLB 1: The first PLB consists of two S3 gates and a 2:1 MUX. This cell has nine unique primary inputs and three primary outputs. As shown in Figure 3

each of the MUX inputs can connect either to a primary input or the output of an S3 gate by placing or removing the appropriate vias on the corresponding jumper wires. Since each of the S3 gates can implement all 2-input functions the entire PLB can be utilized as a 3-LUT. Alternatively, since each S3 gate can implement several 3-input functions, this PLB can simultaneously implement two 3-input functions and a 2-input function (in the MUX). Since a large % of functions in typical 3-LUT mapped netlists can be implemented by a single S3 gate, PLB 1 is more likely to be utilized in the latter configuration.

PLB 2: The second PLB architecture consists of one 3-LUT and three ND3WI gates. Each of the LUT inputs can be connected either to a primary input, or one of the ND3WI gates by patterning the vias on the jumper wires. This PLB has twelve unique inputs and four primary outputs, enabling it to simultaneously implement four 3-input functions.

PLB 3: [2] showed that a heterogeneous PLB consisting of two 3-LUTs and one 2-LUT is 30% more area-efficient than a homogeneous 4-LUT based PLB. Based on this result, we propose a PLB consisting of one 2-LUT, two 3-LUTs, and two ND2WI gates. As shown in Figure 3, the 2-LUT can connect either to primary inputs or the outputs of the 3-LUTs. Furthermore, the 3-LUTs can be driven either by three primary inputs, or two primary inputs and a ND2WI gate. With five primary outputs, this PLB can implement up to three 2-input functions, and two three input functions simultaneously.

PLB 4: [5] showed that an architecture with one 2-LUT, and two 4-LUTs requires 22% fewer bits and 10% fewer pins than a homogeneous 4-LUT based PLB. Based on this result, and the high functional coverage of ND3WI gates, we propose a PLB with one 2-LUT, two 4-LUTs, and two ND3WI gates. As shown in Figure 3, each of the logic elements in this PLB can be driven by primary inputs of the cell, or by other PLB logic elements by patterning the vias on the appropriate jumpers. With five primary outputs, this PLB can implement up to two 4-input functions, two 3-input functions, and one 2-input function simultaneously.

3   Experimental Methodolody

To compare performance and density, we first map and pack a set of standard benchmarks to each of the proposed PLBs with a fabric-specific approach [2] that captures and exploits the benefits of heterogeneity. Next, we place and route the netlists with VPR [7]. To emulate the VPGA routing architecture, we make some modifications to the VPR PLB and delay models. In this section, we discuss the area, delay, and VPR models for our heterogeneous PLBs.

3.1   Area and Delay Models

In the VPGA fabric, the LUTs in the logic blocks are also via-patterned. As shown in Figure 4, the LUT is constructed as a K-1 level tree with a complementary pull up and pull down network.

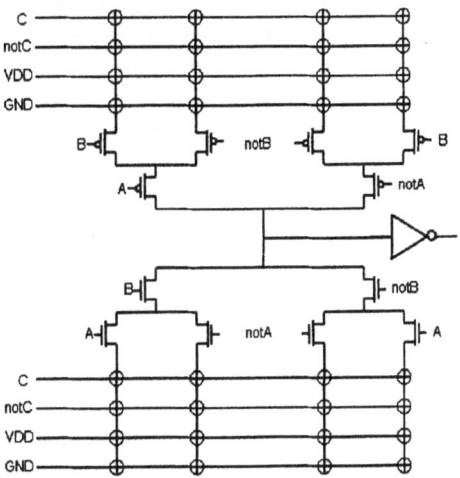

Fig. 4. Via-patterned 3-LUT

Each of the leaf nodes in the LUT can directly connect to VDD, GND, and another input or its complement. We estimate the layout area of the 3-LUT shown in Figure 4 as the area required for three 2:1 MUXes. The layout area of the MUX, and other logic gates in the heterogeneous PLBs, is estimated from the technology specifications for the corresponding minimimum sized gates in our commercial 0.13 μ process. The total PLB area is the sum of the areas for all its logic elements including I/O buffers at each of the PLB inputs, and a single D flip flop.

To determine the delay of the logic elements, we ran HSPICE simulations with an output load comparable to an output buffer. To remain consistent with our area model and control the complexity of the experiment, we kept all transistors minimum sized. For simplicity, the LUTs were configured as NANDs for these simulations. Table 1 summarizes the area and delay values for the logic elements.

3.2   VPR Logic Block Model

As illustrated in Figure 5, VPR assumes that the logic blocks consist of a cluster of N basic logic elements (BLE), each containing a K-input LUT and a register. Furthermore, VPR assumes a fully connected local routing architecture [6] in which each BLE input can connect to a primary input, or the output of any

Table 1. Area and Delay Values

Logic Element	Area (μ sq.)	Delay (ps)
Inverter	6.052	15
ND2WI	24.208	45
ND3WI	32.277	55
S3	64.554	65
2:1 MUX	16.138	20
2-LUT	16.138	50
3-LUT	48.414	70
4-LUT	112.966	90

of the N BLEs via a multiplexer. To evaluate the proposed PLBs with VPR we model each of the PLB logic elements as a K-input BLE, with K set to the number of inputs for the largest logic element. As the VPGA fabric is via-patterned, the connection block and local routing MUXes in Figure 5 are replaced by vias. Hence, the corresponding delay values are set to zero. Similarly, the resistance and capacitance values for the switchboxes are modified to reflect the via-patterned routing, and the combinational delays for the logic elements are set to the values shown in Table 1.

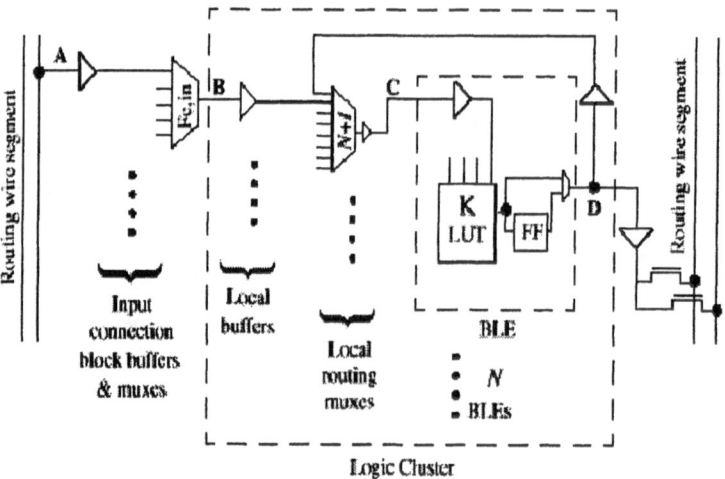

Fig. 5. VPR Logic Block and Delay Model

4 Experimental Results

With the assumptions outlined in the previous section, we place and route the packed netlists with VPR. Then, we compare the total layout area and critical path delay for the benchmark set on each of the architectures. Since the global routing is on top of the CLBs, the total layout area is computed as: max(Routing area, PLB Area)* #PLBs. Furthermore the routing area is estimated as $[(N + (2 * \sqrt{N})) * 0.4]^2$, where:

N = Channel width required by VPR to successfully route the packed netlist
0.4 = Pitch width + Minimum spacing for our 0.13μ process

Since one of the key objectives of the VPGA fabric is to achieve high performance for a *domain of applications*, we present the results in an application-specific manner. Table 2 presents the area, delay, and area-delay product for each of the PLBs across the two application domains. Results for each benchmark are averages over five VPR runs with different placement seeds. The values in Table 2 represent the sum across all the benchmarks in the corresponding application domain.

Table 2. Area, Delay, and Area-Delay Product Results

PLB	Logic Circuits			Datapath Circuits		
	Total Area	Delay	Area*Delay	Total Area	Delay	Area*Delay
	(μ sq.)	(ns)	(*E-03)	(μ sq.)	(ns)	(*E-03)
1	83498	12.26	1.02	62909	14.46	0.91
2	61083	11.06	0.68	47577	12.94	0.62
3	111303	12.78	1.42	84483	14.42	1.22
4	175428	12.56	2.20	131128	14.06	1.84

From Table 2 we observe that PLB 2 has the lowest area, critical path delay, and area-delay product for both logic and datapath circuits. For the logic circuits, PLB 1 has the second best area, delay and area-delay product. However, PLB 1 has the worst delay across the datapath applications. For further analysis of this result, Figure 6 presents the critical path delay for each of the datapath applications. The difference in performance is particularly distinctive for the last five circuits in Figure 6. Across these applications, PLB 2 has a total critical path delay of 11.1 ns, as opposed to 12.3 ns for PLB 4, 12.5 ns for PLB 1 and 12.6 ns for PLB 3. This represents roughly a 10% performance advantage for PLB 2. To explain this result, we examine the percentage utilization (ratio of utilized to available resources) of the ND3WI gates and 3LUT for each of these applications in Figure 7. From Figure 7 we observe that the ND3WI gate utilization is noticeably higher for the ALU, error correction and priority decoder circuits than for the other applications.

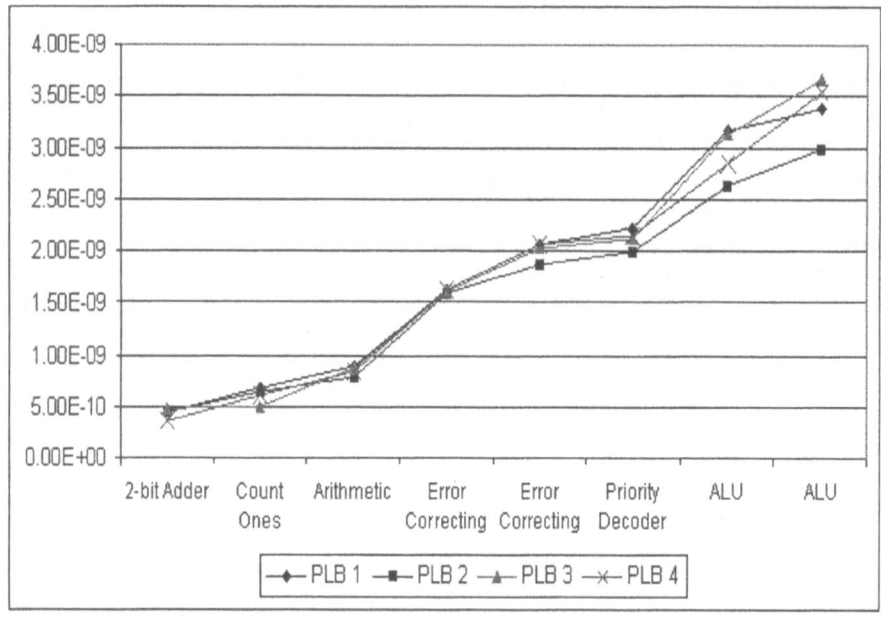

Fig. 6. Delay Comparison for Datapath Applications

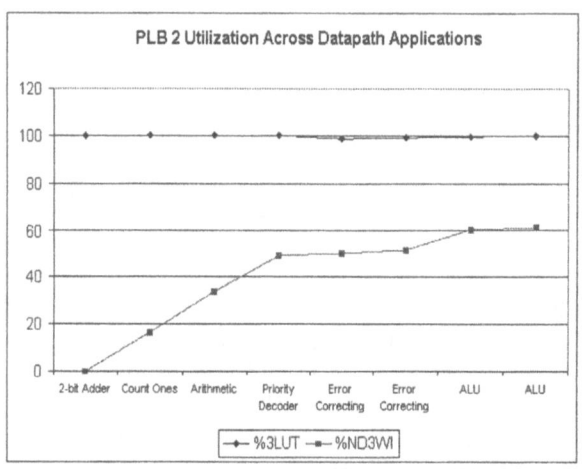

Fig. 7. PLB 2 Fabric Utilization Profile

In [8], the authors presented a physically heterogeneous programmable fabric with a logic array consisting of two kinds of PLBs: a homogeneous PLB with four 4-LUTs, and a CPLD style PAL block. Although we cannot include these results due to space limitations, current work in [9] shows that the heterogeneous PLBs

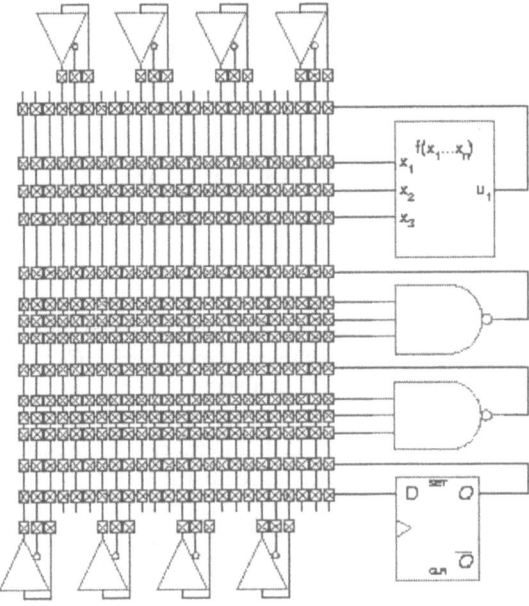

Fig. 8. Selected PLB for the VPGA Fabric

discussed in this paper are more area-efficient than the physically heterogeneous fabric presented in [8] as well.

Based on the results of this section, we select the PLB shown in Figure 8 for the VPGA fabric in this process technology. This cell contains only two ND3WI gates in order to reduce the number of input and output pins. Results presented in [10] show that the VPGA fabric with this logic block is quite competitive with standard cells.

5   Conclusions

In this paper we proposed four new heterogeneous PLBs for via-patterned fabrics and evaluated their performance, density and fabric utilization characteristics for different application domains. Our results suggest employing a heterogeneous PLB consisting of 3-input LUTs and ND3WI gates. Although this analysis was conducted in the context of the VPGA architecture, the proposed heterogeneous PLBs can be employed in any embedded programmable fabric. In future work, we propose to study other via-patterned global routing architectures. We expect the results of these studies to be influenced primarily by variations in delay, since the density benefits of the proposed VPGA PLB are dictated by the high functional coverage of its logic elements. We also propose to explore heterogeneous PLB architectures for more application domains.

Acknowledgements

We would like to thank Chetan Patel and Kim Yaw Tong for the delay simulations and Anthony Cozzie for the VPGA routing area model.

References

1. L. Pileggi, H. Schmit, J. T. Shah, K. Y. Tong, C. Patel and V. Chandra: A Via Patterned Gate Array (VPGA), Technical Report Series, Center for Silicon System Implementation, No. CSSI 02-15, March 2002
2. A. Koorapaty and L. Pileggi: Modular, Fabric-specific Synthesis for Programmable Architectures, Proceedings of the 12th International Conference on Field Programmable Logic and Applications, September 2002
3. A. Koorapaty, V. Chandra, K. Y. Tong, C. Patel, L. Pileggi, and H. Schmit : Heterogeneous Programmable Logic Block Architectures, Proceedings of Design Automation and Test in Europe, March 2003
4. C. Patel, A. Cozzie, H. Schmit, and L. Pileggi: An Architectural Exploration of Via Patterned Gate Arrays, ACM/SIGDA International Symposium on Physical Design, 2003
5. J. He, and J. Rose: Advantages of Heterogeneous Logic Block Architectures for FPGAs, IEEE Custom Integrated Circuits Conference, pp. 7.4.1-7.4.5, May 1993
6. V. Betz, and J. Rose: Cluster-Based Logic Blocks for FPGAs: Area-Efficiency vs. Input Sharing and Size, IEEE Custom Integrated Circuits Conference, pp. 551-554, 1997
7. V. Betz, J. Rose, and A. Marquardt: Architecture and CAD for Deep-Submicron FPGAs, Kluwer Academic Publishers, 1999
8. A. Kaviani, and S. Brown: The hybrid field-programmable architecture, IEEE Design & Test of Computers, Vol. 16, Issue 2, pp. 74-83, April-June 1999
9. A. Koorapaty: Modular, Fabric-specific Synthesis and Novel Logic Block Architectures for Regular Fabrics, Ph.D. Thesis, Center for Silicon System Implementation, Carnegie Mellon University, 2003
10. L. Pileggi, H. Schmit, A. J. Strojwas, P. Gopalakrishnan, V. Kheterpal, A. Koorapaty, C. Patel, V. Rovner, and K. Y. Tong: Exploring Regular Fabrics to Optimize the Performance-Cost Tradeoff, Proceedings of Design Automation Conference, June 2003

A Real-Time Visualization System for PIV

Toshihito Fujiwara, Kenji Fujimoto, and Tsutomu Maruyama

Institute of Engineering Mechanics and Systems, University of Tsukuba,
1-1-1 Ten-ou-dai Tsukuba Ibaraki 305-8573 Japan,
fujiwara@darwin.esys.tsukuba.ac.jp

Abstract. Particle image velocimetry (PIV) is a method of imaging and analyzing field of flows. In the PIV method, small windows in an image of the field (time t) are compared with areas around the windows in the another image of the field (time $t + \Delta t$), and the most similar part to the windows are searched using two dimensional cross-correlation function. The computational complexity of the function is very huge, and can not be processed in real-time by micro-processors. In this paper, we describe a real-time visualization system for the PIV method. In the system, an improved direct computation method is used to reduce the computational complexity. The system consists of only one off-the-shelf Virtex-II FPGA board and a host computer, and calculates the complex function without reducing data bit-width, which becomes possible with one latest FPGA.

1 Introduction

Particle image velocimetry (PIV) is a method for visualizing and analyzing field of flows. In the PIV method, a pair of images (time $= t$ and $t + \Delta t$) of the field is compared, and the flows in Δt are visualized and analyzed. In order to compare two images of the field, many small particles with the same specific gravity of the liquid or air in the field are added and the speed and direction of movement of the particles are measured by finding out pairs of small areas (windows) which are the most similar to each other in the two images. In order to find out the pairs, two dimensional cross-correlation function is used in general. The amount of computations in the function is very huge, and can not be processed in real-time by micro-processors.

We have already proposed a basic idea for efficient computation of the cross-correlation function with FPGA[2]. In that implementation, however, data width of pixels in the images and the intermediate data of the function had to be reduced because of the limitation of the hardware resources and memory width (internal and external), which caused errors in the final results. Because of the recent progress of the size of FPGAs and multipliers built in FPGAs, it became possible to compute the function without reducing the data width, which generates the same results with micro-processors. However, the implementation to obtain the correct results in real-time is not straightforward. In our system, an improved direct computation method of the two dimensional cross-correlation function is used to achieve real-time processing.

P.Y.K. Cheung et al. (Eds.): FPL 2003, LNCS 2778, pp. 437–447, 2003.

In this paper, we first discuss computation methods of the cross-correlation function for real-time processing (the best method depends on the requirements by the application and the performance of the FPGA), and then describe a real-time visualization system which consists of one Virtex-II FPGA board and a host computer.

2   Particle Image Velocimetry (PIV) System

Figure 1 shows an overview of a particle image velocimetry (PIV) system. In PIV method, pairs of images of the field (time t and $t + \Delta t$) with many particles are taken by a CCD camera. Then, the speed and direction of movement of the particles in the field are measured by comparing the pairs of images.

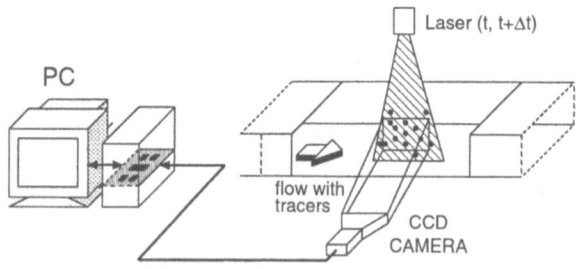

Fig. 1. System Overview

Figure 2 shows a pair of images of the field (t and $t + \Delta t$). Suppose that we want to know where particles in a window (x, y) (a rectangular from (x, y) to $(x+n-1, y+m-1)$) at time t are moving to, and how fast they are moving. Then, the window is compared with all windows in a target area (light gray square in Figure 2(B)), and the most similar window (dark gray square) is searched using cross-correlation function, and the direction of movement and the speed of the particles are obtained.

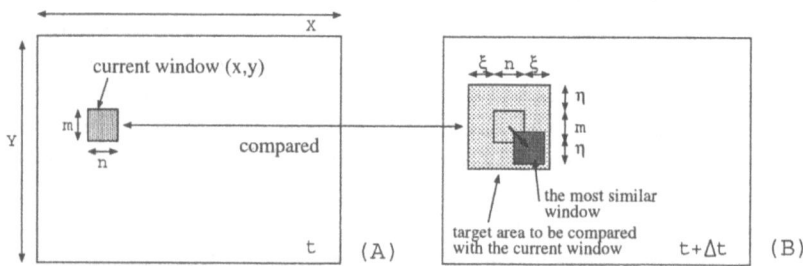

Fig. 2. Particle Image Velocimetry (PIV) Method

The cross-correlation function between two windows (window (x, y) at time t and window $(x + \xi, y + \eta)$ at time $t + \Delta t$) can be described as follows. In this function, I_1 and I_2 show values of pixels of images at time t and $t + \Delta t$ respectively.

$$R(x, y, \xi, \eta) = \frac{\sum_0^{n-1} \sum_0^{m-1} \{I_1(x+i,y+j) - \bar{I}_1\}\{I_2(x+i+\xi,y+j+\eta) - \bar{I}_2\}}{\sqrt{\sum_0^{n-1} \sum_0^{m-1} \{I_1(x+i,y+j) - \bar{I}_1\}^2} \sqrt{\sum_0^{n-1} \sum_0^{m-1} \{I_2(x+i+\xi,y+j+\eta) - \bar{I}_2\}^2}}$$

$$\text{where} \quad \bar{I}_1 = \sum_0^{n-1} \sum_0^{m-1} I_1(x+i, y+j)/nm$$

$$\bar{I}_2 = \sum_0^{n-1} \sum_0^{m-1} I_2(x+i+\xi, y+j+\eta)/nm$$

The computational complexity of the cross-correlation is very huge, and can not be processed in real-time by micro-processors. In our system, an FPGA board (ADC XRC-II by Alpha Data with one XC2V6000) with one additional SRAM board is used to support eight memory banks (each bank has 4MB (or 8MB) memory with 32 bit width). These eight memory banks are necessary to realize real-time processing by an improved direct computation method described below without reducing data-width of the cross-correlation function.

3   Computation of the Cross-Correlation Function

3.1   Requirement for Real-Time Processing

In our target PIV system[1], the size of images is 1008×1008, and 20 pairs of images are compared in one second. The size of windows is up to 30×30, and the range of ξ and η is $[-10$ to $10]$. Suppose that the window size is 24×24, then the number of windows which have to be searched in one pair of images becomes $84(1008/24 * 2) \times 84$ (the position of the windows is moved to right and down by the half of the window size). Under these assumptions, we need to find a pair of the most similar windows in 470 clock cycles in 66 MHz operation frequency in order to realize real-time processing.

3.2   Computation Based on FFT

FFT is often used to compute cross-correlation functions. In FFT, size of the windows has to be $2^k \times 2^l$. Thus, in order to satisfy the requirements for real-time processing, window size must be 64×64 (because the size of the window + the range of ξ and η is larger than 32 and less than 64). As for windows at time t, enlarged area (from 24×24 to 64×64) is filled with 0, while 64×64 pixels are clipped off from images as for windows at time $t + \Delta t$. Figure 3 shows the outline of the computation based on FFT. In Figure 3, FFT values of two windows (time $= t$ and $t + \Delta t$) are first calculated, and the FFT values are multiplied. Then, IFFT values of the multiplied values are calculated and the maximum value is selected to find out the most similar window. In order to discuss the computation time, we focus on the number of multiply operations which is the most time exhaustive (and hardware resource exhaustive) part in the function. As show in Figure 3, the number of multiply operations required is

$$(((32 \times 4) \times log_2 64) \times 64) \times 2 \times 3 + 64 \times 64 \times 4$$

when the row-column two dimensional FFT is used, because one multiply operation of complex numbers requires four integer multiply operations. The key point in the computation based on FFT is that the number of the operations does not change while the size of windows + the range of ξ and η is less than 64 (and larger than 32).

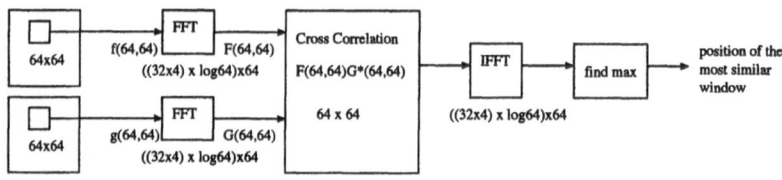

Fig. 3. Outline of the Computation of Cross Correlation Function based on FFT

Our current target FPGA is XC2V6000 (one of the largest FPGAs which is available now). With this FPGA,

1. we can execute 128 multiply operations in parallel because XC2V6000 has 144 built-in multipliers,
2. other add/sub operations can be executed in parallel and in pipeline with the multiply operations,
3. FFT and IFFT can be executed on the same butterfly circuit (with a connection based on omega multi-stage network) by changing only the constant values used for the multiply operations (these values are stored in Block RAMs, and can be easily changed by changing the addresses to the Block RAMs), and
4. 64×64 multiply operations for cross-correlation can be executed on the same butterfly circuit.

Thus, the minimum computation time becomes 2432 clock cycles ignoring any overhead for pipeline processing.

3.3 Direct Computation

Direct computation is often used when window size is small. The total number of multiply operations in this method becomes

$$n \times n \times r \times r$$

where $n \times n$ is the size of windows, and r is the range of ξ and η. When the n is 24 and r is 21, the minimum computation time becomes 1764 clock cycles on XC2V6000 because 144 multiply operations can be executed in parallel, and other add/sub operations can be executed in parallel and pipeline with the multiply operations. This is faster than the computation based on FFT, but still slower than the requirement for real-time processing. The important point in the direct computation is that the computation time depends on the size of windows and the range of ξ and η. If ξ and η vary from -16 to 16, the computation time becomes 4356 clock cycles, which is much slower than the computation based on FFT.

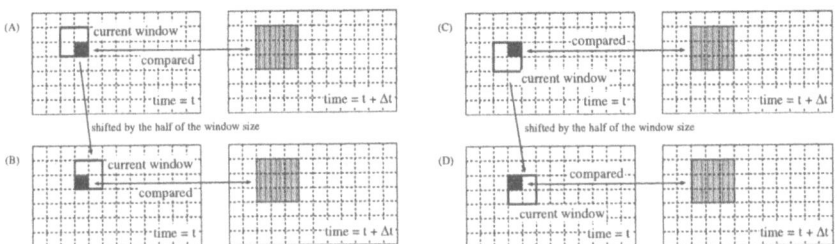

Fig. 4. Redundant Computation in the Direct Computation Method

In case of the direct computation, we can reduce the computation time to 25% of the clock cycles described above, when the CCD camera of the PIV system is placed just above the field of flows and positions of windows are moved to right and down by the half of window size. In Figure 4(A), dark gray part of the current window is compared with light gray part of the image of $t + \Delta t$. Then, the current window is shifted to right by $n/2$, and dark gray part in Figure 4(B) (same with the dark gray part in Figure 4(A)) is compared with the same light gray part again. This dark gray part is compared two more times as shown in Figure 4(C) and (D). In order to avoid these redundant computations, current windows are divided to four sub-windows in our implementation. These sub-windows are calculated first, and their results (intermediate data) are stored in memory. Then, the intermediate data are read back from memory and the cross-correlation for original current windows are calculated. With this optimization, we can reduce the computation clock cycles to 25% of the non-optimized method. Thus, the clock cycles are reduced to 441 clock cycles which satisfy the requirement for real-time processing.

3.4   FPGA Size and Better Computation Method

Because of the reason described above, we used the improved direct computation method for our real-time system. The continuous progress of the size of FPGAs will, however, make it possible to achieve real-time processing by both computation methods in a few years. Then, the computation based on FFT is superior to the direct computation, because in the computation based on FFT, we do not have to care about window size and the range of ξ and η unless sum of them exceed 64 (which is not common case in PIV applications). Users can change window size and the range of the variables at any time of their experiments without changing circuit on the FPGA.

4   Details of the PIV System

Eight memory banks supported in our system, however, is not enough for naive implementation of the optimized direct computation. In this section, we describe the implementation of the optimized direct computation method.

4.1   Details of the Improved Direct Computation

From the view point of the optimization described above, the cross-correlation function for window (x, y) and window $(x+\xi, y+\eta)$ can be transformed as follows (first term of the denominator in the previous definition is deleted because it is a constant).

$$R(x,y,\xi,\eta) = \frac{MAC12(x,y,\xi,\eta) - SUM1(x,y) \times SUM2(x,y,\xi,\eta)/nm}{\sqrt{MAC22(x,y,\xi,\eta) - SUM2(x,y,\xi,\eta) \times SUM2(x,y,\xi,\eta)/nm}}$$

where

$$MAC12(x,y,\xi,\eta) = mac12(x,y,\xi,\eta) + mac12(x,y+m/2,\xi,\eta)$$
$$+ mac12(x+n/2,y,\xi,\eta) + mac12(x+n/2,y+m/2,\xi,\eta)$$
$$MAC22(x,y,\xi,\eta) = mac22(x,y,\xi,\eta) + mac22(x,y+m/2,\xi,\eta)$$
$$+ mac22(x+n/2,y,\xi,\eta) + mac22(x+n/2,y+m/2,\xi,\eta)$$
$$SUM1(x,y) = sum1(x,y) + sum1(x,y+m/2) + sum1(x+n/2,y)$$
$$+ sum1(x+m/2,y+n/2)$$
$$SUM2(x,y,\xi,\eta) = sum2(x,y,\xi,\eta) + sum2(x,y+m/2,\xi,\eta)$$
$$+ sum2(x+n/2,y,\xi,\eta) + sum2(x+m/2,y+n/2,\xi,\eta)$$
$$mac12(x,y,\xi,\eta) = \sum_{0}^{n/2-1} \sum_{0}^{m/2-1} \{I_1(x+i+\xi,y+j+\eta) \times I_2(x+i+\xi,y+j+\eta)\}$$
$$mac22(x,y,\xi,\eta) = \sum_{0}^{n/2-1} \sum_{0}^{m/2-1} \{I_2(x+i+\xi,y+j+\eta) \times I_2(x+i+\xi,y+j+\eta)\}$$
$$sum1(x,y) = \sum_{0}^{n/2-1} \sum_{0}^{m/2-1} I_1(x+i,y+j)$$
$$sum2(x,y,\xi,\eta) = \sum_{0}^{n/2-1} \sum_{0}^{m/2-1} I_2(x+i+\xi,y+j+\eta)$$

Therefore, by storing four kinds of values ($mac12$, $mac22$, $sum1$ and $sum2$) in memory banks (one value for $sum1$ and 21×21 values for $mac12$, $mac22$ and $sum2$ in each sub-window because ξ and η vary from -10 to 10) and reading them out afterward, we can compute the cross-correlation of original windows. If we can store the results of four sub-windows in different memory banks, they can be read out at the same time for the computation of original windows. However, eight memory banks in our system is not enough to store them in different memory banks.

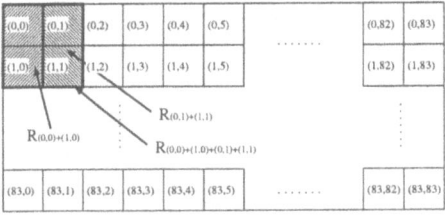

Fig. 5. An Image divided to Sub-windows

Figure 5 shows an image of the field of flows (1008×1008) divided to 84×84 sub-windows (size is 12×12). Sub-windows are processed from $(0, 0)$ to $(83, 83)$, and results of sub-windows $(n, k)_{n\%2=0}$ are stored in memory banks (A), while results of sub-windows $(n, k)_{n\%2=1}$ are stored in memory banks (B). In our implementation, results of only two sub-window rows are stored in memory banks (A) and (B) (results of sub-window (n, k) are overwritten by the results of sub-window $(n + 2, k)$).

In Figure 5, suppose that computation of sub-windows $(0, k)_{k=0,83}$ are finished. In the computation of the next sub-window $(1, 0)$,

1. $R_{(1,0)}$ (results of sub-window $(1,0)$) are calculated and written to memory banks (B),
2. $R_{(0,0)}$ (results of sub-window $(0,0)$) are read from memory banks (A),
3. $R_{(0,0)}$ and $R_{(1,0)}$ are added ($R_{(0,0)+(1,0)}$ are obtained) and stored in a temporal memory in the FPGA.

Then, in the computation of the next sub-window $(1, 1)$,

1. $R_{(1,1)}$ are calculated and written to memory banks (B),
2. $R_{(0,1)}$ are read from memory banks (A),
3. $R_{(0,1)}$ and $R_{(1,1)}$ are added ($R_{(0,1)+(1,1)}$),
4. $R_{(0,1)+(1,1)}$ are added to $R_{(0,0)+(1,0)}$ in the temporal memory ($R_{(0,0)+(1,1)+(0,1)+(1,1)}$ are obtained), and $R_{(0,1)+(1,1)}$ are stored in the temporal memory instead of $R_{(0,0)+(1,0)}$,
5. the cross-correlation function of the original window is calculated using $R_{(0,0)+(1,0)+(0,1)+(1,1)}$

By repeating this procedure to all sub-windows, we can calculate cross-correlation function of all windows. In this processing, only three banks of memory (memory banks (A), memory banks (B) and one internal temporary memory) are used, which can be supported using eight external memory banks and the FPGA in our system.

4.2   Details of the Circuit

Figure 6 shows the block diagram of the circuit implemented on the FPGA. Pairs of images of the field of flows are sent from the CCD camera and stored in SRAM bank0 and bank1. Two bank of SRAM are used to avoid memory access conflicts by the FPGA and the camera. When one pair of images in bank0(bank1) is processed by the FPGA, next pair of images is transferred to bank1(bank0).

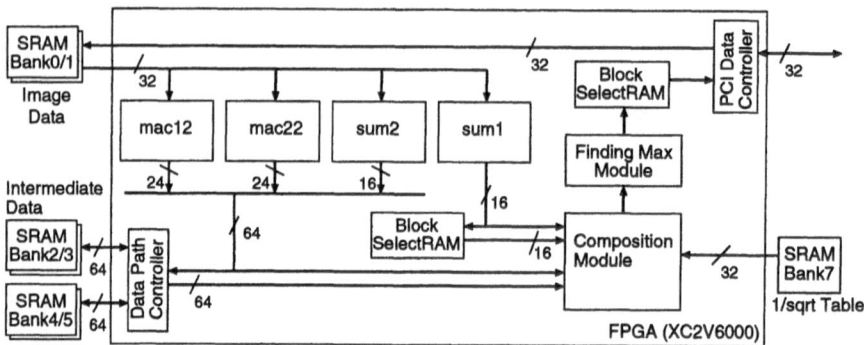

Fig. 6. Block Diagram of the Circuit

The circuit on the FPGA consists of six modules; $mac12$ module, $mac22$ module, $sum1$ module, $sum2$ module, composition module and finding max module. In $mac12$, $mac22$, $sum1$ and $sum2$ modules, values of $mac12$, $mac22$, $sum1$ and $sum2$ of sub-windows are calculated respectively. Outputs of $sum1$ module are stored in one Block RAM, because the number of the outputs for 84 sub-windows $((n, k)_{k=0,83})$ is 84 with 16 bit-width, and Block RAMs support dual-port access. Two sets of two memory banks (bank2/3 and bank4/5) are used for memory banks (A) and (B) because the number of outputs by $mac12$, $mac22$ and $sum2$ for 84 sub-windows is $84 \times 21 \times 21$ with 64 bit-width in total. The cross-correlation function for windows is calculated in the composition module using values stored in the Block RAM and the memory banks (bank 2/3 and 4/5). Bank7 is used to compute $1/sqrt$ in the function. In finding max module, the maximum value of the cross-correlation function is selected and stored into the Block RAM, and the values in the Block RAM are sent back to the host computer.

Figure 7 shows the structure of the composition module. Inputs to this module are $sum1$ from the Block RAM and $sum1$ module, and $mac12/mac22/sum2$

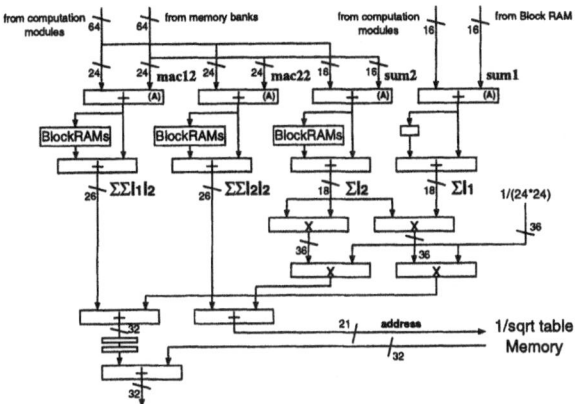

Fig. 7. Composition Module

from the memory banks and $mac12/mac22/sum2$ modules. The values from the memories and the computation modules are added (by adders (A)) to obtain $R_{(n,k)+(n+1,k)}$. Then, the results are stored in Block RAMs, and added with the direct outputs of the adders (A) to obtain $R_{(n,k)+(n+1,k)+(n,k+1)+(n+1,k+1)}$. Then, the cross-correlation function of window is calculated. The only difference from the computation by micro-processors is that the address to the 1/sqrt table is reduce to 21 bits because the memory size of the external memory bank (this cause no difference in the results).

Figure 8 shows the basic idea of the implementation of $mac12$ module[2]. In Figure 8, the size of sub-windows and the range of ξ and η are reduced to 4×4 to simplify the figure. Data of a sub-window (at time $= t$) are stored in a shift register array and data of its target area (at time $= t + \Delta t$) are stored in a distributed RAM array. Data on these two arrays are multiplied in parallel using built-in multipliers (in real $mac12$ module, 144 multiply operations are executed in every clock cycle), and added by pipelined binary tree adders.

Each element in the distributed RAM array stores nine pixel data (for example, LUTs (8 bit-width) labeled 0 (gray part in Figure 8(1)) store nine pixel data labeled 0 (gray parts in target area)). By shifting data in the shift register array to right and down, and by changing addresses to each row and column of the distributed RAM array, we can compute $mac12$ efficiently. In Figure 8(2), multiply-and-accumulate of the sub-window on the shift register array and the gray part in the target area is computed. For this computation, data on the shift register array are shifted five times to right and once to down from Figure 8(1), and addresses to each row and column of the distributed RAM array are given as shown in Figure 8(2) to read out the gray part.

In our implementation, two sets of shift register array and distributed RAM array are used in order to update one of them while the other is used to compute cross-correlation function.

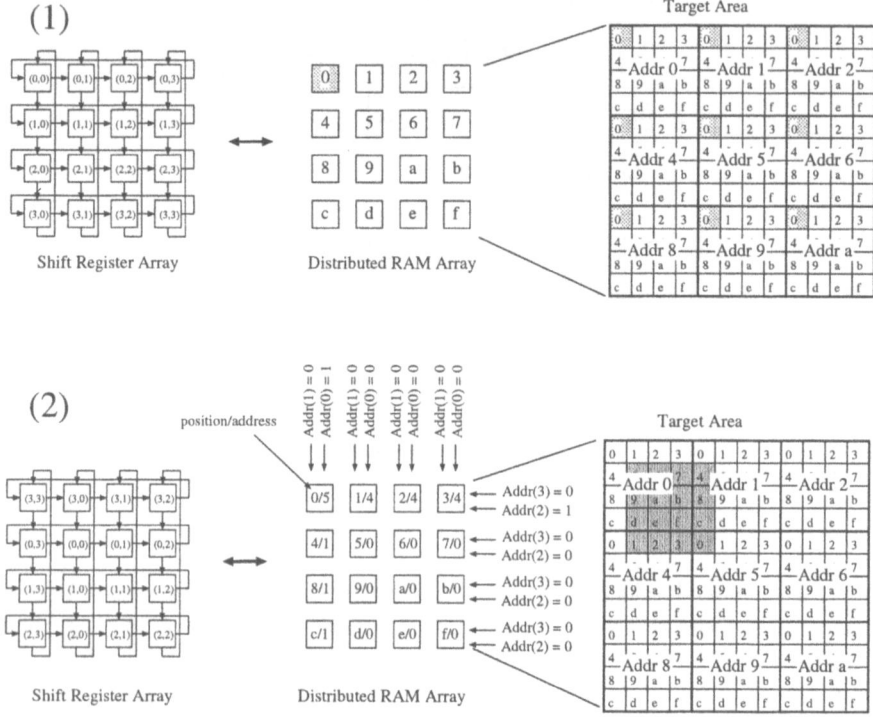

Fig. 8. Basic Idea of $mac12$ module

Figure 9 shows the structure of $mac22$ and $sum2$ modules. In $mac22$ and $sum2$ modules, multiply-and-accumulate and sum of all columns in all sub-windows in the target area ($\sum_{k=y}^{y+11} I_2(x,k) \times I_2(x,k)$ and $\sum_{k=y}^{y+11} I_2(x,k)$) are calculated when data of the target area are loaded to the distributed RAM arrays, and stored in Block RAMs. During the computation of the cross-correlation

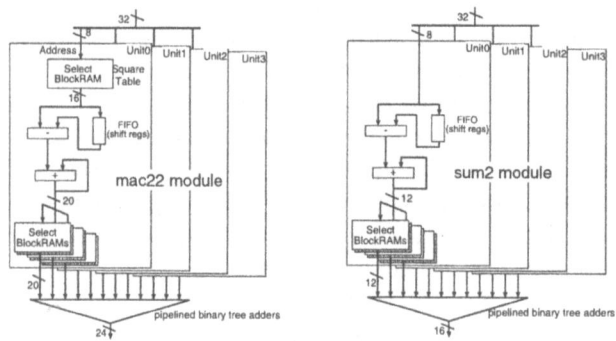

Fig. 9. $mac22$ and $sum2$ modules

function, 12 column data for $mac22$ and $sum2$ are read out in parallel from Block RAMs respectively, and added by binary tree adders to calculate $mac22$ and $sum2$ of the sub-windows.

5 Performance

In our current implementation on XC2V6000, the number of clock cycles to calculate cross-correlations of a sub-window at time t and all sub-windows in its target area (time $t + \Delta t$) is 916 for the first sub-window (sub-window $(0,0)$ in Figure 5) to full-fill all pipeline stages of the circuit, and 441 for other $84 \times 84 - 1$ sub-windows. Therefore, the average clock cycles to find a pair of the most similar windows is almost 441 (less than 470 clock cycles; requirement for real-time processing). Because of this clock cycles, the required operational frequency for real-time processing is relaxed to 62.3MHz. The circuit on XC2V6000 runs at 66 MHz (the maximum operational frequency reported by the CAD is 66.5 MHz). This performance satisfies the requirements for real-time processing.

In the current implementation, all multipliers, 32% of slices and 64 Block RAMs (out of 144) in XC2V6000 are used.

6 Conclusions

In this paper, we described a real-time visualization system of PIV method, which consists of one off-the-shelf FPGA board with one Virtex-II FPGA and a host computer. By using one latest FPGA and an improved direct computation method of the two dimensional cross-correlation function, we could realize real-time processing without reducing data bit-width of the function. Because of the built-in multiplier, the circuit which includes parallel and pipelined multiply-and-accumulate operations for 144 data occupied only 32% of the XC2V6000.

The assumptions for the improvement method, however, can not be always satisfied. We are now developing a circuit to transform a view from a camera which is not placed just above the field of flows to a view from just above the field using the rest area of XC2V6000.

References

1. J. Sakakibara and T. Anzai, "Chain-link-fence structures produced in a plane jet", Physics of Fluids 2001
2. T.Maruyama, Y.Yamaguchi and A.Kawase, "An Approach to Read-time Visualization of PIV Method with FPGA", FPL2001

A Real-Time Stereo Vision System with FPGA

Yosuke Miyajima and Tsutomu Maruyama

Institute of Engineering Mechanics and Systems, University of Tsukuba,
1-1-1 Ten-ou-dai Tsukuba Ibaraki 305-8573 Japan,
miyajima@darwin.esys.tsukuba.ac.jp

Abstract. In this paper, we describe a compact stereo vision system which consists of one off-the-shelf FPGA board with one FPGA. This system supports (1) camera calibration for easy use and for simplifying the circuit, and (2) left-right consistency check for reconstructing correct 3-D geometry from the images taken by the cameras. The performance of the system is limited by the calibration (which is, however, a must for practical use) because only one pixel data can be allowed to read in owing to the calibration. The performance is, however, 20 frame per second (when the size of images is 640 × 480, and 80 frames per second when the size of images is 320 × 240), which is fast enough for practical use such as vision systems for autonomous robots. This high performance can be realized by the recent progress of FPGAs and wide memory access to external RAMs (eight memory banks) on the FPGA board.

1 Introduction

The aim of stereo vision systems is to reconstruct the 3-D geometry of a scene from two (or more) images, which we call *left* and *right*, taken by cameras. Many dedicated hardware systems have been developed for real-time processing, and a stereo vision system with FPGA[1] has achieved real-time processing because of the recent progress of the size and the performance of FPGAs.

Compact systems for stereo vision are especially important for autonomous robots. FPGAs are ideal devices for the compact systems. Depending on situations, a robot may try to reconstruct the 3-D geometry, to find out moving objects which are coming to it, and to find out marker objects to check its position. FPGAs can support all these functions by reconfiguration.

In this paper, as the first step toward a vision system for autonomous robots, we describe a compact real-time stereo vision system which

1. supports camera calibration for easily obtaining correct results with simple circuit, and
2. checks *Left-Right Consistency* to find out occlusions without duplicating the circuit (by only adding another circuit for finding minimum value).

These functions are very imporant to obtain correct 3-D geometry. This system also supports filters for smoothing and eliminating noises to improve the system performance. In order to achieve these functions while exploiting maximum

P.Y.K. Cheung et al. (Eds.): FPL 2003, LNCS 2778, pp. 448–457, 2003.

performance of FPGAs (avoiding memory access conflicts), we need one latest FPGA and eight memory banks on the FPGA board which are just supported on latest off-the-shelf FPGA boards.

The performance of the system is more than 20 frames per second (640×480 inputs and 640×480 output with disparity up to 200), which is much faster than previous works (more than 80 frames if the size of images is 320×240).

This paper is organized as follows. Section 2 describes the overview of stereo vision systems, and details of our system is given in section 3. The performance of the system is discussed in section 4. In section 5, conclusions are given.

2   Overview of Stereo Vision Systems

In order to reconstruct the 3-D geometry of a scene from two images (left and right) taken by two cameras, it is searched which pixels in the two images are projections of the same locations in the scene in the stereo vision systems. In this section, we first discuss the calibration of cameras, which is a very important step to simplify matching computation which is the most time exhaustive part in stereo vision systems, and then discuss matching algorithms and *left-right consistency* to suppress infeasible matches.

2.1   Calibration

Even if the same type of cameras are used to obtain left and right images, the characteristics of the cameras are different, and horizontal (vertical) lines in real-world may not be horizontal (vertical) in the images taken by cemeras. The aim of the calibration is to find out a realtionship (perspective projection) between the 3-D points in real-world and their different camera images. This is a crucial stage in order to simplify the following stages in the stereo vision systems and to obtain correct matching.

2.2   Matching Algorithm

Area-based (or correlation-based) algorithms match small windows centered at a given pixel to find corresponding points between the two images. They yield dense depth maps, but fail within occluded areas. Feature-based algorithms match local cues (e.g., edges, lines, corners) and can provide robust, but sparse disparity maps which requires interpolation. In hardware systems, area-based algorithms are widely used, because the operations required in those algorithms are very regular and simple.

In the area-based algorithms, *epipolar restriction* is used in order to decrease computational complexity. As shown in Figure 1, the corresponding point of a given point lies on its epipolar line in the other image, when the two cameras are arranged so that their principal axes are parallel. Corresponding points can then be found by comparing with every points on the epipolar line on the other image. To use this restriction, calibration of cameras is necessary to guarantee

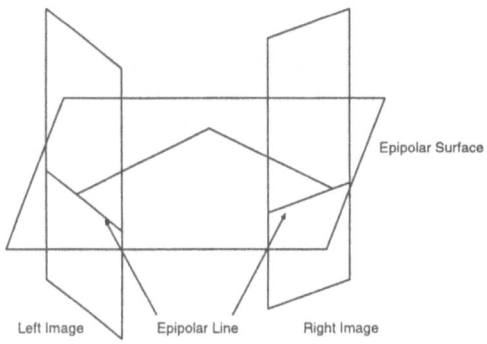

Fig. 1. Epiploar Geometry

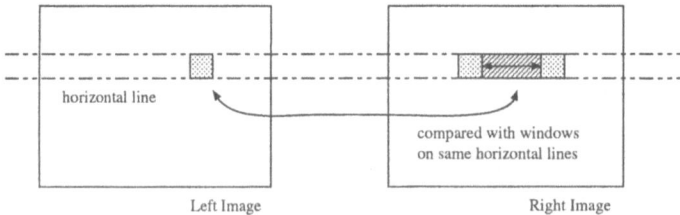

Fig. 2. Stereo Matching on Epilolar Constraint

that objects on a horizontal line in real-world also lie in the same horizontal lines in left and light images taken by the cameras. Then, we need to only compare windows on same horizontal lines in left and right images as shown in Figure 2.

The most traditional area-based matching algorithm is normalized cross-correlation [3], which requires more computation time than the following simplified algorithms. The most common pixel-based matching algorithm are squared intensity differences (SSD)[2] and absolute intensity differences (SAD)[4]. We used the SAD (Sum of Absolute Difference) algorithm because the algorithm is the simplest among them, and the results obtained by the algorithm is alomst same with other algorithms[4]. In SAD algorithm, the value of d which minimizes the following equation is searched.

$$\sum_{i=-n}^{n} \sum_{j=-m}^{m} |I_r(x+i, y+j) - I_l(x+i+d, y+j)|$$

In the equation, I_r and I_l are the right and left image respectively, n and m are the size of the window centered at a given pixel (its position is x and y). d is the disparity, and its range decides how many pixels on the other image are compared with the given pixel. In order to find the corresponding points for an object which is closer to the vision system, larger range for d becomes necessary, though it requires more hardware resources.

2.3   Occlusion and Left-Right Consistency

When pairs of images of objects are taken by two cameras (left and right), some parts of the objects appear in left (right) images and may not apper in right (left) images, depending on the positions and angles between the cameras and the objects. These occlusions are major source of errors in computational stero vision systems, though it has been reported that these occlusions help the human visual system in detecting object boundaries[5].

In many computational systems, one of left and right images is chosen as the base of the matching. Then, windows which include target pixels are selected in the base image, and the most similar windows are searched in another image. If the role of left and right images is reversed, different pairs of the windows may be selected. The so-called *left-right consistency constraint*[6] states that feasible window pairs are those found with both direct and reverse matching.

In our system, occlusions are detected by checking the left-right consistency. Figure 3 shows left image based matching and right image based matching. These matching can be executed in our system without duplicating whole circuit (by adding only another module which consists of comparators and selectors).

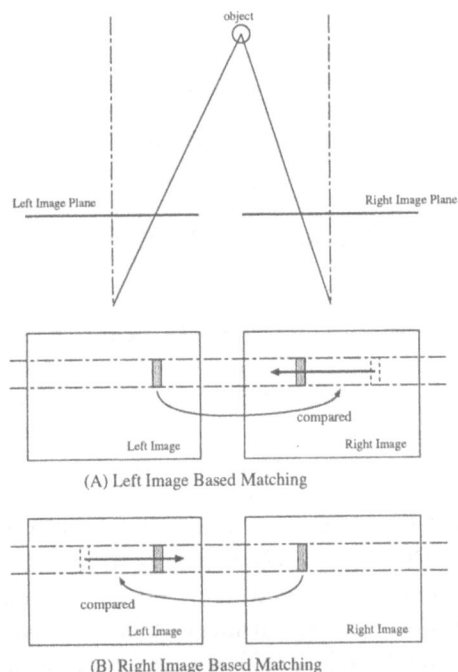

Fig. 3. Left Based / Right Based Matching

2.4   Filters

Filters are often used in the stereo matching for smoothing and eliminating noises to improve the system performance. We prepared Laplacian of Gaussian (LoG) filter for that purpose. Dual-port block RAMs make it possible to implement filters efficiently.

3   Details of the System

3.1   Overview

Figure 4 shows the system overview. Our system consists of a host computer, two cameras and one off-the-shelf FPGA board (ADM-XRC-II by Alpha Data with one additional SSRAM board). Left and right images taken by the two cameras are sent to external RAMs on the FPGA board.

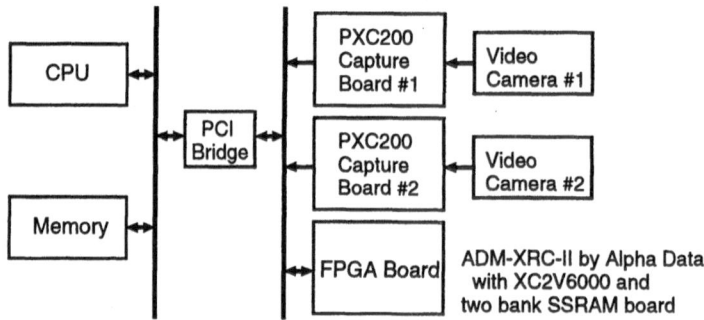

Fig. 4. System Overview

Figure 5 shows the structure of the FPGA board. The board has eight external memory banks (including two memory banks by the additional SSRAM board) which can be accessed independently. The first pair of images (left and right) sent from cameras are stored in bank0 and bank2 respectively, and next pair of images are stored in bank1 and bank3 while the images in bank0 and bank2 are processed by the circuit on the FPGA. The results by the circuit are written back to bank4 and bank5. When the data in bank4 is being sent to the host computer, FPGA writes new results to bank5. In order to exploit maximum performance of FPGA by avoiding memory access conflicts on these external memory banks, we need six memory banks for stereo matching its self. The rest two banks (bank6 and bank7) are used for the calibration described below. Thus, we need at least eight memory banks for our stereo vision system.

3.2   Calibration

In our system, calibration is performed using images of a predefined calibration grid. Before starting the system, the grid is given to left and right cameras (po-

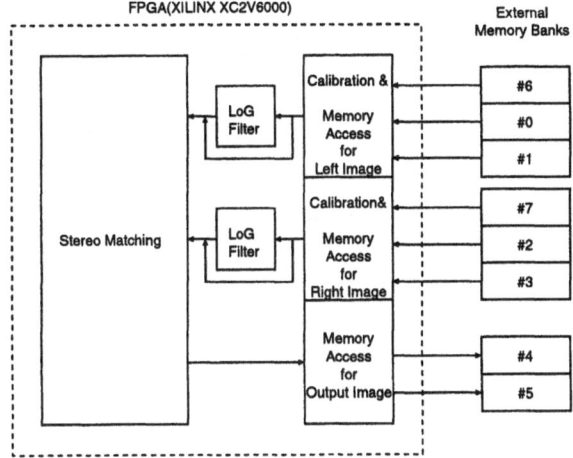

Fig. 5. FPGA Board

sitions of the cameras and the grid are fixed in advance), and the images of the grid taken by both cameras are sent to the host computer. Then, the host computer calculates which pixels on left and right images should be compared, and the positions of the pixels which should be compared are sent back to external RAMs on the FPGA board. These pixel position informations for left and right image are stored in bank6 and bank7, respectively (the size of the information is same with the size of images). In the later matching stages, FPGA first reads out the positions of the pixels which should be compared from the bank6 and bank7, and then the pixel data from bank0/1 and bank2/3 are read out using the positions.

This function is very important to obtain correct 3-D geometry with simple matching circuit, but this allows us to read only one pixel data in each clock cycle from bank0/1 and bank2/3, because the next pixel data which should be compared may not lies in the next address of the image data (when horizontal line in the image do not correspond to the true horizontal lines in real-world owing to the distortion of the lens of the camera and so on). Because of this restriction, the system performance is limited by the access time to the external RAMs on the FPGA board, and can not be improved by providing wider access to the external RAMs.

3.3 Matching and Left-Right Consisteny

Figure 6 shows the outline of the matching circuit. In Figure 6, suppose that window size is $n \times n$, and column data of windows (n pixel data which are shown as Li and Ri in Figure 6) are read out at once in order to simplify the figure (in the actual system, only one pixel data is read out at once as described above).

Column data of windows in left image are broadcasted to all column modules, while column data of windows in right image are delayed by registers and then

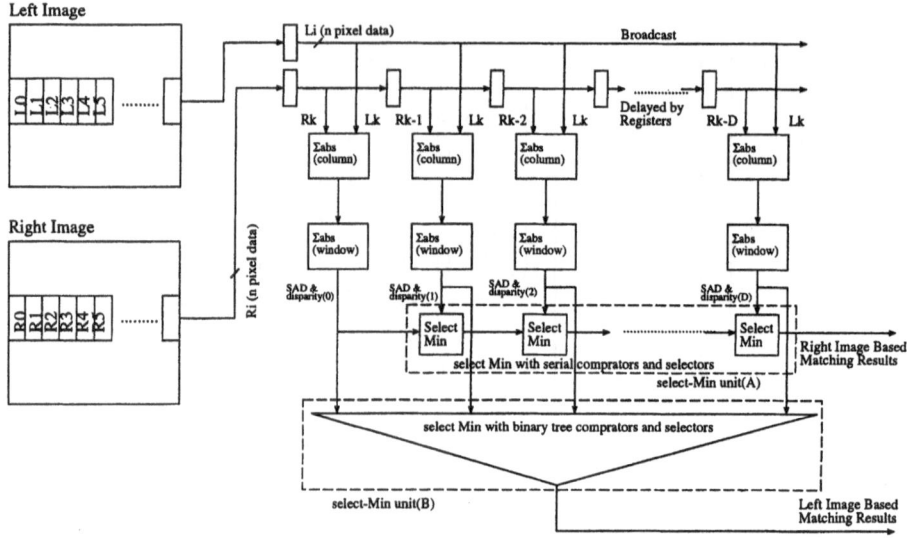

Fig. 6. Outline of the Matching Circuit

given to the column modules. In the column modules, sum of absolute difference of one column data is calculated. The outputs by the column modules are sent to window modules to sum up n values (thus, sum of absolute difference of one window is calculated).

Outputs by the window modules are compared and minimum value and its disparity are selected by two kinds of units. In the select-minimum unit(A), outputs by the window modules are shifted (with daley), and compared with the outputs by the next window module. The smaller value and its disparity are selected and shifted to the next compare module. Then, the output by the last select-min module gives the minimum of sum of absolute difference and its disparity when the right image is chosen as base for the matching. In another select-minimum unit(B), all outputs by the window modules are compared by binary tree comparators and selectors, and minimum value and its disparity are selected. The output by this unit gives the minimum of sum of absolute difference and its disparity when the left image is chosen as base for the matching.

Figure 7 shows the outputs by the window modules when the window size is 5×5. In Figure 7, by comparing outputs of the window modules with shifting and delaying (parts covered by slanting lines in Figure 7), one window in the right image (window {R6-R10}) is compared with windows in the left image({L6-L10},{L7-L11},{L8-L12}...), while one window in left image({L11-L15}) is comapred with windows in the right image({R11-R15},{R10-R14},{R9-R13},...) by comparing outputs of window module at the same time (gray parts in Figure 7).

As described above, left-right consistency check (left image based matching and right image based matching) can be executed by only adding another com-

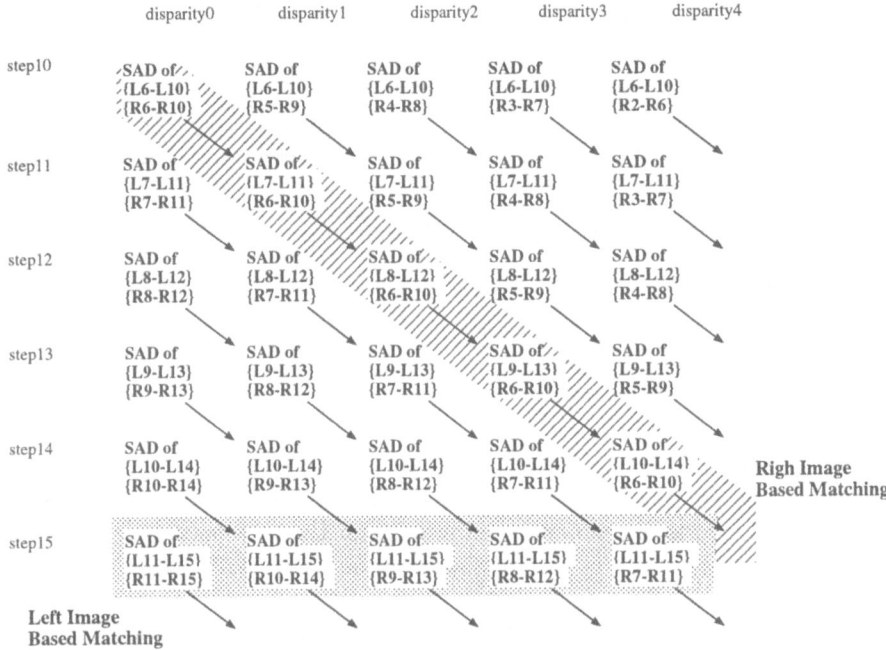

Fig. 7. Left-Right Consistency

pare unit which requires only D-1 comparators and selectors when D (number of window modules, namely maximum disparity) is 2^k.

3.4   Details of the Modules

Figure 8 shows the details of the column module and the window module. In the column module, absolute difference of inputs from left image and right image (one pixel in each clock cycle as described above) is calculated, and summed up for n times when the window size is $n \times n$. The outputs of the column module are sent to the window module, and n outputs are summed up again to caluculate the SAD(sum of absolute difference) of $n \times n$ window. In the window module, new input value is accumulated, and input at (current_step - n) step is subtracted instead of adding n previous values in order to reduce circuit size (previous values are stored in shift registers)[7]. As shown in Figure 7, the output of the window module is, for example, SAD of window {L6-L10}{R6-R10} at step 10, and SAD of {L7-L11}{R7-R11} at step 11. In this case, SAD of window {L7-L11}{R7-R11} can be calculated by the following equation.

SAD of {L7-L11}{R7-R11} =
   SAD of {L6-L10}{R6-R10} – SAD of {L6}{R6} + SAD of {L11}{R11}

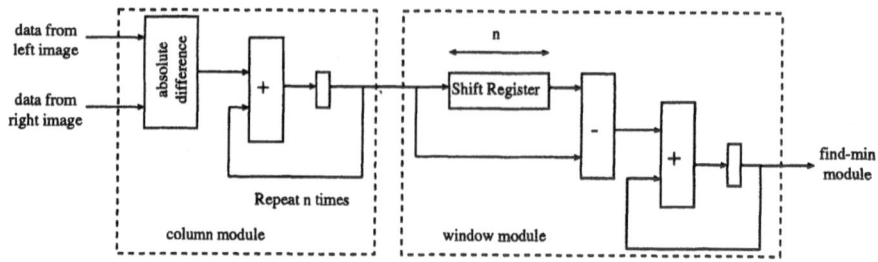

Fig. 8. Details of the Column Module and the Window Module

4   Performance

In our system, major factor that decreases system peformance is window size (because only one pixel is read from external memory owing to calibration, it takes n clock cycles to read one column data for the window), and factors that increase the circuit size is the maximum value of disparity which decide the number of modules.

Table 1 shows the system performance against the window size. The image size used for the evaluation is 640 × 480. As described above, the performance becomes worse as the window size becomes larger. The most often used window size in stereo vision systems is 7 × 7 or 9 × 9. The performace in Table 1 is calculated based on the maximum frequency reported by CAD. In practice, the system could process 20 frames per second in those window sizes, which is fast enough for autonomous robots. When, we need more performance, we can reduce the image size to 320 × 240 (which is widely used in other stereo vision systems). Then, the performace becomes four times faster without changing the circuit.

Table 2 shows the performance when we changed the maximum disparity. In this case, the circuit size becomes larger as the maximum disparity becomes larger, though the performance does not change as described above. Maximum disparity 200 is quite large compared with other stereo vision systems.

Table 1. Window Size and the Performance

Window Size		15 × 15	13 × 13	11 × 11	9 × 9	7 × 7	5 × 5
Performance (frames per second)	8.7		10.0	11.8	14.5	18.9	26.0

The size of left and right image is 640 × 480, and maximum disparity is 80.

Table 2. Maximum Disparity and Circuit Size

Maximum Disparity		80	160	200
Circuit Size		21%	43%	54%
Performance (FPS)		18.9	18.9	18.9
Operation Frequency (MHz)	40		40	40

The size of left and right image is 640 × 480, and window size is 7 × 7.

5   Conclusions

In this paper, we described a compact stereo vision system with one off-the-shelf FPGA board with one FPGA. This system supports (1) camera calibration for easy use and for simplifying the circuit, and (2) left-right consistency check for reconstructing correct 3-D geometry. The performance of the system is limited by the calibration (which is a must for practical use) because only one pixel data can be allowed to read in owing to the calibration. The performance is, however, 20 frame per second (when the size of images is 640 × 480), which is fast enough for practical use such as vision systems for autonomous robots. The operation frequency of the system is still very slow. We are now improving the details of the circuit. We think that we can process more than 30 frames per second by this improvement.

This system became possible because of the continuous progress of FPGAs. We needed at least eight memory banks on the FPGA board to exploit maximum performance of FPGA avoiding memory access conflicts on the memory banks on the FPGA board while supporting calibration. We also needed the latest FPGA to support very large maximum disparity.

References

1. M.Arias-Estrada, J.M.Xicotencatl, "Multiple Stereo Matching Using an Extended Architecture", FPL:2003-212, 2001.
2. P.Anandan, "A computational framework and an algorithm for the measurement of visual motion", IJCV, 2(3):283-310,1989.
3. T.W.Ryan, R.T.Gray, and B.R.Hunt, "Prediction of correlation errors in stereo-pair images", Optical Engineering, 19(3):312-322, 1980.
4. T.Kanade, "Development of a video-rate stereo machine" in IUW, pp. 549-557. 1994.
5. K.Nakayama and S.Shimojo, "Da Vinci stereopsis: Depth and subjective occluding contours from unpaired image points", Vision Research, 30:1811-1825, 1990
6. P.Fua, "Combining stereo and monocular information to compute dense depth maps that preserve depth discontinuities", IJCAI, 1991.
7. O.Faugeras, B.Hotz, H.Mathieu, T.Viville, Z.Zhang, P.Fua, E.Thron, L.Moll, G.Berry, J.Vuillemin, P.Bertin, and C.Proy, "Real time correlation-based stereo: algorithm, implementations and application,", Tech. Rep. 2013, INRIA, August 1993.

Synthesizing on a Reconfigurable Chip an Autonomous Robot Image Processing System*

Jose Antonio Boluda and Fernando Pardo

Departament d'Informàtica, Universitat de València,
Avda. Vicent Andrés Estellés S/N 46100 Burjassot, Spain,
{Jose.A.Boluda,Fernando.Pardo}@uv.es, http://tapec.uv.es

Abstract. This paper deals with the implementation, in a high density reconfigurable device, of an entire log-polar image processing system. The log-polar vision reduces the amount of data to be stored and processed, simplifying several vision algorithms and making it possible the implementation of a complete processing system on a single chip. This image processing system is specially appropriated for autonomous robotic navigation, since these platforms have typically power consumption, size and weight restrictions. Furthermore, the image processing algorithms involved are time consuming and many times they have also real-time restrictions. A reconfigurable approach on a single chip combines hardware performance and software flexibility and appears as specially suited to autonomous robotic navigation. The implementation of log-polar image processing algorithms as a pipeline of differential processing stages is a feasible approach, since the chip incorporates RAM memory enough for storing several full log-polar images as intermediate computations. Two different algorithms have been synthesized into the reconfigurable device showing the chip capabilities.

1 Introduction

The incorporation of visual sensing into an autonomous robot is a desirable objective, since the visual information is accurate, and there can be many different image processing algorithms useful for helping its navigation. Unfortunately, a traditional image processing system based on standard board cards needs a lot of hardware resources. Moreover, an autonomous platform may need the hardware implementation of several vision algorithms due to the real-time nature of its own navigation. On the other hand, the platform carries its own resources, giving as result systems with power consumption, size or weight limitations. In this way, the reconfigurable hardware appears as a new trend in autonomous robot design, since it is possible to maintain software flexibility while keeping hardware performance. Furthermore, the increasing density and performance of

---

* This work has been supported by the Generalitat Valenciana under project CTIDIA/2002/142

reconfigurable devices, specially for the system-on-a-programmable chip families, makes a reconfigurable approach on a single chip as specially suited for a system with hardware restrictions.

Space-variant vision emerges as an attractive image representation, since information reduction is interesting for a system with hardware restrictions. The log-polar mapping shows, as a particular case of space-variant vision, useful properties in addition to the selective reduction of information.

This paper is divided into four sections. Section 2 gives an introduction to log-polar vision and the two differential algorithms useful for robotic navigation. Section 3 gives a brief state-of-the-art about implementation of image processing systems into reconfigurable architectures and explains the advantages that a single chip approach offers. Additionally, section 3 shows how both algorithms have been synthesized into the complex FPGA employed, presenting how they benefit from the use of the on chip RAM for constructing intermediate internal frame-grabbers. Finally, section 4 draws several conclusions.

2   Log-Polar Differential Image Processing Algorithms

A space variant visual sensor distribution presents a selective reduction of information and, in several cases, simplifies geometric computations . The log-polar representation has interesting properties that have been widely studied [1] [2] [3] [4]. As an example of particular computational simplifications, rotations around the sensor center are converted to simple translations along the angular coordinate, and homotheties with respect to the center in the sensor plane become translations along the radial coordinate.

The log-polar transformation can be made directly through a sensor which has the log-polar sensor distribution. Fig. 1 shows the log-polar transformation that is directly performed at the sensor level. The sensor plane is called retinal plane, and the computational plane is called cortical plane. The equations included into the Fig. 1 show the image transformation performed by a log-polar

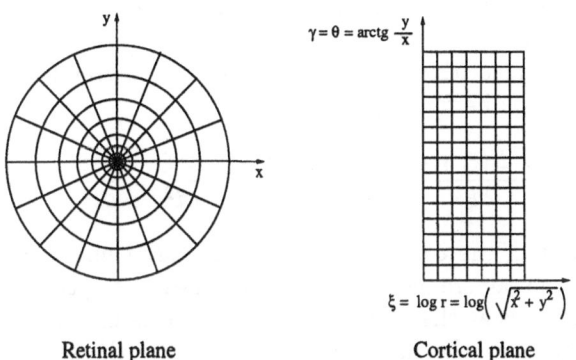

Retinal plane   Cortical plane

Fig. 1. The log-polar transformation

sensor. As an example of log-polar sensor there is a CMOS log-polar visual sensor with a resolution of 76 rings with 128 cells per ring [5]. This sensor has two different areas: the outer 56 rings, or retina, that follows exactly the log-polar equations, and the inner area, or fovea, that follows a linear growth law.

The log-polar conversion, performed at the presented system, consists on arithmetically calculate the transformation from cartesian to log-polar coordinates, for every pixel coming from the camera by means of the CORDIC algorithm. This conversion stage has been integrated into the chip as a first module previous to the processing stage. The chosen implementation for the log-polar transformation has two stages: the first calculates the polar coordinates (radius and angle) of the cartesian coordinates (x, y). The second stage then calculates the logarithm of the radius giving the final log-polar coordinates. This approach gives an image resolution of 56 rings with 128 cells per ring (57.344 bits) following exactly the log-polar transformation equations.

Differential algorithms, developed in log-polar coordinates, extract dynamic information from the scene and are useful for a visual-guided platform. Differential approaches use the temporal and spatial derivatives of the image sequence. The operations involved are simple and systematically applied to all the image pixels, thus suitable for hardware implementation. These algorithms are computationally intensive due to the image size, but they benefit from log-polar data reduction. In this way, the implemented algorithms into the reconfigurable board must optimize temporal and spatial differential computations.

The objective is to have a library of algorithms for programming the reconfigurable chip.

Motion Detection Independent of the Log-Polar Camera Movement: Originally developed in Cartesian coordinates [6], and adapted to log-polar coordinates [7], its experimental effectiveness has been already proved [8]. This algorithm detects moving objects with respect to the static background. In a moving platform, there are image variations due to the self camera movement that may appear as moving objects. This algorithm is able to filter the image displacement due to the camera self movement. The constrains are related to several smoothness conditions in the grey level image and in the camera motion. Theoretically, only objects which are moving with respect to the background are detected. The algorithm constrains are related to grey level and movement smoothness, and can be formulated as follows:

$$\frac{\partial^2 E(\xi, \gamma, t)}{\partial \xi^2} = \frac{\partial^2 E(\xi, \gamma, t)}{\partial \gamma^2} = \frac{\partial^2 \xi}{\partial t^2} = \frac{\partial^2 \gamma}{\partial t^2} = 0 \qquad (1)$$

For any point of the cortical plane the grey level image $E(\xi, \gamma, t)$ must be smoothed, and the camera movement must be linear along the optical axis. If the focus of expansion is at the center of the sensor, this movement is transformed in a translation along the radial coordinate. Under these constrains, the second temporal derivative of the image becomes a small integer number near zero except for the self-moving objects. Therefore, the algorithm is summarized as follows: First, the image must be smoothed, next the first order temporal derivative must be calculated from two consecutive images, and finally the second temporal derivative must be computed, selecting the zero values for binaryzing the image and marking self-moving objects. In this way, the algorithm implementation into programmable logic must compute efficiently image temporal differences. In fact, due to the non-exactly accomplishment of the algorithm constrains, the condition for a point that belongs to a self-moving object is that the second temporal derivative of the log-polar images must be larger than a threshold.

Time to Impact Computation: A second differential algorithm based on log-polar vision, that can be useful for helping the navigation of a robot, is the time to impact computation of a camera to an approaching surface [9]. The time to impact (τ) in the case of polar images, supposing an approaching movement along the optical axis at the sensor center is:

$$\tau = K \frac{-\frac{\partial E}{\partial \xi}}{\frac{\partial E}{\partial t}} \qquad (2)$$

where K is a constant that depends on the log-polar transformation geometric parameters, $E(\xi, \gamma, t)$ is the log-polar image sequence, $\frac{\partial E}{\partial \xi}$ is the radial gradient and $\frac{\partial E}{\partial t}$ is the first order temporal derivative.

Equation (2) shows that the time to impact can be computed as a division of two differential magnitudes: the radial gradient and the first order temporal derivative. A previous stage for smoothing the original log-polar images is required for avoiding non-sense derivatives, like derivatives at the edges (step functions). A detailed discussion about the accuracy of this log-polar algorithm can be found at [10].

3 Synthesizing Differential Algorithms into the Reconfigurable Device

Reconfigurable architectures have been widely used for implementing image processing algorithms. Modular multiboard architectures are formed of several boards, with various reprogrammable devices into each board [11] [12]. The advantages of this approach are flexibility, throughput and computation power. Unfortunately, these good features have the disadvantage of size, weight and power consumption. There are also other simpler approaches or architectures oriented to autonomous platforms implemented in single boards [13], programmable devices [14], or implementations that use few medium PLDs [15]. This last architecture is a pipeline of small Processing Elements (PEs) oriented to processing log-polar vision algorithms. This machine is modular and scalable as the multiboard architectures and small enough for an autonomous robot. Each PE has a medium PLD, SRAM memory and a small PAL. All these elements are not fully used in each algorithm implemented wasting in this way hardware resources

In the range of single board reconfigurable systems, there are some boards that incorporate high-density and high-performance programmable devices. Moreover, these reprogrammable devices offer complete system integration on a single device called system-on-a-programmable-chip (SOPC). These boards and devices seem particularly useful for applications where there are hardware restrictions, and any gate and bit inclusion must be fully justified. In this way, a single high-density SOPC device, which incorporates the possibility of including exactly the memory size and the modules needed at each algorithm will optimize the hardware utilization.

The APEX PCI board from Altera includes a SOPC APEX 20KC device (EP20K1000C) which has 38.400 Logic Elements (LEs) equivalent to 10^6 gates (or $1,7{\cdot}10^6$ system gates) and 320 Kbits of RAM. Furthermore, the board follows the mechanical and electrical PCI interface specifications and it is designed for the integration of a PCI mega-core as input/output interface. A 64 bits master PCI interface fills around 1.400 LEs, which is less than 4% of the resources. This board and this device has been employed for synthesizing the image processing module for the autonomous robot.

It is possible to extract a common structure from the algorithms presented at section 2 and from other similar differential image processing algorithms. A generic log-polar vision algorithm can be split into different well-balanced stages (similar processing delay) for avoiding a slow stage that becomes the pipeline bottleneck. The input of the first stage is a sequence of log-polar images, but the output of this stage, and therefore the inputs to any other stage, can be an ordered data structure.

Any algorithm implemented into the reconfigurable device can be divided as a pipeline of processing stages. In this way, the image processing algorithms that could be efficiently split into such stages are algorithms which have an unidirectional data flow between all the stages of the pipeline. Moreover, the computation of temporal derivatives are implemented using RAM bits as double port memories between stages, so any stage can simultaneously access the information currently processed by the precedent stage and the data stored from the previous computation. The image storage and the reading of the next stage is made simultaneously, but in different memory blocks. Moreover, there is a switching between memories for each new image. The use of Library Parameterizable Macros (LPMs) available with Quartus II allows the easy definition of the inter-stage double port memories and the local-stage memory for computing differential magnitudes. It must be noticed that the vision algorithms are applied to the retina (the outer area in the log-polar images) since this area follows exactly the log-polar growth law, so the algorithms are accomplished and the total data size is 57.344 bits per log-polar image. The APEX20KC device used has 327.680 RAM bits, so it is shown how this approach, that includes double-port RAM memories between stages is a feasible model.

All the stages have been designed with synthetisable VHDL as autonomous and independent modules, with a Finite State Machine (FSM) for implementing its processing functionality. The communication protocol, among all the pipeline

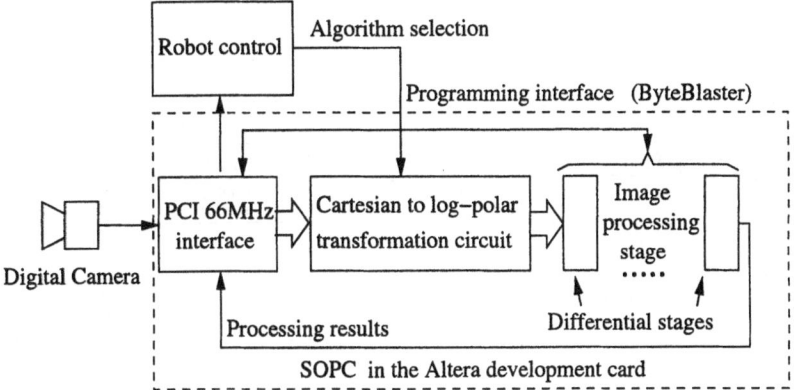

Fig. 2. Overall system

stages, is a simple asynchronous protocol, useful for defining a library of stages that can be mixed for developing different algorithms.

Fig. 2 shows a schema of the overall system organization. A digital cartesian camera acquires the images and send them to the Altera development card by way of the PCI bus. A PCI master/slave megacore is included into the SOPC as image processing system interface as has been already appointed. The chip incorporates a CORDIC module for performing the cartesian to log-polar transformation. Afterwards, the differential pipeline stage processes the images, extracting the information the algorithm delivers, for the robot navigation system. The robot can reconfigure the SOPC selecting the algorithm that is employed for processing the images. The algorithm chosen each time will be the most adequate for helping its navigation.

3.1   Motion Detection Algorithm Synthesis

The algorithm explained in section 2 describes a differential algorithm for detecting objects that move to respect the background, discarding automatically the image displacement due to the camera self movement when this movement is uniform along the camera optical axis. The algorithm has been divided into three stages that work simultaneously as a data pipeline as shown in Fig. 3. Double port memories of exactly the retinal image size (7.168 bytes) are placed between the stages in order to accelerate differential computations. Moreover, the first stage has 128 bytes of local memory for storing a log-polar ring. The ring values are shifted systolically with the aim of computing a grey level average value for accomplishing the smoothed image condition of the algorithm.

1. The first stage smoothes the original log-polar image storing the smoothed image in a double port memory and simultaneously giving it to the next stage. This smooth is made with a simple convolution mask in order to reduce the computations, avoiding a bottleneck at this stage.

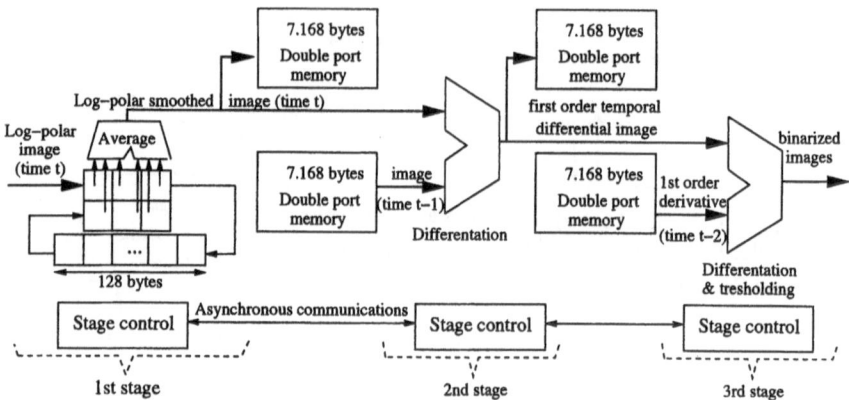

Fig. 3. Motion detection algorithm implementation

2. The second stage computes the first temporal derivative as an image subtraction, accepting the smoothed pixel from the precedent stage. It also reads the double port memory where the precedent smoothed image is stored. Subsequently, first order differences are calculated as a simple pixel subtraction. Finally, the differential pixel is sent to the next stage, being simultaneously stored in a double port memory

3. The third stage computes the second order temporal derivative and binarizes the image. This stage receives first order differences from the preceding one and simultaneously reads the differences previously stored in the double port memories. The second temporal derivative image is computed as differences between two first order temporal derivative images. Moreover, this value is compared to a threshold that is basically related to the robot movement and scene illumination. The final result is a sequence of binarized images which have marks for the points that belong to self-moving objects.

The algorithm has been successfully synthesized into the APEX20K device, occupying 230.400 RAM bits (70% of the total available RAM resources) and less than 1.000 LEs. It is feasible to increase the algorithm accuracy improving the smoothing stage and the derivation computations. The chip clock frequency can reach 200 MHz, but it is limited to 66 MHz that is the system clock frequency. This algorithm has a well balanced number of clock cycles, consuming 4 cycles for computing 1 pixel at each stage. All the stages are working in parallel and the pipeline only needs an additional cycle for the input of each byte and another for the output. In this way, the pipeline yields a processed byte every 6 clock cycles. Taking into account that a complete retinal log-polar image occupies 7 Kbytes it is possible to compute a theoretical processing ratio of more that 250 frames per second. This large high ratio must be limited due to the differential nature of the algorithm. Thus, in order to compute differential magnitudes as are the temporal or spatial derivatives, differences between images must be guaranteed, so a minimum acquisition interval between images must be ensured [8].

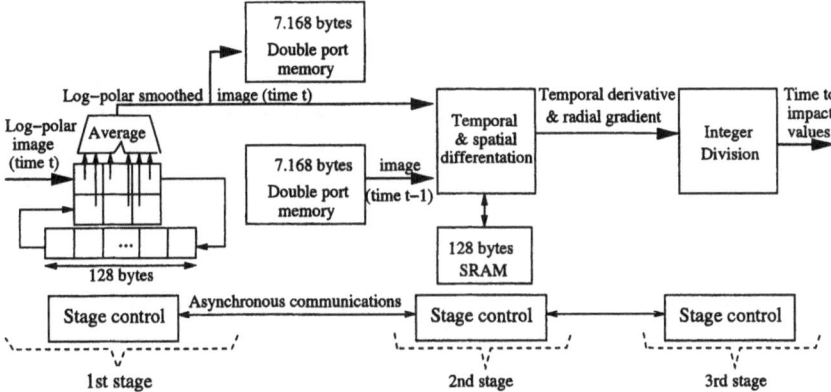

Fig. 4. Time to impact algorithm implementation

3.2   Time to Impact Computation Algorithm Synthesis

The algorithm for time to impact computation has been also implemented with the same methodology of splitting the overall task into a pipeline of stages. Double port memories have been employed for accelerating the computation of the first temporal derivative. Again, the algorithm has been divided into three stages and there are two double port memories as library modules. Fig. 4 shows the algorithm implementation.

1. The first stage is exactly the same smoothing block designed for the previous algorithm. Each stage has been designed with its own control and identical protocol communications. Therefore, any pipeline stage already designed can be re-utilized in any other implementation as a standard library module.
2. The second stage computes the first temporal derivative and the radial gradient. The first order differentiation is computed through the same image subtraction policy described previously. Simultaneously, the radial gradient is computed with the smoothed pixel supplied by the previous stage, and the pixel corresponding to the inferior ring stored in a local small memory of 128 bytes.
3. Finally, the third stage computes the time to impact map for each pixel making an integer division of both values, with a cost of 8 clock cycles.

The algorithm has been also successfully synthesized into the APEX20K device, occupying near 115.000 RAM bits (35% of the total available RAM resources) and less than 1.100 LEs. So, it is also feasible to increase the algorithm accuracy improving the division stage. The clock frequency can be up to 166 MHz, but it is limited to 66 MHz that is the system clock frequency. This algorithm yields a processed byte every 10 clock cycles.

4   Conclusions

Image processing algorithms, involved in visual real-time navigation, benefit from hardware speed. Furthermore, it is not a good solution to have a custom hardware system in an autonomous platform for each desired algorithm due to size, weight and power consumption reasons. Reconfigurable systems on-a-chip appear as a technology that combines hardware performance, software reconfigurability and low resources consumption.

In the other hand, space-variant vision has been employed to reduce the total amount of data to be processed, reducing the memory size for making it possible the implementation of several local frame grabbers inside the chip. Since the involved algorithms are differential, a hardware parallel implementation for computing the temporal differences is a feasible approach that benefits from information reduction. Moreover, the log-polar scheme reduces the computation complexity of the selected algorithms.

A methodology developed previously for splitting differential algorithms into stages for accelerating the temporal difference computations has been applied. All the stages have identical control scheme and communication interface for simplifying the design of algorithms planning, a policy of stage re-usability. Following these ideas, two different algorithms have been implemented in the reconfigurable device, showing its flexibility and performance. Both algorithms take advantage of image reduction and computation simplification, allowing a high rate of processed images per second. The combination of the log-polar formalism, differential algorithms and pipelined architecture shows its good performance for real-time image processing. The synthesis results show that a complete image processing system can be synthesized in a high-density SOPC chip, including a complete PCI interface.

References

1. Bernardino, A., Santos-Victor, J. Vergence Control for Robotic Heads using Log-polar Images. In IEEE/RSJ Int. Conf. on Intelligent Robots and Systems, IROS'96, Osaka, Japan, (1996)
2. Capurro, C,. Panerai F., Sandini, G. Dynamic Vergence using Log-polar Images. International Journal of Computer Vision **24-1** (1997) 79-94
3. Daniilidis, K. Attentive Visual Motion Processing: Computations in the Log-polar Plane. Computing **11** (1996) 1-20
4. Ferrari, F., Nielsen, J., Questa, P., Sandini, G. Space Variant Imaging. Sensor Review **15-2** (1995) 17-20
5. Pardo, F., Dierickx, B., Scheffer, D. Space-Variant Non-Orthogonal Structure CMOS Image Sensor Design. IEEE Journal of Solid State Circuits, **33-6** (1998) 842-849
6. Chen W.G., Nandhakumar, N. A Simple Scheme for Motion Boundary Detection. In Proceedings of the IEEE Intl. Conf. on Systems, Man and Cybernetics (1994)
7. Boluda, J.A., Domingo, J., Pardo, F., Pelechano, J. Detecting Motion Independent of the Camera Movement through a Log-polar Differential Approach. In Computer Analysis of Images and Patterns. 7th International Conference CAIP'97, Lecture

Notes in Computer Science, Vol. 1296. Springer-Verlag Berlin Heidelberg New York (1997) 702-710

8. Boluda, J.A., Domingo, J. On the Advantages of Combining Differential Algorithms and Log-polar Vision for Detection of Self-motion from a Mobile Robot. Robotics and Autonomous Systems **37-4** (2001) 283-296

9. Tistarelli, M., Sandini, G. On the Advantages of Polar and Log-polar Mapping for Direct Estimation of Time-to-impact from Optical Flow. IEEE Trans. on Pattern Analysis and Machine Intelligence, **15-4** (1993) 401-410

10. Pardo, F., Boluda, J.A., Coma, I., Mico, F. High Speed Log-polar Time to Crash Calculation for Mobile Vehicles. Image Processing and Communications, **8-2** (2002) 23-32

11. Duncan Buell, Jeffrey Arnold and Walter Kleinfelder. Splash2: FPGAs in a Custom Computing Machine. IEEE Computer Society Press, (1996)

12. Demigny, D., Kessal, L., Bourguiba, R., Boudouani, N. How to Use High Speed Reconfigurable FPGA for Real Time Image Processing. In Proceedings of the Fifth IEEE International Workshop on Computer Architecture for Machine perception (2000) 240-246

13. Fross, B., Donaldson, R., Palmer, D. PCI-Based WILDFIRE Reconfigurable Computing Engines, In High-Speed Computing, Digital Signal Processing, and Filtering using Reconfigurable Logic. Proceedings of the SPIE Vol. 2914 (1996) 170-179

14. Arias-Estrada, M., Rodriguez-Palacio, E. An FPGA Co-processor for Real-Time Visual Tracking. In 12th International Conference on Field-Programmable Logic and Applications. FPL'02, Lecture Notes in Computer Science, Vol. 2438. Springer-Verlag Berlin Heidelberg New York (2002) 710-710

15. Boluda, J.A., Pardo, F. A Reconfigurable Architecture for Autonomous Visual Navigation. Machine Vision and Applications **13-5** (2003) 322-331

Reconfigurable Hardware SAT Solvers: A Survey of Systems

Iouliia Skliarova and António B. Ferrari

University of Aveiro, Department of Electronics and Telecommunications, IEETA
3810-193 Aveiro, Portugal
{iouliia, ferrari}@det.ua.pt

Abstract. By adapting to computations that are not so well supported by general-purpose processors, reconfigurable systems achieve significant increases in performance. Such computational systems use high-capacity programmable logic devices and are based on processing units customized to the requirements of a particular application. A great deal of research effort in this area is aimed at accelerating the solution of combinatorial optimization problems. Special attention was given to the Boolean satisfiability (SAT) problem resulting in a considerable number of different architectures being proposed. This paper presents the state-of-the-art in reconfigurable hardware SAT satisfiers. The analysis of existing systems has been performed according to such criteria as reconfiguration modes, the execution model, the programming model, etc.

1 Introduction

Although the concept of reconfigurable computing has been known since the early 1960s [1], it is only recently that technologies that allow it to be put into practice became available. The interest started at the beginning of the 1990s as FPGA densities broke the 10K logic gate barrier. Since then, reconfigurable computing became a subject of intensive research. For some classes of applications reconfigurable systems allow very good performance to be achieved compared to general-purpose computers. Other types of applications were mapped to reconfigurable hardware because it offers innovative opportunities to explore. According to the primary objective to be achieved, all these applications can be broadly divided into three categories: hardware emulation and rapid prototyping, evolvable hardware, and the acceleration of computationally intensive tasks. The last category is without doubt the prevalent one.

Recently, a series of attempts have been made to accelerate applications that involve rather complex control flow. In this context special attention was given to problems in the area of combinatorial optimization. Among them, the Boolean satisfiability (SAT) problem stands out. This may be partially explained by the extremely wide range of practical applications in a variety of engineering areas, including the testing of electronic circuits, pattern recognition, logic synthesis, etc. [2]. In addition, SAT has the honor of being the first problem shown to be NP-complete [3]. This means that existing algorithms have an exponential worst-case

P.Y.K. Cheung et al. (Eds.): FPL 2003, LNCS 2778, pp. 468–477, 2003.

complexity. Implementations based on reconfigurable hardware enable the primary operations of the respective algorithms to be executed in parallel. Consequently, the effect of exponential growth in the computation time can be delayed, thus allowing larger size instances of SAT to be solved [2].

SAT is a very well known combinatorial problem that consists of determining whether a given Boolean formula can be satisfied by some truth assignment. The search variant of this problem requires at least one satisfying assignment to be found. Usually, the formula is presented in conjunctive normal form, which is composed of a conjunction of a number of clauses, where a clause is a disjunction of a number of literals. Each literal represents either a Boolean variable or its negation. A survey of algorithmic methods of solving the SAT problem can be found in [2].

In this paper we present the current status of reconfigurable hardware SAT solvers and give an overview of the existing approaches and their tradeoffs. The remaining part of the paper is organized as follows. Section 2 is devoted to the description of the most well known architectures of reconfigurable hardware SAT satisfiers. Analysis and classification of these architectures according to different criteria is performed in section 3. Finally, concluding remarks are given in section 4.

2   Architectures of SAT Solvers

Recently, several research groups have explored different approaches to solve the SAT problem with the aid of reconfigurable hardware [4-5], [9-12], [14-22]. Since names have not typically been given to hardware SAT satisfiers, we will refer to them according to the first author's names of the respective publications.

Suyama et al. [4-5] suggested an architecture of an *instance-specific* SAT solver capable of finding all the solutions (or a fixed number of them) of a given problem instance. The employed algorithm is characterized by the fact that at any moment a full variable assignment is evaluated. A *dynamic decision strategy* based on both experimental unit propagation and a maximum-occurrence-in-clauses-of-minimum-size heuristic has been adopted. A number of circuits have been implemented on an Altera FLEX10K250 FPGA clocked at 10 MHz. *Suyama et al.* were able to achieve a small acceleration compared to the POSIT algorithm [6] executed on an UltraSPARC-II/296 MHz over some instances from DIMACS benchmark suite [7]. However, the time spent in hardware compilation and configuration was not taken into account.

Zhong et al. implemented a version of the well-known Davis-Putnam (DP) algorithm [8]. In their early work [9] they constructed an implication circuit and a state machine for each variable in the formula, all the state machines being connected in a serial chain. As a preprocessing step, all the variables are sorted taking into account the number of their appearances in a given formula. In [10] hardware implementation of *non-chronological backtracking* was proposed. The resulting hardware execution time was quite good but the design had two distinct drawbacks. First, the clock frequency was low (ranging from 700KHz to 2MHz for different formulae). Second, the hardware compilation time took several hours (on a Sun 5/110MHz/64MB) thus canceling all the advantages of fast hardware execution.

In more recent work [11], [12] the basic design decisions were revised. As a result, a regular ring-based interconnecting structure was employed instead of irregular global lines, essentially reducing the compilation time in this way (to an order of seconds) and increasing the clock rate (to 20-30 MHz) [12]. In addition, a technique enabling conflict clauses to be generated and added was proposed. The experimental results are based on both hardware implementation (on an IKOS emulator containing a number of FPGA array boards) and simulation. The speedups achieved over the software satisfier GRASP [13] executing on a Sun5/110MHz/64MB (in a restricted mode), including the hardware compilation and configuration time, are of an order of magnitude [12] for a subset of the DIMACS SAT benchmarks [7].

Abramovici et al. [15] employed the technique of modeling a formula by a 2-level circuit. The SAT solver proposed in [14] is based on the PODEM algorithm. In [15] an improved architecture is suggested that employs the DP algorithm and implements an enhanced variable selection strategy. For hardware implementation Abramovici et al. suggest creating a library of basic modules that are to be used for any formula. The modules have predefined internal placement and routing. In this case the solver circuit will be built from modules, which allows the compilation time to be reduced (to the order of minutes). The authors implemented simple circuits on XC6264 FPGA and simulated the bigger ones. For a circuit occupying the whole area of the XC6264 FPGA the clock frequency is about 3.5 MHz. In [15] Abramovici et al. report speedups from 0.01 to 7000 (after time unit justification) achieved over GRASP [13] for a subset of DIMACS SAT benchmarks [7]. In [15] a virtual logic system was proposed allowing circuits to be constructed for solving SAT problems that are larger than the available hardware resources. This is achieved by decomposing a formula into independent sub-formulae that can be processed in separate FPGAs either concurrently or sequentially.

The SAT solver proposed by Platzner et al. [16], [17] is similar to that of Zhong [9]. It consists of a column of finite state machines, deduction logic and a global control unit. The deduction logic computes the result of the formula based on the current partial variable assignment. All variable assignments are tried in a fixed order. The authors implemented an accelerator prototype on the base of a Pamette board containing 4 Xilinx XC4028 FPGAs. The speedups obtained for hole6...hole10 SAT benchmarks from DIMACS [7], including hardware compilation and configuration time, range from 0.003 to 7.408 compared to GRASP executing on a PII/300MHz/128MB [16]. The designs for the holex problems run at 20 MHz [17].

More recent work in this direction is targeted at avoiding instance-specific layout compilation. Boyd et al. [18] proposed an architecture for a SAT-specific programmable logic device that excludes instance-specific placement and routing. The suggested design consumes polynomial hardware resources (with respect to the number of variables and clauses) and requires polynomial time to configure. The authors implemented a small version of their SAT satisfier for a problem having 8 variables and 8 clauses on a Xilinx XC4005XL running at 12 MHz. However, no results on large benchmark problems were reported.

Sousa et al. [19], [20] were the first to propose partitioning the job between software and reconfigurable hardware with the most computationally intensive tasks (such as computing implications and choosing the next decision variable) assigned to hardware, while the control-oriented tasks (such as conflict analysis, backtrack control and clause database management) are performed in software. The suggested SAT solver has an application-specific architecture that uses configuration registers for SAT formula instantiation [20]. In order to deal with instances that exceed the available hardware capacity, a virtual hardware scheme with context switching has been proposed. The results reported in [19] are based on a software simulator of the system under an estimated clock frequency of 80 MHz, assuming that the context-switching device can swap pages in one clock cycle.

Skliarova et al. [21], [22] proposed an application-specific SAT solver realizing a DP-based algorithm. The problem was formulated over a ternary matrix by setting a correspondence between clauses and variables of a formula and rows and columns of the matrix. In order to solve various problem instances it is only necessary to download the respective matrix data. All the other components of the satisfier remain unchanged. This allows local reconfigurability to be used and reduces the configuration overhead. The problem is partitioned between software and reconfigurable hardware in such a way that an FPGA is only responsible for processing sub-problems that appear at various levels of the decision tree and satisfy the imposed hardware constraints (such as the maximum allowed number of rows and columns in the matrix). This technique permits problems to be solved that exceed the resources of the available reconfigurable hardware. The SAT satisfier was implemented on an ADM-XRC PCI board containing one XCV812E Virtex-EM FPGA (running at 40MHz). The results of experiments on some of DIMACS benchmarks [7] have shown that it is possible to achieve a significant speedup compared to GRASP (up to 111x, including the FPGA configuration time, with GRASP executed on an AMD Athlon/1GHz/256MB).

3 Analysis of Hardware SAT Solvers

In this section we attempt to analyze the reconfigurable hardware SAT solvers according to such criteria as algorithmic issues, programming model, execution model, reconfiguration modes, logic capacity and performance. Table 1 summarizes the respective characteristics of the architectures considered in the previous section.

3.1 Algorithmic Issues

The majority of the existing reconfigurable hardware SAT solvers employs some variation of the Davis-Putnam algorithm [8]. An exception to this is the SAT satisfier of *Abramovici et al.*, which implements a PODEM-based algorithm [14].

The search process in the DP algorithm is usually organized with the aid of a *decision tree*, whose nodes are characterized by the respective partial variable assignments, and arcs represent the *decisions* taken. There exist two basic approaches to the selection of the decision variables: *static* and *dynamic*. Although dynamic selection has been considered to be a difficult task for hardware implementation, it was realized in a number of architectures [5], [19-22].

In the present-day software SAT solvers a lot of advanced techniques (such as non-chronological backtracking [13]) are employed that enable those regions of the search space that do not contain any solution to be identified and avoided. However, up to now these techniques have been largely ignored by hardware SAT solvers. The few exceptions to this rule are the SAT satisfiers of *Zhong et al.* [12] and *Sousa et al.* [19, 20] (the latter implements them in software).

Table 1. Principal characteristics of the reconfigurable hardware SAT solvers

SAT solver	Algorithmic issues	Programming model	Execution model	Reconfiguration mode	Logic capacity	Performance (t_{total})
Suyama et al.	DP-like algorithm with dynamic selection	instance-specific	hardware only	static	multi-FPGA system	$t_{comp} + t_{conf} + t_{ex\_h}$
Zhong et al. [12]	DP-based algorithm with static selection, non-chronological backtracking, conflict analysis	instance-specific	hardware only	dynamic (global)	multi-FPGA system	$t_{comp} + t_{conf} + t_{comm} + t_{ex\_h}$
Platzner et al.	DP-based algorithm with static selection	instance-specific	hardware only	static	use larger device	$t_{comp} + t_{conf} + t_{ex\_h}$
Abramovici et al. [15]	DP-based algorithm with optimized static selection	instance-specific	hardware only	dynamic (global)	logic partitioning in sub-formulae	$t_{comp} + t_{conf} + t_{comm} + t_{ex\_h}$
Sousa et al.	DP-based algorithm with dynamic selection and conflict analysis (in software)	application-specific	software/hardware partitioning according to computational complexity	dynamic (partial)	virtual hardware scheme	$t_{conf} + t_{comm} + t_{ex\_s} + t_{ex\_h}$
Skliarova et al.	DP-based algorithm with dynamic selection	application-specific	software/hardware partitioning according to logic capacity	dynamic (partial)	software/hardware partitioning	$t_{conf} + t_{comm} + t_{ex\_s} + t_{ex\_h}$

3.2 Programming Model

There are two basic approaches to mapping a SAT formula to a reconfigurable system: *instance-specific* and *application-specific*. The first approach has been extensively explored by the SAT research community [4-5], [9-12], [14-17] and assumes the generation of an individual hardware configuration for each problem instance. In this case, a typical design flow is used to describe and implement either a whole instance-specific circuit or a number of primary modules, which are further customized (at compile time) by specially developed software tools to match the respective formula.

In an *application-specific approach* the circuit is designed and optimized only once, after which it can be used for different problem instances [18-22]. This can be achieved with the aid of a hardware template, which is also developed using a typical design flow but is customized with data for a particular problem at run-time (instead of compile-time). It should be noted that in this case a hardware compilation step is completely avoided.

3.3 Execution Model

A SAT problem can be either entirely mapped to reconfigurable hardware (leaving just the tasks of preprocessing and initialization to the host processor) [4-5], [9-12], [14-18] or partitioned between hardware and software [19-22]. There exist different methods of software/hardware partitioning. In the domain of SAT solvers, two are usually employed: *partitioning according to computational complexity* and *partitioning with respect to logic capacity.*

The first method assigns computationally intensive portions of an application to hardware, while the remaining portions that exhibit little parallelism are handled by the host processor [19], [20]. Reconfigurable systems of this type are based on the 90/10 rule, which states that 90% of execution time of an application is spent by 10% of its code. Thus, in order to increase performance it is attempted to accelerate this small portion of an application with the aid of programmable logic devices.

The second method performs partitioning according to the available logic capacity of hardware employed [21], [22]. In this case, if a problem instance does not "fit" to a chosen device (or a number of interconnected devices), it has first to be processed by software up to the point at which it can be transferred to hardware.

3.4 Reconfiguration Modes

In the domain of reconfigurable computing it is common to distinguish between two configuration modes: *static mode* (also known as *compile-time configuration* or *design-time binding*) and *dynamic mode* (frequently referenced as *run-time configuration* or *implementation-time binding*).

Static configuration assumes fixed functionality of the device once it has been programmed [4-5], [9-10], [16-17]. Dynamic reconfiguration allows the functionality of the system to be changed during the execution of an application. Dynamic reconfiguration can in turn be *partial* or *global*. Global reconfiguration reserves all the hardware resources for each step of execution. After a step has been concluded, the device may be reprogrammed for the next step [15]. Partial reconfiguration implies the selective modification of hardware resources [19-22]. This opportunity allows the hardware to be adapted to better suit the actual needs of the application. Since only selected portions are reconfigured, the configuration overhead is less than in the previous case.

A variety of reprogrammable devices can be employed to carry out dynamic reconfiguration. *Single-context* devices require complete reprogramming in order to introduce even a small change. Although many commercially available FPGAs are single-context, there exist techniques (based on hardware templates) that allow partial reconfiguration to take place [19-22]. *Multi-context* devices possess various planes of configuration information with just one of them active at any given moment [19]. The main advantage of such devices is the ability to switch the context very fast. *Partially reconfigurable* devices permit small portions of their resources to be modified without disturbing the remaining parts. Although this kind of devices (such as the XC6200 family of Xilinx) was employed for some SAT solvers [15], the potential for partial reconfigurability has not been explored.

3.5 Logic Capacity

The logic capacity of the employed hardware device is always limited. Thus, efficient techniques are needed to deal with the situation when a problem instance exceeds the available hardware resources. The answers to this issue differ accordingly to the programming and execution models adapted. Basically, four possibilities have been explored.

The first is the expansion of the logic capacity by interconnecting a number of programmable devices and partitioning the circuit between them. It should be noted that fast and efficient multi-device partitioning and routing is quite a difficult task (of course modular design styles [12] can alleviate it). Moreover, the working frequency of such multi-device systems is usually quite limited.

The second method is to partition the problem into a series of configurations to be run either sequentially or in parallel. The partitioning is performed by decomposing an initial formula into a set of independent sub-formulae [15]. Each sub-formula must satisfy the imposed hardware constraints. The main limitation of this method is that the efficiency of the decomposition greatly depends on the characteristics of the formula. As a result, for some problem instances the partitioning time may increase to unacceptable levels.

The third method is based on software/hardware partitioning according to the available logic capacity of hardware that is employed (see section 3.3). In this case just those sub-problems that appear at different levels of the decision tree and respect the capacity limitations are assigned to hardware, the remaining portion of the problem being processed by a software application [21-22].

The last method is based on a virtual hardware scheme proposed in [19-20], which relies on dividing the circuit into a series of hardware pages that are successively run being the intermediate results stored in external memory blocks. Since all the hardware pages have the same structure with only a number of registers being reconfigured, the page switching is performed very fast.

3.6 Performance

The total time (t_{total}) spent by a reconfigurable hardware SAT satisfier to solve a particular problem instance comprises four components: hardware compilation time (t_{comp}), hardware configuration time (t_{conf}), time required for communication between software and hardware (t_{comm}) and actual execution time (t_{ex}). If a problem solution is partitioned between software and hardware then the execution time t_{ex} is composed of software execution time (t_{ex_s}) and hardware execution time (t_{ex_h}). It should be noted that the values of these components depend on the programming and execution models employed and some of them may be zero. For example, if a problem instance is entirely mapped to hardware, usually there is no communication (except for notifying the final result) between the host processor and the programmable device. In the same manner, if an application-specific approach is followed, the hardware compilation time is zero. Actually, the compilation time may constitute a large portion of the total solving time. For easy problem instances it even dominates and cancels out all the benefits of fast hardware execution [4-5], [9-10], [16-17]. That is why a number of techniques targeted at reducing the hardware compilation time have been proposed. They are based on exploiting modular design styles and developing customized software tools instead of using commercially available ones [11-12], [15].

One characteristic inherent in reconfigurable hardware SAT solvers is that it is very difficult to analyze and compare their performance accurately. As a rule, the designers present the results achieved in the light of the software SAT satisfier GRASP [13]. However, GRASP is run on different platforms and with dissimilar parameters that heavily influence its performance. Moreover, the parameters set are frequently not published. The majority of the SAT solvers considered involve a hardware compilation step, which is sometimes ignored (or hidden) when presenting the results. It is also difficult to estimate the exact impact of compilation on the total execution time because of the variety of software platforms used. Nevertheless, in all recent designs a clear intention to reduce and even to avoid the hardware compilation step is apparent [12], [15], [18-22].

As shown by the results of the 2002 software SAT competition [23], GRASP has been surpassed by more recent SAT satisfiers such as zChaff [24] and BerkMin [25]. Consequently novel algorithmic and architectural techniques need to be explored in order to put the reconfigurable hardware SAT solvers in a more favorable light comparing to a software solution.

4 Conclusion

This paper is dedicated to the description and comparison of reconfigurable hardware SAT solvers. The analysis leads to the following conclusions:

- The majority of designers implement complete search algorithms derived from the DP algorithm. Conflict analysis is usually not performed and just chronological backtracking is executed (with a few exceptions).

- Practically all the proposed SAT solvers are based on the instance-specific approach. However, the hardware compilation time restricts the range of problems for which a reconfigurable hardware solution is more effective than the software-based approach. That is why all recent efforts have been focused on avoiding instance-specific placement and routing.

- All the reconfigurable SAT solvers considered are loosely coupled systems with the programmable device (usually, a commercially available FPGA) being attached to the host processor via an external interface.

- It is quite difficult to compare the results that have been achieved. First, the hardware compilation and configuration times are not always clearly exposed. Second, the results are usually compared to GRASP, executed on different platforms with dissimilar parameters, which can lead to variations in the solving time of up to an order of magnitude.

- Real-world SAT formulae are quite large, however how instances that do not fit into an available device can be handled efficiently is not always discussed. Recently, in what seems to be a promising solution, it was suggested that the problem should be partitioned between software and reconfigurable hardware. It should also be noted that due to the rapid evolution in FPGA capacity, many challenging problem instances can now either be fitted into a single FPGA, or at least partitioned more efficiently.

- The speedups achieved by reconfigurable hardware compared to a software solution are significant just for certain classes of SAT instances, for which the optimization techniques proposed and implemented by software SAT satisfiers are not very efficient. Some examples of these techniques are: different decision strategies, exploiting problem symmetry, careful conflict analysis, etc. Consequently, although many interesting and worthwhile architectures have already been proposed, innovative approaches still need to be explored in the reconfigurable hardware domain.

Acknowledgment

This work was supported by the Portuguese Foundation of Science and Technology under grant No. FCT-PRAXIS XXI/BD/21353/99.

References

1. Estrin, G.: Reconfigurable Computer Origins: The UCLA Fixed-Plus-Variable (F+V) Structure Computer. IEEE Annals of the History of Computing. Oct.-Dec. (2002) 3-9
2. Gu, J., Purdom, P.W., Franco, J., Wah, B.W.: Algorithms for the Satisfiability (SAT) Problem: A Survey. DIMACS Series in Discrete Mathematics and Theoretical Computer Science, vol. 35 (1997) 19-151
3. Garey, M.R., Johnson, D.S.: Computers and Intractability: A Guide to the Theory of NP-Completeness. W.H. Freeman and Company. San Francisco (1979)
4. Yokoo, M., Suyama, T., Sawada, H.: Solving Satisfiability Problems Using Field Programmable Gate Arrays: First Results. In: Proc. of 2nd Int. Conf. on Principles and Practice of Constraint Programming (1996) 497-509
5. Suyama, T., Yokoo, M., Sawada, H., Nagoya, A.: Solving Satisfiability Problems Using Reconfigurable Computing. IEEE Trans. on VLSI Systems, vol. 9, no. 1 (2001) 109-116
6. Freeman, J.W.: Improvements to Propositional Satisfiability Search Algorithms. Ph.D. dissertation. Univ. Pennsylvania (1995)
7. DIMACS challenge benchmarks. [Online]. Available: http://www.intellektik.informatik.tu-darmstadt.de/SATLIB/benchm.html
8. Davis, M., Logemann, G., Loveland, D.: A machine program for theorem proving. Communications of the ACM n. 5 (1962) 394-397
9. Zhong, P., Martonosi, M., Ashar, P., Malik, S.: Using Configurable Computing to Accelerate Boolean Satisfiability. IEEE Trans. CAD of Integrated Circuits and Systems, vol. 18, n. 6 (1999) 861-868
10. Zhong, P., Ashar, P., Malik, S., Martonosi, M.: Using reconfigurable computing techniques to accelerate problems in the CAD domain: a case study with Boolean satisfiability. In: Proc. Design Automation Conf. (1998) 194-199
11. Zhong, P., Martonosi, M., Ashar, P., Malik, S.: Solving Boolean satisfiability with dynamic hardware configurations. In Hartenstein, R.W., Keevallik, A. (eds). Field-Programmable Logic: From FPGAs to Computing Paradigm (1998). Springer-Verlag. 326-235
12. Zhong, P.: Using Configurable Computing to Accelerate Boolean Satisfiability. Ph.D. dissertation. Department of Electrical Engineering. Princeton University (1999)

13. Silva, L.M., Sakallah, K.A.: GRASP: a search algorithm for propositional satisfiability. IEEE Trans. Computers, vol. 48, n. 5 (1999) 506-521
14. Abramovici, M., Saab, D.: Satisfiability on Reconfigurable Hardware. In: Proc. 7th Int. Workshop on Field-Programmable Logic and Applications (1997), 448-456
15. Abramovici, M., de Sousa, J.T.: A SAT solver using reconfigurable hardware and virtual logic. Journal of Automated Reasoning, vol. 24, n. 1-2 (2000) 5-36
16. Platzner, M.: Reconfigurable accelerators for combinatorial problems. IEEE Computer. Apr. (2000) 58-60
17. Platzner, M., De Micheli, G.: Acceleration of satisfiability algorithms by reconfigurable hardware. In: Hartenstein, R.W., Keevallik, A. (eds.) Field-Programmable Logic: From FPGAs to Computing Paradigm. Springer-Verlag (1998) 69-78
18. Boyd, M., Larrabee, T.: ELVIS – a scalable, loadable custom programmable logic device for solving Boolean satisfiability problems. In: Proc. 8th IEEE Int. Symp. on Field-Programmable Custom Computing Machines - FCCM (2000)
19. de Sousa, J., Marques-Silva, J.P., Abramovici, M.: A configware/software approach to SAT solving". In: Proc. of 9th IEEE Int. Symp. on Field-Programmable Custom Computing Machines (2001)
20. Reis, N.A., de Sousa, J.T.: On Implementing a Configware/Software SAT Solver. In: Proc. of 10th IEEE Int. Symp. Field-Programmable Custom Computing Machines (2002) 282-283
21. Skliarova, I., Ferrari, A.B.: A SAT Solver Using Software and Reconfigurable Hardware. In: Proc. of the Design, Automation and Test in Europe Conference (2002) 1094
22. Skliarova, I., Ferrari, A.B.: A hardware/software approach to accelerate Boolean satisfiability. In: Proc. of IEEE Design and Diagnostics of Electronic Circuits and Systems Workshop (2002) 270-277
23. Simon, L., Le Berre, D., Hirsch, E.: The SAT2002 Competition. Technical Report (preliminary draft) [Online]. Available: http://www.satlive.org/SATCompetition/onlinereport.pdf (2002)
24. Moskewicz, M.W., Madigan, C.F., Zhao, Y., Zhang, L., Malik, S.: Chaff: Engineering an Efficient SAT Solver. In: Proc. of the 38th Design Automation Conference (2001) 530-535
25. Goldberg, E., Novikov, Y.: BerkMin: a Fast and Robust SAT-solver. In: Proc. Design, Automation and Test in Europe Conference (2002) 142-149

Fault Tolerance Analysis
of Distributed Reconfigurable Systems
Using SAT-Based Techniques*

Rainer Feldmann[1], Christian Haubelt[2], Burkhard Monien[1], and Jürgen Teich[2]

[1] AG Monien, Faculty of CS, EE, and Mathematics,
University of Paderborn, Germany,
{obelix,bm}@upb.de
[2] Department of Computer Science 12, Hardware-Software-Co-Design,
University of Erlangen-Nuremberg, Germany,
{haubelt,teich}@cs.fau.de

Abstract. The ability to migrate tasks from one reconfigurable node to another improves the fault tolerance of distributed reconfigurable systems. The degree of fault tolerance is inherent to the system and can be optimized during system design. Therefore, an efficient way of calculating the degree of fault tolerance is needed. This paper presents an approach based on satisfiability testing (SAT) which regards the question: How many resources may fail in a distributed reconfigurable system without losing any functionality? We will show by experiment that our new approach can easily be applied to systems of reasonable size as we will find in the future in the field of body area networks and ambient intelligence.

1 Introduction

Distributed reconfigurable systems [1, 2] are becoming more and more important for applications in the area of automotive, body area networks, ambient intelligence, etc. The most outstanding property of these systems is the ability of reconfiguration. In terms of system synthesis, this means that the binding of tasks to resources is not static, i.e., the binding changes over time. Recent research was focused on the OS support for FPGAs [3] by dynamically assigning hardware tasks to an FPGA.

In a network of connected FPGAs it is possible to migrate hardware tasks from one node to another. Thus, resource faults can be compensated by *rebinding* tasks to fully functional nodes of the network. The process of rebinding is also called *repartitioning*. Distributed reconfigurable systems that support repartitioning possess an inherent fault tolerance. The degree of fault tolerance is a static property of the system and ,hence, can be optimized during system design. In order to evaluate the degree of fault tolerance, we define a new objective called *k-bindability*. A system is called *k-bindable* iff any set of k resources is redundant. Note, it may be possible that more than k resources are redundant but the k-bindability determines that k such that any set of k arbitrary resources can be removed from the system without losing any functionality.

* Supported in part by the German Science Foundation (DFG), SFB 376 (Massive Parallelität) and SPP 1148 (Rekonfigurierbare Rechensysteme).

P.Y.K. Cheung et al. (Eds.): FPL 2003, LNCS 2778, pp. 478–487, 2003.

The main contribution of this paper is to provide an efficient way based on SAT techniques to determine the k-bindability during system design. This problem is twofold: In a first step, we will reduce the well known binding problem from system synthesis to the satisfiability problem for boolean formulas. Next, we show how to calculate the k-bindability of a system using quantified boolean formulas (QBFs). Therefore, we focus on two particular system synthesis problems:

1. Does there exist a feasible binding for a given specification of a distributed reconfigurable system that supports repartitioning?
2. How many resources may fail in a distributed reconfigurable system that supports repartitioning without losing any functionality?

With this novel approach, we can optimize the fault tolerance of distributed reconfigurable system in an early design phase. In other words, we can maximize the k-bindability of such a system for a limited number of reconfigurable nodes and connections during design space exploration.

The problem to decide the satisfiability of QBFs is an important research issue in Artificial Intelligence, since QBF is the prototypical PSPACE-complete problem. Other PSPACE-hard problems from, e.g., conditional planning [4], non monotonic reasoning [5], and hardware verification [6] have been polynomially reduced to QBF. In the past several decision procedures for QBFs have been proposed in the literature [7–10].

This paper is structured as follows: In Section 2 we introduce the formal specification model of distributed reconfigurable systems used in this paper. The following section shows how to reduce the binding problem to the satisfiability problem of boolean formulas. In Section 4 a QBF-based approach to determine the k-bindability of a distributed reconfigurable systems that supports repartitioning is proposed. Finally, we will show by experiment (Section 5) that problem instances of reasonable size are easily solved by the Davis-Putnam based QBF solver QSOLVE [9].

2   Preliminaries

In order to specify distributed reconfigurable systems, we use a graph-based approach. First, we model the behavior of a system using a directed graph, called *task graph*. The vertices of the task graph represent tasks $t \in T$ where T is a finite set. The edges of the task graph model data dependencies $d \in D$ between the tasks, i.e., $D \subseteq T \times T$.

On the other hand, we model the architecture of our distributed reconfigurable system by a so-called *architecture graph*. An architecture graph is also a directed graph, where vertices correspond to reconfigurable nodes $r \in R$ of the network. Edges of the architecture graph model directed connections $c \in C \subseteq R \times R$ between the nodes.

To relate tasks $t \in T$ and reconfigurable nodes $r \in R$, mapping edges $m \in M$ map tasks to nodes. A mapping edge $m = (t, r)$ indicates that t may be executed on r. Note that more than one mapping edge could be associated with a task t or a reconfigurable node r, modeling possible bindings and resource sharing, respectively. Such graph-based models are also used in commercial systems like VCC [11].

Example 1. Figure 1 shows a specification of a distributed reconfigurable system. The set of tasks T and data dependencies D are given by $T = \{t_0, t_1, t_2\}$ and $D =$

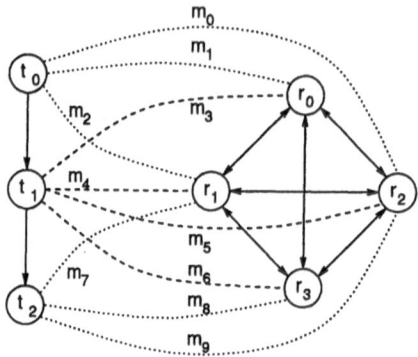

Fig. 1. Distributed control system consisting of a sample task (t_0), a control task (t_1), and a driver task (t_2). The architecture is composed of four reconfigurable nodes ($r_0, \ldots, r_3$). The additional mapping edges ($m_0, \ldots, m_9$) describe possible bindings.

$\{(t_0, t_1), (t_1, t_2)\}$, respectively. This task graph models the coarse grain behavior of a distributed control system, where t_0 corresponds to a sample task sampling a sensor, t_1 corresponds to the control task implementing the actual control, and t_2 models the driver task driving an actuator.

The architecture graph in Figure 1 consists of the reconfigurable nodes $R = \{r_0, r_1, r_2, r_3\}$. Each reconfigurable node could directly communicate with each other, i.e., the architecture graph is a clique. The mapping edges $M = \{m_0, \ldots, m_9\}$ indicate that the sample task t_0 may be executed on any reconfigurable node r_0 to r_2 and the driver task t_2 may be performed on any of the reconfigurable node r_1 to r_3. The controller task t_1 could be bound to any of the reconfigurable nodes in the architecture graph.

3   Binding

With the model introduced previously, the task of system synthesis could be formulated as: "Find a feasible *binding* of the tasks $t \in T$ to reconfigurable nodes $r \in R$, i.e., a subset of mapping edges." Here, a binding is said to be feasible if:

1. each task $t \in T$ is bound to exactly one reconfigurable node $r \in R$ and
2. required communications given by the data dependencies $d \in D$ can be handled by the given architecture graph, i.e., if there is a directed edge $d = (t_i, t_j)$ between task t_i and task t_j then either t_i and t_j have to be performed on the same reconfigurable node r (intra-node communication) or on reconfigurable nodes r_i and r_j which are directly connected via an edge $c = (r_i, r_j)$ (inter-node communication).

Blickle et al. [12] have reduced the problem of finding a feasible binding to the boolean satisfiability problem which is NP-complete. In this paper, we show how to derive boolean functions from specifications given by a task graph, an architecture graph, and the mapping edges as described in Section 2 such that the boolean function is satisfiable iff the specified distributed reconfigurable system has a feasible binding. These function

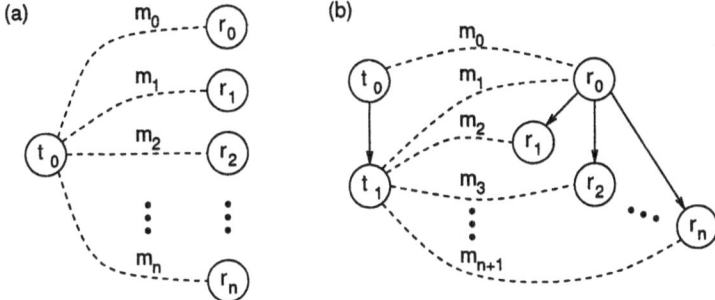

Fig. 2. (a) For each task $t \in T$ exactly one outgoing mapping edge has to be activated. (b) In order to establish the required communication $(d = (t_0, t_1))$, we have to execute the tasks t_0 and t_1 on the same reconfigurable node r_0 or on adjacent reconfigurable nodes.

could be tested by QBF solvers. Later, we extend this idea in order to analyze aspects such as whether a distributed reconfigurable system is fault tolerant and to what degree.

First, we consider the problem of checking the feasibility of a given binding. Therefore, we introduce some notations: Let m_i be a boolean variable, indicating if the mapping edge m_i is part of the binding ($m_i = 1$), or not ($m_i = 0$). The assignment of all variables m_i is denoted by (m). Note, that (m) is a binary coding of the binding. The set of all possible assignments of (m) is denoted by (M). With these new notations, we check the feasibility of a given binding represented by the coding (m) by solving a boolean equation. Therefore, we test both criteria of the feasibility as given above.

Example 2. First, we test if there is exactly one outgoing mapping edge for each task $t \in T$ in the binding. As an example consider Figure 2(a). There is a single task t_0 and $n+1$ mapping edges $m_0, \ldots, m_n$. A boolean function that indicates if there is exactly one outgoing mapping edge for t_0 in conjunctive normal form (cnf) is: $(m_0 + m_1 + m_2 + \cdots + m_n) \cdot (\overline{m}_0 + \overline{m}_1)(\overline{m}_0 + \overline{m}_2) \ldots (\overline{m}_0 + \overline{m}_n) \cdot (\overline{m}_1 + \overline{m}_2) \ldots (\overline{m}_1 + \overline{m}_n) \ldots (\overline{m}_{n-1} + \overline{m}_n)$ Here, $+$ denotes the boolean OR and $\cdot$ is the boolean AND. The first clause ensures that at least one of the mapping edges is activated ($m_i = 1$). The remainder guarantees that at most one mapping edge is part of the binding. The conjunction of both parts results in the required property.

For each task $t \in T$ we have to establish a formula similar to the one given in Example 2. The logical product results in a boolean function $b_1 : (M) \to \{0, 1\}$ with $b_1((m)) = 1$ iff (m) contains exactly one mapping edge per task. Hence, we obtain Equation (1) where $\prod$ denotes the boolean AND and $\sum$ denotes the boolean OR.

$$b_1((m)) = \prod_{t \in T} \left[\left(\sum_{\substack{m \in M: \\ m=(t,r)}} m \right) \cdot \left(\prod_{\substack{m_i, m_j \in M: \\ m_i=(t,r_x) \wedge m_j=(t,r_y) \wedge r_x \neq r_y}} (\overline{m}_i + \overline{m}_j) \right) \right] \quad (1)$$

Now, that we are sure that the first criteria is fulfilled, we check the second property of feasible bindings. All data dependencies $d \in D$ must be provided by the architecture of the implementation. Therefore, let $d = (t_i, t_j)$. If t_i is executed on r_x then t_j has to be performed on r_x also or on an adjacent reconfigurable node r_j, i.e., $(r_i, r_j) \in C$.

Example 3. Consider the example in Figure 2(b). The task t_0 is bound to r_0 by m_0. The execution of task t_1 must take place on node r_0 itself (by m_1) or on any adjacent reconfigurable node $r_1, \ldots, r_n$. An implication that assures this property is given by $m_0 \mapsto (m_1 + m_2 + \cdots + m_{n+1})$. This is equivalent to the clause: $(\overline{m}_0 + m_1 + m_2 + m_3 + \cdots + m_{n+1})$. This equation needs to be satisfied for each mapping edge $m \in M$.

A boolean function $b_2 : (M) \to \{0, 1\}$ to check the second property of feasibility is therefore:

$$
b_2((m)) = \prod_{\substack{m=(t,r)\in M, \\ t_i \in T:(t,t_i)\in D}} \left[\overline{m} + \sum_{\substack{m_j \in M:m_j=(t_i,r_x)\wedge \\ (r=r_x \vee (r,r_x)\in C)}} m_j \right] \tag{2}
$$

With Equation (1) and (2), we formulate a boolean function $b : (M) \to \{0, 1\}$ to check the feasibility of a given binding coded by (m):

$$
b((m)) = b_1((m)) \cdot b_2((m)) \tag{3}
$$

b is given in cnf and $b((m)) = 1$ iff the system has a feasible binding. In order to check if there is at least one feasible binding for a given specification, a SAT solver may be used to solve the following problem

$$
\exists (m) : b((m)) \tag{4}
$$

Checking whether a binding is feasible or whether a partial binding may be completed can be an important task during synthesis, but also in dynamically reconfigurable distributed systems. One application of the above SAT-techniques is therefore the domain of fault tolerance.

4   Fault Tolerance

In reconfigurable systems, the binding may change over time. Therefore, it may be possible to compensate resource faults by *rebinding* tasks to fully functional resources. The process of rebinding is called *repartitioning*. Recent research is focused on the OS support for single FPGA architectures [3]. In this section, we show how to model resource faults and how to measure the robustness of a given distributed reconfigurable system that supports repartitioning, also using SAT-based techniques. Therefore, we define the so-called *k-bindability* which quantifies the number of redundant resources in such a system.

4.1   Modeling Resource Faults

If a reconfigurable nodes fails, i.e., we cannot use this node for task execution any longer, the *allocation* of resources nodes may change. The allocation is the set of used resources

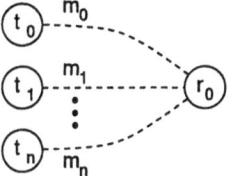

Fig. 3. In case of a resource defect, all incoming mapping edges must be deactivated.

in our implementation. Furthermore, an allocation is said to be feasible if there exists at least one feasible binding for this allocation. As in the case of the binding, we use the term (r) as the coding of an allocation where $r_i = 1$ indicates that the reconfigurable node r_i is part of the allocation. Furthermore, the term (R) describes the set of all possible allocation codings. If a reconfigurable node r_i fails, we have to set the associated binary variable r_i to zero ($r_i = 0$). All adjacent mapping edges m_j to r_i become meaningless, and should not be used in the binding, i.e., $m_j = 0$.

Example 4. Figure 3 shows a single reconfigurable node r_0 and n mapping edges. If r_0 fails $m_0, \ldots, m_n$ must not be used in the binding. We express this fact by n implications in the form $\overline{r}_0 \mapsto \overline{m}_j$ with $j = 0, \ldots, n$. In cnf we get: $(r_0 + \overline{m}_0)(r_0 + \overline{m}_1)(r_0 + \overline{m}_2) \ldots (r_0 + \overline{m}_n)$.

Again, we propose a boolean function to deactivate all mapping edges adjacent to a defect reconfigurable node. This boolean function $e : (M) \times (R) \to \{0, 1\}$ is satisfiable iff no reconfigurable nodes fail or there exists a feasible binding not using any of the mapping edges to the defect node.

$$e\left((m), (r)\right) = \prod_{r \in R, m \in M : m = (t, r)} (r + \overline{m}) \tag{5}$$

With this formula, we can check if a given reconfigurable node is redundant. For example, if we want to test if r_0 is redundant we solve the following SAT formula:

$$\exists (m), (r) : \overline{r}_0 \cdot e\left((m), (r)\right) \cdot b\left((m)\right)$$

4.2   k-Bindability

A frequent question is how many resources could fail in a distributed reconfigurable system that supports repartitioning without losing the desired functionality. Therefore, we define the number of nodes that may fail as *k-bindability*, i.e., k is the maximum number such that any set of k reconfigurable nodes is redundant. Note that we can remove any $n < k$ nodes of our distributed system without losing the specified functionality.

Example 5. Figure 1 shows the specification for a distributed control system. Let us remove one of the reconfigurable nodes $r_0, \ldots, r_3$ in Figure 1. Whatever node we choose, by rebinding the tasks we retain a running system, i.e., our system is at least 1-bindable. If we simultaneously remove any two of the reconfigurable nodes, again, our system remains working through rebinding the tasks. Now, our system is at least 2-bindable.

Let us check for 3-bindability. If we remove the reconfigurable nodes r_0, r_1, r_3 (or they fail simultaneously), we could bind all tasks t_0, t_1, and t_2 to node r_2. But the system is not 3-bindable, since if the nodes r_0, r_1, and r_2 fail simultaneously, we can not find any reconfigurable node to bind task t_0 to. That is: the system is 2-bindable.

In order to check for k-bindability using SAT-based techniques, we formulate a boolean function which encodes all system errors with exactly k reconfigurable node defects:

$$f^{(k)} = (A) \times (R) \to \{0,1\} \tag{6}$$

This function depends on the auxiliary variables a_i with $i = 0, \ldots, a_{|R|-1}$ where $|R|$ denotes the cardinality of R. If exactly k auxiliary variables are set to zero, we set the k corresponding allocation variables r_i to zero, otherwise all allocation variables may be set to one (i.e., no node fails).

Example 6. For the reconfigurable nodes in Figure 1, we encode the single resource defect of r_0 as: $a_3 a_2 a_1 \bar{a}_0$. The implication $(a_3 a_2 a_1 \bar{a}_0) \mapsto \bar{r}_0$ forces the allocation variable r_0 to zero. In cnf this corresponds to: $(\bar{a}_3 + \bar{a}_2 + \bar{a}_1 + a_0 + \bar{r}_0)$.

For all possible faults with exactly one single resource defect, we have to encode $|R|$ different cases (for each resource):

$$f^{(1)}\left((a),(r)\right) = \prod_{j=0}^{|R|-1} \left(\bar{r}_j + a_j + \sum_{i=0, i \neq j}^{|R|-1} \bar{a}_i \right)$$

$f^{(1)}\left((a),(r)\right)$ does not impose any constraints on (r) if more than one variable a_i is set to false. With this boolean function, we check if for all of these faults there is at least one feasible binding. This is done by the following quantified boolean formula:

$$\forall(a)\exists(r),(m) : f^{(1)}\left((a),(r)\right) \cdot e\left((m),(r)\right) \cdot b\left((m)\right)$$

We extend this approach by encoding all faults with exactly k resource defects. This is again an implication and can be written in cnf as:

$$f^{(k)}\left((a),(r)\right) = \prod_{i_1=0}^{|R|-1} \cdots \prod_{i_k=i_{k-1}}^{|R|-1} \prod_{l=1}^{k} \left(\bar{r}_{i_l} + \sum_{\substack{n=0 \\ n=i_1 \vee \cdots \vee n=i_k}}^{|R|-1} a_n + \sum_{\substack{n=0 \\ n \neq i_1 \wedge \cdots \wedge n \neq i_k}}^{|R|-1} \bar{a}_n \right)$$

Note again, that $f^{(k)}\left((a),(r)\right)$ does not impose any constraints on (r) if $p > k$ variables a_i are set to false.

Now, that we know how to code resource defects in a boolean function, we formulate the general form of the QBF solving the k-bindability problem:

$$\forall(a)\exists(r),(m) : f^{(k)}\left((a),(r)\right) \cdot e\left((m),(r)\right) \cdot b\left((m)\right) \tag{7}$$

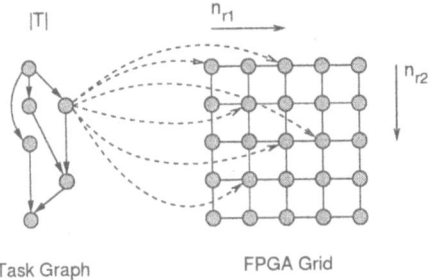

Task Graph   FPGA Grid

Fig. 4. Example specification of an FPGA grid and an application consisting of $|T| = 6$ tasks. The grid is composed of $n_{r1} \times n_{r2} = 5 \times 5$ FPGAs. The number of mapping edges per task is given by $n_m = 6$ and is shown only for one task.

5   Experimental Results

The k-bindability as defined above specifies the degree of fault tolerance of a distributed reconfigurable system. This fault tolerance can be optimized during system design. In order to compare different implementations during design space exploration, we must evaluate these implementations. In this section, we present first results of our new approach by using QSOLVE [9]. For this purpose, we design a benchmark. Our goal is to evaluate the runtime in dependence of the problem size (e.g., number of tasks, number of resources) which could be easily solved and, hence, could be investigated during automated design space exploration.

In a first step, an array of $n_{r1} \times n_{r2}$ reconfigurable nodes is defined. Communications are established such that the array is a 2-dimensional grid. Here, we use a grid in order to construct scalable architectures of distributed reconfigurable systems. Furthermore, a grid is typical for reconfigurable architectures as FPGAs and coarse grain architectures like PACT [13], Chameleon [14], etc. Next, we define a weakly connected random task graph with $|T| = n_t$ tasks. The probability that there is a data dependency between task t_i and task t_j with $j > i$ is given by another parameter called pb. In a last step, n_m mapping edges are randomly drawn from each task $t \in T$ to reconfigurable nodes $r \in R$, i.e., there are $|M| = |T| \cdot n_m$ mapping edges. Figure 4 shows an example of such a specification. The most meaningful results are presented in the following.

5.1   Feasibility of Binding

In a first test, we solve Equation (4) for randomly generated specifications. Here, three different $n_{r1} \times n_{r2}$ grids of reconfigurable nodes are investigated ($5 \times 5, 10 \times 10$, and 15×15). We map randomly generated task graphs onto each of these distributed architectures, while varying the number of tasks ($n_t = 50, 100, 150$). The tasks are connected with a probability of $pb = 0.5$. The number of mapping edges is chosen in a way, that feasible as well as infeasible systems are constructed. Only the cases of infeasible bindings are documented here (Finding a feasible binding by Equation (4) is the easier case). The average results (100 samples each) using QSOLVE [9] obtained on a PC system with a 1.8 GHz processor are shown in Table 1.

Table 1. Number of recursions recur, number of assignments assign, and computation time time required for solving (unsatisfiable) Equation (4).

	$|T| = 50$			$|T| = 100$			$|T| = 150$		
	5×5	10×10	15×15	5×5	10×10	15×15	5×5	10×10	15×15
n_{m}	17	65	140	19	75	160	20	80	–
recur	21	70	180	23	116	237	29	176	–
assign	2149	30285	136675	6527	120421	421960	12809	266205	–
time/s	0.07	1.80	14.05	0.46	14.33	69.67	1.57	43.11	–

The number of recursions *recur* corresponds to the number of nodes in the search tree. We see that systems with 225 reconfigurable nodes and 16,000 mapping edges be checked in a reasonable amount of time (≈ 1 min). Note: We only construct weakly connected task graphs. If we were using n weakly connected task subgraphs that are not connected, our binding problem consists of n independent binding problems. A QBF solver solves these (sub)problems independent of each other, which is much easier.

5.2 k-Bindability

With the results above, we consider the k-bindability problem as described in Section 4. Table 2 shows the average results (100 samples each) obtained from solving the boolean functions with the QBF-solver QSOLVE. We have chosen a 4×4 grid of reconfigurable nodes. Different numbers n_t of tasks are mapped onto this architecture. With parameters $n_{\mathrm{m}} = 13$ and $pb = 0.5$, we check the k-bindability for $k = 4, \ldots, 1$. Table 2 shows that systems with 50 tasks are still solvable in a reasonable amount of time.

As mentioned above, our approach is not limited to grids of reconfigurable nodes but supports arbitrary topologies. Thus, it is possible to optimize the architecture of distributed reconfigurable systems by using SAT-based techniques during design space exploration.

Table 2. Number of recursions recur, number of assignments assign, and computation time time to solve the k-bindability equation (satisfiable) Equation (7).

$|T|$		$k=1$	$k=2$	$k=3$	$k=4$	$|T|$		$k=1$	$k=2$	$k=3$	$k=4$
25	recur	532	2759	10230	30106	40	recur	587	4972	19136	49858
	assign	8569	43856	163972	487057		assign	10179	77695	293135	827385
	time/s	0.10	0.60	2.22	6.72		time/s	0.23	1.78	7.06	20.26
30	recur	434	3235	11931	35222	45	recur	649	4524	17107	51307
	assign	7474	54488	201263	594378		assign	12109	86569	318774	933571
	time/s	0.12	0.94	3.62	10.69		time/s	0.35	2.47	9.15	26.28
35	recur	488	3593	13403	39787	50	recur	633	4893	17818	53272
	assign	9031	65010	240039	705878		assign	12573	93306	342441	1007054
	time/s	0.19	1.45	5.89	16.63		time/s	0.37	2.74	10.02	29.90

6   Conclusions

Distributed reconfigurable systems, e.g., arrays of reconfigurable hardware elements including FPGAs or medium and coarse granular reconfigurable systems such as PACT [13] and Chameleon [14], possess an inherent fault tolerance which can be optimized during system design. The main contribution of this paper is to provide an efficient method to determine the degree of fault tolerance of a system, the so-called k-bindability. Two particular problems were considered in this paper: (i) Does there exist a feasible binding for a given specification of a distributed reconfigurable system that supports repartitioning? (ii) How many resources may fail in a distributed reconfigurable system that supports repartitioning without losing any functionality? Both problems were solved by reducing the binding problem to quantified boolean formulas and applying the QBF solver QSOLVE in order to test the satisfiability of these formulas. We have shown by experiment that our new approach can easily be applied to systems of reasonable size as we will find in the future in the field of body area networks and ambient intelligence. Hence, this approach provides a way to optimize the architecture of distributed reconfigurable systems during design space exploration.

References

1. Dick, R., Jha, N.: CORDS: Hardware-Software Co-Synthesis of Reconfigurable Real-Time Distributed Embedded Systems. In: Proceedigns of ICCAD'98. (1998) 62–68
2. Ouaiss, I., Govindarajan, S., Srinivasan, V., Kaul, M., Vemuri, R.: An Integrated Partitioning and Synthesis System for Dynamically Reconfigurable Multi-FPGA Architectures. In: IPPS/SPDP Workshops. (1998) 31–36
3. Walder, H., Platzner, M.: Online Scheduling for Block-partitioned Reconfigurable Devices. In: Proceedings of Design, Automation and Test in Europe (DATE03). (2003) 290–295
4. Rintanen, J.: Constructing Conditional Plans by a Theorem-Prover. Journal of Artificial Intelligence **10** (1999) 323–352
5. Egly, U., Eiter, T., Tompits, H., Woltran, S.: Solving Advanced Reasoning Tasks Using Quantified Boolean Formulas. In: Proc. of the 17th Nat. Conf. on Artificial Intelligence. (2000) 417–422
6. Scholl, C., Becker, B.: Checking Equivalence for Partial Implementations. In: Proceedings of 38th Design Automation Conference, Las Vegas, USA (2001) 238–243
7. Kleine-Büning, H., Karpinski, M., Flögel, A.: Resolution for Quantified Boolean Formulas. Information and Computation **117** (1995) 12–18
8. Cadoli, M., Giovanardi, A., Schaerf, M.: An Algorithm to Evaluate Quantified Boolean Formulae. In: Proc. of the 15th Nat. Conf. on Artificial Intelligence. (1998) 262–267
9. Feldmann, R., Monien, B., Schamberger, S.: A Distributed Algorithm to Evaluate Quantified Boolean Formulas. In: Proc. of the 17th Nat. Conf. on Artificial Intelligence. (2000) 285–290
10. Giunchiglia, E., Narizzano, M., Tacchella, A.: Backjumping for Quantified Boolean Formulas. In: Proc. of the 17th Int. Joint Conf. on Artificial Intelligence. (2001) 275–281
11. Cadence: Virtual Component Co-design (VCC). (2001) http://www.cadence.com.
12. Blickle, T., Teich, J., Thiele, L.: System-Level Synthesis Using Evolutionary Algorithms. In Gupta, R., ed.: Design Automation for Embedded Systems. 3. Kluwer Academic Publishers, Boston (1998) 23–62
13. Baumgarte, V., May, F., Nückel, A., Vorbach, M., Weinhardt, M.: PACT XPP - A Self-Reconfigurable Data Processing Architecture. In: ERSA, Las Vegas, Nevada (2001)
14. Chameleon Systems: CS2000 Reconfigurable Communications Processor. (2000)

Hardware Implementations
of Real-Time Reconfigurable WSAT Variants

Roland H.C. Yap, Stella Z.Q. Wang, and Martin J. Henz

School of Computing, National University of Singapore
Singapore
{ryap,wangzhan,henz}@comp.nus.edu.sg

Abstract. Local search methods such as WSAT have proven to be successful for solving SAT problems. In this paper, we propose two host-FPGA (Field Programmable Gate Array) co-implementations, which use modified WSAT algorithms to solve SAT problems. Our implementations are reconfigurable in real-time for different problem instances. On an XCV1000 FPGA chip, SAT problems up to 100 variables and 220 clauses can be solved. The first implementation is based on a random strategy and achieves one flip per clock cycle through the use of pipelining. The second uses a greedy heuristic at the expense of FPGA space consumption, which precludes pipelining. Both of the two implementations avoid re-synthesis, placement, routing for different SAT problems, and show improved performance over previously published reconfigurable SAT implementations on FPGAs.

1 Introduction

Stochastic local search (SLS) algorithms have been successful for solving prepositional satisfiability problems (SAT). The WalkSAT family (WSAT) of algorithms [1, 2] contains some of the best performing SLS algorithms. SLS algorithms like WSAT have a very simple structure and are composed of essentially three steps which are iterated until a satisfiable solution is found: (i) evaluate clauses; (ii) choose a variable; and (iii) flip the variable's boolean value.

Since each of the steps is simple, and as the SAT clauses can be directly represented in hardware, it is tempting to build a hardware-based SLS solver. There are a number of such hardware designs and implementations [3, 4, 5, 6] using reconfigurable FPGA hardware. Hardware approaches to systematic search procedures for SAT problems are beyond the scope of this paper; see [7] for an overview.

The use of a hardware SAT solver only makes sense if there is a significant performance advantage compared to software. Software can make use of state of the art processors built with the latest processor technology. A hardware SAT solver, on the other hand, is less likely to have the same level of process technology, and hence longer cycle times. Earlier hardware implementations like [3, 4] did not outperform optimized software. For example, a reimplementation of the design in Hamadi and Merceron [3] which was done in Henz et al. [6] had flip rates between 98 − 962 K flips/s. In some problems, this was a bit faster than software and in other cases slower.

P.Y.K. Cheung et al. (Eds.): FPL 2003, LNCS 2778, pp. 488–496, 2003.
© Springer-Verlag Berlin Heidelberg 2003

In [6], it was shown that GSAT SLS solvers running at one flip per clock cycle was achievable with performance gains of about two orders of magnitude over software. That implementation makes use of the reconfigurable nature of FPGAs to build a custom design specific to a particular SAT problem instance. The contribution of [6] is to show that large speedups are feasible. This approach, however, is not practical as a general SAT problem solver because the time needed to re-synthesize, place and route the specific FPGA design is likely to exceed the runtime improvement from the faster solver.

This paper explores hardware designs for WSAT, which are not instance-specific and thus do not require re-synthesis. In addition to this requirement, a hardware implementation faces interesting design tradeoffs due to the inherently limited logic resources on the chip. We propose two versions of WSAT, which allow real-time reconfiguration. The differences of the WSAT versions lead to different design choices for maximal performance. The first design emphasizes fast cycle times (one flip per clock cycle), employing random variable selection to allow for a pipelined design. The second uses a greedy variable selection heuristic, which precludes pipelining, exemplifying a tradeoff between flip rate and effectiveness of variable selection. Both designs have improved performance over other published non-re-synthesis SLS FPGA implementations.

2 Hardware Implementation Issues

2.1 Cost of Re-synthesis FPGA Implementations

SLS SAT algorithms exhibit large amounts of parallelism and hence are a good match for a hardware solver, which can use the large amounts of parallelism available in the hardware. We focus here on WSAT implementations using Field Programmable Gate Arrays (FPGAs), which provide the benefits of customized hardware but avoid fabrication cost, and thus allow for convenient prototyping of the hardware design. Unlike software, a hardware implementation has to deal with the inherent resource limitations for combinatory logic, memory and routing on an FPGA.

One approach is to maximize performance by making full use of parallelism, exemplified in [6], where clause evaluation and variable selection are parallelized for a GSAT SLS implementation. However, such a high degree of parallelism is expensive in terms of hardware resources. That implementation optimizes the hardware design specifically for a given SAT problem instance, taking advantage of the reconfigurability of FPGAs. This *instance-specific* approach enabled a performance of one flip per clock cycle, more than two orders of magnitude faster than software. The drawback, however, is that a new solver has to be re-synthesized for each SAT problem instance. With current CAD tools, the synthesis, placement and routing for SAT instances with 200 variables can take several hours, while the resulting SAT solver may only take seconds or minutes to find a solution to the instance. Thus, while instance-specific hardware implementations demonstrate the feasibility of very high performance hardware approaches, they are impractical as general-purpose SAT solvers.

A general-purpose hardware SAT solver should instead not require re-synthesis, and be able to handle different SAT instances with only small overheads. One non-re-synthesis approach is given in [4], which takes advantage of the fact that the FPGA configuration file can be altered directly to modify the design. This provides a shortcut to re-synthesis since only small modifications to the definitions of the SAT clauses are necessary. However, this implementation is mostly sequential and does not outperform optimized software. A more serious issue is that current FPGA chips do not have an open architecture. The configuration file for these chips is a black box, which renders this approach unfeasible.

Leong et al [5] achieved a bitstream reconfigurable FPGA implementation for a WSAT variant. Their implementation stores clauses for a SAT problem in the 16x1-bit ROM available in the Logic Cells (LC) of the Xilinx FPGA. A different SAT instance requires various ROM definitions to be modified. Normally, this would require re-synthesis of the FPGA to generate a new bitstream configuration for downloading. Leong et al were able to achieve a non re-synthesis implementation, using a tool to extract the locations of the relevant LCs in the bitstream, and then directly modify the corresponding data for the ROM values in the bitstream file. This approach requires analysis of the bitstream file to figure out how to rebuild the configuration without re-synthesis.

Both of these implementations [4, 5] simulate re-synthesis in a very efficient fashion. However, they are dependent on the ability to modify the FPGA configuration.

The aim of this paper is to obtain a more portable reconfigurable implementation, which nevertheless is capable of providing good search performance, and which exhibits short reconfiguration times.

2.2 A Clause Evaluator without Re-synthesis

The key to avoid re-synthesis is to be able to handle any SAT instance. Hence the clause evaluator in WSAT must be general rather than instance-specific. Our goal is a general clause evaluator, which fits well within an FPGA architecture and can be reconfigured quickly in a portable fashion.

We will focus on the Xilinx Virtex FPGA chips. The basic building block of Virtex FPGA [8] is a LC, which includes a 4-input function generator, carry logic and a storage element. The 4-input function generator is implemented as 4-input look-up table (LUT). Each Virtex CLB (Configurable Logic Block) contains four LCs, organized in two slices. Two LUTs in a slice can be combined to create a 16x1-bit dual port RAM. Our clause evaluator represents the clauses in the SAT instance in a 16x1-bit dual port RAM array, which can be generated from the Xilinx RAM16x1D primitive. The Xilinx RAM16x1D primitive is a 16-word by 1-bit static dual port random access memory with synchronous write capability. The device has two separate address ports; the read address port (DPRA3-DPRA0) and the write address port (A3-A0).

We describe the clause evaluator by example. Consider a SAT clause, c_3, of the form, $x_1 \lor x_2 \lor \overline{x_5}$, and let us assume that c_3 is a clause of a SAT problem over 8 variables. The clause can be written as a disjunction of two simpler functions,

$$f_{3,1}(x_1, x_2, x_3, x_4) \lor f_{3,2}(x_5, x_6, x_7, x_8)$$

where $f_{3,1}(x_1, x_2, x_3, x_4) = x_1 \lor x_2$ and $f_{3,2}(x_5, x_6, x_7, x_8) = \overline{x_5}$. Thus each SAT clause, c_i, can be decomposed into a disjunction of boolean functions on fewer variables. We map each $f_{i,j}$ arising from the *j-th* part of clause i to a RAM16x1D primitive, treating the four variables as the address to the read port (DPRA3-DPRA0). The function $f_{i,j}$ is configured by using the write port (A3-A0) to define its truth table.

One advantage of this representation is that negated variables are handled automatically inside the $f_{i,j}$ block. Figure 1(a) shows an overall block diagram of the reconfigurable clause evaluator for 100 variables and 220 clauses. Figure 1(b) shows each $f_{i,j}$ block, which is configured using the controller in Figure 1(c). The result of each RAM primitive is ORed and stored in the array *all_clause[]*.The clause evaluator evaluates all clauses in parallel in one cycle.

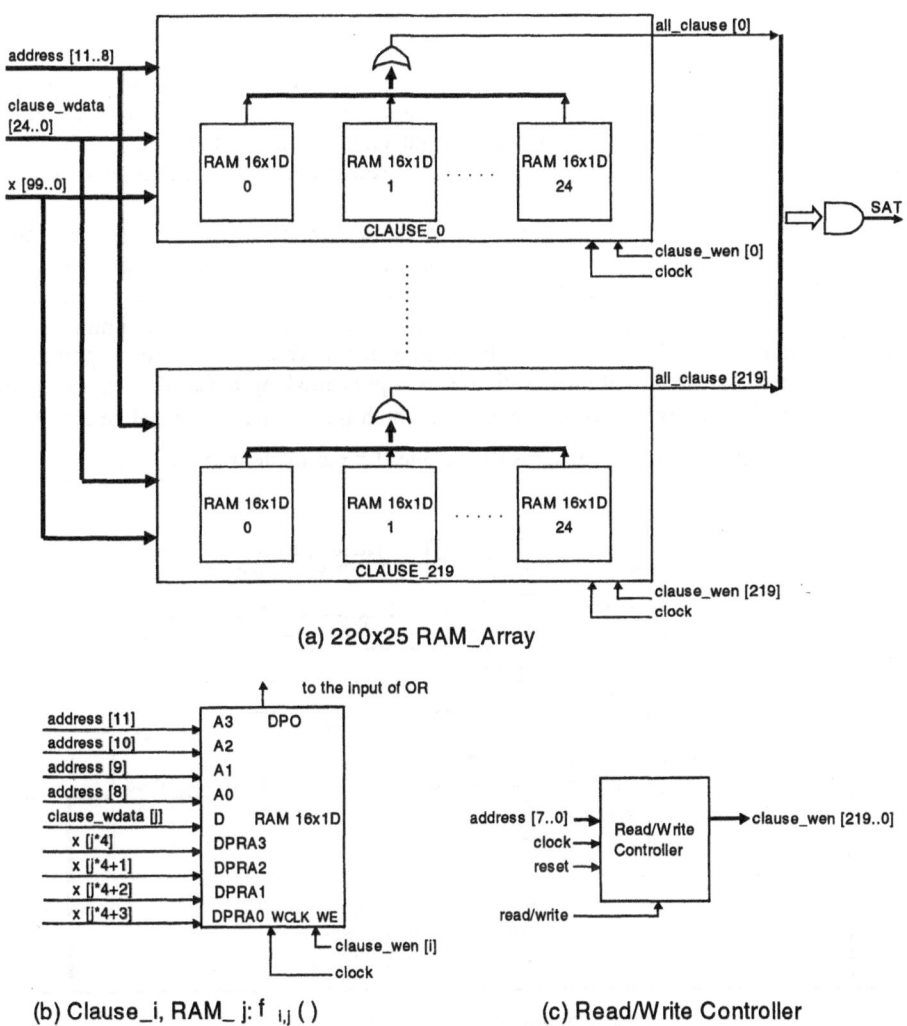

(a) 220x25 RAM_Array

(b) Clause_i, RAM_ j: f $_{i,j}$ ( )   (c) Read/Write Controller

Fig. 1. Block Diagram of the Reconfigurable Clause Evaluator

3 Two FPGA Implementations without Re-synthesis

The reconfigurable clause evaluator requires $O(mn)$ CLBs for an implementation with m clauses and n variables. This component consumes a significant fraction of the available CLBs (as much as 80%). As we would like to be able to handle as large a problem as feasible within the constraints of the FPGA, it is impractical to consider implementations that require multiple clause evaluators. This would consume too much of the chip real estate, even if there is considerable parallelism gain. We present two implementations of WSAT for 3-SAT problems, which represent different tradeoffs in using a single reconfigurable clause evaluator.

3.1 A Pipelined FPGA Implementation Using a Random Selection Heuristic

One strategy is to produce an implementation with a fast cycle time. Given that we are constrained to a single clause evaluator, we are left with pipelining as the only option for increasing the flip rate. For maximal reuse of the clause evaluator, it is important that the pipeline be well balanced with simple pipeline stages. Given that we already have a fully parallel clause evaluator, the most expensive step in WSAT is variable selection. A particularly simple WSAT variant chooses the variable randomly in a selected unsatisfied clause. This strategy is also used in the WSAT implementation of Leong et al [5].

Figure 2 depicts a five-stage pipelined implementation. Stage 1 finds a random unsatisfied clause (this checks all clauses in parallel). Stage 2 generates three variable indices for the selected clause. Stage 3 implements the random selection heuristic, flipping of its input variables. Stage 5 checks for satisfiability. There are a number of storage buffers used. Buffer 1 stores the clause table which gives the mapping of clause to variables used within that clause as represented by a variable index. The SAT problem is initially loaded into buffer 2, which then is used to initialize the $f_{i,j}$

blocks in the clause evaluator. The result is a one flip per cycle implementation.

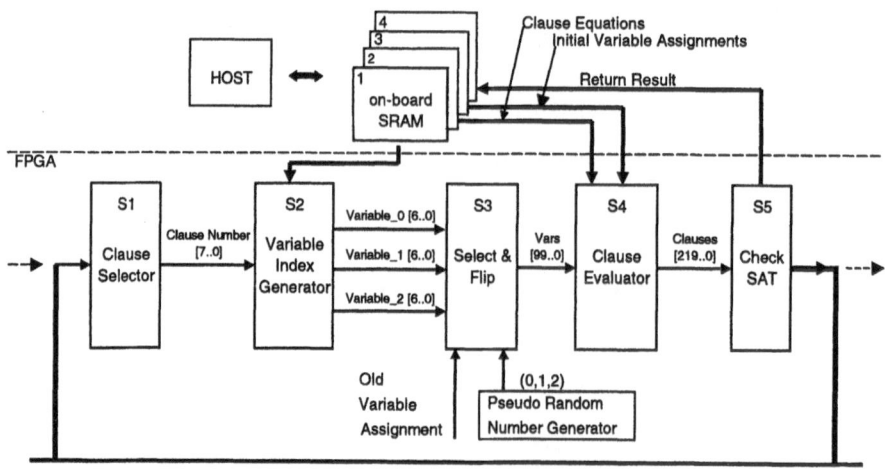

Fig. 2. Pipelined Random WSAT

3.2 A FPGA Implementation Using a Greedy Selection

A more typical WSAT variable selection heuristic is to select the variable, which best improves the score. In terms of the constraints of the hardware, this corresponds to a design with more complex operations. We have chosen to use a pure greedy heuristic without noise (but a noise component can be easily added).

Figure 3 shows the block diagram of a sequential implementation. Since we are dealing with 3-SAT, it is only necessary to determine at most which of the three variables in a clause to select. However, any kind of parallel implementation of this step would require computing the score of each of the three possibilities. This would require three clause evaluator units, which we deem too space consuming for the targeted SAT problem size. Thus, we are restricted to a sequential implementation for the variable selection (Stages 4-6), which reduces the flip rate. Our current implementation performs one flip in nine cycles, as opposed to one cycle achieved by the design for random selection heuristic.

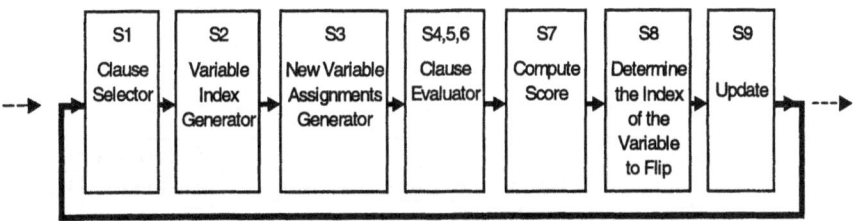

Fig. 3. Sequential Greedy WSAT

4 Results

Our hardware SAT solver is implemented on Celoxica's RC1000-PP standard PCI bus board, which is equipped with a Xilinx XCV1000 FPGA. This board has 8Mb of SRAM directly connected to the FPGA in four 32-bit wide memory banks. Each of the four banks may be granted to either the host CPU or the FPGA at any time. Data can therefore be shared between the FPGA and the host CPU by placing it in the SRAM. It is accessible to the host CPU by DMA transfer across the PCI bus.

As host we use a PC with an AMD Athlon 1.2GHz CPU. Our prototype generates the clause configuration for a new SAT instance in software in about 7ms (this is unoptimized and is probably dominated by file I/O and hence could possibly be faster). Transferring the clause configuration from the host PC to the on-board SRAM takes 0.6ms. The FPGA takes 220 x 16 clock cycles to read the SRAM. With an FPGA clock frequency of 20MHz, this corresponds to 0.176ms. Thus the configuration overhead for solving a new SAT instance is 7.776ms. In contrast, the time to download a new bitstream to the FPGA is around 0.14s.

The prototype implementations investigate the two designs on two SAT problem sizes; a 50 variable/170 clause format and a 100 variable/220 clause format, the latter chosen to such that its reconfigurable clause evaluator fits on the FPGA used. Table 1 gives the hardware costs in terms of slices for the various implementations. The minimum gate delay is as reported by the Xilinx place and route tools. There is only a small difference in gate delay between the two implementations. The larger influence is the increased delay due to larger problem sizes.

Table 1. Time/Space Cost Comparison of FPGA-based Implementation

	Random-Strategy WSAT		Greedy-Strategy WSAT	
System Size	Delay (ns)	Cost of Slices	Delay (ns)	Cost of Slices
50-var/170-c	24.097	4946 (40%)	24.842	6408 (52%)
100-var/220-c	31.005	10396 (85%)	31.639	11834 (96%)

Table 2 shows the flip rate performance comparison given in number of flips per second (fps). We compare FPGA-based hardware implementations versus software for various 3-SAT benchmarks. The benchmarks used are simply those, which fit within the required problem sizes. As the main purpose of the benchmarks is to measure flip rate performance, the difficulty of the benchmarks is not so relevant, as such a mix of more difficult problems and the easier AIM benchmarks are used. Our FPGA implementations were clocked at 20Mhz. The software WSAT implementation is WalkSAT35 by Kautz and Selman [11] running on a Pentium4 1500Mhz PC.

Table 2. Flip Rate Speedups: FPGA-based Hardware versus Pure Software

SAT Problems	Random-Strategy			Greedy-Strategy	
	Software Flip Rate (Kfps)	Hardware Speedup Ours, Pipelined	Leong et al. [5]	Software Flip Rate (Kfps)	Hardware Speedup
Uf20-9	407.6	49.91	0.89	265.8	8.38
Uf20-31	390.9	52.06	0.93	251.8	9.84
Uf20-37	405.8	50.11	0.90	303.2	7.34
Uf50-01	536.2	37.95	-	409.4	5.45
Uf50-010	466.2	43.70	-	459.6	4.84
aim-50-2_0-yes1-1	865.4	23.49	0.41	775.4	2.87
aim-50-2_0-yes1-2	859.8	23.73	0.45	775.3	2.89
aim-50-3_4-yes1-1	618.6	33.36	0.98	609.2	3.66
aim-50-3_4-yes1-2	612.6	33.56	1.58	596.0	3.74
aim-50-3_4-yes1-3	613.1	33.55	0.65	572.4	3.89
aim-50-3_4-yes1-4	609.3	33.83	0.42	561.2	3.97
aim-100-1_6-yes1-1	962.5	21.37	-	814.2	2.74
aim-100-1_6-yes1-2	968.4	21.09	-	787.4	2.86
aim-100-1_6-yes1-3	972.9	21.05	-	805.4	2.76
aim-100-1_6-yes1-4	1014.4	20.29	-	872.2	2.55
aim-100-2_0-yes1-1	838.5	24.07	-	744.9	3.00
aim-100-2_0-yes1-2	814.3	24.92	-	747.7	3.00
aim-100-2_0-yes1-3	812.6	24.93	-	731.4	3.06
aim-100-2_0-yes1-4	834.4	24.25	-	744.6	3.00

The flip rate for the random and greedy variable selection heuristics is constant throughout the problems – 20M flips for random, and 2.2M flips for the greedy heuristics, due to its 9-stage implementation. We also measured actual timings as a reality check. The "Hardware Speedup" columns represent the ratio of measured flip rate versus the software flip rate. Note that the software flip rate varies with the problem, while it is constant in our implementations.

The fourth column compares our pipelined random strategy with the WSAT reconfigurable FPGA implementation from Leong et al. [5], which also uses a random strategy. Their implementation uses a smaller FPGA with problems of up to 50 variables and hence could be clocked at a faster speed of 33Mhz. The speedup has been recomputed using the average timing results in their paper. Where timings or benchmarks are not available, this is indicated by a (-). A major difference between their implementation and the greedy pipelined one here is that our implementation is based on a constant flip rate. Their implementation, on the other hand, has a variable flip rate, because of the use of sequential clause selection and is bounded by a maximum flip rate of 364Kfps.

With the random variable selection heuristic, the preliminary results show that our reconfigurable FPGA implementation is significantly faster than software and previous hardware implementations. This implementation achieves one flip per clock cycle at 20Mhz. The greedy variable selection implementation has more modest speedups. The speedup is likely comparable to software or slightly faster, if the fastest state of art microprocessors are used, since performance scales at a lower rate with clock speed for microprocessors. However, the reduced flip rate may be offset by the increased effectiveness of the variable selection strategy. The greedy heuristic typically gives a better success rate than a random heuristic for WSAT. A detailed analysis of the effect of different variable selection heuristics is given in [9].

5 Conclusion

We demonstrate two prototype hardware solvers implemented on the Xilinx Virtex XCV1000 FPGA with significantly better performance than software and previous hardware WSAT solvers. Furthermore, the solvers are reconfigurable in real-time, with a reconfiguration time of a few milliseconds for problems with 100 variables. Our two implementations illustrate the tradeoff between time, space and effectiveness of the SLS algorithm. The random solver achieves an optimal flip rate at the cost of a simple variable selection strategy, while the greedy solver uses the more expensive and effective strategy but is not amenable to pipelining and is hence slower.

Both implementations are limited by the size of the Xilinx Vertex XCV1000 chip used, which can accommodate a reconfigurable clause checker only for problems with 100 variables and 220 clauses. This chip, dating from 1999, is fabricated using a 5-layer metal 0.22µm CMOS process. In comparison, the current Virtex-II generation uses an 8-layer 0.15µm CMOS process. The XC2V10000 has about 10 times more system gates than the XCV1000 and has significantly faster clock speeds. For example, a 100 variable/600 clause evaluator requires about 30K slices and fits in a XC2V6000 which has 6M system gates.

An FPGA implementation will have more limitations on problem sizes even when larger FPGAs are used. A fast hardware based solver can however still be useful for general SAT solving. One approach is with hybrid search and stochastic solvers. For example, Zhang et al. [10] combine Davis Putnam with stochastic search. Their approach uses Davis Putnam to generate smaller sub-problems which are then solved with WSAT.

Another route to deal with larger problems is to use ASICs rather than FPGAs. Our implementation is not restricted to FPGAs since the reconfiguration for different SAT instances is not dependent on the reconfigurable logic of FPGAs. The prototype uses FPGAs simply because they are more cost effective for development. Given the real-time reconfiguration capability, this may be a promising candidate for direct ASIC implementation, which means higher clock speeds and much more resources for dealing with larger problems.

References

[1]   B. Selman, H. Kautz, and B. Cohen. Noise strategies for improving local search. *Proc. National Conference on Artificial Intelligence*, 337-343, 1994.

[2]   David McAllester, Bart Selman, and Henry Kautz. Evidence for invariants in local search. *Proc. Fourteenth National Conference on Artificial Intelligence (AAAI-97)*, 1997.

[3]   Youssef Hamadi and David Merceron. Reconfigurable architectures: A new vision for optimization problems. *Principles and Practice of Constraint Programming*, 209-221, 1997.

[4]   Wong Hiu Yung, Yuen Wing Seung, Kin Hong Lee, and Philip Heng Wai Leong. A runtime reconfigurable implementation of the GSAT algorithm. *Field-Programmable Logic and Applications*, 526-531, 1999.

[5]   P. H. W. Leong, C. W. Sham, W. C. Wong, H. Y. Wong, W. S. Yuen, and M. P. Leong. A bistream reconfigurable FPGA implementation of the WSAT algorithm. *IEEE Trans. on Very Large Scale Integration (VLSI) Systems*, 9(1): 197-200, 2001.

[6]   Martin Henz, Edgar Tan, Roland Yap. One flip per clock cycle. *Proc. of the Seventh International Conference on Principles and Practice of Constraint Programming*, 509-523, 2001.

[7]   M. Abramovici and A. Sousa. A SAT solver using reconfigurable hardware and virtual logic, *Journal of Automated Reasoning* 24(1/2): 5-36, 2000.

[8]   Xilinx. *Virtex 2.5 Field programmable gate arrays*, 1999.

[9]   H. Hoos and T. Stützle. Local search algorithms for SAT: An empirical evaluation *Journal of Automated Reasoning*, 24:421-481, 2000.

[10] Wenhui Zhang, Zhuo Huang, Jian Zhang. Parallel Execution of Stochastic Search, Procedures on Reduced SAT Instances. *Proc. of the Seventh Pacific Rim International Conference on Artificial Intelligence*, 108-117, 2002.

[11] H. Kautz and B. Selman, Walksat homepage,
      http://www.cs.washington.edu/homes/kautz/walksat/

Core-Based Reusable Architecture for Slave Circuits with Extensive Data Exchange Requirements

Unai Bidarte, Armando Astarloa, Aitzol Zuloaga,
Jaime Jimenez, and Iñigo Martinez de Alegría

University of the Basque Country, E.T.S. Ingenieros,
Department of Electronics and Telecommunications,
Urquijo s/n, E-48013 Bilbao, Spain,
{jtpbipeu,jtpascua,jtpzuiza,jtpjivej,jtpmamai}@bi.ehu.es

Abstract. Many digital circuit's functionality is strongly dependant on high speed data exchange between data source and sink elements. In order to alleviate the main processor's work, it is usually interesting to isolate high speed data exchange from all other control tasks. A generic architecture, based on configurable cores, has been achieved for slave circuits controlled by an external host and with extensive data exchange requirements. Design reuse has been improved by means of a software application that helps on configuration and simulation tasks. Two applications implemented on FPGA technology are presented to validate the proposed architecture.

1   Introduction

When analyzing the data path of a generic digital system, three main elements can be distinguished:

- Data source, processor or sink. Any digital system needs some data source, which can be some kind of sensor or an external system. Process units transform data and finally data clients or sinks make some use of it.
- Data buffer blocks or memories and data exchange control units. The described data units cannot usually be directly connected and an intermediate storage element is needed. On the other hand, when transferring from one device to another, a communication channel and a predefined data transfer protocol must be followed. Data exchange control can be a very time consuming task. In order to liberate the main control unit, it is often adequate to use a data exchange control specific unit.
- High level control unit. It is the digital system master that generates all the control signals needed by the previously noted blocks.

This work studies slave digital systems with much data exchange, that is to say, circuits controlled by a host and with high volume data transfers. The following features summarize the system under study:

P.Y.K. Cheung et al. (Eds.): FPL 2003, LNCS 2778, pp. 497–506, 2003.

- There is a communication channel with the host, which sends control commands. The host can also be a data source or sink.
- Data exchange requires complex and high speed control, which makes a specific data exchange module necessary.
- It is needed to attend several data transfer requirements in parallel.
- Data processing is performed on data terminal units, so, for design purposes, data units are supposed to be source or sink. These elements are not synchronized, so an intermediate data storage is needed.

Fig. 1 represents the block diagram corresponding to the described slave system. Several circuits square with the specifications above: industrial machinery like filling or milling machines, polyphonic audio, generation of three dimensional images, video servers, PC equipment like plotters and printers,...

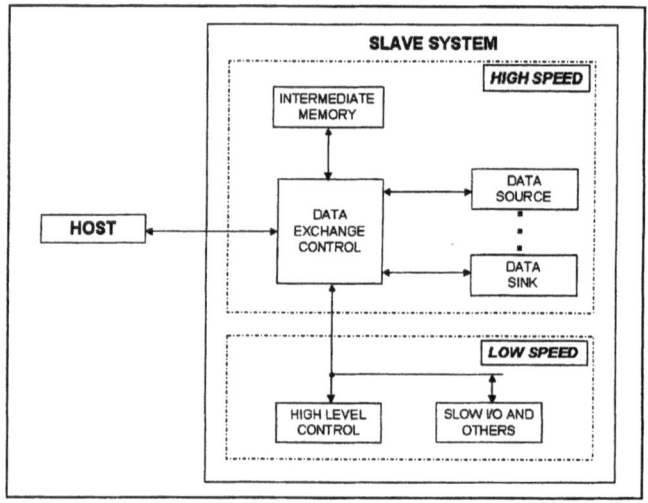

Fig. 1. Slave system's block diagram

The research team's main objective is to achieve a reusable architecture for the presented system. It must be independent of the implementation technology, so a standard hardware description language will be used. In order to facilitate the design reuse, a user friendly software application will be programmed to make the configuration of the architecture parameters and the simulation and verification easier.

2   Slave Digital Systems with Very Time Consuming Data Exchange. Design Alternatives

Traditionally, embedded systems like the one under study have been successfully developed with complex 16 or 32 bits microcontrollers. These process machines perform millions of instructions per second, and include some communication channels, memory interfaces, direct memory access controllers, ... On the other hand, they present many disadvantages that make impossible to fulfill our design goals:

- As they are general purpose integrated circuits, no feature can be adapted to the application. Sometimes software patches will substitute hardware requirements.
- Although they have different communication interfaces, no frame protocol codification or decodification is usually available, so these tasks become software work.
- No data source or sink interfaces are available, so software and general purpose input and output ports are used.
- The mentioned features force the high level control machine to perform data exchange control tasks, so low speed work is seriously limited. An external circuit dedicated to these tasks can be used to alleviate the data exchange control bottleneck.

An optimum solution requires an architecture focused on high speed data exchange performed in an asynchronous mode between source and sink elements [1]. The complete slave system has been achieved on one chip [2]. The design is modular and based in parameterizable cores to facilitate future reuse [3].

There is an intermediate solution between general purpose microcontroller based solution and one chip solution. High level and low speed tasks can be performed by a microcontroller and high speed data exchange left for an autonomous hardware system. This solution can be adequate when features not available on the FPGA but on the microcontroller are needed, such as A/D or D/A data converters, FLASH or EEPROM memory, ... This is a less integrated and slower solution and it is dependant on the chosen microcontroller.

3   System on a Reprogrammable Chip Design Methodology

With today's deep sub-micron technology, it is possible to deliver over two million usable system gates in a FPGA. The availability of FPGAs in the one million system gate range has started a shift of System on Chip (SoC) designs towards using reprogrammable FPGAs, thereby starting a new era of System on a Reprogrammable Chip (SoRC) [4].

Nowadays market expects better and cheaper designs. The only way electronics industry can achieve these needs in a reasonable amount of time is with design reuse. Reusable modules are essential to design complex circuits [5].

So the goal of this work is to achieve a modular, configurable and reusable architecture that performs very high speed data exchange without damaging the low speed tasks. Hardware and software co-design and co-verification is also one of the objectives [6].

Traditionally IP cores used non-standard interconnection schemes that made them difficult to integrate. This required the creation of custom glue logic to connect each of the cores together. By adopting a standard interconnection scheme, the cores can be integrated more quickly and easily by the end user. A standard data exchange protocol is needed in order to facilitate SoRC design and reuse. Excluding external system buses such as PCI, VME, USB and so forth, there are many SoC interconnection buses. Most of them are proprietary: Advanced Microcontroller Bus Architecture (AMBA, from ARM), CoreConnect (from IBM) [7], FISPbus (from Mentor Graphics and Inventra Business Unit), IP interface (from Motorola), and many more. We looked for an open option, that is to say, a bus which does not need any license agreement and with no need to pay any kind of royalty.

The solution is the Wishbone SoC interconnection architecture for portable IP Cores. Wishbone standard defines the data exchange among IP Core modules, and it does not regulate the application specific functions of the IP Core [8]. It offers a flexible integration solution, a variety of bus cycles and data path widths to solve various system problems, and allows cores to be designed by a variety of designers. It is based on a Master / Slave architecture for very flexible system designs. All Wishbone cycles use a handshaking protocol between Master and Slave interfaces.

4   SoRC Core-Based Architecture

The SoRC architecture shown in Fig. 2 complies with the specifications noted. The main objective of the architecture is to isolate high speed data exchange from any other control tasks in the system. That is why the design has been divided into three blocks, each one with its own specific bus:

- The "Data Exchange" block is responsible of all data transfers and uses the high speed Wishbone SoC interconnection Architecture for Portable IP Cores, which makes possible high speed data exchange. A shared bus interconnection with only one Master has been chosen, which controls all transfers in the bus.
- The "Control" block, usually a microcontroller, is the system high level manager, and performs all other tasks. It uses its specific bus to read and write on input and output blocks and any other devices, as well as to communicate with the Wishbone bus.
- The "Host Communication" block is the communication interface with the host part. It can not directly access to the high speed bus and it exchanges information with the frame receiver and transmitter, which is a module on the Wishbone bus.

On the other hand, the SoRC has these connections with the outside: the host communication channel, the memory bus, the data source and / or sink devices interface and the microcontroller side devices interface.

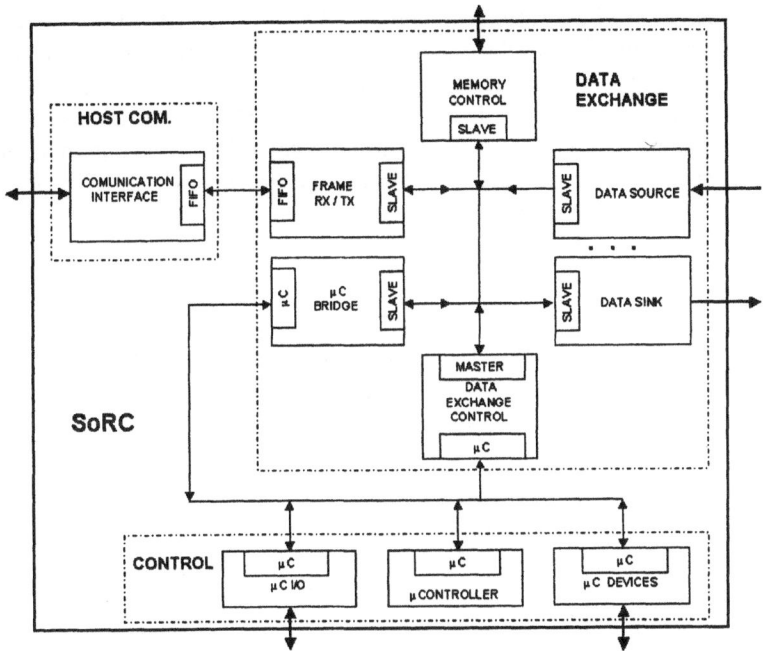

Fig. 2. SoRC architecture

4.1   Communication IP Cores

This core must interface the communication channel. After dealing with a number of different communication channels, the solution we have chosen is:

- If the communication interface needs a very complex controller (Bluetooth, USB) and there is an adequate solution available in ASSP or ASIC format, it is useful and practical to use it. In those cases only the channel interface is implemented into the SoRC.
- For non-complex ones (UARTs, parallel port buses such as IDE or EPP) both the controller and the interface are embedded into the SoC architecture leaving only the physical drivers and protection circuits outside the FPGA.

All the developed cores allow a full duplex communication. These cores have a common interface to the frame receiver/transmitter, which consists of two FIFO memories, one for reception and the other one for transmission. The communication interface presents two control signals to the frame controller to show the status of the FIFOs. The frame controller reads or writes the memories whenever it is needed.

4.2   Data Exchange IP Cores

DATA EXCHANGE CONTROL (DEC): this core allows data transfers between any two Wishbone compatible modules. It is the unique Wishbone master module, so it controls all operations on the bus. The system critical task is high speed data exchange, which must be performed in parallel between different origin and destination pairs. To complete any transfer, the DEC must read the data from the origin and then write it in the destination. Many transfer request can be activated concurrently, so the DEC must be capable of serving them. In order to guaranty that no request is blocked by another one, the DEC priories them following a round robin scheme.

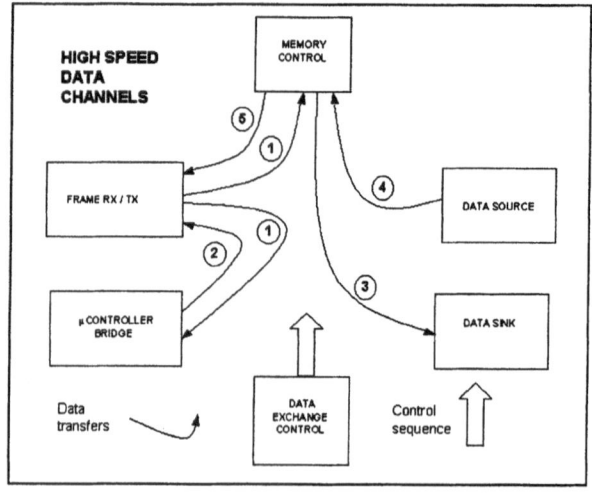

Fig. 3. High speed bus data exchange channels

The key to the control is to manage the right number of data channels, which must be exactly the number of concurrent data movements that can be accepted. Fig. 3 summarizes the solution adopted for hypothetical case with one data source and one data sink. The data channels are as follows:

- From the frame controller to the microcontroller bridge, in case of commands transmission, or to the memory, if data transmission to a sink.
- From the microcontroller bridge to the frame controller.
- From the memory to a data sink.
- From a data source to the memory.
- From the memory to the frame controller.

The number of channels is three plus one additional channel for each data source or sink. Each channel has three control registers: origin, destination and transfer length. Some of them are fixed and others must be configured by the microcontroller before starting data exchange. Additionally, there are two common registers which contain one bit associated to each channel: the control register to enable or disable transfers, and the start register, which must be asserted by the microcontroller, after correctly configuring the three registers of the channel, to start the data transference.

An interruption register is used to acknowledge the termination of the data exchange to the microcontroller. Once the requested data transfer is accomplished, the DEC asserts the interruption register bit associated with the channel. There is only one interruption line, so whenever an interruption occurs, the microcontroller must read the interruption register, process it and deactivate it.

Partial address decoding has been used, so each slave decodes only the range of addresses that it uses. This is accomplished using an address decoder element, which generates all chip select signals. The advantages introduced are: it facilitates high speed address decoding, uses less redundant address decoding logic, supports variable address sizing and supports variable interconnection scheme.

FRAME RX/TX: frame information contains a header, commands for the microcontroller, file configuration, data and a check sequence. The receiver part of this module decodes data frames and sends data to correct destination under the DEC control. The transmitter part is responsible of packaging the outgoing information. This core permits full duplex communication, so receiver and transmitter parts are completely independent.

The high level control block must know about command reception because it configures all transfers. This core generates two interruptions, one in case a command is received and another one whenever a communication error is detected. These are the two only interruptions not generated by the DEC.

When data intensive communication is performed, some kind of data correctness check must be performed. The frame controller is able to perform different kinds of checksum coding and decoding.

MICROCONTROLLER BRIDGE: the microcontroller can not access data on the Wishbone bus directly, so an intermediate bridge between the high speed bus and the microcontroller low speed bus is needed. It must adapt data and control interfaces. Usually the data bus on the high speed side is wider than on the microcontroller side, so one data transfer on the Wishbone side corresponds to more than one operations on the microcontroller side.

MEMORY INTERFACE: data exchange between data source and sink elements is supposed to be performed in asynchronous mode. This is possible using an intermediate data buffer. Large block of RAM memory is needed in data exchange oriented systems and stand alone memories provide good design solutions. The design must be capable of buffering several megabytes, so dynamic memory is needed, and in order to optimize memory access, it must also be synchronous. So synchronous and dynamic memory (SDRAM) controller has been developed [9].

High speed systems like the one presented here must follow synchronous methodology rules. The generation, synchronization and distribution of clock signals is essential. FPGAs designed for SoRC provide high speed, low skew clock distributions through dedicated global routing resources and Delay Locked Loop (DLL) circuits. A DLL works by inserting delay between the input clock and the feedback clock until the two rising edges align. It is not possible to use one DLL to provide both the FPGA and SDRAM clocks. Using two DLLs with the same clock input and separate feedback singnals achieves zero delay between input clock, the FPGA clock, and the SDRAM clock.

DATA SOURCE / SINK: data from a source is written to memory and then transmitted to a sink. The interface to external data source or sink is application dependant.

4.3   High Level Control Unit

This is the slave system central process unit. We have assumed that it is orientated to data exchange, which means that this task is very time consuming and it justifies the specific data exchange control unit. All other tasks can be controlled by a general purpose machine, usually a microcontroller, and it will be chosen in accordance to the application. This multifunction machine uses its own bus to access memory, input/outputs, user interface, other devices, and the high speed bus as well. This bus must be coincident with the one on the bridge core.

A command reception interruption from the frame controller tells the microcontroller about the request from the host. The microcontroller reads it from the bridge, through the DEC, and processes it. If it is a control command, it will send back the answer command. If it is a data command, it will configure the corresponding data channel on the DEC and after this it will send back the acknowledge or answer command to the host. The DEC core will generate the data transfer end interruption when this operation is finished.

Whenever it is detected that data coming from the host is corrupted, the frame controller activates the error interruption. The microcontroller will tell back the host about the failure.

5   Configuration and Verification User Interface

In order to make the reuse and verification of the proposed architecture easier, a user interface application has been developed [10].

The use of a hardware description language like VHDL has allowed doing a parameterizable design. The specifications can be kept open and design alternatives can be evaluated, due to the fact that the design parameters can be modified. Design modularity and cores parameterization greatly improve future reuse possibilities.

To do a generic design, the effort needed at the beginning of the project is bigger than to do a closed design, in which component functionality is fully fixed. But this technique, apart from the advantages mentioned above, could greatly alleviate the unexpected problems that arise in the final stages of the design process.

Some hard coded values in the design have been replaced with constants or generics. In this way, even if the parameter is not going to be changed in the future, code readability is increased. A global package containing the definition of all parameters has been used. The designer can configure the application specific architecture writing on the global package or using the software interface.

Once the desired architecture is configured, and after designing the application specific cores, the complete system functionality must be validated. All the cores, as well as the high level control machine code, must be co-simulated and co-verified. Modelsim from Mentor Graphics is the simulation tool used. It provides a Tool Command Language and Toolkit (Tcl/Tk) environment offering script programmability. A custom testbench has been created to facilitate the visualization of simulation results. A Tcl/Tk program creates a new display based on the simulator's output data, where a selection of the signals can be visualized with data extracted from the simulation results. Some buttons have been added so that new functionality is accessible. Dataflow can be graphically analyzed and design depuration is much easier. Host, communication channel, SDRAM and data source and sink functionality have been described using VHDL behavioural architectures. Data transfers on Wishbone bus are automatically analyzed by a supervisor core, which dramatically simplifies simulation.

6   Results and Conclusions

The following lines describe the application of the proposed architecture to the design of two digital systems.

The first one is a video system connected to a host via ethernet communication channel. The slave system controls two motors related to the camera movement and processes incoming control commands. The DEC core performs image data exchange between the analog to digital converter and the host. The SDRAM is used as a ping-pong memory: while a video frame is being captured, the previous one is being transmitted to the host. A general purpose evaluation board from Altera containing the 20K200EFC484 device has been used for prototyping. System features are summarized in Table 1.

The second one corresponds to an industrial plotter that provides high efficiency on continuously working environments. The microcontroller manages one stepping motor, three dc motors, many input/output signals and a simple user interface. The DEC module is dedicated to high volume data exchange from the host to the printer device using a parallel communication channel. Printing information is buffered in the SDRAM. This circuit is based on a Spartan II

Table 1. Implementation results

Features	Video Processor	Industrial Plotter
Equivalent gates	127.000	182.000
Internal RAM bits	16 Kbits	28 Kbits
User I/O pins	85	94
Max. DEC freq.	47 MHz	65 MHz
High level Proc.	Nios 32 bits	MicroBlaze 32 bits

family device from Xilinx which offers densities up to 200.000 equivalent gates. System features are summarized in Table 1.

The size, speed, and board requirements of today's state-of-the-art FPGAs make it nearly impossible to debug designs using traditional logic analysis methods. Flip-chip and ball grid array packaging do not have exposed leads that can be physically probed. Embedded logic analysis cores have been used for system debugging [11].

The results show that our SoRC architecture is suitable for generating hardware/software designs for slave digital systems with much data exchange. The multicore architecture with configuration parameters is oriented to reuse and a user friendly software application has been developed to help on application specific configuration and system hardware/software coverification. The use of a standard and open SoC interconnection architecture on the high speed bus improves the portability and reliability of the system and results in faster time to market.

References

1. Cesrio, W., Baghdadi, A.: Component-Based Design Approach for Multicore SoCs. Design Automation Conference. Proceedings of the 39th conference. New Orleans, Louisiana (2002)
2. Bergamaschi, R., Lee, W.: Designing Systems on Chip Using Cores. Design Automation Conference. Proceedings of the 37th conference (2000)
3. Gupta, R., Zorian, Y.: Introducing core-based system design. IEEE Design and Test of Computers (October-December 1997) 15-25
4. Xilinx Design Reuse Methodology for ASIC and FPGA Designers. http://www.xilinx.com/ipcenter/designreuse/docs/Xilinx-Design-Reuse-Methodology.pdf
5. Bursky, D.: Core-based design leads the way to flexible system solutions. Electronic Design (May 1997)
6. Balarin, F.: Hardware-Software Co-design of Embedded Systems: The POLIS approach. Kluwer Academic Press (1997)
7. The Connect TM Bus Architecture. http://www-3.ibm.com/chips/techlib/techlib.nsf/productfamilies/CoreConnect-Bus-Architecture
8. Wishbone System-on-Chip (SoC) Interconnection Architecture for Portable IP Cores (rev. B.2), Silicore Corporation, Corcoran (USA), October 2001
9. Gleerup, T.: Memory Architecture for Efficient Utilization of SDRAM: A Case Study of the Computation/Memory Access Trade Off. Int'l Workshop on Hardware-software Codesign (2000) 51-55
10. Quinnell, R.: Development tool suits core-based design. Electronic Design (August 1996)
11. Chakrabarty, K.: Optimal test access architectures for system-on-a-chip. Transactions on Design Automation of Electronic Systems (January 2001) 26-49

Time and Energy Efficient Matrix Factorization Using FPGAs

Seonil Choi and Viktor K. Prasanna

Electrical Engineering-Systems, University of Southern California, Los Angeles, USA,
{seonilch,prasanna}@usc.edu, http://ceng.usc.edu/~prasanna

Abstract. In this paper, new algorithms and architectures for matrix factorization are presented. Two fully-parallel and block-based designs for LU decomposition on configurable devices are proposed. A linear array architecture is employed to minimize the usage of long interconnects, leading to lower energy dissipation. The designs are made scalable by using a fixed I/O bandwidth independent of the problem size. High level models for energy profiling are built and the energy performance of many possible designs is predicted. Through the analysis of design tradeoffs, the block size that minimizes the total energy dissipation is identified. A set of candidate designs was implemented on the Xilinx Virtex-II to verify the estimates. Also, the performance of our designs is compared with that of state-of-the-art DSP based designs and with the performance of designs obtained using a state-of-the-art commercial compilation tool such as Celoxica DK1. Our designs on the FPGAs are significantly more time and energy efficient in both cases.

1 Introduction

FPGAs have become an attractive option for implementing digital signal processing applications because of their high processing power and customizability [7]. The inclusion of new features in the FPGA fabric, such as a large number of embedded multipliers, further enhance their suitability. Recent FPGAs such as Xilinx Virtex-II(pro) [15] and Altera Stratix [1] offer hundreds of multipliers and large memory on a single chip. FPGAs can now be considered for implementing massively parallel and computationally demanding applications [12]. Also, with the proliferation of portable and mobile devices [2], it has become increasingly important that systems are not only fast, but also energy efficient. Even though state-of-the-art configurable devices offer very few features for power control, we show how to effectively use them to improve energy performance.

In this paper, we consider one of the important signal processing kernels: matrix factorization. For example, matrix factorization is a fundamental kernel in adaptive beamforming [9]. Approaches to future wireless communications such as software defined radio (SDR), require the mapping of such signal processing kernels onto reconfigurable hardware like FPGAs [14]. Moreover, the implementations have to be time and energy efficient. First, we develop a linear array

P.Y.K. Cheung et al. (Eds.): FPL 2003, LNCS 2778, pp. 507–519, 2003.

architecture based design for matrix factorization. Then we investigate and apply algorithmic techniques that use a block based approach to obtain time and energy efficient designs in FPGAs. Performance estimation (based on the time and energy performance models) is used for rapid design space exploration. Since block matrix factorization can be realized using various block sizes, we identify an optimal block size that minimizes the total energy dissipation based on the estimation. Candidate designs are implemented. To the best of our knowledge, there are no FPGA based designs for LU decomposition. Hence for the sake of comparison, we implement the matrix factorization on FPGAs using the state-of-the-art compilation tool, Celoxica DK1, with Handel-C [4] and also implement it in software on TI DSP devices. The performance of our designs is compared with that of the DK1 based design and the TI DSP benchmarks.

The remainder of this paper is organized as follows. In Section 2, we present our algorithms and architectures for matrix factorization. In Section 3, time and energy performance is estimated for the proposed algorithms and architectures. Section 4 presents the implementation details and the performance of these synthesized designs. Also, a comparison with Handel-C based and TI DSP-based implementations is made. Finally, Section 5 summarizes our work and discusses possible areas for future work.

2   Time and Energy Efficient Designs for Matrix Factorization

Several methods are known for factoring a given matrix [5]. In this paper, we choose to implement LU decomposition on FPGAs. Essentially, LU decomposition factors a $b \times b$ matrix into a $b \times b$ lower triangular matrix L (the diagonal entries are all 1) and a $b \times b$ upper triangular matrix U.

We propose two designs using two theorems. In Theorem 1, a new algorithm and architecture for LU decomposition is developed for a linear array of processing elements (PEs). Each PE performs computations on the input or intermediate matrix and the results are fed to the neighboring PE. Data dependencies between input and intermediate matrices are solved by efficient and regular scheduling. Each PE uses only two input ports: one for feeding input or intermediate matrices and the other for outputting the decomposed matrix. With this fixed I/O bandwidth regardless of problem size, we achieve an optimal latency of $b^2 + b - 1$ with leading coefficient of 1. The best latency of previously proposed designs [3] is $2b(b+1)$. In Theorem 2, a new parallel design on FPGAs for block LU decomposition is proposed. The design partitions a large matrix into multiple smaller blocks. To perform a computation for the smaller blocks, the architecture/algorithm in Theorem 1 is re-used. By varying the block size, we achieve time and energy efficient designs.

2.1   LU Decomposition

Let A be a $b \times b$ matrix. $a_{x,y}$ denotes an element of matrix A, where x is the row index and y is the column index. Similarly, $l_{x,y}$ ($u_{x,y}$) denotes an element of matrix L (U). We assume that matrix A is a non-singular matrix and, further, we do not consider pivoting. The sequential algorithm in [8] consists of three main steps:

Step 1: The column vector $a_{x,1}$ where $2 \leq x \leq b$ is multiplied by the reciprocal of $a_{1,1}$. The resulting column vector is denoted $l_{x,1}$.

Step 2: $l_{x,1}$ is multiplied by the row vector $a_{1,y}(= u_{1,y})$ where $2 \leq y \leq b$. The product $l_{x,1} \times u_{1,y}$ is computed and subtracted from the submatrix $a_{x,y}$ where $2 \leq x, y \leq b$.

Step 3: Step 1 and 2 are recursively applied to the new submatrix formed in Step 2. An *iteration* denotes an execution of Step 1 and 2. During the k-th iteration, the column vector $l_{x,k}$ and the row vector $u_{k,y}$ where $k+1 \leq x, y \leq b$ are generated. The product $l_{x,k} \times u_{k,y}$ is subtracted from the submatrix $a_{x,y}$ where $k \leq x, y \leq b$ obtained during the $(k-1)$-th iteration.

The time complexity of the sequential algorithm is $\Theta(b^3)$. We propose an architecture and algorithm on a linear array shown in Figure 1 using b PEs. The number of PEs p is the same as the problem size b. Essentially, PE_j performs computations for the j-th column of matrices L and U. Each PE consists of an adder/subtracter, a multiplier, a division lookup table, and a storage LU (p entries per PE). The storage LU of PE_j is used to store the j-th column of matrices L and U. Each PE has two input ports (a_{in}, LU_{in}) and two output ports (a_{out}, LU_{out}). a_{in} and a_{out} are used to feed in and out $a_{x,y}$ or $l_{x,y}$. LU_{in} and LU_{out} are used to output resulting matrices L and U to PE_b in a pipelined manner. Figure 1 (c) shows our algorithm by describing the operations in each PE during each cycle.

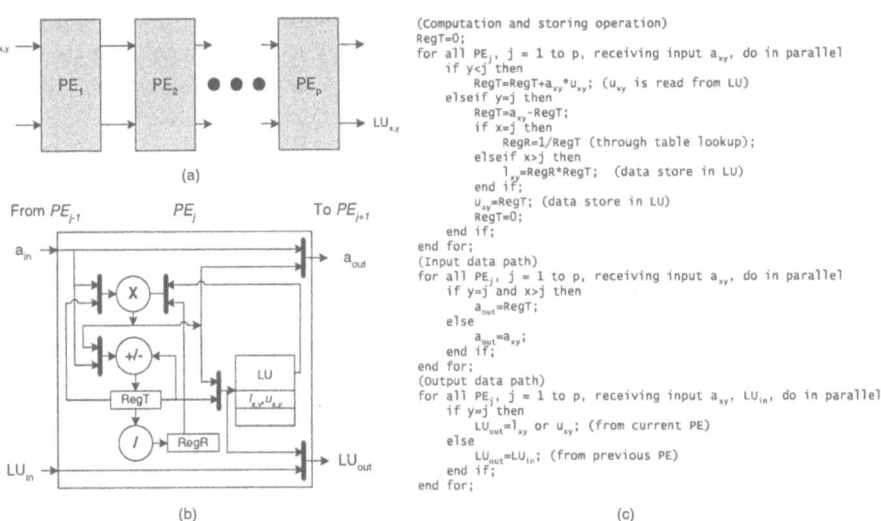

Fig. 1. (a) Overall architecture, (b) architecture of PE, and (c) algorithm for LU decomposition

Theorem 1. *LU decomposition without pivoting of a non-singular $b \times b$ matrix can be performed in $b^2 + b - 1$ cycles using the architecture and the algorithm in Figure 1 using b PEs.*

Proof. The elements $a_{x,y}$ of matrix A are fed in row major order ($a_{1,1}, a_{1,2}, a_{1,3}$, ..., $a_{1,b}, a_{2,1}, ..., a_{b,b}$) to $\mathsf{a_{in}}$ of PE_1. All data are fed from left to right. $a_{x,y}$ arrives at PE_j at cycle $b(x-1) + y + j - 1$ where $1 \leq j \leq b$. Seven operations are performed based on indices x, y and index j of PE_j. The indices x and y can be realized using counters in each PE. They also can be fed to PE_1 and propagated in a pipelined manner.

Op 1) Data propagation: $a_{x,y}$ is passed from PE_{j-1} to PE_{j+1} via PE_j except when $y = j$ and $x > j$. If $y = j$ and $x > j$, $l_{x,y}$ is generated at PE_j and is passed to PE_{j+1} via port $\mathsf{a_{out}}$. $l_{x,y}$ is also stored in the LU of PE_j.

Op 2) Multiplication/Accumulation: If $y < j$, a multiplication and an accumulation are performed in PE_j. $a_{x,y}$ ($= l_{x,y}$ generated at PE_{j-1}) is fed via port $\mathsf{a_{in}}$. During the k-th iteration, PE_j computes the product of the column vector $l_{x,k}$ and the j-th entry from $u_{k,y}$, where $l_{x,k}$ is a column vector generated in PE_k and $u_{k,y}$ is a row vector generated from PE_{k+1} to PE_b during the k-th iteration ($k + 1 \leq x, y \leq b$). $u_{k,y}$ are stored in the LUs of PE_{k+1} to PE_b. An accumulation, $a_{x,y}^{(k)} = l_{x,k} \times u_{k,y} + a_{x,y}^{(k-1)}$, is performed after the multiplication during the same clock cycle. $a_{x,y}^{(k)}$ denotes the intermediate element of submatrix generated during the k-th iteration. $a_{x,y}^{(k)}$ is used either for another accumulation or for normalization (Op 6) and is stored in RegT. RegT is a temporary storage to hold $a_{x,y}^{(k)}$ during the accumulation. Note that accumulation and subtraction share one adder/subtracter since they do not occur simultaneously.

Op 3) Subtraction: If $y = j$, a subtraction is performed after all accumulations are complete. This ensures that $a_{x,y}^{(k)}$ is subtracted from the submatrix $a_{x,y}$ where $k \leq x, y \leq b$ during k-th iteration. For example, $u_{3,3}$ is computed as $\{-(l_{3,1}u_{1,3} + l_{3,2}u_{2,3}) + a_{3,3}\}$. In Step 2 of the sequential algorithm, the subtraction is performed after multiplication and the result is stored for the next subtraction. These operations are done repeatedly. For example, $u_{3,3}$ is computed as $\{(a_{3,3} - l_{3,1}u_{1,3}) - l_{3,2}u_{2,3}\}$.

Op 4) Storing: If $y = j$, $l_{x,y}$ or $u_{x,y}$ is generated in PE_j. If $x \leq j$, $u_{x,y}$ is stored in LU. If $x > j$, $l_{x,y}$ is stored in LU after normalization (Op 6). This operation ensures that the j-th column of the decomposed matrices L and U is stored in PE_j.

Op 5) Reciprocal: Division is required since the normalization is performed by $u_{k,k}$ for the column vector $a_{x,k}^{(k-1)} = (a_{k+1,k}, ..., a_{b,k})$ during the k-th iteration ($1 \leq k \leq b$). $u_{k,k}$ is stored in RegT after subtraction (Op 3) and the reciprocal value of $u_{k,k}$ is stored in RegR. The reciprocal operation occurs if $x = y = j$.

Op 6) Normalization: After the subtraction (Op 3), the value is stored in RegT. If $y = j$ and $x > j$, the values in RegT and RegR are multiplied. This operation generates the column vector $l_{x,k}$ where $k + 1 \leq x \leq b$ in PE_k during the k-th iteration.

Op 7) Output: This operation sends out the results $l_{x,y}$ and $u_{x,y}$ in LU in a pipelined manner. If $y = j$, $l_{x,y}$ or $u_{x,y}$ is sent to port $\mathrm{LU_{out}}$. Otherwise, $l_{x,y}$ or $u_{x,y}$ from PE_{j-1} is passed to PE_{j+1} via port $\mathrm{LU_{out}}$.

To satisfy the data dependency of $l_{x,k}$ being generated during the k-th iteration and used during the $(k+1)$-th iteration and to obtain the minimum latency, two conditions have to be satisfied. Note that the column vector $l_{x,k}$ $(k+1 \leq x \leq b)$ is produced in PE_k during the k-th iteration. The first condition is that $l_{x,k}$ has to propagate from PE_k to PE_{k+1} after $l_{x,k-1}$ (generated during the $(k-1)$-th iteration) propagates to PE_{k+1} and before $l_{x,k+1}$ is generated in PE_{k+1} during the $(k+1)$-th iteration. Let $T_k(l_{x,j})$ be the sum of the time when $l_{x,j}$ is generated in PE_j and the propagation time when it reaches PE_k, which is $T_k(l_{x,j}) = b(x-1) + 2(j-1) + 1 + k - j$. Then, $T_{k+1}(l_{x,k-1}) = b(x-1) + 2k - 1$, $T_{k+1}(l_{x,k}) = b(x-1) + 2k$, and $T_{k+1}(l_{x,k+1}) = b(x-1) + 2k + 1$. Since $T_{k+1}(l_{x,k-1}) < T_{k+1}(l_{x,k}) < T_{k+1}(l_{x,k+1}) \rightarrow -1 < 0 < 1$, the condition is satisfied for all x where $k+1 \leq x \leq b$. To define the second condition, let $lu_{x,y}$ be $l_{x,y}$ if $x > j$, or $u_{x,y}$ if $x \leq j$ in PE_j. Note that $lu_{x,j}$ is computed in PE_j every b cycles. To output the resulting matrices L and U without delay, the second condition that $lu_{x,j}$ arrives at PE_k via port $\mathrm{LU_{in}}$ and $\mathrm{LU_{out}}$ before PE_k produces any $lu_{x,k}$ is required to satisfy. We assume $j < k$. Then, $T_k(lu_{x,j}) = b(x-1) + j + k - 1$ and $T_k(lu_{x,k}) = b(x-1) + 2k - 1$. Since $T_k(lu_{x,j}) < T_k(lu_{x,k}) \rightarrow j < k$, the second condition is satisfied for all x where $1 \leq x \leq b$. Total latency is calculated as the time taken for the last result $lu_{b,b}$ to be available as output: $T_b(lu_{b,b}) = b(b-1) + 2(b-1) + 1 = b^2 + b - 1$. $\square$

Since our design is a pipelined architecture, the first b cycles of the computations on the next matrix can be overlapped with the last b cycles of the computations on the current matrix. For a stream of matrices, one matrix can be decomposed every b^2 cycles. Thus the *effective latency* becomes b^2, which is the time taken to obtain the first output data to the last output data during the current computation.

2.2 Block LU Decomposition

For large matrices, block LU decomposition can be performed. The sequential algorithm is given in [5]. An $n \times n$ matrix A is partitioned into four matrices: A_{11}, A_{12}, A_{21}, and A_{22}. A_{11} is a $b \times b$ matrix, A_{12} is a $b \times (n-b)$ matrix, A_{21} is an $(n-b) \times b$ matrix, and A_{22} is an $(n-b) \times (n-b)$ matrix. The algorithm is to decompose A into two $n \times n$ matrices, L and U, such that $\begin{pmatrix} A_{11} & A_{12} \\ A_{21} & A_{22} \end{pmatrix} =$ $\begin{pmatrix} L'_{11} & 0 \\ L'_{21} & L'_{22} \end{pmatrix} \begin{pmatrix} U'_{11} & U'_{12} \\ 0 & U'_{22} \end{pmatrix}$. The steps of the algorithm are as follows:

Step 1: Perform a sequence of Gaussian eliminations on the $n \times b$ matrix formed by A_{11} and A_{21} in order to calculate the entries of L'_{11}, L'_{21}, and U'_{11}.

Step 2: Calculate U'_{12} as the product of $(L'_{11})^{-1}$ and A_{12}.

Step 3: Evaluate $A'_{22} \leftarrow A_{22} - L'_{21} U'_{12}$.

Step 4: Apply Step 1 to 3 recursively to matrix A'_{22}. During the k-th iteration, the resulting submatrices $L_{11}^{(k)}$, $U_{11}^{(k)}$ $L_{21}^{(k)}$, $U_{12}^{(k)}$, and $A_{22}^{(k)}$ are obtained. An *iteration* denotes an execution of Step 1 to 3.

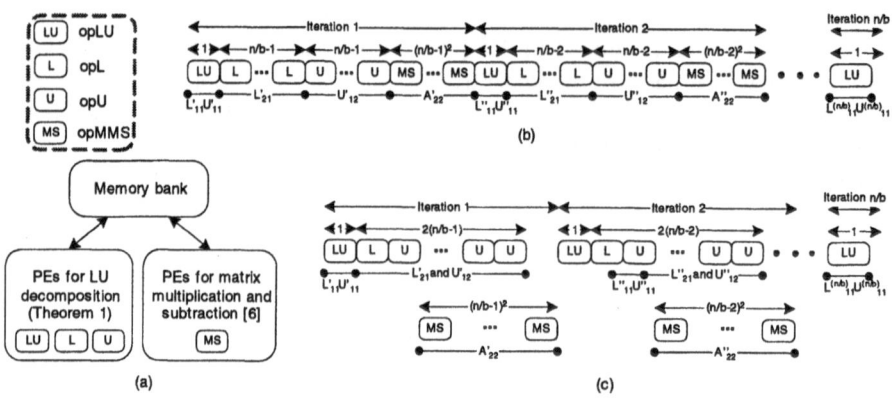

Fig. 2. (a) Overall architecture for block LU decomposition, (b) a schedule for Theorem 2 and (c) a schedule for Corollary 1

By utilizing the architecture and algorithm in Theorem 1 in combination with a matrix multiplication/subtraction architecture, we propose an architecture for block LU decomposition on FPGAs as shown in Figure 2. The block size b is later used as the parameter to realize time and energy efficient designs. There are two sets of PEs: one set performing a $b{\times}b$ LU decomposition and the other performing a $b \times b$ matrix multiplication/subtraction. Each set of PEs is linearly pipelined and both sets are connected to a memory bank. The input matrix is stored in the memory bank and fed to both sets of PEs. After computation, the results are stored back to the memory bank and used for next computation. Four different operations are identified: *opLU*, *opL*, *opU*, and *opMMS*. *opLU* is performed to obtain $L_{11}^{(k)}, U_{11}^{(k)}$. *opLU* from Step 1 is realized by using the algorithm and architecture proposed in Theorem 1. *opL* from Step 1 is performed to obtain $L_{21}^{(k)}$. The same architecture in Theorem 1 is used. However, the matrix $U_{11}^{(k)}$ and the reciprocal of its diagonal entries are required to perform *opL*. Since *opL* is performed after *opLU*, all PEs already hold the reciprocals in RegRs. $U_{11}^{(k)}$ is fed via port LU$_{in}$. We add one more data path that feeds the data from port LU$_{in}$ to storage LU. *opU* from Step 2 is performed to obtain $U_{12}^{(k)}$. *opU* also uses the same architecture. It requires $L_{11}^{(k)}$ from *opLU*. $L_{11}^{(k)}$ are fed via port LU$_{in}$ to storage LU. In Step 3, matrix multiplication/subtraction (*opMMS*) is performed. Once *opL* and *opU* are complete, $L_{21}^{(k)}$ and $U_{12}^{(k)}$ are available for *opMMS*. We have proposed an architecture for this operation [7]. Since there is matrix subtraction after matrix multiplication, additional subtraction logic is added. The matrix multiplication algorithm takes two $b \times b$ submatrices C from $L_{21}^{(k)}$ and D from

$U_{12}^{(k)}$ and computes the product $E \leftarrow C \times D$. Another $b \times b$ submatrix F is taken from $A_{22}^{(k)}$. Then the final values are obtained by $E \leftarrow F - E$. If b is large, the $b \times b$ matrix multiplication can be decomposed into $(\frac{b}{r})^3$ $r \times r$ matrix multiplications, where r is the sub-block size.

Theorem 2. *LU decomposition of $n \times n$ matrix can be performed in $n^2 + \frac{1}{6}\frac{nb^2}{r}\left(\frac{n}{b} - 1\right) \times \left(\frac{2n}{b} - 1\right) + b - 1$ cycles using the architecture in Figure 2 (a) using b PEs for $b \times b$ LU decomposition and r PEs for $r \times r$ matrix multiplication/subtraction, where b is block size and r is sub-block size.*

Proof. At a given time, only one operation is performed and the schedule is shown in Figure 2 (b). As all matrices are fed as streaming input, the computation on the current matrix and the next matrix can be overlapped. Therefore, each of $opLU$, opL, and opU has an effective latency of b^2. $opMMS$ has an effective latency of $\frac{b^3}{r}$ [7]. There are $\frac{n}{b}$ iterations to complete the block LU decomposition. During each iteration, only one $b \times b$ $opLU$ is performed. The effective latency of all $opLU$ is $(\frac{n}{b})b^2$. During the k-th iteration, $(\frac{n}{b} - k)$ opL and opU for $b \times b$ block size are performed. The effective latency of all opL and opU is $b^2 \sum_{k=1}^{n/b}(\frac{n}{b} - k)$. During the k-th iteration, $(\frac{n}{b} - k)^2$ $opMMS$ for $b \times b$ block size are performed. Since $b \times b$ matrix multiplication can be decomposed to $(\frac{b}{r})^3$ $r \times r$ matrix multiplications, the effective latency of all $opMMS$ is $\frac{b^3}{r}\sum_{k=1}^{n/b}(\frac{n}{b} - k)^2$. The total latency is $(\frac{n}{b})b^2 + 2b^2\sum_{k=1}^{n/b}(\frac{n}{b} - k) + \frac{b^3}{r}\sum_{k=1}^{n/b}(\frac{n}{b} - k)^2 + b - 1$, which includes the time to fill the pipeline stages. □

Theorem 2 uses a straightforward schedule since only one set of PEs performs computations at a given time. We can utilize the two sets of PEs in parallel to reduce the total latency.

Corollary 1. *LU decomposition of $n \times n$ matrix can be performed in $3bn - 2b^2 + \frac{1}{6}\frac{nb^2}{r}\left(\frac{n}{b} - 1\right)\left(\frac{2n}{b} - 1\right) + b - 1$ cycles using the schedule in Figure 2 (c).*

Proof. After opL and opU for the first blocks are performed, the input matrices for $opMMS$ are ready. Thus, opL and opU can be performed in parallel with $opMMS$. The effective latency to complete the k-th iteration is $3b^2 + (\frac{n}{b} - k)^2 \cdot \frac{b^3}{r}$. During the $\frac{n}{b}$-th iteration, only one $opLU$ is performed. Thus, the total latency is $\sum_{k=1}^{n/b-1}\{3b^2 + (\frac{n}{b} - k)^2 \cdot \frac{b^3}{r}\} + b^2 + b - 1 = 3bn - 2b^2 + \frac{1}{6}\frac{nb^2}{r}\left(\frac{n}{b} - 1\right)\left(\frac{2n}{b} - 1\right) + b - 1$, which includes the time to fill the pipeline stages. □

While Corollary 1 reduces the total latency compared with Theorem 2, it does not reduce the amount of computation. Total energy is the sum of the energy used for computation and quiescent energy (the energy for configuration memory, static energy, etc.) used by the device even when the logic is idle. The quiescent energy depends only on the total latency. Since Corollary 1 reduces the latency, the quiescent energy and hence the total energy are reduced. Thus we use the architecture and algorithm in Corollary 1 to obtain both time and energy efficient designs.

3   Performance Estimation and Design Trade-Offs

For a given problem size n, varying the parameters such as block size b and sub-block size r creates a large design space. Before implementing the designs and performing low level simulation, we estimate the performance of possible designs, prune the design space, and finally identify "good" candidate designs for time and energy efficiency. The candidate designs were implemented using VHDL (See Section 4).

3.1   High Level Performance Model

To estimate the performance of our designs, we have employed domain-specific modeling proposed in [6]. Domain-specific modeling is a hybrid (top-down plus bottom-up) approach to performance modeling that allows the designer to rapidly evaluate candidate algorithms and architectures in order to determine the design that best meets criteria such as energy, latency, and area. An architecture is divided into *RModules* and *Interconnects*. RModules are hardware elements that are assumed to dissipate the same amount of power no matter where they are instantiated on the device and Interconnects are the wires connecting the RModules. From the algorithm, we know when and for how long each RModule is active. With this knowledge, we can calculate the latency of the design. Additionally, with estimates for the power dissipated by each RModule and the Interconnect, we can estimate the energy dissipated by the design. In the top-down portion of the hybrid approach, the designer's knowledge of the architecture and the algorithm is incorporated, by deriving the performance models to estimate energy, area, and latency. The bottom-up portion is the power estimation of RModules and Interconnects from low level simulations. In our designs, the RModules are multipliers, adders, multiplexers, RAM, reciprocal lookup tables, and registers. The power values of each RModule are as follows. $P_{Mult}(= 11.25 \text{ mW})$ is the power dissipation for a 16×16 multiplier, $P_{Add}(= 1.34 \text{ mW})$ for a 16-bit adder/subtracter, $P_{Div}(= 7.31 \text{ mW})$ for a division lookup table (1024×16 bit), $P_{BSRAM}(= 7.31 \lceil x/1024 \rceil \text{ mW})$ for an on-chip memory where x is the number of entries, $P_R(= 1.17 \text{ mW})$ for a 16-bit register, and $P_{Store}(= 0.126 \lceil x/16 \rceil + 2.18 \text{ mW})$ for a storage LU where x is the number of entries. Table 1 lists the performance models of our designs. The latencies are converted to seconds by dividing them by the clock frequency.

3.2   Design Trade-Offs for Time and Energy Efficiency

To achieve time and energy efficient designs, we explore the various design parameters such as frequency, block size, precision, and number of PEs. All parameters contribute to energy dissipation, latency, and area of a design. For example, the latency and energy of Corollary 1 are a function of the block size b and the sub-block size r. By choosing $b = r = n$, the minimum latency of n^2 (546.1 μsec at 120 MHz for $n = 256$) can be achieved. However, this design does not necessarily have minimum energy dissipation. We explore the parameters,

Table 1. Time and energy performance models

Design	Metric	Performance model
Theorem 1	Latency	$L_{Thm1} = b^2$
	Power	$P_{Thm1} = 6bP_R + bP_{Add} + bP_{Mult} + bP_{Store} + bP_{Div} + 2P_{BSRAM}$
	Energy	$E_{Thm} = L_{Thm1} \cdot P_{Thm1}$
Theorem 2	Latency	$L_{opLU} = (\frac{n}{b})b^2, \ L_{opL} = L_{opU} = \frac{1}{2}(\frac{n}{b})(\frac{n}{b}-1)b^2$
		$L_{opMMS} = \frac{1}{6}(\frac{n}{b})(\frac{n}{b}-1)(2\frac{n}{b}-1) \cdot \frac{b^3}{r}$
		$L_{Thm2} = L_{opLU} + L_{opL} + L_{opU} + L_{opMMS}$
	Power	$P_{opLU} = P_{opL} = P_{Thm1}$
		$P_{opU} = 6bP_R + bP_{Add} + bP_{Mult} + bP_{Store} + 2P_{BSRAM}$
		$P_{opMMS} = 8rP_R + rP_{Add} + rP_{Mult} + 2rP_{Store} + 3P_{BSRAM}$
	Energy	$E_{Thm2} = L_{opLU} \cdot P_{opLU} + L_{opL} \cdot P_{opL} + L_{opU} \cdot P_{opU} + L_{opMMS} \cdot P_{opMMS}$
Corollary 1	Latency	$L_{Cor1} = L_{opLU} + 2(\frac{n}{b}-1)b^2 + L_{opMMS}$
	Power	the same as the ones in Theorem 2
	Energy	$E_{Cor1} = E_{Thm2}$

b and r, and determine their values that minimize the energy dissipation. The estimates are based on 120 MHz designs by considering the operating frequency that can be achieved after implementation (See Section 4). Figure 3 (a) shows the energy dissipation as a function of b and r for $n = 256$. The minimum energy is obtained around $r = 16$ and $b = 16$ while the latency is 2743.5 μsec. Note that the energy optimal design runs 5 times longer than the latency optimal design. Figure 3 (b) shows the energy distribution over four operations when $n = 256$ and b varies. When $b = 16$, we achieve the minimum energy design and *opMMS* is the dominant source of energy dissipation. Through design space exploration, we found that the energy efficient designs are obtained when $b = r = 16$ for $n \geq 32$. Another interesting results are the energy distribution on the core operation, MAC (multiply-and-accumulate) and the rest of operations. In Figure 3 (c), approximately 50% of the total energy is used by MAC, which is relatively high compared with the a general purpose processor or DSP processor.

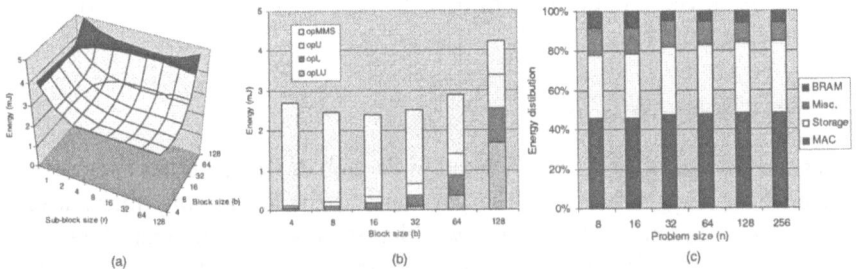

(a)   (b)   (c)

Fig. 3. (a) Energy dissipation as function of b and r for $n = 256$, (b) energy distribution as function of b for $n = 256$, and (c) energy distribution as function of n

4   Design Synthesis and Simulation Results

To obtain time and energy efficient designs, we briefly discuss the optimization techniques used in our designs. Then the synthesized designs for various problem sizes and the results from low level simulations are presented.

4.1   Optimizations for Time and Energy Efficiency

In this section, we summarize the energy efficient design techniques [7] employed in our designs. First, we have chosen a linear array of PEs. In FPGAs, long wires dissipate a significant amount of power [10]. For energy efficient designs, it is beneficial to minimize the number of long wires using a linear array since each PE communicates only with its nearest neighbors. Additionally, the linear array architecture facilitates the use of *parallel processing* and *pipelining*. Both techniques decrease the effective latency of a design and can lead to lower energy dissipation. Another technique is *block disabling*. We design the algorithm such that it utilizes the clock gating technique [15] to disable modules that are not in use during the computation. In our designs, since *opMMS* takes longer time that other operations, a set of PEs for *opLU*, *opL*, and *opU* becomes idle and is disabled to save energy. Another technique is choosing the appropriate *bindings*. In the Xilinx Virtex-II, the storage LU can be implemented as registers, distributed RAM, or embedded Block RAM. When the number of entries > 64, Block RAM is used since it is energy efficient for large memory; otherwise, distributed RAM is used. Similar decisions can be made such as choosing between (embedded) Block multiplier or configured multiplier. We choose Block multiplier since it is energy efficient when both inputs are not constant. To implement the division unit, a lookup table approach is used. This technique is faster and uses less energy compared with other division algorithms [11]. To calculate a/b, we first obtain $1/b$ via a lookup table and perform the multiplication $a \times (1/b)$. The approach is effective if the multiplication is fast. Using Block multipliers, fast multiplication (within one cycle) can be performed. The lookup table for reciprocal is generated as $Inv(b) = Round(2^m/b)$ where b is the value to be inverted, m is the number of bits used to represent the output, and *Round* is the rounding function.

4.2   Simulation Results

Using the performance models defined in Section 3, we identified the energy and time efficient designs based on the parameters. By considering different criteria such as area, latency, and energy, we identified several designs. The minimal energy designs are chosen as candidate designs and are implemented in VHDL. The precision of all designs was 16 bits. These designs were synthesized using XST in Xilinx ISE 4.2i and the frequency achieved was 120 MHz. The place-and-route file as an .ncd file was obtained for the Virtex-II XC2V1500 bg575-6 device. The input test vectors for the simulation were randomly generated such that their average switching activity was 50%. Mentor Graphics ModelSim 5.6b was used to simulate the designs and generate the simulation results as a .vcd file. These .vcd and .ncd files were then used by the Xilinx XPower tool to evaluate the average power dissipation. Energy dissipation was obtained by multiplying the average power by latency. We also compared estimates from Section 3 against actual values based on implemented designs to test the accuracy of the performance estimation. We observed that the energy estimates (See Table 2) were within 10% of the simulation results. The the average power dissipation of the designs on Virtex-II included the quiescent power of 150 mW (from XPower).

Table 2. Performance of the designs based on Corollary 1 (from low-level simulation)

	TI DSP (600MHz)		DK1 based design (50MHz)				Our design (120MHz)					
n	b	L (usec)	E (uJ)	b	L (usec)	Slice/Mult	E_m (uJ)	b	L (usec)	Slice/Mult	E_{est} (uJ)	E_m (uJ)	Err in E_{est}
8	8	2.9	4.2	4	9.6	810/6	2.52	8	0.5	835/8	0.27	0.29	6.8%
16	16	7.2	10.6	4	35.0	810/6	10.27	8	2.1	1955/16	1.3	1.4	7.1%
32	16	31.5	46.6	4	218.2	810/6	42.86	16	10.7	3550/32	8.6	8.2	6.8%
64	16	152.7	229.7	8	1004.8	1318/10	342.7	16	51.2	3550/32	52.3	50.0	5.3%
128	16	836.4	1282.1	8	7087.9	1318/10	1485.7	16	345.6	3550/32	368.8	355.5	4.3%
256	16	5171.3	8068.0	8	56157	1318/10	6863.5	16	2743.5	3550/32	2801.6	2717.2	3.7%
512	16	35335	55857	8	453583	1318/10	34825	16	22421	3550/32	21927	21330	3.3%
1024	16	258619	412251	8	3658982	1318/10	198275	16	182473	3550/32	173710	169236	3.2%

* n is problem size. b is block size. L is latency. Slice is area. Mult is the number of embedded multipliers.
E_{est} is the estimated energy using the performance model. E_m is the measured energy from low level simulation.

We were not aware of any prior FPGA based designs for LU decomposition. Hence, for the sake of comparison, we implemented two baseline designs: one on FPGAs using a state-of-the-art commercial compilation tool and the other, a software implementation on state-of-the-art TI DSPs. The FPGA design was implemented using Handel-C and synthesized using Celoxica DK1.1 [4]. The compilation tool can automatically exploit the parallelism of the algorithm. The synthesized design, with a frequency of 50 MHz, was then implemented with the Xilinx ISE 4.2i. Our designs dissipated 8.8x to 1.2x less energy than the Handel-C based designs.

We also compared the performance of LU decomposition on FPGAs and DSPs. FPGAs are known to be better than DSPs in terms of time and energy performance. Since many target applications for DSP devices and FPGAs are similar, comparing their time and energy performance is beneficial to designers. We chose the TI TMS320C6415 running at 600 MHz as a representative DSP. TMS320C6415 is a high performance DSP and has eight 16-bit MAC units. The LU decomposition was implemented in C and its precision was 16 bits. The matrix multiplication was performed using the function call *DSP_mat_mul* from the TI DSP library. The latency was obtained by using the TI Code Composer 2.1. To compute the energy dissipation, we assumed the 75% high / 25% low activity category of power dissipation for the function call *DSP_mat_mul* since it is a hand-optimized code [13]. The power dissipation for the rest of C code is based on the 50% high / 50% low activity category since the code is optimized by the TI compiler. For the DSP, we chose the block size b, $0 < b < \min(n, 16)$ so as to minimize the energy dissipation. As seen from the results in Table 2, our FPGA implementations perform LU decomposition faster using less energy. While we used the high performance DSP processor, TI also provides low power devices, namely the TMS320VC55xx series. Based on the datasheets, the 55xx series dissipate 150 mW at 300 MHz while the 64xx series dissipate 1500 mW at 600 MHz. The 55xx series have two MACs (600 MIPS) while the 64xx series have eight MACs (4800 MIPS). Thus the scaling factor from 64xx series to 55xx series for energy dissipation can be defined as: $s_e = \frac{P_{55xx}}{P_{64xx}} \times \frac{MIPS_{64xx}}{MIPS_{55xx}} = 0.78$. By applying this scaling factor, the energy dissipation of our designs was determined to be 12.1x to 1.8x less than the TI 55xx series.

5   Conclusion

We developed time and energy efficient designs for LU decomposition on FPGAs. Before implementing the designs, we analyzed the architecture and algorithm to understand the design trade-offs. After pruning the design space, selected designs were implemented using VHDL in the Xilinx ISE design environment. Currently, state-of-the-art FPGAs (e.g., Virtex-II/pro) do not provide low power features such as control for multiple power states or lower static power. The proposed architectures and algorithms are parameterized based on several design parameters. Hence, when more features such as dynamic voltage scaling with dynamic frequency scaling are available, the operations opL and opU can be executed slower than the operation $opMMS$. This might provide the opportunity to use a lower frequency and lower voltage.

Acknowledgements

This work is supported by the DARPA Power Aware Computing and Communication Program under contract F33615-C-00-1633 monitored by Wright Patterson Air Force Base and in part by the National Science Foundation under award No. 99000613. The authors wish to thank Gokul Govindu, Ronald Scrofano, and Zack Baker for helpful discussions and contributions to the low level simulation results.

References

1. Altera Corporation. http://www.altera.com. 2002.
2. J. Becker, T. Pionteck, M. Glesner. DReAM: A Dynamically Reconfigurable Architecture for Future Mobile Communication Applications. *FPL*, 2002.
3. E. Casseau, D. Degrugillier. A Linear Systolic Array for LU Decomposition. *VLSI Design*, 1994.
4. Celoxica Corporation. DK1.1 Design Suite. http://www.celoxica.com. 2003.
5. J. Choi, J. J. Dongarra, L. S. Ostrouchov, A. P. Petitet, D. W. Walker, R. C. Whaley. The Design and Implementation of the Scalapack LU, QR, and Cholesky Factorization Routines. *Scientific Programming*, 5:173–184, 1996.
6. S. Choi, J. Jang, S. Mohanty, V. K. Prasanna. Domain-Specific Modeling for Rapid System-Wide Energy Estimation of Reconfigurable Architectures. *ERSA*, 2002.

7. S. Choi, R. Scrofano, V. K. Prasanna, J.-W. Jang. Energy Efficient Signal Processing using FPGAs. *Field Programmable Gate Array*, 2003.
8. T. H. Cormen, C. E. Leiserson, R. L. Rivest. *Introduction to Algorithms*, McGraw-Hill, 2nd edition, 2001.
9. S. Haykin. *Adaptive Filter Theory*, Prentice Hall, 4th edition, 2002.
10. L. Shang, A. Kaviani, K. Bathala. Dynamic Power Consumption in Virtex-II FPGA Family. *Field Programmable Gate Arrays*, 2001.
11. N. Shirazi, A. Walters, P. Athanas. Quantitative Analysis of Floating Point Arithmetic on FPGA Based Custom Computing Machines. *FCCM*, 1995.
12. H. Styles, W. Luk. Customising Graphics Application: Techniques and Programming Interface. *Field Programmable Custom Computing Machines*, 2000.
13. Texas Instruments. TMS320C64xx Power Consumption Summary, http://www.ti.com.
14. W. Tuttlebee. *Software Defined Radio: Enabling Technologies*, J. Wiley, 2002.
15. Xilinx Incorporated. http://www.xilinx.com.

Improving DSP Performance
with a Small Amount of Field Programmable Logic

John Oliver and Venkatesh Akella

Department of Electrical & Computer Engineering
University of California, Davis
{jyoliver, akella} @ece.ucdavis.edu

Abstract. We show a systematic methodology to create DSP + field-programmable logic hybrid architectures by viewing it as a hardware/software codesign problem. This enables an embedded processor architect to evaluate the trade-offs in the increase in die area due to the field programmable logic and the resultant improvement in performance or code size. We demonstrate our methodology with the implementation of a Viterbi decoder. A key result of the paper is that the addition of a field-programmable data alignment unit (FPDAU) between the register-file and the computational blocks provides 15%-22% improvement in the performance of a Viterbi decoder on the state-of-the-art TigerSHARC DSP. The area overhead of the FPDAU is small relative to the DSP die size and does not require any changes to the programming model or the instruction set architecture.

1 Introduction

Can we improve the performance and power or memory requirements of a state-of-the-art DSP with programmable logic? Many researchers have addressed this question in the past and many solutions have been proposed including customized instructions, loops [1], reconfigurable functional units, [2] and co-processor [3,4,5,6,7,8]. However most of the existing approaches do not factor the cost of the programmable logic in their evaluation - they tacitly assume that the die size penalty of adding programmable logic is not important. However, in embedded applications where DSPs are used, cost is a very critical factor. So, we would like to find a sweet spot for the programmable logic where a small addition to the die size in the form of programmable logic realizes maximum return in terms of improvements to performance (throughput), power, or memory requirements. For this we believe that the integration of programmable logic with a DSP should be viewed as a hardware/software co-design problem.

We illustrate our proposal using a Viterbi decoder as an example. First we analyze the optimized assembly code for Viterbi decoding on state-of-the-art DSPs and show that it is not the functional units that are the problem but the restrictions on the connection between the register file and the computational units that are the bottleneck which can be elegantly overcome by using a flexible interconnect network that can be realized using field-programmable logic. We call this new hardware block - FPDAU (Field Programmable Data Alignment Unit). This is situated between the register file of a processor and the computational units. This block dynamically re-configures the dataflow between the register file and the functional unit and hence

P.Y.K. Cheung et al. (Eds.): FPL 2003, LNCS 2778, pp. 520–532, 2003.

eliminates a significant fraction of the instructions in the kernels of many important signal-processing algorithms. In order to determine the configuration of the FPDAU the implementation has to be approached as a hardware/software co-design problem. We will show the details of our implementation again using the Viterbi decoder on a TigerSHARC DSP as an example.

The techniques presented are general enough to be used with any other DSP that supports SIMD style processing such as the AltiVec and TI's TMS320c62xx. Also, we show that this approach is not just meant for Viterbi decoding, it can be used with other algorithms as well. In fact, a variety of DSP oriented computations like vector and matrix operations like transposing a matrix, finding the determinant of a matrix can benefit with the proposed architecture.

1.1 Organization of This Paper

First we will introduce Viterbi decoding and how it is efficiently implemented in assembly language on the TigerSharc that already has support for ACS computation. Then we will show what the bottleneck of the implementation is and its impact on the execution time and memory for K=5, 7, and 9 Viterbi decoding. We then propose a simple scheduler and programmable interconnect to rectify the problem. The design of the scheduler is described and its cost in terms of equivalent look-up-tables and die size is estimated. We show how the field programmable interconnect is used by the DSP programmer. The improvements in performance are then presented. We then describe other algorithms that can benefit by the proposed solution to demonstrate that this is not just for Viterbi Decoding. Finally, we compare our approach to related solutions in the area of DSP+PL hybrids and show why our approach is more promising.

2 Overview of Viterbi Decoding and Its DSP Implementation

Viterbi decoding is a critical application in embedded communication systems like 802.11-based wireless LAN; CDMA based cellular technologies and host of other applications that require data communication over noisy channels. It is part of the EEMBC benchmark suite. In spite of special support to execute Viterbi algorithm efficiently modern DSP are unable to meet the high data rate Viterbi decoding requirements imposed by standards such as the 3G and 802.11(a). So, we use Viterbi algorithm as an example in this paper to illustrate our technique.

First, to understand the computational requirements of a Viterbi decoder, it is useful to start with a convolutional encoder. Fig. 1 shows a ½ rate convolutional encoder for constraint length K=3. In this encoder, for every input bit, two output bits are transmitted. Each input is convolved through XOR operations with the previous two bits. The circuit in Fig. 1 can also be represented as a state-machine shown in Figure 2.

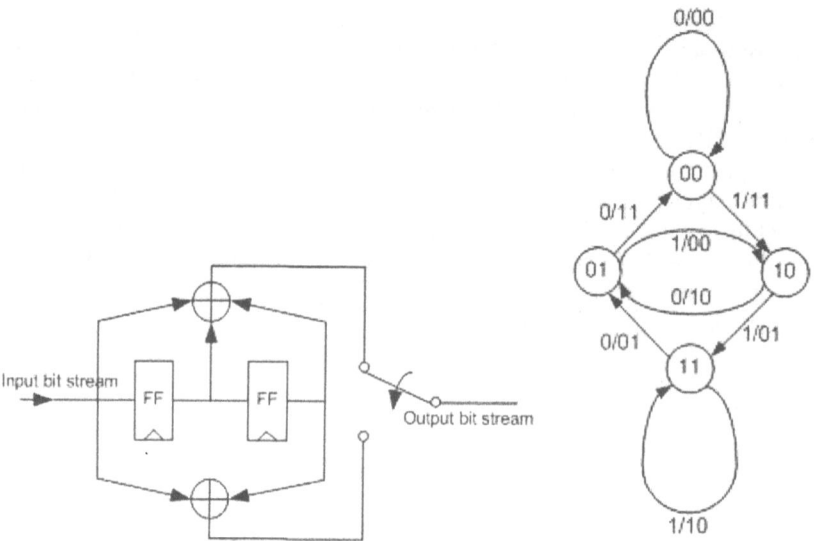

Fig. 1. K=3 Convolutional Encoder

Fig. 2.   State Diagram of K=3 Convolutional Encoder

The goal of a Viterbi decoder is to determine what the *most likely* inputs were, given an output data stream corrupted by a noisy transmission channel. A trellis is a map of all of the states from the encoder, drawn out to show each step in time. For the K=3 encoder shown in Figure 2, there would be four states in each time instance of the trellis.   Fig. 3 shows a trellis used for Viterbi decoding for the K=3 convolutional encoder. Viterbi decoding consists of two tasks - the population of the trellis and the trace back through the trellis to find the path that yields the most likely sequence of states. Population of the trellis works as the follows.   For each pair of input bits, the distance between the input bits and the expected output for each transition between states is calculated for each of the possible state transitions.   In the

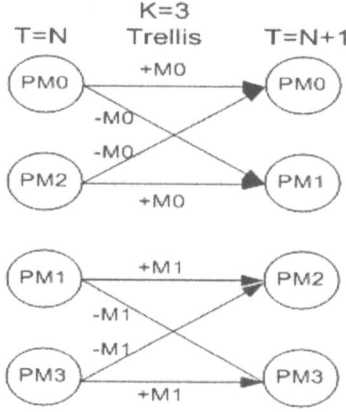

Fig. 3. K=3 Trellis Diagram

K=3 state machine shown in Figure 2, there are a total of 8 possible transitions, two to each state. This distance is represented by the +/- M0 and M1 in the trellis diagram in Fig. 3. The smallest distance to each state is chosen and saved for each state. For the next time instance, the same procedure is used except the chosen smallest distance to each state is added to the previous metric saved for that state. These accumulated distances are referred to as *path metrics*. This process of adding the input bits against the local path value, comparing the two local path values to find the smallest distance, and selection of the smallest path distance is often referred to as an Add-Compare-Select, or ACS. Many DSPs have custom ACS instructions to accelerate this process.

The traceback of a Viterbi decoder is simply the selection of the smallest accumulated state metric for each of states of the trellis. This computation is mostly serial, and relatively inexpensive in terms of instructions for Viterbi decoders of constraint lengths of 7 or more as a fraction of the total time.

2.1 AltiVec Implementation

The Altivec DSP co-processor[9] is a vector processor that operates on 128-bit vectors in 8, 16, or 32 bit SIMD mode. Assuming that path metrics are 32-bit values, we could store the four path-metrics PM0 to PM3 for the the ACS kernel for K=3 trellis (shown in figure 3) in one 128bit vector. Figure 4 shows the pseudo-assembly code for the implementation of the ACS kernel for K=3 where V0, V1, V2, V3, V4 and V5 are 128 bit vector registers. PM(x) denotes a 32-bit value that holds the accumulated path metric of state x. M0 and M1 represent the magnitude of the two different possible distances that may be generated for any input pair of bits. Figure 5 illustrates the flow of data between the registers and the result of the computation. For example, it shows that the least significant 32 bits of register V0 are obtained by adding PM(3) and M1 and so on. Now, in order to compute the new value PM(0) we need to find the minimum of PM(0)+M0 and PM(2)-M0 (please refer to Figure 3), but the vector-min instruction expects the two operands to be in adjacent locations in the vector register. This is an alignment restriction in SIMD processing and is results in simplification of the hardware.

So, the data needs to re-ordered so that the pairs of candidate path metrics are in adjacent sub-word locations in a vector register. This necessitates the need for the two *vec_merge* instructions shown in the pseudo code in Fig. 4.

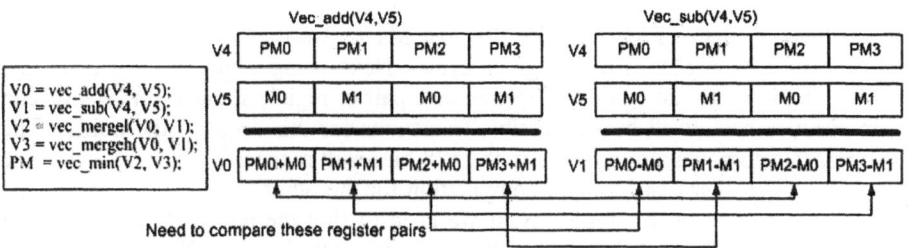

Fig. 4. Altivec Register Mapping of ACS and Pseudo Code

2.2 TigerSHARC Implementation

Is this restriction just a limitation of the AltiVec processor or is it more general? To investigate this we looked at other DSP architectures (with a completely different architecture style), namely, the TigerSHARC [10] from Analog Devices, which is a statically scheduled superscalar with various SIMD modes of computation and two independent functional units, that operate on 64-bit data.   A block diagram of the TigerSHARC computational block is shown in Fig. 7. Each computational block is fed by a 32 entry, 32-bit register file with 4 read ports and 4 write ports. Within each computational block, there are 3 different SIMD modes, allowing for sub-word computations on 32-bit, 16-bit or 8-bit boundaries similar to the Altivec. There are restrictions on which registers may be used in a SIMD instruction. 32-bit SIMD calculations may be completed only on adjacent 32-bit registers. Similar restrictions are placed on 16-bit and 8-bit SIMD computations.

The K=3 trellis (presented in Fig. 3) can be mapped to one of the TigerSHARC computational blocks. Again the pseudo assembly code is shown in Figure 7 and the register dataflow is shown in Figure 8.   PM(x) denotes a 32-bit value that holds the accumulated path metric of state x. M0 and M1 represent the magnitude of the two different possible distances that may be generated for any input pair of bits. PM3and PM2 are stored in register pair R5:4, PM1 and PM0 are stored in register pair R3:2 and M0 and M1 are assumed stored in register pair R1:0. This grouping of registers allows the TigherSHARC to use its 32-bit SIMD mode and represents an efficient implementation of Viterbi on TigerSHARC.

```
R15:12 = Add/Subtract(R5:4, R1:0);

R11:8  = Add/Subtract(R3:2, R1:0);

R15:12 = Merge(R15:14, R11:10);

R11:8  = Merge(R13:12, R9:8);

R15:12 = VectorMin(R15:14, R11:10);

R11:8  = VectorMin(R13:12, R9:8);
```

Fig. 5. TigerSHARC Viterbi ACS Pseudo-Code

On the right half of Fig. 6, both M0 and M1 as well as PM0 and PM1 can be fetched from the register file in a given cycle. Next, using a special instruction that allows addition and subtraction to operate on a pair of registers, the TigerSHARC can then produce half of all of the possible transitions for this stage of the trellis. The result of this computation is shown in the 128-bit result register R11:8. Likewise, the left half of Fig. 6 shows a similar computation for the other four possible transitions for the same stage of the trellis. Next we need to find the path with the minimum metric for each of the pairs of transitions to each state in the trellis, which can be done by the special *vector_min* instruction which also supports SIMD mode However, to utilize the SIMD mode, the vector minimum instruction expects the data to be compared in the same bit locations in both operands. The overlapping arrows in the

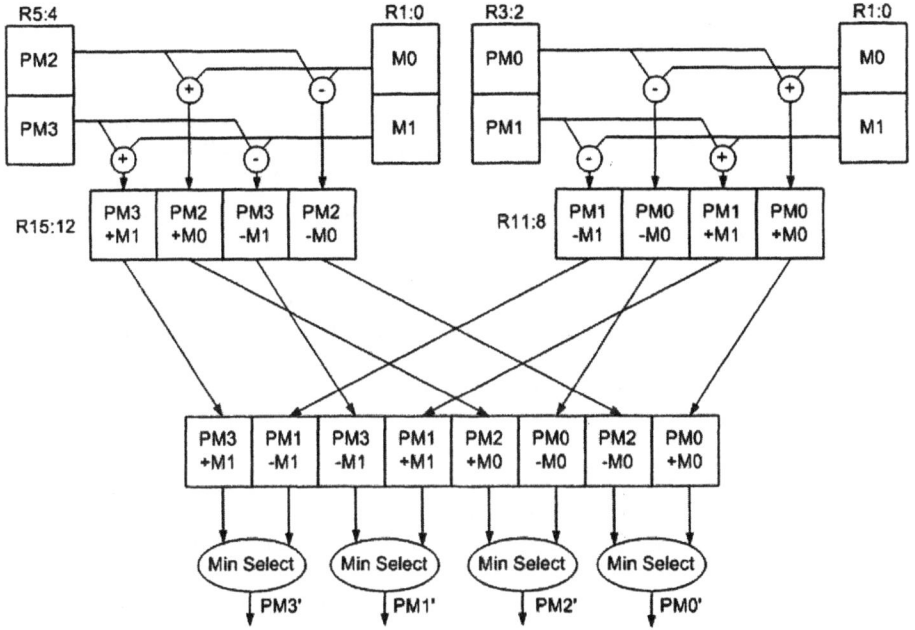

Fig. 6. Mapping of Viterbi ACS Dataflow to the TigerSHARC DSP

middle of Fig. 6 indicate the required data movement in order to utilize the SIMD vector minimum instruction. The overlapping arrows are realized by the *permutation* instructions that are similar in spirit to the *vec*_merge instructions in Altivec, i.e., they rearrange data in the register file. Finally, we analyzed the Texas Instrument'sC62xx DSP and found that a similar permute instructions are needed to overcome the SIMD restriction [11]. Table 1 shows the performance of the TigerSHARC on various different Viterbi decoders. The %ACS row indicates the fraction of the total cycles the TigerSHARC DSP spends on the trellis population and the %Permutes row indicates the fraction of the total cycles spent on permutations. These cycles are for the entire implementation of the GSM decoder. The fraction of the total cycles spent on ACS and permutations is similar for TI C6x DSP [12]and the AltiVec vector processor. From here on out, we will focus on the TigerSHARC architecture as we had access to the simulation tools for this platform.

Is there a more efficient way to address this problem? To investigate this we decided to profile the TigerSHARC implementation of a Viterbi decoder developed for the GSM wireless handset standard, which requires K=5, 16-bit data and 189 bit data frame.

Table 1. TigerSHARC Viterbi ACS Performance

For ½ Rate Viterbi Decoder, L=190 Bits	K=5	K=7	K=9
ACS Cycles	1960	4191	8459
Traceback Cycles	960	1245	1625
% Execution Cycles in ACS	67.1%	77.1%	83.9%
% of ACS Instructions which are Permutes	23.3%	25.0%	26.9%

3 HW/SW Co-design and the FPDAU

The data in Table 1 shows that a significant fraction of the computation cycles in the Viterbi decoder are spent in permute instruction, which are actually not doing anything useful in terms of the Viterbi algorithm. They are merely there to overcome the data flow restrictions to the function units in a typical DSP. So, the problem is not that the DSP do not have the appropriate instructions or the memory bandwidth (as shown in the previous section, most DSP do have special instructions to support Viterbi), but it is the data alignment restriction.

We propose a Field Programmable Data Alignment Unit (FPDAU) to circumvent the need for these permutation instructions. So, the data rearrangement will be done in hardware instead of software as it is being done now. This gives us two key benefits. It eliminates the instructions from the critical kernel of the computation and thereby provides improvements in performance and memory requirements and possibly reduces power and instruction cache pollution. It also gives us additional flexibility, because with a field-programmable hardware unit we can customize the dataflow to the specific algorithm being implemented.

Next we describe the details of the FPDAU and its integration with the DSP architecture and its programming model. We will illustrate this with the TigerSHARC DSP because we have access to their simulation tools. As noted before, a similar structure would work with other DSP as well; the programming model and the interface will differ.

The FP-DAU consists of two parts - a flexible interconnect that connects the register file to the ALU, Shifter and MAC units and a dynamically programmable state-machine to control the configuration of the flexible interconnect. The controller has configuration register that is mapped into the TigerSHARC's memory space. The placement of the FP-DAU is shown in Fig. 8.

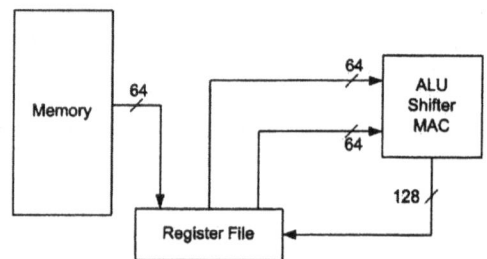

Fig. 7. Block Diagram of a TigerSHARC Computational Block

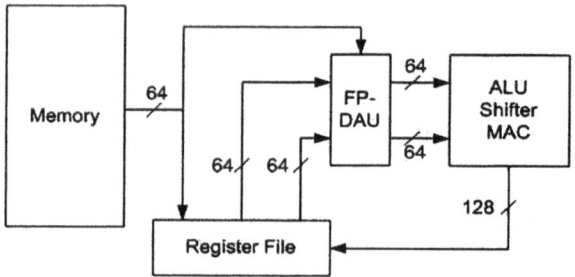

Fig. 8. Block Diagram of a TigerSHARC Computational Block with FPDAU

The detailed block diagram of the FPDAU is shown in Fig. 9. To support the data alignment required for Viterbi (the overlapping arrows in the middle of Fig. 6), we need an interconnect that is flexible only on word i.e. 32-bit boundaries. However, since the TigerSHARC does supports operations on bytes, we will design the FPDAU to support byte-wide granularity. The TigerSHARC register file has two 64-bit read ports, as shown in Fig. 8. The FPDAU needs to select one of 16 bytes from the register file and connect each of those bytes to a byte input of the computational unit. This interconnect can be built with 128 16-to-1 multiplexers. The dynamically programmable state machine inside the FPDAU controls the configurations of the multiplexers. As far as the impact of the FPDAU on the DSP critical path goes, there is a delay of an additional 16:1 multiplexer which does endanger the 300 MHz operating frequency of the TigerSHARC DSP. In the future as we move to finer geometries, we expect this to be less of an issue. We propose an identical FPDAU in both of the independent computational blocks of the TigerSHARC DSP.

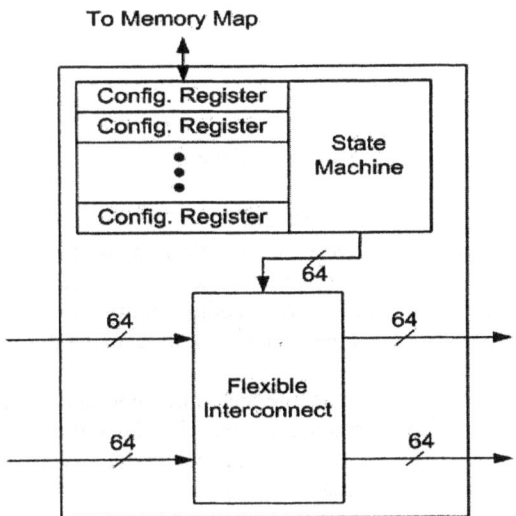

Fig. 9. FP-DAU Block Diagram

Next, the design of the dynamically programmable controller or the state machine shown at the top of Fig. 9 is described. The purpose of the controller is to define the configuration of the flexible interconnect of the FPDAU. This controller will be realized on traditional LUT-based fabric to give it maximum flexibility. This state machine will be clocked by the *read enable* signal of the TigerSHARC register file. In order to minimize the impact of the FPDAU on the instruction set architecture we require that the state machine does not have any additional inputs. Therefore, every time that the read enable is clocked and the FPDAU is active, the state machine will proceed to the next state. This has two consequences. First, we need as many states as there are register reads in the inner most loop of the algorithms that utilize the FPDAU. Secondly, it precludes us from using the FP-DAU in inner loops that have non-linear flow, such as branches or jumps.

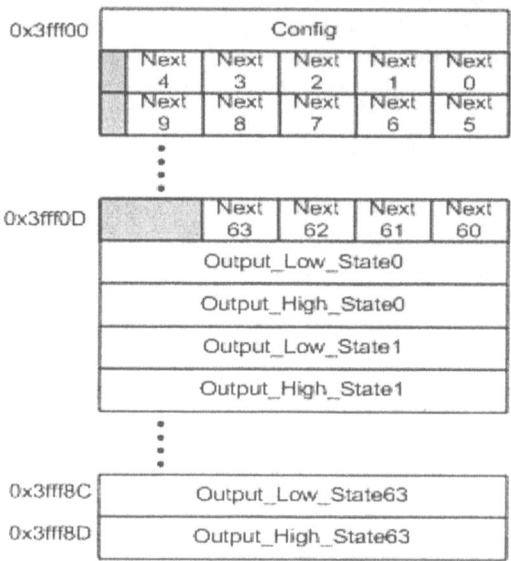

Fig. 10. FP-DAU memory map

However, with predicated execution and the tight loops in DSP kernels this is not much of a restriction. Note that this is a design decision to minimize the impact of the FPDAU on the instruction set architecture. If one has the ability to slightly modify the instruction set architecture (define new opcodes) more efficient and more general-purpose programmable state machines can be realized inside the FPDAU, without the restrictions listed above.

The configuration of the state machine has to be generated during the compilation of the application to the DSP processor. This will allow the data flow between the register file and the ALU to change (in customized way) every clock. The configuration space is memory mapped into the TigerSHARC's internal memory address space, as shown in Fig. 10. This allows the state of the programmable logic to be saved to memory, and also allows new states to be saved and restored by the TigerSHARC. In addition, the FPDAU needs a control register (one bit) that defines whether the FPDAU is active or not.

As noted in the beginning of the paper, the main objective of our work is to minimize the amount of programmable logic to achieve a certain level of performance improvement. That is why we did not advocate a new functional unit or a coprocessor to execution. So, how much area is required for the FPDAU? This requires the estimation of the area for the programmable state machine that is implemented in LUTs. For flexible interconnect structure (that gives us byte-wide data realignment), we need 128 16:1 multiplexers that results in 64 bits for each state of the state machine. Let us assume we have 64 states for the state machine, which should be sufficient to cover a wide range of applications (the inner loops for most applications have fewer than 64 instructions). Each state must have 6 bits to indicate which is the next state. Fig. 10 shows how the state machine of the FP-DAU is memory mapped into the TigerSHARC architecture. Table 2 summarizes the overhead of the FPDAU. The maximum initialization overhead should only be incurred if all 64 states of the

FPDAU's controller are used. The start overhead is the overhead of writing to the configuration register of the FPDAU's controller to start or stop the operation of the FPDAU. The hardware overhead cost is relatively minor when compared to a typical DSP die area of about 1 sq. cm^2. In the hardware overhead, we assumed that the FPDAU's controller is resident inside 4-LUTs. However, the controller could also use configuration SRAM, which would decrease the number of 4-LUTs needed dramatically. Finally, other functions could be included in the FPDAU, like zero/one insertion, bit reversal or any other simple operation. These added functions could marginally increase the needed hardware for the FPDAU, but would allow us to leverage the strengths of programmable logic on a DSP platform.

Table 2. FPDAU Overhead

Initialization Overhead	72 cycles, Maximum
Start Overhead	2 cycles per ACS Trellis Frame
Hardware Overhead	128 16:1 Muxes and 200 4-LUTs

3.1 Programming the FPDAU

Next we will describe how the programmer or the compiler uses the FPDAU, using the ACS computation of Viterbi as an example. The configuration for the FPDAU is generated from the register-transfer level assembly code. From the pseudo code shown in Fig. 5 (Viterbi decoder with K=3) we can see that if we omit the permute instructions; we only need four cycles to complete a single stage of the ACS trellis update. Since each of these four instructions accesses the register file, we will need a state-machine with four states to control the flexible interconnect. Note that typically a branch would be executed at the bottom of the loop, but it is omitted from the pseudo code in Fig. 5 for simplicity. Since the branch does not access the register file, it will not clock the state machine so we can ignore it from the perspective of configuring the FPDAU. Figure 11 shows the resultant state machine for Viterbi ACS derived from Figure 7. The register reads for the two add/subtract instructions are done in normal i.e. without any permutation. They are indicated by state A and state B in Figure 11. The two *vector_min* instructions are executed in states C and D of the state machine which requires the FPDAU to program the flexible interconnect to permute the data corresponding to the pattern shown at the bottom of Fig. 6. The new pseudo code required to complete a single stage of the ACS trellis update is also shown in Figure 11, as expected it eliminates the two permute instructions.

Finally, it is important to note that the FP-DAU should be disabled and the configuration of the FP-DAU is saved and restored upon entering interrupt routines. If the FP-DAU is to be utilized inside an interrupt service routine, the states of the FP-DAU must be saved and restored to the TigerSHARC's on-chip memory upon entering and exiting the interrupt, respectively. In most cases the entire configuration memory is not utilized, so the overhead of saving the FPDAU is typically a few cycles, especially given that the TigerSHARC has be ability to read/write 128-bits to memory in a given cycle. However, if the entire configuration space of the FP-DAU does indeed have to be saved, the maximum penalty is around 72 cycles to save the entire FP-DAU state.

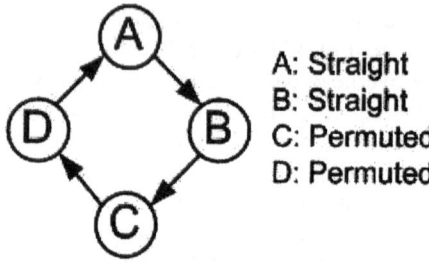

A: Straight
B: Straight
C: Permuted
D: Permuted

R15:12 = Add/Subtract(R5:4, R1:0);   // A
R11:8 = Add/Subtract(R3:2, R1:0);   // B
R15:12 = VectorMin(R15:14, R11:10); // C
R11:8 = VectorMin(R13:12, R9:8);   // D

Fig. 11. Example FP-DAU Configuration for Viterbi ACS

4 Results from Viterbi Implementation

In this section we will summarize the results of the implementation of the Viterbi decoder on the TigerSHARC enhanced with the FPDAU. As noted before, the programmable logic is configured at compile time i.e. statically by analyzing the kernel of the computation, which in the case of this decoder has 20 instructions. Using the simple compilation scheme described above would translate into 20 states for the FPDAU programmable state machine. The one time overhead of writing to the FP-DAU configuration register and programming the states in the FP-DAU is 16 cycles. Two additional cycles are needed to turn on/off the FP-DAU when entering/exiting the ACS inner loop.   Table 3 shows the performance improvements of the TigerSHARC DSP with the FP-DAU on Viterbi decoders of different lengths. Note that the improvement in terms of cycles saved is quite impressive (15 % to 23%) given that the TigerSHARC is already optimized to implement Viterbi efficiently. Also, note improvement also results in improvements to code density, which is quite useful in embedded applications. It may also result in power savings but the FPDAU itself will consume some power but we do not have access to the gate-level netlists of the TigerSHARC to evaluate exactly what the savings would be.

The additional area required for 64 16-to-1 Muxes is Y, incurring a total delay of Z in W process technology.   The state machine in the FP-DAU requires the equivalent of X number of CLBs, at an area estimate of A um2 in W process technology.

Table 3. TigerSHARC Viterbi Performance with FP-DAU

L=190 Bits	K=5	K=7	K=9
ACS	1506	3146	6183
Traceback	960	1245	1625
Total Cycles	2466	4391	7808
% Speed Up	15.5%	19.2%	22.6%

5   Other Applications of FPDAU

Even though the focus of this paper was the implementation of a Viterbi decoder, it should be pointed out that the FPDAU concept is quite general and it has many applications. Basically, the FPDAU restores some flexibility of a Vector, VLIW, or SIMD mode processor by allowing the functional units to operate on any data in the register file. Without the FPDAU one has to waste valuable CPU cycles and power in rearranging the data so that a given instruction can execute properly. We have found applications for FPDAU in a variety of DSP applications especially those that involve matrix operations like Reed-Solomon decoding, finding the minimum or maximum in a vector, data interleaving and de-interleaving and matrix transpose. In each of these applications the FPDAU can be used, but exactly how it is used is determined by the hardware/software co-design of the application, as illustrated in this example. The interface and the programming model of the FPDAU will be the same but the configuration of the state machine will be different in each case and depending on the application the amount of improvement will also vary. For example, in an experiment with matrix transpose on the AltiVec we found that only half the merge instructions in the inner loop can be eliminated with the FPDAU. So, it is important to note that *not all permute (or data rearrangement) operations* can be eliminated with the FPDAU; this is the trade-off between the amount of configurable logic inside the FPDAU and its interface and the amount of flexibility. We deliberate restrict the inputs to the FPDAU to two and byte-level reconfigurability to minimize the area overhead of the FPDAU and its impact on the critical path of the processor.

6   Related Work and Conclusions

The idea of utilizing field programmable logic to accelerate computations is not new. Starting with the PRISC project in Harvard [13] and the work in BYU[3] on integration of DSP and reconfigurable logic and more recently the reconfigurable functional unit idea in the Chimera project in Northwestern University[2], there have been numerous efforts at integrating programmable logic with a processor. The key difference between those efforts and the proposed solution is in two areas (a) we treat DSP + programmable logic integration as a hardware/software co-design problem, hence what we propose is a methodology rather than a specific solution. So, it can be applied to any processor and any application (b) unlike the previous efforts we focus on the cost issue, which precludes us from using a co-processor or a new functional unit because that would add to the cost and change the instruction set architecture of the underlying processor – which poses problems in terms of adoption in embedded processors especially in the commercial arena. We believe that the solution proposed here finds a sweet spot in terms of return on investment in terms of the amount of programmable logic and the improvement in performance achieved. Also, it has minimal impact on the instruction set architecture and the programming model of a DSP, so it can be ignored without significant penalty if the application domain does not required it.

Also, if one has more chip area to spend on the programmable logic the FPDAU can be expanded to include other operations in the LUT area that is currently being used to only implement the programmable state machine. For example, one could have a bit-level operations support that could help in encryption algorithms like DES and AES. So, again the proposal here is a co-design methodology for the DSP + programmable logic platform, where the architect can choose how much chip area to spend on programmable logic and what operations to implement there with the FPDAU providing a general framework for programming and interface. If it is expanded further it will resemble the RFU idea in Chimera or the co-processor concept in the BYU project or Riverside project[1].

References

1   Stitt, G. and Vahid, F.: Energy Advantages of Microprocessor Platforms with On-chip Configurable Logic. IEEE Design and Test, 2002
2   Ye, Z., Moshovos, A., Hauck, S., Banerjee, P.: CHIMAERA: A High-Performance Architecture with a Tightly-Coupled Reconfigurable Functional Unit, Computer Architecture News, (2000).
3   Graham, P., Nelson, B.: Reconfigurable Processors for High-Performance, Embedded Digital Signal Processing, Field Programmable Logic and Applications. (1999)
4   Fisher, J., Faraboschi, P., Desoli, G.: Custom-Fit Processors: Letting Applications Define Architectures. Hewlett-Packard Laboratories Cambridge, Cambridge, MA, (1996)
5   Compton, K., Hauck, S.: Reconfigurable Computing: A Survey of Systems and Software, http://www.ee.washington.edu/faculty/hauck/publications/ConfigCompute.pdf
6   Dehon, A.: The Density Advantage of Configurable Computing. IEEE Computer Magazine, (2000)
7   Tessier, R., Burleson, R.: Reconfigurable Computing for Digital Signal Processing: A Survey. Journal of VLSI Signal Processing Systems for Signal, Image, and Video Technology, (2001)
8   Hartenstein, R.: Reconfigurable Computing: A New Business Model – and it's Impact on SoC Design. Proceedings Euromicro Symposium on Digital Systems Design. IEEE Comput. Soc. (2001)
9   Ollmann, I.: Altivec. http://www.simdtech.org/apps/group_public/documents.php
10 Analog Devices: TigerSHARC DSP Hardware Specification. http://www.analog.com/Analog_Root/static/library/dspManuals/Tigersharc_hardware.html
11 Fridman, J.: Data Alignment for Sub-Word Parallelism in DSP. IEEE Signal Processing Magazine, IEEE, (2000). p.27-35.
12 Texas Instruments: TMS320C6000 CPU and Instruction Set Reference Guide. (2000), http://www-s.ti.com/sc/psheets/spru189f/spru189f.pdf
13 Razdan, R., Smith, M.: High-Performance Microarchitectures with Hardware-Programmable Functional Units, Proc. 27th Annual IEEE/ACM Intl. Symp. on Microarchitecture, pp. 172-180, November (1994)

Fully Parameterized Discrete Wavelet Packet Transform Architecture Oriented to FPGA

Guillermo Payá, Marcos M. Peiró, Francisco Ballester, and Francisco Mora

Universidad Politécnica de Valencia. Department of Electronic Engineering.
Camino de Vera s/n, 46022, SPAIN.
guipava@doctor.upv.es, {mpeiro, fballest, fmora}@eln.upv.es

Abstract. The present paper describes a fully parameterized Discrete Wavelet Packet Transform (DWPT) architecture based on a folded Distributed Arithmetic implementation, which makes possible to design any kind of wavelet bases. The proposed parameterized architecture allows different CDF wavelet coefficient with variable bit precision (data input and output size, and coefficient length). Moreover, by combining different blocks in cascade, we can expand as many complete stages (wavelet packet levels) as we require. Our architecture need only two FIR filters to calculate various wavelet stages simultaneously, and specific VIRTEX family resources (SRL16E) have been instantiated to reduce area and increase frequency operation. Finally, a DWPT implementation for CDF(9,7) wavelet coefficients is synthesized on VIRTEX-II 3000-6 FPGA for different precisions.

1   Introduction

For years, the Discrete Wavelet Transform (DWT) has been used in a wide range of applications, including signal analysis and coding, data compression, image and video compression, numerical analysis, statistics, physics... Recently, new studies propose opening the high-pass branches in the DWT (see Fig. 1). This fact trades new data compression capability when the information is distributed in low and higher frequency ranges. New designs with this structure have been successfully applied in 1D data processing, such as in medical audio processing [1], in ECG compression [2], and in digital modulations systems like CDMA or OFDM [3]. Two-dimensional applications of these structures are employed when images have strong high-pass components, e.g., the fingerprint images and high-contrast medical images. This wavelet structure is called Wavelet Packet Transform (WPT) due to the frequency 'packets' obtained in the output of its binary tree. In [4], Coifman introduced the adaptive tree structure concept using packets as an evolution of wavelet for signal and image compression. He proposed to expand the wavelet tree selecting the best wavelet bases.

This work describes a new metodology to implement differents DWPT structures, depending on bit precision (data input, coefficients and data output) and wavelet bases selected. Distributed Arithmetic (DA) technique has been applied to implement

P.Y.K. Cheung et al. (Eds.): FPL 2003, LNCS 2778, pp. 533–542, 2003.
© Springer-Verlag Berlin Heidelberg 2003

easily-parameterized structures and gives optimum results on FPGA devices. This work is organized as follows: next section introduces the DA by using a finite-impulse response (FIR) filter implementation. In section three, we explain the DWPT architecture based on DA and Polyphase Decomposition. Fourth and fifth sections present the modular DWPT architecture and results. Finally, the conclusions are exposed.

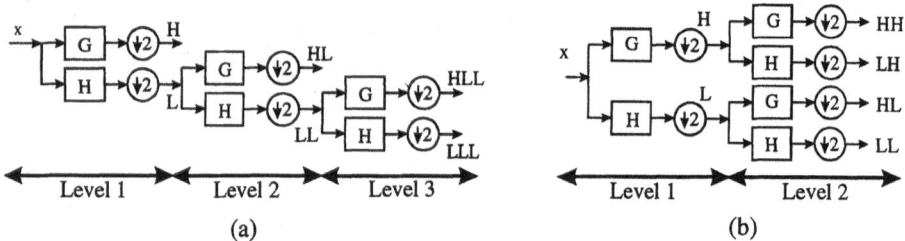

Fig. 1. (a) Three stages Discrete Wavelet analysis filter bank and (b) Two stages Packet Wavelet analysis filter bank

2   Distributed Arithmetic Technique on FPGA Device

The Distributed Arithmetic technique is an efficient procedure for computing sum-of-products (inner products) between a fixed and a variable data vector. The basic principle is owed to Croisier et al. [5], but Peled and Liu [6] have independently presented a similar method. This arithmetic trades memory for combinatory elements, resulting ideal to implement custom digital signal processors in look-up table (LUT-based) FPGA [7]. In addition to a DA implementation, the designer also can select from a bit-serial to a full-parallel implementation [8].

DWT and its packet version are based on a cascade of FIR filters. The operation of these filters involves inner products of the equation 1 type. The inner product can be rewritten as equation 2 with two's complement representation for coefficient (α_i) and data input (x_i). The data input are scaled so that $|x_i| \leq 1$.

$$y = \sum_{i=1}^{N} \alpha_i \cdot x_i \qquad (1)$$

$$y = \sum_{i=1}^{N} \alpha_i \cdot \left[-x_{i0} + \sum_{k=1}^{Wd-1} x_{ik} \cdot 2^k \right] \qquad (2)$$

The symbols x_{ik} represent the kth bit in x_i data input, and Wd is the number of bits of the data. By modifying the order of the summations we get

$$y = -\left[\sum_{i=1}^{N} \alpha_i \cdot x_{i0} \right] + \sum_{k=1}^{Wd-1} \left[\sum_{i=1}^{N} \alpha_i \cdot x_{ik} \right] \cdot 2^{-k} \qquad (3)$$

The elements in brackets take only a finite number of values, 2^N, so we compute and store these values in a look-up table (LUT).

Fig. 2 shows a block diagram for computing an inner product according to previous equation. Since the output is divided by 2, by the inherent shift, the circuit is called *shift-accumulator*. Bits x_{ik} are the address of the LUT which store the binary coefficient additions. Data inputs x_i are shifted one-bit at a time (1BAAT) generating a bit-serial DA structure. Of course, we can implement a parallel form of DA by allocating a LUT to each term in brackets in equation 3. Our work computes DA in bit-serial fashion using the minimum resources into the FPGA. On the other hand, we save half of the area resources in DA implementations, when a FIR filter has symmetric coefficients.

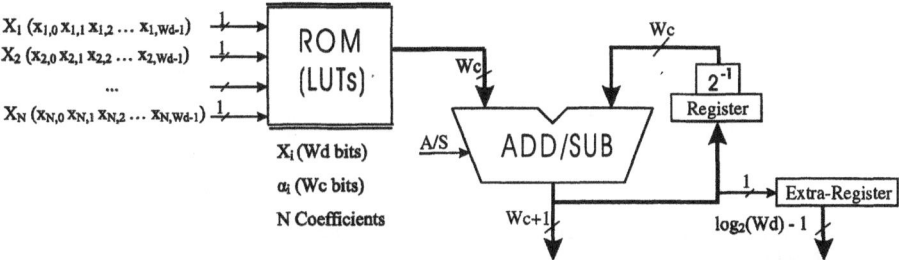

Fig. 2. Bit serial Distributed Arithmetic basic cell

Moreover, using shift registers LUT mode (SRL16E and SRLC16E primitives, see Fig. 3.) we can reduce a 50% on area when original design has a higher number of flip-flops. Virtex family can configure any LUT as a 16-bit shift register without using the flip-flops available in each slice. Shift-in operations are synchronous with the clock, and output length is dynamically selectable. A dedicated output allows the cascading of any number of 16-bit shift registers to create whatever size shift register is needed. Nevertheless, the configurable 16-bit shift register cannot be set or reset.

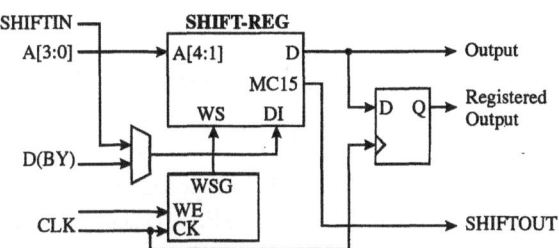

Fig. 3. Shift Register mode LUT Configuration

3   Wavelet Packet and Polyphase Decomposition

The theory of wavelet signal decomposition was firstly introduced by S.G. Mallat [9]. In his work, he computes the wavelet representation with a Pyramidal Algorithm (PA) based on convolutions with Quadrature Mirror Filters (QMF) filters (the basic wavelet cell showed in Fig. 4a) and decimators. In this figure, G represents the high-pass filter that obtains the signal details whereas H obtains the coarser resolution (low-pass component) of the input signal. In addition, the decimator can be translated to the input of the filters obtaining the commutator model [10] for the QMF cell (Fig. 4b).

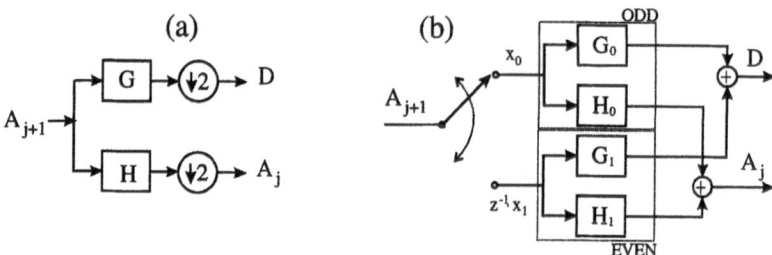

Fig. 4. (a) The QMF of the wavelet decomposition and (b) the commutator model of the Polyphase QMF bank

By repeating in cascade this algorithm the wavelet representation of a signal A on J resolution levels (or stages) is computed. Since the filter outputs are decimated in the basic DWT elementary cell, we apply the polyphase decomposition of the filter banks. For example, the algorithm expresses the symmetric 9-tap filter $H(z)$,

$$H(z) = \alpha_0 \cdot z^{-4} + \alpha_1 \cdot \left(z^{-3} + z^{-5}\right) + \alpha_2 \cdot \left(z^{-2} + z^{-6}\right) + \alpha_3 \cdot \left(z^{-1} + z^{-7}\right) + \alpha_4 \cdot \left(1 + z^{-8}\right) \qquad (4)$$

using the biphase decomposition in the form

$$H(z) = H_0(z^2) + z^{-1} \cdot H_1(z^2) \qquad (5)$$

where

$$H_0(z) = \alpha_0 \cdot z^{-2} + \alpha_2 \cdot \left(z^{-1} + z^{-3}\right) + \alpha_4 \cdot \left(1 + z^{-4}\right) \qquad (6)$$
$$H_1(z) = \alpha_1 \cdot \left(z^{-1} + z^{-2}\right) + \alpha_3 \cdot \left(1 + z^{-3}\right)$$

To compute the DWT, M.Vishwanath [11] proposed an alternative to the PA algorithm called Recursive Pyramidal Algorithm (RPA). Basically, RPA consists of rearranging the order of the outputs such that an output is scheduled at the earliest instance that it can be scheduled. To compute the 2-level DWPT, we propose an RPA-modified algorithm. It consists of rearranging the low-pass and high-pass outputs in order to open 2-levels DWPT (see Fig. 5).

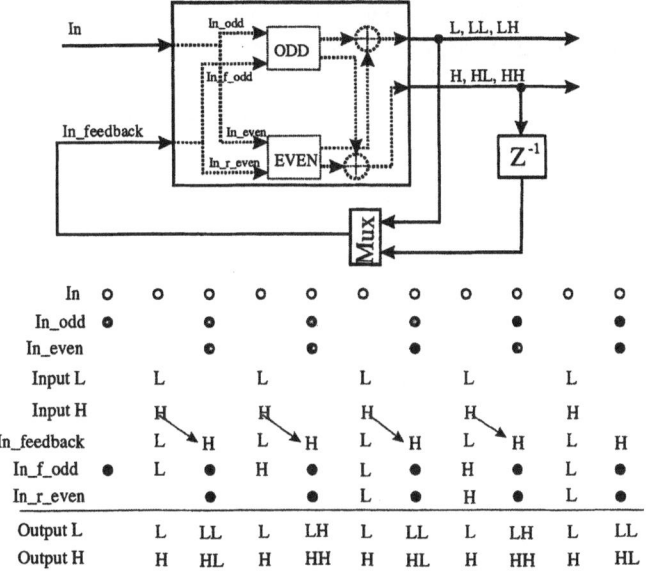

In	o	o	o	o	o	o	o	o	o	o o
In_odd	●		◉		◉		◉		●	●
In_even		◉		◉		●		◉		●
Input L		L		L		L		L		L
Input H	H		H		H		H		H	
In_feedback	L	H	L	H	L	H	L	H	L	H
In_f_odd	● L	●	H ●	L	● H	●	L ●	H	● L	●
In_r_even		●		●	L ●		H ●		L ●	●
Output L	L	LL	L	LH	L	LL	L	LH	L	LL
Output H	H	HL	H	HH	H	HL	H	HH	H	HL

Fig. 5. 2-levels DWPT structure based on the Recursive Pyramid Algorithm

The sampling grid for the DWPT obtains the output schedule by push down all the horizontal lines of samples until they form a single line. The order of the outputs obtained gives us the output schedule.

Taking profit of bit rate, we can expand as many complete levels as we need, by replying blocks in cascade (see Fig. 6). Using the free time slots, we can expand two additional stages by means of 4-Stages block (4-S). This 4-Stages block has similar structure than 2-Stages block (2-S) described before. The differences are located on the number and length of the registers, as we will see in next section.

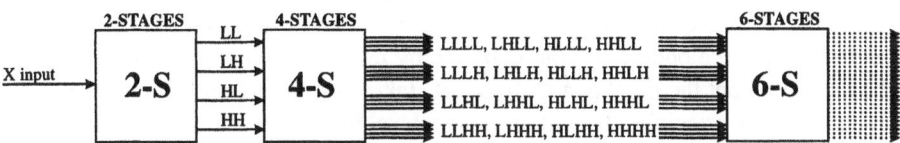

Fig. 6. J-Stages DWPT structure.

4   FPGA-Based Fully Parameterized DWPT Architecture

Our proposed architecture is fully parameterized and modular. We can select the different modules to implement different discrete wavelet packets. In Fig. 7. we present a DWPT structure for a CDF(9,7) wavelet bases. CDF is chosen because it has been applied in most of the previously mentioned applications. These fixed coefficients can be referenced as CDF(G,H), representing the number of taps in the high-pass and the low-pass filter coefficients.

In Fig. 7 the parallel input sequence for even (x_o) and odd (x_1) paths are serialized by the Parallel to Serial converter (P/S) and then introduced to the z^{-1} filter registers of Wd bits. The adders perform both the symmetry coefficients in each filter and the shared additions between two sub-filters. Next, the LUT and the scaling-accumulator are performed for each sub-filter.

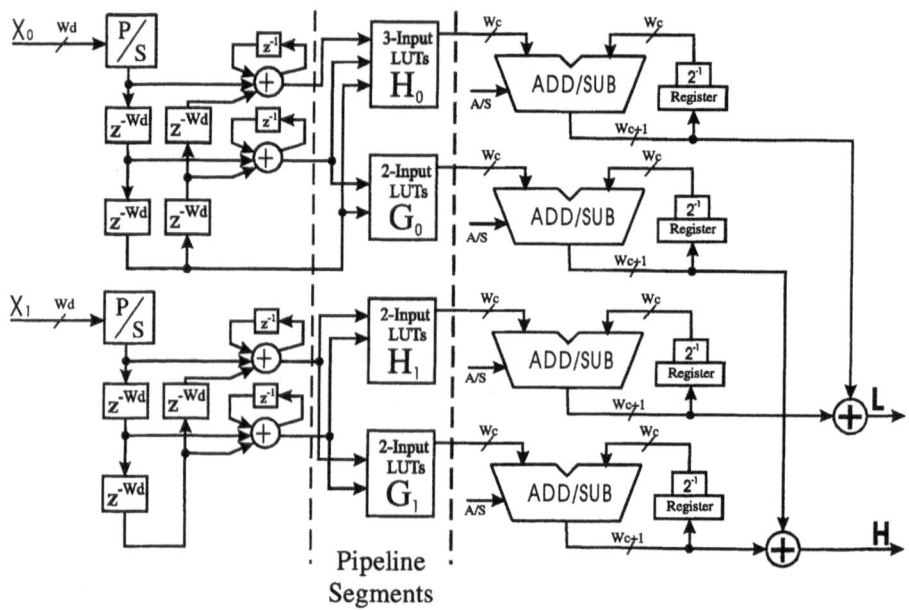

Fig. 7. DWPT structure for a CDF(9,7) wavelet bases

Our basic cell can be RPA characterized by increasing the registers in both the input feedback data from the previous stage and the registers in the scaling-accumulator of the DA structure (see Fig. 5 scheduling). The idea is that filters work at double frequency than the odd/even input data-rate by using free time slots. These registers allow us to accept the input at a uniform rate taking profit of decimation by two. Applying the RPA principles, the output to next level is connected to feedback input (x_i). The scheduling generated from these additional registers is described in next figures.

4.1   Parallel to Serial Converters

The Parallel to Serial Converters have a special functionality as Fig. 8 shows. In 2-Stages block, it converts parallel data (Xi inputs with T·Wd period, really 2·T·Wd because we select the odd or even inputs) and Li and Hi with 2·T·Wd period (4·T·Wd odd/even period) to bit serial data with T period. Fig. 8 shows a detailed version of the P/S converter with its synthesis results.

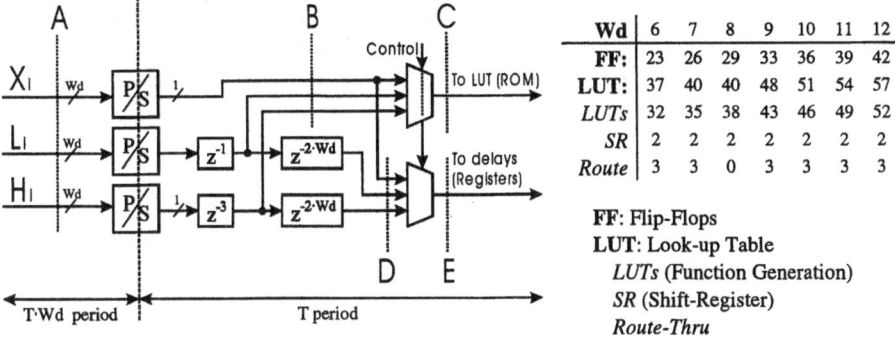

Wd	6	7	8	9	10	11	12
FF:	23	26	29	33	36	39	42
LUT:	37	40	40	48	51	54	57
LUTs	32	35	38	43	46	49	52
SR	2	2	2	2	2	2	2
Route	3	3	0	3	3	3	3

FF: Flip-Flops
LUT: Look-up Table
   LUTs (Function Generation)
   SR (Shift-Register)
   Route-Thru

Fig. 8. Parallel to Serial Converter with scheduling registers for 2-Stages block. Notation (for Wd = 4 bits): $X_i = (X_{3,i}X_{2,i}X_{1,i}X_{0,i})$

The z^{-1} and z^{-3} are included to fix the low-pass and high-pass feedback inputs (L_i and H_i, respectively) between x_i bits of the data. The vertical dotted lines represent the scheduling points (A, B, C, D, E) explained in Fig. 9. The cutset D represents the z^{-2wd} delays needed to synchronize new data inputs x_i with the inputs from the previous stages.

Time	15	14	13	12	11	10	9	8	7	6	5	4	3	2	1	0
Cutset B		$X_{3,j+2}$	$L_{3,j}$	$X_{2,j+2}$		$X_{1,j+2}$	$L_{2,j}$	$X_{0,j+2}$		$X_{3,j}$	$L_{1,j}$	$X_{2,j}$		$X_{1,j}$	$L_{0,j}$	$X_{0,j}$
	$H_{3,i}$				$H_{2,i}$				$H_{1,i}$				$H_{0,i}$			
Cutset C	$H_{3,j}$	$X_{3,j+2}$	$L_{3,j}$	$X_{2,j+2}$	$H_{2,j}$	$X_{1,j+2}$	$L_{2,j}$	$X_{0,j+2}$	$H_{1,j}$	$X_{3,j}$	$L_{1,j}$	$X_{2,j}$	$H_{0,j}$	$X_{1,j}$	$L_{0,j}$	$X_{0,j}$
Cutset D		$X_{3,j+2}$	$L_{1,j}$	$X_{2,j+2}$		$X_{1,j+2}$	$L_{0,j}$	$X_{0,j+2}$		$X_{3,j}$		$X_{2,j}$		$X_{1,j}$		$X_{0,j}$
	$H_{1,i}$				$H_{0,i}$											
Cutset E	$H_{1,i}$	$X_{3,j+2}$	$L_{1,j}$	$X_{2,j+2}$	$H_{0,i}$	$X_{1,j+2}$	$L_{0,j}$	$X_{0,j+2}$		$X_{3,j}$		$X_{2,j}$		$X_{1,j}$		$X_{0,j}$

Fig. 9. Scheduling of P/S Converter for Wd = 4 bits. Notation: $X_i = (X_{3,i}X_{2,i}X_{1,i}X_{0,i})$, $L_i = (L_{3,i}L_{2,i}L_{1,i}L_{0,i})$ and $H_i = (H_{3,i}H_{2,i}H_{1,i}H_{0,i})$

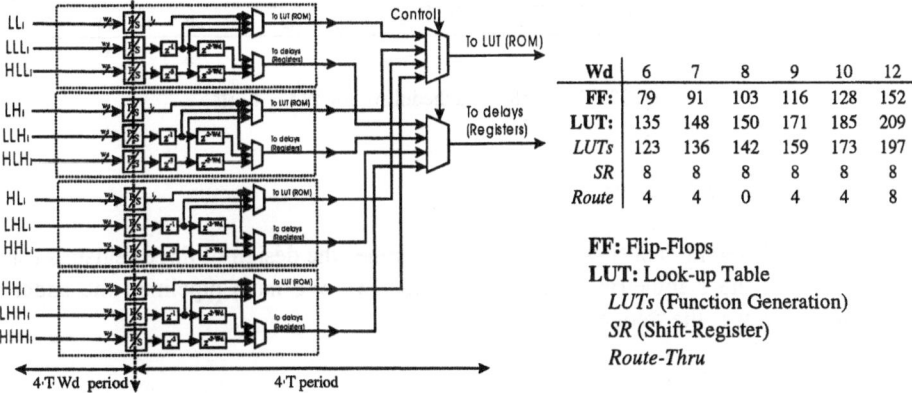

Wd	6	7	8	9	10	12
FF:	79	91	103	116	128	152
LUT:	135	148	150	171	185	209
LUTs	123	136	142	159	173	197
SR	8	8	8	8	8	8
Route	4	4	0	4	4	8

FF: Flip-Flops
LUT: Look-up Table
   LUTs (Function Generation)
   SR (Shift-Register)
   Route-Thru

Fig. 10. Parallel to Serial Converter for 4-Stages block

The P/S Converters for the 4-Stages block have 4 inputs and 8 feedbacks inputs. We have to replicate four times the Fig. 8 structure. The input rates are divided by 4 because the decimators. Fig. 10 represents the structure with the synthesis results.

4.2   Delay Blocks

The delay blocks are detailed in Fig. 11a, and Fig. 12 describes its scheduling. Paying attention on the synchronization between F outputs and the previous C outputs, the delay blocks for the 4-Stages block only differ in the register length (see Fig. 11b).

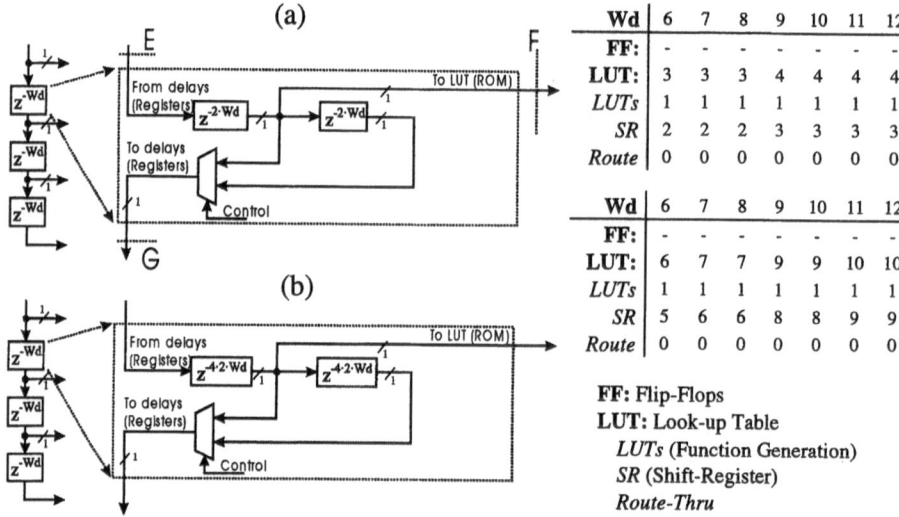

Wd	6	7	8	9	10	11	12
FF:	-	-	-	-	-	-	-
LUT:	3	3	3	4	4	4	4
LUTs	1	1	1	1	1	1	1
SR	2	2	2	3	3	3	3
Route	0	0	0	0	0	0	0

Wd	6	7	8	9	10	11	12
FF:	-	-	-	-	-	-	-
LUT:	6	7	7	9	9	10	10
LUTs	1	1	1	1	1	1	1
SR	5	6	6	8	8	9	9
Route	0	0	0	0	0	0	0

FF: Flip-Flops
LUT: Look-up Table
    LUTs (Function Generation)
    SR (Shift-Register)
    Route-Thru

Fig. 11. (a) Delay blocks for the 2-Stages block and (b) Delay blocks for the 4-Stages block

Time	15	14	13	12	11	10	9	8	7	6	5	4	3	2	1	0
Cutset E	$H_{1,j}$	$X_{3,j+2}$	$L_{1,j}$	$X_{2,j+2}$	$H_{0,j}$	$X_{1,j+2}$	$L_{0,j}$	$X_{0,j+2}$		$X_{3,j}$		$X_{2,j}$		$X_{1,j}$		$X_{0,j}$
Time	23	22	21	20	19	18	17	16	15	14	13	12	11	10	9	8
Cutset F	$H_{1,j}$	$X_{3,j+2}$	$L_{1,j}$	$X_{2,j+2}$	$H_{0,j}$	$X_{1,j+2}$	$L_{0,j}$	$X_{0,j+2}$		$X_{3,j}$		$X_{2,j}$		$X_{1,j}$		$X_{0,j}$
Time	31	30	29	28	27	26	25	24	23	22	21	20	19	18	17	16
Cutset G	$H_{1,j}$	$X_{3,j+4}$	$L_{1,j}$	$X_{2,j+4}$	$H_{0,j}$	$X_{1,j+4}$	$L_{0,j}$	$X_{0,j+4}$		$X_{3,j+2}$		$X_{2,j+2}$		$X_{1,j+2}$		$X_{0,j+2}$

Fig. 12. Delay blocks scheduling for Wd = 4 bits

4.3   Symmetrical Adders

The use of different data flows (X, L and H) implies the design of three registers (one per flow) in the symmetrical adders (see Fig. 13). We have pipelined the adder's output to increase the frequency operation.

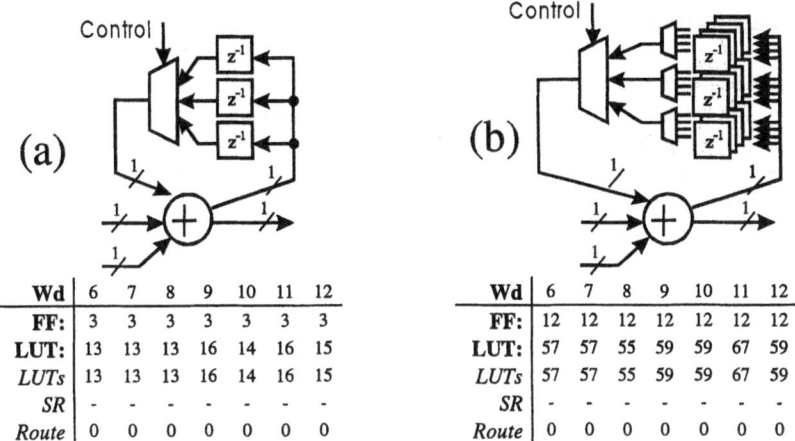

Wd	6	7	8	9	10	11	12
FF:	3	3	3	3	3	3	3
LUT:	13	13	13	16	14	16	15
LUTs	13	13	13	16	14	16	15
SR	-	-	-	-	-	-	-
Route	0	0	0	0	0	0	0

Wd	6	7	8	9	10	11	12
FF:	12	12	12	12	12	12	12
LUT:	57	57	55	59	59	67	59
LUTs	57	57	55	59	59	67	59
SR	-	-	-	-	-	-	-
Route	0	0	0	0	0	0	0

Fig. 13. (a) Symmetrical-adders structure for the 2-Stages block. (b) Symmetrical-adders structure for the 4-Stages block.

4.4 Memory ROMs (LUTs)

As usual, in DA technique, the symmetry of G_0, G_1, H_0 and H_1 halves the LUT size in their physical implementation. LUT-area depends on coefficient length (Wc), not on number of stages. Only four memories of Wc LUT are needed in the overall design.

4.5 Adder/Subtract Accumulators

Finally, we have to use the accumulator registers to keep the inner products results of the different subbands. The structure (see Fig. 14) is similar then the Fig. 13.

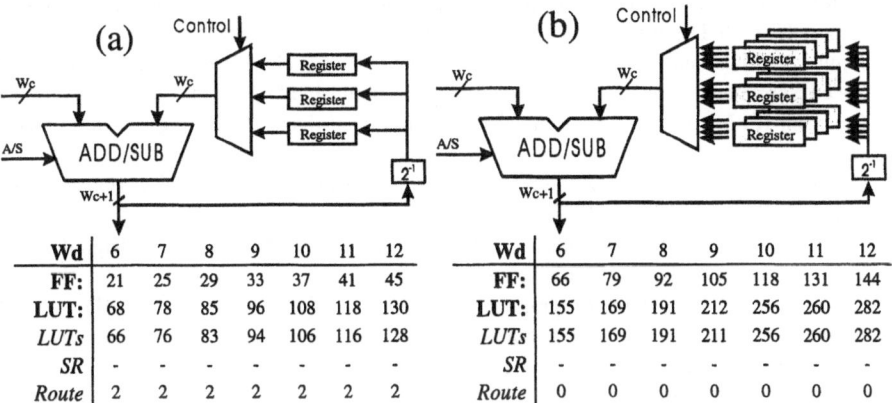

Wd	6	7	8	9	10	11	12
FF:	21	25	29	33	37	41	45
LUT:	68	78	85	96	108	118	130
LUTs	66	76	83	94	106	116	128
SR	-	-	-	-	-	-	-
Route	2	2	2	2	2	2	2

Wd	6	7	8	9	10	11	12
FF:	66	79	92	105	118	131	144
LUT:	155	169	191	212	256	260	282
LUTs	155	169	191	211	256	260	282
SR	-	-	-	-	-	-	-
Route	0	0	0	0	0	0	0

Fig. 14. (a) Adder/Subtract Accumulator registers structure for 2-Stages blocks. (b) Adder/Subtract Accumulator registers structure for 4-Stages blocks

5   Implementation Results and Conclusions

We have synthesized and implemented a CDF(9,7) DWPT into a XILINX Virtex-II 3000-6 FPGA device. For 2-Stages DWPT and Wd=Wc=8, the clock frequency reaches 130 MHz with a hardware cost of 255 Flip-flops and 495 LUTs.

In Fig. 15, we have estimated the occupation area in Slice FF and LUTs for different CDF DWPT structure (CDF(2,2) with 5 and 3 coefficients and CDF(9,7) with 9 and 7 coefficients). The maximum clock frequency results always over 100 MHz.

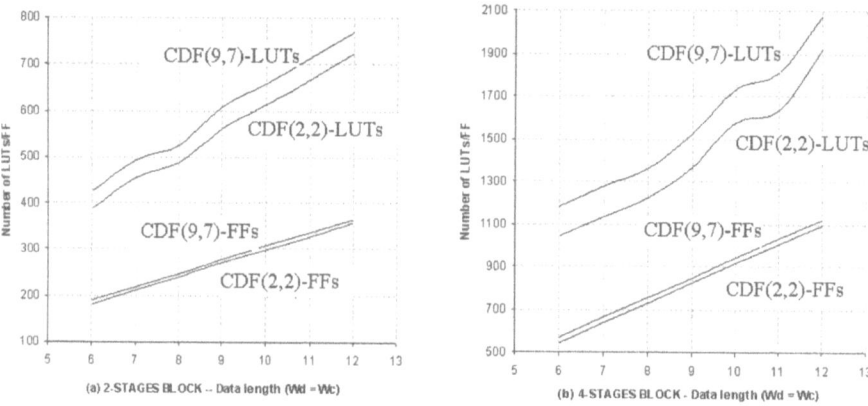

Fig. 15. Data length vs. Number of LUTs and Flip-flops in (a) 2-Stages blocks and (b) 4-Stages blocks

References

1. Trenas, M..A, Lopez, J., and Hongyi, C.: A Configurable Architecture for the Wavelet Packet Transform. IEE Electron Letters (1999) 499-500
2. Hilton, M.L.: Wavelet and Wavelet Packet Compression of Electrocardiograms. Technical Report TR9505, Department of Computer Science, University of South Carolina (1997)
3. Jamin, A., Mähönen, P.: FPGA Implementation of the Wavelet Packet Transform for High Speed Communications. Conference on Field-Programmable Logic and Applications (2002)
4. Coifman, R., Meyer, Y., Quake, S., Wickerhauser, V.: Signal Processing an Compression with Wave Packets. Numerical Algorithms Research Group, Yale University (1990).
5. Croiser, A., Esteban, D.J., Levilion, M.E., Rizo, V.: Digital Filter for PCM Encoded Signals. U.S. Patent 3 777 130 (1973).
6. Peled, A., Liu, B.: A New Approach to the Realization of NonRecursive Digital Filter. IEEE Trans. On Audio and Electroacoustic, vol. 21, no. 6 (1973) 477-485.
7. LB_2DFDWT – Line-Based Programmable Forward DWT, AllianceCore™ (2001)
8. White, S.A.: Applications of Distributed Arithmetic to Digital Signal Processing: a Tutorial Review. ASSP Magazine, vol. 6, Issue. 3 (1989) 4-9
9. Mallat, S.: Multifrequency Channel Decompositions. IEEE Trans. On Acoustics, Speech and Signal Processing, Vol.37, no. 12 (1989)
10. Vaidyanathan, P.P.: Multirate Systems and Filters Banks, Prencitce-Hall Inc (1993)
11. Vishwanath, M.: The Recursive Pyramid Algorithm for the Discrete Wavelet Transform. IEEE Trans. On Signal Processing, vol. 42, no. 3 (1994) 673-677

An FPGA System for the High Speed Extraction, Normalization and Classification of Moment Descriptors

Stavros Paschalakis[1], Peter Lee[2], and Miroslaw Bober[1]

[1] Mitsubishi Electric ITE-VIL, The Surrey Research Park
20 Frederick Sanger Road, Guildford, Surrey GU2 7YD, UK
{Stavros.Paschalakis, Miroslaw.Bober}@vil.ite.mee.com
[2] Department of Electronics, University of Kent at Canterbury
Canterbury, Kent CT2 7NT, UK
P.Lee@kent.ac.uk

Abstract. We propose a new FPGA system for the high speed extraction, normalization and classification of moment descriptors. Moments are extensively used in computer vision, most recently in the MPEG-7 standard for the region shape descriptor. The computational complexity of such methods has been partially addressed by the proposal of custom hardware architectures for the fast computation of moments. However, a complete system for the extraction, normalization and classification of moment descriptors has not yet been suggested. Our system is a hybrid, relying partly on a very fast parallel processing structure and partly on a custom built, low cost, reprogrammable processing unit. Within the latter, we also propose FPGA circuits for low cost double precision floating-point arithmetic. Our system achieves the extraction and classification of invariant descriptors for hundreds or even thousands of intensity or color images per second and is ideal for high speed and/or volume applications.

1 Introduction

The theory of moments is among the most commonly used methodological frameworks in computer vision applications such as document analysis and OCR [1], object recognition [2] and, recently, for the specification of the MPEG-7 region shape descriptor [3]. Moments are projections of the image function onto a basis function and different basis functions have given rise to different types of moments, such as geometric, Zernike and Legendre [4], each type with distinct characteristics with regards to its information content, ease of normalization to image transformations, etc. Geometric moments are among the most commonly used, mainly due to their ease of calculation in relation to other types of moments. Nevertheless, even geometric moments are computationally demanding in terms of their extraction, in spite of the increasing performance of computer systems. This is especially troublesome for high speed and/or volume systems. A lot of effort has been devoted to the development of algorithmic modifications and custom hardware structures to alleviate this computational complexity problem. Thus, there are techniques which allow the fast calculation of moments based solely on simple integer arithmetic operations. Nevertheless, in order for moment descriptors to be used in image processing applications, e.g. object

P.Y.K. Cheung et al. (Eds.): FPL 2003, LNCS 2778, pp. 543–552, 2003.

recognition and image retrieval, they usually require a set of normalization procedures and a classification framework. Such processes are usually assigned to "host" microprocessors, such as CPUs, due to their extended arithmetic processing capabilities and their reprogrammability, which facilitates algorithmic enhancements over time. This partly negates the benefit of the custom hardware implementation, e.g. for applications where a CPU does not actually exist, such as intelligent sensors.

In this paper we propose a new FPGA architecture for the high speed extraction, normalization and classification of moment descriptors. We have chosen an FPGA as our implementation vehicle because it combines the reprogrammability advantage of general purpose processors with the parallel processing and speed advantages of custom hardware. The proposed system has a hybrid form, comprising a parallel processing structure working alongside a low cost, custom built, reprogrammable processing unit. The latter can be reprogrammed for different normalization and classification functions, or even different image processing problems. In the context of this general processing unit we also propose low cost FPGA circuits for 64-bit double precision floating-point arithmetic operations, i.e. addition/subtraction, multiplication, division and square root. Such circuits have been investigated by researchers but only for 32-bit single precision or, more commonly, lower precision custom formats. Our system achieves the extraction and classification of invariant moment descriptors for hundreds or even thousands of intensity or color images per second, making it ideal for high speed and/or volume applications.

2   Algorithmic Framework

The theory and normalization procedures of moments are only briefly presented here to place the subsequent designs in context. A more detailed analysis can be found in [2]. For a grayscale image $g(x,y)$, with $x = 0, 1, ..., M$ and $y = 0, 1, ..., N$, the geometric moments m_{pq} of order $p+q$ are defined as

$$m_{pq} = \sum_{x=0}^{M}\sum_{y=0}^{N} x^p \cdot y^q \cdot g\,(x,y) \tag{1}$$

Translation invariance is achieved by calculating the central moments given by

$$\mu_{pq} = \sum_{r=0}^{p}\sum_{s=0}^{q}\binom{p}{r}\cdot\binom{q}{s}\cdot\left(-\bar{x}\right)^r\cdot\left(-\bar{y}\right)^s\cdot m_{p-r,q-s} \quad \text{with} \quad \bar{x}=\frac{m_{10}}{m_{00}}, \bar{y}=\frac{m_{01}}{m_{00}} \tag{2}$$

Invariance with respect to isometric scale and to scalar intensity changes (which arise from uniform illumination intensity changes) is achieved by n_{pq}, using

$$n_{pq} = \frac{\mu_{pq}}{\mu_{00}}\cdot\left(\frac{\mu_{00}}{\mu_{20}+\mu_{02}}\right)^{\frac{p+q}{2}} \tag{3}$$

Invariance with regards to in-plane rotations and/or reflections is more involved, and a number of feature sets exist. The complex moment magnitudes give rise to a concise yet powerful rotation and reflection invariant descriptor, and will be considered here. The complex moments C_{pq} can be calculated from n_{pq} of using

$$C_{pq} = \sum_{r=0}^{p} \sum_{s=0}^{q} \binom{p}{r} \cdot \binom{q}{s} \cdot i^{p+q-r-s} \cdot (-1)^{q-s} \cdot n_{r+s,\,p+q-r-s} \tag{4}$$

The complex moment magnitudes are then calculated as

$$\left| C_{pq} \right| = \sqrt{\left(C_{pq}^{real} \right)^2 + \left(C_{pq}^{imag} \right)^2} \tag{5}$$

Thus, the above descriptor is normalized with respect to translation, isometric scale, illumination intensity changes, in-plane rotation and reflection. A classification function is also required to compare it to the descriptors of the given templates, e.g. for image retrieval. With moment-based methods, distance-based classification functions are most commonly used. An example is the weighted Euclidean metric, defined between an unknown sample descriptor X and the descriptor of the ith template Y_i as

$$d_i = \sqrt{\sum_{j=1}^{n} w_{ij} \left(X_j - Y_{ij} \right)^2} \tag{6}$$

where n is the dimensionality of the descriptors. The weight w is commonly given by an expression which takes into account the variance for each feature and for the different templates when multiple samples represent each template, or it can be 1 for single sample templates, giving rise to the simple Euclidean metric.

The algorithms presented here have been used mostly in the processing of binary and grayscale images. In [2], a framework is proposed for the processing of color images based on the above methodology or, indeed, using any type of moments and normalization procedures. This entails the treatment of each color plane as an isolated grayscale image for the derivation of invariant descriptors followed by a fusion stage. Different fusion schemes are examined in [2], such as descriptor aggregation followed by classification or single plane classification followed by decision fusion. Furthermore, because for RGB images changes in the intensity or spectral power distribution of the incident illumination result in an independent scalar change of each color plane, this framework achieves invariance not only to geometric and illumination intensity changes but also to changes in the color of the incident illumination.

3   Parallel Moment Computation Circuit

The fast calculation of the geometric moments of an intensity image relies on a parallel computation structure. Our module is based on Hatamian's work [5] towards the implementation of a VLSI moment calculation chip and is, effectively, a cascaded accumulator structure. The organization of the moment computation module for the calculation of moments up to the fifth order can be seen in Figure 1. The row processing elements (RPEs) process each pixel of each image row y and produce the outputs $Y_0(y)...Y_5(y)$. The column processing elements (CPEs) process the results produced by the RPEs for each row y and produce outputs for the entire image.

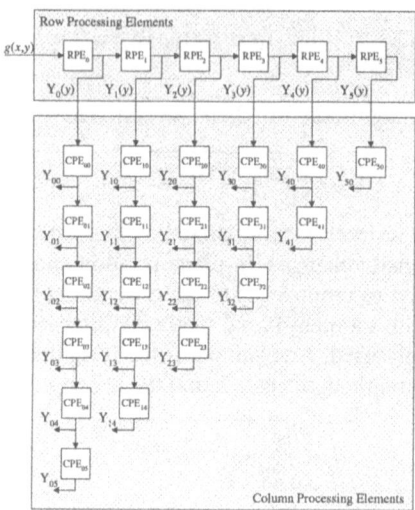

Fig. 1. Organization of the parallel moment computation circuit

The RPEs have been implemented as parallel accumulators, to achieve a pixel per cycle processing rate. Each RPE has a different size, from 17 bits for RPE_0 up to 53 bits for RPE_5. These values were chosen so that an overflow is guaranteed not to occur for an 8-bit intensity image with rows of up to 512 pixels. The CPEs have been implemented as identical 64-bit serial accumulators, giving rise to a CPE systolic array. This facilitates a straightforward implementation and simple control logic. The serial implementation is the preferred design choice, since the CPEs operate at the image row level and need not be as fast as the RPEs. The serial accumulators have been implemented using the FPGA's function generator RAMs (LUTRAMs) instead of registers, drastically reducing the circuit size. The 64-bit accumulator size has been chosen so that an overflow is guaranteed not to occur for 512×512 pixel 8-bit intensity images. As for the calculation of moments of higher than fifth order, this can be easily achieved by extending the RPE chain and the CPE systolic array. The reconfigurability of the FPGA device allows the redeployment of such modified modules. The CPE structure considered here is more efficient than the one proposed by Hatamian, which requires almost twice the resources because it calculates not only the moments up to the required order, but also incomplete sets of higher order moments.

Note that the values produced by this circuit are not, in fact, the geometric moments of the image. The moments m_{pq} are derived through the simple algebraic combination of the Y_{pq} outputs. These equations are not included here, but they are can be found in [2]. Of relevance here is the fact that the calculation of m_{pq} from Y_{pq} involves a total of 45 multiplications of Y_{pq} values with 8-bit integers followed by 45 additions of these products. Although this processing is quite simple, no custom component was created because it will performed by the processing units described next.

Table 1 shows the implementation statistics of the complete moment computation module on a XILINX® XCV1000 Virtex™ FPGA of –6 speed grade [6]. At 4.56% usage, the circuit is quite small. These figures also include 53 I/O synchronization

Table 1. Circuit statistics for the moment computation module

Slices	535 (4.35%)
Slice flip-flops	628
4-input LUTs	664
Dual Port LUTRAMs	84
GCLKs	1 (25%)
Total equivalent gate count	21,287

registers. The circuit can operate at up to 50MHz, the critical path lying on the RPE chain and comprised of 62.3% and 37.7% logic and routing delays respectively. Thus, for an image of an $M{\times}N$ pixels, the frame rate is $(50{\times}10^6)/(M{\times}N)$ frames/sec., e.g. ~3051 frames/sec. for 128×128 pixel images and ~190 frames/sec. for 512×512 pixel images. This is ~750 times faster than a direct implementation and ~100 faster than a recursive addition implementation on a SUN UltraSPARC10 server.

4   Floating-Point Arithmetic Unit

The design and implementation of FPGA floating-point arithmetic circuits has been examined by various researchers [7-11]. Although such circuits cannot match the performance of the floating-point units of state-of-the-art microprocessors, they are extremely useful in systems and applications which benefit from custom parallel processing structures, but also require floating-point capabilities, such as the system considered in this paper. Thus, this section presents FPGA circuits for all the common floating-point operations, i.e. addition/subtraction, multiplication, division and square root, the last two being the least investigated in the literature. While previous work addressed the implementation of operators in the 32-bit single precision or even lower precision custom formats, in order to reduce the circuit costs and increase their speed, our aim was the creation of low cost operators that provide 64-bit double precision arithmetic. Furthermore, our circuits are IEEE-754 [12] compliant, implementing round-to-nearest-even rounding, able to handle infinities and NaNs, etc. Because the detailed treatment of this material is not feasible here, this section will focus only on the circuit statistics and performance of the units, while a very detailed description of these designs and implementations can be found in [2]. The implementation statistics of all four operators on the aforementioned device are shown in Table 2.

Addition and subtraction are, in general, the most frequent floating-point operations in scientific computing. Our double precision floating-point adder aims at a low implementation cost combined with a low latency. A non-pipelined design was adopted, so that key components may be reused, with a fixed latency of three clock cycles. At 5.49% usage, the circuit is quite small. These figures also include 194 I/O synchronization registers The circuit can operate at up to 25MHz, the critical path lying on the significand processing path and comprised of 41.1% and 58.9% logic and routing delays respectively. Since the design is not pipelined and has a fixed latency of three clock cycles, this gives rise to a performance in the order of 8.33MFLOPS. Obviously, the implementation considered here is small enough to allow multiple instances to be incorporated in a single FPGA device if needed.

Table 2. Double precision floating-point operator circuit statistics

	Adder	Multiplier	Divider	Square Root
Slices	675 (5.49%)	495 (4.03%)	343 (2.79%)	347 (2.82%)
Slice flip-flops	336	460	400	316
4-input LUTs	1,118	604	463	399
GCLKs	1 (25%)	2 (50%)	1 (25%)	1 (25%)
Total equiv. gate count	10,334	8,426	6,464	5,366

The double precision floating-point multiplier also aims at a low implementation cost while maintaining a relatively low latency, considering the scale of the significand multiplication involved. A non-pipelined design was adopted with a fixed latency of ten cycles. The circuit operates on two clocks, a primary or global clock (CLK_1), to which the ten clock cycle latency corresponds, and an internal secondary clock (CLK_2), which is twice as fast as the primary clock and is used by the significand multiplier. The overall circuit is quite small, occupying only 4.03% of the device. These figures also include 193 I/O synchronization registers. The primary clock CLK_1 can be set to a frequency of up to 40MHz, its critical path comprised of 36.4% and 63.6% logic and routing delays respectively, while the secondary clock CLK_2 can be set to a frequency of up to 75MHz, its critical path comprised of 36.8% and 63.2% logic and routing delays respectively. Since the circuit is not pipelined with a fixed latency of ten CLK_1 cycles, a frequency of 33MHz and 66MHz for CLK_1 and CLK_2 respectively gives rise to a performance in the order of 3.3MFLOPS.

In general, division is much less frequent than the previous operations. Because of this, our double precision floating-point divider aims mainly at a low implementation cost. A non-pipelined design was adopted, incorporating an economic significand divider, with a fixed latency of 60 clock cycles. The circuit is very small, occupying only 2.73% of the device, which also includes 193 I/O synchronization registers. This circuit can operate at up to 60MHz, the critical path comprised of 42.8% and 57.2% logic and routing delays respectively. Since the design is not pipelined and has a fixed latency of 60 clock cycles, this gives rise to a performance in the order of 1MFLOPS.

The square root function is the least frequent of the operations considered here. As for the divider, our double precision floating-point square root circuit aims at a low implementation cost. A non-pipelined design was adopted with a fixed latency of 59 cycles. The circuit is very small, occupying only 2.83% of the device, which also includes 129 I/O synchronization registers. The circuit can operate at up to 80MHz, the critical path comprised of 53.0% and 47.0% logic and routing delays respectively. Since the implementation considered here is not pipelined and has a fixed latency of 59 clock cycles, this gives rise to a performance in the order of 1.36MFLOPS.

The above discussion and the associated implementation statistics illustrate that our aim was the implementation of high precision operators at a low cost rather than with a very high performance. Our floating-point unit can perform a few million double precision operations per second and is meant to complement custom FPGA architectures, such as the one of Section 3, eliminating the need for integrating a microprocessor into the system. Furthermore, the circuits sizes allow the integration of multiple operators on the same device if necessary, while their self-contained implementation allows one to choose only the required operators for a given system.

5 System Architecture

The system that we have implemented for the processing of intensity images based on the components of the previous sections is illustrated in Figure 2(a). The first main stage in our architecture is the parallel processing structure of Section 3, the moment computation module. The input to this module can be a video signal directly from a camera, some other image delivery mechanism, or some other pre-processing module, e.g. a noise filter. The second main stage is a custom FPGA general processing unit which, in turn, comprises two main parts. The first main part is a double precision floating-point unit (FPU) which includes the circuits of the previous section. Two additional modules, which we did not describe earlier due to the simplicity of their implementation, are an integer to floating-point converter and a floating point comparator. The converter is very small, occupying ~1% of the device, has a variable latency of 1 to 15 clock cycles and a maximum clock speed of 145MHz. The comparator also occupies ~1% of the device, has a fixed latency of 1 cycle (input buffering) and a maximum clock speed of 135MHz. The second main part of the general purpose processing unit is a controller module, the main components of which are a control logic unit, an instruction RAM and a data RAM. The data RAM is an 8-bit address 64-bit data memory, implemented using the dedicated RAM blocks of the FPGA (BRAMs). This memory is used by the FPU, as well as for externally provided constants and statistics. Clearly, external RAM modules may very easily be used if large storage is required or if no RAM is available in the FPGA device. The instruction RAM is a 10-bit address 28-bit data memory, also implemented using BRAMs. This memory stores all the operations required for the calculation and classification of the invariant descriptors in the form of FPU commands.

In order to assess how the custom processing unit compares to the fast and parallel moment computation circuit, consider the following scenario. Assume that all the normalization procedures of Section 2 are desired. Obviously, these equations are not stored in the closed forms examined earlier, but must be expressed in an explicit form. Once expressed in such a form, one also discovers that there exists a great amount of redundancy and duplication of calculations, which can be eliminated. Note that the system is not restricted to these normalization and classification functions. By changing the contents of the instruction and/or data RAMs, different functions may be implemented or even different types of moments based on geometric moments. Furthermore, assume that the target application is a 10-way classification problem, i.e. assigning an unknown image to one of ten different templates. This affects the complexity of the classification function. The entire general processing unit has a fixed latency, with the floating-point converter being the only exception, for which we can assume a constant worst case latency. With the above scenario, all the processing, i.e. from the delivery of the results of the moment computation circuit up to a classification decision being delivered, requires 6806 clock cycles. The derivation of this figure is not included here but may be found in [2]. During all of this time, the moment computation module can carry on with the processing of the next image. Now, compare that with the processing time of the moment computation module, which has a pixel per cycle processing rate, e.g. 16384 cycles for 128×128 pixel images and 262144 cycles for 512×512 pixel images. In both cases, the latency of the moment

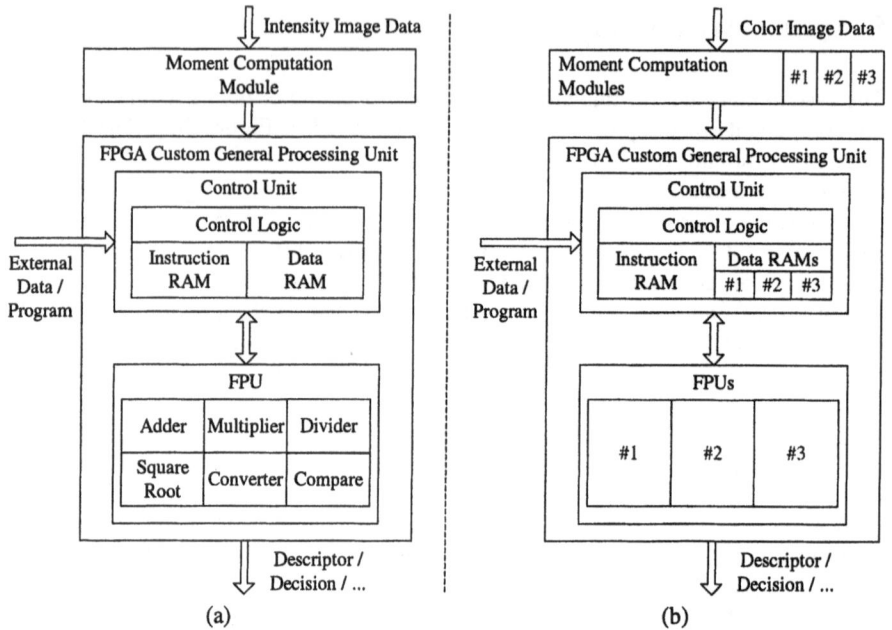

Fig. 2. (a) Binary/grayscale descriptor system (b) Color descriptor system

Table 3. Implementation statistics for the intensity and color descriptor systems

	Intensity descriptor system	Color descriptor system
Slices	3,437 (27.97%)	9,825 (79.96%)
Slice flip-flops	3,108	8,821
4-input LUTs	3,775	10,975
Dual port LUTRAMs	84	252
Bonded IOBs	77 (19.06%)	79 (19.55%)
TBUFs	1,600 (12.76%)	4,800 (38.27%)
BRAMs	11 (34.38%)	19 (59.38%)
GCLKs	2 (50%)	2 (50%)
GCLKIOBs	2 (50%)	2 (50%)
Total equiv. gate count	248,531	509,541

computation module completely overshadows the latency of all the other circuits. The timing analysis indicates that the entire system can be clocked at up to 25MHz, the speed of the floating point adder being the determining factor. However, one can adopt a dual clock strategy and clock the moment computation circuit at 50MHz, which is feasible based on the results of Section 3. In that case, the entire system has the same processing rates with the moment computation module, i.e. ~3051 frames/sec. for 128×128 pixel images and ~190 frames/sec. for 512×512 pixel images. These speeds certainly satisfy the requirements for real-time systems or for multiple camera systems or for other high volume applications. The implementation statistics of this system can be seen in Table 3. Occupying 27.97% of the aforementioned XCV1000 device, the design is not negligible, but it is certainly feasible.

With regards to the processing of color images based on the information fusion framework briefly described in Section 2, a number of choices are available. Thus, one can use the hardware described above (its speed certainly allows that, even for real-time processing) for the sequential processing of the color planes, storing different sets of statistics for each plane for use by the classifier. Alternatively, one can employ three moment computation modules and a single general processing unit, and so on. The system outlined in Figure 2(b) is yet another possibility. There, three moment computation modules are used along with three complete FPUs, the FPUs sharing a common control logic and instruction register but each having its own data RAM. The sharing of the control and instruction logic is possible because all the color planes are processed in the same manner and the different components of the system have a fixed latency (the worst case latency must be enforced on the floating-point converters). Assuming a descriptor aggregation scheme for the fusion step, the actions performed by this system are basically the same as before, with some additional steps in the instruction RAM to implement the fusion framework. With a dual clock strategy, this system can also process ~3051 frames/sec. and ~190 frames/sec. for 128×128 pixel and 512×512 pixel 24-bit RGB images respectively. Alternatively, the same organization can be used for the parallel processing of grayscale images, achieving ~9153 frames/sec. and ~570 frames/sec. for 128×128 pixel and 512×512 pixel grayscale images respectively. The statistics of this system are also shown in Table 3. Occupying 79.96% of the device, the system requires the best part of the chosen FPGA, but is still feasible within a single chip.

6 Discussion and Conclusions

We have presented a new FPGA architecture for the high speed extraction, normalization and classification of moment descriptors. FPGAs are ideal for the implementation of such systems because they combine the reprogrammability advantage of general purpose processors with the parallel processing and speed advantages of custom hardware. Our system relies on a custom parallel moment computation module working alongside a custom low cost general processing unit. The specific characteristics of the moment computation module, such as highest moment order and maximum image dimensions, can be easily changed and the module redeployed by reconfiguring the FPGA. The reprogrammable general processing unit, on the other hand, allows the easy deployment of different normalization and classification functions and can be programmed for different recognition problems without any design modifications. Even if such changes are required, the reconfigurability of the FPGA makes such a task feasible. In this manner, one can always implement the exact processing unit required for a system. It is interesting to note that the design of this general processing unit was actually straightforward once all the modules had been created. In terms of design effort, the design and implementation of the double precision floating-point operators was probably the most involved task, these circuits also involving the most laborious testing procedures. It is for this reason that we have implemented each operator as a low cost standalone unit instead of implementing an FPU that shares hardware components between the operators. That is, so that one can choose only the operators needed for a given system, image processing or not, and also have multiple instances of individual operators if necessary. The performance of our overall architecture, in the order of hundreds or thousands of binary, grayscale or color images per second, makes it suitable for high speed and/or volume applications, e.g. for image recognition, detection and retrieval, while its self-contained implementation also allows its deployment in small standalone systems.

References

1. Rahman, A.F.R.: Study of Multiple Expert Decision Combination Strategies for Handwritten and Printed Character Recognition, Ph.D. Thesis, Electronic Engineering Laboratory, University of Kent at Canterbury, UK (1997)
2. Paschalakis, S.: Moment Methods and Hardware Architectures for High Speed Binary, Greyscale and Colour Pattern Recognition, Ph.D. Thesis, Department of Electronics, University of Kent at Canterbury, UK (2001)
3. Bober, M., Preteux, F., Kim, W.Y.: Shape Descriptors. In Manjunath, B.S., Salembier, P., Sikora, T. (eds.): Introduction to MPEG-7, John Wiley & Sons (2002) 231-260
4. Teh, C.H., Chin, R.T.,: On Image Analysis by the Method of Moments, IEEE Transactions on Pattern Analysis and Machine Intelligence, vol. 10, no. 4 (1988) 496-513
5. Hatamian, M.: A Real-Time Two-Dimensional Moment Generating Algorithm and Its Single Chip Implementation, IEEE Transactions on Acoustics, Speech, and Signal Processing, vol. ASSP-34, no. 3 (1986) 546-553
6. Virtex™ 2.5V Field Programmable Gate Arrays, XILINX® Corporation (2000)
7. Ligon, W.B., McMillan, S., Monn, G., Schoonover, K., Stivers, F., Underwood, K.D.: A Re-Evaluation of the Practicality of Floating Point Operations on FPGAs, In Proc. IEEE Symposium on FPGAs for Custom Computing Machines (1998) 206-215
8. Louca, L., Cook, T.A., Johnson, W.H.: Implementation of IEEE Single Precision Floating Point Addition and Multiplication on FPGAs, In Proc. IEEE Symposium on FPGAs for Custom Computing Machines (1996) 107-116
9. Li, Y., Chu, W.: Implementation of Single Precision Floating Point Square Root on FPGAs, In Proc. Fifth IEEE Symposium on Field Programmable Custom Computing Machines (1997) 226-232
10. Tangtrakul, A., Yeung, B., Cook, T.A.: Signed-Digit On-Line Floating-Point Arithmetic for FPGAs, In Proc. High-Speed Computing, Digital Signal Processing, and Filtering Using Reconfigurable Logic (1996) 2-13
11. Shirazi, N., Walters, A., Athanas, P.: Quantitative Analysis of Floating Point Arithmetic on FPGA Based Custom Computing Machines, In Proc. IEEE Symposium on FPGAs for Custom Computing Machines (1995) 155-162
12. ANSI/IEEE Std 754-1985: IEEE Standard for Binary Floating-Point Arithmetic, IEEE (1985)

Design and Implementation
of a Novel FIR Filter Architecture
with Boundary Handling on Xilinx VIRTEX FPGAs

A. Benkrid, K. Benkrid, and D. Crookes

School of Computer Science, The Queen's University of Belfast, UK
a.benkrid@qub.ac.uk

Abstract. This paper presents the design and implementation of a novel architecture for FIR filters on Xilinx Virtex FPGAs. The architecture is particularly useful for handling the problem of signal boundaries filtering, which occurs in finite length signal processing (e.g. image processing). It cleverly exploits the Shift Register Logic (SRL) component of the Virtex family in order to implement the necessary complex data scheduling, leading to considerable area savings compared to the conventional implementation (based on a hard router), with no speed penalty. Our architecture uses bit parallel arithmetic and is fully scalable and parameterisable. A case study based on the implementation of the standard low filter of the Daubechies-8 wavelet on Xilinx Virtex-E FPGAs is presented.

1 Introduction

Finite Impulse Response (FIR) filters are widely used in digital signal processing. An N-tap FIR filter is defined by the following input-output equation [1]:

$$out(n) = \sum_{i=0}^{N-1} x(n-i)\,h(i) \qquad (1)$$

where $\{h(i): i = 0, \ldots, N-1\}$ are the filter coefficients.

Figure 1 shows the two conventional structures (the direct and the inverse form) of an FIR filter [2]. Both structures of Fig 1 seek to align the products $x(n-i)h(i)$ of equation (1) in time before accumulation.

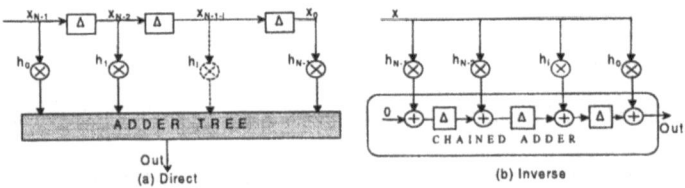

Fig. 1. Two conventional FIR filter structures

P.Y.K. Cheung et al. (Eds.): FPL 2003, LNCS 2778, pp. 553–564, 2003.

An FIR filter implements a convolution operation [1], which is often built on the assumption of infinite length signals e.g. continuous audio signal. Finite length signals (e.g. images) on the other hand, have discontinuities at the boundaries (see Fig 2). Thus emerges the problem of which values to use at these regions.

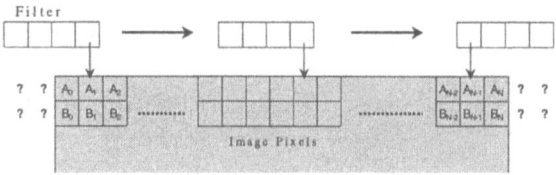

Fig. 2. Boundary problem when filtering an image

Although, this problem could be ignored for a one-stage convolution, it cannot be discarded when implementing a multi-stage convolution as in Discrete Wavelet Transform [3]. A usually recommended solution to this problem is to extend each row by reflection at the signal boundary [4] as shown in Fig 3.

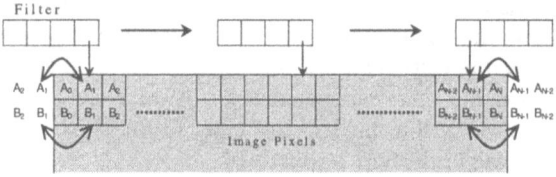

Fig. 3. A finite 2-D signal (image) filtering with boundary extension by reflection

For an N-Tap FIR, the minimum number of extra samples to be introduced is constant and equal to N-1. This is because P+(N-1) input samples are required to generate P output samples. However, the number of samples to be added at the left border of the input signal (referred to by α) or the right one (referred to by μ) can be variable, i.e. not a constant.

To handle the problem of boundary processing in hardware, Chakrabati [5] proposed the use of a router (or switcher) to feed the appropriate data, in parallel, to the multipliers (see Fig 4).

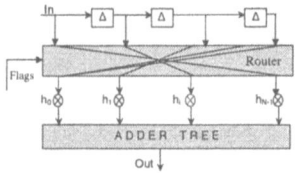

Fig. 4. A conventional FIR architecture with a hard-router to handle the boundary processing

The hard router is implemented using multiplexers. A controller is needed to drive the appropriate multiplexers' selection signals. A minimum of $W(\sum_{i=2}^{\alpha+1}\lceil i/2\rceil + \sum_{j=2}^{N-\alpha}\lceil j/2\rceil)$

LUTs are required to implement the router for an N-tap FIR filter[6] if its input signal is extended by α samples at its left side. This represents an $O(N^2W)$ hardware complexity[1] and therefore requires considerable area and routing resources which will have a negative effect on the speed performance of the implementation.

To overcome this high area cost of the Hard Router, we have presented in [6] a novel architecture for symmetric FIR filter family. The suggested structure is parameterised and scalable. It led to considerable area saving but with speed penalty as it requires the use of clock doubler. The actual paper overcomes this problem and addresses basically a general FIR filter. Nonetheless, the results provided in this paper can be tailored for the symmetric FIR filter type. Unfortunately, because of the paper size limit, this will not be detailed in this paper. The following will explain our approach for a general FIR filter.

Our suggested architecture is tailored to the Xilinx Virtex FPGA family. It exploits mainly the Virtex Shift Register Logic component: SRL16 [7]. SRL16 is implemented by the Virtex slices' LUT (see Fig 5). There are two LUTs in every Virtex slice. Each one can be configured to create a shift register (SRL16) that varies in length from 1 to 16 bits. Longer shift register length can be implemented with multiple chained SRL16. As shown in Fig 5, the SRL16 configuration consists of a chained delay with a multiplexer at the output. The input address Addr[3:0] selects which bit in the chained delay to be output hence controlling the length of the shift register from 1 to 16. Note that each SRL16 can be immediately pipelined by using the flip-flop available on the same slice logic cell [7].

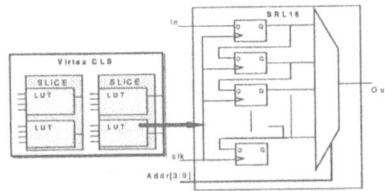

Fig. 5. Virtex CLB and SRL16 configuration.

The remaining of this paper will first present the basic architecture of our novel FIR architecture regardless of the signal boundaries filtering. It then shows how this basic architecture can be easily extended to handle signal boundaries processing with little hardware penalty. A detailed approach will be given. Then, area measurements of our novel architecture will be presented and compared with the corresponding results from a conventional hard-router based FIR implementation. Timing and area results of a case study implementation will be provided. Finally conclusions will be drawn.

Throughout the remaining of the paper, we will assume the use of bit parallel arithmetic. The term SRL will correspond to either one SRL16 component or a chained SRL16 components if required. The abbreviation 'cc' denotes the term clock cycle and the term $SRL_{(i)}$ refers to the SRL associated with the filter coefficient h_i.

---

1 $\sum_{i=1}^{N} i = N(N+1)/2 = O(N^2)$

2 Our Novel FIR Filter Architecture

In order to handle the boundary filtering efficiently, we suggest the novel architecture shown in Fig 6. The input data samples (X) in this figure structure are first multiplied in parallel with the filter coefficients. Then, the multiplication results $x(n-i).h(i)$ of equation (1) are skewed and aligned in time properly using the SRL to produce the filter output.

The SRL layer in Fig 6 structure aims to skew the products and align them properly in time before parallel accumulation as shown in Fig 7. In fact, unlike Fig 1.a structure, our structure does not include an input samples chained delay and therefore the supply of the products onto the adder tree needs to be synchronised by the SRL layer before *parallel* accumulation.

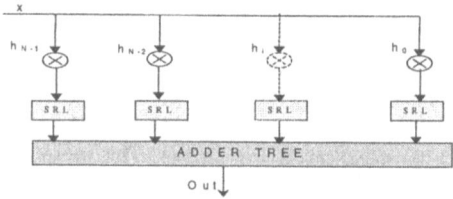

Fig. 6. Our novel FIR filter structure for signal borders processing using symmetry extension

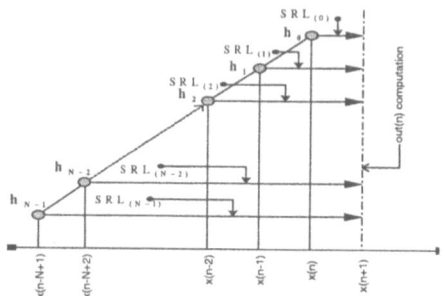

Fig. 7. Data Dependence Graph (DDG) of N-tap FIR filter. The horizontal arrows show the SRLs' delays length (the filled circles represent the relevant products $x(n-i).h(i)$)

Table 1 gives the SRLs delay length. Each SRL delay length is equal to the projection length of its associated product instant on the time axis abscise "n+1"(see Fig 7).

Table 1. The SRLs' delay length for an FIR filter with no boundary processing

$SRL_{(0)}$	$SRL_{(1)}$	$SRL_{(2)}$	....	$SRL_{(N-2)}$	$SRL_{(N-1)}$
1	2	3		N-1	N

It is worth noting that the SRLs in Fig 6 structure can be placed before or after the multipliers (thus skewing the product $x(n-i)h(i)$ or the multiplicands x). If the filter coefficients are fractional and truncation is carried out at the output of the multipliers, the multipliers' output word lengths could be smaller than their input ones. Thus, for area optimisation, the SRLs should be placed at the multipliers output. On the other hand, if the filter coefficients values are greater than 1, the multipliers' output word lengths will be greater than their inputs. Therefore, placing the SRLs before the multipliers will lead to a more compact implementation.

Besides from being able to implement a convolution operation, our structure seeks primarily to handle signal boundaries processing. This is handled efficiently with very little hardware overhead, thanks to the SRLs dynamic skewing feature. The following sections will detail our own approach to achieve this goal.

Throughout, the term P_i denotes the FIR DDG's product associated with the multiplier h_i. In particularly, P_{N-1} (P_0) denotes the first (last) FIR DDG's node.

3   An Extension to Handle Signal Boundaries

To handle *signal boundary filtering*, no alteration on the architecture of Fig 6 is necessary. In fact, it is handled efficiently by a proper skewing of the products P_i through *a dynamic SRLs addressing*. For this purpose, we will present an algorithm, which allows us to find the required SRLs delay length values according to the filter length, N, and the symmetry-filtering axis. The latter is defined through the number of input samples, α, which should be added at the left side of the signal, and to the number of input samples, μ, which should be introduced at its right side, where $\alpha+\mu=N-1$.

Fig 8 shows a 4-tap FIR filter DDG at the boundary regions. The dashed arrows show where the signal extension using symmetry reflection is applied. The deduction of such graphs will be detailed later in the section. In this figure, the filled circles refer to the filter products whereas the shaded rectangle shows the boundary region between the two sequences I and (I-1). Sequence-I refers to the actual sequence of samples, whereas Sequence-(I-1) to its previous one. The negative values (-1, -2, -3...) will denote all instants before instant 0 of sequence I. The boundary DDGs associated with sequence-I represent actually the DDGs at the left side signal boundary, whereas the ones of sequence–(I-1) represent the ones at its right side edge.

From Fig 8, and as a general rule we can see that the *regular* DDG (a straight line as depicted in Fig 7) of a generic N-tap filter ends at the P_{N-1} of instant "–N" (= -4 in Fig 8), and starts from the P_0 of instant N-1 (= 3 in Fig 8). Between those two values, the DDG becomes irregular.

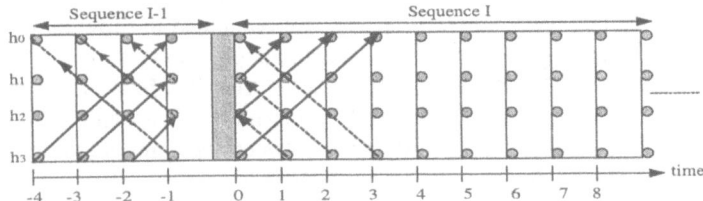

Fig. 8. Data dependence graph of a 4-tap FIR filter (N=4) using boundary processing with symmetry extension

We will refer to the irregular DDG *deflection point* by the term **Hub**. The term P_{Hub} will represent its associated product whereas P_{Hub-i} (P_{Hub+i}) represent the ith DDG's product that comes before (after) the Hub (i is a positive integer).

Because of the DDG irregularity at the signal boundary, the SRLs delay length given in table 1 should then be updated. We therefore suggest the following approach.

An Approach to Determine the SRLs Delay Length when Using the Symmetry Signal Extension

Once the filtering symmetry axis (i.e. α and μ, where $\alpha+\mu=N-1$) is determined, the updated SRLs' delays length could be deduced using two main steps associated respectively with the left side and the right side input signal.

Fig 9.a (Fig 9.b) shows the DDG at the left (right) side boundary of sequence-I input signal, where α (β) samples are introduced through symmetry reflection (see the arced arrows). This was needed since the P_{Hub-i} (P_{Hub+i}) multiplicands at this boundary region correspond to sequence–(I-1) (sequence–(I+1)) samples (see the dashed line).

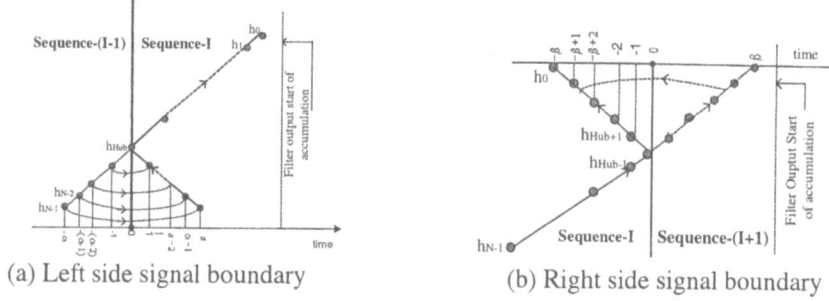

(a) Left side signal boundary   (b) Right side signal boundary

Fig. 9. The irregular DDG for an FIR filter when using symmetry extension.

For the proper functionality of the Fig 6 structure, all the DDG products should feed the adder tree at the same time. In fact, the parallel accumulation can't start till all the products have been computed. Table 1 gives the SRLs delay length when no boundary processing is handled. The values given assume implicitly the instant of computation of P_0 as a reference (see Fig 7). The question that arises: will this reference needs to be changed if we process the border filtering?

Fig 10 shows two cases where P_0 and P_{N-1} are computed in different relative order. We can see from Fig 10.b that the accumulation can't start one cc after the computation of P_0 if this latter is calculated earlier than P_{N-1}. Instead, the accumulation can start one cc after the computation of P_{N-1} (instant t_2 rather than t_1, see Fig 9.b). The change in time (t_2-t_1) denotes the difference between the computation instants of P_{N-1} and P_0. We therefore define a variable ξ such that:

$$\xi=\max[(t[P_{N-1}]-t[P_0]),0]$$

This variable value should then be added to the SRLs delay length values given in table 1. In fact, when $t[P_{N-1}]-t[P_0]<0$ (see Fig 10.a), ξ is equal to zero, since no update is needed (the accumulation starts one cc after the computation of P_0 as assumed implicitly in table 1). However, when $t[P_{N-1}]-t[P_0]>0$ (see Fig 10.b), $\xi= t[P_{N-1}]-t[P_0]$ which is equal to the required SRLs delays length increment as explained in the last paragraph.

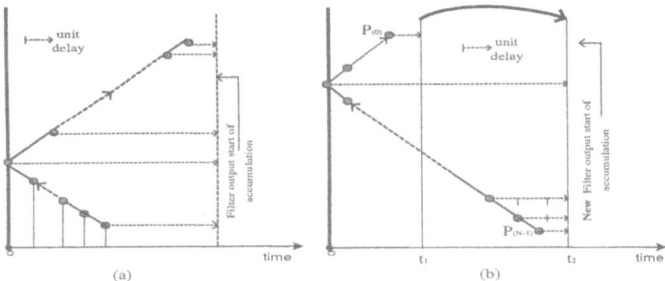

Fig. 10. The effect of P_{N-1} and P_0 relative time order on the start of accumulation instant

It is worth noting that if we extend the input signal by α samples at its left side, $\xi=\max[2\alpha+1-N,0]$ since $t[P_{N-1}]= \alpha$ and $t[P_0]= N-(\alpha+1)$.This will correspond to the SRLs delay length update value for the first filter output. However, since all the filter outputs are produced regularly in time (delayed by one cc), the same update value (ξ) needs to be added to the SRLs delays length associated with the second, third etc filter outputs, and in particularly to the SRLs delays length associated with the non-boundary regions. Therefore, we can represent the delays length of the SRLs list in the *non-boundary* region by a *RegSkew* list where:

$$RegSkew=[\lambda,\lambda-1, \lambda-2,\ldots,\xi+3, \xi+2, \xi+1], \text{ where } \lambda=N+\xi \quad (2)$$

This list values are applied on tap-(N-1) to tap-0.

Nonetheless, the individual SRLs delay length at the boundary regions need still to be determined because of the DDGs irregulaity. For this sake, we consider in the following the left side of the signal and then its right side.

Left Side Signal Boundary Extension

Taking into account Fig 7 structure as a reference, we can see from Fig 9.a that the P_{hub-i} products are advanced in time relatively to the accumulation instant. Thus, their associated regular delays length should be *decreased*. From this figure, we can see that P_{N-1} is computed at instant α instead of $(-\alpha)$. This represents a change of 2α cc's delay. Logically, the associated SRL delay length should be *decremented* by this value. Similarly P_{N-2} is computed at instant $\alpha-1$ instead of $(-\alpha+1)$. This represents a change of $2(\alpha-1)$ cc's delay. The associated SRL delay length should be then *decremented* by this value. Following the same reasoning, we can conclude easily that if α samples are introduced at the left side of the input signal, the P_{hub-i} ($i \in [\alpha,\alpha-1,\ldots,0]$) delays length value for the first filter output should be *updated* by the L_1 list values:

$$L_1= [-2\alpha,-2(\alpha-1), \ldots\ldots,-2,0]$$

The same reasoning can be applied on the second boundary filter output. The latter corresponds to P_{N-1} of instant $(\alpha-1)$. The reader can verify easily that the update list L_2 for the P_{hub-i} SRLs delay length is:

$$L_2=[-2(\alpha-1),-2(\alpha-2), \ldots.,-2,0]$$

These updates lists L_1, L_2,...should be added to the regular skew list RegSkew (equation 2) to deduce the final SRLs delay length.

By applying the same reasoning on the remaining α boundary outputs, the SRLs delay length at the boundary region can be represented with a *Left-Matrix* matrix expression such that:

Left-Matrix=[α_Matrix|Tail(α,N-α-1)], where:

o   **Tail(α,N-α)** is a matrix of size [α,N-α-1], where each of its rows is equal to:
[λ-α-1, λ-α-2, λ-α-3,..... ,ξ+2, ξ+1]

o α_**Matrix** is a matrix of size [α,α+1], where:

$$\alpha\_Matrix=\begin{bmatrix} \lambda-2\alpha & \lambda-2\alpha+1 & \lambda-2\alpha+2 & . & \ldots & \lambda-\alpha-2 & \lambda-\alpha-1 & \underline{\lambda-\alpha} \\ \lambda-2\alpha+2 & \lambda-2\alpha+3 & \lambda-2\alpha+4 & . & \ldots & \lambda-\alpha+2 & \underline{\lambda-\alpha+1} & \lambda-\alpha \\ \lambda-2\alpha+4 & \lambda-2\alpha+5 & \lambda-2\alpha+6 & . & \ldots & \underline{\lambda-\alpha+2} & \lambda-\alpha+1 & \lambda-\alpha \\ . & . & . & . & \ldots & . & . & . \\ \lambda-6 & \lambda-5 & \lambda-4 & \underline{\lambda-3} & \ldots & \lambda-\alpha+2 & \lambda-\alpha+1 & \lambda-\alpha \\ \lambda-4 & \lambda-3 & \underline{\lambda-2} & \lambda-3 & \ldots & \lambda-\alpha+2 & \lambda-\alpha+1 & \lambda-\alpha \\ \lambda-2 & \underline{\lambda-1} & \lambda-2 & \lambda-3 & \ldots & \lambda-\alpha+2 & \lambda-\alpha+1 & \lambda-\alpha \end{bmatrix}$$

The underlined elements in this matrix expression refer to the delay length of the SRL_{Hub}. All the matrix elements put on bold refer to the pre-hub SRLs delay length. They form an upper triangular matrix.

Right Side Signal Boundary Extension

We can see from Fig 9.b that because of the irregularity of the DDG, P_0 is computed at instant $(-\beta)$ instead of β. This represents a change of 2β cc's delay. Logically the associated SRL delay length should be increased by this value. Similarly P_1 is computed at instant $(-\beta+1)$ instead of $(\beta-1)$. This represents a change of $2(\beta-1)$ cc's delay. Logically the associated SRL delay length should be incremented by this value. Following the same reasoning we can conclude easily that if β samples are introduced at the right side of a signal, the SRL_{hub+i} $(0\leq i \leq\beta)$ delays length of the first output should be *updated* by the following list, R_1,

$$R_1=[0,2,4,6,....,2\beta]$$

This update list should be added to the RegSkew list expression (equation 2) to deduce the final SRLs delays length. By doing so for every β value in the [1,μ] interval, the SRLs delay length at the boundary region can be represented by a *Right-Matrix* matrix expression such that:

Right-Matrix=[Head(μ,N-μ-1)|μ_Matrix] where:

Head(μ,N-μ-1) is a matrix of size [μ,N-μ-1], such that each of its row is equal to:

$[\lambda, \lambda-1, \ldots, \xi+\mu+2]$

μ_Matrix a matrix of size $[\mu, \mu+1]$, where:

$$\mu\text{-Matrix} = \begin{bmatrix}
\xi+\mu+1 & \xi+\mu & \xi+\mu-1 & \xi+\mu-2 & \cdots & \xi+3 & \underline{\zeta+2} & \xi+3 \\
\xi+\mu+1 & \xi+\mu & \xi+\mu-1 & \xi+\mu-2 & \cdots & \underline{\zeta+3} & \xi+4 & \xi+5 \\
\xi+\mu+1 & \xi+\mu & \xi+\mu-1 & \xi+\mu-2 & \cdots & \xi+5 & \xi+6 & \xi+7 \\
\cdot & \cdot & \cdot & \cdot & \cdots & \cdot & \cdot & \cdot \\
\cdot & \cdot & \cdot & \cdot & \cdots & \cdot & \cdot & \cdot \\
\xi+\mu+1 & \xi+\mu & \underline{\zeta+\mu-1} & \xi+\mu & \cdots & \xi+2\mu-5 & \xi+2\mu-4 & \xi+2\mu-3 \\
\xi+\mu+1 & \underline{\zeta+\mu} & \xi+\mu+1 & \xi+\mu+2 & \cdots & \xi+2\mu-3 & \xi+2\mu-2 & \xi+2\mu-1 \\
\underline{\zeta+\mu+1} & \xi+\mu+2 & \xi+\mu+3 & \xi+\mu+4 & \cdots & \xi+2\mu-1 & \xi+2\mu & \xi+2\mu+1
\end{bmatrix}$$

The underlined elements in this matrix expression refer to the delay length of the SRL_{Hub}. All the matrix elements put in bold refer to the post-hub SRLs delay length. They form a lower triangular matrix. By using the *Left_Matrix*, *Right_Matrix* matrices and the *RegSkew* list expressions, all the required delays length for the SRL layer of Fig 6 structure are determined.

4 Area Measurements

In this section, we will compare the area cost of our structure and Fig 4's architecture. Table 2 lists the resources used by those two structures where N is the filter length and W is the input wordlength.

Table 2. Logic resources consumed by different FIR structures

	Number of Multipliers	Number of Word Delays	Number of Adders	Router Hardware (LUTs)
Fig 4 Architecture	N	N-1	AdderTree(N)	$W(\sum_{i=2}^{\alpha+1}\lceil i/2\rceil + \sum_{j=2}^{N-\alpha}\lceil j/2\rceil)$
BenKrid Architecture	N	0	AdderTree(N)	SRLs Layer

From table 2, we can see clearly that our architecture consumes as many multipliers as Fig-4 structure. Our structure does not infer input samples delays unlike with Fig-4 structure saving thus a (N-1) parallel word delays resource. However, it consumes as much hardware as Fig 4 structure for the accumulation task. The last remaining resource in table 2 is related to the router functionality. As explained in section 1, the hard router consumes $W(\sum_{i=2}^{\alpha+1}\lceil i/2\rceil + \sum_{j=2}^{N-\alpha}\lceil j/2\rceil)$ LUTs. With our structure, the routing functionality is implemented through the SRL layer. For each SRL, we need to determine its maximum delay length (DL) value. This can be deduced easily from the expression of the Left_Matrix, Right_Matrix and RegSkew expression. If this value is less or equal to 16, one SRL16 (one LUT) can be used. However, if more depth is needed, more SRL16 should be chained, thus increasing the required number of LUTs. The number of the required LUTs is simply equal to $\lceil \frac{DL}{16} \rceil$. If we omit the area cost of the SRLs' address generators (which indeed can be considered negligible in front of the final filter area), our SRL layer cost area will depend solely on the number of SRLs used.

For different α, μ and N values, Fig 11 depicts the router functionality area cost evolution. It shows clearly that the hard router consumes much more area than our suggested structure.

Therefore, we can conclude that our novel architecture consumes fewer resources than the conventional structure of Fig 4. The next section presents the real hardware implementation results for the standard low filter of the Daubechies-8 wavelet (8 taps) [3] on the Xilinx Virtex family FPGAs. These will be compared to an implementation of Fig 4 architecture (i.e. a conventional FIR architecture with a hard router).

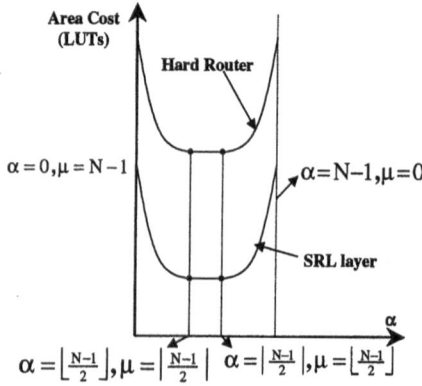

Fig. 11. Router area cost evolution when implemented using an SRL layer and a multiplexer layer (Hard Router),W=1

5   Case Study

In the following, we assume a bit parallel arithmetic. The FIR filters were implemented using 9-bit input word length, 8-bit coded coefficients and a 2-bit intermediate and final precision results. Our adders and multipliers were designed using the dedicated carry logic of the Virtex Xilinx CLBs. Since the filter coefficients are constant, the Canonic Signed Digit (CSD) representation based approach was used to design the multipliers [8]. Timing constraints were applied to determine the maximum achievable frequency.

Table 3 shows the performances achieved from implementing the Daubechies-8 FIR filter with no boundary processing. The structures of Fig 1 are used as well as ours (Fig 6). We can see that when using Fig 1.a structure, the implementation delivers higher speed but requiring more area comparing to the inverse form structure of Fig1.b. Those two effects are due to the input chained delay of Fig 1.a structure, which will increase its area (since Fig 1.b structure does not include such resources) and will avoid using long routing line to feed the multipliers as in Fig 1.b structure. Our structure with no boundary processing delivers the same speed as for the conventional inverse FIR structure. However, it consumes more area because of the SRL layer.

Table 3. Performance of a low Daubechies-8 FIR filter implementation using two different architectures on Xilinx XCVE50-8 FPGA with no boundary processing

	Fig1-a architecture	Fig 1-b architecture	Benkrid architecture
Area(Slices)	147	113	146
Speed(Mhz)	~167	~159	~159

When handling the signal boundaries processing, the architecture of Fig 4 is implemented and compared to our structure. The SRLs Layer in our structure get addressed dynamically by the values of the *Left_Matrix*, *Right_Matrix* and *RegSkew* expression values. To implement the SRLs' address generators, we can conclude from section 3 that the SRL addressing can be subdivided into two categories:

Boundary regions: that will corresponds to N-1 states (corresponding to *Left_Matrix*, *RightMatrix* rows). These states can be stored in the slices distributed RAMs.

Non-boundary regions: the address will be constant. It corresponds to a *RegSkew* element value.

A counter line and multiplexer are needed to flag the boundary transition instant.

Table 4 lists the performances achieved. The second column of table 3 and 4 shows the effect of the Hard Router on Fig 1.a structure. It involves the use of 78 extra slices with ~6 Mhz speed penalty. On the other hand, by comparing the performance of our structure with and without boundary processing, we can see clearly that the dynamic skewing of the SRL layer does introduce a slight area penalty (12 slices) with no speed penalty.

Table 4 shows clearly that our architecture is more compact in area (almost 70 slices less) and run at the same speed range. It is worth noting that the Hard Router has been implemented at word level therefore seeking its highest speed implementation.

Table 4. Performance of a low Daubechies-8 FIR filter implementation using two different architectures on Xilinx XCVE50-8 FPGA with boundary processing

	Fig 4 architecture	Benkrid architecture
Area(Slices)	225	158
Speed(Mhz)	~161	~159

6 Conclusion

In this paper, we have presented a novel architecture for FIR filters with signal boundary handling. This architecture is tailored to Xilinx Virtex FPGAs family. It cleverly exploits the SRL16 component to implement the FIR filters. The problem of signal boundary processing, which occurs in finite length signals filtering, is smartly and efficiently handled by an appropriate skewing of input data, rather than using the conventional brute force hard-router. The architecture is fully scalable and parameterisable. Its real hardware implementation on FPGAs leads to a very compact configuration compared to a hard-router based implementation, with no speed penalty. Moreover, unlike the conventional architecture, our architecture allows the use of multiplier blocks or sub-expression sharing, leading therefore to more compact implementation.

References

[1] Proakis. J.G., Manolakis .D.G, 'Introduction to Digital Signal Processing', McMillan Publishing, USA, 1989.
[2] Peter Pirsch, 'Architectures for Digital Signal Processing', John Wiley & Sons, 1999.
[3] Vetterli M, Kovacevic M, 'Wavelets and Subband Coding' Prentice Hall, New Jersey, USA, 1995
[4] Smith M.J.T, Eddins S 'Analysis/Synthesis techniques for subband image coding', IEEE Trans On Acoustics, Speech and Signal Processing, 1446-1456, August 1990
[5] Chakrabarti C, 'A DWT based encoder architecture for symmetrically extended images', Proceedings of the International Symposium on Circuits and Systems, 1999.
[6] Benkrid A, Benkrid K, Crookes D, 'Design and Implementation of a Novel Architecture for Symmetric FIR filters with Boundary Handling on Xilinx Virtex FPGAs', IEEE Conference on Field-Programmable Technology, FPT'2002, December 16
[7] http://www.xilinx.com/partinfo/ds022.pdf
[8] Keshab K. Parhi, 'VLSI Digital Signal Processing Systems: Design and Implementation', John Wiley & Sons, 1999

A Self-reconfiguring Platform

Brandon Blodget[1], Philip James-Roxby[2], Eric Keller[2],
Scott McMillan[1], and Prasanna Sundararajan[1]

[1] Xilinx, 2100 Logic Drive, San Jose, CA, 95124, USA,
{brandonb,mcmillan,prasanna}@xilinx.com
[2] Xilinx, 3100 Logic Drive, Longmont, CO, 80503, USA,
{jamespb,keller}@xilinx.com

Abstract. A self-reconfiguring platform is reported that enables an FPGA to dynamically reconfigure itself under the control of an embedded microprocessor. This platform has been implemented on Xilinx Virtex II^{tm} and Virtex II Pro^{tm} devices. The platform's hardware architecture has been designed to be lightweight. Two APIs (Application Program Interface) are described which abstract the low level configuration interface. The Xilinx Partial Reconfiguration Toolkit (XPART), the higher level of the two APIs, provides methods for reading and modifying select FPGA resources. It also provides support for relocatable partial bitstreams. The presented self-reconfiguring platform enables embedded applications to take advantage of dynamic partial reconfiguration without requiring external circuitry.

1 Introduction

This paper presents a self-reconfiguring platform (*SRP*) for Xilinx Virtex II^{tm} and Virtex II Pro^{tm} devices[1]. It begins by reviewing the motivation for developing the *SRP*, before presenting the first detailed description of the two core software components, the ICAP API and the Xilinx Partial Reconfiguration Toolkit(XPART). Recent improvements and revisions to the *SRP* hardware architecture are also described in more detail.

Dynamic reconfiguration and self-reconfiguration are two of the more advanced forms of FPGA reconfigurability. Dynamic reconfiguration implies that an active array may be partially reconfigured, while ensuring the correct operation of those active circuits that are not being changed. Self-reconfiguration extends the concept of dynamic reconfigurability. It assumes that specific circuits on the logic array are used to control the reconfiguration of other parts of the FPGA. Clearly the integrity of the control circuits must be guaranteed during reconfiguration, so by definition self-control is a specialized form of dynamic reconfiguration.

Both dynamic reconfiguration and self-reconfiguration rely on an external reconfiguration control interface to boot an FPGA when power is first applied or the device is reset. Once initially configured, self-control requires an internal reconfiguration interface that can be driven by the logic configured on the logic array. On Xilinx Virtex II and Virtex II Pro parts, this interface is called the internal configuration access port (ICAP)[2].

The hardware component of *SRP* is composed of the ICAP, control logic, a small configuration cache, and an embedded processor. The embedded processor can be Xilinx's MicroBlaze^{tm}, which is a 32-bit RISC soft microprocessor core[3]. The hardcore

P.Y.K. Cheung et al. (Eds.): FPL 2003, LNCS 2778, pp. 565–574, 2003.

PowerPC on the Virtex II Pro can also be used as the embedded processor. The embedded processor provides intelligent control of device reconfiguration at runtime. The integration of this functionality is especially attractive for embedded systems. This lightweight approach maximizes flexibility while minimizing additional external circuitry.

The software component of *SRP* defines two APIs. The lower level one is the ICAP API. The ICAP API provides access to the configuration cache and controls reading and writing the cache to the device. The higher level API is Xilinx Partial Reconfiguration Toolkit(XPART). XPART is derived from the JBits API work[4]. Like the JBits API it abstracts the bitstream details providing seemingly random access to select FPGA resources. XPART also contains methods for relocating partial bitstreams.

SRP opens up a number of interesting possibilities. Firstly it gives more reconfiguration options to the designer. Adding *SRP* to a system allows an application to get reconfiguration data from any peripheral and partially reconfigure the device. For example if the system has a network connection partial bitstreams could be pulled off the network. Secondly the embedded processor could manipulate the data before reconfiguring the device. This could permit custom encryption or compression of bitstreams. Thirdly the XPART API enables fine grain reconfiguration control over select FPGA resources. This allows tuning of MGTs (multi-gigabit transceiver) , or constant folding achieved by modifying LUTs. Finally it is envisioned that a subsystem like *SRP* could play an important role in an embedded operating system running on a platform FPGA. *SRP* could help the OS manage hardware tasks. It could provide the capability to swap tasks in and out of hardware. It would also allow these tasks to be relocated to different regions on the device[5].

The paper is arranged as follows: Section 2 reviews previous work relating to *SRP*. Section 3 looks at the details of the ICAP and the reconfiguration mechanisms of the Virtex line of FPGAs. This provides the background necessary for an appreciation of the *SRP* hardware and software infrastructure that are described in sections 4 and 5. Sections 5.1 and 5.2 describe the APIs in the software layers. Section 6 presents future work and concludes the paper.

2   Related Work

Dynamic reconfiguration implies that an active array may be partially reconfigured, while ensuring the correct operation of those active circuits that are not being changed. Much research has been done on dynamic reconfiguration which shows it can be used to reduce circuit complexity, increase performance and simplify system design[6].

Self-reconfiguration is a special case of dynamic reconfiguration where the configuration control is hosted within the logic array that is being dynamically reconfigured. The part of the array containing the configuration control remains unmodified throughout execution. A formal definition of self-reconfiguration is presented in [7]. There are several desirable features of such an arrangement. Firstly the control logic is as close to the logic array as possible, thus minimizing the latencies associated with accessing the configuration port. Secondly, fewer discrete devices are required, reducing the overall system complexity[8].

For a device to support self-reconfiguration it must be dynamically reconfigurable. A second desirable, but not required, characteristic is the device provides internal access to the configuration port. This way, the configuration stream does not have to exit the chip and use board level resources to gain access to the configuration port. The Xilinx XC6200 family of devices first provided both of these features, newer devices from Xilinx such as the Virtex II and Virtex II Pro families are both dynamically reconfigurable and provide access to the configuration port to internal logic.

The first references to self-reconfiguration using FPGAs in the literature was in [9] where a small amount of static logic is added to a reconfigurable device in order to produce a self-reconfiguring processor. The Flexible URISC[10] defined an abstract model of virtual circuitry that had self-configuring capability. A pattern-matching algorithm was used to investigate the viability of systems that exhibit self-control of reconfiguration management[8]. A number of applications mapped using self-reconfiguration have been shown to improve performance over existing approaches[9][11].

Xilinx application note 662 describes a method for using self-reconfiguration on a Virtex II Pro device to update MGT (multi-gigabit transceiver) parameters[12]. These attributes must be modified to optimize the MGT signal transmission prior to and after a system has been deployed in the field. The work presented in this paper extends the framework described in Xilinx application note 662 to enable a general purpose self-reconfiguring platform.

The XPART software layer is derived from Xilinx's JBits API work. XPART provides a lightweight, minimal set of JBits API features implemented in the C language. XPART also provides some basic functionality for supporting relocatable modules. A relocatable module is a partial bitstream that can be relocated to multiple places on the FPGA. This functionality has also been called Dynamic Hardware Plugins(DHP) and was used for implementing multiple networking applications in hardware for high-performance programmable routers[13]. The methods that XPART provide are very similar to PARBIT (PARtial BItfile Transformer)[14]. The tool JBitsDiff also provides the ability to create relocatable modules directly from bitstreams[15]. The added benefit XPART provides is these functions can run on an embedded processor.

3   Virtex I/II/II Pro Configuration Architecture

SRAM based FPGAs are configured by loading application specific data into configuration memory. All Virtex^{tm} series devices (Virtex I, Virtex II and Virtex II Pro) have their configuration memory segmented into *frames*. These devices are partially reconfigurable and a frame is the smallest unit of reconfiguration. There are multiple frames per CLB column. For example Virtex devices have 48 frames per CLB column. Frames are one bit wide and differ in length depending on the number of CLB rows in the device. For example an XC2V40 has 832 bits per frame and an XC2V1000 has 3392 bits per frame.

Virtex II and Virtex II Pro devices have an internal reconfiguration access port (ICAP) which can be controlled by internal FPGA logic. The ICAP interface is a subset of the SelectMAP^{tm} interface. Figure 1 shows a comparison between the two interfaces. The ICAP interface has fewer signals than the SelectMAP interface because it does not have to do full configurations and it does not have to support different configuration modes.

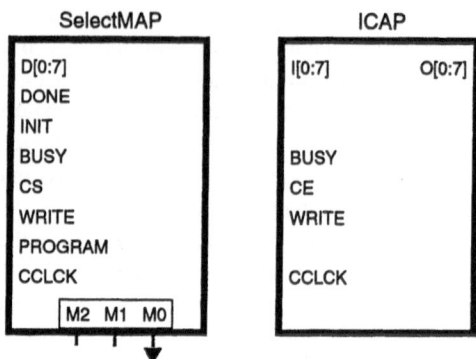

Fig. 1. SelectMap vs ICAP

It also differs in that the SelectMAP bi-directional *D* port is split into an *I* data port and an *O* data port. One other item to note is the functionality provided by the ICAP *CE* pin is equivalent to the SelectMAP *CS* pin.

The eight bit data *I* port on the ICAP allows faster reconfiguration than serial modes of reconfiguration. The maximum frequency that the SelectMAP and ICAP interfaces can be clocked at without checking the BUSY signal is 66MHz[16].

Virtex I/II/II Pro FPGAs require a pad frame be added to the end of the configuration data. This pad frame flushes out the reconfiguration pipeline. Therefore to write one frame to the device it is necessary to clock in two frames, the data frame plus a pad frame. Thus the minimal time to reconfigure over ICAP a single XC2V40 frame at 66MHz is 3.2us. It would take 13us to reconfigure a single XC2V1000 frame at 66MHz.

4   The SRP Hardware Architecture

SRP's hardware component is composed of the ICAP, some control logic, a small con-figuration cache, and an embedded processor. These peripherals communicate over the CoreConnect^{tm} Open Peripheral Bus (OPB)[17]. Figure 2A shows a block diagram of the current hardware subsystem implementation.

In this implementation a 32 bit register is used to interface to the ICAP port. Figure 2A shows how this register is mapped to the ICAP signals. The control logic for reading and writing data to the ICAP is implemented in a low level software driver. This driver defines methods for reading and writing to the ICAP interface register. There are also methods for reading and writing blocks of data to the ICAP. This methodology is similar to the XVPI register JBits SDK used to access the SelectMAP interface[18].

The BlockRAM (BRAM) shown is Figure 2A is used to cache configuration data. To keep the *SRP* hardware as lightweight as possible only one BRAM is used. One Virtex II BRAM can store 16K bits of data. Since the largest Virtex II Pro device, the XC2VP125, has a frame length of 11K bits, one BRAM can easily store a whole frame of even this largest Virtex II Pro device.

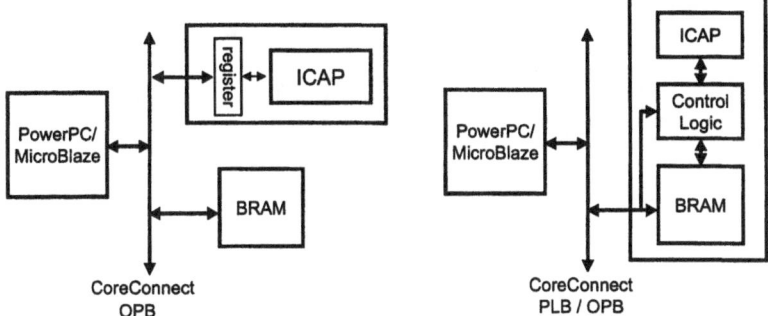

A. Current SRP Hardware Architecture   B. Planned SRP Hardware Architecture

Fig. 2. Curent and Planned SRP Hardware Architecture

This hardware system has been implemented on both Virtex II and Virtex II Pro devices. On the Virtex II device the MicroBlaze soft processor was used. On Virtex II Pro the embedded PowerPC was targeted. The Xilinx Embedded Developer Kit (EDK) and Xilinx 5.1i ISE tools were used to build these hardware platforms.

The next generation of the hardware system will provide much better performance. Figure 2B shows this system. This will be achieved via a few changes. Firstly the ICAP control logic will move from being in software (hardware drivers) to being implemented directly in hardware. This move also means there will be less communication over the system bus. Secondly, the configuration cache BRAM will be moved inside the ICAP control peripheral. This will allow the dual ported nature of the BRAM to be exploited. One port of the BRAM will be exposed to the system bus. The other port will be accessed by the ICAP control logic. This way the ICAP control logic will not take up system bus cycles to transfer data to and from the configuration cache. The embedded processor will still be able to access the configuration cache BRAM over the system bus. Thirdly, the ICAP control peripheral will be a master peripheral. This way the peripheral will be able to fetch configuration data directly from external memory without requiring the processor be involved. Lastly, we plan to implement the peripheral as a Xilinx IPIF (IP Interface) peripheral. IPIF peripherals can interface to both the OPB and the faster PLB (Processor Local Bus).

5   The SRP Software Architecture

The software components of *SRP* are designed in layers. Figure 3 shows the different software layers. The software layers are divided into hardware dependent and hardware independent parts. This division is enabled by the ICAP API. This API defines methods for transferring data between the configuration cache implemented in BRAM and the active configuration memory. It also provides methods for accessing the configuration cache.

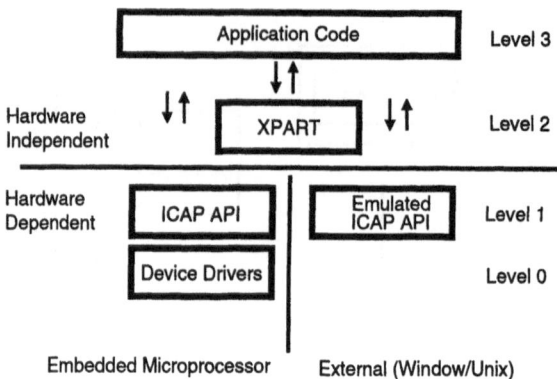

Fig. 3. SRP Software Layers

We have created two implementations of the ICAP API. The first uses the SRP hardware described in the previous section to enable embedded applications to readback or modify the active configuration of the FPGA. The second implementation emulates the active configuration store and the configuration cache entirely in software. This emulated version of the ICAP API can then be compiled for a Windows or Unix workstation.

The Xilinx Partial Reconfiguration Toolkit (XPART) is built on top of the ICAP API. Thus XPART is hardware independent and can be compiled for an embedded system, or for a Windows or Unix system. This portability also applies to applications that use XPART and/or the ICAP API.

One advantage of this portability is one can do a degree of debug on a standard workstation before moving to the target embedded platform. It is also possible to use XPART on a workstation, manipulate partial bitstreams, then write them to the computer's hard drive instead of the physical FPGA device.

5.1 The ICAP API

The ICAP API defines methods for accessing configuration logic through the ICAP port. The main methods move data between the configuration cache (BRAM) and the active configuration memory (the device). Other methods allow the processor to read and write to the configuration cache. Finally the setDevice() method indicates which device is being targeted. This method allows the designer to retarget their application to a different device by changing one line in the code. All Virtex II and Virtex II Pro family members are supported. Table 1 gives an outline of the methods in the ICAP API.

5.2 XPART – Xilinx Partial Reconfiguration Toolkit

The Xilinx Partial Reconfiguration Toolkit (XPART) has been built on top of the ICAP API. The purpose of this toolkit is to allow embedded processors to modify resources and relocate modules. Currently this toolkit has four methods. Table 2 gives an overview of these methods.

Table 1. The ICAP API

Routines	Description
setDevice()	Indicates the target device.
storageBufferWrite()	Writes data to the configuration storage buffer
storageBufferRead()	Reads data from the configuration storage buffer
deviceWrite()	Transfers specified number of bytes from storage buffer to ICAP_IN
deviceRead()	Transfers specified number of bytes from ICAP_OUT to storage buffer
deviceAbort()	Aborts the current operation
deviceReadFrame()	Reads one or more frames from the device into the storage buffer
deviceWriteFrame()	Writes one or more frames to the device from the storage buffer
setConfiguration()	Loads a configuration from memory
getConfiguration()	Writes current configuration to memory

Table 2. XPART Methods

Routines	Description
getCLBBits()	Reads back the state of a selected CLB resource.
setCLBBits()	Reconfigures the state of a selected CLB resource.
setCLBModule()	Place the module at a particular location on the device
copyCLBModule()	Given a bounding box copy it to a new location on the device

XPART provides functions for on the fly resource modification. This is done through the getCLBBits() and setCLBBits() methods. These methods abstract the bitstream and provides seemingly random access to selected resources. Currently XPART has the capability to read or modify all Virtex II CLB logic and routing resource. It currently does not have access to IOB, DCM, MGT, BRAM or other non CLB resources.

The strategy employed is to do a read/modify/write of configuration data. The following lists the steps that setCLBBits() would perform to update a resource like a LUT:

1. Calculate the target frame
2. Find LUT bits in target frame
.3. read target frame from device and put in storage buffer using deviceReadFrame()
4. Modify the LUT bits in the storage buffer using writeStorageBuffer()
5. Reconfigure the device with new LUT bits using deviceWriteFrame()

The memory elements that define the content of a LUT in Virtex II are all located in one frame. This is not true with all resources. Some resources have their configuration bits scattered across multiple frames. The setCLBBits() and getCLBBits() methods are optimized so they will not have to read or write the same frame twice during the same call. Figure 4 shows sample code that updates the content of a LUT.

XPART provides some basic functionality for supporting relocatable modules. A relocatable module is a partial bitstream that can be relocated to multiple places on the FPGA. XPART provides two methods for dealing with relocatable modules. The two methods are setCLBModule() and copyCLBModule(). The setCLBModule() method works on regular partial bitstreams that contains information about all of the rows in the

```
#include <XPART.h>
#include <LUT.h> /* Bitstream resource library for LUTs */

int main(int argc, char *args[]) {
    char* value;
    int error, i, row, col, slice;
    setDevice(XC2VP7); // Set the device type

    ... [initialize row,col,slice and value to desired values] ...

    error = setCLBBits(row, col, LUT.RES[slice][LE_F_LUT], value, 16);
    return error;
} /* end main() */
```

Fig. 4. Code to update LUT contents

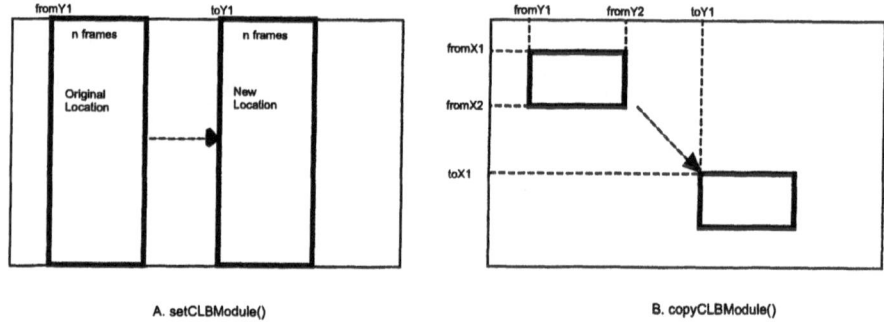

Fig. 5. Illustration of Module Methods

included frames. It works by modifying the header of the partial reconfiguration packet. Figure 5A illustrates the setCLBModule() command.

The copyCLBModule() function copies any sized rectangular region of configuration memory and writes it to another location. The copied region contains just a subset of the rows in a frame. This allows the designer to define dynamic regions that have static regions above or below it. Figure 5B shows an illustration of copyCLBModule().

The copyModule() function employees a read/modify/write strategy like the resource modification functions. This technique enables changing select bits in a frame and leaving the others bits to their current configured state. The Virtex II provides glitchless configuration logic, meaning if a bit stays the same between two configurations no glitch will occur. For example a frame may be modified that contains a routing resource. If the value of the bits controlling that routing resource remain the same between the old and new configurations, no glitch will occur on that routing resource.

6   Future Work and Conclusions

Several extensions and applications of the system are in progress. As noted earlier we are working on a more efficient implementation of our *SRP* hardware architecture. In the current implementation the ICAP control logic is implemented in software via device drivers. This implementation requires 10ms to modify a LUT value on an XC2V1000 device with MicroBlaze running at 50MHz. The next version of the *SRP* hardware should allow us to clock configuration data into and out of the device at the maximum no handshake frequency of 66MHz. Modifying one LUT on the XC2V1000 will take 13us for the read, negligible time for the modify, and 13us for the write for a total of approximately 26us. This is over two orders of magnitude faster than the existing *SRP* hardware architecture.

We are currently using *SRP* to develop a reconfiguration controller for a high I/O reconfigurable crossbar switch[19]. The original reconfiguration controller was implemented using external logic and memory. The data for the reconfiguration controller was generated offline using JBits SDK. *SRP* should allow us to implement the reconfigurable crossbar and controller on the same chip.

In conclusion, we have described a self-reconfigurable platform. *SRP* is an intelligent subsystem for lightweight reconfiguration of Xilinx Virtex II and Virtex II Pro FPGAs in embedded systems. The system enables self-reconfiguration under software control within a single FPGA. *SRP* has a layered hardware and software architecture that permits a variety of different interfaces to maximize flexibility and ease-of-use.

References

1. Blodget, B., McMillan, S., Lysaght, P.: A lightweight approach for embedded reconfiguration of fpgas. In: Design, Automation and Test in Europe (DATE03), IEEE (2003) 399–400
2. Xilinx, Inc.: Xilinx 5.1i Libraries Guide. (2002)
3. : Xilinx web site. http://www.xilinx.com/ipcenter/processorcentral/microblaze (2003)
4. Sundararajan, P., Guccione, S.A., Levi, D.: JBits: Java based interface for reconfigurable computing. In: 2nd Annual Military and Aerospace Applications of Programmable Devices and Technologies (MAPLD99), Laurel, MD (1999)
5. Nollet, V., Coene, P., Verkest, D., Vernalde, S., Lauwereins, R.: Designing an operating system for a heterogeneous reconfigurable soc. In: RAW'03 workshop. (2003) (Accepted)
6. Compton, K., Hauck, S.: Reconfigurable computing: A survey of systems and software. In: ACM Computing Surveys, Vol 34, No. 2. (2002) 171–210
7. Sidhu, R., Prasanna, V.K.: Efficient metacomputation using self-reconfiguration. In: Field Programmable Logic and Applications (FPL02), Springer (2002) 698–709
8. McGregor, G., Lysaght, P.: Self controlling dynamic reconfiguration: A case study. In: Field Programmable Logic and Applications (FPL99), Springer (1999) 144–154
9. French, P.C., Taylor, R.W.: A self-reconfiguring processor. In: IEEE Symposium on File-Programmable Custom Computing Machines (FCCM93), Napa Valley, California (1993) 50–59
10. Donlin, A.: Self modifying circuitry - a platform for tractable virtual circuitry. In: Field Programmable Logic and Applications (FPL98), Springer (1998) 199–208
11. Sidhu, R.P.S., Mei, A., Prasanna, V.K.: Genetic programming using self-reconfigurable fpgas. In: Field Programmable Logic and Applications (FPL99), Springer (1999)

12. Eck, V., Kalra, P., LeBlanc, R., McManus, J.: In-circuit partial reconfiguration of RocketIO attributes. Xilinx Application Note XAPP662, version 1.0, Xilinx, Inc. (2003)
13. David E. Taylor, Jonathan S. Turner, J.W.L.: Dynamic hardware plugins (DHP): exploiting reconfigurable hardware for high-performance programmable routers. In: Open architectures and Network Programming Proceedings, IEEE (2001) 25–34
14. Horta, E., Lockwood, J.W.: PARBIT: a tool to transform bitfiles to implement partial reconfiguration of field programmable gate arrays (FPGAs). Technical Report WUCS-01-13, Washington University in Saint Louis, Department of Computer Science (July 6, 2001)
15. James-Roxby, P., Guccione, S.A.: Automated extraction of run-time parameterisable cores from programmable device configurations. In: IEEE Symposium on File-Programmable Custom Computing Machines (FCCM00), Napa Valley, California, IEEE Computer Society (2000) 153–161
16. Xilinx, Inc.: Virtex-II Platform FPGA User Guide. (2002)
17. : IBM web site. http://www.chips.ibm.com/products/coreconnect (2003)
18. Sundararajan, P., Guccione, S.A.: XVPI: A portable hardware/software interface for virtex. In: Reconfigurable Technology: FPGAs for Computing and Applications II, Proc. SPIE 4212, SPIE – The International Society for Optical Engineering (2000) 90–95
19. Young, S., Alfke, P., Fewer, C., McMillan, S., Blodget, B., Levi, D.: A high i/o reconfigurable crossbar switch. In: IEEE Symposium on File-Programmable Custom Computing Machines (FCCM03), Napa Valley, California, IEEE Computer Society (2003)
20. Carmichael, C.: Virtex FPGA series configuration and readback. Xilinx Application Note XAPP138, version 1.1, Xilinx, Inc. (1999)
21. Lim, D., Peattie, M.: Two flows for partial reconfiguration: Module based or small bit manipulations. Xilinx Application Note XAPP290, version 1.0, Xilinx, Inc. (2002)
22. McMillan, S., Guccione, S.A.: Partial run-time reconfiguration using JRTR. In: Field Programmable Logic and Applications (FPL00), Springer (2000) 352–360
23. Sidhu, R., Prasanna, V.K.: Fast regular expression matching using fpgas. In: IEEE Symposium on File-Programmable Custom Computing Machines (FCCM01), Rohnert Park, California (2001)
24. Horta, E.L., Lockwood, J.W., Taylor, D.E., Parlour, D.: Dynamic hardware plugins in an FPGA with partial run-time reconfiguration. In: Design Automation Conference (DAC), New Orleans, LA (2002)

Heuristics for Online Scheduling Real-Time Tasks to Partially Reconfigurable Devices*

Christoph Steiger, Herbert Walder, and Marco Platzner

Swiss Federal Institute of Technology (ETH) Zurich, Switzerland,
platzner@tik.ee.ethz.ch

Abstract. Partially reconfigurable devices allow to configure and execute tasks in a true multitasking manner. The main characteristics of mapping tasks to such devices is the strong nexus between scheduling and placement. In this paper, we formulate a new online real-time scheduling problem and present two heuristics, the *horizon* and the *stuffing* technique, to tackle it. Simulation experiments evaluate the performance and the runtime efficiency of the schedulers. Finally, we discuss our prototyping work toward an integration of scheduling and placement into an operating system for reconfigurable devices.

1 Introduction

Todays reconfigurable devices provide millions of gates capacity and partial reconfiguration. This allows for true multitasking, i.e., configuring and executing tasks without affecting other, currently running tasks. Multitasking of dynamic task sets can lead to complex allocation situations which clearly asks for well-defined system services that help to efficiently operate the system. Such a set of system services forms a *reconfigurable operating system* [1] [2]. This paper deals with one aspect of such an operating system (OS), the problem of online scheduling hard real-time tasks.

Formally, a task T_i is modeled as rectangular area of reconfigurable logic blocks given by its width and height, $w_i \times h_i$. Tasks arrive at arbitrary times a_i, require execution times e_i, and carry deadlines d_i, $d_i \geq a_i + e_i$. The reconfigurable device is also modeled as rectangular area $W \times H$ of logic blocks. We focus on non-preemptive systems – once a task is loaded onto the device it runs to completion.

The complexity for mapping tasks to such devices depends heavily on the used *area model*. We use two different area models, a 1D and a 2D model as shown in Figure 1. In the simpler 1D area model, tasks can be allocated along the horizontal device dimension; the vertical dimension is fixed. The 1D area model suffers badly from two types of fragmentation. The first type is the area wasted when a task does not utilize the full device height. The second type occurs when the remaining free area is split into several unconnected vertical

---

* This work was supported by the Swiss National Science Foundation (SNF) under grant number 2100-59274.99.

P.Y.K. Cheung et al. (Eds.): FPL 2003, LNCS 2778, pp. 575–584, 2003.

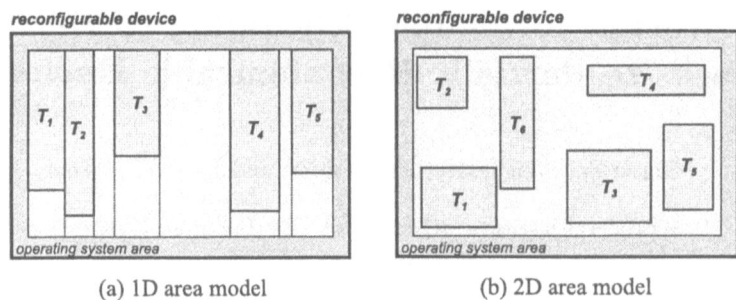

(a) 1D area model   (b) 2D area model

Fig. 1. Reconfigurable resource models

stripes. Fragmentation can prevent the placement of a further task although a sufficient amount of free area exists. The more complex 2D area model allows to allocate tasks anywhere on the 2D reconfigurable surface and suffers less from fragmentation.

The main contribution of this paper is the development of online hard real-time scheduling heuristics that work for both the 1D and the 2D area model. The limitations of the models and related work are discussed in Section 2. Section 3 states the online scheduling problem and presents the two heuristics. An experimental evaluation is done in Section 4. Section 5 shows our work toward a prototype implementation and, finally, Section 6 summarizes the paper.

2 Limitations and Related Work

Our task and device models are consistent with related work in the field [1] [3] [4] [5] [6]. However, we also have to discuss the limitations when it comes to practical realization on currently available technology. The main abstraction is that tasks are modeled as relocatable rectangles. While the latest design tools allow to constrain tasks to rectangular areas, the relocatability rises questions concerning the i) device homogeneity, ii) task communication and timing, and iii) partial reconfigurability.

We assume surface uniformity which is in contrast with modern FPGAs that contain special resources such as block memories and embedded multipliers. However, a reconfigurable OS takes many of these resources (e.g., block RAM) away from the user task area. Tasks must use predefined communication channels to access such special resources [7][2]. Further, the algorithms in this paper can easily be extended to handle additional placement constraints for tasks, e.g., to relocate tasks at different levels of granularity or even to place tasks at fixed positions. The basic problems and approaches will not change, but the resulting performance.

Arbitrarily relocated tasks that communicate with each other and with I/O devices would require online routing and delay estimation of their external signals, neither of which is sufficiently supported by current tools. The state-of-the-art in reconfigurable OS prototypes [7] [2] overcomes this problem by using a slightly different area model that partitions the reconfigurable surface into fixed-size blocks. These OSs provide predefined communication interfaces to tasks and asynchronous intertask communication. The same technique can be applied to our 1D area model. For the 2D model, communication is an unresolved issue. Related work mostly assumes that sufficient resources for communication are available [4].

The partial reconfiguration capabilities of the Xilinx Virtex family, which reconfigures a device in vertical chip-spanning columns, fits perfectly the 1D area model. While the implementation of a somewhat limited 2D area model on the same technology seems to be within reach, ensuring the integrity of non-reconfigured device areas during task reconfiguration is tricky.

In summary, given current technology the 1D area model is realistic whereas the 2D model faces unresolved issues. Most of the related work on 2D models targets the (meanwhile withdrawn) FPGA series Xilinx XC6200 that is reconfigurable on the logic block level and has a publicly available bitstream architecture. Requirements for future devices supporting the 2D model include block-based reconfiguration and a built-in communication network that is not affected by user logic reconfigurations. As we will show in this paper, the 2D model has great advantages over the 1D model in terms of scheduling performance. For these reasons, we believe that it is worthwhile to investigate and develop algorithms for both the 1D and 2D area models.

3 Scheduling Real-Time Tasks

3.1 The Online Scheduling Problem

The online scheduler tries to find a placement and a starting time for a newly arrived task such that its deadline is met. In the 1D area model a *placement* for a task T_i is given by the x coordinate of the left-most task cell, x_i, with $x_i + w_i \leq W$. The *starting time* for T_i is denoted by s_i. The main characteristics of scheduling to dynamically reconfigurable devices is that a scheduled task has to satisfy intertwined time and placement constraints:

Definition 1 (Scheduled Task). *A scheduled task T_i is a task with a placement x_i and a starting time s_i such that:*

i) $[(x_i + w_i) \leq x_j] \vee [(s_i + e_i) \leq s_j] \vee [x_i \geq (x_j + w_j)] \vee [s_i \geq (s_j + e_j)]$
   $\forall T_j \in \mathcal{T}, T_j \neq T_i$ *(scheduled tasks must not overlap in space and time,*
   $\mathcal{T}$ *denotes the set of scheduled tasks)*
ii) $s_i + e_i \leq d_i$ *(deadline must be met)*

We consider an online *hard real-time* system that runs an acceptance test for each arriving task. A task passing this test is guaranteed to meet its deadline. A task failing the test is rejected by the scheduler in order to preserve the schedulability of the currently guaranteed task set. The scheduling goal is to minimize the number of rejected tasks. Accept/reject mechanisms are typically adopted in dynamic real-time systems [8] and assume that the scheduler's environment can react properly on a task rejection, e.g., by migrating the task to a different computing resource.

Our scheduling problem shares some similarity with orthogonal placement problems, so-called strip packing problems. Strip packing tries to place a set of two dimensional boxes into a vertical strip of width W by minimizing the total height of the strip. Translated to our scheduling problem, the width of the strip corresponds to the device width and the vertical dimension corresponds to time. The offline strip packing problem is NP-hard [9] and many approximation algorithms have been developed for it. There are also some online algorithms with known competitive ratios. However, to the best of our knowledge, there is no published online algorithm with a proven competitive ratio for the problem described in this paper which differs in following characteristics: First, our optimization goal is to minimize the number of rejected tasks, based on an acceptance test that is run at task arrival. Second, time proceeds as tasks are arriving. We cannot schedule tasks beyond the current timeline, i.e., into the past. Finally, tasks must not be rotated or otherwise modified.

The simplest online method is to check whether a newly arrived task finds an immediate placement. If there is none, the task is rejected. This crude technique needs to know only about the current allocation situation but will show a low performance. We include this method as a reference point in our experimentation. Sophisticated online methods increase the acceptance ratio by *planning*, i.e., looking into the future. We may delay starting a task for its laxity (until $s_{i-latest} = d_i - e_i$) and still meet its deadline. The time interval $[a_i, s_{i-latest}]$ is the planning period for a task. In the following sections we discuss two online methods, the *horizon* technique and *stuffing* technique. These planning methods are runtime efficient as they do not reschedule previously guaranteed tasks.

3.2   The Horizon Technique

The horizon technique implements scheduling and placement by maintaining two lists, the *scheduling horizon* and the *reservation list*. The scheduling horizon is a set of intervals that fully partition the spatial resource dimension. Each horizon interval is written as $[x_1, x_2]@t_r$, where $[x_1, x_2]$ denotes the interval in x-dimension and t_r gives the last release time for the corresponding reconfigurable resources. The set of intervals is sorted according to increasing release times.

Figure 2 shows an example for a device of width $W = 10$. At time $t = 2$, two tasks (T_1, T_2) are running on the device and further four tasks (T_3, T_4, T_5, T_6) are scheduled. The resulting scheduling horizon is given by the four intervals shown in Figure 3a) and indicated as dotted lines in Figure 2.

When a new task arrives, the scheduler walks through the list of intervals and checks whether the task can be appended to the horizon. When a horizon interval $[x_1, x_2] @ t_r$ is hit that is large enough to accommodate the task, the task is scheduled to placement x_1 and starting time t_r and the planning process stops. Otherwise, the scheduler tries to merge the interval with already released and adjacent horizon intervals to form a larger interval. If this larger interval does not fit either, the next interval in the horizon is considered. Should several

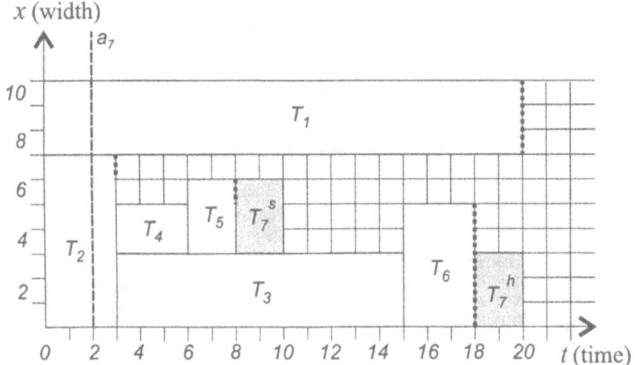

Fig. 2. 1D allocation with scheduling horizon (*dotted lines*) before accepting T_7 and placements for T_7 in the horizon (T_7^h) and stuffing methods (T_7^s)

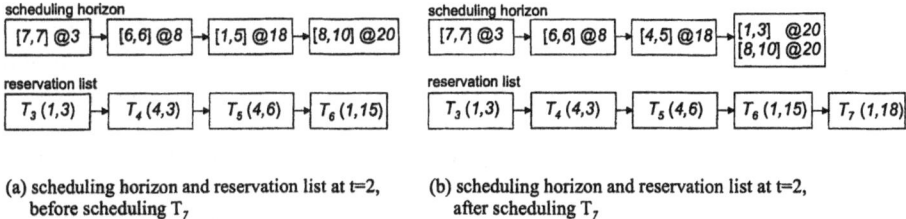

(a) scheduling horizon and reservation list at t=2, before scheduling T_7

(b) scheduling horizon and reservation list at t=2, after scheduling T_7

Fig. 3. Horizon method: scheduling horizon and reservation list

intervals fit at some point, the scheduler applies the best-fit rule to select the interval with the smallest width.

In the example of Figure 2, a new task T_7 arrives with $(a_7, w_7, e_7, d_7) = (2, 3, 2, 20)$ which gives a planning period of $[a_7, (d_7 - e_7)] = [2, 18]$. The first horizon interval to be checked is $[7, 7]$ @ 3, which is too small to fit T_7. The scheduler proceeds to $[6, 6]$@8, which is too small again. Then, these two intervals are temporarily merged to $[6, 7]$ @ 8 which is still insufficient. The next interval is $[1, 5]$ @ 18 which allows to schedule T_7 to $(x_7, s_7) = (1, 18)$.

The *reservation list* stores all scheduled but not yet executing tasks. The list entries are denoted as $T_i(x_i, s_i)$ and hold the placement and starting time. The reservation list is sorted in order of increasing starting times. The horizon technique ensures that new tasks are only inserted into the reservation list when they do not overlap in time or space with other tasks in the list. The scheduler is activated whenever a new task arrives, a running task terminates, or a scheduled task is to be started. On each event, the horizon is updated. For the example of Figure 2, the reservation list at time $t = 2$ is displayed in Figure 3a). Figure 3b) shows the updated horizon and reservation lists after scheduling and accepting T_7 at time $t = 2$.

The central point of the horizon scheduler is that tasks can only be *appended* to the horizon. Particularly, it is not possible to schedule tasks before the horizon, as the procedure maintains no knowledge about the time-varying allocation between the current time and the horizon. The advantage of this technique is that maintaining the horizon is simple compared to maintaining the complete future.

3.3   The Stuffing Technique

The stuffing technique schedules tasks into arbitrary free rectangles that will exist in the future. The implementation uses again two lists, the *free space list* and the reservation list. The free space list is a set of intervals $[x_1, x_2]$ that denote currently unused resource rectangles with height H. The free spaces are ordered according to their x-coordinates.

On arrival of T_i, we assume n tasks are executing on the device and m tasks are waiting in the reservation list. Two types of events occur during the planning period of T_i. First, $n', n' \leq n$ tasks terminate which generates new free areas. Second, $m', m' \leq m$ previously guaranteed tasks are started which reduces the free area. When a new task T_i arrives, the scheduler starts walking through the task's planning period, simulating all future allocations of the device by mimicking task terminations and placements together with the underlying free space management. On arrival of T_i and the termination of the n' placed tasks, the placer is called to find a feasible interval. If one is found, the scheduler accepts and adds a new reservation for T_i, and planning stops. If no sufficient interval is found, the scheduler proceeds until $s_{i-latest}$. During the planning process, the scheduler merges adjacent free spaces. Should several intervals fit, the best-fit rule is applied.

For T_7 in the example of Figure 2 planning proceeds until time $t = 8$, where the free space list contains the interval $[4, 7]$ which fits T_7. The stuffing technique leads to improved performance over the horizon method. The drawback is the increased complexity as we need to simulate future task terminations and planned starts to identify free space. Both schedulers use two lists. They differ, however, in the planning process for an arriving task. While the horizon method updates the horizon list only once per task arrival, the stuffing method updates the free space list for $n' + m'$ times.

3.4   Extension to the 2D Area Model

The concepts and methods discussed so far extend naturally to the 2D area model. A 2D placement is given by the coordinates of the task's bottom-left cell, (x_i, y_i), with $x_i + w_i \leq W$ and $y_i + h_i \leq H$, and the resulting scheduling problem relates to 3D strip packing problems [6]. The main difference compared to the 1D model lies in the placer. Instead of keeping lists of intervals, we need to manage lists of rectangles. The 2D scheduling horizon is a set of rectangles that fully partition the device rectangle, together with the last release time for each rectangle. The 2D stuffing method maintains the free space as a list of free rectangles. The placers we use to implement the 2D horizon and stuffing methods are based on the approach presented by Bazargan et al. [5] and have been described in [10].

3.5 Runtime Efficiency

The runtime efficiency of the schedulers depends largely on the underlying placer implementation. Both our 1D free space list and the 2D Bazargan placer [5] keep $\mathcal{O}(n)$ free areas for n currently placed tasks. Thus, the asymptotic worst-case runtime complexity is the same for 1D and 2D area models. The reference scheduler that considers only immediate placements has a complexity of $\mathcal{O}(n)$. It can be shown that the horizon scheduler's complexity is given by $\mathcal{O}(n + m)$, where m denotes the number of guaranteed but not yet scheduled tasks. The complexity of the stuffing method amounts to $\mathcal{O}(n^2 m)$.

4 Evaluation

4.1 Simulation Setup

To evaluate the online schedulers, we have devised a discrete-time simulation framework. Tasks are randomly generated. We have simulated a wide range of parameter settings. The results presented in this section are typical and are based on following settings: The simulated device consists of 96 × 64 reconfigurable units (Xilinx XCV1000). The task areas are uniformly distributed in [50, 500] reconfigurable units; task execution times are uniformly distributed in [5, 100] time units. The aspect ratios are distributed between [5, 0.2]. We have defined three task classes A, B, C, with laxities uniformly distributed in [1, 50], [50, 100], and [100, 200] time units, respectively. Runtime measurements have been conducted on a Pentium-III 1000MHz, taking advantage of Visual C++'s profiling facilities. All the simulations presented below use 95 % confidence level with an error range of ±3 percent.

4.2 Results

Scheduling to the 1D Area Model: Figure 4a) compares the performance of the reference scheduler with the horizon and stuffing techniques. The aspect ratios are distributed such that 50 % of the tasks are taller than wide (standing tasks) and 50% are wider than tall (lying tasks). The reference scheduler does not plan into the future. Hence, its performance is independent of the laxity class. As expected, the stuffing method performs better than the horizon method which in turn is superior to the reference scheduler. The differences between the methods grow with increasing laxity because longer planning periods provide more opportunity for scheduling. For laxity class C, the horizon scheduler outperforms the reference by 14.46 %; the stuffing scheduler outperforms the reference by 23.56 %.

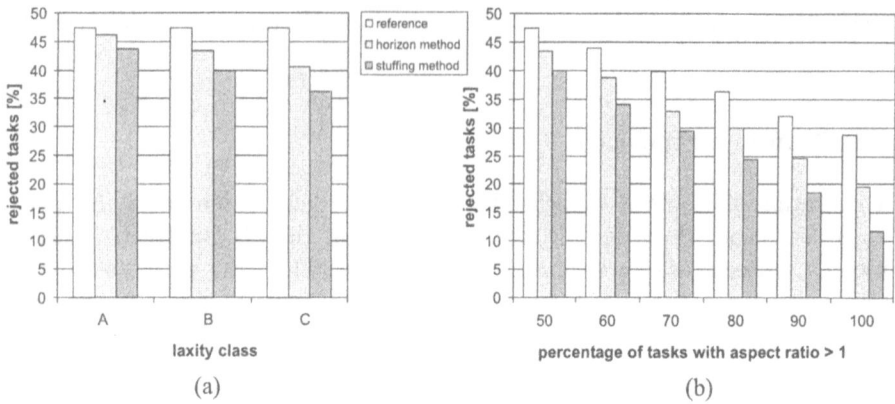

Fig. 4. Performance of the scheduling heuristics for the 1D area model

Figure 4b) shows the number of rejected tasks as function of the aspect ratio, using laxity class B. For the 1D area model standing tasks are clearly preferable. The generation of such tasks can be facilitated by providing placement and routing constraints. In Figure 4b), a percentage of tasks with aspect ratio > 1 of 100 % denotes an all standing task set. The results demonstrate that all schedulers benefit from standing tasks. The differences again grow with the aspect ratio. For 100 % standing tasks, the horizon method results in a performance improvement over the reference of 32%, the stuffing method even in 58.84 %.

Comparison of 1D and 2D Area Models: Figure 5a) compares the performance between the 1D and 2D area models for the stuffing technique. The aspect ratios are distributed such that 50 % of the tasks are standing. The results clearly show the superiority of the 2D area model. For laxity class A, the performance

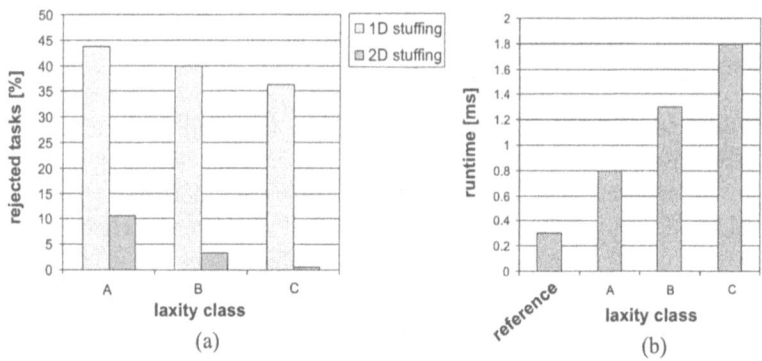

Fig. 5. (a) performance of the 1D and 2D stuffing methods; (b) runtimes for scheduling one task for the reference scheduler and the 2D stuffing method

improvement in going from 1D to 2D is 75.57 %; for laxity class C it is 98.35 %. An interesting result (not shown in the figures) is that the performance for the 2D model depends only weakly on the aspect ratio distribution. Due to the 2D resource management, both mixed and standing task sets are handled well.

Runtime Efficiency: Figure 5b) presents the average runtime required to schedule one task for the 2D stuffing method, the most complex of all implemented techniques. The runtime increases with the length of the planning period. However, with $1.8\,ms$ at most the absolute values are small. Assuming a time unit to be $10\,ms$, which gives us tasks running from $50\,ms$ to $1\,s$, the runtime overheads for the online scheduler and for reconfiguration (in the order of a few ms) are negligible. This justifies our simulation setup which neglects these overheads.

5   Toward a Reconfigurable OS Prototype

Figure 6(a) shows the 1D area model we use in our current reconfigurable OS prototype. The reconfigurable surface splits into an OS area and a user area which is partitioned into a number of fixed-size task slots. Tasks connect to predefined interfaces and communicate via FIFO buffers in the OS area. The hardware implementation is done using Xilinx Modular Design [11]. The prototype runs an embedded networking application and has been described elsewhere in more detail [2]. The application is packet-based audio streaming of encoded audio data (12kHz, 16bit, mono) with an optional AES decoding. A receiver task checks incoming Ethernet packets and extracts the payload to FIFOs. Then, AES decryption and audio decoding tasks are started to decrypt, decode and stream the audio samples. The task deadlines depend on the minimal packet inter-arrival time and the FIFO lengths. Our prototype implements rather small FIFOs and guarantees to handle packets with a minimal inter-arrival time of $20\,ms$.

Our prototype proves the feasibility of multitasking on partially reconfigurable devices and its applicability to real-time embedded systems. However,

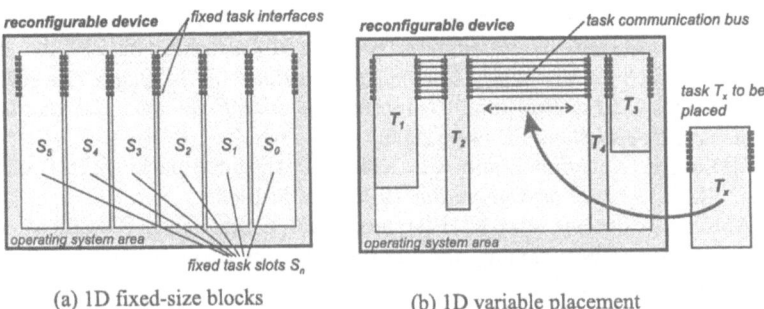

(a) 1D fixed-size blocks   (b) 1D variable placement

Fig. 6. Prototype OS structure

the 1D block-partitioned area model differs from the model used in this paper which requires task placements to arbitrary horizontal positions as shown in Figure 6(b). While task relocation is solved, prototyping a communication infrastructure for variably-positioned tasks, as proposed by [3], remains to be done.

6   Conclusion and Further Work

We discussed the problem of online scheduling hard real-time tasks to partially reconfigurable devices and developed two online scheduling heuristics for 1D and 2D area models. Simulations show that the heuristics are effective in reducing the number of rejected tasks. While the 1D schedulers depend on the tasks' aspect ratios, the 2D schedulers do not. In all cases, the 2D model dramatically outperforms the 1D model. Finally, the scheduler runtimes are so small that the more complex stuffing technique will be the method of choice for most application scenarios.

References

1. G. Brebner. A Virtual Hardware Operating System for the Xilinx XC6200. In *Int'l Workshop on Field-Programmable Logic and Applications (FPL)*, pages 327–336, 1996.
2. H. Walder and M. Platzner. Reconfigurable Hardware Operating Systems: From Concepts to Realizations. In *Int'l Conf. on Engineering of Reconfigurable Systems and Architectures (ERSA)*, 2003.
3. G. Brebner and O. Diessel. Chip-Based Reconfigurable Task Management. In *Int'l Conf. on Field Programmable Logic and Applications (FPL)*, pages 182–191, 2001.
4. O. Diessel, H. ElGindy, M. Middendorf, H. Schmeck, and B. Schmidt. Dynamic scheduling of tasks on partially reconfigurable FPGAs. 147(3):181–188, 2000.
5. K. Bazargan, R. Kastner, and M. Sarrafzadeh. Fast Template Placement for Reconfigurable Computing Systems. 17(1):68–83, 2000.
6. S. Fekete, E. Köhler, and J. Teich. Optimal FPGA Module Placement with Temporal Precedence Constraints. In *Design Automation and Test in Europe (DATE)*, pages 658–665, 2001.
7. T. Marescaux, A. Bartic, Verkest D., S. Vernalde, and R. Lauwereins. Interconnection Networks Enable Fine-Grain Dynamic Multi-tasking on FPGAs. In *Int'l Conf. on Field-Programmable Logic and Applications (FPL)*, pages 795–805, 2002.
8. G.C. Buttazzo. *Hard Real-time Computing Systems: Predictable Scheduling Algorithms and Applications*. Kluwer, 2000.
9. B.S. Baker, E.G. Coffman, and R.L. Rivest. Orthogonal packings in two dimensions. *SIAM Journal on Computing*, (9):846–855, 1980.
10. H. Walder, C. Steiger, and M. Platzner. Fast Online Task Placement on FPGAs: Free Space Partitioning and 2D-Hashing. In *Reconfigurable Architectures Workshop (RAW)*, 2003.
11. D. Lim and M. Peattie. Two Flows for Partial Reconfiguration: Module Based or Small Bit Manipulations. XAPP 290, Xilinx, 2002.

Run-Time Minimization of Reconfiguration Overhead in Dynamically Reconfigurable Systems

Javier Resano[1], Daniel Mozos[1], Diederik Verkest[2,3,4],
Serge Vernalde[2], and Francky Catthoor[2,4]

[1] Dept. Arquitectura de Computadores, Universidad Complutense de Madrid, Spain.
`javier1@fdi.ucm.es`, `mozos@dacya.ucm.es`

[2] IMEC vzw, Kapeldreef 75, 3001, Leuven, Belgium
`{Verkest, Vernalde, Catthoor}@imec.be`

[3] Professor at Vrije Universiteit Brussel, Belgium

[4] Professor at Katholieke Universiteit Leuven., Belgium.

Abstract. Dynamically Reconfigurable Hardware (DRHW) can take advantage of its reconfiguration capability to adapt at run-time its performance and its energy consumption. However, due to the lack of programming support for dynamic task placement on these platforms, little previous work has been presented studying these run-time performance/power trade-offs. To cope with the task placement problem we have adopted an interconnection-network-based DRHW model with specific support for reallocating tasks at run-time. On top of it, we have applied an emerging task concurrency management (TCM) methodology previously applied to multiprocessor platforms. We have identified that the reconfiguration overhead can drastically affect both the system performance and energy consumption. Hence, we have developed two new modules for the TCM run-time scheduler that minimize these effects. The first module reuses previously loaded configurations, whereas the second minimizes the impact of the reconfiguration latency by applying a configuration prefetching technique. With these techniques reconfiguration overhead is reduced by a factor of 4.

1 Introduction and Related Work

Dynamically Reconfigurable Hardware (DRHW), that allows partial reconfiguration at run-time, represents a powerful and flexible way to deal with the dynamism of current multimedia applications. However, compared with application specific integrated circuits (ASICs), DRHW systems are less power efficient. Since power consumption is one of the most important design concerns, this problem must be addressed at every possible level; thus, we propose to use a task concurrency management (TCM) approach, specially designed to deal with current dynamic multimedia applications, which attempts to reduce the energy consumption at task-level.

Other research groups have addressed the power consumption of DRHW, proposing a technique to allocate configurations [1], optimizing the data allocation both for improving the execution time and the energy consumption [2], presenting an energy-

P.Y.K. Cheung et al. (Eds.): FPL 2003, LNCS 2778, pp. 585–594, 2003.

conscious architectural exploration [3], introducing a methodology to decrease the voltage requirements [4], or carrying out a static scheduling [5]. However, all these approaches are applied at design-time, so they cannot tackle efficiently dynamic applications, whereas our approach selects at run-time among different power/performance trade-offs. Hence, we can achieve larger energy savings since our approach prevents the use of static mappings based on worst-case conditions.

The rest of the paper is organized as follows: sections 2 and 3 provide a brief overview of the ICN-based DRHW model and the TCM methodology. Sect. 4 discusses the problem of the reconfiguration overhead. Sect. 5 presents the two modules developed to tackle this problem. Sect. 6 introduces some energy considerations. Sect. 7 analyses the experimental results and Sect. 8 presents the conclusions.

2 ICN-Based DRHW Model

The Interconnection-Network (ICN) DRHW model [6] partitions an FPGA platform into an array of identical tiles. The tiles are interconnected using a packet-switched ICN implemented using the FPGA fabric. At run-time, tasks are assigned to these tiles using partial dynamic reconfiguration. Communication between the tasks is achieved by sending messages over the ICN using a fixed network interface implemented inside each tile. As explained in [6], this approach avoids the huge Place & Route overhead that would be incurred when directly interconnecting tasks. In the latter case, when a new task with a different interface is mapped on a tile, the whole FPGA needs to be rerouted in order to interconnect the interface of the new task to the other tasks. Also the communication interfaces contain some Operating System (OS) support like storage space and routing tables, which allow run-time migration of tasks from one tile to another tile or even from an FPGA tile to an embedded processor.

Applying the ICN model to a DRHW platform greatly simplifies the dynamic task allocation problem, providing a software-like approach, where tasks can be assigned to HW resources in the same way that threads are assigned to processors. Thus, this model enables the use of the emerging Matador TCM methodology [7].

3 Matador Task Concurrency Management Methodology

The matador TCM methodology [7, 9] proposes a task scheduling technique for heterogeneous multiprocessor embedded systems. The different steps of the methodology are presented in figure 1. It starts from an application specification, composed of several tasks, called Thread Frames (TF). These TFs are dynamically created and deleted and can even be non-deterministically triggered. Nevertheless, inside each TF only deterministic and limited dynamic behavior is allowed.

The whole application is represented using the grey-box model, which combines Control–Data Flow Graphs (CDFG) with another model (called MTG*) specifically designed to tackle dynamic non-deterministic behavior. CDFGs are used to model the

TFs, whereas MTG* models the inter-TF behavior. Each node of the CDFG is called a Thread Node (TN). TNs are the atomic scheduling units.

TCM accomplishes the scheduling in two phases. The first phase generates at design-time a set of near-optimal scheduling solutions for each TF called a Pareto curve. Each solution represents a schedule and an assignment of the TNs over the available processing elements with a different performance/energy tradeoff. Whereas this first step accomplishes separately the design-space exploration of each TF, the second phase tackles their run-time behavior, selecting at run-time the most suitable Pareto point for each TF. The goal of the methodology is to minimize the energy consumption while meeting the timing constraints of the applications (typically, highly-dynamic multimedia applications). TCM has been successfully applied to schedule several current multimedia applications on multiprocessor systems [8, 9].

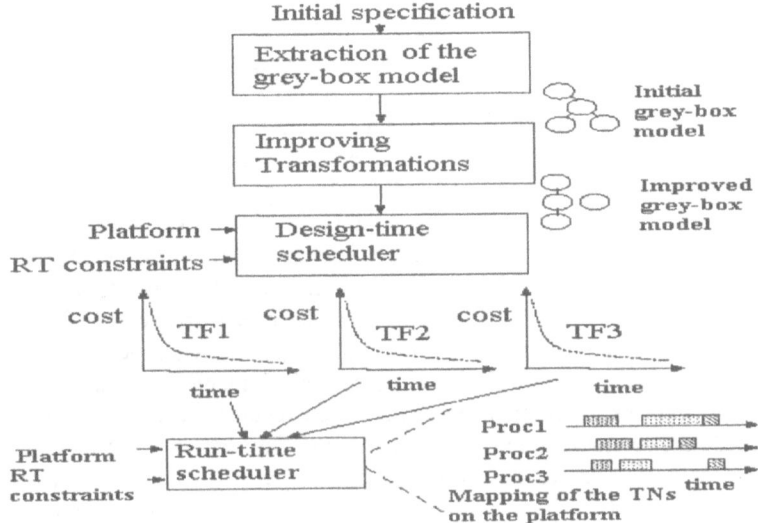

Fig. 1. Matador Task Concurrency Management Methodology

4 Dynamic Reconfiguration Overhead

Fig. 2 represents the Pareto curve obtained using the existing TCM design scheduler for a system with a SA-1110 processor coupled with a Virtex2 v6000 FPGA. The TF corresponds to a motion JPEG application. In the figure optimal, and non-optimal schedules are depicted, but only the optimal ones are included in the Pareto curve. This curve is one of the inputs for the run-time scheduler. Thus, the scheduler can select at run-time between different energy/performance trade-offs. For instance, if there is not a tight timing constraint, it will select the least energy consuming solution, whereas, if the timing constraint changes (for instance if a new task starts), the run-time scheduler will look for a faster solution which meets the new constraint (with the consequent energy penalty).

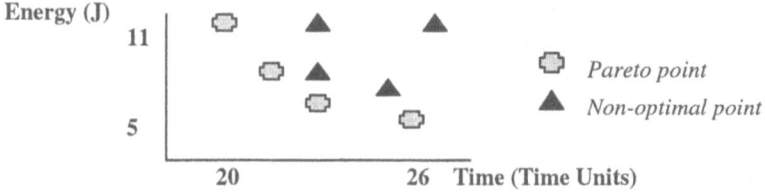

Fig. 2. Pareto curve of the motion JPEG

In this example, assuming that 80% of the time the most energy-efficient solution can be selected TCM can achieve a 47% energy saving (on average 11*0.2+5*0.8= 5.78 J/iteration) than a static worst-case approach while providing the same peak performance (the worst-case consumes 11 J/iteration). Hence, the TCM approach can drastically reduce the overall energy consumption while still meeting hard real-time constraints.   However, current TCM scheduling tools neglect the task context-switching overhead, since for many existing processors it is very low. However, the overhead due to the load of a configuration on a FPGA is much greater than the afore-mentioned context-switching overhead, e.g. reconfiguring a tile of our ICN-based FPGA consumes 4 ms (assuming that a tile occupies one tenth of a XC2V6000 FPGA and the configuration frequency is 50 MHz). The impact of this overhead on the system performance greatly depends on the granularity of the TNs. However, for current multimedia applications, TN average execution time is likely to be in the order of magnitude of milliseconds (a motion JPEG application must decode a frames in 40 ms). If this is the case, the reconfiguration overhead can drastically affect both the performance and the energy consumption of the system, moving the Pareto curve to a more energy and time consuming area. Moreover, in many cases, the shape of the Pareto curve changes when this overhead is added. Therefore, if it is not included the TCM schedulers cannot take the optimal decisions. To address this problem, we have added two new modules to the TCM schedulers, namely: configuration reuse, and configuration prefetch. These modules are not only used to accurately estimate the reconfiguration overhead, but also to minimize it. Configuration reuse attempts to reuse previously loaded configurations. Thus, if a TF is being executed periodically, at the beginning of each iteration the scheduler checks if the TNs loaded in the previous iteration are still there, if so, they are reused preventing unnecessary reconfigurations.

Configuration prefetch [10] attempts to overlap the configurations of a TN with the computation of other TNs in order to hide the configuration latency. A very simple example is presented in figure 3, where 4 TNs must be loaded and executed on an FPGA with 3 tiles. Since current FPGAs do not support multiple simultaneous recon-figurations, configurations must be load sequentially. Without prefetching, the best on-demand schedule result is depicted in 3(a) because TN3 cannot be loaded before the loading of TN2 is finished and TN4 must wait until the execution of both TN2 and TN3 is finished. However, applying prefetch (3b), the loads of TN2 and TN4 overlap with the execution of TN1 and TN3. Hence, only the loads of TN1 and TN3 penalize the system execution time.

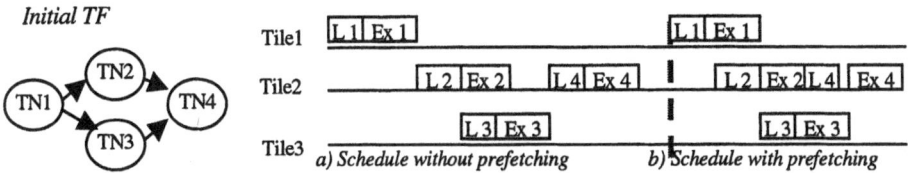

Fig. 3. Configuration prefetch on a platform with 3 FPGA tiles. L: loading, Ex: executing

Clearly, this powerful technique can lead to significant execution time savings. Unfortunately, deciding the best order to load the configurations is a NP-complete problem. Moreover, in order to apply this technique in conjunction with the configuration reuse technique, the schedule of the reconfiguration must be established at run-time, since the number of configurations that must be loaded depends on the number of configurations that can be reused and typically this number will differ from one execution to another if the system behavior is non-deterministic. Therefore, we need to introduce these techniques in the TCM run-time scheduler while attempting to keep the resulting overhead to a minimum.

5 Run-Time Configuration Prefetch and Reuse

The run-time scheduler receives as an input a set of Pareto curve and selects a point on them for each TF according to timing and energy considerations. Currently, at run-time we never alter the scheduling order imposed by the design-time scheduler, thus, in order to check if a previous configuration can be reused we just look for the first TN assigned to every FPGA tile in the given schedule (we use the term *initial* for those TNs). Since all tiles created by the use of the ICN in the FPGA are identical, the actual tile in which a TN is executed is irrelevant. When the run-time scheduler needs to execute a given TF, it will first check whether the *initial* TNs are still present in any of the FPGA tiles. If so, they are reused (avoiding the costly reconfiguration). Otherwise the TNs are assigned to an available tile. For instance in the example of Fig. 3, the run-time scheduler will initially check if the configurations of TN1, TN2, and TN3 are still loaded in the FPGA, in this case TN1, and TN3 will be reused. The run-time scheduler will not check if it can reuse the configuration of TN4 since it knows that even when this TN could remain loaded in the FPGA it will be overwritten by TN2.

The run-time configuration reuse algorithm is presented in figure 4.a, its complexity is $O(NT*I)$, where NT is the number of tiles and I the number of *initial* TNs. Typically I and NT are small numbers and the overhead of this module is negligible.

Once the run-time scheduler knows which TN configurations can be reused, it has to decide when the remaining configurations are going to be loaded. The pseudo-code of the heuristic developed for this step is presented in figure 4.b. It starts from a given design-time schedule that does not include the reconfiguration overhead and updates it according to the number of reconfigurations needed. Then, it schedules the reconfigurations by applying prefetching to try to minimize the execution time.

```
a) for (i=0; i < Number of initial TNs; i++){
      while (not found)and(j < Number of FPGA tiles){
          found = look_for_reuse(i, j);
          if(found){assign the TN i to actual tile j; j++;}}
   Assign the remaining TNs to actual tiles;
```

```
b) If there are TN configurations to load{
      schedule the TNs that do not need to be loaded
      for (i=0; i < Number of configurations to load; i++){
      select&schedule a configuration;
          schedule its successors that do not need to
          be loaded on to the FPGA;}}
```

Fig. 4. a) Configuration reuse pseudo-code. **b)** Configuration prefetch pseudo-code

We have developed a simple heuristic based on an enhanced list-scheduling. It starts by scheduling all the nodes that do not need to be configured on the FPGA, i.e. those assigned to SW or those assigned to the FPGA whose configuration can be reused. After scheduling a TN, the heuristic attempts to schedule its successors. If they do not need to be loaded on to the FPGA the process continues, otherwise, a reconfiguration request is stored in a list. When it is impossible to continue scheduling TNs, one of the requests is selected according to the following criteria:

- If, at a given time **t,** there is just one configuration ready for loading this configuration is selected. A configuration is ready for loading if the previous TN assigned to the same FPGA tile has already finished its execution.
- Otherwise, when several configurations are ready, the configuration with the highest weight is selected.

The weight of a configuration is assigned at design-time and represents the maximum time-distance from a TN to the end of the TF. This weight is computed carrying out an ALAP scheduling in the TF. Those configurations corresponding to nodes in the critical path of the TF will be heavier than the other ones. The complexity of this module is $O(N*C)$ where N is the number of TNs and C the number of configurations to load.

Since these two techniques must be applied at run-time, we are very concerned about their execution time. This time depends on the number of TN, FPGA tiles and configurations that must be loaded. The configuration-prefetch module is much more time-demanding that the configuration-reuse module. However we believe that the overhead is acceptable for a run-time scheduler. For instance a TF with 20 nodes, 4 FPGA tiles and 13 configurations to be loaded is scheduled in less than 2.5 µs using a Pentium-II running at 350MHz. This number has been obtained starting from a C++ initial code, and disabling all the compiler optimizations. We expect that this time will be significantly reduced when starting from C code and applying optimization compiler techniques. But even if is not reduced, it will be compensated by the prefetching time-savings if the heuristic hides the latency of one reconfiguration at least once every 1600 executions (assuming that the reconfiguration overhead is 4ms). In our experiments the heuristic has exhibited much better results than this minimum requirement. Although we believe that the overhead due to the prefetch module is acceptable, it is not always worthwhile to execute it. For instance, when all the configurations of a design are already loaded in the FPGA there will be no gains applying prefetch. Thus, in this case the module will not be executed, preventing unnecessary computation. There is another common case where this run-time computation can be substituted by design-time computation; this is when all the configurations must be loaded. This case happens at least once (when the TF is loaded for the first time), and it can be very frequent if there is a great amount of TNs competing for the resources. Hence, we can save some run-time computation analyzing this case at design-time. However, this last optimization duplicates the storage space needed for a Pareto point, since now for each point in the Pareto curve two schedules are stored. Hence, if the system has a very limited storage space, this optimization should be disabled. The run-time scheduling will follow the steps depicted in figure 5.

```
for each Pareto point to evaluate {
  apply configuration reuse
  If there are not configurations to load {
   read energy and execution time of scheduling 1}
  Else if all the configurations must be loaded{
   read energy and execution time of scheduling 2 }
  Else {
  actualize scheduling 1 applying configuration prefetching}}
```

Fig. 5. Run-time evaluation process pseudo-code. Schedules 1 and 2 are computed at design time. Both of them share the same allocation of the TNs on the system processing elements, but they have different execution time and energy consumption since schedule 1 assumes that all the TNs assigned to the FPGA have been previously loaded, whereas schedule 2 includes the reconfiguration overhead of these TNs

6 Energy Considerations

TCM is an energy-aware scheduling technique. Thus, these two new modules should not only reduce the execution time but also the energy consumption. Clearly, the configuration reuse technique can generate energy savings, since loading a configuration to an FPGA involves both an execution time and an energy overhead. According to [5], when a FPGA is frequently reconfigured, up to 50% of the FPGA energy consumption is due to the reconfiguration circuitry. Hence, reducing the number of reconfigurations is a powerful way to achieve energy savings.

Configuration prefetch can also indirectly lead to energy savings. If we assume that loading a configuration on to a tile has a constant overhead Ec, and a given schedule involves 4 reconfigurations, the energy overhead due to the reconfigurations will be 4*Ec independently of the order of the loads. However, configuration prefetch reduces the execution time of the TFs and the run-time scheduler can take advantage of this extra time to select a slower and less energy consuming Pareto point. Fig. 6 illustrates this idea with one TF.

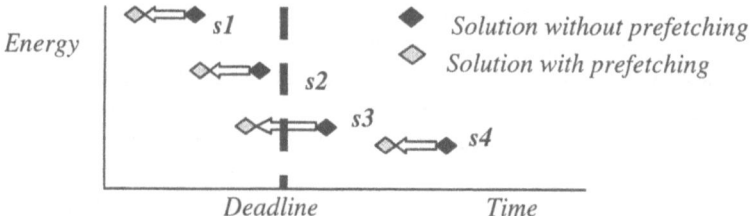

Fig. 6. Configuration prefetch technique for energy savings. Before applying prefetching, s2 was the solution that consumes less energy meeting the timing constraint. After applying prefetching the time saved can be used to select a more energy efficient solution (s3)

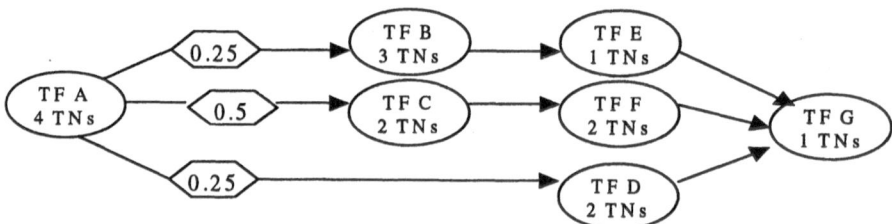

Fig. 7. Set of TFs generated using the Task Graph For Free (TGFF) system [11]. Inside each node its name (a letter from A to G), and the number of TNs assigned to HW are depicted. When there are different execution paths, each one is tagged with the probability of being selected. We assume that this set of TF is executed periodically

7 Results and Analysis

The efficiency of the reuse module depends on the number of FPGA tiles, the TNs assigned to these tiles and on the run time events. Figure 8 presents the average reuse percentage for the set of TFs depicted in figure 7 for different number of FPGA tiles. This example contains 15 TNs assigned to HW, although not all of them are executed every iteration. The reuse percentage is significant even with just 5 FPGA tiles (29%). Most of the reuse is due to the TFs A and G, since they are executed every iteration. For instance, when there are 8 tiles, 48 % of the configurations are reused and 30% are due to TF A and TF G (5 TNs). When there are 17 tiles, only the 71 % of configurations are reused, this is not an optimal result since as long as there are more tiles than TNs it should be possible to reuse 100% of the configurations. However, we are just applying a local reuse algorithm to each TF instead of applying a global one to the whole set of TFs. We have adopted a local policy because it creates almost no runtime overhead, and can be easily applied to systems with non-deterministic behavior. However, we are currently analyzing how a more complex approach could lead to higher percentages of reuse with an affordable overhead.

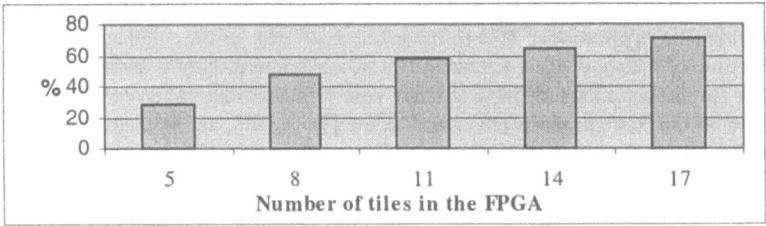

Fig. 8. Percentage of configurations reused vs Number of tiles in the FPGA

We have performed two experiments to analyze the prefetch module performance. Firstly, we have studied how good the schedules computed by our heuristic are. To this end we have generated 100 pseudo-random TFs using the TGFF system, and we have scheduled the configuration loads both with our heuristic, and with a branch&bound (b&b) algorithm that accomplishes a full design space exploration. This experiment shows that the b&b scheduler finds better solutions (on average 10% better) than our heuristic. However, for TFs with 20 TNs, it needs 800 times more computational time to find these solutions. Hence, our heuristic generates almost optimal schedules, in an affordable time.

The second experiment presents the time-savings achieved due to the prefetch module for three multimedia applications (table 1). It assumes that the whole application is executed on an ICN-like FPGA, with 4 tiles that can be reconfigured in 4 ms, which is currently the fastest possible speed. However, even with this fast reconfiguration assumption, it is remarkable how the reconfiguration overhead affects the system performance, increasing up to the 35% the overall execution time. This overhead is drastically reduced (in average a factor of 4) when the prefetch module is applied.

8   Conclusions

The ICN-based DRHW model provides the dynamic task reallocation support needed to apply a TCM approach to heterogeneous systems with DRHW resources. With this model both the DRHW and the SW resources can be handled in the same way, simplifying the tasks of the TCM schedulers. We have identified that the DRHW reconfiguration overhead significantly decreases the system performance and increases the energy consumption. Hence, we have developed two new modules for the TCM run-time scheduler (namely configuration reuse and configuration prefetch) that can reduce this problem while improving at run-time the accuracy of the execution time and energy consumption estimations.

Configuration reuse attempts to reuse previously loaded configurations, leading to energy and execution time savings. Its efficiency depends on the number of FPGA tiles, and TNs assigned to these tiles. However, the module exhibits significant reuse percentage even with 15 TNs competing for 5 FPGA tiles. Configuration prefetch schedules the reconfigurations to minimize the overall execution time. The results show that it reduces the execution time reconfiguration overhead by a factor of 4.

Table 1. Reconfiguration overhead with and without applying the prefetch module for three actual multimedia applications. **TNs** is the number of TNs in the application, **Init T** is the execution time of the application assuming that no reconfigurations are required. **Prefetch** and **Overhead** are the percentage of the execution time increased due to the reconfiguration overhead, when all the configurations are loaded in the FPGA, with and without the prefetch module. The first two applications are different implementations of a JPEG decoder, the third application computes the Hough transform of a given image in order to look for certain patterns

	TNs	Init T	Overhead	Prefetch
JPEG decoder	4	81ms	+25%	+5%
Enhanced JPEG decoder	8	57ms	+35%	+7%
Pattern Recognition	6	94ms	+17%	+4%

Acknowledgements

The authors would like to acknowledge all our colleagues from the T-Recs and Matador groups at IMEC, for all their comments, help and support. This work has been partially supported by TIC 2002-00160 and the Marie Curie Host Fellowship HPMT-CT-2000-00031.

References

1. R. Maestre et al., "Configuration Management in Multi-Context Reconfigurable Systems for Simultaneous Performance and Power Optimizations", ISSS'00, pp 107-113, 2000.
2. M. Sánchez-Elez et al "Low-Energy Data Management for Different On-Chip Memory Levels in Multi-Context Reconfigurable Architectures". DATE'03, pp. 36-41, 2003.
3. M. Wan et al. "Design Methodology of a Low-Energy Reconfigurable Single-Chip DSP System", Journal of VLSI Signal Processing 28, pp. 47-61, 2001.
4. A. D. Garcia et al., "Reducing the power Consumption in FPGAs with keeping a high Performance Level", WVLSI00, pp 47-52, 2002.
5. Li Shang et al., "Hw/Sw Co-synthesis of Low Power Real-Time Distributed Embedded Systems with Dynamically Reconfigurable FPGAs", ASP-DAC'02, pp. 345-360, 2002.
6. T. Marescaux et al.,"Interconnection Network enable Fine-Grain Dynamic Multi-Tasking on FPGAs", FPL'02, pp. 795-805, 2002.
7. P. Yang et al., "Energy-Aware Runtime Scheduling for Embedded-Multiprocessors SOCs", IEEE Journal on Design&Test of Computers, pp. 46-58, 2001.
8. P. Marchal et al, "Matador: an Exploration Environment for System-Design", Journal of Circuits, Systems and Computers, Vol. 11, No. 5, pp. 503-535, 2002.
9. P. Yang et al, "Managing Dynamic Concurrent Tasks in Embedded Real-Time Multimedia systems", ISSS'02, pp. 112-119, 2002.
10. Z. Li and S. Hauck, "Configuration prefetching techniques for partial reconfigurable coprocessor with relocation and defragmentation" Int'l Symp. FPGAs, pp. 187-195, 2002.
11. R.P. Dick et al, "TGFF: Task Graphs for Free", Int'l Workshop HW/SW Codesign, pp. 97-101, 1998.